MAJOR TOPICS OF APPLICATIONS

SELECTED ACTUAL APPLICATIONS

Functional Calculus

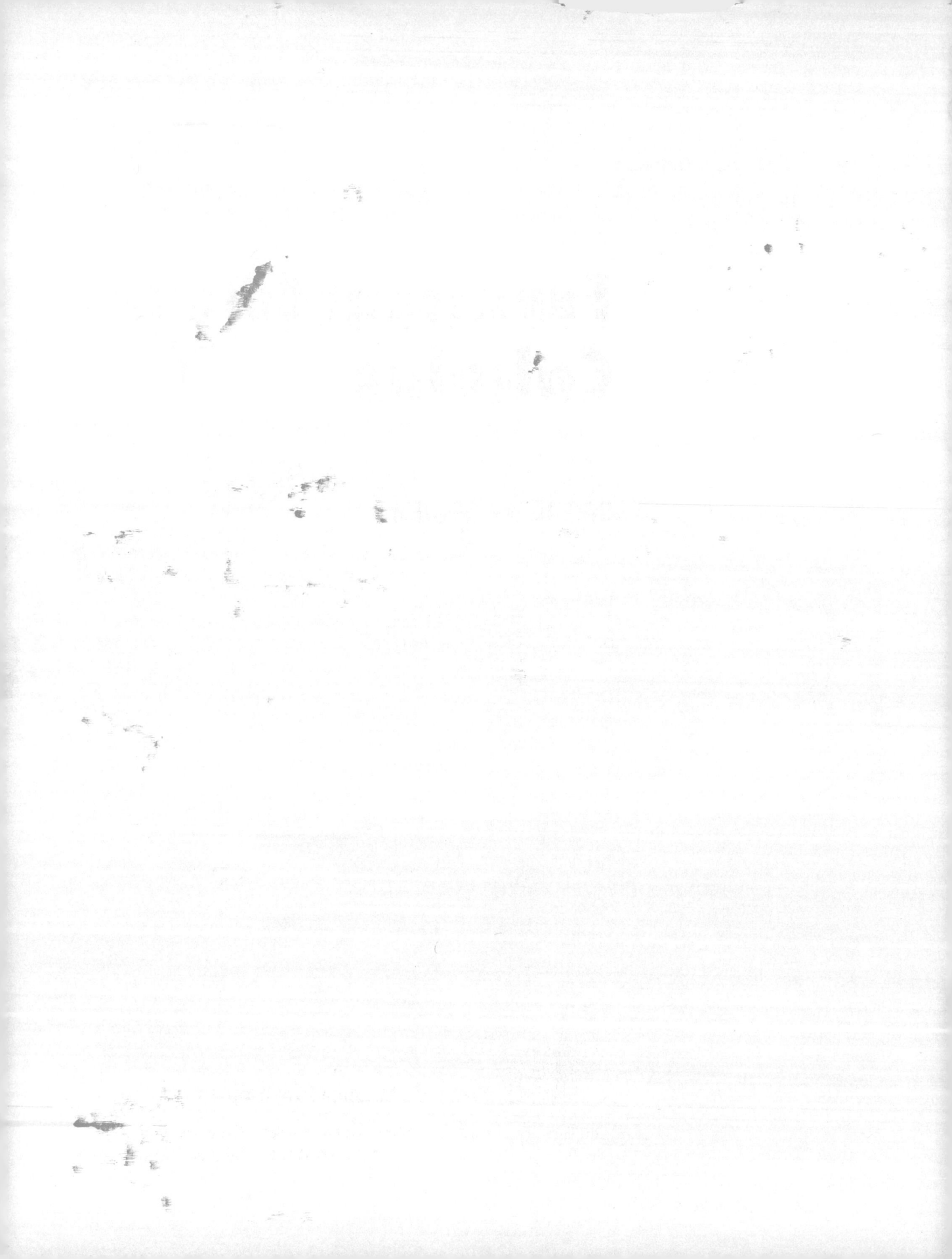

Functional Calculus

Brief Calculus for Management, Life, and Social Sciences

William C. Ramaley
Fort Lewis College, Durango, Colorado

WCB **Wm. C. Brown Publishers**

Dubuque, IA Bogota Boston Buenos Aires Caracas Chicago
Guilford, CT London Madrid Mexico City Sydney Toronto

Book Team

Developmental Editor   *Rachel Riley*
Publishing Services Coordinator   *Julie Avery Kennedy*

WCB

Wm. C. Brown Publishers
A Division of Wm. C. Brown Communications, Inc.

Vice President and General Manager   *Beverly Kolz*
Vice President, Publisher   *Earl McPeek*
Vice President, Director of Sales and Marketing   *Virginia S. Moffat*
Vice President, Director of Production   *Colleen A. Yonda*
National Sales Manager   *Douglas J. DiNardo*
Marketing Manager   *Julie Joyce Keck*
Advertising Manager   *Janelle Keeffer*
Production Editorial Manager   *Renée Menne*
Publishing Services Manager   *Karen J. Slaght*
Royalty/Permissions Manager   *Connie Allendorf*

Wm. C. Brown Communications, Inc.

President and Chief Executive Officer   *G. Franklin Lewis*
Senior Vice President, Operations   *James H. Higby*
Corporate Senior Vice President, President of WCB Manufacturing   *Roger Meyer*
Corporate Senior Vice President and Chief Financial Officer   *Robert Chesterman*

Cover Image © by Index Stock Photography, Inc.

Copyediting, Design, and Production by GTS Graphics, Inc.
Composition and Prepress by GTS Graphics, Inc.

The credits section for this book begins on page C-1 and is considered an
extension of this copyright page.

to my wife, Annette

CONTENTS

The * denotes an optional section.

PREFACE

To the Student and Instructor

Isaac Newton created calculus, in part, because he was fascinated by its applications. Today's students are also interested in finding such applications.

Goals of the Book. This book provides a brief calculus course that focuses on applications. All decisions about what material to include in this book were guided by asking: Suppose a student is majoring in business, economics, management science, or a biological or social science. What aspects of calculus should this student know? What are the applications of calculus to this student's area of interest? This book attempts to arouse students' *mathematical interests* while developing *mathematical abilities.*

Every effort has been made to present realistic, practical, and interesting problems as they might arise in the world of business meetings, boardrooms, and laboratories. Enabling students to use calculus to discuss and resolve problems is the essential goal of this book.

Goals of Students. Students want to be able to find, within themselves, the confidence to proceed in a situation in which mathematical formulations and approaches may be helpful. They want to know some worthwhile and useful ideas about how to begin a problem and where they can find additional help. This book is written to help them achieve their goals.

Examples. Each example and application has been selected carefully to provide just enough detail to make the problem clear and meaningful, and to avoid making it overwhelmingly complex. Examples are offered in a lively form, which students can understand, handle, and become successful at doing. It is important for students to realize that many of life's complications can be made simpler and more manageable by a gradual analytic approach.

As the topics of each section are gradually introduced, they are illustrated by two types of examples. First, there is the traditional "fully worked out" example. Following that example is often an additional example, called a "Practice Exercise," whose detailed solution is postponed to the end of the section. This provides students the opportunity to work on the practice exercise before being shown all the steps. Involvement improves learning and retention.

Informal Mathematics. Mathematical formulas and results are presented in an informal manner, but never in a misleading or incorrect form. In some situations, a more formal discussion may occur in an (optional) appendix to the section.

Complex Problems and Calculators.   Problems arising from actual situations rarely have "nice" answers, with small and simple numbers. Naturally, any textbook must contain many "nice" problems when a new concept is being introduced so that the concept itself is not obscured. However, if all the problems were of this sort, students might get an incorrect view of the applicability of mathematics. If we consider a real-world application, the chance that it is "nice" is probably less than the chance of our finding a twenty-dollar bill while walking to lunch. Of course, we should keep our eyes open, but we should be prepared for some problems that have "messy" answers. To do these problems, we need to have a calculator and to learn how to use it. A graphing calculator is optional, but there are many examples and some problems that use them.

Section Problems.   Problem sets begin with "Foundations" problems, which illustrate the basic skills required to do the exercises. These, in turn, begin with questions to develop students' skills and confidence. Following these problems are more interesting applications and situations. At the end of most sections are problems that ask students to write and discuss some mathematical ideas, and there are problems using graphing calculators. Furthermore, there are "Enrichment Problems," intended for students more interested in the topic and willing to explore the material at a richer level.

Functions and Other Introductory Topics.   Chapter 1 covers topics some students may have encountered before, but this chapter introduces some terminology used in business and other applications. The chapter may be omitted because the essential terms are reintroduced later. However, the presence of these definitions and their use in applications keeps Chapter 1 from being merely a boring review of previous work. Furthermore, most students need a gradual reintroduction to the variety of functions in that chapter. An understanding of functions forms the bedrock of calculus.

Word Problems.   A special comment on word problems is warranted: Most instructors and students approach word or story problems with an attitude that might be expressed as, "Don't pay any attention to the words at all. Get to the essentials. Look for key words and phrases. The words are just window dressing." On one hand, that attitude is exactly right, and we do need to develop the skills to dive right into the problem, see through the setting, and firmly grasp the essentials. On the other hand, students need to realize that the mathematics they are learning does have real-world applications. Under all the complex, full, and beautiful things we see is a hidden mathematical framework, a scaffold to support the applications. Many instructors of mathematics take for granted the existence and intrinsic interest of this mathematical scaffold, but students may not.

Historical Comments.   This book contains many historical comments. In these comments, a special effort has been made to point out those contributions made by mathematicians to solve applied problems of interest to society. We may not find it easier to find a derivative just because we learn the two creators of calculus, Newton and Leibniz, both directed their countries' mints for coining money, but we may see mathematicians as being more human. Furthermore, such knowledge allows us to feel less removed from mathematicians and doing mathematics ourselves. Lastly, historical comments are often just plain fun.

Especially to the Instructor

This book may be used in several ways. For a three or four credit-hour semester course, you may use the first six chapters (omitting the starred sections), plus sections 7.1–7.3. If that seems to be too much material, I would urge at least a mention of the idea of optimization of a function of two variables. A longer or more intensive course could cover all of the first six chapters, plus most of Chapter 7.

For a class well versed in functions, Chapter 1 may be covered swiftly as a review. However, the many applications there are of interest even to students who have been exposed to the mathematical material previously.

Especially to the Student

The Book.   This text is meant to be *used by you*. It is more than merely a collection of topics and problems. It is an essential resource. The effort that you and your instructor put into this course will be connected together by this book, for it contains explanations of ideas as well as worked examples and problems. All of these will help you understand calculus and the many uses of calculus concepts.

Answers.   Students are often intimidated by the fact that most problems at this level of mathematics have just one answer. However—and this "however" is very important—very often there are different approaches to arrive at that answer. Thus, your way of solving a problem may seem (or even actually be) different from the way shown in the book or by your instructor. This does not mean that one particular way is always the only correct way. In fact, if you can work a problem by several methods, then you have a check on your work.

The answers given in the back of the book are often presented in both exact and approximate forms. The reason for this is that the world outside the classroom uses decimal approximations. We should also note that there are often several equivalent forms for an answer. For example, you might obtain an answer of $\sqrt{20}$ and find $2\sqrt{5}$ as the answer in the back. If your answer differs from the answer given, do not abandon your answer immediately. Take a moment to see if the answers are equivalent.

Evaluating Answers.   When at last you arrive at an answer, ask yourself, "Does this answer make sense?" Try to understand what you have done. In the real world, you do not want to be telling the filling station operator that you have just pumped 137 gallons of gas into a car that has a 20-gallon tank, when you meant 13.7 gallons. It is very important to try to get some "feel" for reasonable answers. That can only be done by working many, many problems.

Success. I expect that as you work through this book, there will be times when you ask, "What does he mean by this?" or "How did he get that?" I hope those times are few and I wish I could be there to try to explain what I did mean, or to work through another example, or to try a different explanation. But please do have some patience with yourself. Thousands of students have learned this material successfully and *you can too.*

Acknowledgments

My thanks go to my family, my colleagues, and the staffs of Wm. C. Brown and GTS Graphics for their support and encouragement. Over the years some particularly key people have been: Lisa Moller, Cindy Jones, Jane Parrigin, Rachel Riley, Pat Foard, Greg Bell, Heather Stratton, and Hannelore Hanks. Thanks also to those students at Fort Lewis College who used and commented on earlier versions of the text.

I welcome further comments or correction.

I am grateful to the following reviewers for their help in preparing this book.

Burla J. Sims
University of Arkansas at Little Rock

Arlene M. Sherburne
Montgomery College

Diane M. Spresser
James Madison University

Thomas Metzger
University of Pittsburgh

Oiyin Pauline Chow
Harrisburg Area Community College

M. B. Elgindi
University of Wisconsin at Eau Claire

Reza Abbasian
Texas Lutheran College

George Witt
Glendale Community College

Cindy Jones
Indiana University and Purdue University at Indianapolis

John Pommersheim
Bowling Green State University

Bill Ramaley

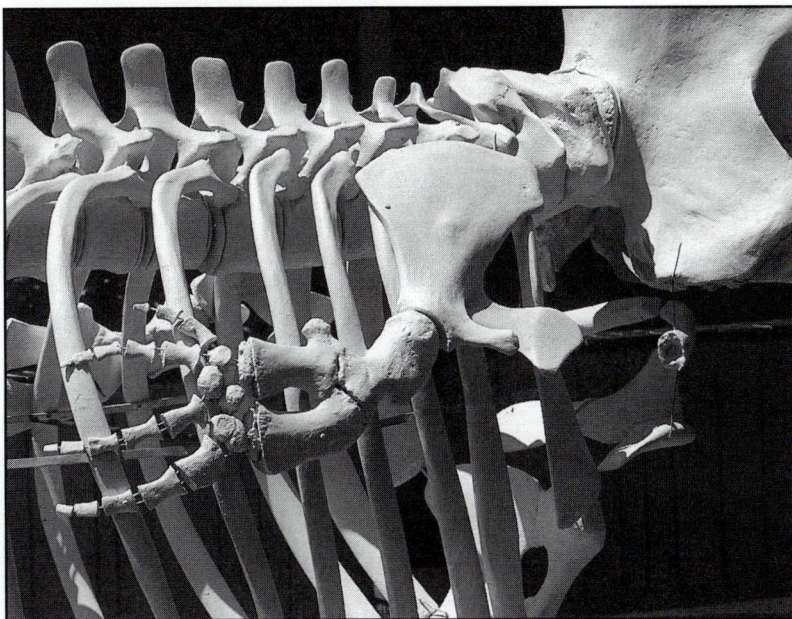

CHAPTER 1

NUMBERS AND FUNCTIONS

THE TOPICS OF THIS CHAPTER PROVIDE THE NECESSARY FOUNDATION FOR building calculus. Much of the material may be familiar, but its review will serve two purposes. First, its study provides a common background and momentum for studying calculus. Second, we will see and use these mathematical ideas to create models of the real world. Our mathematical models will apply to income taxes, whale skeletons, cardiac output, the Golden Gate Bridge, and beyond.

1.1   REAL NUMBERS, SETS, AND INEQUALITIES

In this section we discuss real numbers, inequalities, and absolute values, and then present specifications for sets in several ways. In particular, we introduce sets that are intervals of numbers. One use of intervals is to describe the solutions of inequalities.

1

MATHEMATICAL INNOVATORS
Fibonacci of Pisa (1175–1250)

In our everyday computations we are comfortable using both positive and negative numbers. However, as recently as the thirteenth century only positive numbers were allowed as solutions to problems. Writers and thinkers would have asked, "What could be less than nothing?"

The use of negative numbers first appeared in the book *Flos,* written about 1225 by Fibonacci. He was explaining a business problem and interpreted a negative number as representing a financial loss.

Fibonacci, the son of a merchant, was born about 1175 in Pisa, Italy. He studied while traveling throughout Northern Africa and countries bordering the eastern Mediterranean Sea. There he learned the positional system of notation, which we use today, with its concept of "zero." His fellow merchants found it difficult to accept the idea that "zero" was something that could stand for nothing. However, gradually they saw the advantages of writing a number such as 1202 in that form, rather than in the Roman numeral form of MCCII.

Real Numbers

The only numbers we are going to discuss in this book are real numbers, so everywhere in this book the word *number* means "real number." Sometimes these numbers are referred to as "the reals."

The term *system of real numbers* refers not only to the numbers themselves, but also to the arithmetical properties of addition, subtraction, multiplication, and division, all of which allow us to operate on numbers to create new numbers.

The system of real numbers is founded on the *integers:*

$$\cdots, -4, -3, -2, -1, 0, 1, 2, 3, 4, \cdots$$

Real numbers that can be expressed as ratios of integers, such as $-5/2$ or $3/4$, are called *rational numbers.* A rational number expressed in decimal form either terminates or produces a repeating block of digits. Furthermore, any terminating decimal or repeating decimal can be expressed as the quotient of two integers. For example, both the terminating decimal 0.75 and the repeating decimal $0.121212\cdots$, are rational numbers. The first equals $3/4$ and the second equals $4/33$.

In addition, there are real numbers that are not rational numbers. For example, the number represented by $\sqrt{2}$ is an *irrational number.* Calculators may lead us to believe that all numbers are decimal numbers that terminate. For example, by entering $\boxed{2}$ and then pressing the $\boxed{\sqrt{}}$ key on a calculator, we may be tempted to write $\sqrt{2} = 1.414213562$. However, $\sqrt{2}$ is not exactly equal to 1.414213562. The calculator is simply rounding off the value of $\sqrt{2}$ to nine decimal places.

There are situations in which we need an exact value and other situations for which approximations are satisfactory. In fact, many times the approximate solution is more useful. For example, when asked whether $\sqrt{2}$ or 13/9 is larger, most people decide by using decimal approximations for each. Furthermore, the world beyond the classroom almost always uses decimal approximations.

Numbers that are quite large or quite small are expressed in *scientific notation.* On a calculator, part of the display of the number will be the power of 10 by which the displayed value should be multiplied. For instance, check that on your calculator the number 1/1996 is displayed as $\boxed{5.01002^{-04}}$ or $\boxed{5.01002\,\text{E}-04}$ or something equivalent. This display means the number is approximately $5.01002 \cdot 10^{-4}$ $= 0.000501002$.

In this book, we usually use four places to the right of the decimal point as our approximation. We say that π is about 3.1416, written as $\pi \simeq 3.1416$, or that $\sqrt{2}$ is about 1.4142. If the number is expressed in scientific notation, then we use four places to the right of the displayed decimal point. This rule is only a guideline; some situations require more precision than others.

Using all of the digits shown by our calculator may give quite an incorrect impression. For example, suppose someone is told that a swimming pool is 25 meters long and they want to know how many feet that is. A standard formula used for the conversion of meters to feet is to multiply the number of meters by 3.281. Doing so would give a length of $25 \cdot 3.281 = 82.025$ feet. However, it is quite misleading to say the pool is 82.025 feet long, because to do so implies that we know the length to the nearest thousandth of a foot. Quite likely, the length in meters is only known to the nearest tenth, or maybe hundredth, of a meter.

When we are performing calculations, we should not use any approximation until we need to do so. For example, suppose we were to use 4.4 as an approximation for $\sqrt{19}$ and 1.7 as an approximation for $\sqrt{3}$. Because our calculator displays 4.4/1.7 as being 2.588235294, we might be tempted to use 2.6 as a one-place approximation for $\sqrt{19}/\sqrt{3}$. Actually, to find the correct one-place approximation for $\sqrt{19}/\sqrt{3}$ we should find 4.358898944 divided by 1.7320508, which the calculator displays as 2.516611478, and which we would round off to 2.5.

Often we wish to transform a number or answer into an equivalent form. For example, the process known as *rationalization* changes $1/\sqrt{2}$ into $\sqrt{2}/2$ by multiplying the numerator and denominator by $\sqrt{2}$.

$$\frac{1}{\sqrt{2}} \cdot \frac{\sqrt{2}}{\sqrt{2}} = \frac{\sqrt{2}}{2}$$

The knowledge and skills needed to transform different forms of mathematical expressions are precisely some of the most valuable aspects of mathematics, and we devote a considerable amount of effort to acquiring this knowledge. However, sometimes such transformations become distracting. We should always consider the time and effort involved in finding equivalent exact answers and whether such an effort is appropriate.

To illustrate the situation, consider the expressions $1/\sqrt{2}$ and $\sqrt{2}/2$. The form $\sqrt{2}/2$ was preferred at a time when long divisions were performed by hand because it was far easier to divide 1.4142 by 2 than to divide 1 by 1.4142. However, using calculators to obtain an approximation, we use *fewer* keystrokes to evaluate $1/\sqrt{2}$ than to evaluate $\sqrt{2}/2$. As a consequence, if a decimal approximation is desired, we should not transform $1/\sqrt{2}$ into $\sqrt{2}/2$ before finding the approximation.

The Real Number Line

A convenient representation of all of the real numbers is to associate each number with exactly one point on a horizontal straight line. To do this, we choose an arbitrary point and assign it to the number 0. That point is called the *origin* or the *zero-point.* Choose another point, to the right of the zero-point, and assign it to the number 1. Using the distance between the 0 and the 1 as our measure of one unit of distance, we assign all the points to the right of 0 to the positive numbers and all the points to the left to the negative numbers. A line so marked is a *real number line* (figure 1.1.1), and the numbers are the *coordinates* of the points.

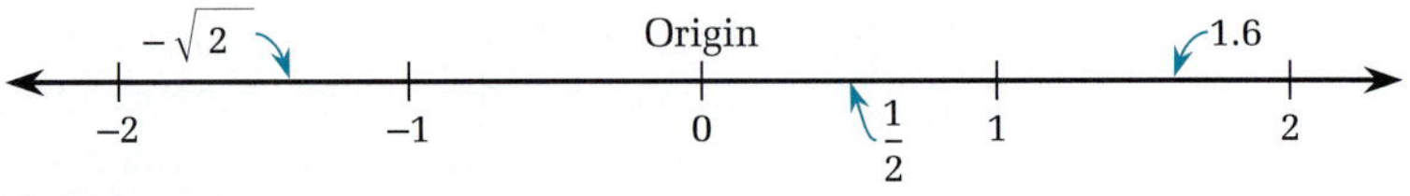

FIGURE 1.1.1

Inequalities

DEFINITION

Greater than and less than

If the point on the line that corresponds to the number a is to the left of the point that corresponds to the number b, then we say a *is less than* b and write: $a < b$. If the point for a is to the right of the point for b, we say that a *is greater than* b and we write: $a > b$. We write $a \le b$ when a *is less than or equal to* b. Similarly, $a \ge b$ means a *is greater than or equal to* b.

Expressions such as $3 \leq 5$, $6 > 2$, and $-3 \leq -1$ are examples of *inequalities.* At our convenience, we write inequalities in either direction. For example, to express the relationship between 3 and 5 we may think of 3 as being less than 5 and so write $3 < 5$. On the other hand, it is equivalently true that 5 is greater than 3, and so we can write $5 > 3$.

Shown in figure 1.1.2 are $5 < 7$, $3 \leq 7$, $3 \geq -1$, and $4 \leq 4$.

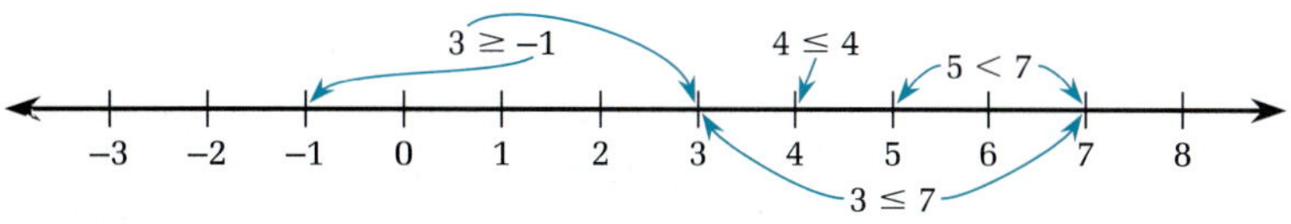

FIGURE 1.1.2

If we write a double inequality, such as $a < b \leq c$, we mean that *both* $a < b$ and $b \leq c$ have to be true. For example, $3 < x \leq 5$ means that x must be greater than 3 but no greater than 5.

WARNING

When writing double inequalities be careful with notation. For example, consider the double inequality $1 < x < -2$. This says that x is a number such that *both* $1 < x$ and $x < -2$ have to be true. There are no such values for x. If x was meant to be a real number that could be greater than 1 or could be less than -2, then one correct way of writing this would be $1 < x$ or $x < -2$.

The following properties of inequalities allow us to perform arithmetic with inequalities. Similar rules hold if "$<$" is replaced by "$\leq$" and "$>$" is replaced by "$\geq$."

INEQUALITY PROPERTIES

For any real numbers a, b, and c:

(i) If $a < b$, then $a + c < b + c$ and $a - c < b - c$

(ii) If $a < b$ and $c > 0$, then $ac < bc$ and $a/c < b/c$

(iii) If $a < b$ and $c < 0$, then $ac > bc$ and $a/c > b/c$

Notice that properties (ii) and (iii) state that the direction of an inequality upon multiplication or division is kept the same when multiplying or dividing by a positive number and the direction is reversed when multiplying or dividing by a negative number.

EXAMPLE 1   Inequality properties

As illustrations of inequality properties, consider the following examples. In each case we check the correctness of the resulting inequality by actually doing the arithmetic and comparing the resulting numbers. The particular numbers used are chosen as random representatives. Each example illustrates the inequality property with the corresponding number.

Property (i). From knowing $2 < 5$, we have $2 + 3 < 5 + 3$. As a check, we calculate $2 + 3 = 5$ and $5 + 3 = 8$. Indeed, $5 < 8$. From knowing $3 < 7$, we have $3 - 2 < 7 - 2$. Again, we can check by finding $3 - 2 = 1$ and $7 - 2 = 5$ and seeing $1 < 5$.

Property (ii). From $1 < 3$ and $2 > 0$, by multiplying we have $1 \cdot 2 < 3 \cdot 2$ (checking, $2 < 6$) and by dividing we have $1/2 < 3/2$ (checking, $0.5 < 1.5$).

Property (iii). Because $2 < 4$ and $-3 < 0$, we have $2(-3) > 4(-3)$ (checking, $-6 > -12$).

We may think geometrically about any of these examples. For instance, in figure 1.1.3 we can imagine 2 being to the left of 4. Upon multiplication by -3 we have $2(-3) = -6$ and $4(-3) = -12$. The point for -6 is to the right of -12, so $-6 > -12$.

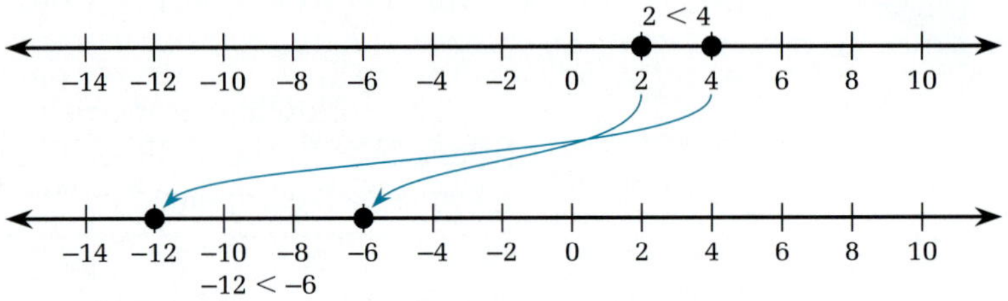

FIGURE 1.1.3

Absolute Value

Sometimes we only want, or need, to know the size of a number. This size is determined by the distance between its associated point and the zero point and it is the *absolute value* of the number.

DEFINITION

The absolute value

The *absolute value* of a number a is written as $|a|$ and is equal to a if a is positive or zero and is equal to $-a$ when a is negative. In symbols,

$$|a| = \begin{cases} a \text{ if } a \geq 0 \\ -a \text{ if } a < 0 \end{cases}$$

The second part of the definition, which says that $|a|$ is equal to $-a$ when a is negative, may seem difficult to understand because we always want the absolute value to be positive. But keep in mind that if we are finding the absolute value of a negative number such as -3.5, the a in the definition is the entire negative number, -3.5, not just the 3.5.

Hence, with $a = -3.5$, then $|a| = -a$ gives us $-(a) = -(-3.5) = 3.5$.

EXAMPLE 2   Absolute values

Evaluate $|3|$, $|5.4|$, $|-3|$, $|-0.7|$, and $|x - 2|$.

SOLUTION

$|3| = 3$, $|5.4| = 5.4$, $|-3| = -(-3) = 3$, and $|-0.7| = 0.7$. To evaluate the expression $|x - 2|$ we need to know more about x. If $x - 2 \geq 0$, which is equivalent to $x \geq 2$, then $|x - 2| = x - 2$. However, if $x - 2 < 0$, which is equivalent to $x < 2$, then $|x - 2| = -(x - 2) = -x + 2$. Graphically, we can represent this in figure 1.1.4.

$|x|$ is the distance from x to 0.

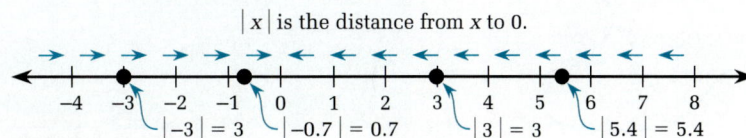

$|-3| = 3$ $|-0.7| = 0.7$ $|3| = 3$ $|5.4| = 5.4$

FIGURE 1.1.4

PRACTICE EXERCISE 1

Find the absolute values of the following: 5, $\sqrt{3}$, -2, -0.4, $x + 3$.

The following example is an application of the concept of absolute value that is of interest to economists and sociologists. It uses absolute values to determine a comparative distribution of two resources.

EXAMPLE 3   Application to economics

Coefficient of Concentration

Economists define the *coefficient of concentration, r,* to be equal to the sum of all the absolute differences between percentages, written decimally, of two resources.

For example, suppose there are three regions, A, B, and C, as shown in figure 1.1.5. Region A contains 25% of all the doctors and 40% of all the people; region B contains 5% of all the doctors and 35% of all the people; and region C contains 70% of all the doctors and 25% of all the people. The coefficient of concentration is:

$$r = |0.25 - 0.40| + |0.05 - 0.35| + |0.70 - 0.25|$$
$$= |-0.15| + |-0.30| + |0.45| = 0.9$$

A coefficient of concentration that is near zero means that the resources under comparison are distributed comparably equally. That is, continuing with the situation in Example 3, if r was near zero, then the doctors would be located where the people were.

The following properties of absolute values can be proved.

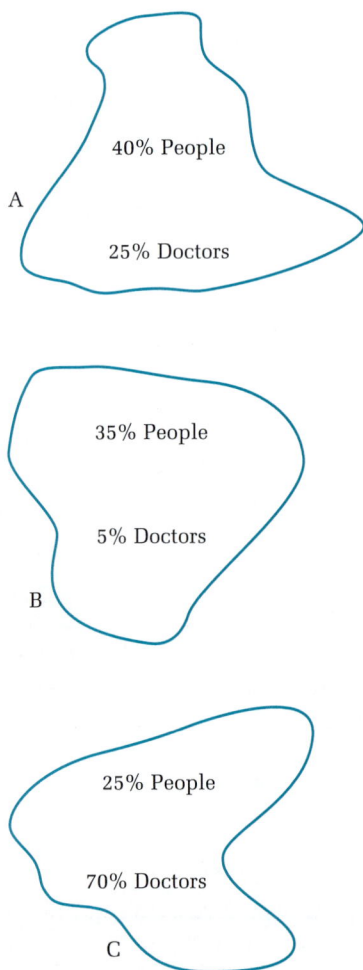

40% People

A

25% Doctors

35% People

5% Doctors

B

25% People

70% Doctors

C

FIGURE 1.1.5

PROPERTIES OF ABSOLUTE VALUES

For all real numbers a and b

(i) $|a| \geq 0$ **(iii)** $|a| = |-a|$

(ii) $|a \cdot b| = |a| \cdot |b|$ **(iv)** $|a + b| \leq |a| + |b|$

It is often convenient to change an inequality that involves an absolute value into an inequality that does not. For example, if we wish to find values of x for which $|x| < 6$, we may prefer to rewrite the condition to read: find values of x for which $-6 < x < 6$. These are equivalent conditions on the possible values for x.

The following are equivalent conditions on the possible values for the number x. (Remember, we are using the phrase "any number" to mean "any real number.")

EQUIVALENT CONDITIONS BETWEEN INEQUALITIES AND ABSOLUTE VALUES

For any number x and for a positive number b

(i) $|x| < b$ is equivalent to: $-b < x < b$

(ii) $|x| > b$ is equivalent to: $x > b$ or $x < -b$

(iii) $|x| = b$ is equivalent to: $x = b$ or $x = -b$

Similar results can be stated for $|x| \leq b$ and $|x| \geq b$ for any $b \geq 0$.

WARNING

Remember that the expression $-b < x < b$ means that both $-b < x$ and $x < b$.

EXAMPLE 4   Equivalent conditions using inequalities and absolute values

$|x| > 7$ is equivalent to: $x > 7$ or $x < -7$.
$|x| \leq 4$ is equivalent to: $-4 \leq x \leq 4$.
$|x| = 5$ is equivalent to: $x = -5$ or $x = 5$.
If $x > 2$ or $x < -2$, then $|x| > 2$.
If $-1.5 \leq x \leq 1.5$, then $|x| \leq 1.5$.

The next definition is an application of absolute values.

DEFINITION   Distance between points

Suppose A and B are two points on the line in figure 1.1.6 and have coordinates a and b, respectively. The *distance between A and B*, written as $d(A, B)$, is given by

$$d(A, B) = |b - a|$$

The distance is demonstrated on the real number line in figure 1.1.6.

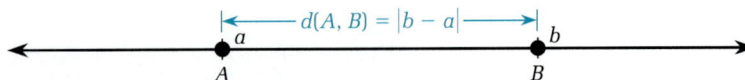

FIGURE 1.1.6

EXAMPLE 5   Distance between points

In figure 1.1.7 we see the distance between the points that correspond to 1 and 5 is $|5 - 1| = |4| = 4$

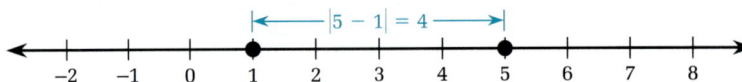

FIGURE 1.1.7

Notice the distance between the points 1 and 5 is the same as the distance between the points 5 and 1. This is because $|5 - 1| = |4| = 4$ and $|1 - 5| = |-4| = 4$. Thus, the use of absolute values to measure distance makes the order of the points unimportant.

EXAMPLE 6   Distance between points

Suppose A, B, C, and O (which stands for the origin) have coordinates -2, 2, 5, and 0, respectively. Find $d(A, B)$, $d(A, C)$, $d(B, C)$, $d(C, B)$, $d(A, O)$ and $d(B, O)$.

SOLUTION

The points are represented on figure 1.1.8.

$$d(A, B) = |2 - (-2)| = |2 + 2| = |4| = 4$$
$$d(A, C) = |5 - (-2)| = |5 + 2| = |7| = 7$$
$$d(B, C) = |5 - 2| = |3| = 3$$
$$d(C, B) = |2 - 5| = |-3| = 3$$
$$d(A, O) = |0 - (-2)| = |2| = 2$$
$$d(B, O) = |0 - 2| = |-2| = 2$$

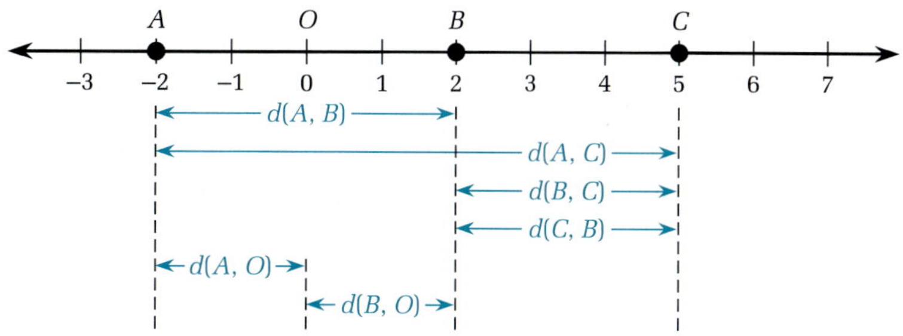

FIGURE 1.1.8

PRACTICE EXERCISE 2

Suppose *A, B,* and *O* have coordinates -1, 3, and 0, respectively. Find $d(A, B)$, $d(A, O)$, $d(B, A)$ and $d(O, B)$.

Sets

Sets are collections of objects. A set is specified by listing everything in it or by stating some property that lets us decide whether some particular object is in the set or not. A member of a set is said to be *an element of* that set.

One way of writing a set is to use the brace symbols, { and }. For example, the set of the first ten letters in the alphabet can be indicated by writing {a, b, c, d, e, f, g, h, i, j}.

A listing of all members of a set is impossible if the set has an infinite number of members, such as the set of all positive integers. For large sets or sets whose members are unknown, we can write something about members of the set. That is, we may write {$x : x$ satisfies some property}. In this expression, the colon (:) is read as "such that." For instance, the set {$x : x$ subscribes to *Newsweek* magazine} is read as "the set of all x such that x subscribes to *Newsweek* magazine." We certainly do not speak that way in ordinary conversation; instead we talk about "the set of all subscribers of *Newsweek*" or just "all *Newsweek* subscribers." Nonetheless, to avoid possible confusion, there does exist a standard form that uses a symbol as a variable to represent the potential members of the set and then describes what must be true about that variable for actual membership.

As another example, the set of all numbers greater than 1 can be written as {$x : x > 1$}. This expression is read as "the set of all numbers x such that x is greater than 1."

Just as some numbers are represented by special symbols, such as π, so are some sets. The set of all real numbers is represented by the special symbol $\mathbb{R}$. Frequently this set is called the set of all *reals*.

DEFINITION   Equality for sets, subset of a set

Two *sets are equal* if they contain exactly the same elements.
One set is a *subset* of another set if everything in the first set is in the second set.

EXAMPLE 7   Sets and subsets

The set of all integers is a subset of the set of all real numbers. The set of all readers of *Newsweek* magazine is a subset of the set of readers of all magazines.

PRACTICE EXERCISE 3

(a) Write in set notation the set of all numbers x such that x is greater than 5.

(b) Describe in words the set $\{x : -2 \leq x\}$.

(c) List the elements of the set of even positive integers less than 9.

Intervals

The *graph* of any set of real numbers consists of all the points on the real number line that correspond to the numbers of the set.

If $a < b$, then the set $\{x : a < x < b\}$ of all numbers between a and b is called an *open interval* and is represented by (a, b). The values of a and b are the *endpoints* of the interval.

The term "interval" is often used to mean the "graph of an interval" and we will use this term where there can be no confusion.

EXAMPLE 8   Graphs of open intervals

Figure 1.1.9 contains the graphs of $(1, 3)$, $(-1, 2)$, and a general open interval (a, b). To indicate that the endpoints are not included in these graphs, an open circle is placed at each end of the line segment.

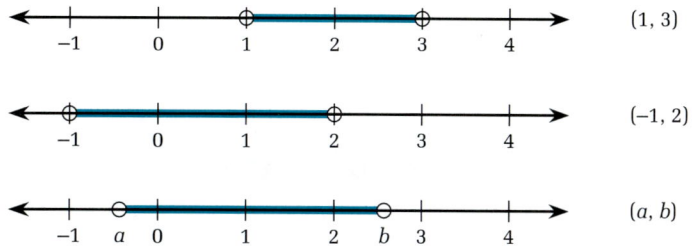

(1, 3)

(−1, 2)

(a, b)

FIGURE 1.1.9

PRACTICE EXERCISE 4

Draw the graphs for (−2, 1) and (1, 4).

If the endpoints of an interval are to be included, we have the set $\{x : a \leq x \leq b\}$, which is a *closed interval* and is represented as $[a, b]$.

In figure 1.1.10 of the following example, a solid dot is placed at the ends of the graph for each interval to emphasize that the endpoint is included.

EXAMPLE 9   Graphs of closed intervals

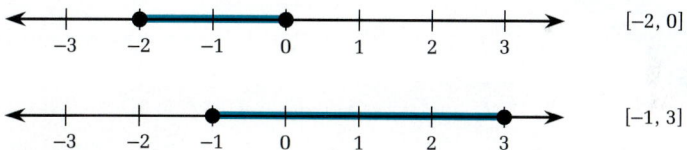

[−2, 0]

[−1, 3]

FIGURE 1.1.10

Not all intervals are either open or closed. There are other possibilities, such as including just one of the endpoints. Furthermore, an interval could be unbounded in either a positive or negative direction, or both.

EXAMPLE 10   Intervals that are neither open nor closed

The interval [−1, 2) includes only one endpoint because −1 is included but 2 is not. The graph is shown in figure 1.1.11.

[−1, 2)

FIGURE 1.1.11

The graph of (2, 5] is shown in figure 1.1.12.

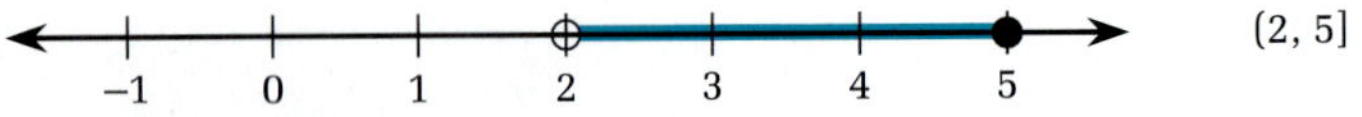

(2, 5]

FIGURE 1.1.12

The interval $[1, \infty)$ is the set $\{x : 1 \leq x\}$. This is an unbounded interval, which we graph as shown in figure 1.1.13.

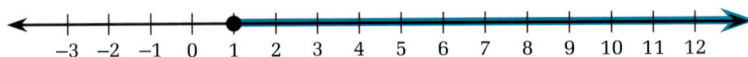

FIGURE 1.1.13

The set $\{x : x < 3\}$ is represented as $(-\infty, 3)$ and graphed as shown in figure 1.1.14.

FIGURE 1.1.14

WARNING

The symbols "∞" and "$-\infty$" are not real numbers. They are only a notation to tell us the interval does not end. We should never assert ∞ or $-\infty$ is a part of an interval. In particular, do *not* write expressions such as $[1, \infty]$ or $[-\infty, 5]$.

PRACTICE EXERCISE 5

Graph the intervals

(a) $[2, 5]$ (b) $[-2, 1)$ (c) $(3, \infty)$ (d) $(-\infty, 1)$

Solutions of Inequalities

Consider the expressions $3x \geq 6$ or $x^2 - 5 < x + 8$. These inequalities involve variables and so their truth depends on the values of x. The values of x for which a particular inequality is true are the *solution set* of the inequality.

For instance, the solution set for the inequality $3x \geq 6$ is $\{x : x \geq 2\}$, which is the interval $[2, \infty)$ shown in figure 1.1.15.

FIGURE 1.1.15

Generally, solving an inequality consists of finding sets that are equal to the original set, but are each described in a progressively simpler, more explicit fashion. In our example, the set {x : 3x ≥ 6} is equal to the set {x : x ≥ 2}. This follows from properties about inequalities, which allow us to start with $3x \geq 6$ and divide each side by the positive number 3 to obtain the equivalent inequality $x \geq 2$.

When we find equivalent inequalities we need to keep in mind that addition or subtraction of the same value on each side always preserves the direction of the inequality. Multiplication or division of both sides by a positive number also preserves the direction of the inequality. However, multiplication or division of both sides by a negative number reverses the direction of the inequality. In determining a solution, often the greatest difficulty arises because we are dividing or multiplying by a variable without considering the sign of that variable.

EXAMPLE 11   Solution of an inequality and its graph

Solve the inequality $3x - 2 \leq x + 4$ and represent the solution graphically.

SOLUTION

From $3x - 2 \leq x + 4$ we can add 2 to each side to have

$3x \leq x + 6$ Subtracting x gives

$2x \leq 6$ Finally, dividing each side by 2 yields

$x \leq 3$ Thus the solution set is $(-\infty, 3]$, shown in figure 1.1.16.

FIGURE 1.1.16

In the next example we have a double inequality. When such occurs, we need to be certain that the solution consists of values for x that satisfy both of the inequalities. This can be done by working on each inequality separately or by simultaneously solving both of them, as is done in the example.

EXAMPLE 12   Solution of a double inequality and its graph

Solve the inequality $-2 < -3x + 1 \leq 9$ and represent the solution graphically.

SOLUTION

Subtracting 1 throughout gives us $-2 - 1 < (-3x + 1) - 1 \leq 9 - 1$, so we have

$-3 < -3x \leq 8$ Dividing by -3 reverses the directions of the inequalities to produce

$1 > x \geq -\dfrac{8}{3}$ This interval can be represented by $\left[-\dfrac{8}{3}, 1 \right)$

as shown in figure 1.1.17.

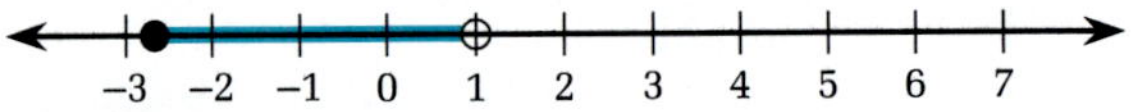

FIGURE 1.1.17

An inequality may be expressed in terms of an absolute value, as is done in the next example.

EXAMPLE 13   Solution of an inequality given by an absolute value

Solve the inequality $|2x - 3| < 4$ and represent the solution graphically.

SOLUTION

We saw that $|2x - 3| < 4$ was equivalent to $-4 < 2x - 3 < 4$, so by adding 3 throughout we have $-4 + 3 < (2x - 3) + 3 < 4 + 3$, which gives us

$-1 < 2x < 7$ Dividing by 2 we have

$-\dfrac{1}{2} < x < \dfrac{7}{2}$ which can be represented by $\left(-\dfrac{1}{2}, \dfrac{7}{2} \right)$

as shown in figure 1.1.18

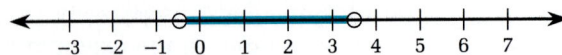

FIGURE 1.1.18

Notice in this last example that we could have transformed the inequality

$$-4 < 2x - 3 < 4 \text{ into}$$

$$-2 < x - \frac{3}{2} < 2, \text{ which is equivalent to } \left| x - \frac{3}{2} \right| < 2.$$

This last form, which can be interpreted as requiring x to be within 2 of 3/2, is one of the most common ways we encounter the need to solve inequalities. The usual formulation of this type of problem is in words, such as is shown in the following example.

EXAMPLE 14   An interval defined by "within"

Find the interval for which x is within 0.5 of 3.3.

SOLUTION

We have $|x - 3.3| < 0.5$. This is equivalent to $-0.5 < x - 3.3 < 0.5$. Adding 3.3 throughout gives $2.8 < x < 3.8$, which is the interval $(2.8, 3.8)$ and is shown in figure 1.1.19.

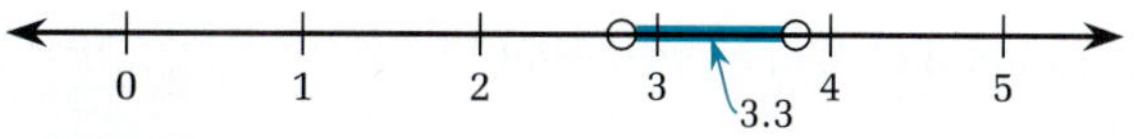

FIGURE 1.1.19

PRACTICE EXERCISE 6

Solve the following inequalities and represent the solutions graphically:

(a) $2x + 3 < 5x - 2$ (b) $|x - 1.5| < 2$

1.1 PROBLEMS

Foundations

The exercises of this section require the basic skills illustrated by Problems 1–5. Evaluate the following.

1. $-4 + \dfrac{22}{3}$ **2.** $\dfrac{5}{16} - \dfrac{3}{4}$ **3.** $7x - 3x$

4. Find the decimal equivalent of 60%.

5. What percentage is 21.7 of the sum of 21.7, 21.5, and 56.8?

Exercises

Evaluate each of the following absolute values by rewriting them so as to eliminate the absolute value sign.

6. (a) $|3|$ (b) $\left| \dfrac{2}{3} \right|$ (c) $|3 + 5|$ (d) $|2.3|$

(e) $|2 - 4|$ (f) $\left| -4 + \dfrac{22}{3} \right|$ (g) $\left| \sqrt{2} - \dfrac{13}{9} \right|$

(h) $|-2a + 5a|$ (for $a > 0$)

7. (a) $|2|$ (b) $\left|\dfrac{2}{5}\right|$ (c) $|1 + 3|$ (d) $|1 - 4|$

(e) $|-5|$ (f) $\left|\dfrac{5}{16} - \dfrac{3}{4}\right|$ (g) $\left|\sqrt{3} - \dfrac{17}{10}\right|$

(h) $|a - 3a|$ (for $a < 0$)

8. Use interval notation to express the set $\{x : 2 < x \le 5\}$.

9. Use interval notation to express the set $\{u : -1 < u < 3\}$.

10. Use interval notation to express the set $\{x : x \le 2\}$.

11. Use interval notation to express the set $\{x : x > -2\}$.

12. Express in words the set $\{x : x = 3\}$.

13. Express in words the set $\{x : x = 5 \text{ or } x = -2\}$.

14. Express in set notation the set of countries in North America.

15. Express in set notation the set of all positive odd integers less than 10.

16. Express in set notation the set of all readers of *Time* magazine.

17. Express in set notation the set of all real numbers greater than or equal to 8.

18. Suppose A, B, and C are points with coordinates -1, 5, and 2, respectively. Find each distance.
(a) $d(A, B)$ (b) $d(A, C)$ (c) $d(B, C)$
(d) $d(C, A)$

19. Suppose A, B, and C are points with coordinates 2, -1, and 4, respectively. Find each distance.
(a) $d(A, B)$ (b) $d(A, C)$ (c) $d(B, C)$
(d) $d(C, A)$

Graph the following intervals and state whether the interval is open, closed, or neither.

20. $(-2, 5)$ **21.** $(2, 3)$ **22.** $[-1, 2]$

23. $[-2, 1]$ **24.** $(0, 3)$ **25.** $(-1, 2)$

26. $(-2, 3]$ **27.** $(-3, -2]$ **28.** $(2, \infty)$

29. $(-1, \infty)$ **30.** $[0.5, \infty)$ **31.** $[1.5, \infty)$

32. $(-\infty, 1]$ **33.** $(-\infty, -2]$

Find the solution set of each inequality and express the set in terms of intervals.

34. $2x + 1 > 5$ **35.** $2x + 3 > 2$

36. $3x \le 5 - x$ **37.** $2x \le 1 + x$

38. $7x > 3x + 1$ **39.** $3x > -x - 3$

40. $-1 < 2x < 5$ **41.** $2 \le 2x \le 7$

42. $-2 \le 2x - 1 \le 3$ **43.** $-1 \le 3x + 2 \le 2$

44. $|x - 2| < 2$ **45.** $|x - 1| \le 3$

46. $|2x - 8| < 7$ **47.** $|3x - 1| < 5$

48. $|x - 2| \le -1$ **49.** $|3x - 1| < -2$

50. $|8 - 2x| \le 1$ **51.** $|2 - 3x| \le 2$

52. Find the interval for which x is within 1.5 of 3.75.

53. Find the interval for which x is within 0.8 of 2.3.

54. *(Cardiovascular Fitness)* To gain cardiovascular benefit from exercise, the American College of Sports Medicine recommends individuals should maintain a target heartbeat rate within a range of 60% to 90% of their maximum heart rate. This maximum heart rate is found by subtracting the individual's age from 220.*
(a) Find the range of target heartbeat rates for 20 year olds.
(b) Those just starting an exercise program should maintain a heartbeat rate that is approximately in the range of 60% to 75% of their maximum heart rate. Find a range of heartbeat rates for such a 50-year-old person.

55. *(Coefficient of Concentration)* Suppose there are three countries producing speedboats, and the percentages of production and population are shown in the following chart. Find the coefficient of concentration.

COUNTRY	SPEEDBOAT PRODUCTION	POPULATION
Red	10%	20%
Green	50%	10%
Blue	40%	70%

56. *(Coefficient of Concentration)* See the table at the top of the following page. Suppose there are three factories producing trucks and cars. Find the coefficient of concentration.

Vitality Magazine, February 2, 1992, p. 20.

FACTORY	TRUCK PRODUCTION	TRUCK AND CAR PRODUCTION
Times Square	5%	25%
Central Park	15%	30%
Towers	80%	45%

57. *(Coefficient of Concentration)*   Suppose we consider the populations, the number of federal employees, and the number of workers employed in the manufacture of motor vehicles. If we do this for the states of California, Ohio, New York, and New Mexico, then by using the percentages based on the totals of these categories in just these four states we have the data shown in the table at the bottom of this page.
(a) Find the coefficient of concentration between population and the workers in the manufacture of motor vehicles.
(b) Find the coefficient of concentration between population and the federal workers.

Writing and Discussion Problems

58. What are the advantages of using positional notation rather than Roman numerals? In particular, discuss the process of multiplication using each notation. For example, what is the product of XXXVI (36) and XCI (91)? One of the reasons positional notation was disliked by merchants was the chance of making errors in locating or transposing digits. What would you respond to such an objection?

59. *(Coefficient of Concentration)*
(a) What is the possible range of values for the coefficient of concentration?
(b) Under what conditions is the coefficient of concentration 0?
(c) Under what conditions is the coefficient of concentration its maximum value?
(d) Choose three states of the United States and find both their populations and areas. Determine the coefficient of concentration.

60. The definition of $\sqrt{x^2}$ is $|x|$, rather than simply x.
(a) Show that this definition avoids the apparent paradox in the following: We know $3^2 = 9$, so $\sqrt{9} = 3$. However, we also know $(-3)^2 = 9$ and so $\sqrt{9} = -3$. Hence $-3 = 3$.
(b) Use the definition of $\sqrt{x^2}$ to give an alternate statement of the distance between two points, x_1 and x_2, which does not directly use absolute values.

Enrichment Problems

61. *(Coefficient of Concentration)*   The question "Are we getting our fair share?" is frequently discussed by comparing percentages of population with allocations of some sort. For instance, the chart on the following page gives the percentage of U.S. population in four geographic regions, together with the amount of Federal Research and Development money spent in each of the regions.

IN 1980	PERCENTAGE OF WORKERS IN MANUFACTURE OF MOTOR VEHICLES	PERCENTAGE OF FEDERAL WORKERS	PERCENTAGE OF TOTAL POPULATION
California	21.7	53.4	44.4
New Mexico	0.0	4.9	2.5
New York	21.5	26.8	32.9
Ohio	56.8	14.9	20.2

"Employment Hours and Earnings, States and Areas 1939–1982," U.S. Department of Labor Statistics, Jan. 1984, Bulletin 1370-17.

IN 1986	PERCENTAGE OF TOTAL POPULATION	FEDERAL R&D (IN MILLIONS OF DOLLARS)
Northeast	20.6	1,806
Midwest	24.4	908
South	34.4	7,020
West	20.6	2,785

Statistical Abstracts, 1990, p. 19 and p. 584.

Find the coefficient of concentration for this allocation and determine which region is getting an amount closest to its "fair share."

62. A car going x miles per hour is going $(22/15)x$ feet per second. Suppose a driver is keeping her speed within 5 miles per hour of 60 miles per hour. Measured in feet per second, how near to 88 feet per second is she keeping her speed?

63. One meter is equal to approximately 3.281 feet. Suppose a swimming pool was supposed to be 25 meters long. If its actual length is within 0.1 meter of 25 meters, give the possible variation on the number of feet for the pool's length.

64. Show that $|a + b| \leq |a| + |b|$ for any real numbers a and b by supplying reasons to go with the following argument.

$$(a + b)^2 = a^2 + 2ab + b^2$$
$$\leq |a|^2 + 2 \cdot |a| \cdot |b| + |b|^2$$
$$= (|a| + |b|)^2$$

Since $\sqrt{x^2} = |x|$ (as we saw earlier), and for positive values of r and s it is true that $|r + s| = |r| + |s|$, we have the final result.

65. Find values for a and b for which it can be that $|a + b| \neq |a| + |b|$.

SOLUTIONS TO PRACTICE EXERCISES

1. $5, \sqrt{3}, 2, 0.4, x + 3$ if $x \geq -3$ and $-(x + 3)$ if $x < -3$

2.

FIGURE 1.1.20

$$d(A, B) = |3 - (-1)| = |3 + 1| = |4| = 4$$
$$d(A, O) = |0 - (-1)| = |1| = 1$$
$$d(B, A) = |-1 - 3| = |-4| = 4$$
$$d(O, B) = |3 - 0| = |3| = 3$$

3. (a) The set can be written as $\{x : 5 < x\}$ or as $\{x : x > 5\}$.
  (b) The set of all numbers such that the number is greater than or equal to -2. (Another acceptable answer would be "the numbers greater than or equal to -2.")
  (c) $\{2, 4, 6, 8\}$

4.

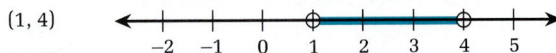

FIGURE 1.1.21

5.

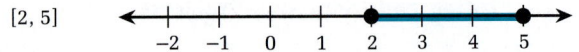

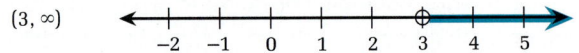
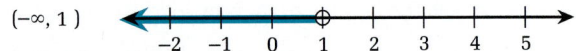

FIGURE 1.1.22

6. $2x + 3 < 5x - 2$ is equivalent to $5 < 3x$, which gives $5/3 < x$. Thus the interval is $(5/3, \infty)$ and the graph is shown in figure 1.1.23.

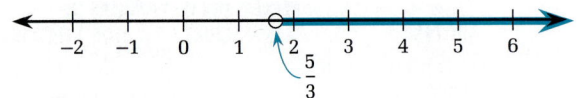

FIGURE 1.1.23

$|x - 1.5| < 2$ is equivalent to $-2 < x - 1.5 < 2$, so $-0.5 < x < 3.5$. The interval is $(-0.5, 3.5)$ and its graph is shown in figure 1.1.24.

FIGURE 1.1.24

FUNCTIONS AND GRAPHS

In everyday conversation we frequently say that one thing is a function of something else. A simple situation may involve only a function of one variable. For example, the cost of mailing a first-class letter within the United States is a function of only the weight of the letter. More complicated situations involve functions of several variables. As an example of a function of several variables, consider automobile insurance. The insurance premium is a function of many things, such as our age, the car's age, our driving record, where we register the car, and so on. In this chapter we study functions of just one variable. In a later chapter, we will study functions of several variables.

Functions

Consider hand-held calculators, which we can think of as being function machines. The typical hand-held calculator has four kinds of keys. There are the number keys, which we can press and cause a display of that number, such as key $\boxed{2}$ or $\boxed{5}$. The second kind of key is the arithmetic key. These include $\boxed{+}$, $\boxed{\times}$, and even $\boxed{x^y}$. These are used to perform operations on numbers. The third kind of key is the function key. Such keys as $\boxed{1/x}$, $\boxed{\pm}$ and $\boxed{\ln}$ will operate on whatever number is being displayed and produce a new number. Last are those keys that let us choose an alternate role for another key. For example, there may be a key that enables the calculator to display the value of π. However, to obtain $\boxed{3.14159265}$ as a display for $\boxed{\pi}$ may involve a key that is perhaps labeled $\boxed{\text{inv}}$ or $\boxed{\text{2nd}}$.

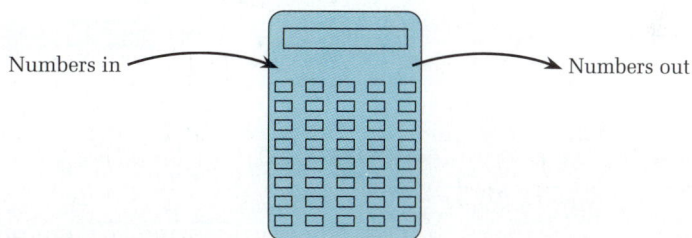

Numbers in ⟶   ⟶ Numbers out

FIGURE 1.2.1

In addition to the keys on an ordinary calculator, graphing calculators have keys such as $\boxed{\text{graph}}$ and $\boxed{\text{trace}}$, which perform other operations on numbers and expressions. However, let us only consider non-graphing calculators for the moment.

If we turn a calculator on, press the $\boxed{9}$ key, and then press the $\boxed{\sqrt{x}}$ key, we get $\boxed{3.0}$. The *square root function* key determines the square root of whatever number was displayed before the $\boxed{\sqrt{x}}$ key was pressed.

Two important things to notice are: (1) the $\boxed{\sqrt{x}}$ key cannot take a square root unless it has some number to work with, and (2) the

$\boxed{\sqrt{x}}$ may not work on some of the numbers we try. For example, if we enter -9 and then press the $\boxed{\sqrt{x}}$ key, our calculators will give us an error message of some sort because there is no real number whose square is negative.

On the left side of figure 1.2.2 is shown a set of numbers and on the right is shown the corresponding response of the calculator upon our pressing the $\boxed{\sqrt{x}}$ key. Only four digits to the right of the decimal point are displayed.

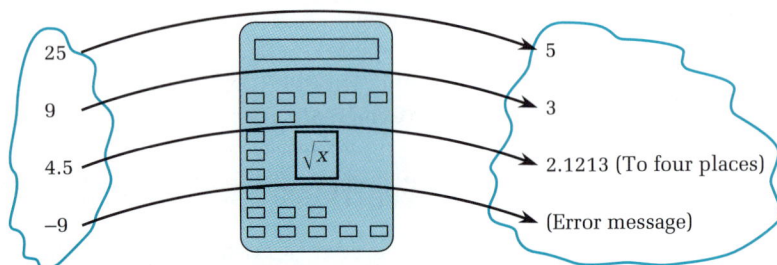

FIGURE 1.2.2

Another function key on our calculator is $\boxed{1/x}$. Its operation on several numbers is illustrated in figure 1.2.3.

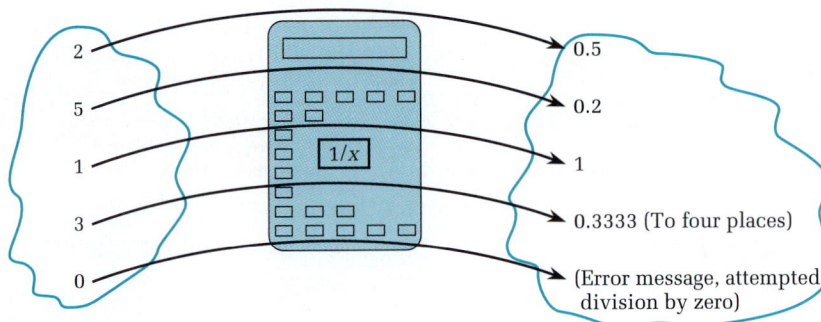

FIGURE 1.2.3

The keys $\boxed{\sqrt{x}}$ and $\boxed{1/x}$ are only two of the many function keys on a calculator. In addition to such standard functions built into the machine by the manufacturer, some calculators allow us to create special functions. Let us imagine that we build a calculator with many function keys. For instance, suppose we wish to take the value x as input and produce the value $2x + 1$ as output. We set up the calculator so that when a special function key is pressed, the desired operation is performed. If we call this the $\boxed{2x + 1}$ key, then we can watch its use in figure 1.2.4 on some sample inputs.

As we think of a function transforming an input into an output, we may be imagining a time relationship between input and output. In fact, it is very convenient to do so. However, functions need not have any cause and effect relationship over time. The only thing that

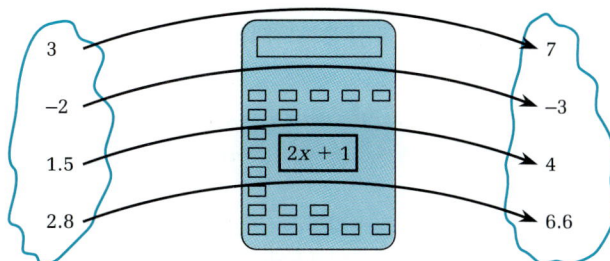

FIGURE 1.2.4

must happen is that for every element in the input set there corresponds just one element in the output set.

Although each input gives a unique output, it is quite possible that different inputs produce the same output. For instance, we know both 5^2 and $(-5)^2$ are 25. Thus, after the function key $\boxed{x^2}$ has been pressed, the same display of $\boxed{25}$ appears whether the previous display was $\boxed{5}$ or $\boxed{-5}$.

The set of all possible inputs is said to be the *domain* of the function, and the set of all of the outputs is said to be the *range* of the function.

A summary of our discussion follows.

DEFINITION

Function, domain, range

A *function* is a rule that assigns to each member of one set, called the *domain*, a unique member of a second set, called the *range*.

In figure 1.2.5 we have a schematic illustration of a function. The set A is the domain and B is the range. The symbol x represents a typical member of A and y is the element of B that corresponds to x.

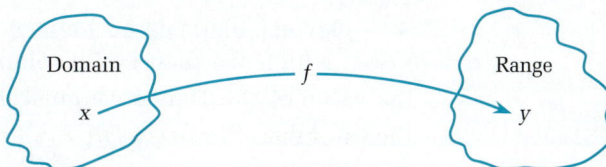

FIGURE 1.2.5

Let us represent the function by f. Then the expression $f(x)$ represents the value y, which the function associates with x. This is called the *"value of the function f at x"* or more simply *"f of x"* or *"f at x."*

WARNING

$f(x)$ is *not* "f times x." We are *not* multiplying two things together.

If we write the expression $y = f(x)$, we are thinking of x as an *independent variable* and y as a *dependent variable* because y depends on the value of x used as input.

Functions can be described by an algebraic formula, such as $f(x) = \sqrt{x}$ or $f(x) = x + 1$; or in words, such as "the area of a circle is the product of π times the square of the radius"; or in some graphical or tabular form, such as the sales tax tables used by retail businesses or the income tax tables.

In whatever form the function is described, for it truly to be a function a particular value from the domain (the input) must always produce the same value in the range set (the output) each time that domain value goes into the function.

The commonest forms of functions are algebraic formulas. These can be as simple as $f(x) = 2x$ or they can be quite complicated. To evaluate an algebraic formula for some particular value of the domain we replace the variable by that domain value. The domain value may be a number, but sometimes it is an expression. This next example shows evaluation.

EXAMPLE 1   Evaluating an algebraic function

Suppose $f(x) = \sqrt{x}$. Find the value of $f(x)$ at each of the input values 4, 5, 9, a, $a + 1$, and a^2.

SOLUTION

For this function, we may use our knowledge of the squares of numbers to find special values of $f(x)$. Since $2^2 = 4$, we have $f(4) = \sqrt{4} = \sqrt{2^2} = 2$ and similarly $f(9) = 3$. For numbers that are not known squares, we can use our calculator. Doing so, we find $f(5)$ to be $\sqrt{5} \approx 2.2361$.

Our calculator is no help in finding $\sqrt{a}$. There is no $\boxed{a}$ key to press, which we then could follow by pressing the $\boxed{\sqrt{x}}$ key to find the value of $\sqrt{a}$. Hence we must leave $f(a)$ as just $\sqrt{a}$, some unspecified number. Similarly $f(a + 1) = \sqrt{a + 1}$.

Finally, $f(a^2) = \sqrt{a^2} = |a|$. Notice that $\sqrt{a^2} = |a|$ instead of just a, because whether a is positive or negative, it is always true that $a^2 > 0$ and $\sqrt{a}$ is defined as the positive number whose square is a.

If we have a positive constant a and we consider the expression $\sqrt{a}$, we think of this as being some specific (but unspecified) number. In contrast, we think of the expression $\sqrt{x}$ as defining a function of x.

We look at $f(x) = ax + b$ and think of a and b as constants and x as a variable. The a and b are unspecified in the formula and so is x. However, even when a and b are given specific values, the x will not be fixed, or $f(x)$ would not be a function.

Both $f(x) = 2x + 3$ and $g(x) = 4x - 5$ are functions, but $2 \cdot 4 + 3$ and $4 \cdot 4 - 5$ are merely values.

EXAMPLE 2   Evaluating a function

Suppose $f(x) = 3x^2 + 4x - 1$. Find $f(2)$, $f(-1)$, and $f(a + 1)$.

SOLUTION

To find $f(2)$ substitute 2 for each occurrence of x in $f(x)$ to get

$$f(2) = 3 \cdot 2^2 + 4 \cdot 2 - 1.$$

This is $f(2)$, but it is not a usable form of $f(2)$. We want as simplified a form of $f(2)$ as we can find. Thus, we go on to find

$$f(2) = 3 \cdot 4 + 8 - 1 = 12 + 8 - 1 = 19.$$

Similarly,

$$f(-1) = 3(-1)^2 + 4 \cdot (-1) - 1$$
$$= 3 \cdot 1 - 4 - 1 = 3 - 4 - 1 = -2.$$
$$f(a + 1) = 3(a + 1)^2 + 4(a + 1) - 1$$
$$= 3(a^2 + 2a + 1) + 4a + 4 - 1$$
$$= 3a^2 + 6a + 3 + 4a + 3 = 3a^2 + 10a + 6.$$

PRACTICE EXERCISE 1

For $f(x) = x^2 - 2x + 1$, find $f(-1)$, $f(2)$, $f(3)$, $f(3 + a)$.

In calculus, it is very important to be able to evaluate $[f(x + h) - f(x)]/h$ (for $h \neq 0$) and to find a simplified form for this expression. The next example illustrates the process.

EXAMPLE 3   Simplifying $\dfrac{f(x + h) - f(x)}{h}$

Suppose $f(x) = 3x^2 + 4x - 1$. Find a simplified form for $[f(x + h) - f(x)]/h$, $h \neq 0$.

SOLUTION

By substituting $x + h$ for the input variable, we have $f(x + h) = 3(x + h)^2 + 4(x + h) - 1$ so that

$$f(x + h) - f(x) = [3(x + h)^2 + 4(x + h) - 1] - [3x^2 + 4x - 1]$$

Multiplying this out gives us

$$= 3x^2 + 6xh + 3h^2 + 4x + 4h - 1 - 3x^2 - 4x + 1$$
$$= 6xh + 3h^2 + 4h \text{ by collecting together similar terms.}$$

Each of these terms has an h in it, so the h can be factored out to give

$$= (6x + 3h + 4)h.$$

Because $h \neq 0$, we may cancel h out of the numerator and denominator and find a simplified expression for

$$\frac{f(x + h) - f(x)}{h} \quad \text{to be} \quad \frac{(6x + 3h + 4)h}{h} = 6x + 3h + 4$$

Notice in the last example we said we were substituting $x + h$ for the input variable. We did *not* say we were letting $x + h = x$.

Piecewise Functions

A function may be defined using different rules over different parts of its domain. We have already seen one such function, the absolute value function. If $f(x) = |x|$, then $f(x) = x$ if $x \geq 0$ and $f(x) = -x$ if $x < 0$.

EXAMPLE 4   A function defined piecewise

Suppose $f(x) = \begin{cases} x^2 + 1 & \text{if } x \geq 1 \\ -x + 2 & \text{if } x < 1 \end{cases}$. Find $f(2)$, $f(0)$, and $f(-1)$.

SOLUTION

Because $2 \geq 1$, we find $f(2)$ using the first part of the definition. Hence, $f(2) = 2^2 + 1 = 5$. Because $0 < 1$, we find $f(0)$ using the second part of the definition. With this, $f(0) = -0 + 2 = 2$. Similarly, $f(-1) = -(-1) + 2 = 1 + 2 = 3$.

So far we have used only "x" as a variable name, but we may use any convenient name for a variable.

The functions $f(x) = \sqrt{x + 1}$ and $f(u) = \sqrt{u + 1}$ are the same function, provided they have the same domain. The x used in the first function is a place-holder, merely a dummy variable, to stand for domain numbers. In the second function, u is also used as a place-holder. However, in either description of the function, $f(3) = \sqrt{3 + 1} = \sqrt{4} = 2$ and $f(-1) = \sqrt{-1 + 1} = \sqrt{0} = 0$.

In fact, $f(x) = \sqrt{x + 1}$ and $g(u) = \sqrt{u + 1}$ are the same function because both produce the same output from the same particular input. We have achieved a truly important success along our road of understanding functions when we realize these represent the same function. The function is the relationship between numbers in two sets.

The use of a variable in defining a function is similar to the use of a variable to describe a set. The set $\{s : s > 1\}$ is the same set as $\{x : x > 1\}$. Both sets describe the same real numbers. The "s" in the first set and the "x" in the second are merely place-holding names.

MATHEMATICAL INNOVATORS
René Descartes (1596–1650)

We use letters near the end of the alphabet, such as *x, y,* and *z,* for variables, and we use letters near the front of the alphabet for constants. This mathematical convention dates from the writings of René Descartes. His book *Discours de la Méthode,* published in 1637, announced to the world the concept of analytic geometry, a subject that weds geometry to algebra and is sometimes called "Cartesian geometry" in his honor. The development of calculus depended on the existence of analytic geometry. Outside of the world of mathematics, Descartes is best known for the phrase, "I think, therefore I am."

Finding Domains

Suppose a function is described by an algebraic formula. Sometimes its domain is given explicitly. However, if no domain is specified, then we assume that the domain is the largest subset (of the real numbers) that can be used as input values for the function. Recall that the entire set of real numbers is abbreviated by the symbol $\mathbb{R}$.

EXAMPLE 5   Finding domains

Find the largest possible domain for each of the following functions.

(a) $f(x) = \sqrt{x - 2}$ (b) $g(w) = \dfrac{1}{w}$ (c) $h(s) = \dfrac{1}{\sqrt{s - 1}}$

SOLUTION

(a) The square root is defined only for non-negative numbers; therefore, for $f(x)$ to exist we need to have $x - 2 \geq 0$. This is equivalent to needing $x \geq 2$, so the domain is the interval $[2, \infty)$.

(b) Division by zero is not possible, so the domain of $g(w)$ is the set of all real numbers except 0. This can be written as $\{w : w \neq 0\}$.

(c) This function combines the types of restrictions found in the preceding two functions: $s - 1$ has to be non-negative for the square root to exist, and $s - 1 \neq 0$ to avoid division by zero. Combining these restrictions, we need to insist that $s - 1 > 0$. The domain of $h(s)$ is therefore $\{s : s > 1\}$. This is the interval $(1, \infty)$.

These last functions illustrate two very common restrictions on domains. Square roots must be taken only of non-negative numbers, and division by zero is not possible. In fact, until we become familiar with a greater variety of functions, we will encounter only these two restrictions on unspecified domains.

PRACTICE EXERCISE 2

Find the largest possible domain for each of the following:

(a) $f(x) = 2x + 1$ (b) $g(y) = \dfrac{1}{(y - 1)}$ (c) $h(t) = \sqrt{2t - 1}$

Graphs

In order to understand a function, it is helpful to have a graphical representation of it. To do this, we associate points on the plane with values of the domain and range.

Locating points on a plane is achieved by placing two number lines at right angles to each other and identifying each point on the plane with an ordered pair of numbers. Usually, the horizontal number line is called the *x-axis* and the vertical number line is the *y-axis*. The ordered pair of numbers, (x, y), determines a unique point, $P(x, y)$, which is said to have the *coordinates* (x, y). The first coordinate is the *x-coordinate,* which gives the horizontal location, with increasingly positive values to the right. The second coordinate is the *y-coordinate,* which gives the vertical location, with increasingly positive values going up. The *origin* is assigned the point $P(0, 0)$.

These pairs of numbers are said to be *ordered pairs,* for indeed their order is important. The two points, $A(1, 3)$ and $B(3, 1)$, are shown in figure 1.2.6.

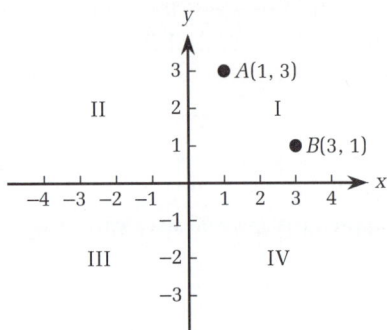

FIGURE 1.2.6

The two axes divide the plane into four *quadrants,* numbered with Roman numerals in a counterclockwise fashion, as illustrated in figure 1.2.6.

DEFINITION

Graph of a function

The *graph* of the function f is the set of all points $P(x, y)$ where x is any value of the domain and $y = f(x)$.

Usually, only a portion of a graph is displayed because whether we use a graphing calculator or are determining values of $f(x)$ ourselves and plotting the results, we are unable to use infinite domains because our sheet of paper has edges and our electronic screen has boundaries. Thus, even if the graph of a particular function is unbounded, we can display only some limited part of the graph.

DEFINITION

Screen of a graph

The displayed portion of a graph is the *screen of the graph.*

WARNING

The boundaries used by the graphing calculator are called "ranges" for both the x and y values. Keep in mind that this use of the word "range" for screen size is different from the use of the word "range" in connection with the outputs of a function.

The graphing of functions is a topic we will discuss many times in this book. This chapter presents an overview of a variety of functions. In Chapter 3 the topic of graphing is a focus for investigating those points on graphs that are of special interest—points of maximum or minimum values. At this stage of development, we want to avoid becoming mired in how we choose which points to graph. Hence, let us simply display a few screens of graphs and postpone any detailed discussion of how those screens were chosen.

EXAMPLE 6   Screens of graphs

Show the screens of the graph of $y = 3x^2 + 4x - 1$ determined by the following values of x and y:

(a) $0 \leq x \leq 1$ $\qquad -2 \leq y \leq 7$

(b) $-2 \leq x \leq 1$ $\qquad -3 \leq y \leq 7$

SOLUTION

Figures 1.2.7 and 1.2.8 show the screens.

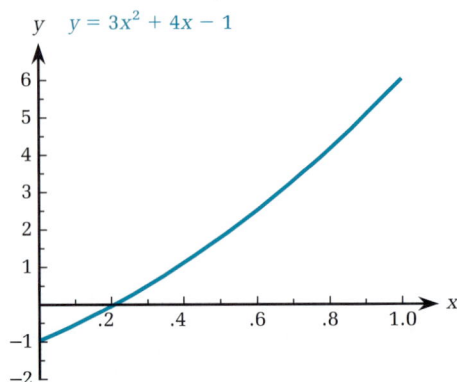

FIGURE 1.2.7

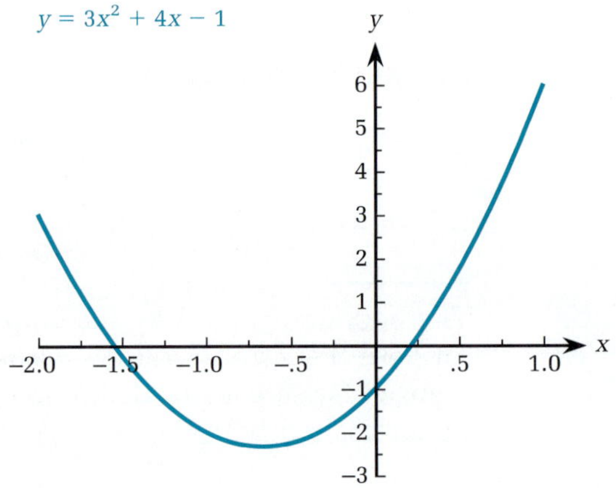

FIGURE 1.2.8

Notice how important it is to choose axes with values of x and y that provide an interesting screen of the graph of the function. It is possible for a screen to completely miss any of the graph, or scales can be so big or so small that important aspects of the graph are obscured or omitted altogether. A major goal of this book is to help us become better at understanding which points are essential to a good sketch. In fact, finding appropriate values is a major application of calculus.

Vertical Line Test

A vertical line will intersect the graph of a function only once, if at all. Geometrically, this is equivalent to saying that if x is some particular value of the domain $x = a$, then the vertical line with that x-coordinate will have only the one point $(a, f(a))$ in common with the graph. The graph in figure 1.2.9 shows a function. The graph in figure 1.2.10 is not a function.

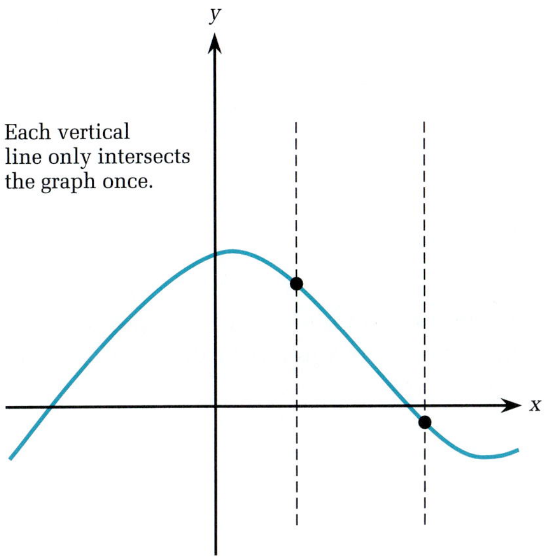
y
x
Each vertical
line only intersects
the graph once.

FIGURE 1.2.9

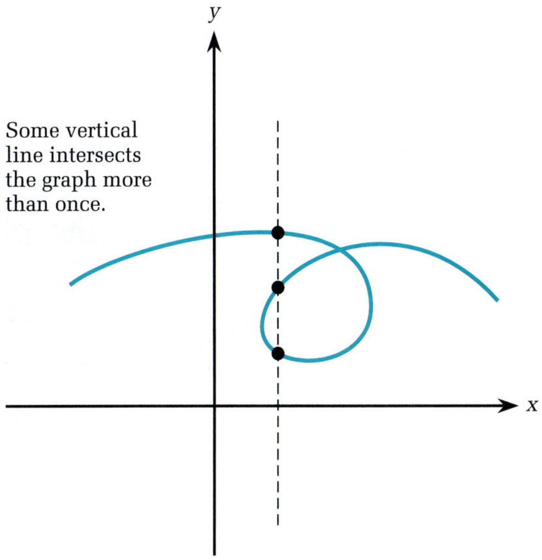
y
x
Some vertical
line intersects
the graph more
than once.

FIGURE 1.2.10

A second type of graph occurs when we simply plot available data and use the points themselves to determine the function. For example, the Dow Industrial Index has been kept on a regular basis since 1896. Figure 1.2.11 is a graphical representation of the Dow Index for the years 1987 through 1989. This graph determines a function, whereby we select the dates at the bottom as the domain and the values of the Dow on those dates as the value of the function for each date. Usually such graphing is done in the hope that the display of past data will yield a clue to future performance.

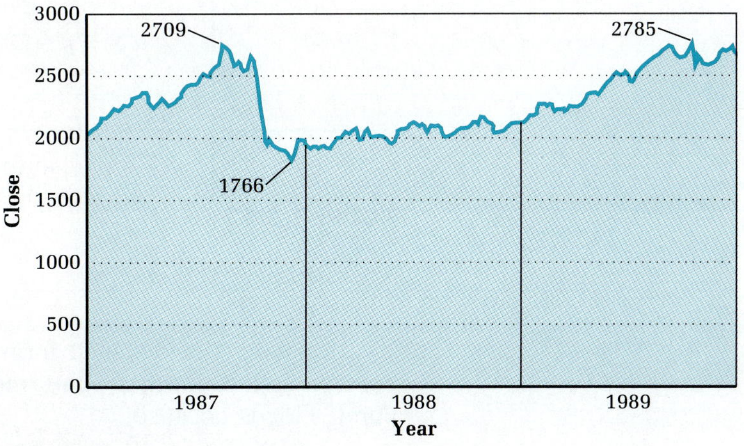
3000
2500
2000
1500
1000
500
0
Close
2709
2785
1766
1987
1988
1989
Year

FIGURE 1.2.11

Subscripted Variables

It might seem there are quite enough letters to stand for variables, but in fact it is convenient to introduce subscripting. For example, x_1 (read as "x sub 1") might stand for the x-coordinate of a first point under discussion, whereas y_1 would stand for the y-coordinate of the same point. As a result, we can represent two points by $P_1(x_1, y_1)$ and $P_2(x_2, y_2)$ and understand the coordinates are of a first point and a second point.

The Distance Formula

The following *distance formula* follows from the Pythagorean Theorem, which states that the square of the length of the hypotenuse of a right triangle is equal to the sum of the squares of the lengths of the other two sides.

DEFINITION   ## Distance between two points

The distance between the points $P_1(x_1, y_1)$ and $P_2(x_2, y_2)$ is given by $d(P_1, P_2) = \sqrt{(x_2 - x_1)^2 + (y_2 - y_1)^2}$ as shown in figure 1.2.12.

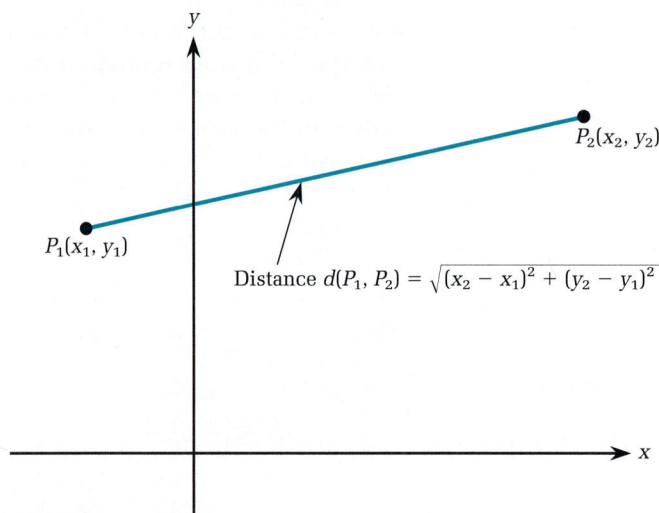

Distance $d(P_1, P_2) = \sqrt{(x_2 - x_1)^2 + (y_2 - y_1)^2}$

FIGURE 1.2.12

In using the distance formula, it makes no difference which point is called the first point and which point is called the second point. This is because

$$(x_2 - x_1)^2 = (x_1 - x_2)^2 \quad \text{and} \quad (y_2 - y_1)^2 = (y_1 - y_2)^2$$

EXAMPLE 7   Distance between points

Find the distance between $P_1(1, 3)$ and $P_2(5, 2)$.

SOLUTION

These points are shown in figure 1.2.13. The distance between the points is $d(P_1, P_2) = \sqrt{(5 - 1)^2 + (2 - 3)^2} = \sqrt{4^2 + (-1)^2} = \sqrt{17} \approx 4.1231$.

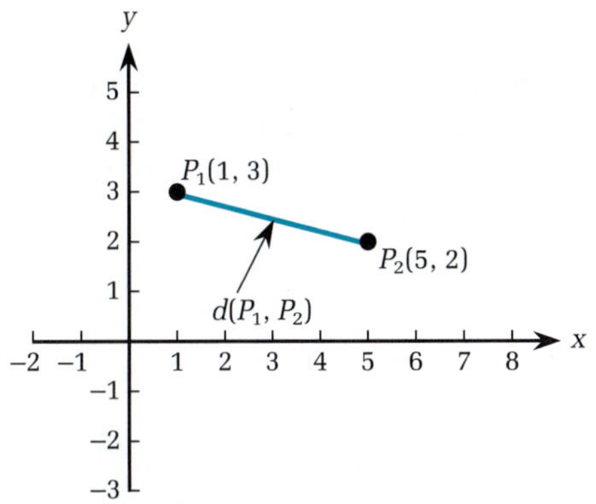

FIGURE 1.2.13

In the next example we use the distance formula to determine the distance between points on a graph.

EXAMPLE 8   Distance between points on a graph

Find the distance between the points on the graph of $f(x) = x^2 + 1$ determined by $x = -1$ and $x = 3$, as shown in figure 1.2.14.

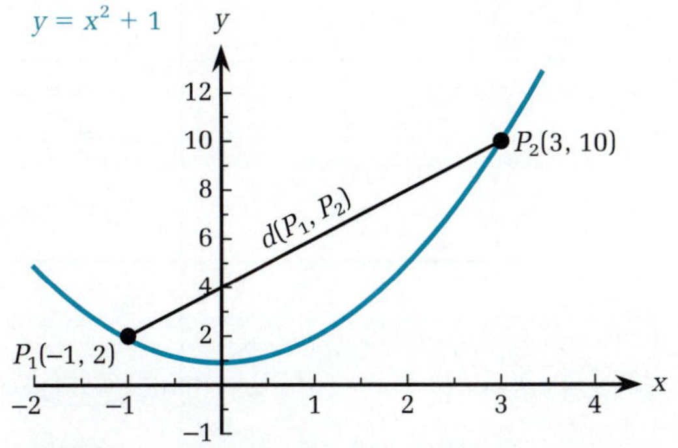

FIGURE 1.2.14

SOLUTION

$f(-1) = (-1)^2 + 1 = 2$ and $f(3) = 3^2 + 1 = 10$. Thus the two points are $P_1(-1, 2)$ and $P_2(3, 10)$. The distance between these points is $\sqrt{(3 - (-1))^2 + (10 - 2)^2} = \sqrt{4^2 + 8^2} = \sqrt{80} \approx 8.9443$. ●

PRACTICE EXERCISE 3

Find the distance between the points on the graph of $g(x) = x^2 + x$ determined by $x = 1$ and $x = 3$.

Equation of a Circle

A circle consists of all points that are the same distance from some fixed point. The distance is the *radius* of the circle and the fixed point is the *center* of the circle. By using the distance formula we find the following equation to describe the points of the circle.

EQUATION OF A CIRCLE

Suppose a circle has center at the point $C(a, b)$ and has a radius r. The points $P(x, y)$ of the circle satisfy $(x - a)^2 + (y - b)^2 = r^2$.

In figure 1.2.15 we see r is the distance from a point on the circle to the center. By the distance formula, we have $r = d(P(x, y), C(a, b)) = \sqrt{(x - a)^2 + (y - b)^2}$. Squaring this equation gives the equation given above.

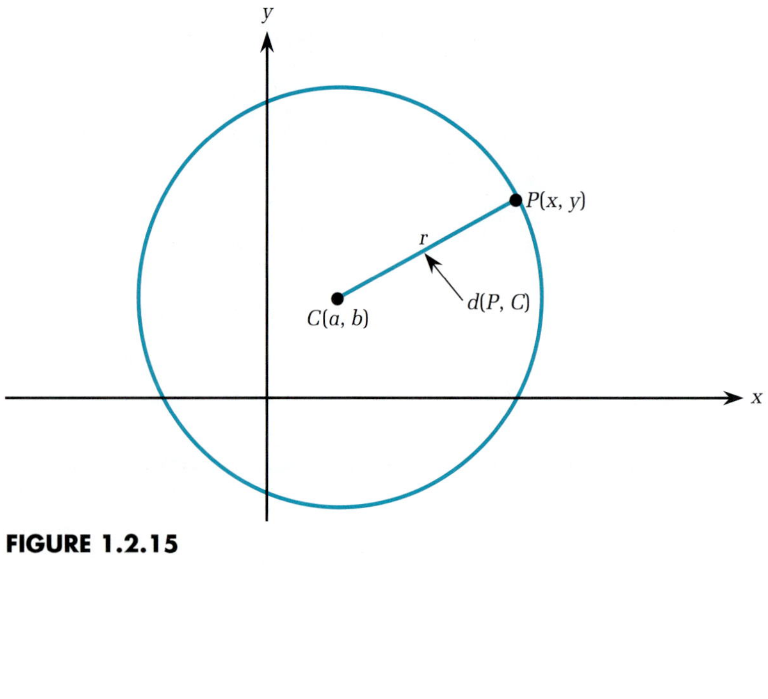

FIGURE 1.2.15

EXAMPLE 9 The equation of a circle

Find the equation for a circle with a radius of 2 and a center at $C(1, 3)$.

SOLUTION

The radius is $r = 2$. The center is at $C(1, 3)$, so we have $a = 1$, $b = 3$. Using the formula, we find

$$(x - 1)^2 + (y - 3)^2 = 2^2 = 4.$$

Another application of the distance formula is found in laying pipeline or electrical lines, in which a common goal is to use the least amount of pipe or cable. The next example illustrates such a use.

EXAMPLE 10 Gasfield pipeline

The San Juan Basin of New Mexico has one of America's largest fields of natural gas. Suppose three wells can be located by the coordinates $A(0, 0)$, $B(3, 4)$, and $C(-5, 12)$. The three wells are to be connected by pipeline, laid in straight lines from well to well. In order to minimize the total amount of pipe, the two wells farthest apart will be connected only by pipelines through the third well. Find which two wells are farthest apart.

SOLUTION

The locations of the wells are plotted in figure 1.2.16.

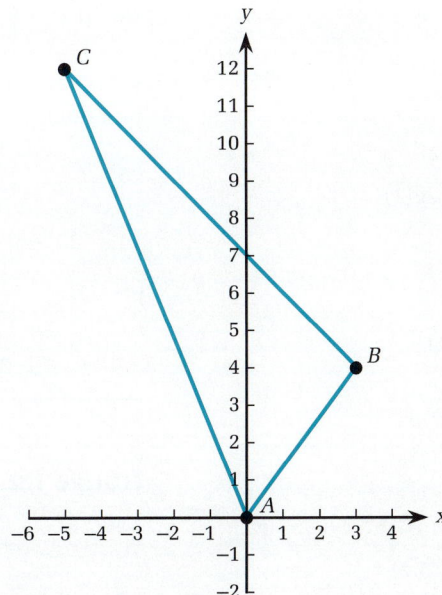

FIGURE 1.2.16

$$d(A, B) = \sqrt{(3 - 0)^2 + (4 - 0)^2} = \sqrt{9 + 16} = \sqrt{25} = 5$$
$$d(B, C) = \sqrt{(-5 - 3)^2 + (12 - 4)^2} = \sqrt{(-8)^2 + 8^2}$$
$$= \sqrt{64 + 64} = \sqrt{128} \approx 11.3137$$
$$d(C, A) = \sqrt{(0 - (-5))^2 + (0 - 12)^2} = \sqrt{5^2 + (-12)^2}$$
$$= \sqrt{25 + 144} = \sqrt{169} = 13$$

Because the distance between C and A is the greatest, it should not have a direct pipeline connection. ●

1.2 PROBLEMS

Foundations

The problems of this section require the basic skills illustrated by the following:

1. Find the interval on which $x - 2 \geq 0$.

2. Find the interval on which $4 - x^2 \leq 0$.

3. Find $(2 + h)^2$.

4. Find $|3 - 2|$.

Exercises

Find the largest possible domain for each function and evaluate the functions as indicated.

5. $f(x) = 3x - 2, f(0), f(2)$

6. $f(x) = 2x + 5, f(0), f(-2)$

7. $g(u) = \dfrac{1}{2}u + 1, g(2), g(-1)$

8. $f(u) = |u - 1|, f(0), f(1), f(2)$

9. $g(x) = |2x + 5|, g(-4), g(0), g(1)$

10. $f(x) = \sqrt{x - 2}, f(2), f(3)$

11. $h(x) = \sqrt{2x + 1}, h(1), h(3)$

12. $g(x) = \dfrac{5}{1 - x}, g(3), g(0), g(1)$

13. $g(t) = \dfrac{1}{2 + t}, g(2), g(0), g(-2)$

14. $f(t) = \dfrac{2}{t} - 3t, f(1), f(0), f(-1)$

15. $f(u) = 2u + \dfrac{1}{u}, f(3), f(0), f(5.2)$

16. $h(x) = \sqrt{5 - x}, h(5), h(0), h(9)$

17. $g(y) = \sqrt{2 - 3y}, g(0), g\left(\dfrac{1}{2}\right), g(1)$

18. $f(x) = |-x + 1|, f(2), f(0), f(-2)$

19. $f(x) = \sqrt{4 - x^2}, f(0), f(2)$

20. $g(t) = \dfrac{1}{2 - t^2}, g(0), g(-2)$

21. $f(x) = \begin{cases} 2x + 3 & \text{if } x \geq 0 \\ -x^2 & \text{if } x < 0 \end{cases}, f(2), f(-1)$

22. $g(u) = \begin{cases} u - 2 & \text{if } u > 1 \\ -2u & \text{if } u \leq 1 \end{cases}, g(0), g(3)$

23. $f(x) = \begin{cases} x - 3 & \text{if } x \geq 0 \\ 2x^2 & \text{if } x < 0 \end{cases}, f(-2), f(1)$

24. $g(u) = \begin{cases} u^2 - 2 & \text{if } u > 1 \\ 2u & \text{if } u \leq 1 \end{cases}, g(-1), g(1), g(2)$

Which of the following graphs are graphs of functions?

25.

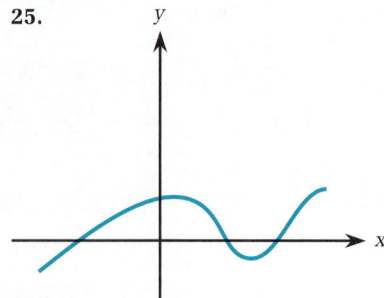

FIGURE 1.2.17

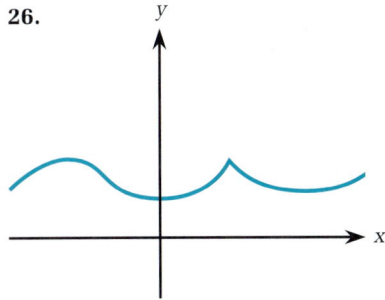

26.

FIGURE 1.2.18

27.

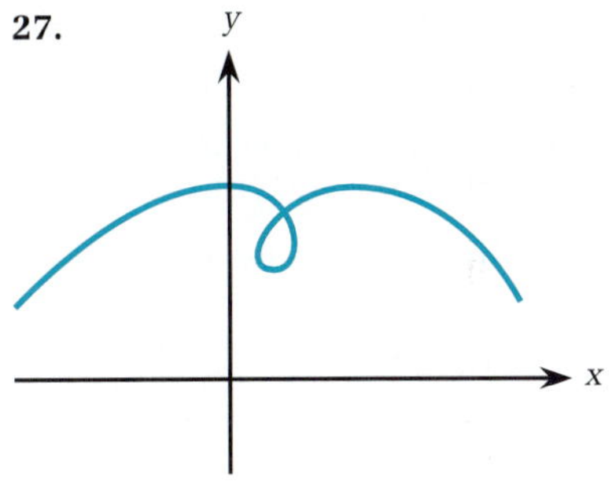

FIGURE 1.2.19

28.

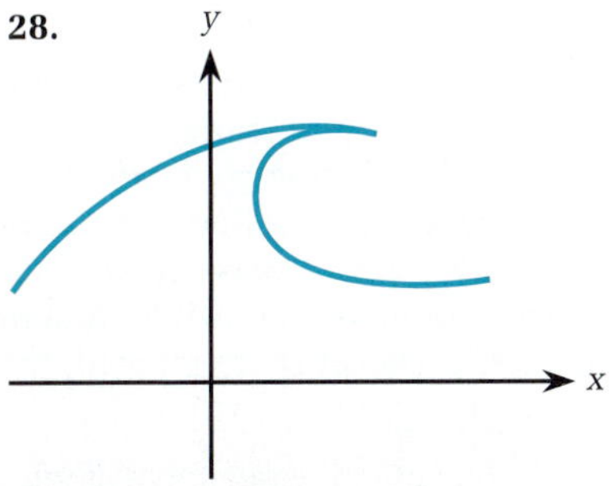

FIGURE 1.2.20

29. Find the distance between $P_1(1, 3)$ and $P_2(5, 3)$.

30. Find the distance between $P_1(0, 5)$ and $P_2(1, 4)$.

31. Suppose $f(x) = x^2 - 1$.
(a) Find each of the following: $f(1)$, $f(3)$, $f(a)$, $f(3 + h)$, $f(a + h)$.
(b) Find the distance between the points on the graph of $f(x)$ determined by $x = 1$ and $x = 3$.

(c) Find a simplified form for the quotient $[f(a + h) - f(a)]/h$, $h \neq 0$.

32. Suppose $f(x) = x^2 + 2$.
(a) Find each of the following: $f(1)$, $f(2)$, $f(a)$, $f(2 + h)$, $f(a + h)$.
(b) Find the distance between the points on the graph of $f(x)$ determined by $x = 1$ and $x = 2$.
(c) Find a simplified form for the quotient $[f(a + h) - f(a)]/h$, $h \neq 0$.

DEFINITION

The *greatest integer* of a number x, also known as the *floor integer* of x and written as $\lfloor x \rfloor$, is the one integer r such that $r \leq x < r + 1$. The *ceiling integer* of a number x, written as $\lceil x \rceil$, is the one integer s such that $s - 1 < x \leq s$.

Examples of finding floor and ceiling integer functions are: $\lceil 2.3 \rceil = 3$, $\lceil 2.8 \rceil = 3$, $\lceil 1.98 \rceil = 2$, $\lfloor 2.3 \rfloor = 2$, $\lfloor 1.2 \rfloor = 1$, $\lfloor -3.1 \rfloor = -4$, and $\lceil -3.1 \rceil = -3$. Evaluate the following:

33. $\lfloor 1.8 \rfloor$ **34.** $\lfloor 2.05 \rfloor$ **35.** $\lceil 8.1 \rceil$

36. $\lceil 2.05 \rceil$ **37.** $\lceil -1.2 \rceil$ **38.** $\lceil -2.3 \rceil$

39. *(Postage Stamp Function)*   The ceiling integer function is what the post office uses to find the postage due. For example, if you have an envelope that weighs 2.3 ounces, they will charge you the rate for 3 ounces. In 1993, it cost 29 cents for the first ounce (or fraction thereof) and 23 cents for each additional ounce (or fraction thereof) to mail a first-class letter in the United States. Express this in terms of the ceiling integer function. (Hint: The function may have to be defined in several pieces.)

40. *(Machine Rentals)*   The cost of renting a particular electric generator is $100 plus $50 per day or any fraction of a day. Express this in terms of the ceiling integer function.

The next two problems depend on the fact that if a triangle is such that the sum of the squares of the distances of two of the sides is equal to the square of the distance of the third side, then the triangle is a right triangle. For example, a triangle with sides of lengths 3, 4, and 5 is a right triangle because $3^2 + 4^2 = 5^2$.

41. Plot the points $A(6, 2)$, $B(2, -4)$, and $C(-1, -2)$ and show by the distance formula that these points are the vertices of a right triangle.

42. Plot the points $A(-3, 2)$, $B(9, 4)$, and $C(2, -3)$ and show by the distance formula that these points are the vertices of a right triangle.

43. Suppose three pump stations are located at $A(0, 0)$, $B(3, 1)$, and $C(2, 3)$. It is desired to connect all three with pipe, which must be straight and can go only from one pump station to another (that is, no branching off part way along the path between stations). There does not need to be a pipe from each pump station to every other, but there must be some pathway that connects all the pump stations together. What is the minimum amount of pipe necessary?

For problems 44–47, find equations of circles with the given conditions.

44. Center $C(1, -2)$ and radius 3.

45. Center $C(0, 3)$ and radius 2.

46. Center $C(-1, -1)$ with the point $P(2, 3)$ on the circle.

47. Center $C(2, 1)$ and diameter 6.

▨ Graphing Calculator Problems

48. Show the screens of $y = x^2 - 1$ over each of the following ranges:
   (a) $-1 \le x \le 5$, $0 \le y \le 2$
   (b) $-1 \le x \le 5$, $-2 \le y \le 25$
   (c) Explain why the second of these screens is so much more informative than the first.

49. Suppose $f(x) = \begin{cases} 2x + 3 & \text{if } x \ge 0 \\ -x^2 & \text{if } x < 0 \end{cases}$. There are two ways of finding the value of $f(a)$ for a particular value of a. One method is to plot both $y = 2x + 3$ and $y = -x^2$ with a continuation instruction that then allows you to use the $\boxed{\text{Trace}}$ key and the arrow keys to switch back and forth from one graph to the other, depending on whether a is positive or negative.

   (a) Do this, even though it clutters the screen up with points that are not really on the graph of $y = f(x)$.
   (b) Graph $y = 2(\sqrt{x})^2 + 3$.
   (c) Use the graph from part (b) and a continuation instruction for a second function so that all that is displayed on the screen is the actual graph of $y = f(x)$.

Writing and Discussion Problems

50. What is a function?

51. Which of the two graph screens shown in Example 6 is more useful? Why?

52. A circle cannot be a graph of just one function. Why? Can a circle be made from the graphs of two functions? Justify your answer.

53. There are situations in which we use the word "function" in everyday conversation but without a number attached to its use. An example would be to say, "Whether I will study tonight is a function of how tired I feel then."

   There seems to be a tendency to attempt to attach numbers to everything: business enterprise risk, ice skating competition, feelings of "coldness," drug effectiveness, and many other matters. Is it reasonable to do this? For instance, would you be using "function" in the mathematical sense if you said, "On a scale of 1 to 10, my tiredness is a 7.5."?

54. From the mathematical perspective of what a function should be, discuss which of the following statements are correct:
   (a) The area of a circle is a function of its radius.
   (b) The taste of a pizza is a function of its price.
   (c) The height of a car is a function of its age.
   (d) The volume of a 12-ounce can is a function of its height.
   (e) The cost of a bunch of bananas is a function of the bunch's weight.
   (f) Whether water is frozen or not is a function of its temperature.
   (g) The worth of advice is a function of its cost.
   (h) The volume of a sphere is a function of its surface area.

Enrichment Problems

55. The governor of a state wants to visit four cities by helicopter. The coordinates of the cities are given by $A(0, 0)$, $B(20, 10)$, $C(50, 20)$, and $D(10, 30)$. The governor starts in city A and does not care in which city the tour finishes. Flying in straight lines from city to city, what is the smallest number of miles for the tour?

56. The following problems use the "floor" and "ceiling" functions defined above before problem #33.
(a) Find values of a and b such that $\lceil a + b \rceil < \lceil a \rceil + \lceil b \rceil$.
(b) Show that $\lceil a + b \rceil \leq \lceil a \rceil + \lceil b \rceil$ and $\lfloor a + b \rfloor \geq \lfloor a \rfloor + \lfloor b \rfloor$ for any real numbers a and b.

SOLUTIONS TO PRACTICE EXERCISES

1. $f(-1) = (-1)^2 - 2(-1) + 1 = 1 + 2 + 1 = 4$
$f(2) = 2^2 - 2 \cdot 2 + 1 = 4 - 4 + 1 = 1$
$f(3) = 3^2 - 2 \cdot 3 + 1 = 9 - 6 + 1 = 4$
$f(3 + a) = (3 + a)^2 - 2(3 + a) + 1 = 9 + 6a + a^2 - 6 - 2a + 1 = a^2 + 4a + 4$

2. (a) The set of all real numbers $\mathbb{R}$.
(b) The set of all numbers except 1, which we can write as $\{y: y \neq 1\}$.

(c) Because we need $2t - 1 \geq 0$, the domain is $\{t: t \geq 1/2\}$, which is the interval $[1/2, \infty)$.

3. $g(1) = 1^2 + 1 = 2$ and $g(3) = 3^2 + 3 = 9 + 3 = 12$. Hence the two points are $P_1(1, 2)$ and $P_2(3, 12)$ and their distance apart is $\sqrt{(3 - 1)^2 + (12 - 2)^2} = \sqrt{2^2 + 10^2} = \sqrt{104}$.

1.3

THE ALGEBRA OF FUNCTIONS

This section is about functions that can be represented in terms of other functions. As an example, suppose we describe the worth of a company that begins with \$10,000 in start-up capital and has costs of \$500 each week. As a result, without considering income, the amount of original capital after x weeks is given by $g(x) = \$10,000 - 500x$, whose graph is shown in figure 1.3.1. However, suppose the company estimates its income will be $h(x) = \$7x^2$, whose graph is shown in figure 1.3.2.

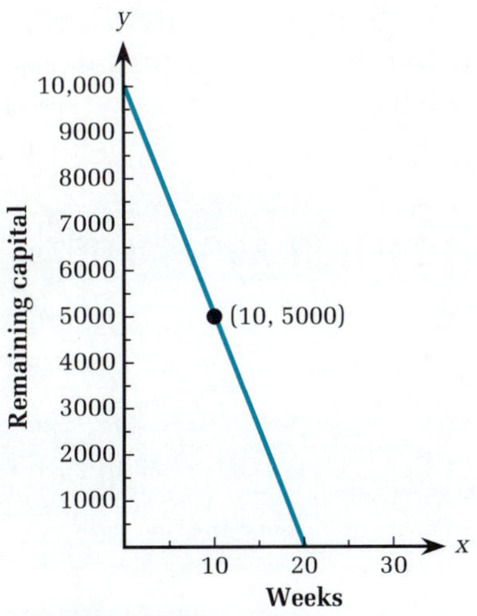

FIGURE 1.3.1

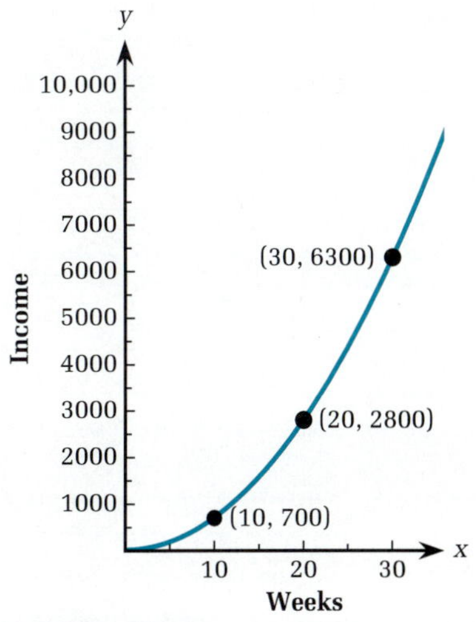

FIGURE 1.3.2

We may combine these two functions to create a function, $W(x)$, which represents the total worth in week x.

$$W(x) = g(x) + h(x)$$
$$= 10{,}000 - 500x + 7x^2$$

The graph of $y = W(x)$ is shown in figure 1.3.3.

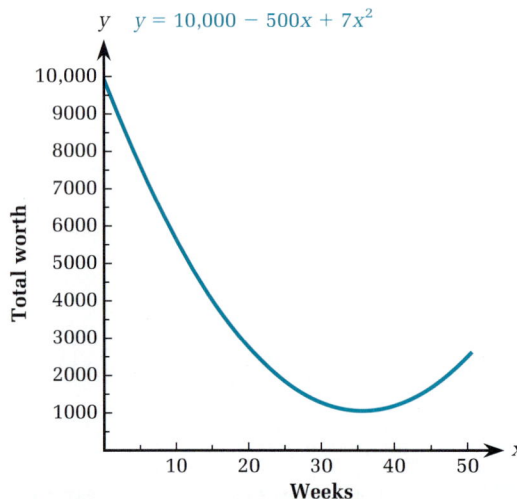

FIGURE 1.3.3

In this section we will be combining simple functions into more complex ones, as well as (even more importantly!) beginning with a complex function and breaking it down into simpler functions. The other topic of this section is inverse functions.

The Arithmetic of Functions

Functions can be created by any arithmetical process that adds, subtracts, multiplies, or divides the values of other functions. In the following, each function is defined by specifying its value over its domain.

FUNCTION ARITHMETIC

Given functions f and g, and any real number a, we define the following:

(i) $(a \cdot f)(x) = a \cdot f(x)$

(ii) $(f + g)(x) = f(x) + g(x)$

(iii) $(f - g)(x) = f(x) - g(x)$

(iv) $(f \cdot g)(x) = f(x) \cdot g(x)$

$$\textbf{(v)} \qquad \left(\frac{f}{g}\right)(x) \; = \frac{f(x)}{g(x)} \; \text{provided } g(x) \neq 0$$

These rules are illustrated by the next example.

EXAMPLE 1   Function arithmetic

Suppose $f(x) = x + 1$ and $g(x) = x^2 - 2$. Find each of the following functions: $(3 \cdot f)(x)$, $(f + g)(x)$, $(f - g)(x)$, $(f \cdot g)(x)$, and $(f/g)(x)$.

SOLUTION

$(3 \cdot f)(x) = 3 \cdot f(x)$ by rule (i)

$\qquad\qquad\qquad = 3 \cdot (x + 1) = 3x + 3$

$(f + g)(x) = f(x) + g(x)$ by rule (ii)

$\qquad\qquad\qquad = (x + 1) + (x^2 - 2)$

$\qquad\qquad\qquad = x^2 + x - 1$

$(f - g)(x) = f(x) - g(x)$ by rule (iii)

$\qquad\qquad\qquad = (x + 1) - (x^2 - 2)$

$\qquad\qquad\qquad = x + 1 - x^2 + 2 = -x^2 + x + 3$

$(f \cdot g)(x) = f(x) \cdot g(x)$ by rule (iv)

$\qquad\qquad\qquad = (x + 1)(x^2 - 2)$

$\qquad\qquad\qquad = x^3 + x^2 - 2x - 2$

$\left(\dfrac{f}{g}\right)(x) = \dfrac{f(x)}{g(x)}$ by rule (v)

$\qquad\qquad\qquad = \dfrac{x + 1}{x^2 - 2}$ (provided $x^2 - 2 \neq 0$, that is, $x \neq \pm\sqrt{2}$)

PRACTICE EXERCISE 1

For the functions in Example 1, find $(f + 3g)(x)$.

Composition of Functions

Functions may have input values that are themselves outputs from other functions. That is, the domain of one function may consist of values in the range of another.

For example, to determine $\sqrt{1/x}$ first press the $\boxed{1/x}$ key, and follow it by pressing the $\boxed{\sqrt{x}}$ key. Therefore, to find $\sqrt{1/5}$ we can enter $\boxed{5}$, then press the $\boxed{1/x}$ key, so that $\boxed{0.2}$ is displayed, then press the $\boxed{\sqrt{x}}$ key and see that $\boxed{0.447213595}$ is displayed. This process is called *function composition.*

DEFINITION

Composite function

Suppose f and g are functions, and the range of g lies in the domain of f. The *composite function* of f and g is the function h defined by $h(x) = f(g(x))$. The domain of $f(g(x))$ consists of that part of the domain of g that produces values in the domain of f.

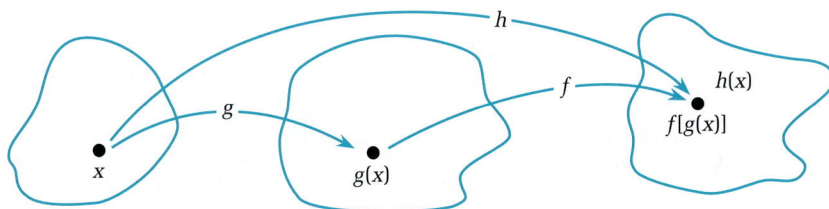

FIGURE 1.3.4

In order to show more clearly that $f(g(x))$ is the result of replacing the variable x in $f(x)$ by the value of $g(x)$, it is sometimes desirable to rewrite the function $f(x)$ with the x being replaced by a new variable name.

EXAMPLE 2   Changing variable names during composition

Suppose $f(x) = 2/x$ and $g(x) = x^2 - 3$. Find $f(g(x))$.

SOLUTION

$f(x) = 2/x$ is the same function as $f(t) = 2/t$, so if we let $t = g(x)$, then we have $f(g(x)) = [2/(g(x))] = 2/(x^2 - 3)$.

EXAMPLE 3   Composition of functions

Suppose $f(x) = 2x + 1$ and $g(x) = x^2$. Find each of the following: $f(g(2))$, $f(g(x))$, $g(f(2))$, $g(f(x))$, $f(f(x))$.

SOLUTION

$$f(g(2)) = f(2)^2 = f(4) = 2 \cdot 4 + 1 = 8 + 1 = 9$$
$$f(g(x)) = f(x^2) = 2 \cdot x^2 + 1$$
$$g(f(2)) = g(2 \cdot 2 + 1) = g(5) = 5^2 = 25$$
$$g(f(x)) = g(2x + 1) = (2x+ 1)^2$$
$$f(f(x)) = f(2x + 1) = 2(2x + 1) + 1 = 4x + 2 + 1 = 4x + 3$$

PRACTICE EXERCISE 2

Find $g(f(1))$ for the functions in Example 3.

WARNING

The order of composition is very important. Usually, as we found in Example 3, $f(g(x)) \neq g(f(x))$.

Decomposition of Functions

Creating a complicated function from simpler functions is important, but the real power of the concept of function composition comes from the ability to *decompose* a function into simpler functions. Thus, when we realize that $f(x) = \sqrt{x + 2}$ can be decomposed into a square root function and an addition function, then we discover there are fewer functions we need to study.

A function may be decomposed into simpler functions in several different ways.

EXAMPLE 4   Alternate decompositions of a function

Show that each of the following provides decompositions of $f(x) = (x + 1)^2$.

1. First decomposition. Express $f(x)$ as $g(h(x))$, where $h(x) = x + 1$ and $g(x) = x^2$.

2. An alternate decomposition. Express $f(x)$ as $r(s(x))$, where $s(x) = x^2 + 2x$ and $r(x) = x + 1$.

SOLUTION

To check the first decomposition, $g(h(x)) = g(x + 1) = (x + 1)^2$. For the alternate decomposition, $r(s(x)) = r(x^2 + 2x) = (x^2 + 2x) + 1 = (x + 1)^2$.

The first decomposition seems more natural, but nonetheless the second one is also valid.

EXAMPLE 5   Decomposition

Find functions $g(x)$ and $h(x)$ such that $1/\sqrt{2x + 1} = g(h(x))$.

SOLUTION

One possibility is to let $h(x) = 2x + 1$ and $g(x) = 1/\sqrt{x}$.

PRACTICE EXERCISE 3

Find functions $f(x)$ and $g(x)$ such that $(x^2 + 1)^3 = f(g(x))$.

WARNING

Be very careful to realize $f(g(x))$ is *not* $f(x) \cdot g(x)$. The composition of $f(x)$ and $g(x)$ as $f(g(x))$ means the output of $g(x)$ is used as an input for $f(x)$, while $f(x) \cdot g(x)$ means that the output of $f(x)$ is multiplied by the output of $g(x)$. As an example, if we let $f(x) = 2x + 1$ and $g(x) = 1/\sqrt{x}$ then $f(g(9)) = f(1/\sqrt{9}) = f(1/3) = 2 \cdot 1/3 + 1 = 5/3$, whereas $f(9) \cdot g(9) = (2 \cdot 9 + 1) \cdot 1/\sqrt{9} = 19 \cdot 1/3 = 19/3$.

EXAMPLE 6   Composition, mail orders

Suppose you order $\$x$ worth of merchandise and must pay sales tax of 3% and a handling and shipping fee of \$3 plus 5% of your order cost (including the tax). Find a function to express your total cost when you order $\$x$ worth of merchandise.

SOLUTION

Let $g(x) = 0.03x$ represent the tax and $h(x) = 3 + 0.05x$ represent the handling and shipping fee. The total cost, $T(x)$, is the sum of the order cost plus tax and also the shipping and handling fee on that order cost plus tax. In symbols, $T(x)$ can be expressed as

$$T(x) = (x + g(x)) + h(x + g(x))$$

Because $x + g(x) = x + 0.03x = 1.03x$, we have

$$
\begin{aligned}
T(x) &= 1.03x + h(1.03x) \\
&= 1.03x + (3 + 0.05(1.03x)) \\
&= 1.03x + 3 + 0.0515x \\
&= 3 + 1.0815x
\end{aligned}
$$

Inverse Functions

In Section 1.2 we noted that any vertical line intersects the graph of the function in no more than one point. This is not true for horizontal lines. However, there are some functions for which any horizontal line also intersects the graph in no more than one point.

The graphs of $f(x) = 2x + 1$ and $g(x) = x^2 - x$ are shown in figures 1.3.5 and 1.3.6.

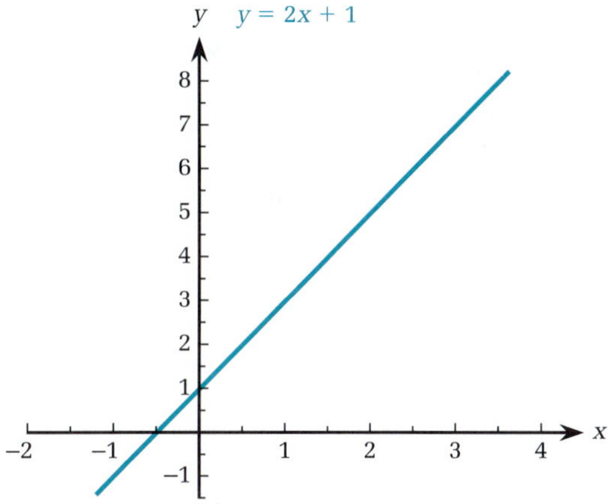

FIGURE 1.3.5

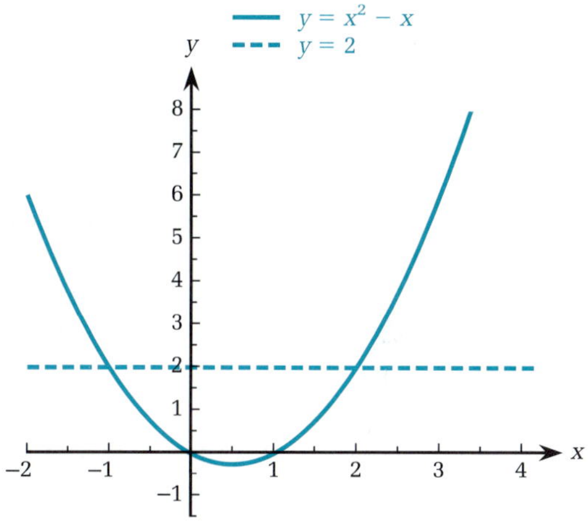

FIGURE 1.3.6

Every horizontal line intersects the graph of $f(x)$ in exactly one point, but for $g(x)$ that is not the situation. For instance, the horizontal line $y = 2$ intersects the graph of $g(x)$ at $P(2, 2)$ and at $Q(-1, 2)$. This is because $g(2) = 2^2 - 2 = 4 - 2 = 2$ and $g(-1) = (-1)^2 - (-1) = 1 + 1 = 2$.

DEFINITION

One-to-one function

Suppose a function $y = f(x)$ is such that each value of y is produced by a unique value of x. That is, each value in the range is the image of exactly one value in the domain. Such a function is called *one-to-one*.

Any horizontal line drawn through the graph of a one-to-one function will intersect the graph in at most only one point. In figure 1.3.7 we illustrate a function that is one-to-one. Because each value for y is associated with only one value of x, we can define a function g as follows: Let y be any value in the range of $f(x)$. Then $g(y) = x$. This function g is an *inverse function*.

FIGURE 1.3.7

DEFINITION

Inverse function

Suppose f and g are functions such that $g(f(x)) = x$ for all values in the domain of f and $f(g(x)) = x$ for all values in the domain of g. Then f and g are *inverses* of each other. If g is the inverse of f, we may denote g by f^{-1}, which is read "f-inverse." The domain of the inverse function, f^{-1}, is the range of f.

WARNING

The notation for an inverse, f^{-1}, does *not* mean $1/f$.

If *any* horizontal line intersects the graph of the function in more than one point, then the function does not have an inverse. An example of a graph of a function that has no inverse is shown in figure 1.3.8.

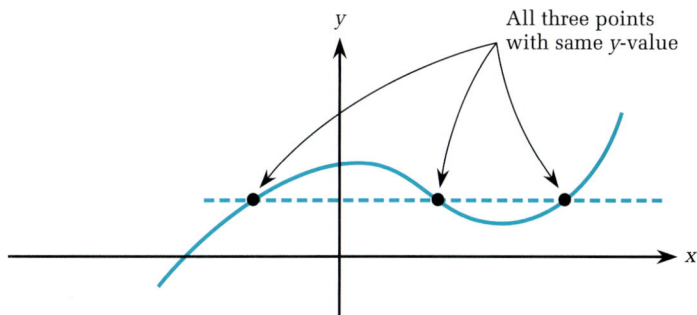

All three points with same *y*-value

FIGURE 1.3.8

Rarely are we given two functions and asked to check that they are inverses. The more typical situation is that we know one function and we wish to find its inverse. This may sometimes be done by solving an algebraic equation, as we do in the next example. First, we state the procedure we follow to find an inverse function.

FINDING AN INVERSE FUNCTION

The key idea is that because the point $(x, f(x))$ is on the graph of $y = f(x)$, and the inverse function g is such that $g(f(x)) = x$, then the point $(f(x), x)$ must be on the graph of g. Thus, if we solve $y = f(x)$ for x in terms of y, we have the inverse function, expressed as a function of y. To express the inverse function $g(y)$ as a function of x, replace all occurrences of y in $g(y)$ by x.

EXAMPLE 7   Finding an inverse function

Find the inverse function of $f(x) = 2x - 1$.

SOLUTION

Let $y = 2x - 1$. Solving for x, we have, first, $y + 1 = 2x$ and then $(y + 1)/2 = x$. Thus, the inverse function will take a value, y, and produce $g(y) = (y + 1)/2$. Expressing this as a function of x we have $g(x) = (x + 1)/2$.

As an illustration, suppose we want to find a value a such that $f(a) = 7$. Using the inverse function, we have $g(7) = (7 + 1)/2 = 8/2 = 4$. We check that $f(4) = 2 \cdot 4 - 1 = 8 - 1 = 7$, as we wanted.

To verify that $g(x)$ truly is the inverse we show that $g(f(x)) = x$.

Indeed, $g(f(x)) = g(2x - 1) = [(2x - 1) + 1]/2 = 2x/2 = x$, as desired.

PRACTICE EXERCISE 4

Find the inverse function of $f(x) = \dfrac{2}{(x + 1)}$.

EXAMPLE 8   Finding the domain of an inverse function

Find the inverse function for $f(x) = \sqrt{x} + 1$ and determine its domain.

SOLUTION

Letting $y = \sqrt{x} + 1$, we have $y - 1 = \sqrt{x}$.

By squaring both sides, $(y - 1)^2 = x$. Thus, the inverse function is $g(y) = (y - 1)^2$. Rewriting this as a function of x, $g(x) = (x - 1)^2$.

The domain of $g(x)$ is $\{x: x \geq 1\}$ rather than the set of all real numbers $\mathbb{R}$, as might have appeared from the form of $g(x)$. This is because the domain of g is the range of f and since $\sqrt{x} \geq 0$, the range of $f(x) = \sqrt{x} + 1$ consists of the real numbers equal to or greater than 1. ●

1.3 PROBLEMS

Foundations

The problems of this section require the basic skills illustrated by the following:

1. Find those values of x for which $3x + 9 > 0$.

2. Evaluate $3\left(1/\sqrt{9}\right) + 4$.

3. Simplify the expression $3/[(2/x) - 2]$.

Exercises

4. For $f(x) = 3x + 9$ and $g(x) = 1/\sqrt{x}$, find each of the following:
   (a) $(f + g)(x)$ (b) $(f \cdot g)(x)$ (c) $(f/g)(x)$
   (d) $f(g(4))$ (e) $f(g(x))$ (f) $g(f(3))$
   (g) $g(f(x))$ (h) the domain of $f(g(x))$
   (i) the domain of $g(f(x))$.

5. For $f(x) = x^2 - 4$ and $g(x) = x/(x + 3)$, find each of the following:
   (a) $(f - g)(x)$ (b) $(f \cdot g)(x)$ (c) $(f/g)(x)$
   (d) $f(g(2))$ (e) $f(g(x))$ (f) $g(f(2))$
   (g) $g(f(x))$ (h) the domain of $f(g(x))$
   (i) the domain of $g(f(x))$.

For the next two problems, find both (a) $f(g(x))$ and its domain and (b) $g(f(x))$ and its domain.

6. $f(x) = \dfrac{3}{x - 2}$ $g(x) = \dfrac{2}{x}$

7. $f(x) = \dfrac{x - 1}{x}$ $g(x) = \dfrac{2}{x + 1}$

8. Suppose $h(x) = \sqrt{x^2 + 2}$. Find functions $f(x)$ and $g(x)$ so that $h(x) = f(g(x))$.

9. Suppose $r(x) = 1/(\pi x^2 + 1)$. Find functions $f(x)$ and $g(x)$ so that $r(x) = f(g(x))$.

10. El Nopal makes piñon nut candy bars, which are packed 36 to the box. The profit from selling x boxes is $\$10.80 \cdot x$. Express the profit as a function of n, the number of piñon nut bars sold.

11. The Olden Days Postcard Company charges tax at the rate of 5% and shipping and handling of $\$4$ plus 2% of the cost of the order (including tax). Determine a function that gives the total cost of ordering $\$x$ worth of postcards.

For the functions in problems 12–23 find an inverse function for each of those that have inverses.

12. $f(x) = 2x + 1$ **13.** $g(x) = 3 - 5x$

14. $f(x) = \dfrac{1}{2 - x}$ **15.** $f(x) = \dfrac{1}{2x}$

16. $f(x) = \dfrac{x + 1}{2x - 1}$ **17.** $g(x) = \dfrac{2 - x}{x - 1}$

18. $f(x) = x^2 - 2$ **19.** $g(x) = -2x^2 + 1$

20. $f(x) = 2$ **21.** $f(x) = 2x^2 + 1, x \geq 0$

22. $r(x) = -x^2 + 2, x \geq 1$

23. $s(x) = -x^3 + 1$

Show that the next two functions are their own inverses.

24. $g(x) = -x + b$

25. $k(x) = \dfrac{2x + b}{cx - 2}$ for $c \neq 0$

26. The Mirage Lake Water Company determines that the cost of removing $x\%$ of the impurities from a source of water is given by the formula $C(x) = (2x + 3)/(101 - x)$, in thousands of dollars. Find a formula to express the percentage of impurities that can be removed by spending x thousand dollars.

Writing and Discussion Problems

27. Explain the difference between composition and multiplication of functions.

28. Suppose $f(x) = x^2 + 1$ and $g(x) = \sqrt{x}$. Find the domain of $f(g(x))$ and explain why it is not all of the real numbers even though $f(g(x)) = x + 1$. Give a definition of the domain of $f(g(x))$ when $f(x)$ and $g(x)$ are general functions.

29. Graph $f(x) = \sqrt{x} + 1, x \geq 0$ and $g(x) = (x - 1)^2$, $x \geq 1$. Show that these graphs are reflections of each other with respect to the line $y = x$. Explain why the domains of $f(x)$ and $g(x)$ are given as they are. Would the reflection of $y = \sqrt{x} + 1$ be $y = (x - 1)^2$ with no restriction on x?

Enrichment Problems

30. Show that the graphs of f^{-1} and f are symmetric with respect to the line $y = x$ by supplying reasons for the following steps.
  (a) If $P(x, y)$ is a point on the graph of f, then $Q(y, x)$ is a point on the graph of f^{-1}.
  (b) The midpoint of the line segment joining P and Q lies on the line $y = x$.
  (c) The line segment joining P and Q is perpendicular to the line $y = x$.

SOLUTIONS TO PRACTICE EXERCISES

1. $(f + 3g)(x) = f(x) + (3g)(x) = f(x) + 3 \cdot g(x) = x + 1 + 3(x^2 - 2) = x + 1 + 3x^2 - 6 = 3x^2 + x - 5$.

2. $g(f(1)) = g(2 \cdot 1 + 1) = g(3) = 3^2 = 9$.

3. One pair would be $f(x) = x^3$ and $g(x) = x^2 + 1$. Another pair is $f(x) = (x + 1)^3$ and $g(x) = x^2$.

4. Let $y = 2/(x + 1)$. Then $y(x + 1) = 2$ and so $yx + y = 2$. Solving for x gives $x = (2 - y)/y$. Expressed as a function of x, the inverse function is $g(x) = (2 - x)/x$.

1.4   LINEAR FUNCTIONS AND MODELS

This section first discusses mathematical models. Then we review properties of lines and their equations. This material provides a basis for understanding linear functions and their use as mathematical models.

Mathematical Models

Pablo Picasso, the famous twentieth-century artist, once said, "Art is a lie that makes us realize the truth." A good mathematical model, like a great work of art, both simplifies and unifies in order to show the essentials. In doing so, it often does not consider every aspect of a situation. However, a good model reveals the truth of a situation. It helps us understand.

The most important reason for the creation of a model is to enable the model builder to explain and predict future behavior. If predictions are verified, then there is widespread belief that the assumptions used in making the model accurately describe the real situation.

Situations in real life are rarely easy to model, even for experts. Consider the serious financial difficulties of Citicorp Bank in the mid-1980s. In the 1970s Citicorp had begun a program of loans to "developing countries." These loans were based on models that were developed by economists and predicted rapid growth in those countries. However, a mere decade later Citicorp was faced with $15 billion in bad loans. The president of Citicorp, John S. Reed, began an ongoing program of workshops and a research program on economics. He told workshop participants to devise a new model, one that would explain and predict the global economy.

The most common difficulty faced by modelers is deciding which type of function, or functions, to use. There is no guaranteed method that always determines what function is best. For many situations no one knows any functions that fit very well. In this section we consider only linear models, but in subsequent sections of the book we will discover a great variety of functions, and models based on those functions.

Lines and Their Equations

Suppose $P(x_1, y_1)$ and $Q(x_2, y_2)$ are two points and $x_1 \neq x_2$, as shown in figure 1.4.1.

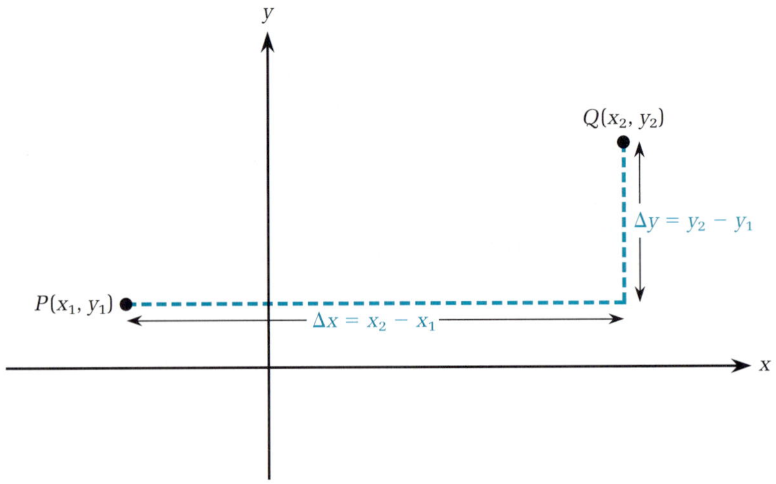

FIGURE 1.4.1

Define $\Delta y = y_2 - y_1$ and $\Delta x = x_2 - x_1$. The symbol Δ, which is read "delta," is the capital form of the letter of the Greek alphabet that corresponds to our letter "d" and it stands for "the change in $\cdots$" or "the difference of $\cdots$" Thus, Δx is the difference in the x-values.

DEFINITION   ## Slope

The *slope, m,* of a nonvertical line through the points $P(x_1, y_1)$ and $Q(x_2, y_2)$ is defined as the ratio

$$m = \frac{\Delta y}{\Delta x} = \frac{y_2 - y_1}{x_2 - x_1}, \; x_2 \neq x_1$$

The slope's value is the same no matter which point we call the first point and label as $P(x_1, y_1)$ and which point we call the second point and label $Q(x_2, y_2)$. To see this, rewrite $(y_2 - y_1)/(x_2 - x_1)$ as $(-y_1 + y_2)/(-x_1 + x_2)$ and then multiply both numerator and denominator by -1, which does not change the value of the quotient. Therefore,

$$\frac{y_2 - y_1}{x_2 - x_1} = \frac{(-y_1 + y_2)}{(-x_1 + x_2)} = \frac{-(-y_1 + y_2)}{-(-x_1 + x_2)} = \frac{y_1 - y_2}{x_1 - x_2}$$

WARNING

Even though it makes no difference which point is called the first and which the second, we must still make certain that if we use the x-value of one of the points as our x_1 then the y-value of the same point must be used for the y_1.

EXAMPLE 1   Finding the slope

Find the slope of the line through the points $P(1, 2)$ and $Q(4, 7)$ as shown in figure 1.4.2.

SOLUTION

By the definition of the slope, m, we have $m = (7 - 2)/(4 - 1) = 5/3$.

A line with a positive slope rises from left to right, as shown in figure 1.4.3, and a line with a negative slope falls from left to right, as shown in figure 1.4.4. If the slope has a large absolute value, then the line is steep. A line with a zero slope is horizontal, as in figure 1.4.5.

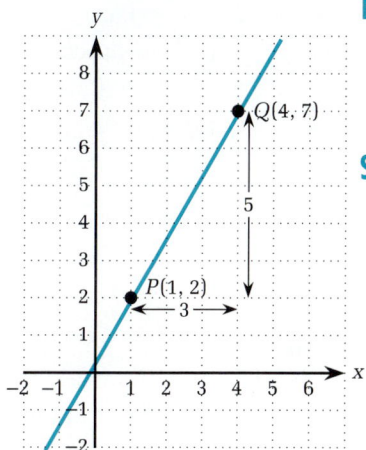

FIGURE 1.4.2

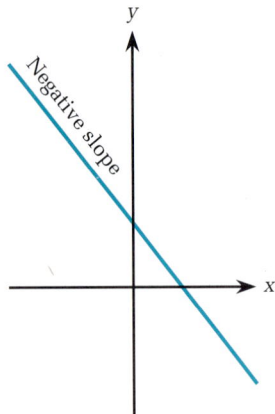

FIGURE 1.4.3   **FIGURE 1.4.4**   **FIGURE 1.4.5**

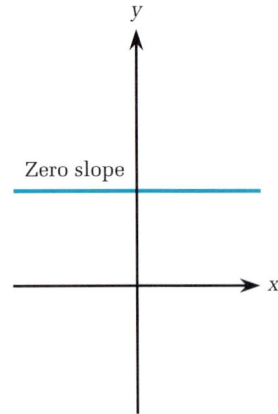

DEFINITION   *y*-intercept

The *y-intercept* of a line is the *y* coordinate of the point at which the line crosses the *y*-axis.

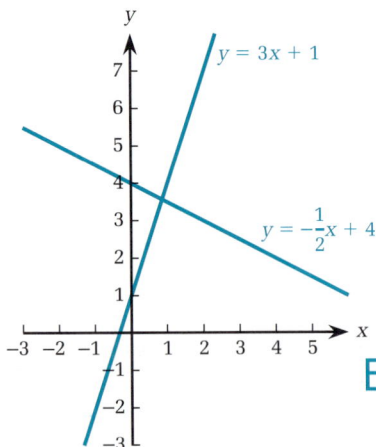

FIGURE 1.4.6

FORMS OF EQUATIONS FOR LINES

Equations for nonvertical lines are often expressed in one of the following forms:

Slope-Intercept $y = mx + b$, where m is the slope and b is the *y*-intercept.

Point-Slope $(y - y_1)/(x - x_1) = m$, where m is the slope and $P(x_1, y_1)$ is a point on the line.

Two-Point $(y - y_1)/(x - x_1) = (y_2 - y_1)/(x_2 - x_1)$, where $P(x_1, y_1)$ and $P(x_2, y_2)$ are two points on the line.

Vertical lines are written in the form $x = a$, where a is a constant. Vertical lines are said to have no slope.

EXAMPLE 2   Slope and *y*-intercept

Find the slope and the *y*-intercept of the lines $y = 3x + 1$ and $y = -(1/2)x + 4$, shown in figure 1.4.6.

SOLUTION

These lines are written in the slope-intercept form. The line $y = 3x + 1$ rises steeply and crosses the y-axis at the point $(0, 1)$. The slope is 3 and the y-intercept is 1. The line $y = -(1/2)x + 4$ falls slowly and crosses the y-axis at the point $(0, 4)$. Its slope is $-1/2$ and the y-intercept is 4.

PRACTICE EXERCISE 1

Find the slope and the y-intercept for the line $y = -2x + 4$ and sketch its graph.

A nonvertical line whose equation is written in some form other than $y = mx + b$ can be rewritten to be in that form. Doing so makes possible a quick sketch of the graph.

EXAMPLE 3   Rewriting in slope-intercept form

Rewrite $3y + 2x = 4$ in slope-intercept form and sketch its graph.

SOLUTION

First subtract $2x$ from both sides of the equation to get $3y = -2x + 4$. Then divide by 3 to get $y = -(2/3)x + 4/3$. In the rewritten form, both the slope and the y-intercept are explicit. A negative slope means the graph is going down. The graph is indicated in figure 1.4.7.

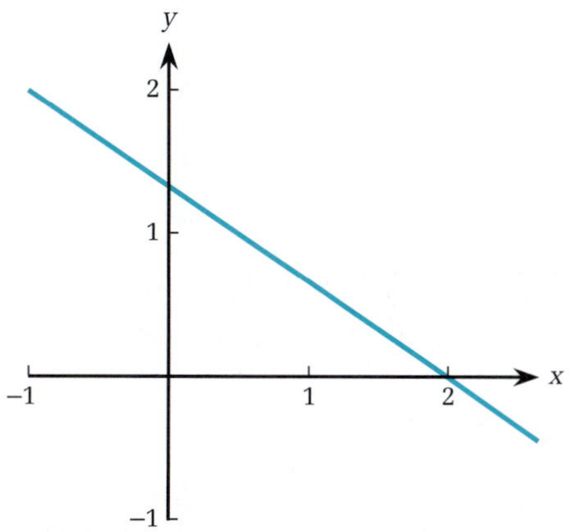

FIGURE 1.4.7

EXAMPLE 4   Forms of equations for a line

Find both the two-point form and the slope-intercept form of the line through $P(1, 2)$ and $Q(4, 7)$ and sketch its graph.

SOLUTION

The two-point form would be $(y - 2)/(x - 1) = (7 - 2)/(4 - 1)$, or $(y - 2)/(x - 1) = 5/3$. To put this into the slope-intercept form, we first multiply each side by $x - 1$ to get

$$y - 2 = \frac{5}{3}(x - 1) = \frac{5}{3}x - \frac{5}{3}$$

Finally we have

$$y = \frac{5}{3}x - \frac{5}{3} + 2 = \frac{5}{3}x + \frac{1}{3}$$

The graph is shown in figure 1.4.8.

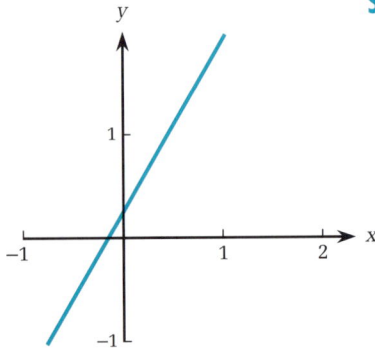

FIGURE 1.4.8

Parallel and Perpendicular Lines

The following theorem can be proved.

SLOPES OF PARALLEL AND PERPENDICULAR LINES

Suppose the line l_1 has slope m_1 and the line l_2 has slope m_2. The lines l_1 and l_2 are parallel to each other if and only if $m_1 = m_2$. The lines l_1 and l_2 are perpendicular to each other if and only if $m_1 \cdot m_2 = -1$.

This result is illustrated in figure 1.4.9.

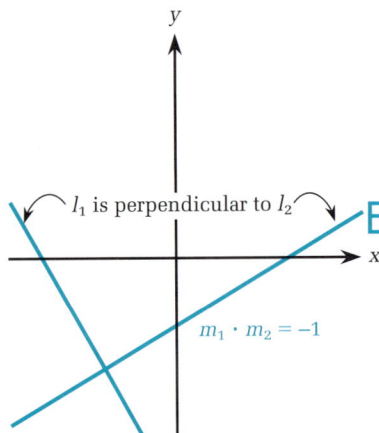

FIGURE 1.4.9

EXAMPLE 5   Parallel and perpendicular lines

Suppose a line l has the equation $3x + 4y = 12$.

(a) Find an equation for a line parallel to l but containing the point $P(1, 3/2)$.

(b) Find an equation for a line perpendicular to l and containing the point $Q(1, 0)$.

SOLUTION

Rewriting the equation for l in the slope-intercept form we have $y = -(3/4)x + 3$, so the slope of l is $-3/4$.

(a) Let k_1 be a line parallel to l. A line parallel to l must have the same slope as l, so we know the equation for k_1 has the form $y = -(3/4)x + b$. It remains for us to evaluate b to get the equation of the line. We do this by substituting the x- and y-coordinates of point $P(1, 3/2)$ and finding $3/2 = -(3/4)1 + b$. Hence, $b = 3/2 + 3/4 = 9/4$, and the parallel line has the equation $y = -(3/4)x + 9/4$.

(b) Let k_2 be a line perpendicular to l. Perpendicular lines have slopes that are negative reciprocals. The slope of k_2 is therefore given by

$$-\frac{1}{(\text{slope of } l)} = -\frac{1}{-\dfrac{3}{4}} = -1\left(-\frac{4}{3}\right) = \frac{4}{3}$$

Thus, starting again with the slope-intercept form, we have $y = (4/3)x + b$, and from this we evaluate b by substituting 1 for x and 0 for y. Doing so makes certain the point $Q(1, 0)$ is on the line.

From $0 = 4/3 \cdot 1 + b$ we find $b = -4/3$ and thus the desired equation is $y = (4/3)x - 4/3$.

The lines k_1 and k_2 are shown in figure 1.4.10.

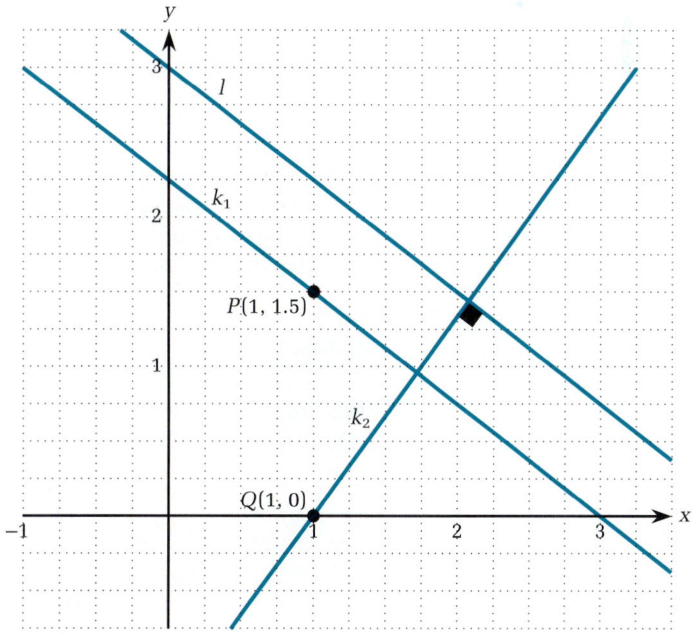

FIGURE 1.4.10

PRACTICE EXERCISE 2

Find an equation of the line parallel to $2x - y = 5$ and containing the point $P(1, 3)$.

Linear Functions

The points on a nonvertical line are the graph of a *linear function*.

DEFINITION   Linear function

The function $f(x) = ax + b$ is a *linear function* and its graph is a line with slope a and with b as its y-intercept.

Linear Models

Suppose we have a collection of data and all of the data points appear to fall on, or close to, a line. Then we may believe a function whose graph is a line will provide a reasonably accurate approximation for the data. Such a model is a *linear model.* As an illustration of the use of a linear model, consider the following example from the world of television and advertising.

EXAMPLE 6   Broadcast television audience share

The percentage of the daytime television audience shared by the broadcast networks ABC, CBS, Fox, and NBC declined in the 1977–1989 period, as reported by the Nielsen poll.* The table below gives the percentage of the total daytime audience watching one of the four broadcast networks.

PROGRAMMING SEASON	NETWORKS' PERCENTAGE OF DAYTIME AUDIENCE
1977/78	78
1979/80	75
1981/82	68
1983/84	65
1985/86	61
1987/88	57

*Denver Post, January 16, 1989.

To better understand the significance of these numbers, let us create the graph in figure 1.4.11. The horizontal axis will be used for the season and the vertical axis for the percentage of the daytime audience.

To use a function with small input values we use a process known as "coding the data" whereby data are given new and simpler values. In this example, we let $x = 0$ represent 1977–1978. Doing so we use $x = 2$ to represent 1979–80, $x = 4$ to represent 1981–1982, and so on up to $x = 10$ to represent 1987–1988. We are increasing x in steps of 2 because the programming seasons are taken in steps of 2 years.

On figure 1.4.11 is also plotted the graph of the linear function $f(x) = 78 - 2x$.

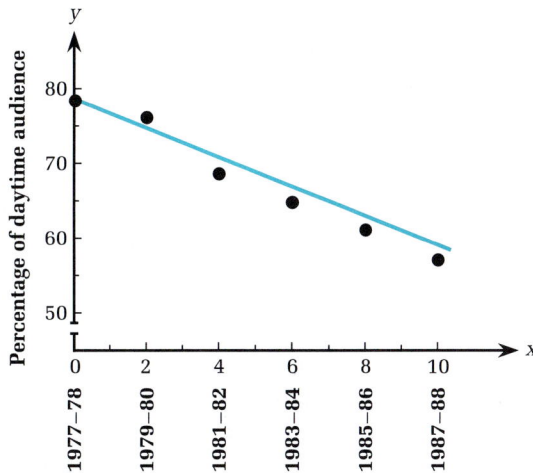

FIGURE 1.4.11

The graph of $f(x)$ comes close to going through the data points. It is not exactly on them, but it is close. Can the function be used to predict the future share of the audience? If so, then for the 1995–1996 season we would use $x = 18$ and find $f(18) = 78 - 2 \cdot 18 = 42$. Thus, the model would predict that in that season only 42 percent of the daytime audience will be watching broadcast network television.

PRACTICE EXERCISE 3

Refer to Example 6, and use $f(x) = 78 - 2x$ as the model for the broadcast network daytime audience.

(a) What is the value of x for the year 2001–2002?

(b) Find $f(24)$ and interpret this value as a percentage of the daytime audience.

EXAMPLE 7   Shoe sizes

Jan Matzeliger (1852–1889) was an African American who lived in Philadelphia and in Lynn, Massachusetts. He revolutionized the shoe manufacturing industry by patenting a machine that mechanized the attaching of the upper part of the shoe to the insole.* Adult shoe sizes start with size 1, having an insole with length 8.875 inches and with each full size being 1/3 inch longer.

(a) Find a formula for adult shoe insole lengths.

(b) How long is the insole of a size 10.5 shoe?

SOLUTION

(a) Let x be the size. Then the insole length is $L(x) = 8.875 + (1/3)(x - 1)$.

(b) $L(10.5) = 8.875 + (1/3)(10.5 - 1) \approx 8.875 + 3.167 \approx 12.04.$

Straight Line Depreciation

There are several standard methods used by accountants to determine the value of an asset over a period of time. One simple method uses a linear function and is called *straight line depreciation.* This model assumes an asset declines in value over time in a linear manner. That is, each year it loses exactly the same fixed amount of its initial value.

Suppose P is initial price, L is the useful lifetime, S is scrap value at the end of that lifetime, and t is a variable used for measuring time. Then the value of the asset at any time t, $V(t)$, is a linear function that passes through the points $(0, P)$ and (L, S), as shown in figure 1.4.12.

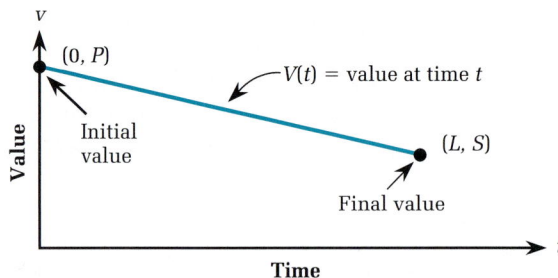

FIGURE 1.4.12

The slope of this line is given by $(S - P)/(L - 0)$, which is more conveniently written $-[(P - S)/L]$.

*D. Karwalka, "Against All Odds," *American Heritage of Invention and Technology* 6:30 (Winter, 1991), pp. 50–55.

The *y*-intercept of the line, *P*, is its price at time $t = 0$. We find the value function, $V(t)$, at time t is thus given by the following formula.

STRAIGHT LINE DEPRECIATION

Suppose P is initial price, L is the useful lifetime, S is scrap value at the end of that lifetime, and t is a variable used for measuring time. Then the value $V(t)$ at time t is given by

$$V(t) = P - \left(\frac{P - S}{L}\right) \cdot t$$

EXAMPLE 8   Straight line depreciation

Consider the value of a car costing $12,000 with a useful lifetime of eight years and a scrap value of $2000. Find $V(t)$, the car's value after t years, using straight line depreciation. Evaluate $V(6)$. Graph $V(t)$ over the useful lifetime of the car.

SOLUTION

$P - S = 12,000 - 2000 = 10,000$, giving us $-[(P - S)/L] = -(10,000/8) = -1250$. Thus, the value function is $V(t) = 12,000 - 1250t$. After six years the value of the car would be $V(6) = 12,000 - (1250) \cdot 6 = 12,000 - 7500 = 4500$.

The graph of $V(t)$ is shown in figure 1.4.13.

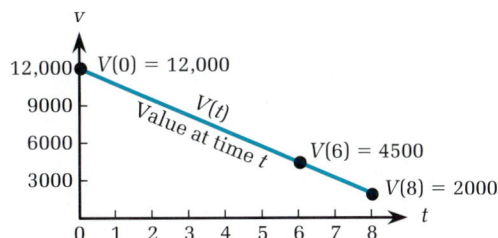

FIGURE 1.4.13

PRACTICE EXERCISE 4

Find the value function, $V(t)$, of a Xerox machine under straight line depreciation. The purchase price is $9000, and after five years the scrap value is $1000.

Piecewise Linear Functions

DEFINITION

Piecewise linear function

A function is *piecewise linear* if its graph is composed of line segments.

The segments may be connected, but that is not always the situation, as is shown by the next example.

EXAMPLE 9   Piecewise linear function

Graph the function $f(x) = \begin{cases} 1 & 0 \leq x \leq 1 \\ -1 & 1 < x \leq 2 \\ 3 & 2 < x < 3 \\ 0 & 3 \leq x \leq 4 \end{cases}$

SOLUTION

The graph of the function is shown in figure 1.4.14.

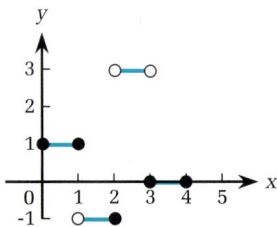

FIGURE 1.4.14

EXAMPLE 10   Piecewise linear function, cost of postage

Consider the function that gave the postage charge for mailing a first-class letter in 1993. The cost of mailing a first-class letter in 1993 was 29 cents for the first ounce and then 23 cents for each additional ounce or fraction thereof. The graph of this function, often called a postage stamp function, is shown in figure 1.4.15.

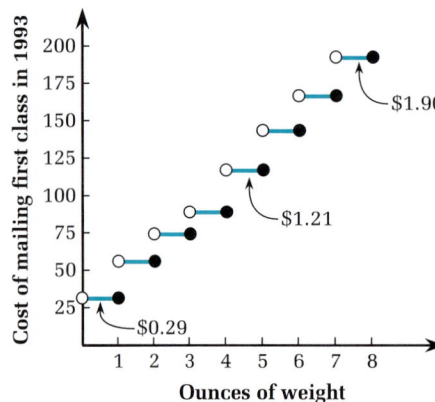

FIGURE 1.4.15

Any function whose graph is similar to the postage stamp function is often called a *step function,* since it consists of unconnected, level pieces, resembling stairsteps.

In some situations our goal is to not have steps, but rather to have pieces of lines that are all connected together. The next example shows such a function.

EXAMPLE 11   Federal income tax

The 1991 United States income tax for a single taxpayer was calculated using a three-part piecewise function, as shown in the table below.

IF TAXABLE INCOME IS OVER	BUT NOT OVER	TAX OWED IS	OF THE AMOUNT OVER
$0	$20,350	15%	$0
$20,350	$49,300	$3052.50 + 28%	$20,350
$49,300		$11,158.50 + 31%	$49,300

Another way of expressing this same information is: If a single taxpayer had a taxable income of x dollars, then the tax due was

$$T(x) = \begin{cases} 0.15x & \text{if } x < 20{,}350 \\ 3052.50 + 0.28(x - 20{,}350) & \text{if } 20{,}350 \leq x < 49{,}300 \\ 11{,}158.50 + 0.31(x - 49{,}300) & \text{if } 49{,}300 \leq x \end{cases}$$

The graph of $T(x)$ is shown in figure 1.4.16.

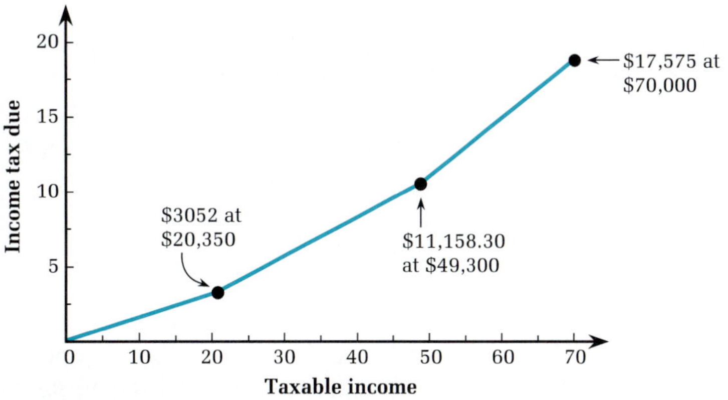

FIGURE 1.4.16

These pieces of lines are connected, as the graph shows. Because of those connections, when taxable income passes one of the "corners" there is not a sudden jump in tax. Notice that $3052.50 is 15% of $20,350. Therefore, the person whose taxable income was $21,000 and in consequence owed $3052.50 + 0.28(21,000 − 20,350) = $3052.50 + 182 = $3234.50 did not pay a lot more than someone whose taxable income was $20,000 and hence owed 0.15(20,000) = $3000.00. Another way of putting this is to say that the person in the 28% tax bracket was only paying that 28% on the amount of taxable income over $20,350, not on the entire income.   ●

Actually, a single person with taxable income of $21,000 in 1991 would have used the tax tables supplied by the government to find that the tax was $3242. Tables with thousands of entries were used because experience had shown that very few people could use even such a simple formula as $T(x)$. While using the tables produced an actual tax-due function that was an unconnected step function, like the postage stamp function, nonetheless the function stayed very close to the graph of $T(x)$. As a final comment, notice that the formula and the tax table values do not give exactly the same tax. This is because of "rounding-off" in the categories.

PRACTICE EXERCISE 5

Using the formula for $T(x)$ from Example 11, find the tax due on a taxable income of $60,000 in 1991.

Direct Proportionality

A special case of a linear model is that of "direct proportionality."

DEFINITION

Directly proportional

The variable y is *directly proportional* to the variable x if there is some nonzero constant k such that

$$y = kx.$$

An example of direct proportionality would be the number of tires used on a motorcycle production line, which is two times the number of motorcycles manufactured. Another example would be the price of bananas, which is directly proportional to the number of pounds bought.

We can find k by knowing some particular values for x and y, as illustrated in Example 12.

EXAMPLE 12   Direct proportionality

Suppose 950 football game fans ate 320 hot dogs. Assume the sale of hot dogs is directly proportional to the number of fans. How many hot dogs will 1500 fans eat?

SOLUTION

In the formula $y = kx$, let $x = 950$ and $y = 320$.

Solving $320 = k \cdot 950$ we have $k \approx 0.3368$ (recall we are usually rounding to four places). Because this decimal is unnecessarily precise, we might use the value $k = 0.34$. Then we would expect 1500 fans to eat approximately $0.34 \cdot 1500 = 510$ hot dogs.   ●

PRACTICE EXERCISE 6

Suppose 500 hot-air balloons use 6000 gallons of propane fuel. If the use of fuel is directly proportional to the number of hot-air balloons, how many gallons of propane fuel will be used by 580 hot-air balloons?

The process of converting units of measurement from one system to another often uses direct proportionality. For example, to change x miles per hour to y kilometers per hour you must know that one mile is equal to 1.609 kilometers. Thus, $y = 1.609x$. Using this formula, 55 miles per hour is $(1.609)55 = 88.5$ kilometers per hour. Another example is the formula $y = (22/15)x$, which converts x miles per hour into y feet per second. By this formula, 60 miles per hour is $(22/15) \cdot 60 = 88$ feet per second.

Notice that for any two variables x and y, if y is directly proportional to x then x is directly proportional to y. This can be seen by realizing if there is some nonzero value k such that $y = kx$, then we can solve this equation for x in terms of y, and have $x = (1/k)y$.

Cost, Revenue, Profit, Breakeven

Let us start with an example of making a publishing decision.

EXAMPLE 13   Economics of publishing

Prudent Press is considering whether or not to publish a new cookbook. The estimated cost of the book is $5000 plus $1.27 for each copy printed. It decides 2000 copies of the book can be sold at $5 per copy. Should it publish the book?

Prudent's cost for 2000 copies would be $5000 + 1.27 \cdot 2000$, which equals $7540. Revenue would be $5 \cdot 2000 = \$10,000$. With a forecast profit of $2460, it decides to publish.

FUNCTIONS OF COST, REVENUE, AND PROFIT

Throughout this book we use the following terminology. If x represents the number of units of a quantity, then

$C(x)$ represents the *cost* of x units,

$R(x)$ represents the *revenue* from x units, and

$P(x)$ represents the *profit* from x units.

DEFINITION

Breakeven

If the cost of a project is equal to the income from the project, then the project is said to be at the *breakeven point* or, simply, to *break even*.

Profit is found by subtracting costs from revenue, or $P(x) = R(x) - C(x)$. Thus breakeven occurs when $P(x) = 0$, which is equivalent to having $C(x) = R(x)$. A graphical representation is shown in figure 1.4.17.

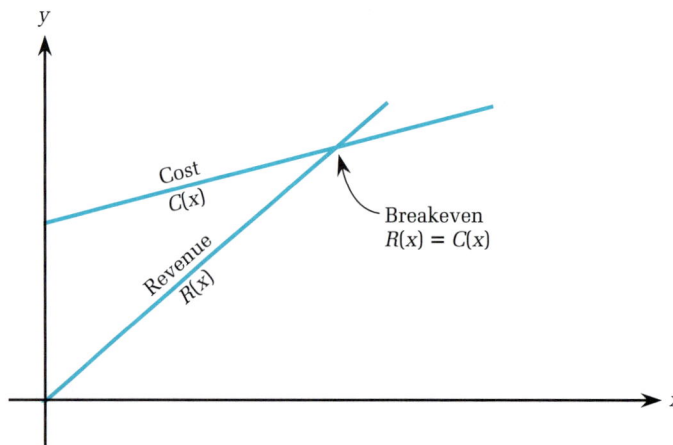

FIGURE 1.4.17

EXAMPLE 14   Breakeven

Suppose in Example 13 that Prudent Press was not certain it could sell 2000 copies. How many would it have to sell to break even?

SOLUTION

To solve this, let us create functions that describe cost, revenue, and profit. Let x stand for the number of books produced, $C(x)$ for their cost, $R(x)$ for the revenue from selling them, and $P(x)$ for the profit realized from selling those x books.

For the Prudent Press cookbook, $C(x) = 5000 + 1.27x$, $R(x) = 5x$, and $P(x) = R(x) - C(x) = 5x - (5000 + 1.27x) = -5000 + 3.73x$.

Breakeven occurs when $R(x) = C(x)$, that is, when $P(x) = -5000 + 3.73x = 0$. Solving for x we have $x = 5000/3.73 = 1340.48$. Because only whole numbers of books are sold, we will say that $x = 1341$ is the breakeven value. Actually, no press is going to print such an odd number of copies, so a more realistic answer would be to say 1400 is the breakeven point.

PRACTICE EXERCISE 7

Suppose you buy a taxicab for $12,000 and it costs you $0.22 per mile to operate the taxi. If you charge $2 a mile for a rider, find the approximate number of miles required to break even.

1.4 PROBLEMS

Foundations

The problems of this section require the basic skills illustrated by the following:

1. Evaluate $\dfrac{3 - (-2)}{-2 + 1}$ **2.** Evaluate $\dfrac{-1}{-\dfrac{3}{2}}$

3. Solve for y if $2x + 3y + 5 = 0$

Exercises

Find the slope of the line containing the following pairs of points.

4. $A(0, 0)$ $B(5, 2)$ **5.** $A(2, 3)$ $B(4, 7)$

6. $A(3, 1)$ $B(4, 0)$ **7.** $A(1, 4)$ $B(3, -1)$

8. $A(2, 3)$ $B(-1, 2)$ **9.** $A(3, 1)$ $B(-2, -3)$

10. $A(0.5, -1)$ $B(0.3, -2)$

11. $A(0, 0)$ $B(-2, 2)$

Graph the line given by the equation and find the slope and the y-intercept.

12. $y = 3$ **13.** $y = -2$

14. $y = 2x - 3$ **15.** $y = \dfrac{1}{2}x + 2$

16. $y = -3x + 2$ **17.** $2x + y = 3$

18. $x - 2y = 3$ **19.** $2x + 3y + 5 = 0$

Find an equation for a line which:

20. has slope 2 and y-intercept -2

21. has slope $-(1/2)$ and y-intercept 3

22. contains the points $P(1, 6)$ and $Q(3, 10)$

23. contains the points $P(2, 5)$ and $Q(4, 1)$

24. contains the point $P(2, 1/2)$ and is parallel to $2x + y = 5$

25. contains the point $P(1, 5)$ and is parallel to $3x + 2y = 1$

26. contains the point $P(2, 3)$ and is perpendicular to $2x + y = 5$

27. contains the point $P(4, 3)$ and is perpendicular to $3x + 2y = 1$

28. Using the formula given in the section, express the speed, measured in feet per second, of a car traveling
    (a) 45 miles per hour   (b) 70 miles per hour.

29. Using the formula given in the text following Practice Exercise 6, express in kilometers the distance from Toledo to Chicago (which is 235 miles).

30. *(Child Shoe Sizes)*   Child shoe sizes start with 0 being an insole length of 3 and 11/12 inches and each full size being 1/3 inch longer.
    (a) Find the equation for the length of an insole for a child's shoe.
    (b) Find the length of the insole of a child's shoe of size 3.

31. *(Straight Line Depreciation)*   The purchase price of a color TV is $450, and over ten years its value drops to a salvage value of $50.
    (a) Find a formula that will give the value of the TV after t years, using straight line depreciation.
    (b) Find the value of the TV at the end of the sixth year.

32. *(Straight Line Depreciation)*   The purchase price of an ice-cream machine is $6000, and after five years it will have a salvage value of $2000.
    (a) Find a formula that will give the value of the ice-cream machine after t years, using straight line depreciation.
    (b) Find the value of the machine at the end of the third year.

33. Temperatures are often expressed in °F, "degrees Fahrenheit," or in °C, "degrees Celsius." Using the equivalencies that 0°C = 32°F and 100°C = 212°F, find a formula to convert a measurement in Fahrenheit to a value on the Celsius scale.

34. *(Retail Discounting)*   A department store sold Red Top socks at a regular price of $1 a pair. As a special sale, they offered a discount of 20% for eight pairs or less and 25% for nine pairs or more. (For example, six pairs would be $4.80 and ten pairs would be $7.50.) Find a piecewise linear function that would describe this pricing, letting x be the number of pairs a customer bought and $p(x)$ the price paid.

35. *(Traffic Fines)*   A county judge assessed fines for speeding based on the following: For speeds from 55 mph to 75 mph, the fine was $30 plus $2 for every mile per hour over 55 mph. For 75 mph and higher, the fine was $100 plus $5 for every mile per hour over 75 mph. Find a piecewise linear function that describes this system of fines, letting x be the miles per hour and $f(x)$ be the amount of the fine.

36. *(Breakeven Analysis, Videos)*   Bill's Video Rentals buys videos for an average price of $46 each and rents them for $3.25 a time. Whenever a video is rented there are costs such as paperwork, building rent, and so on that have to be paid, and Bill estimates those at $0.55 per video rental.
    (a) How many times must a video be rented for Bill to break even?
    (b) What would be the profit (loss) on a video he rented 15 times?
    (c) What would be the profit (loss) on a video he rented 25 times?

37. *(Commissions on Sales)*   Frankie is hired to sell balloons along the parade route, and she can choose between two alternate wage plans. Plan A pays her 30% of the sales. Plan B pays her $3.75 an hour plus 25% of the sales. Frankie believes she can sell $150 worth of balloons an hour.
    (a) Which plan should she choose?
    (b) For what dollar sales per hour would the two plans pay Frankie the same wage?

38. *(Directly Proportional)*   The sale of balloons is directly proportional to the size of the crowd. Last week 325 balloons were sold to a crowd of approximately 4000 people.
    (a) This week the crowd is expected to be about 7000. What is the projected sale of balloons?
    (b) How big would the crowd have to be to produce a sale of 1000 balloons?

39. (Directly Proportional and Breakeven)   The sale of boxes of popcorn is directly proportional to the size of the crowd. For 200 people, 40 boxes are sold. The fixed cost of setting up the popcorn stand is $13.50 and the cost of popcorn is $0.10 per box.
   (a) Suppose the popcorn is sold at $1.00 per box.
      (i)   How many boxes must be sold for the stand to break even?
      (ii)   How large must the crowd be for the popcorn stand to break even?
      (iii)   What is the profit (or loss) on a crowd of 250?
   (b) Answer the three questions of part (a) under the assumption that popcorn is sold for $0.35 per box.

40. (Whales)   Based on a study of 170 whales, it was found that the mass of their skeletons was related to their total body mass by
$M_{skel} = 0.105 M_{body}$.*
   (a) Estimate the mass of a skeleton of a whale whose body mass was 1.8 tons.
   (b) Estimate the total body mass of a whale whose skeleton weighed 0.9 tons.

41. (Plastics Recycling)   In 1993, the Union Carbide Chemicals and Plastics Company operated a plastic recycling plant in Bound Brook, New Jersey. Each month the plant converted 4.5 million pounds of plastic bottles into pellets. Compressed plastic bottles weigh about 5.2 pounds per cubic foot.
   (a) Find the number of cubic feet of compressed plastic bottles that the plant converted each month.
   (b) An acre is 43,560 square feet. Over an area of one acre, how deep would be the compressed bottles that represented 2.5 million pounds?

42. (TV Audience Share)   Nielsen reported that the share of the prime-time television audience held by the four networks (ABC, CBS, Fox, and NBC) was 91% in the 1977–1978 season and 73% in the 1985–1986 season. Assume a linear decline of audience share.
   (a) Find a formula expressing the percentage share of the prime-time television audience held by the networks as a function of t (where $t = 0$ at 1977–1978 and $t = 8$ at 1985–1986).

*K. Schmidt-Nielson, *Scaling: Why Is Animal Size So Important?*, Cambridge: Cambridge University Press, 1984, p. 48.

   (b) What percentage does your formula predict for the 1987–1988 season? (The actual Nielsen figure was 66% for 1987–1988.)

43. (Baseball)   An outstanding baseball pitcher throws a fastball at a rate of about 90 miles per hour. The distance to home plate is about 60 feet.
   (a) Estimate the number of seconds it takes for the pitch to reach home plate.
   (b) A fastball is spinning at a rate of about 1600 revolutions per minute. Estimate the number of revolutions en route to home plate.

44. (Consumer Spending on Poultry)   The amount of money spent annually per capita in the United States on poultry increased (approximately) linearly from $25 in 1975 to $48 in 1987.
   (a) Using $t = 0$ for 1975 and $t = 12$ for 1987, find a formula that will fit a linear model for consumer spending on poultry.
   (b) Estimate the per capita spending on poultry in 1990.

45. (Drug Dosages for Children)   Pharmaceutical manufacturers suggest dosages for children as well as adults. Two suggested models for modifying adult dosages for children are:

$$Cowling's\ Rule: y = \frac{t + 1}{24} a$$

and

$$Friend's\ Rule: y = \frac{2}{25} ta$$

In each formula, t is the child's age, a is the adult dosage, and y is the child's dosage.
   (a) Suppose the adult dosage is 75 mg. Find the dosage for a 4-year-old child suggested by each rule.
   (b) Suppose the adult dosage is 50 mg. Find the dosage for an 8-year-old child suggested by each rule.
   (c) For what aged child (if ever) do Cowling's and Friend's rules suggest the same dosage?
   (d) For each rule, how old would the child have to be in order to be prescribed an adult dosage? (In other words, what are natural domains for each rule?)

46. *(Breakeven on Pharmaceutical Manufacturing)* A particular drug cost $125,000 to develop and bring on the market. It sells for $20 per hundred and costs $15 per hundred to manufacture. How many hundreds must be sold for the manufacturer to break even?

47. *(Caloric Consumption)* The chart shown below comes from information collected by the American Dietetic Association. The chart gives the number of calories burned per hour by a 137-pound person who is performing the listed activity.

ACTIVITY	CALORIES PER HOUR
Jogging or aerobic exercise	500
Playing racquetball	660
Downhill skiing	365
Ballroom dancing	192
Shopping	240
Sitting quietly on the couch	84

  (a) How many calories does the jogger burn in 1.5 hours?
  (b) How many calories are burned by the racquetball player in 45 minutes?
  (c) How long does it take such a person to burn 1000 calories while
      (i)   doing aerobics?
      (ii) shopping?
      (iii) sitting quietly on the couch?
  (d) Find the difference between the total number of calories burned over 2 hours by a ballroom dancer and by someone sitting quietly on the couch.

48. *(Cost, Revenue, Profit, Breakeven)* TopCal Steam Co. spent $200,000 before they were able to generate any electricity with geothermal steam. Once in production, it costs them 3 cents per kilowatt-hour (kwh) to generate power, which they then sell for 7 cents per kilowatt-hour. Using x to stand for the number of kwh, write:
  (a) a revenue function,
  (b) a cost function, and
  (c) a profit function for TopCal.
  (d) How many kilowatt-hours must they sell to break even?

  (e) What is their profit (loss) on 8,000,000 kwh?

49. *(Cost, Revenue, Profit, Breakeven)* Jake has $30 a day in fixed costs in running a hot dog cart. This includes the cost of his license, the fuel to heat the hot dogs, and so forth. His average daily cost for each hot dog (including raw hot dog, bun, catsup, and everything else) is 72 cents.
  (a) Selling them for $1 each, how many hot dogs does he need to sell to break even?
  (b) What would be his profit (loss) on selling 120 hot dogs?

50. Refer to problem 49. At what price should Jake sell his hot dogs if he wants to break even selling about 80 hot dogs?

51. *(Forestry Logging)* The height of a lodgepole pine tree can be closely approximated by $H = 5.5 + 0.6D$, where D is the diameter of a stump cut 15 centimeters from the ground. D is measured in centimeters and H in meters.*
  (a) Estimate the height of a tree of 20 cm diameter.
  (b) Suppose loggers contract to cut only trees at least 20 meters tall. What stump diameters are they permitted to cut?

Graphing Calculator Problems

52. Let $y = 0.05x + 35$.
  (i) Graph this equation using the following ranges of x and y to determine screens
       (a) $-4.7 \le x \le 4.7$, $-3.1 \le y \le 3.1$
           (Why is the screen blank?)
       (b) $-75 \le x \le 75$, $-50 \le y \le 50$
       (c) $-800 \le x \le 100$, $0 \le y \le 50$
       (d) $-800 \le x \le 100$, $30 \le y \le 40$
  (ii) Suggest a method that uses the x- and y-intercepts to determine a preliminary range of screen values.

53. Graph the lines given by the following equations and use the ⎡trace⎤ or ⎡box⎤ function to estimate the point of intersection: $x + 2y = 1.8$ and $4y = 7x$.

---

*P. Koch and J. Schlieter, *Spiral Grain and Annual Ring Width in Natural Unthinned Stands of Lodgepole Pine in North America*, USDA Forest Service Intermountain Research Station, Research Paper INT-499, September, 1991.

Writing and Discussion Questions

54. What is a line?

55. Describe a procedure for finding the x-intercept of the equation of a line.

56. The criterion for perpendicularity is: Lines with slopes m_1 and m_2 are perpendicular to each other if and only if $m_1 \cdot m_2 = -1$. Explain why the criterion does not apply to the situation of a vertical line and a horizontal line. What can be said to cover the situation where one of the lines is horizontal?

57. Go to the library and find the tax schedule rates for the current year. Then construct the function for $T(x)$, the amount of tax due by a single taxpayer, in the same manner as was done in Example 11.

58. *(Misuse of a Linear Model)* In the years 1952–1955, the actual annual appropriations by the U.S. government were as follows (amounts in billions of dollars):

1952: 127.8; 1953: 94.9; 1954: 74.8; 1955: 54.7. These data can be approximated fairly well by the graph of $A(t) = -24t + 126$, where $t = 0$ in 1952 and $t = 3$ in 1955.
  (a) What would this model have predicted as the annual appropriation for 1956?
  (b) In what year would the government have quit spending money?
  (c) Discuss the limitations of linear models.

Enrichment Problems

59. Refer to problem 49. Joan opens a competing hot dog stand and charges 95 cents. Jake knows that he is selling about 200 hot dogs a day and he can keep all his customers if he cuts his price to 90 cents.
  (a) Can he do this and still break even?
  (b) How low a price can he charge and still break even on selling 200 hot dogs?

60. Refer to problem 39. Suppose the crowd is 250 people. What is the lowest price per box the popcorn stand can charge and still earn a profit of $10.00?

SOLUTIONS TO PRACTICE EXERCISES

1. The line $y = -2x + 4$ has slope -2 and a y-intercept of 4, as shown in figure 1.4.18.

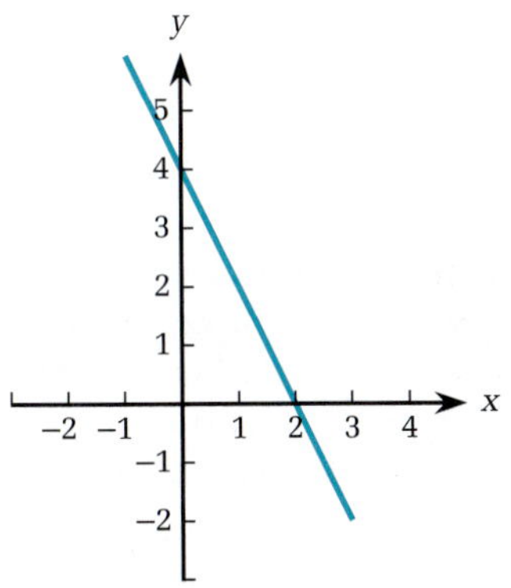

FIGURE 1.4.18

2. $y = 2x - 5$ has a slope of 2. A parallel line would have the same slope. Hence we need to find the value of b such that $y = 2x + b$ contains the point $P(1, 3)$. Substituting $x = 1$ and $y = 3$, we have $3 = 2 \cdot 1 + b$, so $b = 1$ and the desired equation is $y = 2x + 1$.

3. (a) The season 2001–2002 is coded as $x = 24$.
  (b) $f(24) = 78 - 2 \cdot 24 = 30$. In that year the networks would have 30% of the daytime TV audience.

4. For $P = 9000$, $S = 1000$, and $L = 5$, we have $V(t) = 9000 - (8000/5) t = 9000 - 1600t$.

5. The tax on a taxable income of $60,000 was $11,158.50 + 0.31(60,000 - 49,300) = $14,475.50$.

6. From $6000 = k \cdot 500$ we find $k = 12$. Thus, 580 balloons will need approximately $12 \cdot 580 = 6960$ gallons of propane.

7. The total cost of operating the taxi for x miles is $C(x) = 12,000 + 0.22 \cdot x$ and revenue is $R(x) = 2x$. Breakeven occurs if $C(x) = R(x)$, which happens when $12,000 + 0.22 \cdot x = 2x$. Solving this for x, the breakeven is at $x \approx 6742$ miles.

1.5   POLYNOMIAL FUNCTIONS AND THEIR ZEROES

This section begins with factoring quadratic expressions and reviews the use of factoring and the quadratic formula as techniques for solving quadratic equations. The maximum and minimum values of quadratic functions are studied with a view to applications, especially of revenue and profit functions.

As one application, we will find out why the North Carolina legislators made a costly mistake in 1990 when they voted for a $10 increase in the price of vanity license plates, which are vehicle plates with messages on them, such as PAID 4 or MATH 4 ME.

The last part of the section concerns general polynomial functions.

MATHEMATICAL INNOVATORS
Fra Luca Pacioli (1445–1514) and Girolamo Cardan (1501–1576)

Five hundred years ago the European center of mathematical writing and thought was in Italy, the crossroads of all Mediterranean trade and learning. In Venice, an influential encyclopedia of mathematics was published in 1494 by Fra Luca Pacioli, a Franciscan friar. Pacioli was the first to explain to the general public the mechanics of double-entry bookkeeping. He is still honored as being one of the "fathers of accounting."

Shortly after Pacioli's work, methods for solving all cubic (3rd degree) and quartic (4th degree) equations were discovered by several other Italian mathematicians. Most of their work appeared in *Ars Magna,* a book written by Girolamo Cardan and printed in 1545.

Girolamo Cardan was a famous physician, outstanding mathematician, and prolific writer (131 published books). He was also an astrologer and a shameless scoundrel. His biography provides a fascinating glimpse of Renaissance life in the Italian city-states. From scholarly debates, conducted with bitter passion, to poisonings, all too frequent in his day, Cardan experienced it all.

Factoring Quadratic Expressions

On the graph of $y = f(x)$, we usually want to indicate values of x such that $f(x) = 0$; such values are called *zeroes* of the function. To solve $f(x) = 0$, it is often helpful if we can write $f(x)$ as a product of other expressions, called *factors.* The process of finding these factors is called *factoring.*

Factoring is important because if a product of factors is equal to zero, then one of those factors must equal zero. That is, if $r \cdot s = 0$, then $r = 0$ or $s = 0$. Hence, if we are solving $f(x) = 0$, we may be able to solve the easier problem of determining when the factors of $f(x)$ are zero.

EXAMPLE 1   The usefulness of factoring

Solve $x^2 + x - 6 = 0$.

SOLUTION

The expression $x^2 + x - 6$ factors into $(x + 3)(x - 2)$. Hence, solving $x^2 + x - 6 = 0$ is equivalent to solving $(x + 3)(x - 2) = 0$. Using the fact that a product of factors is zero only if one of those factors is zero, we have $x + 3 = 0$ or $x - 2 = 0$. Thus, $x = -3$ or $x = 2$. ●

FACTORING PERFECT SQUARE EXPRESSIONS

Two formulas that help us factor quadratic expressions are:

$$x^2 + 2sx + s^2 = (x + s)^2 \quad \text{and} \quad x^2 - 2sx + s^2 = (x - s)^2$$

These are called *perfect squares.*

EXAMPLE 2   Factoring a perfect square

Factor $x^2 + 6x + 9$.

SOLUTION

Look at the coefficient of x, in this case 6. If the square of one-half of it is the constant term, as it is since $6/2 = 3$ and $3^2 = 9$, then the quadratic is of the form $x^2 + 2sx + s^2$. By the formula for a perfect square, we have $x^2 + 6x + 9 = (x + 3)^2$. ●

FACTORING GENERAL QUADRATIC EXPRESSIONS

To factor $ax^2 + bx + c$, try to find values of r, s, t, and u so that

$$ax^2 + bx + c = (rx + s) \cdot (tx + u)$$
$$= rtx^2 + (ru + st)x + su$$

Finding these values, if indeed any exist, can be complicated. Thus, usually we search only if we have small integer values for a, b, and c. Even then, we only try integers for the other constants.

EXAMPLE 3   Factoring

Factor $x^2 + 8x + 15$.

SOLUTION

With $a = 1$, $b = 8$, and $c = 15$, we are searching for r and t so that $rt = 1$. Since we will check only integer solutions, and $r \cdot t = 1$, we let $r = 1$ and $t = 1$.

Next, we need u and s so that $u \cdot s = 15$ and $u + s = 8$. The factorings of 15 are $1 \cdot 15$, $3 \cdot 5$, $-1 \cdot -15$, and $-3 \cdot -5$. For each choice of u and s, we check the value of $u + s$ and find

$15 = u \cdot s$	$1 \cdot 15$	$3 \cdot 5$	$-1 \cdot -15$	$-3 \cdot -5$
$u =$	1	3	-1	-3
$s =$	15	5	-15	-5
$u + s =$	16	8	-16	-8

We want $u + s = 8$, so we use $u = 3$ and $s = 5$. Hence, we find

$$x^2 + 8x + 15 = (x + 3)(x + 5)$$

Actually, as soon as we found $3 + 5 = 8$, we would quit checking values of $u + s$.

PRACTICE EXERCISE 1

Factor $x^2 + 3x - 10$.

The Quadratic Formula

Suppose we have a quadratic expression, $p(x)$, and do not know a factored form for it. The expression could be one such as $x^2 + 7x - 800$, which in fact does factor into $(x + 32)(x - 25)$, but whose factoring involves much drudgery. Or, we simply may be unsuccessful in a search for factors. In either case, to solve $p(x)$, we can always use the following formula.

THE QUADRATIC FORMULA

For $f(x) = ax^2 + bx + c$, with $a \neq 0$, the solutions of $f(x) = 0$ are given by the *quadratic formula*

$$x = \frac{-b \pm \sqrt{b^2 - 4ac}}{2a}$$

In the quadratic formula the "$\pm$" sign means that one solution is created using the "+" and one solution is created using the "$-$." The value $b^2 - 4ac$ is the *discriminant,* and its value determines the number of solutions. If $b^2 - 4ac > 0$, there are two real solutions. If $b^2 - 4ac < 0$, there are no real solutions. When $b^2 - 4ac = 0$, there is one solution.

Quadratic equations that arise in realistic applications almost never factor, forcing us instead to use the quadratic formula. Nonetheless, although factoring plays a limited role as a method of solving quadratic equations, the ability to factor expressions is quite useful, as will be shown in later sections.

EXAMPLE 4   Using the quadratic formula

Solve $2x^2 - 5x - 4 = 0$.

SOLUTION

Using the quadratic formula with $a = 2$, $b = -5$, and $c = -4$, we have

$$x = \frac{-b \pm \sqrt{b^2 - 4ac}}{2a} = \frac{-(-5) \pm \sqrt{(-5)^2 - 4(2)(-4)}}{2(2)}$$

$$= \frac{5 \pm \sqrt{25 + 32}}{4} = \frac{5 \pm \sqrt{57}}{4}$$

WARNING

When we evaluate expressions such as the quadratic formula, we should be careful about how we keep track of the products and the minus signs. Using parentheses is one good means to do this.

Quadratic Functions

Let us review some terminology and results about quadratic functions.

Quadratic function

A *quadratic function* is a function of the form $f(x) = ax^2 + bx + c$ with $a \neq 0$.

The graph of a quadratic function is a *parabola*. The graph opens up if $a > 0$ (figure 1.5.1) and opens down if $a < 0$ (figure 1.5.2). The *vertex* is the high point of a parabola that opens down and the low point of a parabola that opens up.

The location of the vertex may be found by *completing the square,* which is a technique of rewriting $ax^2 + bx + c$ in the form $a(x - h)^2 + k$. The vertex of $y = a(x - h)^2 + k$ is located at $V(h, k)$. To see this, realize that for any value of x, $(x - h)^2 \geq 0$ and is 0 only for $x = h$. Hence the graph of $y = a(x - h)^2 + k$ is always above the horizontal line $y = k$ for $a > 0$, and it is always below the line $y = k$ for $a < 0$. The situation is shown in figure 1.5.3.

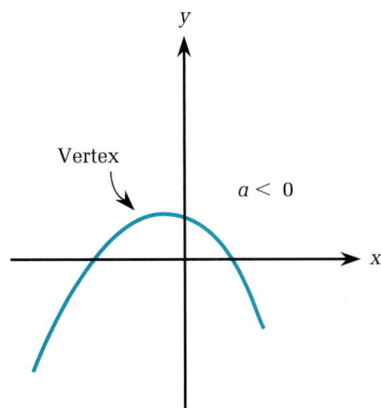

FIGURE 1.5.1

FIGURE 1.5.2

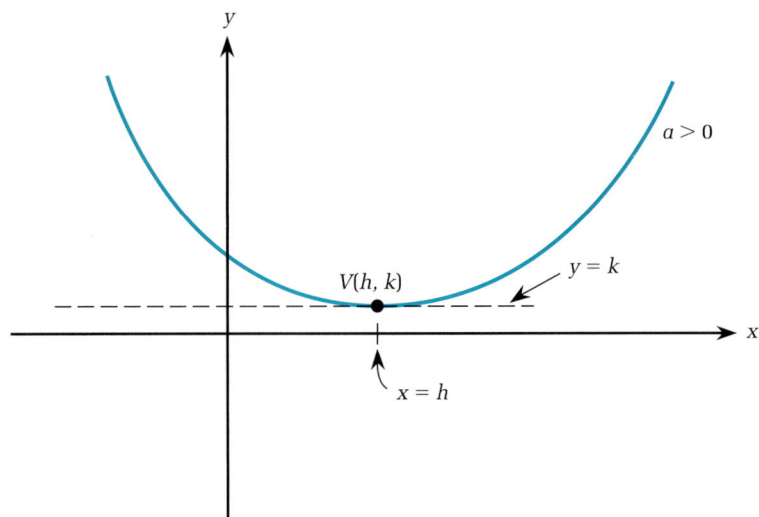

FIGURE 1.5.3

EXAMPLE 5   Finding the vertex by completing the square

Graph $f(x) = 4x^2 - 4x - 1$.

SOLUTION

First, write the coefficient of x^2, in this case the value 4, as a factor of both the term involving x^2 and the term involving x.

$$4x^2 - 4x - 1 = 4(x^2 - x) - 1$$

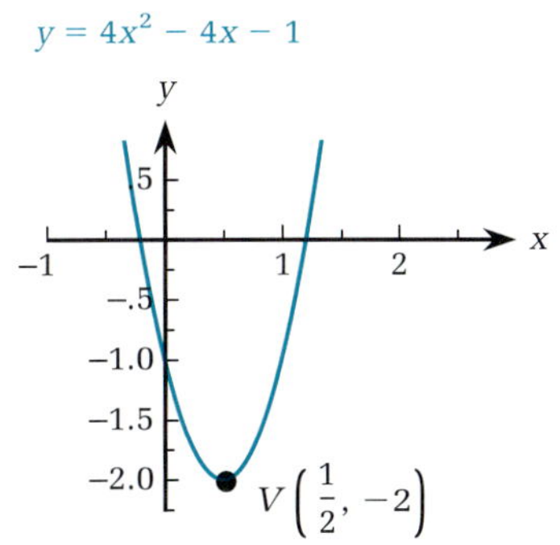

$y = 4x^2 - 4x - 1$

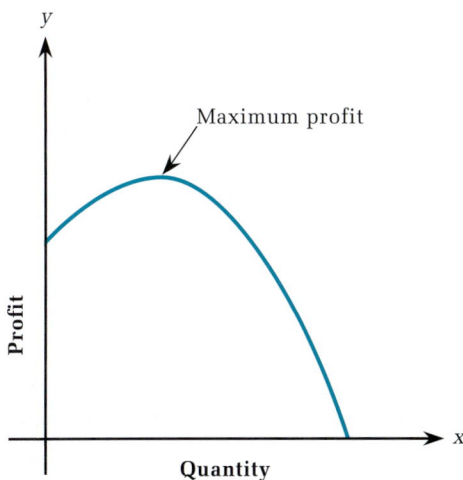

FIGURE 1.5.4

In order to have the expression in the parentheses, $x^2 - x$, be a perfect square, we need to find one-half of the coefficient of x and add its square to that expression. Since $1/2 \cdot 1 = 1/2$ and $(1/2)^2 = 1/4$, we have

$$4x^2 - 4x - 1 = 4(x^2 - x + \boxed{1/4}) - 1 \boxed{-1}$$

The two framed terms compensate for each other. The second one is $-4 \cdot 1/4 = -1$ because we added $4 \cdot 1/4$ in our creation of a square term. Rewritten, we now have

$$4x^2 - 4x - 1 = 4(x - 1/2)^2 - 2$$

The vertex is therefore located at $V(1/2, -2)$ and the parabola opens upward, as shown in figure 1.5.4.

The process of completing the square can be used to show the next result.

THE VERTEX OF A PARABOLA (THE GRAPH OF A QUADRATIC FUNCTION)

The vertex of the graph of $y = f(x) = ax^2 + bx + c$ is located at

$$V\left(-\frac{b}{2a}, f\left(-\frac{b}{2a}\right)\right)$$

The vertex is an extremely important point on a parabola and is usually indicated on the graph. For example, if the quadratic function represents profit, then the vertex of the parabola may give the maximum profit (figure 1.5.5). On the other hand, if it represents cost, then the vertex of the parabola may give the minimum cost (figure 1.5.6).

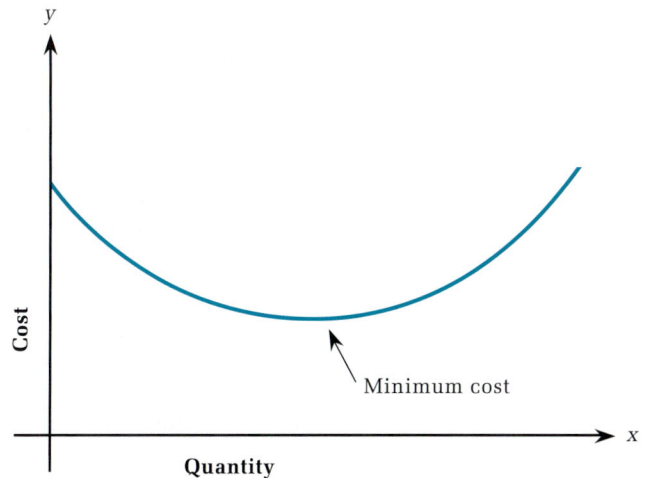

FIGURE 1.5.5

FIGURE 1.5.6

In addition to the vertex, other points that are usually plotted include those at which the graph crosses the x-axis (if it does so) and y-axis. To find the crossing of the y-axis, we substitute 0 for x and find $f(0)$, which is the c value. Any crossings of the x-axis are determined by solving $f(x)$ by the quadratic formula or by factoring.

EXAMPLE 6   The graph of a quadratic function

Sketch a graph of $f(x) = 3x^2 - x - 1$.

SOLUTION

For this function, $a = 3$, $b = -1$, and $c = -1$. Since $f(0) = -1$, the graph crosses the y-axis at -1. The vertex has an x-coordinate of $-b/(2a) = -(-1)/(2 \cdot 3) = 1/6$. By evaluating $f(1/6)$, we find the vertex has a y-coordinate of $f(1/6) = 3 \cdot (1/6)^2 - (1/6) - 1 = 1/12 - 1/6 - 1 = -(13/12) \approx -1.0833$. Hence the vertex is located at about $V(0.1667, -1.0836)$.

By the quadratic formula, the function crosses the x-axis when

$$x = \frac{-b \pm \sqrt{b^2 - 4ac}}{2a}$$

$$= \frac{-(-1) \pm \sqrt{(-1)^2 - 4 \cdot 3 \cdot (-1)}}{2 \cdot 3}$$

$$= \frac{1 \pm \sqrt{1 + 12}}{6}$$

$$= \frac{1 \pm \sqrt{13}}{6} \approx \frac{1 \pm 3.6056}{6}$$

Thus, the x-intercepts are approximately 0.7677 and -0.4343.

In addition to the four points we now know, a few additional points might give us a better picture of the graph. We may choose x values that are near other, interesting values, or x values that would help us fill in a large gap. For example, consider $x = 1$, for which $f(1) = 3 \cdot (1)^2 - (1) - 1 = 1$. Thus, the point $P(1, 1)$ is on the graph.

We often use inputs that are integers because the arithmetic and algebra generally are easier with an integer. However, we have already seen one interesting point that did not have an integer x-value (the vertex), so we know we should not restrict ourselves to only integer x-values.

For instance, even though we have no reason to suspect $x = 1.2$ provides an especially interesting point, let us find $f(1.2) = 3 \cdot (1.2)^2 - 1.2 - 1 = 4.32 - 1.2 - 1 = 2.12$. Thus, $Q(1.2, 2.12)$ is yet another point on the graph.

As we become more familiar with the graphs of different functions, only rarely do we plot more than five or six points. The graph of $f(x)$ for this example is shown in figure 1.5.7.

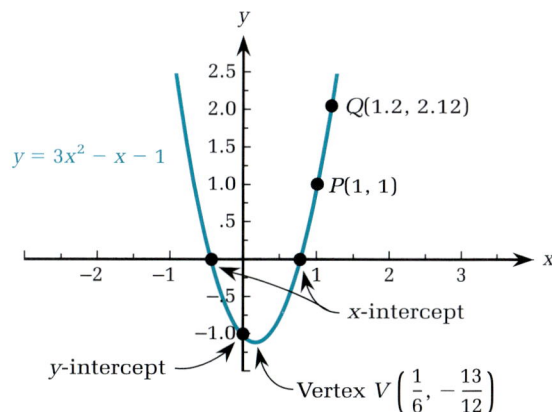

FIGURE 1.5.7

PRACTICE EXERCISE 2

Find the vertex and intercepts for $f(x) = 2x^2 - x - 2$.

Applications of Quadratic Functions

By locating the vertex of the graph of a quadratic function we can solve several types of maximize/minimize problems. These problems ask for the maximum of some quantity (such as production, revenue, or profit) or the minimum of some quantity (such as cost).

EXAMPLE 7   Maximizing oil production

Suppose an oil field currently has five wells per square mile, each well producing 600 barrels of oil a day. A company is thinking of drilling additional wells, but as these would be drilled, the decreased underground pressure would cause the production in each and every well to decline. After tests, the company decides that if it would drill x new wells, the production in each well, both the old and the new wells, would be only $600 - 40x$ barrels of oil a day. Find the number of new wells that should be drilled to maximize total production and find that production.

SOLUTION

Total production is given by the number of wells, multiplied by the production of each well. If x new wells are drilled, then the total number of wells is $5 + x$. Each well would be producing $600 - 40x$ barrels of oil, so that the total production would be

$$P(x) = (5 + x) \cdot (600 - 40x) = 3000 + 400x - 40x^2$$

This quadratic has $a = -40$, so the curve opens down and has a maximum value at the vertex. By the vertex formula, the x-coordinate at the vertex is

$$x = -\frac{b}{2a} = -\frac{400}{2(-40)} = -\frac{400}{-80} = 5$$

The y-coordinate of the vertex is given by substituting 5 in $P(x)$, which gives us

$$P(5) = 3000 + 400 \cdot 5 - 40 \cdot 5^2 = 3000 + 2000 - 1000 = 4000$$

Hence, if the company drills five new wells, there will be a total of ten wells per square mile, and total production will be maximized at 4000 barrels a day. If fewer or more than five new wells are drilled, production will be less than 4000 barrels a day. The graph of $P(x)$ is shown in figure 1.5.8.

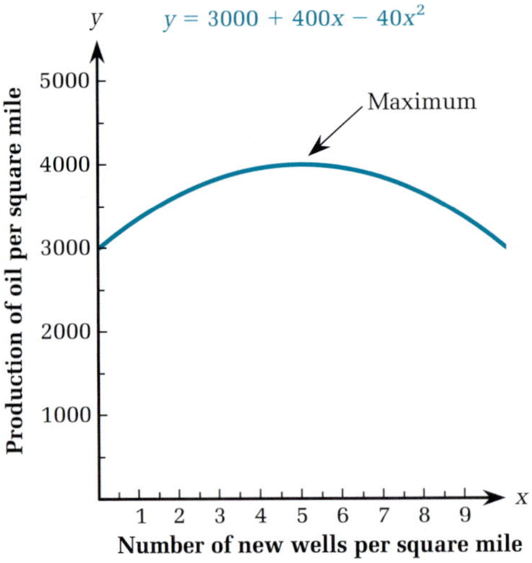

FIGURE 1.5.8

To use the quadratic and vertex formulas, it is helpful to rewrite the quadratic in the form $P(x) = ax^2 + bx + c$. However, notice that the expression $P(x) = (5 + x) \cdot (600 - 40x)$ states the production situation in a clearer way because it gives the production as "the number of wells multiplied by the output per well." Of course, this gives the same total production of

$$P(5) = (5 + 5) \cdot (600 - 40 \cdot 5) = 10 \cdot 400 = 4000$$

PRACTICE EXERCISE 3

Suppose, in Example 7, the production in each well would be $600 - 24x$ barrels of oil a day. Find the number of new wells that should be drilled to maximize total production and find that maximum production.

Demand Functions in Business

The demand for an item is related to its price. As the price increases, the demand decreases; conversely, as the price decreases, the demand increases.

We may think of the demand as a function of price, or we may think of the price as a function of demand. Economists prefer the second format. They reason that price should be a function of the quantity sold, just as revenue, cost, and profit are all functions of the quantity sold. As a result, although many people would rather think of the quantity sold graphed as a function of price, as shown in figure 1.5.9, we will use the standard form used by economists and graph price as a function of quantity, shown in figure 1.5.10. Thus, the demand function will be written as $p = p(x)$, where p is the price and x is the quantity. In some instances, the demand function is named $d(x)$.

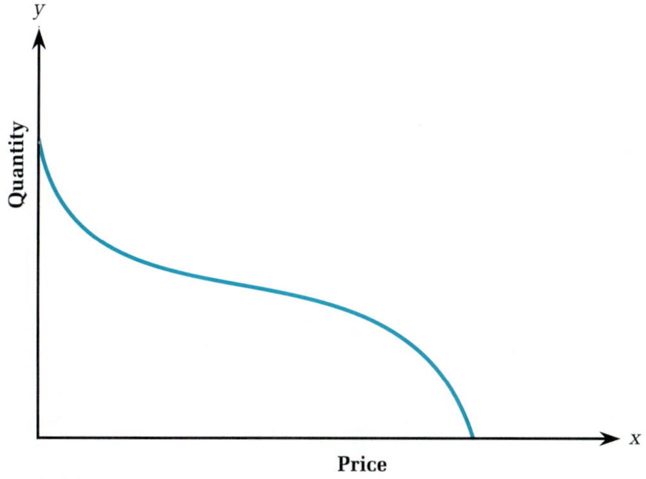

FIGURE 1.5.9

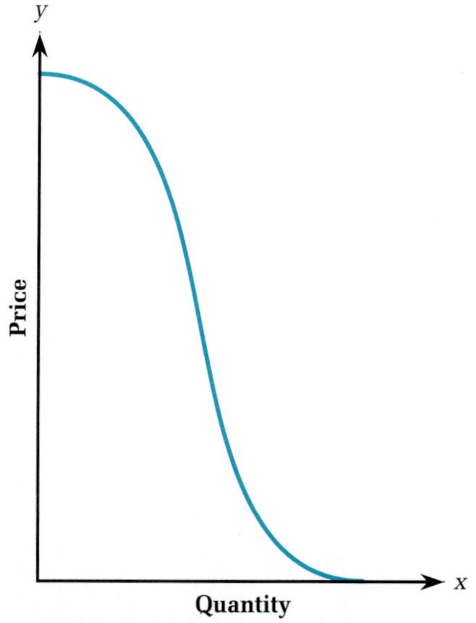

FIGURE 1.5.10

THE REVENUE FUNCTION IN TERMS OF THE DEMAND FUNCTION

Suppose the demand function is $p(x)$ and it gives the price per item as a function of x items. Then the revenue function, $R(x)$, is given by $R(x) = x \cdot p(x)$.

Example 8 is an example of a demand function for a particular product.

EXAMPLE 8   Demand function of vanity license plates

The state of North Carolina planted wildflowers along roadsides using designated revenues earned from the sale of vanity license plates for vehicles. Such plates might read MATH-4-ME or NEVER-L8. During a period in 1989 they sold 60,354 vanity plates at $30 each. Thus, $1,810,620 was raised for roadside beautification. The 1989 legislature decided to raise the fee to $40. They assumed that while some people might not be willing to pay the extra $10, nonetheless most people would do so. They made a serious mistake. In the same period in 1990, only 31,122 plates were sold. Thus, revenue plummeted to a mere $1,244,880.*

Let us simplify the model and use the approximation that 30,000 licenses would be sold at $40 and 60,000 licenses would be sold at $30. Further, let us assume the demand is linear.

(a) Find the equation for a linear demand function.

(b) Find the revenue function.

(c) Determine how many plates would yield the greatest revenue.

(d) Find the price of those plates.

(e) Find the greatest total possible revenue for the sale of vanity license plates.

SOLUTION

(a) On figure 1.5.11 are shown the two known points of the demand function and the line connecting them.

$$y = -\frac{1}{3000}x + 50$$

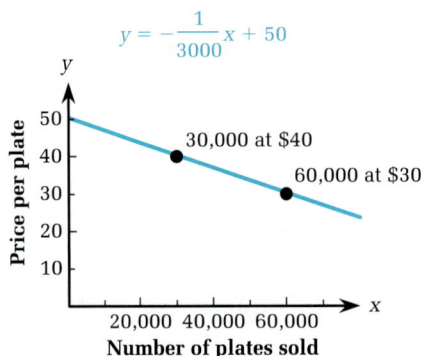

FIGURE 1.5.11

*Time Magazine, July 16, 1990, p. 29.

The slope of the line is $(30 - 40)/(60{,}000 - 30{,}000) = -(10/30{,}000) = -(1/3000)$. Thus, the line has the form $y = -(1/3000)x + b$. We determine b by using the fact that the point $(60{,}000, 30)$ is on the line.

From $30 = -(1/3000)(60{,}000) + b = -20 + b$ we find $b = 50$. Therefore, $p(x) = -(1/3000)x + 50$ is the demand function.

(b) The revenue, $R(x)$, is the product of the quantity of plates sold times the price for each license plate. Hence $R(x) = x \cdot p(x) = x(-(1/3000)x + 50) = -(1/3000)x^2 + 50x$.

$R(x)$ is graphed in figure 1.5.12.

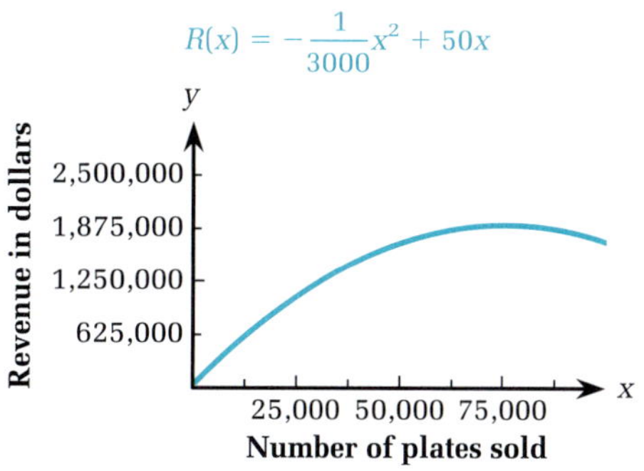

$$R(x) = -\frac{1}{3000}x^2 + 50x$$

FIGURE 1.5.12

(c) The maximum revenue occurs for $x = -b/(2a) = -50/(2(-1/3000)) = 75{,}000$.

(d) The price of plates that generates the maximum revenue is $p(75{,}000) = -(1/3000)(75{,}000) + 50 = -25 + 50 = 25$.

(e) The maximum revenue is $R(75{,}000) = 75{,}000 \cdot 25 = 1{,}875{,}000$.

As a result, if we assume the model to be fairly accurate, North Carolina was already very close to achieving the maximum possible revenue from the sale of vanity license plates, and they could only do better by *reducing* the cost of the plates to $25. We might note that in 1990, the states of Ohio and Colorado charged exactly an additional $25 for a vanity license plate.

The next situation also arises from an actual study. In it we first consider maximizing revenue, but then we go on to consider how to maximize profit. Again, we make the assumption that the demand for the product can be modeled by a linear function. Although that may not be true over a wide range of prices, nonetheless it appears to be reasonable for the data being analyzed.

EXAMPLE 9   Grocery pricing, maximizing profit

Writing in the *Harvard Business Review,* Lodish and Reibstein*
reported data on the sales of Minute Rice by one grocery, which had
deliberately varied the retail price to determine sales. The results are
shown in the following table.

RETAIL PRICE	AVERAGE WEEKLY SALES OF BOXES
0.89	80
0.99	66
1.09	62
1.19	50
1.29	42

A graph for Minute Rice sales is shown in figure 1.5.13.

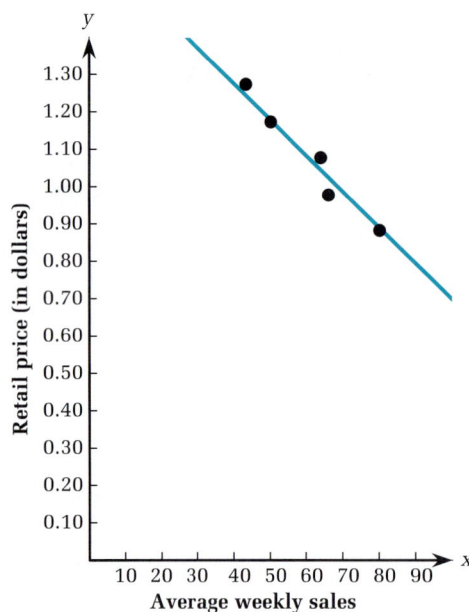

FIGURE 1.5.13

The demand function gives price per item as a function of the
quantity demanded. In this example the demand appears to be fairly
linear and the function $p(x) = -0.01x + 1.69$ approximates the data.
The wholesale price of Minute Rice was $0.69.

Find the pricing that (a) maximizes revenue and (b) maximizes
profit.

*Lodish and Reibstein, "New Goldmines and Minefields in MR," *Harvard Business Review,* January
1986, p. 169.

SOLUTION

(a) The revenue function, $R(x)$, on selling x items at $p(x)$ dollars each is

$$R(x) = x \cdot p(x) = x \cdot (-0.01x + 1.69) = -0.01x^2 + 1.69x$$

This quadratic has a maximum at the vertex, whose x-coordinate is $-b/(2a) = -1.69/(-0.02) = 84.5$. Because a fraction of a box cannot be sold, we use 85 as a value for the number of boxes. The revenue is maximized when the grocery sells 85 boxes. From the demand function, we see 85 boxes corresponds to a price of $-0.01(85) + 1.69 = 84$ cents a box. The revenue is then $85(\$0.84) = \$71.40$. (Interestingly, if we use 84 for the number of boxes, the total revenue is still $\$71.40$ because we would be selling the boxes for 85 cents.)
    The graph of $R(x)$ is shown in figure 1.5.14.

(b) Let us consider profit. The grocery pays 69 cents a box, so 85 boxes will cost the grocer $85 \cdot (0.69) = \$58.65$. Thus, the profit on 85 boxes sold at 84 cents each is $\$71.40 - \$58.65 = \$12.75$. Is that the best profit possible?
    Profit is the difference between cost and revenue, so the profit function would be $P(x) = R(x) - C(x) = x \cdot p(x) - C(x)$, which in our case gives

$$P(x) = (-0.01x^2 + 1.69x) - (0.69x) = -0.01x^2 + x$$

The graph of $P(x)$ is shown in figure 1.5.15. It has a vertex at $x = 50$ and a profit of $P(50) = \$25$. Hence, the greatest profit does not occur at the quantity and price that produce the greatest revenue but rather it occurs when 50 boxes are sold at $\$1.19$ each.

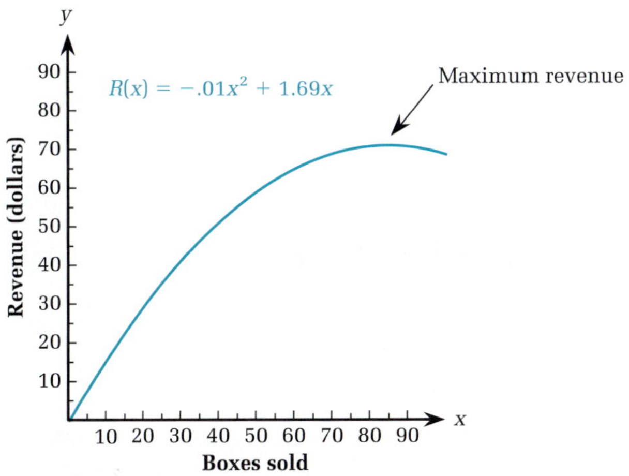

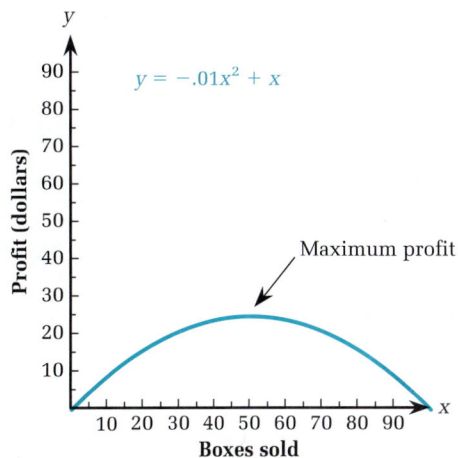

FIGURE 1.5.14

FIGURE 1.5.15

PRACTICE EXERCISE 4

If Batesville Burgers prices their hamburgers at $0.75 each, they sell 150 hamburgers. If the hamburgers are priced at $0.50, they sell 200. Assume the demand curve is linear. Find functions for (a) the demand and (b) the revenue for Batesville Burgers. (c) What is their maximum revenue?

Polynomial Functions

Linear and quadratic functions are examples of polynomial functions. Other examples are $x^3 + 2x - 1$, $3x^5 + 4x^3 - 1$, or $5x^{83} - x^7 + 2x - 6$.

DEFINITION   Polynomial function

A *polynomial function of degree n* has the form

$$f(x) = a_n x^n + a_{n-1}x^{n-1} + \cdots + a_1 x + a_0$$

where the coefficients a_n, a_{n-1}, $\cdots$, a_1, a_0 are all real numbers, $a_n \neq 0$.

If we are trying to create a polynomial function to model data, Minitab or some other statistical computer package is often used to estimate coefficients to produce a "best fit." The higher the degree of the polynomial, the closer will be the fit. However, along with the higher degree also goes a great deal more computer time and work on our part.

Example 10 is an example of an application of a cubic polynomial.

EXAMPLE 10   Fish lengths, polynomial modeling

According to published data on bass fish lengths and weights, a fairly good formula for predicting the weight, in ounces, of a bass that is x inches long is $w(x) = 0.00853x^3$. Estimate the weight of a 14-inch bass.

SOLUTION

$$w(14) = 0.00853(14)^3 = 23.4 \text{ ounces.}$$

Graphing a polynomial function is very difficult unless the function is very simple. For a general function we certainly can try various input values, then examine the output values with the hope that high or low points have been found, but how do we decide which values to input? The answers will come from calculus! But until we study calculus, we are forced to plot away and hope for the best. With that warning, we will sketch some graphs.

In the following examples, a graphing calculator is of considerable assistance, for it graphs many, many points in the intervals given. Even so, in this section we specify the intervals to use.

EXAMPLE 11   Graphing a polynomial function

Graph $f(x) = 2x^3 - 9x^2 + 12x + 2$ where $0 \le x \le 3$.

SOLUTION

When we use the integers 0, 1, 2, and 3 as inputs we have

$$f(0) = 2$$
$$f(1) = 2 - 9 + 12 + 2 = 7$$
$$f(2) = 16 - 36 + 24 + 2 = 6$$
$$f(3) = 54 - 81 + 36 + 2 = 11$$

These points are plotted in figure 1.5.16.

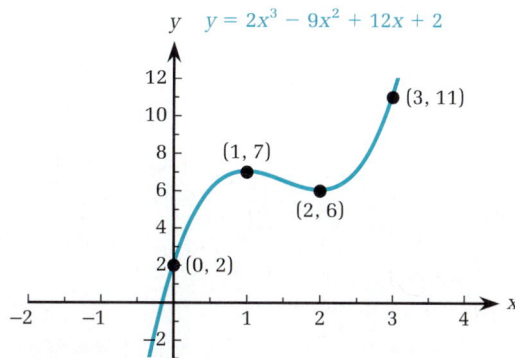

FIGURE 1.5.16

Apparently, there is a high point at $P(1,7)$ and a low point at $Q(2,6)$, but to be certain we would need to check values near 1 and near 2. In general, we cannot depend on always having high and low points occurring at integer x-values.

PRACTICE EXERCISE 5

Graph $f(x) = 2x^3 - 9x^2 + 2$ for $-1 \leq x \leq 4$ (using integer values as inputs).

If we need to solve a polynomial equation, $f(x) = 0$, then, as we saw for a quadratic function, we search for polynomials $g(x)$ and $h(x)$ such that $f(x) = g(x) \cdot h(x)$. We do this because if $f(x) = g(x) \cdot h(x)$, then the only way to have $f(x) = 0$ is to have $g(x) = 0$ or $h(x) = 0$. Although, apart from textbook examples, few polynomials are factorable, some are, and we should attempt a few minutes of investigation. The reason for this is that the lower the degree of the polynomial, the easier it is to solve the equation. We do not want to discuss techniques of factoring in this book, but the following are examples of such factoring.

EXAMPLE 12   Factoring a polynomial expression

Factor $x^3 + x^2 - 2x$.

SOLUTION

First factor out an x to get $x(x^2 + x - 2)$. Then $x^2 + x - 2$ factors as $(x + 2)(x - 1)$ so that a complete factoring is $x^3 + x^2 - 2x = x(x + 2)(x - 1)$.

EXAMPLE 13   Using factors to solve a polynomial equation

Solve $x^3 - 2x^2 + x - 2 = 0$.

SOLUTION

Factor $x^3 - 2x^2 + x - 2 = x^2(x - 2) + x - 2 = (x^2 + 1)(x - 2)$. The term $x^2 + 1$ does not factor into two linear terms. Thus, the only solution is when $x - 2 = 0$, which gives $x = 2$.

PRACTICE EXERCISE 6

Solve $x^3 + 2x^2 - x - 2 = 0$.

If we cannot solve $p(x) = 0$, we may be able to approximate solutions. There are many methods for finding such approximations. The simplest is to find progressively smaller intervals within which the sign of $p(x)$ changes from positive to negative, or from negative to positive. Within any such interval there is at least one value for which the polynomial is zero. This property of polynomials is stated in the next rule.

> ### DETECTING ZEROES BY CHANGES IN SIGN OF A POLYNOMIAL
>
> If $p(x)$ is a polynomial and $p(a)$ and $p(b)$ are of different signs, then between a and b there is at least one value of x for which $p(x) = 0$.

EXAMPLE 14   Changes in sign

Show that $x^3 - 5x^2 + 3 = 0$ for some value of x between 0 and 2.

SOLUTION

Let $p(x) = x^3 - 5x^2 + 3$. By evaluation, $p(0) = 0^3 - 5 \cdot 0^2 + 3 = 3$ and $p(2) = 2^3 - 5 \cdot 2^2 + 3 = 8 - 5 \cdot 4 + 3 = 8 - 20 + 3 = -9$. Since $p(0) > 0$ and $p(2) < 0$, there must be at least one value of x between 0 and 2 for which $p(x)$ is zero.

Graphing Calculators

Access to a graphing calculator does make it much easier to create good graphs because the calculator does so many computations in the process of making the display on the screen. However, it is still true that we will be able to make better graphs, and do them more rapidly, after we have applied calculus to graphing. Having said that, let us do some exploration with a graphing calculator.

EXAMPLE 15   Using the graphing calculator to find zeroes of a polynomial

Approximate the solutions of $x^3 - 12x^2 + 16x - 5 = 0$.

SOLUTION

We are not given any range for the values of x and y, so we must assign some. Suppose we begin with the built-in default values of the graphing calculator. On one brand those values are: $-4.7 \leq x \leq 4.7$ and $-3.1 \leq y \leq 3.1$ with a scale of 1 for both x and y.

Enter the function as $y = x^3 - 12x^2 + 16x - 5$ and press $\boxed{\text{EXE}}$. The screen we see is given in figure 1.5.17. Apparently, two solutions are being shown. One of them appears to be about 0.5 and the other about 1.

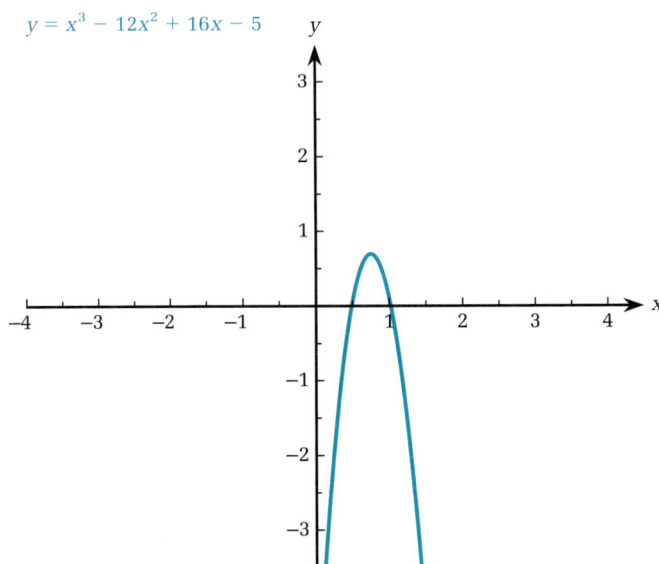

$y = x^3 - 12x^2 + 16x - 5$

FIGURE 1.5.17

Using the $\boxed{\text{Trace}}$ or $\boxed{\text{Zoom}}$ we can examine the area around $x = 0.5$ and find there is a solution near $x = 0.47$. Similarly, we can check the region near $x = 1$ and we find that 1 is actually an exact solution.

Are these the only two solutions? There sometimes can be only two zeroes of a cubic polynomial, but no cubic polynomial can look like the screen in figure 1.5.17. A cubic polynomial has to become unbounded on both sides of the x-axis. That is, if $p(x)$ is a cubic polynomial and it becomes unboundedly positive for large positive values of x, then it must become unboundedly negative for unboundedly negative values of x. On the other hand, if it becomes unboundedly negative for large positive values of x, then it must become unboundedly positive for unboundedly negative values of x. So, there must be another zero of $x^3 - 12x^2 + 16x - 5$.

We start over with the initial ranges of $-4.7 \leq x \leq 4.7$ and $-3.1 \leq y \leq 3.1$ and scale of 1. Draw the first screen of $x^3 - 12x^2 + 16x - 5$ and then $\boxed{\text{Zoom}}$ out until we suddenly see some more of the graph appear. Now the screen might appear as in figure 1.5.18. By using ☐Trace or ☐Box we can find that the x-value of the newly appeared crossing is about $x = 10.5$. By narrowing down, we find that x is about 10.53.

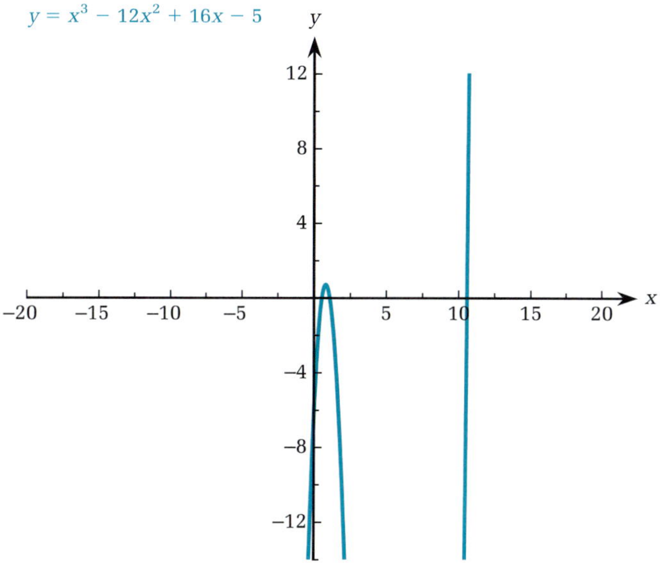

$y = x^3 - 12x^2 + 16x - 5$

FIGURE 1.5.18

Thus, the solutions of $x^3 - 12x^2 + 16x - 5 = 0$ are approximately 0.47, 1, and 10.53.

Once we know where to look, each solution of $p(x) = 0$ can be found to any desired amount of accuracy by zooms or setting ranges with little variation in the values of x and y. We do need to remember to change the scale values if we reset range values or we may find ourselves staring at a blank screen and wondering why no graph is appearing.

The real difficulty is knowing how many solutions exist. Even if we start with a great spread of range values we may miss seeing some of the solutions. Calculus will help.

1.5 PROBLEMS

Foundations

The problems of this section require the basic skills illustrated by the following:

1. Find the equation of the line through the points $A(12, 5)$ and $B(20, 3)$.

2. Find the equation of the line through the points $A(300, 25)$ and $B(250, 30)$.

For problems 3 and 4, find all of the possible products of factors that are both integers.

3. 10

4. 15

For problems 5 and 6, express the polynomials as products of factors.

5. $x^2 - x - 6$

6. $x^3 + x$

For problems 7 and 8, evaluate $b^2 - 4ac$ and determine whether $\sqrt{b^2 - 4ac}$ is a real number.

7. $a = 4, b = 2, c = -1$ **8.** $a = 2, b = 1, c = 2$

Exercises

Solve each of the equations in problems 9–16.

9. $x^2 - 3x - 10 = 0$ **10.** $x^2 - 2x - 15 = 0$

11. $x^2 - 2x = 0$ **12.** $x^2 + x = 0$

13. $x^2 - 3x = 4$ **14.** $x^2 - x = 2$

15. $x^2 + 2x - 2 = 0$ **16.** $x^2 + 3x - 1 = 0$

Sketch graphs of problems 17–26, indicating the vertex and the x- and y-intercepts of each.

17. $f(x) = x^2 - 3x - 10$ (see problem 9)

18. $f(x) = x^2 - 2x$ (see problem 11)

19. $f(x) = -x^2 + x + 2$ (see problem 14)

20. $f(x) = -x^2 - 3x + 1$ (see problem 16)

21. $f(x) = 2x^2 - x - 3$

22. $f(x) = 3x^2 + 2x + 1$

23. $f(x) = 2x^2 + x + 2$

24. $f(x) = 4x^2 + 2x - 1$

25. $f(x) = -x^2 - 3x - 2$

26. $f(x) = -3x^2 + x + 1$

Use factors to find when the polynomial function is zero, then sketch a graph of the function, indicating any x- or y-intercepts.

27. $f(x) = x^3 - 4x$

28. $g(x) = x^3 + 2x^2$

29. $g(x) = x^3 - x^2 - 6x$

30. $f(x) = x^3 + x^2 - 12x$

31. $f(x) = x^3 - x^2 + 2x - 2$

32. $g(x) = 2x^3 - 2x^2 + x - 1$

33. $f(x) = x^4 - 4x^2$

34. $r(x) = 9x^4 - x^2$

Show by changes of sign that each of the following equations has zero in the given interval. Do not find such zeroes.

35. $f(x) = x^5 + 3x^2 - 2$ in $[0, 1]$

36. $f(x) = -3x^3 - 2x + 1$ in $[0, 1]$

37. $f(x) = -x^4 + x^3 - 6x$ in $[-1, 1]$

38. $g(x) = x^3 - 9x^2 + 3x + 5$ in $[2, 10]$

39. *(Biology, Angiocinematography)* In the medical evaluation of a heart, one process is to inject a dye near the heart and film the presence of the dye. A formula stating the concentration of the dye after t seconds is $C(t) = -0.006t^4 + 0.140t^3 - 0.053t^2 + 1.79t$.
(a) Find $C(0)$, $C(5)$, $C(10)$, $C(15)$, and $C(20)$.
(b) Show that $C(t) = 0$ for some value of t between $t = 23$ and $t = 24$.
(c) Sketch a graph of $C(t)$ for $0 \le t \le 23$.

40. *(Fishery Management, Weight Gains)* According to a published formula,* the weight increase, G, measured in grams, when N fish are planted is given by $G = 500 - 50N$. The total fish production $T = NG$.
(a) Find a formula for T expressed in terms of N.
(b) Find the value for N that maximizes T.
(c) Find the maximum value of T.

Problems 41 and 42 concern the Fragrant Flowers company, which operates several pushcarts that sell flowers on street corners.

41. *(Maximizing Revenue)* Suppose that Fragrant Flowers has seven pushcart flower stands. Each stand sells an average of $525 worth of flowers per day. If x new stands are added, then the average revenue per day per stand (both old and new stands) will be $\$(525 - 25x)$.
(a) Find the number of stands that will maximize the total revenue from sales of flowers.
(b) What is that total sales amount?

42. *(Maximizing Revenue)* Refer to problem 41. Suppose the addition (or subtraction by closing) of x flower stands will cause an average revenue per stand (both old and new stands) of $\$(525 - 100x)$ at each of the $7 + x$ stands.
(a) Find the number of stands that will maximize the total revenue from sales of flowers.
(b) What is that total sales amount?

*I. Chaston, *Mathematics for Ecologists,* London: Butterworth, 1971, p. 17.

43. **(Maximizing Revenue, Vanity Car License Plates)**   Many states sell vanity license plates for automobiles and motorcycles. One state has a fee of $55 and sells 5000 plates. In considering a price increase, their Budget Committee has found that when the price goes up by x dollars, the sales will be $5000 - 40x$ plates. The legislature needs to know how the plates should be priced so as to maximize state revenue.
    (a) What price will generate the maximum revenue from the sale of vanity plates?
    (b) What is that maximum revenue?

44. **(Maximizing Revenue)**   When the Little Dinner Theater priced their tickets at $25 they sold 300. When they raised their price to $30, they sold only 250. Assume the demand curve for tickets is linear.
    (a) Find the equation of the demand function.
    (b) Find the ticket price and the number of tickets at that price that will produce the greatest total revenue.
    (c) What is the maximum possible revenue?

45. **(Maximizing Profit)**   Fresh Flowers Cat Food, whose manufacturing cost is $22 a sack, was selling 25 sacks a week at $40 each. They lowered their price to $30 and sold 50 sacks a week. Assume the demand curve is linear.
    (a) Find the equation of the demand function.
    (b) Find the price and quantity of sacks that will provide the greatest profit.
    (c) What is that greatest profit?

46. **(General, St. Louis Arch)**   If you start walking from one end of the St. Louis Gateway Arch and stay under the arch, then when you are x feet from your starting point, the height of the arch above you is approximately $f(x) = 4x - 0.00635x^2$ feet. Estimate the maximum height of the arch and the width of the base.

47. **(General, Cassette Tape Length)**   A cassette tape recorder has a counter reader and you can measure the number of seconds that elapse as the recorder plays. The data shown in the following table were collected from one recorder.

Show that the formula $t = 1.94293c + 0.001045773c^2$ gives a very good approximation to the observed time values by computing t at $c = 100$, 400, and 700.

Graphing Calculator Problems

48. $x^3 - 11x^2 + 14x - 4 = 0$ has three solutions. Approximate each of them to two decimal places.

49. $x^3 - 7x^2 - 12x - 4 = 0$ has three solutions. Approximate each of them to two decimal places.

50. Graph both $f(x) = x^3 - 9x^2 + 1$ and $g(x) = 3x + 6$ on the same scale and estimate the solutions of $f(x) = g(x)$. Find a polynomial equation $p(x)$ that has the same solutions as $f(x) = g(x)$.

Writing and Discussion Problems

51. In Renaissance Italy, debates often determined who would be hired to teach mathematics at the royal courts or in the newly founded universities. Research and write about the debate of 1548 in Milan between Niccolo Tartaglia (1500–1557) and Ludovico Ferrari (1522–1565). The role of Girolamo Cardan in this debate was crucial.

52. Have 25 people fill out data for the chart below. (It may help to have friends collect data.)

During one month, I would attend the following number of movies if the cost were:

COST	$3.50	$4.00	$4.50	$5.00	$5.50
Number of movies					

COUNTER READING	100	200	300	400	500	600	700	800
time (in secs.)	205	430	677	945	1233	1542	1870	2224

(a) Estimate a linear demand function by averaging your data.

(b) Determine a price that would maximize a theater's revenue.

(c) Discuss possible sources of error in your analysis.

53. Discuss how the value of $b^2 - 4ac$ (being positive, zero, or negative) determines the number of crossings of the x-axis by the graph of $y = ax^2 + bx + c$.

54. We have used the fact that if $a \cdot b = 0$, then $a = 0$ or $b = 0$. If $a \cdot b = c \neq 0$, then what, if anything, can we conclude about a and b?

Enrichment Problems

55. Refer to problem 41. Suppose the average sales per stand per day were $\$(525 - rx)$ for some value r. For what values of r should Fragrant Flowers be adding any new stands at all?

56. Refer to problem 43. Suppose sales of vanity plates will be $5000 - rx$ for some value r. For what values of r should the Budget Committee be raising the price of the vanity plates at all?

57. The Compact Turner bought a shipment of compact discs for a total cost of \$4000. The employees found ten of the discs defective. Selling the good discs at \$2 over their original cost, the Compact Turner made a profit of \$900.
   (a) Find the number of discs bought.
   (b) What was the retail price of the discs sold by Compact Turner?

58. A photographer took one picture of each passenger on a commercial riverboat ride. Her cost of doing this was \$80. She offered the photos for sale at a price equal to her cost plus \$3.20. Seventy-five of the passengers declined to buy. Her profit was \$20.
   (a) How many passengers took the ride?
   (b) What was her cost per picture?
   (c) What was her asking price per picture?

SOLUTIONS TO PRACTICE EXERCISES

1. To solve $x^2 + 3x - 10 = (x + s)(x + u)$, we want to find values for s and u such that $su = -10$ and $s + u = 3$. By checking possibilities, we find $s = 5$ and $u = -2$ meet this criterion. Thus $(x + 5)(x - 2)$ is the desired factorization.

2. The x-coordinate of the vertex is at $x = -b/(2a) = -(-1)/(2 \cdot 2) = 1/4 = 0.25$. The y-coordinate of the vertex is at $2(1/4)^2 - (1/4) - 2 = -17/8 = -2.125$. The y-intercept is -2. The x-intercepts are given by the quadratic formula as $x = \left(1 \pm \sqrt{17}\right)/4$ so $x \approx 1.2808$ and $x \approx -0.7808$.

3. $P(x) = (5 + x)(600 - 24x) = 3000 + 600x - 120x - 24x^2 = 3000 + 480x - 24x^2$.

   The x-coordinate of the vertex is $x = -b/(2a) = -480/(2(-24)) = 10$. Because $a < 0$, the vertex will be at a maximum. Hence, the maximum production occurs when 10 new wells are drilled and the total production at that value is $P(10) = (5 + 10)(600 - 24 \cdot 10) = (15)(360) = 5400$ barrels per day.

4. (a) $p(x) = -0.005x + 1.50$
   (b) $R(x) = x \cdot p(x) = -0.005x^2 + 1.50x$
   (c) The vertex is at $x = 150$ and the maximum revenue is $R(150) = \$112.50$.

5. The function $f(x) = 2x^3 - 9x^2 + 2$ for $-1 \leq x \leq 4$ has values $f(-1) = -9, f(0) = 2, f(1) = -5, f(2) = -18, f(3) = -25, f(4) = -14$. The graph is shown in figure 1.5.19.

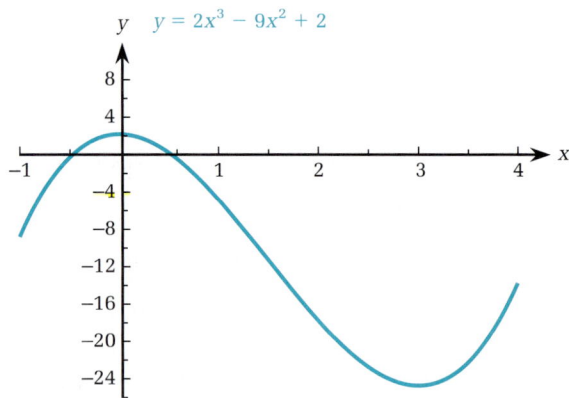

FIGURE 1.5.19

6. $x^3 + 2x^2 - x - 2 = x^2(x + 2) - (x + 2) = (x^2 - 1)(x + 2) = (x - 1)(x + 1)(x + 2)$. Thus, if $x^3 + 2x^2 - x - 2 = 0$, we need $(x - 1) = 0$ or $(x + 1) = 0$ or $(x + 2) = 0$. By solving each of these, the solutions to the original equation are $x = 1, x = -1$, and $x = -2$.

1.6 RATIONAL FUNCTIONS

This section deals with quotients of polynomials. Such functions are used to describe inverse proportionality and the "law of diminishing returns," which concerns situations wherein achieving success requires efforts that increase greatly as the goal is neared.

Consider the case of air quality in the Grand Canyon National Park. There are large, coal-fired electric generating plants at Page, Arizona, and Farmington, New Mexico, as well as at other locations in the southwestern United States. Each of these plants emits gases as well as visible solids. In order to provide visitors to the Grand Canyon an opportunity to see the canyon as it was in bygone years, the federal government has mandated that the power plants limit their emissions.

The elimination of 50% of the emissions is far less costly than the elimination of 90% of the emissions. As the power producers are urged to eliminate the last few percent, the costs become very great.

A graph showing a typical "diminishing returns" curve is shown in figure 1.6.1.

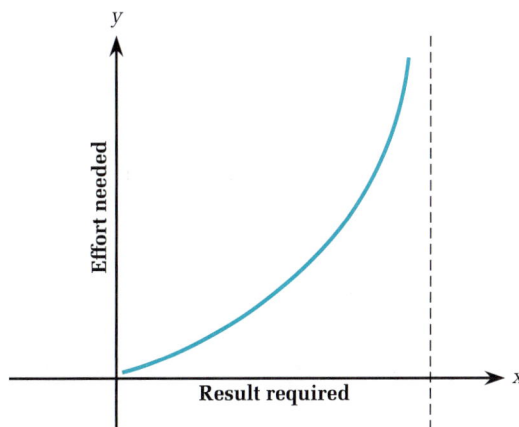

FIGURE 1.6.1

Rational Functions

We begin with a definition of a rational function.

DEFINITION

Rational function

A *rational function* is expressible as the quotient of two polynomials.

Two examples of rational functions are $f(x) = x^2/(x^3 + 1)$ and $g(x) = (3x - 1)/(2x)$. The function $k(x) = 3x + 2/x$ is a rational function because it can be expressed as $(3x^2 + 2)/x$, the quotient of two polynomials. The function $r(x) = (\sqrt{x} - 1)/x$ is not a rational function since the numerator is not a polynomial.

To determine the domain of a rational function, first recall that the domain of any polynomial function is $\mathbb{R}$, the set of all real numbers. Hence, if $f(x) = r(x)/s(x)$ is a rational function, both $r(x)$ and $s(x)$ have $\mathbb{R}$ as their domains. However, $s(x)$ cannot be zero because division by zero is not possible. As a result, the domain of a rational function consists of just those real numbers for which the denominator is not zero.

DOMAIN OF A RATIONAL FUNCTION

The domain of the rational function $f(x) = r(x)/s(x)$ consists of all real numbers x such that $s(x) \neq 0$.

EXAMPLE 1   Domains of rational functions

Find the domains of $f(x) = \dfrac{x^3}{x + 1}$ and $g(x) = \dfrac{x - 1}{x}$.

SOLUTION

The domain of $f(x) = x^3/(x + 1)$ is determined by considering $x + 1 \neq 0$. Solving this for x we have $x \neq -1$. Hence the domain of $f(x)$ is the set $\{x : x \neq -1\}$. Similarly, the domain of $g(x) = (x - 1)/x$ is $\{x : x \neq 0\}$.

Graphing rational functions will be considerably easier when we use calculus, so at this point we will not invest much effort in graphing complicated rational functions. However, we present the graphs of a few functions to illustrate the nature of these graphs. Before doing so, recall that if the graph of a function appears to be straightening out and approaching a line, then that line is called an *asymptote* of the graph.

EXAMPLE 2   Graphing a rational function

Graph $f(x) = \dfrac{1}{x - 1}$ for $0 \leq x$.

SOLUTION

This function has domain $\{x : x \neq 1\}$ because the denominator, $x - 1$, cannot be zero. To determine the graph, we evaluate $f(x)$ for the values of x shown in the following table, being especially careful to select values near $x = 1$.

x	0	0.5	0.75	0.9	0.95	1.05	1.1	1.5	2	3
$\dfrac{1}{x-1}$	-1	-2	-4	-10	-20	20	10	2	1	0.5

Plotting the points determined by these values and connecting them, we have the graph in figure 1.6.2. The dotted vertical line at $x = 1$ represents a vertical asymptote. The x-axis itself is a horizontal asymptote.

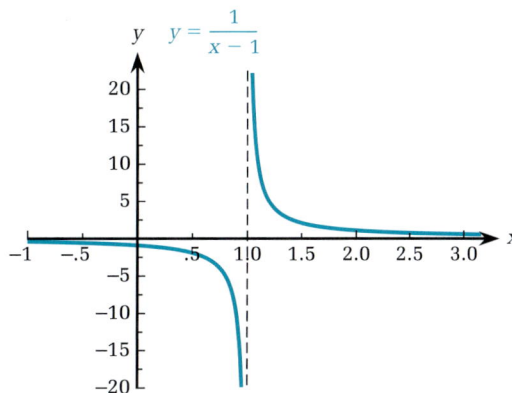

FIGURE 1.6.2

EXAMPLE 3   Graphing a rational function with two vertical asymptotes

Sketch the graph of $f(x) = \dfrac{1}{x^2 - 4}$.

SOLUTION

Because $x^2 - 4 = 0$ if $x = \pm 2$, the domain of $f(x)$ is $\{x : x \neq \pm 2\}$. There are vertical asymptotes at $x = \pm 2$. Finding $f(x)$ for some sample values, including some values of x close to 2, gives us the following table.

x	0	0.5	1.5	1.9	2.1	2.5	3
$\dfrac{1}{x^2 - 4}$	−0.25	−0.27	−0.57	−2.56	2.44	0.44	0.2

There is no need to evaluate $f(x)$ for any negative values of x since $f(-x) = f(x)$ and so the graph is symmetric with respect to the y-axis. The graph of $f(x)$ is shown in figure 1.6.3.

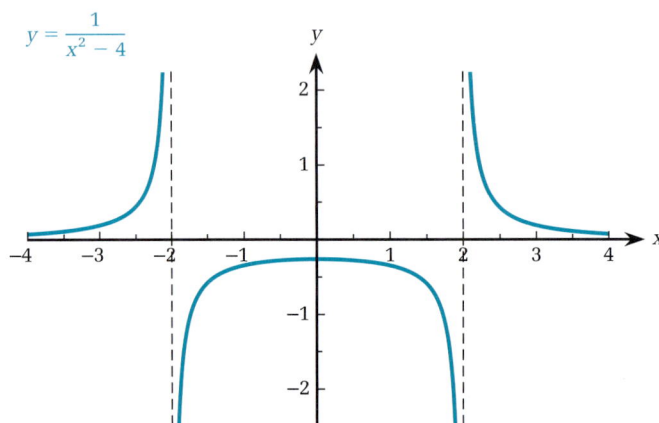

$y = \dfrac{1}{x^2 - 4}$

FIGURE 1.6.3

PRACTICE EXERCISE 1

Determine the domain of $g(x) = (x^3 + 1)/x$. Find the value of $g(x)$ for $x = -2, -1, -0.5, 0.2, 0.5, 1, 1.5, 2$, and 2.5 and sketch the graph of $g(x)$.

The graphs in Examples 2 and 3 have vertical asymptotes, but not all rational functions have graphs with vertical asymptotes. First, it can happen that the denominator is never zero. Second, the graph can have a *puncture point*, that is, there is a "missing point" of the graph. Examples of each situation follow.

EXAMPLE 4   A rational function whose denominator is never zero

Find the graph of $f(x) = \dfrac{x}{x^2 + 1}$.

SOLUTION

The domain of $f(x)$ consists of all the real numbers because $x^2 + 1$ is never zero. The following table presents the results of evaluating $f(x)$ for some values of x.

x	-3	-2	-1	-0.5	0	0.5	1	2	3	4
$\dfrac{x}{x^2 + 1}$	-0.3	-0.4	-0.5	-0.4	0	0.4	0.5	0.4	0.3	0.2353

The x-axis is a horizontal asymptote, but there is no vertical asymptote. The graph is shown in figure 1.6.4.

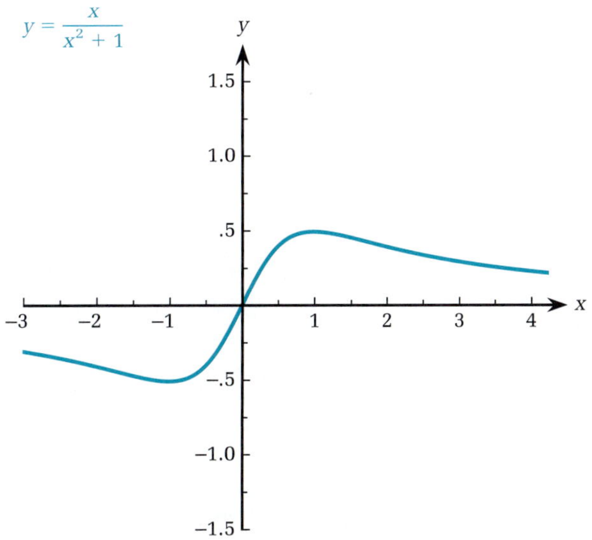

$$y = \frac{x}{x^2 + 1}$$

FIGURE 1.6.4

EXAMPLE 5   A rational function whose graph has a puncture point

Find the graph of $r(x) = \dfrac{x^2 - 4}{x - 2}$.

SOLUTION

The reason that $r(x)$ does not have a vertical asymptote at $x = 2$ is that even though $x - 2$ does equal zero at $x = 2$, the function $r(x)$ is equal to $x + 2$ for all values of x except 2 itself. This follows from factoring $x^2 - 4$ into $(x + 2)(x - 2)$. Hence $(x^2 - 4)/(x - 2) = [(x + 2)(x - 2)]/(x - 2) = x + 2$ if $x \neq 2$. Thus the graph of $r(x)$ is the same as the graph of $x + 2$, except for the puncture point $A(2, 4)$, which is not on the graph of $r(x)$. This is shown in figure 1.6.5.

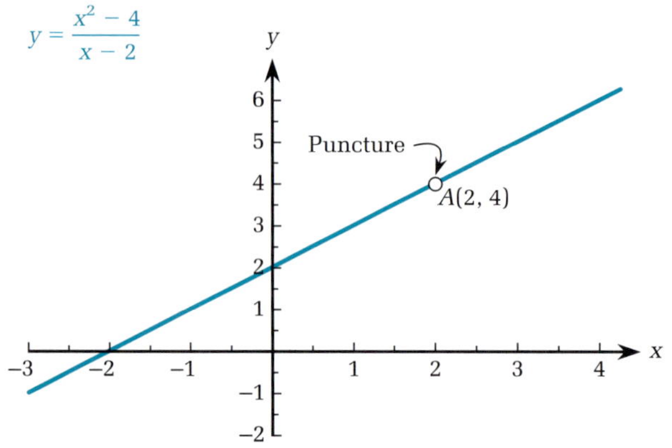

FIGURE 1.6.5

The rule for deciding if $x = c$ is a vertical asymptote of the rational function $f(x)/g(x)$ is given by the following:

VERTICAL ASYMPTOTES OF RATIONAL FUNCTIONS

The rational function $f(x)/g(x)$ has a vertical asymptote at $x = c$ if $(x - c)$ is repeated more times as a factor of $g(x)$ than it is as a factor of $f(x)$.

Graphing Calculators

A graphing calculator quickly sketches screens of the graph. As we have said before, and as we will repeat again, the choice of screen range will be much easier when we use calculus. For now we simply use the Zoom and Box features. There are two situations to be aware of, which are demonstrated nicely in the graphing of rational functions:

1. The calculator may not indicate an asymptote, but rather try to connect both sides of the graph.

2. The calculator may or may not indicate that the "missing point" is truly missing in the graph.

   Figure 1.6.6 shows an incorrect graph of the function $f(x) = 1/(1 - x)$, for it does not show the vertical asymptote that exists at $x = 1$.

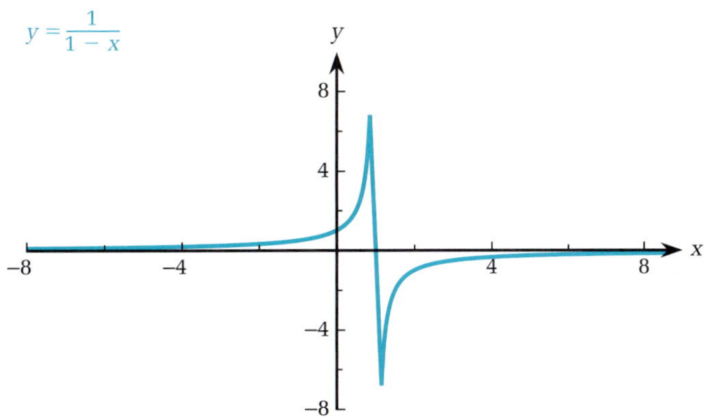

$$y = \frac{1}{1-x}$$

FIGURE 1.6.6

The vertical asymptote is not shown because the calculator is connecting points that it has plotted. Depending on the calculator used and the ranges chosen for x and y, asymptotes and punctures may or may not be seen.

Inverse Proportionality

One thing is said to be inversely proportional to another if as one increases, the other decreases and vice versa. For example, we might say that our enjoyment of an evening at the pizza parlor is inversely proportional to the price of the pizza because as the price increases, our enjoyment decreases.

In mathematical writing, the phrase "inversely proportional" is given a more precise meaning.

DEFINITION

Inversely proportional

The variable y is *inversely proportional* to the variable x if $y = k(1/x)$ for some $k \neq 0$.

EXAMPLE 6   Inversely proportional

The graph of $y = 3/x$ is shown in figure 1.6.7. Notice that as x increases from 1 to 6, the values for y decrease from $y = 3/1 = 3$ to $y = 3/6 = 0.5$.

$y = \dfrac{3}{x}$

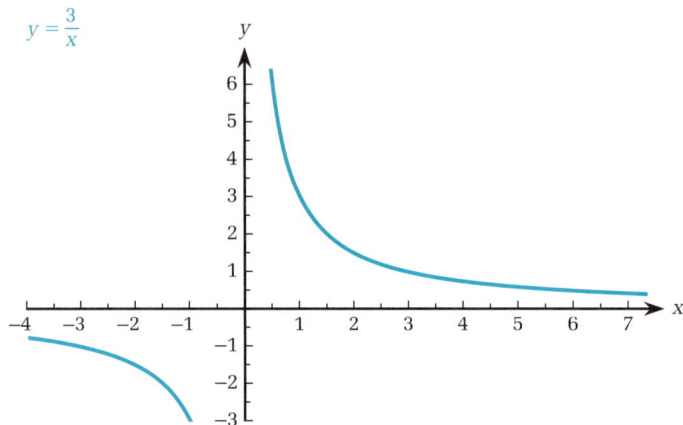

FIGURE 1.6.7

EXAMPLE 7   Application of inverse proportionality to the "prime rate"

The interest rate at which bankers will lend money to their customers with the greatest financial stability is called the "prime rate." The earnings/sales ratio (ESR) of a company often is inversely proportional to the prime rate. This is particularly true of a company that borrows heavily, such as a power company.

The prime rate is inversely proportional to the ESR of the Animas Gas Company. Suppose the ESR of Animas Gas is 0.04 when the prime rate is 9.5%. What would be their ESR if the prime rate rises to 11%?

SOLUTION

Let y be the prime rate and x be the ESR. The assumption that y is inversely proportional to x can be represented by the formula $y = k(1/x)$. Another way of writing $y = k(1/x)$ is $yx = k$, so we can find the constant of inverse proportionality k by multiplying any known y by its corresponding x. For $y = 9.5$ and $x = 0.04$, we have $k = (0.04) \cdot 9.5 = 0.38$. With the new prime rate of 11%, the new ESR must be $0.38(1/11) \approx 0.0345$.

PRACTICE EXERCISE 2

Continuing with Example 7, what would the prime rate have to be so that Animas Gas would have an ESR of 0.05?

Law of Diminishing Returns

The next two examples use rational functions to model that familiar situation in which the amount of work required increases as the goal is neared. In some cases, no matter how much effort is expended, perfection may be impossible.

EXAMPLE 8   Biology, law of diminishing returns on collecting seed by birds

Suppose a bird is collecting seed from a field that contains 100 grams of seed. In the beginning, the bird quickly collects quite a bit of the seed, but as time goes on, the remaining seed is more sparsely distributed and it is harder to collect. A formula that describes this situation may be of the form described below.

Collecting x grams takes the bird $f(x) = (5x + 1)/(100 - x)$ minutes $(0 < x < 100)$. For example, to collect 50 grams takes the bird $(5 \cdot 50 + 1)/(100 - 50) = 5.02$ minutes. To collect 80 grams takes $(5 \cdot 80 + 1)/(100 - 80) = 20.05$ minutes, and to achieve the near-perfection goal of 99 grams collected takes $(5 \cdot 99 + 1)/(100 - 99) = 496$ minutes (or eight hours and 16 minutes!).

A graph of $f(x)$ is shown in figure 1.6.8.

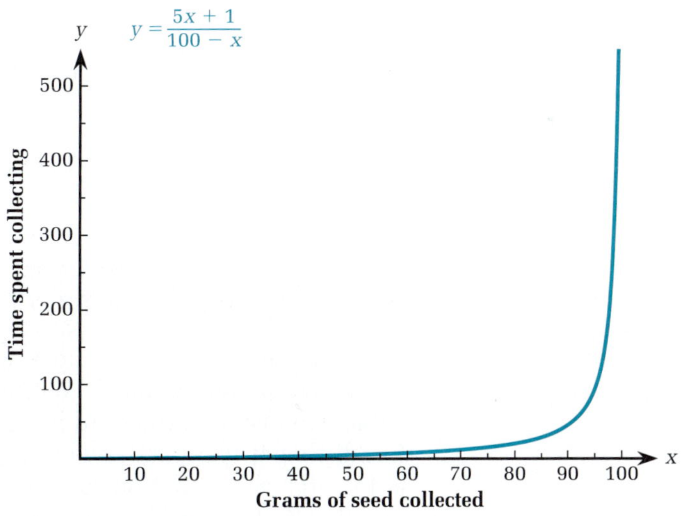

FIGURE 1.6.8

PRACTICE EXERCISE 3

Continuing with the function $f(x)$ from Example 8, how much time would the bird need to collect 90 grams of seed?

In the next example, it appears possible to attain perfection, but the cost of increasing results still follows the law of diminishing returns.

EXAMPLE 9   Cost-benefit analysis, law of diminishing returns

The Environmental Protection Agency (EPA) is requiring the Mirage Lake Water Company to remove impurities from its water. The cost of removing x% of the impurities is $C(x) = (2x + 3)/(101 - x)$, measured in thousands of dollars.

(a) What will it cost to remove 85% of the impurities?

(b) What percentage of the impurities can be removed for a cost of $3000?

SOLUTION

(a) We need to be careful in using this formula and realize that the percentage is *not* given in its decimalized form. Therefore, for 85% we use $x = 85$ rather than $x = 0.85$.

$$C(85) = \frac{2 \cdot 85 + 3}{101 - 85} = \frac{173}{16} = \$10.81 \text{ thousands} = \$10,810$$

(b) We need to solve the equation $C(x) = 3$ (thousand dollars) to obtain a value of x. Beginning with $(2x + 3)/(101 - x) = 3$, we can multiply both sides by $(101 - x)$ to give $3(101 - x) = 2x + 3$. Multiplying and collecting terms, we have $303 - 3 = 2x + 3x$, which simplifies to $300 = 5x$, yielding $x = 300/5 = 60$. We interpret this to mean that for $3000 the company can remove 60% of the impurities.

A graph of $C(x)$ is shown in figure 1.6.9.

The second part of the last example consisted of solving an inverse function problem. While we were solving for a particular value, we just as well could have found the general inverse function. Recall that if $y = f(x)$, then one method for determining an inverse is to solve this equation for x in terms of y and rewrite the inverse function in terms of x.

EXAMPLE 10   Finding an inverse function of a rational function

Using the function of Example 9, find the inverse of $f(x) = (2x + 3)/(101 - x)$.

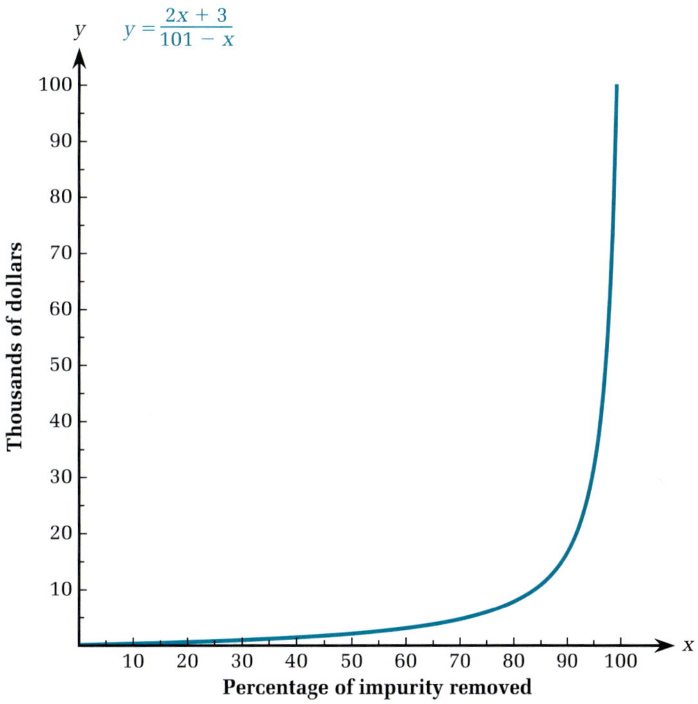

$$y = \frac{2x + 3}{101 - x}$$

FIGURE 1.6.9

SOLUTION

Let $y = (2x + 3)/(101 - x)$ and solve for x in terms of y. Multiplying by $(101 - x)$ gives $y(101 - x) = 2x + 3$. Then $101y - yx = 2x + 3$. Collecting all the terms involving x on the right side and then factoring out an x gives us $101y - 3 = (2 + y)x$. Hence, we have $x = (101y - 3)/(2 + y)$. Expressing this inverse function in the variable x gives us $f^{-1}(x) = (101x - 3)/(2 + x)$.   ●

PRACTICE EXERCISE 4

Refer to Example 8 (of a bird collecting seed).

(a) Find the inverse function $G(y)$, which states the grams of seed the bird collects by working y minutes.

(b) Evaluate $G(10)$ and interpret your answer.

1.6 PROBLEMS

Foundations

The problems of this section require the basic skills illustrated by the following:

Solve the following equations.

1. $2x + 3 = 0$ **2.** $x^2 - 4 = 0$ **3.** $x^2 + 1 = 0$

Factor each of the following polynomials.

4. $x^2 + 2x$ **5.** $4x^2 - 1$ **6.** $x^2 - x - 2$

Exercises

Express the following as rational functions.

7. $3x + \dfrac{1}{x}$

8. $x^2 - \dfrac{2x}{x + 1}$

9. $\dfrac{x + 1}{x - 1} + 2$

Find the domains of the following functions.

10. $f(x) = \dfrac{x + 2}{x^2 - 1}$

11. $f(x) = \dfrac{2x}{x^2 - 2x - 3}$

12. $f(x) = \dfrac{2}{2x^2 - x - 1}$

For problems 13 to 24 indicate any vertical asymptotes and sketch the graph.

13. $f(x) = \dfrac{2}{x + 1}$

14. $f(x) = \dfrac{1}{x - 1}$

15. $g(x) = -\dfrac{1}{3x - 2}$

16. $g(x) = \dfrac{5}{3x + 2}$

17. $f(x) = \dfrac{3}{x^2 - 4}$

18. $f(x) = \dfrac{4}{x^2 - 1}$

19. $f(x) = \dfrac{x + 1}{x - 1}$

20. $h(x) = \dfrac{2x - 1}{3x - 2}$

21. $r(x) = \dfrac{2 - 3x}{2x + 3}$

22. $r(x) = \dfrac{x}{x - 2}$

23. $g(x) = \dfrac{2}{x^2 + 2}$

24. $g(x) = \dfrac{5}{x^2 + 1}$

For problems 25 to 28, find the puncture points and sketch the graph.

25. $y = \dfrac{x^2 - 4}{x + 2}$

26. $y = \dfrac{4x^2 - 1}{2x + 1}$

27. $y = \dfrac{x^2 - x - 6}{x - 3}$

28. $y = \dfrac{x^2 + x - 6}{x + 3}$

For problems 29 to 32, find any vertical asymptotes and puncture points. Sketch the graph.

29. $g(x) = \dfrac{x + 2}{x^2 + 2x}$

30. $f(x) = \dfrac{x}{x^2 + x}$

31. $f(x) = \dfrac{2x + 1}{4x^2 - 1}$

32. $h(x) = \dfrac{x + 1}{x^2 - x - 2}$

33. *(Cost-Benefit, Ecology, Law of Diminishing Returns)*   The Environmental Protection Agency decides to clean up a radioactive waste site in Utah. The cost of removing $x\%$ of the waste is given by $C(x) = 200/(100 - x)$ in millions of dollars.
(a) What is the cost of removing 50% of the waste?
(b) What percentage of the waste can they remove for 20 million dollars?
(c) Sketch a graph of $C(x)$ for $0 \le x \le 100$.

34. *(Diminishing Returns, Sociology)*   The Mirage Lake Water Co. is trying to contact all of its customers by telephone. When a person does not answer, a call-back is made. It is very difficult to make contact with some people, necessitating many call-backs. The cost of making contact with $x\%$ of their customers is $C(x) = 44/(102 - x)$ in thousands of dollars.
(a) What will it cost them to contact 80% of their customers?
(b) What percent of their customers can they contact with a calling budget of $5000?
(c) Sketch a graph of $C(x)$ for $0 \le x \le 100$.

35. *(Inverse Proportionality, Bond Prices and Yields)*   The *current yield* on a bond with a *face value* of $1000 is calculated as being equal to [1000 · (the coupon rate)]/current purchase price. The *coupon rate* is the rate that was set when the bond was first issued. For example, a $1000 bond with a coupon rate of 8% and a purchase price of $850 has a current yield of $(1000 \cdot 8)/850 = 9.41\%$. Because bonds sold on Wall Street have to be competitive with other yields, such as Treasury bills, the current yields are inversely proportional to the price of the bond. A bond sold at below *face value* is said to be selling at a *discount* and a bond selling above the face value is said to be selling at a *premium.*

(a) Mirage Lake Water Co. issued $1000 bonds with a coupon rate of 6.5%. Mary's purchase price was $750. What is the current yield?

(b) Frank knows that the current yield on his Mirage Lake $1000 bonds is 9.2% and he just paid $1150 for them. What is the coupon rate on the bond?

(c) Mirage Lake Series B $1000 bonds were issued with a coupon rate of 4%. What should they sell for if the current yield must be 8.5%?

36. *(Inverse Proportionality, Bond Prices and Yields)*   [See problem 35 for the relation between bond yields and prices.] A $10,000 U.S. Treasury bill has a stated yield (which corresponds to the coupon rate) of 9.1%.

(a) If it is sold at a discount at $9850, what is its current yield?

(b) If the current yield market changes so that 9.7% is the current yield, what is the new price of the bond?

37. *(Inverse Proportionality, Value of Dollar in Buying Power and Consumer Price Index)*   The amount of goods and services that a dollar will buy is inversely proportional to the Consumer Price Index (CPI). As goods and services become more expensive, the value of a dollar declines.

For example, using a CPI of 100 in 1967, when the CPI was 109.8 in 1969, the dollar was "worth" $0.91 in 1969.

(a) What was a dollar "worth" in 1915 when the CPI was 30.4?

(b) What was a dollar "worth" in 1940 when the CPI was 42.0?

(c) What was a dollar "worth" in 1981 when the CPI was 272.3?

(d) What was a dollar "worth" in 1987 when the CPI was 340.4?

(e) Graph the "value of a dollar" as a function of the CPI.

38. *(Yield to Maturity, YTM)*   The YTM of a $1000 bond with an annual coupon interest rate I (expressed decimally), a current market price of M

dollars, and having N years to maturity, is approximately*

$$YTM = \frac{2 \cdot I \cdot 1000 \cdot N + 2000 - 2 \cdot M}{N(1000 + M)}$$

Find the YTM of a $1000 bond with a current market price of $850, a coupon interest rate of 8%, and having ten years to maturity.

39. *(Video Tape Rentals)*   The rustic town of Videoville, population 12,000, has five video tape rental stores. Jack is thinking of opening another. After a market study, he finds that presently each store is averaging 225 rentals per day. Because of near market saturation, the total number of rentals in the entire town will go up only slightly if he opens another store. In fact, each new store that opens only adds about 25 new daily tape rentals to the entire town's total number of video tapes being rented. The rest of a new store's business consists of rentals that had previously been made at other stores. Thus, the average for each store will decline.

(a) What can Jack expect to be his store's average daily number of tapes rented?

(b) Suppose Jack decides to open up n new stores. What will be the average daily number of tapes rented at each of his stores?

Graphing Calculator Problems

40. *(Airplane Engine Maintenance Expense)*   The function $C(t) = (0.7t^2 + 2.75t + 0.28)/(t + 0.53)^2$ approximates the percentage in year t of the cost of the first year's maintenance for the JT3D turbojet engines made by Pratt and Whitney.† Find $C(0)$, $C(0.35)$, $C(1)$, $C(5)$, $C(10)$, and $C(50)$ and sketch the curve.

Sketch the graphs of the following, approximating any asymptotes and x-intercepts.

41. $f(x) = \dfrac{x}{x^2 - 4}$

42. $g(x) = \dfrac{x + 1}{x^2 - 5}$

43. $f(x) = \dfrac{x^2 - x - 6}{2x^2 - 2x}$

44. $f(x) = \dfrac{x^2 - x - 1}{3x^2 - 1200}$

45. $h(x) = \dfrac{x^2 - 2x - 4}{x^2 - 625}$

Business and Investment Almanac. Homewood, Illinois: Dow Jones–Irwin, 1982, p. 430.
†NASA Document CR-134645, Section II, Figure 11-I.

46. **(Handicapping Road Running Races)** Functions have been developed to model the absolute limits for the number of minutes required for a human to run 5 kilometers.* These are:
    (a) For a woman x years old, $f_w(x) = 0.0049x^2 - 0.2629x + 18.2708 + 17.1793/x$
    (b) For a man x years old, $f_m(x) = 0.0023x^2 - 0.0625x + 11.1674 + 64.9426/x$

    Plot and compare these for $5 \le x \le 85$.

Writing and Discussion Questions

47. Is a polynomial itself a rational function?

48. Is the quotient of two rational functions itself a rational function?

Enrichment

The next two problems are suggested by the work of Arthur Laffer, an American economist who argued that lowering a tax rate might increase the total revenue from taxes. For an income tax this might occur because if a worker were keeping a larger percentage of each dollar earned, the worker might be willing to work so many additional hours that the total tax paid would exceed the previous tax. For a sales tax on a product, a lower sales tax might result in increased sales to such an extent that the total sales tax collected would increase.

What came to be known as a *Laffer Curve* has a shape somewhat like figure 1.6.10.

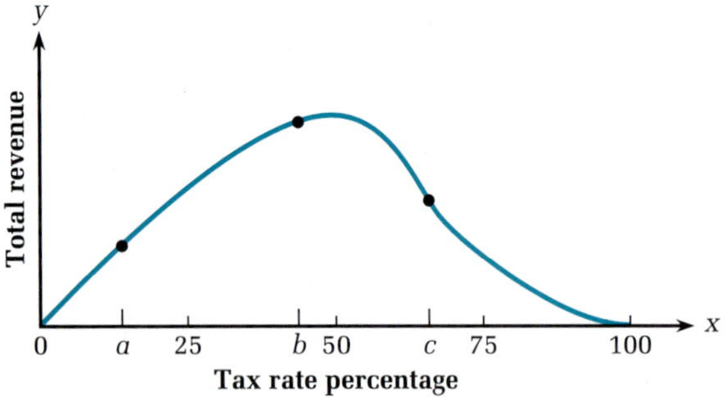

FIGURE 1.6.10

On this curve, if the rate increases from a to b, then the total tax revenue increases. However, if the

rate were to increase from b to c, the total tax revenue would decrease. All economists agree, of course, on the fact that tax rates of 0% and 100% yield no revenue, but the shape of an actual curve is a subject of great controversy.

49. **(Laffer Curve)** Suppose a Laffer Curve is given by $R(x) = (100x - x^2)/(x + 10)$, where x is the tax rate in percent and $R(x)$ is the tax revenue in millions of dollars. A sketch of $R(x)$ is shown in figure 1.6.11.

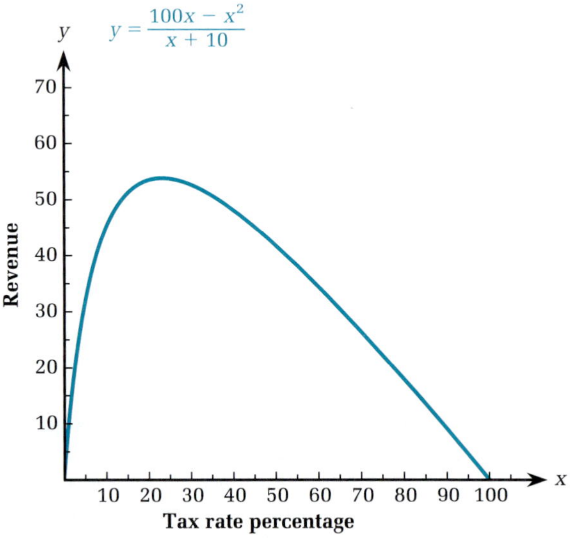

FIGURE 1.6.11

    (a) From the graph of figure 1.6.11, estimate the following: $R(5)$, $R(10)$, $R(15)$, $R(20)$, $R(25)$, $R(30)$, $R(35)$.
    (b) From the graph, estimate the rate that yields the maximum revenue.

50. **(Laffer Curve)** Suppose a Laffer Curve is given by $R(x) = (100x - x^2)/[x(x + 100)]$, where x is the tax rate in percent and $R(x)$ is the tax revenue in millions of dollars. A sketch of $R(x)$ is shown in figure 1.6.12.
    (a) From the graph of figure 1.6.12, estimate the following: $R(5)$, $R(10)$, $R(20)$, $R(40)$, $R(50)$.
    (b) From the graph, estimate the rate that yields the maximum revenue.

*J. D. Camm and T. J. Grogan, "An Application of Frontier Analysis: Handicapping Running Races," *Interfaces* 18:6 (Nov.–Dec. 1988), pp. 52–60.

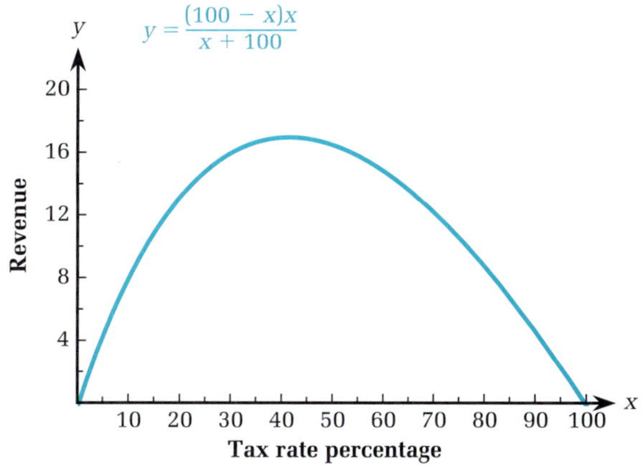

$$y = \frac{(100 - x)x}{x + 100}$$

FIGURE 1.6.12

SOLUTIONS TO PRACTICE EXERCISES

1. The domain of $g(x)$ consists of all reals except zero.

x	-2	-1	-0.5	0.2	0.5	1	1.5	2	2.5
$\dfrac{x^3 + 1}{x}$	3.5	0	-1.75	5.04	2.25	2	2.92	4.5	6.65

These values are graphed in figure 1.6.13.

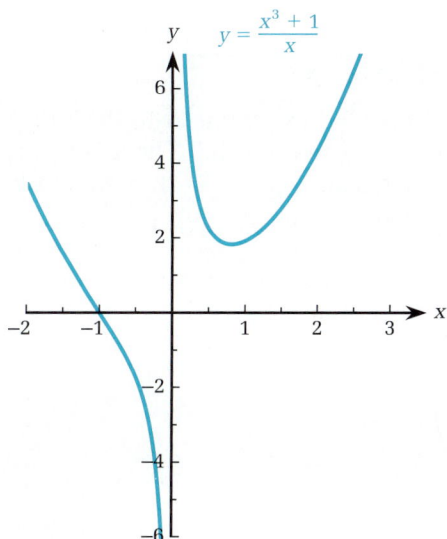

$$y = \frac{x^3 + 1}{x}$$

FIGURE 1.6.13

2. The value of k would still be 0.38, so the prime would be $0.38/0.05 = 7.6\%$.

3. $(5 \cdot 90 + 1)/10 = 451/10 = 45.1$ minutes.

4. (a) Starting from $y = (5x + 1)/(100 - x)$, we have $y(100 - x) = 5x + 1$, which will lead us to $100y - 1 = 5x + yx$, and finally to $(100y - 1)/(5 + y) = x = P(y)$.

   (b) $P(10) = (100 \cdot 10 - 1)/(5 + 10) = 999/15 = 66.6$ grams in ten minutes.

1.7 PROPERTIES OF EXPONENTS AND POWER FUNCTIONS

The functions we have studied so far have been polynomials and rational functions. These functions used only positive integer powers of x. That is, in the expression x^n, we assumed n was a positive integer. In this section we extend this notation considerably and discuss the meanings of x^a for any real number a.

If the variable is being raised to a fixed power, such as $x^{1.4}$, then we have a *power function*. Models using power functions are used to represent a great variety of situations, many in the biological and social sciences. By analyzing various aspects of size, biologists attempt to understand why the largest flying birds weigh about 13 kilograms. To take another situation, suppose we measure the wing area of a bird in square centimeters. It turns out that the wing area is approximately $k \cdot m^{2/3}$, where the constant k depends on the type of bird, and the mass of the bird, m, is measured in grams.

Based on arguments about the possibilities for flight, a paper published in 1959 concluded that the maximum muscle power that a well-conditioned human could generate would just be sufficient to fly an ultralight airplane. Twenty years later, in 1979, the human-powered *Gossamer Albatross* was flown from England to France.

The largest hovering birds weigh about 100 grams. Several groups are currently working on the problem of a human-powered hovering vehicle.

Exponent Notation

Let us first review exponential notation and terminology for positive integer exponents. For a real number x and a positive integer n, x^n represents the product of n factors of x.

$$x^n = \overbrace{x \cdot x \cdot \;\cdots\; \cdot x}^{n \text{ factors}}$$

The expression "x^n" is read as "x raised to the nth power," or "x to the nth power," or just "x to the nth." The x is called the *base* and n is the *power*. For instance, $2^3 = 2 \cdot 2 \cdot 2 = 8$, and 2^3 has 2 as the base and 3 as the power.

EXAMPLE 1

Evaluate 2^4, 7^3 and $\left(\dfrac{1}{2}\right)^3$.

SOLUTION

$$2^4 = 2 \cdot 2 \cdot 2 \cdot 2 = 16, \quad 7^3 = 7 \cdot 7 \cdot 7 = 343,$$
$$(1/2)^3 = 1/2 \cdot 1/2 \cdot 1/2 = 1/8$$

PRACTICE EXERCISE 1

Evaluate 4^3.

The same words, "base" and "power," are used even if the power is not an integer.

DEFINITION

Base and power

In the expression x^y, the *base* is x and the *power* is y.

In Example 1 and Practice Exercise 1, the evaluation of the number can be done using repeated multiplication. However, for even a fairly small power it may be easier to calculate x^y than to calculate repeated multiplications. For instance, $3^6 = 3 \cdot 3 \cdot 3 \cdot 3 \cdot 3 \cdot 3$ may be calculated as a product of six factors, but it is easier to calculate using an exponential key on a calculator. If the power is not an integer, then the exponential key is essential.

On your calculator, one of the keys is labeled $\boxed{x^y}$ (Casio), $\boxed{y^x}$ (Hewlett-Packard), or $\boxed{\wedge}$ (Texas Instruments). To gain access to this function key, your calculator may require that you use a second function key, which may be labeled as $\boxed{\text{inv}}$ or $\boxed{\text{2nd}}$.

EXAMPLE 2   Evaluation by a calculator

Evaluate $(0.45)^4$, $(1.34)^2$, $(-2)^3$, $3^{1.3}$, $(-1.2)^{0.5}$.

SOLUTION

The first three of these can be done by repeated multiplication, although doing so is tedious.

$$(0.45)^4 = 0.04100625 \qquad (1.34)^2 = 1.7956 \qquad (-2)^3 = -8$$

The expression $3^{1.3}$ requires the use of a calculator, which approximates the value to be 4.171167511. We should indicate that this is only an approximation by using the symbol "$\approx$" to write $3^{1.3} \approx 4.171167511$ or, using just four places, $3^{1.3} \approx 4.1712$. In many instances such approximations are not indicated by the "$\approx$" symbol, but simply written as "=." Nonetheless, if it is important to emphasize an approximation is being used, then the "$\approx$" symbol must be written.

Using our calculator in an attempt to evaluate $(-1.2)^{0.5}$, we receive an error message. Perhaps you will find the display of $\boxed{-E-}$. This is because the exponent 0.5 indicates a square root, and the square root of a negative number is not defined.

The last part of Example 2 shows that x^y is not always defined when x is negative. However, if x is positive, then for any y there is a value for x^y, whether y is positive or negative, a whole number, or simply an ordinary (maybe somewhat messy) real number such as 2.356. The fact that $x^{2.356}$ is defined for any positive value of x is a mathematical fact, but we are not going to discuss the reason at this point. Suffice it to say that our calculator gives a good approximation for the value of any number of any exponential form, such as $2^{2.356} \simeq 5.119489655$ or $2^{\sqrt{2}} \simeq 2.665144142$.

To avoid the difficulties of deciding whether x^y is defined or not, let us agree that if y is not an integer, then x^y is considered only for positive values of x.

Properties of Exponents

The following chart summarizes the properties of exponents.

PROPERTIES OF EXPONENTS

For positive real numbers x and y and any real numbers w and z, we have

E1 $\quad x^w \cdot x^z = x^{w+z}$

E2 $\quad x^0 = 1$

E3 $\quad x^{-w} = \dfrac{1}{x^w}$

E4 $\quad (x \cdot y)^z = x^z \cdot y^z$

E5 $\quad \dfrac{x^w}{x^z} = x^{w-z}$

E6 $\quad (x^z)^w = x^{z \cdot w}$

E7 $\quad \left(\dfrac{x}{y}\right)^z = \dfrac{x^z}{y^z}$

EXAMPLE 3 Properties of exponents

The following illustrate each of the properties of exponents.

Property	Examples
E1 $\quad x^w \cdot x^z = x^{w+z}$	$3^4 \cdot 3^2 = 3^{4+2} = 3^6 = 729$
	$x^2 \cdot x^3 = x^{2+3} = x^5$
	$x^4 \cdot x^2 \cdot x^{1.3} = x^{4+2+1.3} = x^{7.3}$
E2 $\quad x^0 = 1$	$5^0 = 1$

The expression 0^0 is not defined.

E3 $x^{-w} = \dfrac{1}{x^w}$ $2^{-5} = \dfrac{1}{2^5} = \dfrac{1}{32} = 0.03125$

$x^{-4} = \dfrac{1}{x^4}$

E4 $(x \cdot y)^z = x^z y^z$ $(2 \cdot 5)^2 = 2^2 \cdot 5^2 = 4 \cdot 25 = 100$

$(x \cdot y)^3 = x^3 \cdot y^3$

$(2x)^3 = 2^3 \cdot x^3 = 8x^3$

E5 $\dfrac{x^w}{x^z} = x^{w-z}$ $\dfrac{2^5}{2^2} = 2^{5-2} = 2^3 = 8$

$\dfrac{x^5}{x^3} = x^{5-3} = x^2$

$\dfrac{x^{2.3}}{x^{1.8}} = x^{2.3-1.8} = x^{0.5}$

$\dfrac{x^3}{x^5} = x^{3-5} = x^{-2} = \dfrac{1}{x^2}$

E6 $(x^z)^w = x^{z \cdot w}$ $(3^2)^4 = 3^{2 \cdot 4} = 3^8 = 6561$

$(x^2)^3 = x^{2 \cdot 3} = x^6$

E7 $\left(\dfrac{x}{y}\right)^z = \dfrac{x^z}{y^z}$ $\left(\dfrac{4}{3}\right)^5 = \dfrac{4^5}{3^5} = \dfrac{1024}{243} \approx 4.2140$

$\left(\dfrac{x}{y}\right)^3 = \dfrac{x^3}{y^3}$

In the last example, suppose we evaluated $(4/3)^5$ by using the four-place approximation for $4/3 = 1.3333$ and then calculated $(1.3333)^5$. Had we done so, the approximate answer (to four places) then would be 4.2135, rather than 4.2140. This demonstrates again the hazard of performing further operations on an approximation. We should not use an approximation until necessary.

PRACTICE EXERCISE 2

Simplify the following. Evaluate the numeric exercises.
(a) $x^r \cdot x^s$ (b) $3^3 \cdot 3^{-1}$ (c) $3^2 \cdot 3^{1.3}$ (d) $x^3/(x^{-1})$
(e) $(x^2)^3$ (f) $((1/2)^3)^{-2}$ (g) $(x^2 \cdot y^{-1})^3$ (h) 5^0
(i) $(3x)^2$ (j) $(4/(3^{-1}))^2$

WARNING

In using exponential notation, as in all our writing, we need to be careful to say what we want to say. For example, notice that the expression $2x^3$ is not equal to $(2x)^3$. Without parentheses to direct us otherwise, exponentiation is done before multiplication. Hence, $2 \cdot 3^2 = 2 \cdot 9 = 18$, not 36.

With a graphing calculator, use parentheses extensively. We know $2^{3 \cdot 2}$ means 2^6, which is 64. But entering $2 \wedge 3 \times 2$ gives 16. We must enter $2 \wedge (3 \times 2)$ to obtain the correct result.

Another frequent mistake occurs in calculating an expression such as -2^4, which may be (mistakenly) said to be $(-2)^4 = 16$, rather than its correct value of $-(2^4) = -16$.

Roots

An alternate terminology exists for exponents of the form $1/n$, where n is a positive integer.

DEFINITION

nth root

For any positive integer n, the *nth root* of x, written $\sqrt[n]{x}$, is $x^{1/n}$.

As a special case, the square root is written omitting the number 2. Therefore, the square root of x is written as $\sqrt{x}$ rather than $\sqrt[2]{x}$. For our convenience, our calculator includes a $\boxed{\sqrt{x}}$ key as well as the $\boxed{x^y}$ key because square roots occur frequently and it is quicker to use the $\boxed{\sqrt{x}}$ key.

Your calculator may also have a $\boxed{x^{1/y}}$ key, which can be used to evaluate roots if you want (or need) to avoid the approximations necessary for decimal exponents. Such a key is extremely handy if you wish to find odd roots of negative numbers. For example, $(-8)^{1/3} = -2$ whereas $(-8)^{0.33333333} = \boxed{-E-}$ (an error message). This will not greatly bother us, for we plan to work only with positive bases, but the example does show another difficulty with rounding off. The error occurs because $1/3$ is not exactly 0.33333333.

EXAMPLE 4   Evaluating roots

Evaluate (a) $\sqrt[3]{8}$ (b) $\sqrt[5]{1/32}$.

SOLUTION

(a) $\sqrt[3]{8} = 8^{1/3} = 2$

(b) $\sqrt[5]{1/32} = (1/32)^{1/5}$. At this point we have several choices for the next step. We can find $1/32 = 0.03125$ and then evaluate that to the 1/5 power, perhaps even converting 1/5 to its decimal equivalent of 0.2. This value is 1/2, or 0.5.

On the other hand, we can rewrite $(1/32)^{1/5}$ as $(1)^{1/5}/(32)^{1/5}$ and then evaluate this as 1/2, the same answer.   ⬤

Both of the roots used in Example 4 had values so that the answer was an integer or a fraction, which could be evaluated without a calculator. However, if we have $\sqrt[5]{3}$, we would need to use the $\boxed{x^y}$ key (with $y = 0.2$) or the $\boxed{x^{1/y}}$ key (with $y = 5$) to find $\sqrt[5]{3} \approx 1.24573094$.

PRACTICE EXERCISE 3

Find (a) $\sqrt[4]{16}$ (b) $\sqrt[4]{10}$.

Power Functions

Next we consider a function that is a power of a variable.

DEFINITION

Power function

A *power function* is of the form $f(x) = ax^b$, where a and b are real numbers.

Examples of power functions are $2 \cdot x^{1/2}$ (which is the same as $2\sqrt{x}$), $4x^{0.145}$, $5x^{\sqrt{3}}$, and $3x^{-4}$.

The domain of a power function depends on b. For example, when b is an integer, the domain is $\mathbb{R}$ (all real numbers), but when $b = 1/2$, then the domain is only the non-negative reals.

Lacking the use of a graphing calculator, we can construct the graphs of power functions by determining several values of the function (often choosing values for x that make the calculations easy) and smoothly connecting the resulting points. In the following table used for the construction of the graph, the values of the function are given only to two decimal places because of the difficulty of plotting points by hand with accuracy.

EXAMPLE 5   Graph of a power function

Graph $y = f(x) = 3\sqrt{x}$, using x values of 0, 1, 2, 3, 4, and 9.

SOLUTION

Calculated values are shown in the following table and the graph in figure 1.7.1.

x	0	1	2	3	4	9
$3\sqrt{x}$	0	3	4.24	5.20	6	9

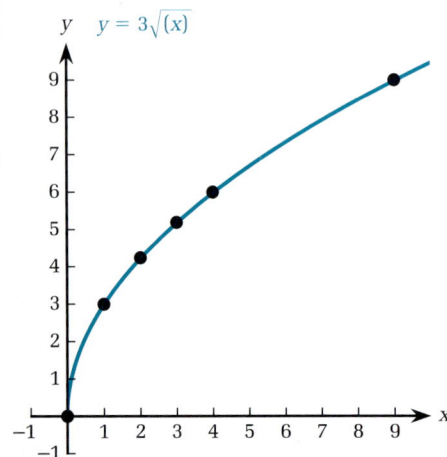

FIGURE 1.7.1

For a function such as $g(x) = 2\sqrt{5x}$, remember that $\sqrt{5x} = \sqrt{5} \cdot \sqrt{x}$. Therefore, we may rewrite $g(x)$ as $2\sqrt{5x} = 2 \cdot \sqrt{5} \cdot \sqrt{x} \simeq 4.4721\sqrt{x}$.

PRACTICE EXERCISE 4

The diameter of a circular pizza pie is found by the formula $d = 2\sqrt{A/\pi}$, where d is the diameter in inches and A is the area of the pizza in square inches. Graph this function, finding d for areas of 50, 75, and 100 square inches. If the area doubles, does the diameter double?

EXAMPLE 6   Application of a power function to motion pictures

Judy, a movie stunt coordinator, is asked to jump off a roof. She knows if she falls x feet, she will hit the air bag on the ground with a velocity of $v(x) = 5.45\sqrt{x}$ miles per hour.

(a) If she falls 25 feet, how fast will she hit the bag?

(b) If she does not want to be going over 40 miles per hour when she hits, what is the tallest roof height for her jump?

SOLUTION

(a) $5.45\sqrt{25} = 27.25$ miles per hour.

(b) Substituting into the equation, we have $40 = 5.45\sqrt{x}$. Hence $\sqrt{x} = 40/5.45$, giving us the maximum height of $x = (40/5.45)^2 = 53.9$ ft.

PRACTICE EXERCISE 5

Use the formula from Example 6.

(a) If Judy falls 30 feet, how fast will she hit the bag?

(b) If she gets a softer bag and can be going 50 mph when she lands, what is the maximum height for her jump?

Models from Biology and Medicine

Allometry is the study of the size of organisms and the consequences of size. In 1637, Galileo was the first writer to point out that large animals had bones that were scaled out of proportion to their linear dimensions. That is, he observed that a large, massive animal had thick bones, whose ratio of length to cross section was smaller than was the corresponding ratio for a lightweight animal.

Over the years, biologists have collected measurements for many variables and sought to fit curves to the measurements. They search for an explanation of each curve and any significant variation of individual data points. Often these explanations have aided the biologists in their attempts to determine and understand the physical principles that affect biological growth and behavior.

EXAMPLE 7   Basal metabolism

For most mammals, the resting metabolic rate, P_m, measured in kilocalories per day, can be expressed as $P_m = 73.3(M_b)^{0.75}$, where M_b is body mass measured in kilograms (kg).* For example, a person weighing 80 kg (about 176 pounds) and not performing strenuous exercise, will burn approximately $73.3(80)^{0.75} \approx 73.3(26.75) \approx 1961$ kilocalories per day. (Note: the food we eat has a calorie value stated

*K. Schmidt-Nelsen, *Scaling: Why Is Animal Size So Important?*, Cambridge: Cambridge University Press, 1984, p. 58.

in Calories (spelled with a capital C), but these values are actually "kilocalories" to the physicist.)

Plot the values of resting metabolic rate for a rat (0.15 kg), a human (70 kg), a bison (250 kg), and a large steer (680 kg).

SOLUTION

$$73.3(0.15)^{0.75} \simeq 73.3(0.24) \simeq 17.6 \text{ kilocalories for a rat}$$
$$73.3(70)^{0.75} \simeq 73.3(24.2) \simeq 1774 \text{ kilocalories for a human}$$
$$73.3(250)^{0.75} \simeq 73.3(62.9) \simeq 4610 \text{ kilocalories for a bison}$$
$$73.3(680)^{0.75} \simeq 73.3(133.2) \simeq 9764 \text{ kilocalories for a large steer}$$

These values are plotted in figure 1.7.2.

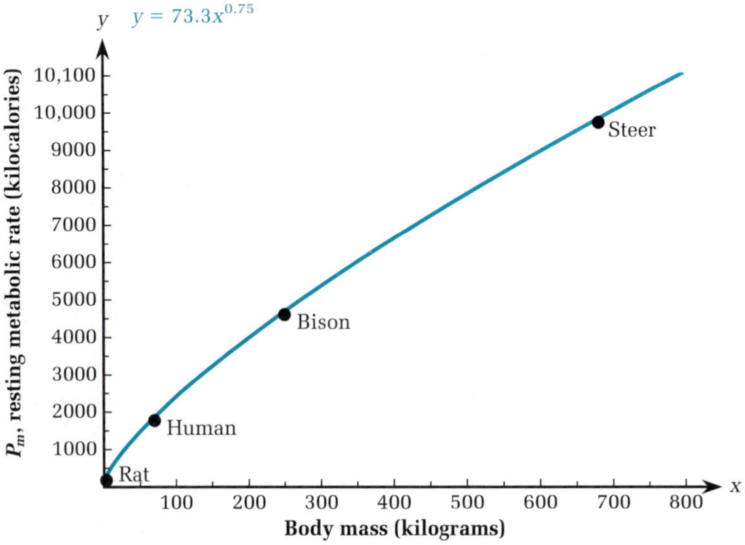

FIGURE 1.7.2

Biologists are interested in understanding why the metabolic rate of so many mammals satisfies this power equation. Of almost equal interest is understanding why the metabolic rates of seals and whales are almost exactly twice as great as predicted by the formula, whereas the rate for desert dwellers is lower than predicted.

PRACTICE EXERCISE 6

For most mammals, the number of heartbeats per minute is approximately given by $f_h = 241(M_b)^{-0.25}$, where M_b is measured in kilograms (kg).* Find f_h for the mammals in Example 7 and sketch the graph of f_h.

*Ibid., p. 101.

EXAMPLE 8   Hibernation of mammals

Some mammals spend part of the year in a physiological state of hibernation. Most of that time is spent sleeping. However, all hibernators have regular periods when they rewarm and maintain a high body temperature for an interval of time before sinking back into sleep. For instance, a marmot (or groundhog), which weighs approximately 2 kg, will stir from his torpor and have an arousal period of about 15.6 hours before sinking back into inaction.

The length of time of arousal for small mammals is approximately $W(m) = 12(m^{0.38})$ hours for an animal of m kilogram size.* (Note: the inactive phase may last several days or even weeks before another arousal occurs.)

Graph this curve and estimate the arousal time for each of the following mammals: (a) small bat (5 grams), (b) jumping mouse (60 grams), (c) ground squirrel (800 grams), (d) a fat marmot (6 kilograms = 13.2 pounds).

SOLUTION

The curve is plotted in figure 1.7.3. From the formula we find (a) $W(0.005) = 1.6$ hours, (b) $W(0.06) = 4.1$ hours, (c) $W(0.8) = 11$ hours, and (d) $W(6) = 23.7$ hours.

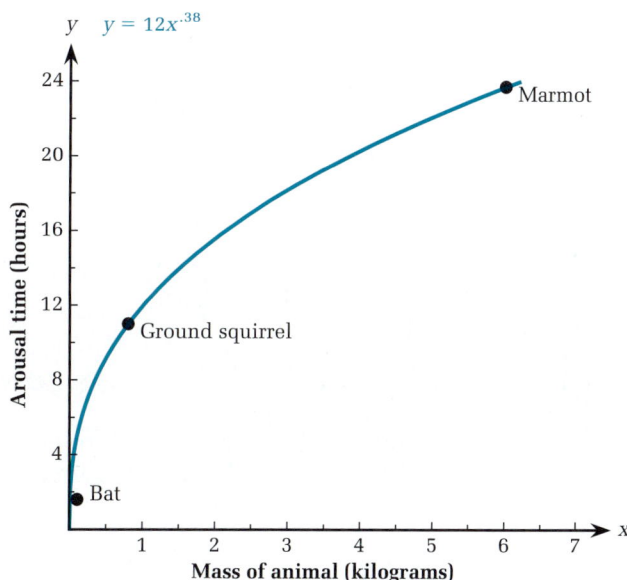

FIGURE 1.7.3

*A. French, "The Patterns of Mammalian Hibernation," *American Scientist* 76:6 (Nov.–Dec. 1988), pp. 568–575.

Stevens's Power Law (a Model from Psychology)

A mathematical part of psychology is "psychophysics." One of the topics studied is the perception of an external stimulus. For instance, if two weights are held, how much heavier does one feel in comparison to the other? In stating the following law, psychologists use the Greek letters ψ and κ.

STEVENS'S POWER LAW

If ψ is the sensation perceived by a particular individual and S is the actual physical stimulus, then for all individuals, $\psi = \kappa S^a$, where κ may vary for different individuals but a depends only on the sensation being investigated.

For instance, for heaviness, $a = 1.45$ for all individuals. Thus, if one individual is offered a two-pound weight and then a four-pound weight, she will perceive the two-pound weight as weighing $\kappa(2)^{1.45} = 2.73\kappa$ and the four-pound weight as weighing $\kappa(4)^{1.45} = 7.46\kappa$. In other words, although the second weight is only twice as heavy, she will perceive it as being almost 2.73 times as heavy.

Generally, if the first weight is any amount, w, and the second weight is twice as great, $2w$, then because $\kappa(2w)^{1.45} = \kappa(2)^{1.45} \cdot (w)^{1.45} = \kappa(2.732)(w)^{1.45}$, she would perceive the second weight as being almost 2.73 times as heavy. The value of κ, because it depends on each individual, cannot be evaluated independently of performing an actual experiment with the individual.

Values for a for selected sensations are listed in the table to the left.*

SENSATION	a
Smell	0.2
Brightness or loudness	0.3
Visual distance	0.67
Visual velocity	1.77
Electric current passed through finger	3.5

EXAMPLE 9   Stevens's Power Law

Suppose John is shown two lights, with the second having twice the actual candlepower of the first. How much brighter will the second light be perceived by John?

SOLUTION

For the first light we can let S be 1 and for the second we can let S be 2. Then $\psi = \kappa \cdot 1$ and $\psi = \kappa \cdot 2^{0.3} = 1.23\kappa$. Hence the second light

*S. S. Stevens, "On the Psychophysical Law," *The Psychological Review* 64:2 (1957), p. 166.

will seem to be about 123% as bright as the first light. Another way of expressing this is to say the second light appears to be 23% brighter than the first light.

PRACTICE EXERCISE 7

Continuing Example 9, suppose John is shown a light that has five times the actual candlepower of the first light. How much brighter will John perceive it to be?

1.7 PROBLEMS

Foundations

The problems of this section require the basic skills illustrated by the following. Solve each of the following for a.

1. $16 = 2^a$ 2. $36 = 6^a$ 3. $27 = 3^a$

4. $\dfrac{1}{8} = 2^{-a}$ 5. $\sqrt{13} = 13^a$ 6. $\sqrt[4]{80} = 80^a$

Exercises

Write problems 7–18 in terms of a base to a power. Then evaluate that expression.

7. $2^3 2^2$ 8. $3^2 3^2$ 9. $(-2)^3 2^2$

10. $3^2(-3)^3$ 11. $2^{-1}2^3$ 12. $5^2 5^{-3}$

13. $2^3 4^2$ 14. $3^3(-9)^2$ 15. $2^2(-3)^{-2}$

16. $2^5 3^{-5}$ 17. $\dfrac{2^{-3}}{2^2}$ 18. $\dfrac{3^2}{3^{-4}}$

For problems 19–26, give a decimal approximation.

19. $16^{3/4}$ 20. $\left(\dfrac{25}{36}\right)^{3/2}$ 21. $\left(\dfrac{25}{4}\right)^{-1/2}$

22. $(27)^{2/3}$ 23. $4^{0.3}$ 24. $(12)^{0.23}$

25. $\sqrt{13}$ 26. $\sqrt{8 + 2^{1.4}}$

Write problems 27–34 as a base to a single power.

27. $3^x 3^{x+1}$ 28. $2^y 2^{y-2}$ 29. $x^3 x^2$

30. $y^{-1}y^4$ 31. $(x^3 \cdot x^{-1})^2$ 32. $(y \cdot y^{-2})^3$

33. $\dfrac{r^3}{r}$ 34. $\dfrac{x^4}{x^2}$

Write the following in exponent form, with positive exponents.

35. $\sqrt{s^2 t^6}$ 36. $\sqrt{r^{-2}s^3}$ 37. $\sqrt[3]{x^6 y^3}$

38. $\sqrt[3]{x^{-3}y^6}$

Approximate problems 39–42.

39. $\sqrt[4]{6}$ 40. $\sqrt[4]{81}$ 41. $\sqrt[4]{80}$ 42. $\sqrt[3]{41}$

43. Find $f(x) = x^{1.5}$ for $x = 0, 1, 2, 4, 9,$ and 25 and use your values to sketch the graph of $f(x)$ for $x \geq 0$.

44. Find $f(x) = x^{1/3}$ for $x = 0, 1, 2, 8,$ and 27 and use your values to sketch the graph of $f(x)$ for $x \geq 0$.

45. **(Free-fall Skydiving)**   Our movie stuntwoman in Example 6 neglected the effect of air resistance because it is so small on a short fall. During a long free fall, an object reaches a maximum speed, called the *terminal velocity*. The terminal velocity of a general class of objects (all of the same general density and shape) is given by $v(w) = k \cdot w^{1/6}$, where the v is in miles per hour and w is the weight in pounds (lb). A 160-lb sky diver reaches about 120 mph.
   (a) What is k for sky divers?
   (b) What is the terminal velocity of a 200-lb sky diver?

46. **(Veterinary Medicine)** After a house cat has dropped about 75 feet it spreads its legs out, which slows its continued fall. Its maximum rate of fall has been estimated to be only 40 mph. Hence, as veterinarians in major cities learned, cats regularly survive falls from tall skyscrapers. A cat weighs about 8 pounds. Using the formula in problem 45, what would be the maximum rate of fall for a 200-pound lion?

47. **(Economies of Scale in Chemical Manufacturing, the Rule of 0.6)** The chemical manufacturing industry has a rule of thumb that states that if a process (such as refining or smelting) has a capacity of C_1 and costs X_1 dollars, then if the capacity is changed to C_2, the new cost will be $X_2 = (C_2/C_1)^{0.6} \cdot X_1$ dollars. For example, an oven that is three times as large as a $3500 oven will cost $3^{0.6} \cdot 3500 = (1.933) \cdot 3500 = \$6766$. Because capacity is tripled without tripling, or even doubling, cost, an *economy of scale* has been achieved.
    (a) Use the rule of 0.6 to find the cost of doubling the oven in our example.
    (b) Suppose a burger broiler costs $30,000. What would be the cost of building a burger broiler three times as large? (Round your answer to the nearest thousand.)

48. **(Biology, Rates of Breathing)** For most mammals, the frequency of breaths per minute is approximated by $f_r = 53.3 \cdot M_b^{-0.26}$, where M_b is measured in kilograms.[*] Find how often a person weighing 50 kg breathes.

49. **(Biology, Human Surface Area)** The total surface area of a person, measured in square centimeters, can be approximated by $S = 11 \cdot M_b^{0.67}$, where M_b is measured in kilograms.[†] Estimate the surface area of a person who weighs 60 kg.

50. **(Biology, Cardiac Medicine)** The total cardiac output of a human heart can be approximated by $C = 187 \cdot M_b^{0.81}$, where C is measured in milliliters per minute and M_b is in kilograms.[‡] Estimate the total cardiac output of a person weighing 60 kilograms.

51. **(Biology, Cardiac Medicine)** When at rest, an average person recirculates his or her total blood supply in about a minute. The heart of a well-trained athlete who is engaged in strenuous exercise may be able to recirculate the total blood supply in a mere 12 seconds. The recirculation time for mammals is approximated by the formula $t = 17.4 \cdot M_b^{0.25}$, where M_b is in kilograms and t is in seconds.

    Sketch the curve for recirculation of mammals by plotting points for an elephant (4000 kg), horse (700 kg), human (70 kg), rat (0.2 kg), and mouse (0.03 kg).

52. **(Biology, Forestry)** Tree rings provide visible evidence of the annual growth of a tree. For lodgepole pine (*Pinus contorta*), the width of the new annual ring is approximately $y = 2x^{-0.6}$ ($0 \le x \le 10$), where y is measured in millimeters and x is the age-class of the tree measured in decades.[§] For example, a tree aged 61 to 70 years would have an x value of 7.
    (a) Estimate the thickness of a new tree ring for a tree 35 years old.
    (b) Sketch the graph of y.

53. **(Psychology, Stevens's Law)** The Stevens's function for estimation of visual distance from the observer is $\psi(S) = \kappa(S)^{0.67}$. Suppose an object is given two placings.
    (a) Determine how many times farther the second placing appears to be than the first if these are 3 and then 6 feet from the observer.
    (b) Determine how many times farther the second placing appears to be than the first if these are 3 and then 9 feet from the observer.

54. **(Psychology, Stevens's Law)** The Stevens's function for estimation of visual velocity of a moving object is $\psi(S) = \kappa(S)^{1.77}$.
    (a) Suppose an airplane flies by going 100 miles per hour and then flies by again at the same distance from the observer but at a speed of 200 miles per hour. How much faster does the observer perceive the plane is flying the second time compared to the first time?
    (b) Suppose an airplane flies by going 200 miles per hour and then flies by again at the same distance from the observer but at

[*]K. Schmidt-Nelsen, *Scaling: Why Is Animal Size So Important?*, Cambridge: Cambridge University Press, 1984, p. 101.
[†]Ibid., p. 81.
[‡]Ibid., p. 127.

[§]P. Koch and J. Schlieter, *Spiral Grain and Annual Ring Width in Natural Unthinned Stands of Lodgepole in North America*, USDA Forest Service, Intermountain Research Station INT-499, September, 1991.

a speed of 300 miles per hour. How much faster does the observer perceive the plane is flying the second time compared to the first time?

Graphing Calculator Problems

55. Graph $y = x^a$ for $a = 0.5, 0.8, 1, 1.5$, and 2. What can be concluded about the shape of the graph, depending on the value of a?

56. Approximating solutions to the equation $x^{2.5} = x^2 + 3$ can be achieved in two ways.
   (a) Graph $f(x) = x^{2.5} - x^2 - 3$ and use the Trace or Box function to approximate solutions of $f(x) = 0$.
   (b) Graph $g(x) = x^{2.5}$ and $h(x) = x^2 + 3$ on the same screen. This should be done so as to allow the swapping of the Trace function between the graphs of the two functions by using the △ and ▽ keys. By moving the * between graphs, approximate the x-value of the point of intersection.

Writing and Discussion Problems

57. The first property for exponents was $x^w \cdot x^z = x^{w+z}$, which we called $E1$. Use this property, with $z = 0$, to show that $x^0 = 1$, and then go on to show why we want $x^{-w} = 1/x^w$.

58. In order to create a meaningful model, care must be taken in all measurements.
   (a) In the model for surface area described in problem 49, suggest how the surface area of a human can be measured.
   (b) Not all animals stand quietly while their surface area is being measured. For cows, the measurements are actually made by using a paint roller. Suggest how a cylindrical paint roller, with length 5 inches and outside diameter 2.5 inches, can be used to determine the surface area of a cow.

59. It was said that Galileo wrote that large animals had bones that were scaled out of proportion to their linear dimensions. Who was Galileo and what were several of his contributions to mathematics and mathematical modeling?

Enrichment Problems

60. Simplify $((25/9)x^4)^{-1/2}$.

61. Show how $E3$ follows from $E1$ and the fact that $x^w \cdot 1/x^w = 1$.

62. ***(Credit Card Use versus Discounting for Cash)*** Prior to 1981 a rebate of more than 5% for cash payment was considered a finance charge levied against credit card users and was illegal in the United States. Such rebates were legalized in 1981, when Congress passed the Cash Discount Act. We offer a formula to determine the cash rebate equivalent to the use of a credit card.

   A retailer incurs two costs in a credit card transaction. There is the *factoring fee,* which is the percentage of the transaction that the credit card company charges to convert the charge to cash. Second, there is the *time lag* between the sale and the collection of the funds. For example, suppose the cost of capital is 20% and an average of six days occurs between the sale and the collection of the proceeds, and suppose the factor fee is 5%. Then a credit sale of $10,000 is equal to $9472 in cash. Hence, the retailer would come out as well on that sale by offering a 5.3% discount for cash. The formula is $x = 1 - [(1 - z)/(1 + r)^{t/365}]$, where x is the discount rate in decimal form, z is the factoring fee in decimal form, r is the annual cost of capital in decimal form, and t is the time lag in number of days.
   (a) Verify that the formula gives the discount for cash claimed in the example.
   (b) Find the cash discount that could be offered if the factor fee was 4%, the time lag was eight days, and the annual cost of capital was 15%.
   (Note: There is considerably more to this problem than simply determining what cash discount could be given. A customer who does not know what the equivalency figure should be may choose a cash discount of 3% when the equivalency figure was 5.3%. For that matter, some customers may be paying cash regardless, and the retailer does not want to offer those customers a cash discount. For a more detailed discussion, consult Levy, Michael, and Ingene, Charles A., "Retailers: Head Off Credit Cards with Cash Discounts," *Harvard Business Review* 3 (May/June 1983), p. 18.)

Solutions to Practice Exercises

1. $4^3 = 4 \cdot 4 \cdot 4 = 64$

2. (a) $x^r \cdot x^s = x^{r+s}$
(b) $3^3 \cdot 3^{-1} = 3^{3-1} = 3^2 = 9$
(c) $3^2 \cdot 3^{1.3} = 3^{3.3} \approx 37.5405$
(d) $\dfrac{x^3}{x^{-1}} = x^{3-(-1)} = x^{3+1} = x^4$
(e) $(x^2)^3 = x^{2 \cdot 3} = x^6$
(f) $\left(\left(\dfrac{1}{2}\right)^3\right)^{-2} = \left(\dfrac{1}{2}\right)^{3 \cdot (-2)} = \left(\dfrac{1}{2}\right)^{-6}$
$= \dfrac{1}{\left(\dfrac{1}{2}\right)^6} = \dfrac{1}{\dfrac{1}{64}} = 64$
(g) $(x^2 \cdot y^{-1})^3 = (x^2)^3 \cdot (y^{-1})^3 = x^6 \cdot y^{-3}$
(h) $5^0 = 1$
(i) $(3x)^2 = 3^2 \cdot x^2 = 9x^2$
(j) $\left(\dfrac{4}{3^{-1}}\right)^2 = (4 \cdot 3)^2 = 12^2 = 144$

3. (a) $\sqrt[4]{16} = 2$
(b) $\sqrt[4]{10} \approx 1.7783$

4. When $A = 50$, $d = 8$; $A = 75$, $d = 9.78$; $A = 100$, $d = 11.28$. The diameter did not double. A graph of d as a function of A is shown in figure 1.7.4.

5. (a) $5.45\sqrt{30} = 29.85$ mph.
(b) $50 = 5.45\sqrt{x}$. Then $\sqrt{x} = 50/5.45 = 9.1734$, so $x = 84.17$ ft.

6. $241(0.15)^{-0.25} \approx 387$ for a rat. $241(70)^{-0.25} \approx 83$ for a human. $241(250)^{-0.25} \approx 61$ for a bison. $241(680)^{-0.25} \approx 47$ for a large steer. These values are plotted on the graph shown in figure 1.7.5.

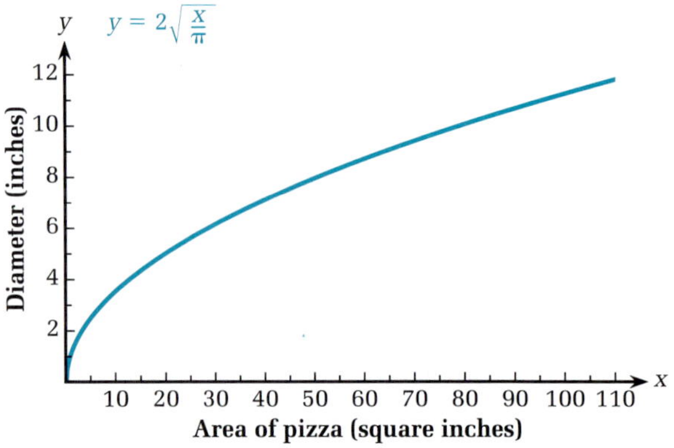

FIGURE 1.7.4

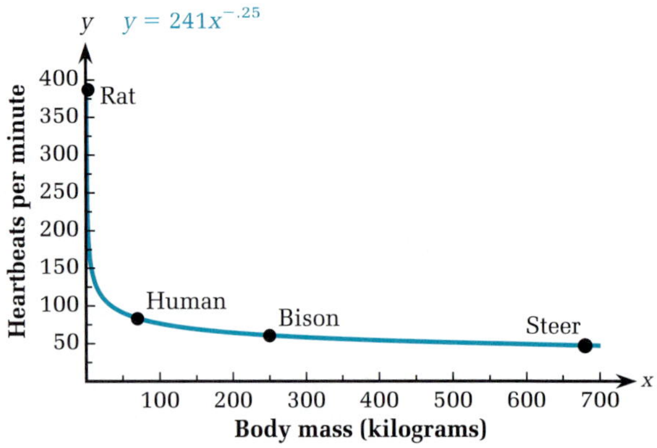

FIGURE 1.7.5

7. Because $\kappa \cdot 5^{0.3} = 1.62\kappa$, John will perceive it as being 62% brighter than the first light.

CHAPTER 1 REVIEW

Discuss or define:

1. Real numbers and the real number line
2. Inequalities and their properties
3. Absolute value
4. Coefficient of concentration
5. Distance between points on the real number line
6. Sets
7. Intervals (open, closed, and other types)

8. Solving inequalities
9. Function (including domain and range)
10. Independent and dependent variables
11. Cartesian plane
12. Graph of a function
13. x- and y-intercepts
14. Distance between points

15. Floor integer and ceiling integer functions

16. Mathematical models

17. Equations of lines (including slope and intercepts)

18. Slopes of parallel and perpendicular lines

19. Straight line depreciation

20. Piecewise linear

21. Directly proportional

22. Cost, revenue, demand, and profit functions

23. Breakeven

24. Factoring quadratic expressions

25. Quadratic formula

26. Quadratic function

27. Parabola

28. Vertex of a parabola

29. Extreme values of a quadratic function

30. Polynomial functions

31. Degree

32. Rational function

33. Vertical asymptote

34. Inversely proportional

35. Inverse function

36. Inverse function test

37. Algebra of functions (including composition)

38. The base and power of x^y

39. Properties of exponents

40. Roots

41. Power function

REVIEW PROBLEMS FOR CHAPTER 1

Evaluate the following.

1. $|3.2|$

2. $|-2.3|$

3. $\left|\sqrt{3} - 2\right|$

4. $\left|4 - 2\sqrt{3}\right|$

5. $\left|\dfrac{1}{3} - \dfrac{2}{5}\right|$

6. $\left|4\left(\dfrac{2}{3} - \dfrac{1}{2}\right)\right|$

7. Express in words the sets:
   (a) $\{x : -3 \le x < 5\}$ (b) $[-2, 5)$
   (c) $\{x : -1 < x < 2.5\}$

8. Represent the following sets as intervals and graph each on the real number line.
   (a) $\{x : x < 2\}$ (b) $\{x : -1 \le x < 3\}$
   (c) $\{x : 1 < x\}$ (d) $\{x : 1 < x \le 3\}$
   (e) $\{x : 0 \le x < -1\}$ (f) $\{x : 3 \ge 2x\}$

9. Find the solution set of each of the following inequalities, expressing the set in terms of an interval.
   (a) $|x - 1| < 5$ (b) $|2x - 1| < 4$
   (c) $|3 + 2x| < 6$ (d) $x + 2 \le 3x - 1$
   (e) $1 < 3x - 1 \le 5$ (f) $0 < 2 - 3x \le 5$

10. Find the domains of the following.
   (a) $f(x) = \dfrac{x}{x^2 - 1}$ (b) $g(x) = \sqrt{x + 4}$
   (c) $g(x) = \dfrac{\sqrt{x}}{\sqrt{x + 1}}$ (d) $\dfrac{x}{x^2 - x}$

11. For $f(x) = \sqrt{4 - x}$, find
   (a) $f(1)$ (b) $f(0)$ (c) $f(a)$
   (d) $f(a + 1)$

12. For $g(x) = 3/x$, find
   (a) $g(1)$ (b) $g(0)$ (c) $g(a)$
   (d) $g(a + 1)$

13. Evaluate the following.
   (a) $\lceil 1.5 \rceil$ (b) $\lfloor 1.5 \rfloor$ (c) $\lceil 2.8 \rceil$
   (d) $\lfloor -2.1 \rfloor$
   (These are the floor integer and ceiling integer functions.)

14. A study of intercollegiate sport teams produced the data shown in the following table. Both the funding and participation for each sport combine the men's and the women's teams' values. Find the coefficient of concentration.

	PARTICIPANTS	BUDGET
Basketball	45	$200,000
Skiing	25	$100,000
Swimming	30	$ 50,000

15. Find the distance between the points $P_1(5, -1)$ and $P_2(1, 3)$.

16. Plot the points $A(2, 0)$, $B(4, -1)$, and $C(5, 6)$ and use the distance formula to show these points form the vertices of a right triangle.

17. Give the standard form of the equation of a circle with center at $P(-1, 4)$ and radius 3.

18. Suppose a line segment joining the points $P_1(1, 3)$ and $P_2(4, 2)$ is the diameter of a circle.
(a) Find the center of the circle.
(b) Find the radius of the circle.
(c) Find the standard form of the equation of the circle.

19. For $f(x) = 20 - 2x^2$:
(a) Find each of the following: $f(0)$, $f(1)$, $f(a)$, and $f(a + h)$.
(b) Find the distance between the points on the graph determined by $x = 0$ and $x = 2$.
(c) Find a simplified form for the quotient $[f(x + h) - f(x)]/h$.

20. The company Just the FAX Please charges $5 for the first 10 minutes and then 50¢ for each additional minute or fraction thereof. For example, 12.5 minutes costs $5 + 3(0.50) = $6.50. Express the charge for x minutes in terms of the ceiling integer function.

21. Graph the line given by the equation $-3x + 2y = 6$ and find:
(a) the slope
(b) the y-intercept

22. (a) Find an equation for the line that contains the points $A(1, 5)$ and $B(4, 7)$.
(b) What is its slope?
(c) What is its y-intercept?

23. Find an equation for the line parallel to the line $3y + 6x = 5$ that contains the point $P(1, 2)$.

24. Find an equation for the line through the point $P(5, 1)$ and perpendicular to a line through the points $A(1, 0)$ and $B(3, 5)$.

25. An office CD system costs $16,000 and over six years its value drops to a salvage value of $10,300. Using straight line depreciation, find a formula that will give the value of the CD system t years after purchase.

26. The Speedy Service Sizzle Steak Co. bought an Acura NSX for $80,000 in 1990. They plan to sell it in five years for a salvage value of $10,000. Using straight line depreciation, find a formula that will give the value of the Acura NSX t years after purchase.

27. Factor
(a) $x^2 + 6x + 8$
(b) $2x^2 - 11x - 6$

28. Solve
(a) $3x^2 - x - 2 = 0$
(b) $x^2 - x - 12 = 0$

29. Solve
(a) $2x^2 - x - 2 = 0$
(b) $x^2 + 5x - 1 = 0$

30. Sketch the graph of $f(x) = 2x^2 + x - 6$. Include on your graph the coordinates of the vertex and both the x- and y-intercepts.

31. Sketch the graph of $f(x) = x^2 - 3x + 1$. Include on your graph the coordinates of the vertex and both the x- and y-intercepts.

32. For $f(x) = ax^2 + bx + c$, state a condition on $b^2 - 4ac$ that guarantees the graph of $f(x)$ does not touch or cross the x-axis.

33. Suppose $R(x) = -x^2 + 10$ is a revenue function and $C(x) = 2x + 2$ is a cost function. Find
(a) the profit function and
(b) the breakeven value for x.

34. Factor
(a) $x^3 - x^2 - 6x$
(b) $x^3 + x^2 - 4x - 4$

35. *(Air Pressure, Brooklyn Bridge)* The normal air pressure at sea level is about 15 pounds per square inch. At a depth in water of d feet, the pressure is approximately $p(d) = 0.45d + 15$ pounds per square inch. When digging the supports for the Brooklyn Bridge (or any structure that extends down into water), workers were in a casing that was pressurized to keep the water out.
(a) What pressure was necessary at 70 feet?
(b) If workers could only stand 45 pounds per square inch pressure, how deep could they work?

36. The sales of soda at a game are directly proportional to the size of the crowd. For 1000 people, sales are 15 gallons.
(a) Find an equation for sales (in gallons) as a function of the size of the crowd.
(b) What are sales for a crowd of 3000?
(c) What was the size of the crowd on a day when sales were 30 gallons?

Suppose $6 is the revenue per gallon sold. The cost of setting up the soda stand is $100 and the cost of the soda is $2 per gallon.

(d) Find an equation for revenue as a function of the number of gallons sold.

(e) Find an equation for revenue as a function of the size of the crowd.

(f) What size does the crowd have to be for the soda stand to break even?

(g) What is the expected profit for a crowd of 5000?

37. The luxury ocean cruiseliner *QE2* carries 4440 tons of oil for fuel. Traveling at its normal cruising speed of 32.8 miles per hour, fuel is burned at the rate of 15.4 tons per hour, whereas slowing to 23 miles per hour cuts consumption to 6.17 tons per hour.* Let y be the tons of fuel consumed per hour when traveling x miles per hour and assume there exist values of a and b such that $y = f(x) = ax + b$.

(a) Determine a and b.

(b) How many hours can the *QE2* travel at 25 miles per hour?

(c) Compare the maximum distance the *QE2* can travel at constant speeds of 25 miles per hour or 32.8 miles per hour.

38. Such-a-Burger sells 1400 burgers a day at 89 cents each. At 99 cents apiece they sell 1200. Assume the demand for burgers is linear.

(a) Find an equation for the demand curve.

(b) Find an equation for the revenue curve.

(c) What price maximizes their revenue?

(d) How many burgers are sold at that price?

(e) What is the maximum revenue?

39. The Glass Table Store bought n glasses at a total cost of $8000. Upon unpacking, they found five broken glasses. By pricing each glass at $1 over their wholesale cost of $8000/(n - 5)$, the profit on the $n - 5$ glasses was $995.

(a) Find n.

(b) Find the wholesale price of each glass.

40. For $f(x) = 1/x^2$ and $g(x) = 3x + 1$, find the following:

(a) $(f + g)(3)$ (b) $(f \cdot g)(3)$ (c) $(f/g)(x)$

(d) $(f \cdot g)(x)$ (e) $f(g(x))$ (f) $g(f(x))$

(g) the domain of $f(g(x))$

41. For $f(x) = 2x^2 + \sqrt{x}$ and $g(x) = 1/x$, find the following:

(a) $(f + g)(x)$ (b) $(f \cdot g)(x)$ (c) $(f/g)(x)$

(d) $f(g(x))$ (e) $g(f(x))$

42. Graph $f(x) = 1/(4x^2 - 1)$, giving the x-coordinates of any vertical asymptotes.

43. Find an inverse function for

(a) $g(x) = 2x + 3$

(b) $f(x) = 2x/(x - 1)$

44. Find functions $g(x)$ and $h(x)$ so that $f(x) = \sqrt{x^2 + 2} = g(h(x))$. [Note: The answer is not unique.]

45. *(Lead Contamination)* Lead sinkers used by fishermen are filling up Lake Mirage. The cost-benefit model for cleaning out x percent of the lake is given by $C(x) = 30/(100 - x)$ thousand dollars.

(a) How much will it cost to clean out 90 percent?

(b) What percent can be cleaned out for $2000?

46. Simplify (and evaluate those that have numeric values).

(a) $x^3 x^{-1}$ (b) $3^4 3^{-2}$ (c) $\dfrac{(x^2 y)^2}{x}$

(d) $(x^2 y^{-2})^3$ (e) $\sqrt{\dfrac{18}{2}}$

47. Evaluate each of the following:

(a) $2^{1.4} \cdot 2^5$ (b) $(25)^{3/2}$ (c) $\sqrt[4]{256}$

(d) $\sqrt[4]{60}$

48. Write $x^{1.4} \cdot (x^{-2}/y^2)^3$ in the form $x^a y^b$.

49. Find $x^{1.6}$ for $x = 0, 0.5, 1, 1.5, 2, 2.5,$ and 3 and sketch a graph of $f(x) = x^{1.6}$.

50. *(Froude's Law, Salmon Speed)* The sustained speed of a sockeye salmon (*Oncorhynchus nerka*) over one hour, as a function of its length, can be approximated by the formula $V(l) = 19.5\sqrt{l}$, where V is measured in centimeters/second and l is measured in centimeters.†

(a) Sketch the graph of $V(l)$ over the domain of lengths $10 \le l \le 50$.

(b) How many times the speed of a 25-centimeter salmon is the speed of a 50-centimeter salmon?

---

*H. Petroski, "Driven by Economics," *American Scientist* 79:6 (Nov.–Dec. 1991), pp. 491–495.

†J. R. Brett, "The Relation of Size to Rate of Oxygen Consumption and Sustained Swimming Speed of Sockeye Salmon (*Oncorhynchus nerka*)," *J. Fish. Res. Bd. Canada* 22 (1965), pp. 1491–1497.

DERIVATIVES

CALCULUS IS ABOUT FUNCTIONS AND THEIR RATES OF CHANGE. CREATED in the seventeenth century, calculus was used at that time to investigate and describe physical changes. As one example, shortly before 1600, Galileo had noticed that a chandelier swinging in the cathedral at Pisa always took the same time to complete one full swing, no matter how large the arc of that swing. Galileo found this to be true of any pendulum, but he did not have an explanation of the fact. By the end of the century, Isaac Newton and Gottfried Leibniz, two of the founders of calculus, could explain not only the swing of a chandelier, but even the path of Earth around the sun and many other physical phenomena.

Although the innovators of calculus were often motivated by the dynamics of physical science, their labors created a mathematical system with far-reaching applicability. Examples of the use of calculus in today's world include an accountant analyzing her company's profits, a farmer trying to decide when to harvest a crop, or a physician studying the spread of an epidemic. For each of these, and many others, mathematical models using calculus provide a valuable tool for investigation and decision making.

Consider the graph in figure 2.0.1, which is based on information from Nintendo of America, Inc. Shown are the sales of video games in the United States for the period 1979–1990.*

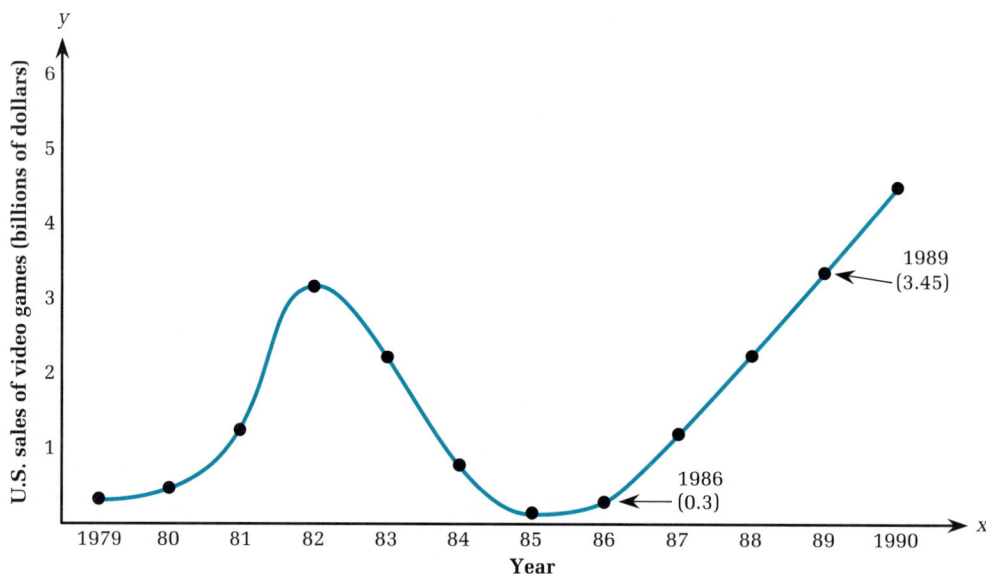

FIGURE 2.0.1

We can see that sales increased slowly during 1979 and 1980, then increased rapidly, peaking in 1982. Following that high point, sales declined until 1985, and then rebounded strongly to the end of the decade. Nintendo might use calculus to develop a model for sales. A reliable model, one that explained changes in sales, could be used by their marketing department to develop a plan for increasing sales, by their production division to plan purchasing of materials and the allocation of equipment, and by their human resources department to forecast staffing requirements.

To study a general function, $f(x)$, and the shape of the graph of $y = f(x)$, we want to investigate the rate of change and determine how rapidly the function is increasing or decreasing. In many cases, we want to find out whether the function has reached a maximum or minimum value. Calculus helps us investigate these situations.

Denver Post (from the *New York Times*), June 14, 1990.

2.1   LIMITS, AN INTUITIVE APPROACH

A fundamental difference between precalculus and calculus is the dependence of calculus on the idea of a limit. The concept of a limit, as it is used in mathematics, is very specific. However, before giving a definition of limit, let us develop our intuitive understanding of the concept of limits through several examples.

Limits of Functions

The following example introduces the concept of a function, $f(x)$, approaching a limit as x approaches a, some real number.

EXAMPLE 1   Limit of a function, selling chocolate

Suppose the candy store El Chocolaté de Oro sells individual pieces of chocolate pecan caramel for $12 a pound. What is the limit of the price of an actual selection of individual pieces as the weight of that selection approaches one-half pound?

SOLUTION

Because the chocolate costs $12 per pound, the cost of x pounds is given by the cost function $C(x) = \$12x$. If pieces could be selected whose total weight was exactly one-half pound, their price would be $C(0.5) = \$12(0.5) = \$6.00$.

If the selection weighs only approximately one-half pound, then its cost would be only approximately $6.00. However, the closer the weight of the selection is to one-half pound, the closer the cost will be to exactly $6.00. Using the terminology of limits, we say, "For $C(x) = 12x$, as x approaches 1/2, the limit of $C(x)$ is 6." This limit is called a *two-sided limit* because the values for x could be more or less than 1/2.

Figure 2.1.1 shows two vertical bands centered on $x = 1/2$. These indicate various selections of weights, as measured on the x-axis. The vertical band on the x-axis from 0.4 pounds to 0.6 pounds represents that variation in weights that corresponds to a variation in price of $4.80 to $7.20. These prices are indicated by a corresponding horizontal band from 4.8 to 7.2 on the y-axis. A second vertical band on the x-axis represents weights from 0.45 pounds to 0.55 pounds and it corresponds to a horizontal band of prices between $5.40 and $6.60 on the y-axis.

We can see that if we want to restrict the possible variation in price from $6, we first create a horizontal "price-band." For each such horizontal "price-band" there is a vertical "weight-band," which contains the value of 1/2. Any selection of chocolate whose weight is within the vertical "weight-band" has an associated price within the "price-band."

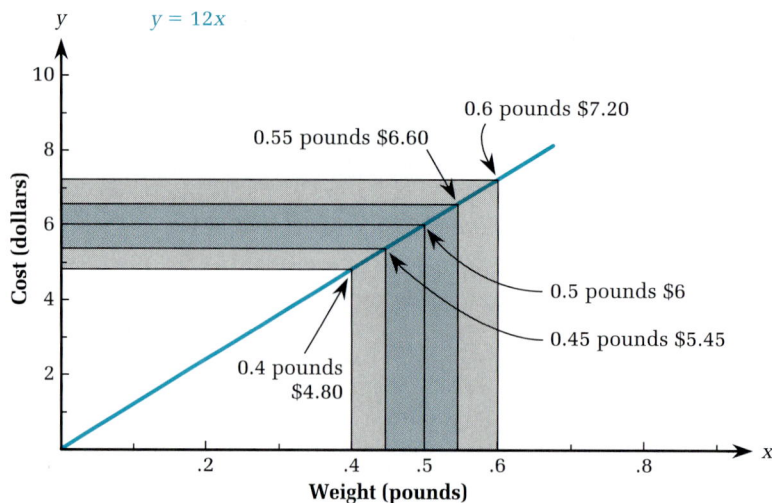

FIGURE 2.1.1

WEIGHT	PRICE
0.47	$5.64
0.48	5.76
0.49 ⟷	5.88
0.50	6.00
0.51	6.12
0.52	6.24
0.53	6.36

Another way to picture this limit is to imagine a band on a weighing scale that has weight on the left side and price on the right side. The schematic in the accompanying table shows the region of the band around 0.50, giving some sample weights near 0.50, and the associated prices. The arrow marker shown at 0.49–5.88 indicates that the selection being weighed is 0.49 pounds and has a cost of $5.88. The closer the arrow marker is to a weight of 0.50, the closer the cost is to $6.00.

PRACTICE EXERCISE 1

Continuing our example of $C(x) = 12x$, what price is the limit as x approaches 3/4 of a pound?

The graph of the function $C(x) = 12x$ can be drawn through the point $P(0.5, 6)$ in a smooth, continuous manner. That is, the value of the limit of $C(x)$ as x approaches 0.5 is the same as $C(0.5)$. Both of them are 6. Whenever this happens for the graph of a function $f(x)$ and a value a, then we say the function is continuous at that value.

DEFINITION

$f(x)$ continuous at a

The function $f(x)$ is *continuous at a* if all three of the following conditions are satisfied.

(i) $f(x)$ is defined for $x = a$

(ii) $f(x)$ has a limit as x approaches a

(iii) that limit is equal to $f(a)$

We have seen that $C(x) = 12x$ is continuous at $x = 6$. In fact, $12x$ is continuous at any value of x. The next three examples illustrate functions that are not continuous at some value.

EXAMPLE 2   ## The limit exists as x approaches a, but the function is not defined at $x = a$

Show that the function $g(x) = (2x^2 - 2)/(x - 1)$ has the limit of 4 as x approaches 1, although $g(1)$ is not defined.

SOLUTION

The graph of $g(x) = (2x^2 - 2)/(x - 1)$ is shown in figure 2.1.2. The graph has a puncture hole at $P(1, 4)$ because division by zero is not possible. Hence $g(1)$ is not defined. However, on both sides of the hole the graph shows the values of the function are close to 4. Hence, as x approaches 1, the values of $g(x)$ approach 4 as a limit, even though $g(1)$ is not defined.

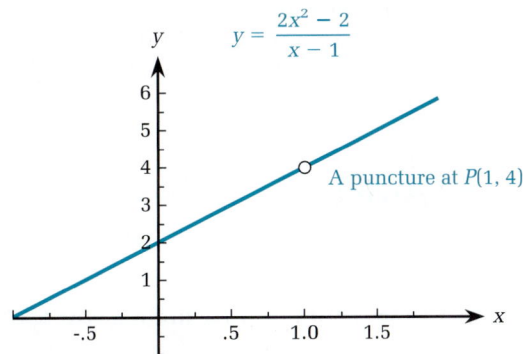

FIGURE 2.1.2

On any graph, we may need to indicate that a particular point is actually on the graph, or is missing. The point $P(1, 4)$ is missing from the graph of $y = g(x)$ in Example 2 because $g(1)$ is not defined. When we want to emphasize the point is missing we put in a "○" and when we want to emphasize that the point is truly present we put in a "●".

Graphing Calculators

If we graph a function by using a graphing calculator or a software package and are using a "connecting mode," we may not be able to tell there is a point "missing." This is due to the limitations of the calculator. Actually, as we know, a true individual "point" is vanishingly small and cannot even be seen. What a calculator does is display a big "blob" and call it a point.

If we ask any calculator to determine the value of $g(1) = (2 \cdot 1^2 - 2)/(1 - 1)$, the calculator will indicate an error has been made in data entry because division by zero is not defined. A graphing calculator may draw a graph of $g(x) = (2x^2 - 2)/(x - 1)$, and show the omission of a puncture point if its calculation involves division by zero. However, because the values for which $g(x)$ is evaluated are values chosen by the scaling package built into the calculator, it may happen that $x = 1$ is never used as input. In this case the calculator does not attempt division by zero and therefore it draws a connected line with no gap in it. In other words, the graphing calculator calculates selected values and then connects the individual points as though the function were continuous.

A similar situation occurs in playing compact discs. There the process known as "oversampling" transforms individual digital values into a smoothed-out sound.

Another possibility for whether or not a limit exists is illustrated by Example 3. This type of behavior is common among functions that test coins for authenticity in coin-operated machines or functions that describe how a combination lock works to open at only one special value.

EXAMPLE 3   The limit of $f(x)$ exists as x approaches a, but it is not $f(a)$

Let a function be defined by $f(x) = \begin{cases} 3 \text{ if } x \neq 1 \\ 2 \text{ if } x = 1 \end{cases}$. Does this function have a limit as x approaches 1?

SOLUTION

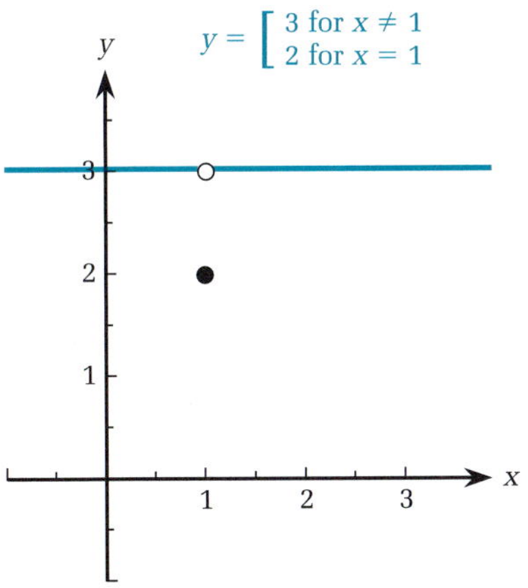

$$y = \begin{bmatrix} 3 \text{ for } x \neq 1 \\ 2 \text{ for } x = 1 \end{bmatrix}$$

The graph of this function is shown in figure 2.1.3.

Even though $f(x)$ is defined for all values of x, the value of $f(1) = 2$ is not close to the values of $f(x)$ for x near 1. Regardless of this, we say the limit exists as x approaches 1 and its value is 3.

Examples 1, 2, and 3 all have functions that are defined for values of x near a, and in each case we understand that there is a limit as x approaches a. These examples show that for a general function, it is possible the limit exists as x approaches a, although $f(x)$ is not defined at a. Even if $f(a)$ is defined, it may not equal the limit of $f(x)$ as x approaches a. In the next example we have a function that does not have a limit as x approaches a.

FIGURE 2.1.3

EXAMPLE 4   The nonexistence of a limit for $f(x)$ as x approaches a

The so-called "ceiling function," written as $h(x) = \lceil x \rceil$, is defined as the integer just larger than or equal to x. For instance, $\lceil 1.5 \rceil = 2$, $\lceil 1.05 \rceil = 2$, $\lceil 2.93 \rceil = 3$, and $\lceil 3 \rceil = 3$. What can we say about the existence of the limit of $h(x)$ as x approaches an integer?

SOLUTION

The graph of $h(x)$ is shown in figure 2.1.4.

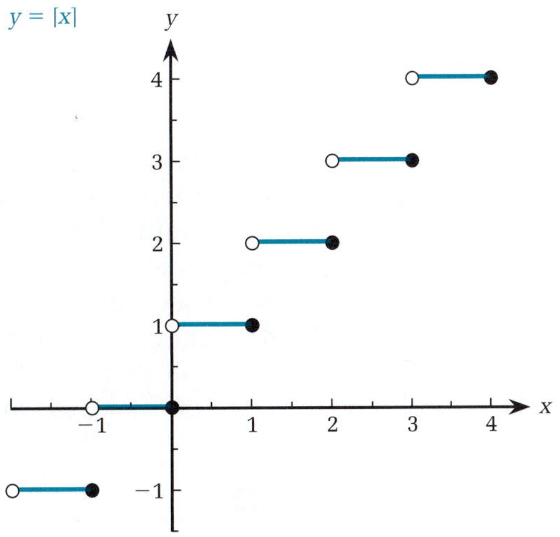

$y = \lceil x \rceil$

FIGURE 2.1.4

On each of the horizontal sections there is a last point on the right side so it is included on the graph and we emphasize its inclusion by filling in a circle there. However, there is no last point on the left side of each horizontal section. That endpoint is "missing." In other words, there is a jump at each left side of each horizontal section, or, in other words, there is a jump at each integer. For instance, when x is slightly less than 1 then the ceiling integer is 1, but when x is slightly more than 1 then $\lceil x \rceil = 2$. At $x = 1$ itself, $\lceil 1 \rceil = 1$.

There is no limit of $h(x)$ as x approaches any integer because the function "jumps," taking on quite different y-values on either side of any integer x-value.

EXAMPLE 5   The circle as a limit

The constant we abbreviate as π occurs in the formulas for the area and circumference of a circle. For a circle of radius r, the area is $A = \pi r^2$ and the circumference is $C = 2\pi r$. Although π is often approximated by the fraction 22/7, its exact value is not a rational number. Approximations of π can be obtained by using regular polygons and letting the number of sides increase. Show that a circle is the limit of the process of letting the number of sides of regular polygons increase.

SOLUTION

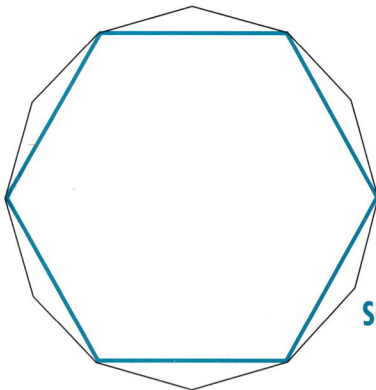

FIGURE 2.1.5

Begin with a six-sided regular polygon, then construct a regular 12-sided polygon around it, as shown in figure 2.1.5.

Continue this process, each time doubling the number of sides by adding new vertices. The limit of this process is a circle. The circle's radius is the distance from the center of the polygons to any vertex. All of the polygons are inside the circle and their vertices are points on the circle, as shown in figure 2.1.6.

Because π is the ratio of the circumference of a circle to its diameter, π actually was estimated by using regular polygons that very closely approximated a circle. In 1610, a prodigious and lifelong effort by Ludolph van Ceulen resulted in an approximation correct to 35 places. He had used a regular polygon of 2^{62} sides. Methods from calculus greatly simplified the calculation of π.

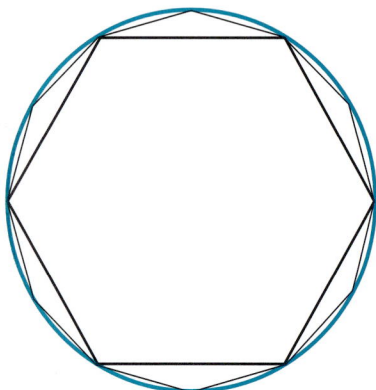

FIGURE 2.1.6

Today, computers are used to calculate better and better approximations to the exact value for π. By 1992, more than one billion digits of the decimal expansion for π were known. Such prodigious calculations are used as tests of the power of new supercomputers and the virtuosity of their programmers.

Limits Defined

We are now ready to introduce notation and summarize our ideas about limits. Let $f(x)$ represent a general function and let x approach a, an arbitrary value. In the following we first consider what happens if x stays on one side of a, and then we consider a two-sided approach.

In the next definition, the expression "$x \to a^+$" is read "x approaches a from the right" and "$x \to a^-$" is read "x approaches a from the left."

DEFINITION

One-sided limit (left-hand and right-hand limits)

Suppose $x < a$ and the values of $f(x)$ are arbitrarily close to some number L for values of x sufficiently close to a. Then we write $\lim\limits_{x \to a^-} f(x) = L$ and say: The *left-hand limit of $f(x)$, as x approaches a, is L.*

Similarly, but with $x > a$, we define the *right-hand limit of $f(x)$, as x approaches a.* For the right-hand limit we write $\lim\limits_{x \to a^+} f(x)$.

Two comments about this definition are in order. First, the phrase "arbitrarily close" means that no matter how small a positive number we choose, we can make the difference between $f(x)$ and L be less than that number by making sure x is very close to a. Second, the values of $f(x)$ can actually be L, or can even cross over the line $y = L$.

Formally, these definitions of limits do not require any sense of time or any sequence of actions, even though we use the phrase "as x approaches a." Actually, it greatly helps us to think of what is happening to $f(x)$ as x is approaching a in some sense of movement over time. We can think of a dot "•" on the x-axis moving toward a from the left-hand side, and at the same time imagine a corresponding "•" moving on the graph of the function, while yet a third related "•" hovers on the y-axis. In order to show this interconnectedness, the corresponding dots are given the same numbers in figure 2.1.7.

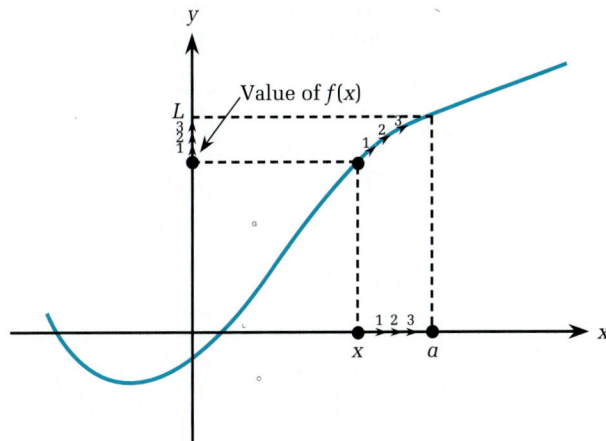

FIGURE 2.1.7

A similar situation exists for $\lim\limits_{x \to a^+} f(x)$, where we think of x as moving toward a from the right-hand side.

In the following definition we want x to be able to move toward a from either the right- or left-hand side.

DEFINITION

Two-sided limit

The limit of $f(x)$, as x approaches a, is L if the left- and right-hand limits are both defined and they are equal. In symbols, if $\lim\limits_{x \to a^-} f(x) = L$ *and* $\lim\limits_{x \to a^+} f(x) = L$*, then* $\lim\limits_{x \to a} f(x) = L$.

The phrase "two-sided limit" emphasizes that the limit being considered needs to exist on both sides of $x = a$. Even when we do not explicitly use the phrase "two-sided," we need to be careful that when we write "limit," we realize we are working with a two-sided limit.

Graphing Calculators

If we use a graphing calculator and have it display the x- and y-coordinates while tracing along a curve, then we can display the related x- and y-values. As the tracing cursor moves and the x-values become close to a, we see the y-values becoming closer to L. In fact, if we do not know L, we may use zooming and tracing to try to determine it.

One disadvantage of the graphing calculator is that the right-hand and left-hand steps cannot be made arbitrarily small but are of fixed sizes that depend on the scaling used. This causes a difficulty if the function takes on quite different values for values of x near a. Generally, however, the functions we study in this course will not exhibit such behavior.

Suppose $f(x)$ is a general function and a is an arbitrary value. The limit as x approaches a does not exist if both of the one-sided limits do exist, but they have different values. We saw that happen in Example 4 as x approached 1 for the function $h(x) = \lceil x \rceil$. Following is another example of such a function.

EXAMPLE 6   Unequal left- and right-hand limits

Show that the left- and right-hand limits are different for $g(x) = |x|/x$, $x \neq 0$.

SOLUTION

For $x > 0$ we define $|x| = x$, so that $|x|/x = x/x = 1$. As a result, as x approaches 0 from the right, the value of $g(x)$ is always 1. Thus, we have $\lim\limits_{x \to 0^+} |x|/x = \lim\limits_{x \to 0^+} 1 = 1$.

However, for $x < 0$ we define $|x| = -x$, so that $|x|/x = -x/x = -1$. Thus, as x approaches 0 from the left, $\lim\limits_{x \to 0^-} |x|/x = -1$.

A graph of $g(x)$ is shown in figure 2.1.8. Notice that $g(0)$ is not defined.

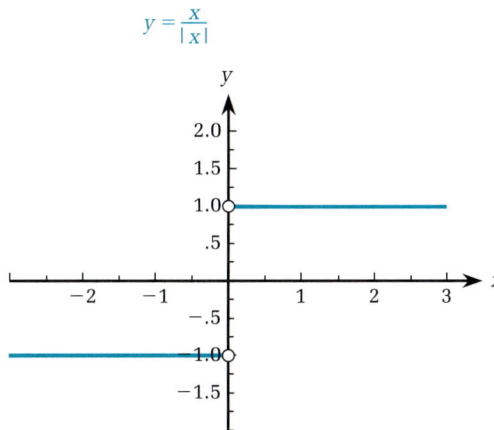

$$y = \frac{x}{|x|}$$

FIGURE 2.1.8

Another reason a limit may not exist is because the function becomes unboundedly large or small as x approaches a, as shown in Example 7.

EXAMPLE 7   Unbounded values prevent a limit from existing

Show that $\lim\limits_{x \to 2} \dfrac{1}{x - 2}$ does not exist.

SOLUTION

The graph of $g(x) = 1/(x - 2)$ is shown in figure 2.1.9. Notice that as x approaches 2 from the right, the function becomes unboundedly large, and as x approaches 2 from the left, the function becomes unboundedly small. Thus, there is no limit for $g(x)$ as x approaches 2.

In order to give further information about the behavior of $g(x)$ and its graph, we may be tempted to write that $\lim\limits_{x \to 2^+} 1/(x - 2) = \infty$ and $\lim\limits_{x \to 2^-} 1/(x - 2) = -\infty$. These expressions are an effort to say in

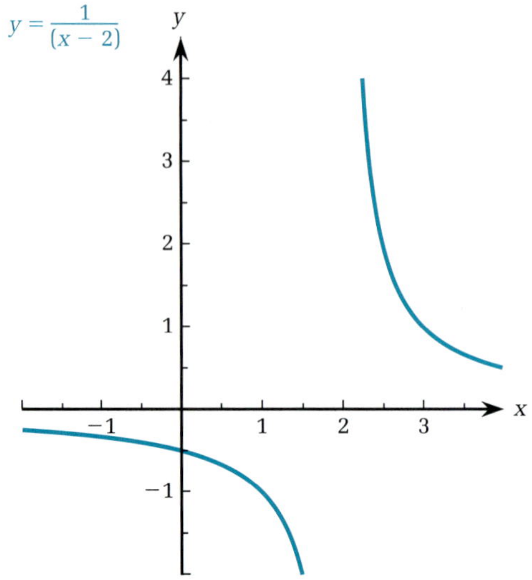

$$y = \frac{1}{(x-2)}$$

FIGURE 2.1.9

symbols that $g(x)$ becomes unboundedly large as x approaches 2 from the right-hand side and $g(x)$ becomes unboundedly small as x approaches 2 from the left-hand side. However, this use of notation may lead us to think of the "∞" symbol as a real number being the value of some limit. However, there is no "infinite" real number. Hence, we will say "a limit exists" only when there is a real number for a limit. Otherwise, we simply say "no limit exists."

Graphing Calculators

The window of a graphing calculator is bounded. If we display a portion of a graph and it goes off the screen, we may not know whether the function is truly unbounded or whether we simply need to change the region of the graph being displayed. For instance, using the display range of $-3 \leq x \leq 3$ and $-5 \leq y \leq 5$, the graph of the function $f(x) = -5x^3 + 13x^2 - 3$ disappears from the screen, and judging from the display shown in figure 2.1.10, $f(x)$ could become unbounded.

Using the reverse zoom feature, or readjusting the scale, may show us that the function is bounded, as indeed is shown in figure 2.1.11, but from our original graph we are not certain *from what we see* whether the limit at 2 exists or fails to exist.

EXAMPLE 8   Finding limits from a graph

Consider the graph of $f(x)$, as shown in figure 2.1.12. Find the one- and two-sided limits as the values of x approach each value 1, 2, 3, and 5.

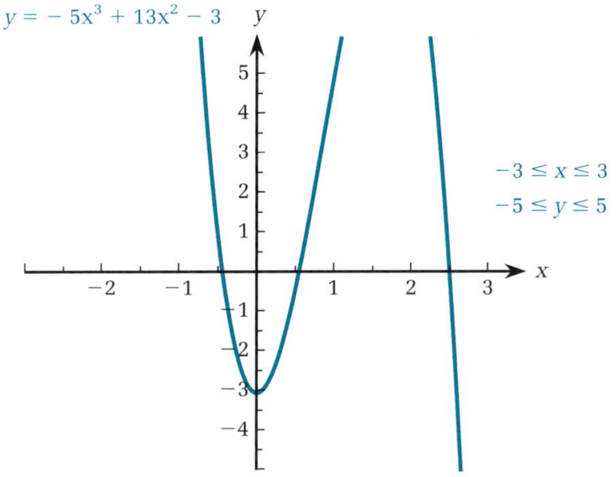

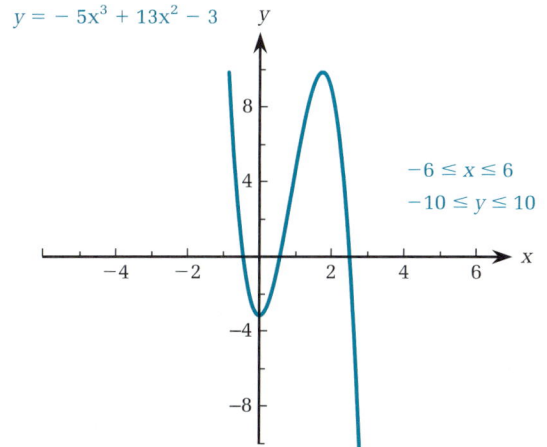

$y = -5x^3 + 13x^2 - 3$

$-3 \le x \le 3$
$-5 \le y \le 5$

FIGURE 2.1.10

$y = -5x^3 + 13x^2 - 3$

$-6 \le x \le 6$
$-10 \le y \le 10$

FIGURE 2.1.11

SOLUTION

Using the graph in figure 2.1.12, we find:

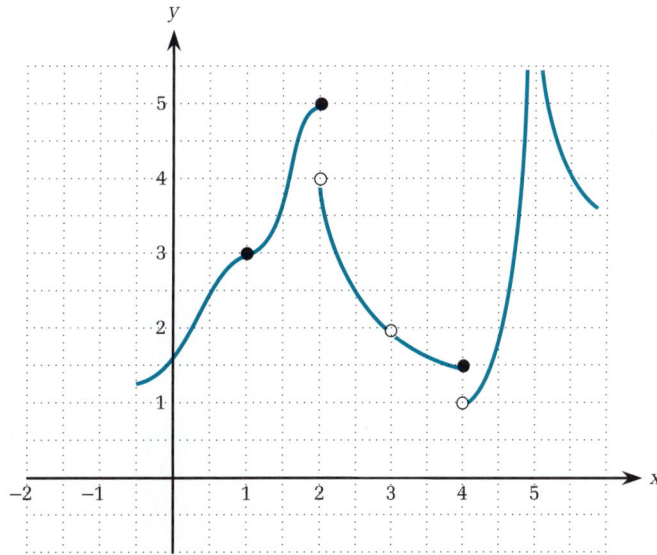

FIGURE 2.1.12

As x approaches 1, the one-sided limits are $\lim\limits_{x \to 1^-} f(x) = 3$ and $\lim\limits_{x \to 1^+} f(x) = 3$. Because both of the one-sided limits exist and they are equal to each other, then the two-sided limit exists and we have $\lim\limits_{x \to 1} f(x) = 3$.

As x approaches 2, the one-sided limits both exist, but they are unequal. The left-hand limit is $\lim\limits_{x \to 2^-} f(x) = 5$, whereas the right-hand limit is $\lim\limits_{x \to 2} f(x) = 4$. We say no limit exists as x approaches 2 because the word "limit," used all by itself, means a two-sided limit, and this exists only when the two one-sided limits exist and are equal.

As x approaches 3, the limit of $f(x)$ is 2, even though $f(x)$ is not defined at $x = 3$. The limit depends on the values of $f(x)$, which are produced by values of x near 3. Thus, because both one-sided limits exist and they are equal, the limit exists. In symbols, $\lim_{x \to 3^-} f(x) = 2$ and $\lim_{x \to 3^+} f(x) = 2$, therefore $\lim_{x \to 3} f(x) = 2$.

As x approaches 5, there is no limit of any sort. The function becomes unboundedly large on each side of $x = 5$. ●

PRACTICE EXERCISE 2

Discuss the limits of $f(x)$ as x approaches 4, as shown in figure 2.1.12.

From our discussion and examples, let us single out three points for further emphasis.

EXISTENCE OF A LIMIT

1. If the right-hand limit is not equal to the left-hand limit, then there is not a (two-sided) limit.

2. The existence and the value of the limit as x approaches a does not depend on the value of $f(a)$. If $\lim_{x \to a} f(x)$ does exist, it does not necessarily equal $f(a)$.

3. Even if $f(a)$ is not defined, the $\lim_{x \to a} f(x)$ may still exist.

2.1 PROBLEMS

Foundations

The problems of this section require the knowledge of the "floor function" and the "ceiling function." The floor function $\lfloor x \rfloor$ gives the integer equal to or just below x. For example, $\lfloor 2.3 \rfloor = 2$. This function is often known as "the greatest integer in x."

1. Find
(a) $\lfloor 2.95 \rfloor$
(b) $\lfloor 5 \rfloor$

The ceiling function $\lceil x \rceil$ gives the integer equal to or just above x. For example, $\lceil 2.3 \rceil = 3$.

2. Find
(a) $\lceil 3.5 \rceil$
(b) $\lceil 3 \rceil$

Exercises

Use the graphs shown in figures 2.1.13−2.1.16 to answer problems 3−6, finding those limits that do exist. If a limit does not exist, indicate the reason.

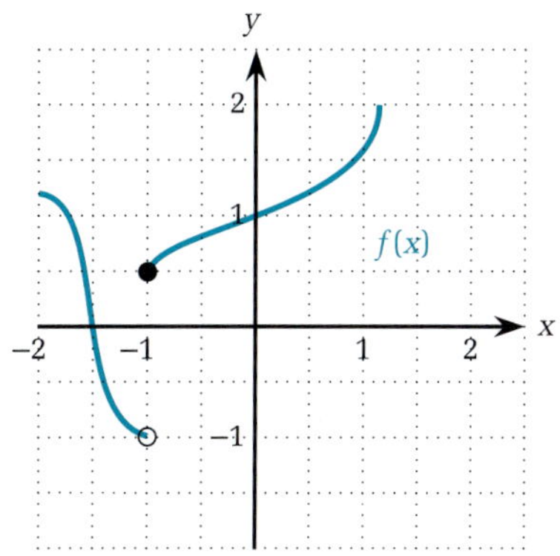

FIGURE 2.1.13

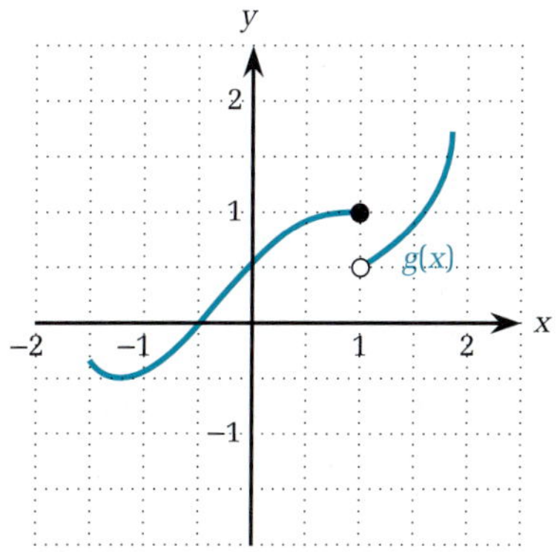

FIGURE 2.1.14

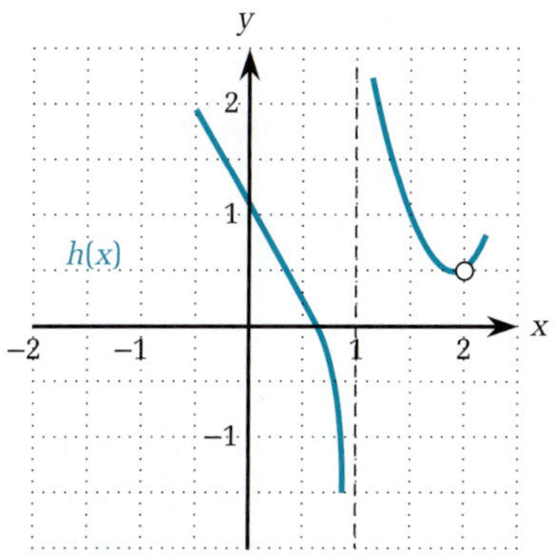

$h(x)$

FIGURE 2.1.15

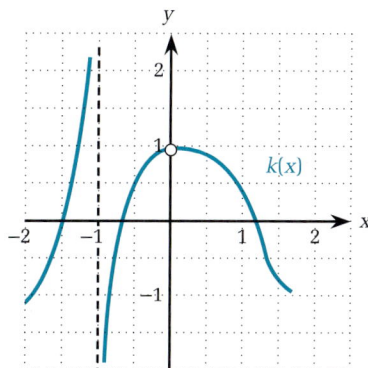

FIGURE 2.1.16

3. (a) Find $f(0)$. (b) Find $\lim\limits_{x\to0} f(x)$.

(c) Find $\lim\limits_{x\to-1^+} f(x)$. (d) Find $\lim\limits_{x\to-1^-} f(x)$.

(e) Find $\lim\limits_{x\to-1} f(x)$.

4. (a) Find $g(0)$. (b) Find $\lim\limits_{x\to0} g(x)$.

(c) Find $\lim\limits_{x\to1^+} g(x)$. (d) Find $\lim\limits_{x\to1^-} g(x)$.

(e) Find $\lim\limits_{x\to1} g(x)$.

5. (a) Find $h(0)$. (b) Find $\lim\limits_{x\to0} h(x)$.

(c) Find $\lim\limits_{x\to2^+} h(x)$. (d) Find $\lim\limits_{x\to2^-} h(x)$.

(e) Find $\lim\limits_{x\to2} h(x)$.

6. (a) Find $k(-1)$. (b) Find $\lim\limits_{x\to-1} k(x)$.

(c) Find $\lim\limits_{x\to0^+} k(x)$. (d) Find $\lim\limits_{x\to0^-} k(x)$.

(e) Find $\lim\limits_{x\to0} k(x)$.

Evaluate the following limits:

7. $\lim\limits_{x\to2^+} 3x$ **8.** $\lim\limits_{x\to3^+} -2x$

9. $\lim\limits_{x\to2} 3x$ **10.** $\lim\limits_{x\to3} -2x$

11. $\lim\limits_{x\to4} (-x + 2)$ **12.** $\lim\limits_{x\to-3} (x + 2)$

13. $\lim\limits_{x\to3} (x + 1)$ **14.** $\lim\limits_{x\to4} (x + 3)$

15. $\lim\limits_{x\to-3} \dfrac{1}{x + 5}$ **16.** $\lim\limits_{x\to2} \dfrac{2}{3x - 1}$

Find each of the following limits that exist:

17. $\lim\limits_{x \to 3} \dfrac{2}{x - 3}$

18. $\lim\limits_{x \to 2} \dfrac{2x}{x - 2}$

19. $\lim\limits_{x \to 1^+} \dfrac{|x - 1|}{x - 1}$

20. $\lim\limits_{x \to 2^-} \dfrac{|x - 2|}{x - 2}$

21. $\lim\limits_{x \to 1^-} \dfrac{|x - 1|}{x - 1}$

22. $\lim\limits_{x \to 2^+} \dfrac{|x - 2|}{x - 2}$

23. $\lim\limits_{x \to 1} \dfrac{|x - 1|}{x - 1}$

24. $\lim\limits_{x \to 2} \dfrac{|x - 2|}{x - 2}$

25. (a) $\lim\limits_{x \to 3^+} \lfloor x \rfloor$ (b) $\lim\limits_{x \to 3^-} \lfloor x \rfloor$ (c) $\lim\limits_{x \to 3} \lfloor x \rfloor$

26. (a) $\lim\limits_{x \to 5^+} \lfloor x \rfloor$ (b) $\lim\limits_{x \to 5^-} \lfloor x \rfloor$ (c) $\lim\limits_{x \to 5} \lfloor x \rfloor$

27. (a) $\lim\limits_{x \to 2^+} \lceil x \rceil$ (b) $\lim\limits_{x \to 2^-} \lceil x \rceil$ (c) $\lim\limits_{x \to 2} \lceil x \rceil$

28. The Parking Spot charges $10 per car for parking up to four hours. Beyond four hours, they charge $1 per half-hour or any fraction thereof. Let $C(x)$ be the cost of parking x hours.
(a) Express $C(x)$ using the ceiling function, $\lceil x \rceil$.
(b) For what values of a does $\lim\limits_{x \to a} C(x)$ exist?
(c) Sketch a graph of $C(x)$ for $0 \le x \le 8$.

29. The Airport Courier charges $10 for any parcel weighing up to one pound. For heavier parcels they charge another $1 for each additional pound or fraction thereof. Let $C(x)$ be the cost of shipping a parcel weighing x pounds.
(a) Express $C(x)$ using the ceiling function, $\lceil x \rceil$.
(b) For what values of a does $\lim\limits_{x \to a} C(x)$ exist?
(c) Sketch a graph of $C(x)$ for $0 \le x \le 6$.

30. El Chocolaté de Oro sells piñon nut brittle for $9 a pound.
(a) Find a function, $C(x)$, for the cost of x pounds of piñon nut brittle.
(b) Find $\lim\limits_{x \to 0.5} C(x)$.
(c) What would be the range of prices on packages whose weights are all within one-quarter of a pound of being exactly one-half pound?

31. Suppose hamburger costs $1.84 per pound.
(a) Find a function, $C(x)$, for the cost of x pounds of hamburger.
(b) Find $\lim\limits_{x \to 1.5} C(x)$.

(c) What would be the range of prices on the packages if each package were within one-tenth of a pound of 1.5 pounds?

32. Suppose shrimp costs $4.96 per pound.
(a) Find a function, $C(x)$, for the cost of x pounds of shrimp.
(b) Find $\lim\limits_{x \to 1} C(x)$.
(c) What would be the range of prices on the packages if each package were within one ounce of being one pound?

Writing and Discussion Problems

33. The approximation for π provided by using 22/7 is $3.\overline{142857}$. Another rational approximation for π was given about 480 A.D. by a Chinese writer on physics, Tsu Ch'ung-chih. His value was $355/113$. To how many decimal places do each of these provide a correct value of π? (For further information on the chronology of π, read Eves, Howard, *An Introduction to the History of Mathematics,* Fifth Edition, CBS College Publishing, 1983, pp. 85–90.)

34. Discuss the concept of limit.

35. Discuss how the pricing of prepackaged items, such as apples, is related to one-sided limits.

Enrichment Problems

36. Referring to problem 30, what would be the range of weights if all packages had to be within:
(a) $0.50 of $4.50?   (b) $0.25 of $4.50?

37. Referring to problem 31, what would be the range of weights if each package had to be within:
(a) fifty cents of $2.76?
(b) ten cents of $2.76?

38. For what values of a does $\lim\limits_{x \to a} \lceil 2x \rceil$ exist?

39. Suppose $f(x) = (2 + x)/(1 + x)$ for $x \ne -1$. Then $f(f(x)) = [2 + f(x)]/[1 + f(x)]$. For instance, $f(1) = (2 + 1)/(1 + 1) = 3/2$ and $f(f(1)) = f(3/2) = [2 + (3/2)]/[1 + (3/2)] = (7/2)/(5/2) = 7/5$. If we let L represent the limit of $f(f(f(\cdots f(1))))$, then $f(L)$ is also L. Use that result to show that $L = \sqrt{2}$.

40. $\lim\limits_{x \to 4} (3x - 5) = 7$. How close to 4 does x have to be to guarantee $3x - 5$ is within 0.01 of 7?

SOLUTIONS TO PRACTICE EXERCISES

1. The limit of $C(x) = 12x$ as x approaches 3/4
   is \$9.

2.2

EVALUATING LIMITS

In Section 2.1, we discussed the existence or non-existence of limits. Now, let us concentrate our efforts on the question of determining $\lim_{x \to a} f(x)$ in some cases when the limit does exist.

For some values of a and some functions, the determination of $\lim_{x \to a} f(x)$ may be quite straightforward, as we have seen in Section 2.1 and as is illustrated again by the next example.

EXAMPLE 1   The limit of a linear function

Find $\lim_{x \to 2} (3x - 2)$.

SOLUTION

To find $\lim_{x \to 2} (3x - 2)$, we are interested in values of x that are near 2. The values in the following table are plotted in figure 2.2.1.

FIGURE 2.2.1

x	1.5	1.9	1.95	2	2.05	2.1	2.5
$f(x)$	2.5	3.7	3.85	4	4.15	4.3	5.5

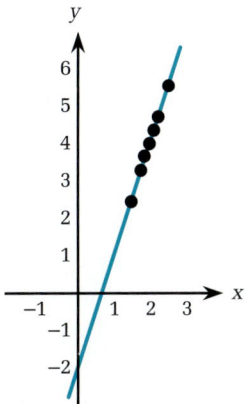

It appears that $\lim_{x \to 2} (3x - 2) = 4$.   ●

Recall that a function is continuous at a if $\lim_{x \to a} f(x) = f(a)$. The function $f(x) = 3x - 2$ is continuous at all values of a, so the determination of $\lim_{x \to 2} (3x - 2)$ can be achieved simply by finding $f(2)$. Quite often we are dealing with a continuous function and are able to determine a limit in this manner.

If $f(x)$ is not continuous at a, however, we need some other means of finding $\lim_{x \to a} f(x)$. Sometimes we are forced to evaluate $f(x)$ for values of x that approach a and then guess at both the existence and value of the limit. This technique is illustrated in the next example. Such "computational guesswork" does not guarantee success. However, from it we may correctly guess that the limit exists and even may guess the value of the limit. Once we have a guess for the limit it may be possible to verify it. That is the case with the function in Example 2.

EXAMPLE 2   A limit determined by computation

Find $\lim\limits_{x \to 1} \dfrac{\sqrt{x} - 1}{x - 1}$.

SOLUTION

The function $f(x) = (\sqrt{x} - 1)/(x - 1)$ is not defined at $x = 1$, yet the following table suggests that $\lim\limits_{x \to 1} f(x) = 1/2$, which indeed is the case.

$f(x)$	0.52786	0.51317	0.50126	undefined	0.49876	0.48809	0.47723
x	0.8	0.9	0.99	1	1.01	1.1	1.2

A graph of $f(x)$, shown in figure 2.2.2, illustrates this. The open circle shown at the point $P(1, 0.5)$ indicates that $f(x)$ is not defined at $x = 1$.

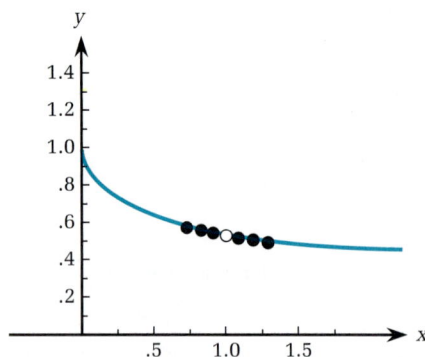

FIGURE 2.2.2

The method of "computational guesswork" is not satisfactory, for some functions have great variations in y-values for x-values near a. Even graphing calculators may not be able to plot enough points for us to see what is happening, or the limit may be some number whose decimal approximations do not seem familiar to us. For instance, if the limit were really $\sqrt{5}/2$, would we recognize the decimal version?

Properties of Limits

In this section we restrict ourselves to functions whose limits can be found by using the properties given in the following summary. These properties can be proven, but we will not attempt to do so. Their use is illustrated in Examples 3–7.

PROPERTIES OF LIMITS

Let M, P, a, r, and c be real numbers and assume

$$\lim_{x \to a} f(x) = M \text{ and } \lim_{x \to a} g(x) = P$$

L_1) If $p(x)$ is any polynomial, then $\lim_{x \to a} p(x) = p(a)$.

L_2) $\lim_{x \to a} [f(x) \pm g(x)] = M \pm P$.

L_3) $\lim_{x \to a} c \cdot f(x) = c \cdot M$ for any real number c.

L_4) $\lim_{x \to a} [f(x) \cdot g(x)] = M \cdot P$.

L_5) $\lim_{x \to a} \left(\dfrac{f(x)}{g(x)} \right) = \dfrac{M}{P}$ provided $P \neq 0$.

L_6) $\lim_{x \to a} (f(x))^r = M^r$ for any real number r.

EXAMPLE 3   Evaluating the limit of a constant function

Find $\lim_{x \to a} c$, where c is any constant.

SOLUTION

$\lim_{x \to a} c = c$ by L_1 because $p(x) = c$ is a special case of a polynomial.

As particular instances of this example, $\lim_{x \to 3} 5 = 5$ and $\lim_{x \to -2} 4 = 4$.

EXAMPLE 4   Evaluating the limit of a linear function

Find $\lim\limits_{x \to a} x$.

SOLUTION

$\lim\limits_{x \to a} x = a$ by L_1 because $p(x) = x$ is a special case of a polynomial.

EXAMPLE 5   Evaluating the limit of a polynomial function

Find $\lim\limits_{x \to 3} (2x^2 - x + 4)$.

SOLUTION

$\lim\limits_{x \to 3} (2x^2 - x + 4) = 2 \cdot 3^2 - 3 + 4$ by L_1

$= 2 \cdot 9 - 3 + 4 = 18 - 3 + 4 = 19$

PRACTICE EXERCISE 1

Find $\lim\limits_{x \to 2} (3x^2 - 2x)$, stating which rules are used.

EXAMPLE 6   Evaluating the limit of a rational function

Find $\lim\limits_{x \to 3} \dfrac{x^2 - 2x}{x + 1}$.

SOLUTION

$$\lim_{x \to 3} \frac{x^2 - 2x}{x + 1} = \frac{\lim\limits_{x \to 3} (x^2 - 2x)}{\lim\limits_{x \to 3} (x + 1)} \text{ by } L_5$$

$$= \frac{3^2 - 2 \cdot 3}{3 + 1} \text{ by } L_1$$

$$= \frac{9 - 6}{4} = \frac{3}{4} = 0.75$$

PRACTICE EXERCISE 2

Find $\lim\limits_{x \to 5} \dfrac{1 + 3x}{2}$, stating which rules are used.

The next example shows that using the limit properties L_1 to L_6 we can find limits of many functions, not simply polynomials and rational functions.

EXAMPLE 7   Evaluating by limit properties

Find $\lim\limits_{x \to 4} \dfrac{x + \sqrt{x}}{x - 1}$.

SOLUTION

$$\lim_{x \to 4} \frac{x + \sqrt{x}}{x - 1} = \frac{\lim\limits_{x \to 4} (x + \sqrt{x})}{\lim\limits_{x \to 4} (x - 1)} \text{ by } L_5$$

$$= \frac{\lim\limits_{x \to 4} x + \lim\limits_{x \to 4} \sqrt{x}}{\lim\limits_{x \to 4} x - \lim\limits_{x \to 4} 1} \text{ by } L_2$$

$$= \frac{4 + \sqrt{\lim\limits_{x \to 4} x}}{4 - 1} \text{ by } L_1 \text{ and } L_6 \text{ (using } r = \frac{1}{2} \text{ for the square root)}$$

$$= \frac{4 + \sqrt{4}}{3} \text{ by } L_1$$

$$= \frac{4 + 2}{3} = \frac{6}{3} = 2$$

PRACTICE EXERCISE 3

Find $\lim\limits_{x \to 3} \dfrac{x^2 - 2x}{\sqrt{x + 1} + x}$.

The Alternate-Function Method

In our evaluation of limits, we have been making extensive use of the continuity of our functions and finding $\lim_{x \to a} f(x)$ by calculating $f(a)$ and using the properties of limits. This is the method of *direct substitution*. This method would not work in Example 2, to find $\lim_{x \to 1} (\sqrt{x} - 1)/(x - 1)$, because the denominator is 0 if $x = 1$.

The method of direct substitution works only for continuous functions, and there are many other functions whose limits we will need to evaluate. The following theorem allows us to determine many of those limits by finding an "alternate" function.

THE ALTERNATE-FUNCTION THEOREM

Let a be a real number and let $f(x) = g(x)$ for all x (except possibly $x = a$). Then $\lim_{x \to a} f(x) = \lim_{x \to a} g(x)$.

This theorem provides us an extremely powerful method of finding $\lim_{x \to a} f(x)$, as is illustrated by Examples 8 and 9.

"ALTERNATE-FUNCTION" METHOD OF DETERMINING $\lim_{x \to a} f(x)$

If there exists a function $g(x)$ such that:

(i) $g(x) = f(x)$ for all x for which $f(x)$ is defined, and

(ii) $g(x)$ is continuous at a,

then $\lim_{x \to a} f(x) = g(a)$.

Geometrically speaking we hope the graph of $f(x)$ actually consists of an unbroken curve or one that has only a simple puncture, such as the missing point shown in figure 2.2.3. Fixing that puncture by filling in the hole, as shown in figure 2.2.4, would give a continuous function that has the same limit as x approaches a, but whose limit we can determine easily because $\lim_{x \to a} g(x) = g(a)$.

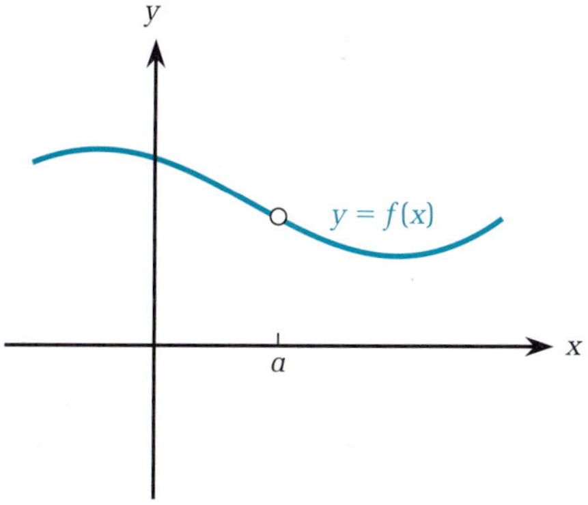

FIGURE 2.2.3

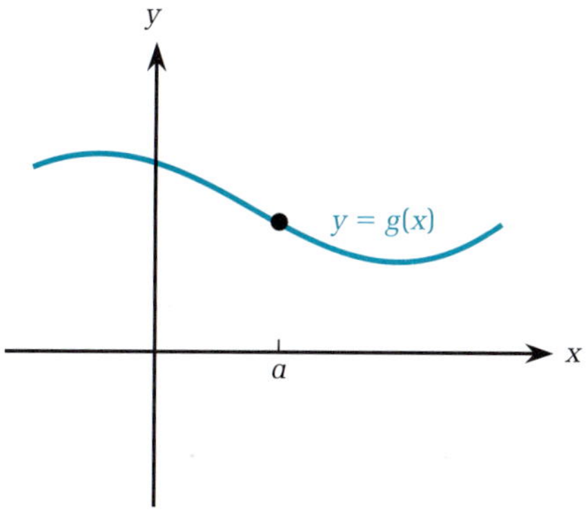

FIGURE 2.2.4

EXAMPLE 8   Using the alternate-function method

Let $f(x) = \dfrac{x^2 - 4}{x - 2}$. Show $f(2)$ does not exist, but $\lim\limits_{x \to 2} f(x) = 4$.

SOLUTION

$f(2) = (2^2 - 4)/(2 - 2) = 0/0$, which does not exist because division by 0 is not defined. However,

$$\frac{x^2 - 4}{x - 2} = \frac{(x - 2)(x + 2)}{x - 2}$$

Hence, if $x \neq 2$, $f(x) = x + 2$. By letting $g(x) = x + 2$, we have $\lim\limits_{x \to 2} g(x) = \lim\limits_{x \to 2} (x + 2) = 4$. Thus, by the alternate-function method, $\lim\limits_{x \to 2} f(x) = 4$.

The graph of $f(x)$ is shown in figure 2.2.5. It is a line with a hole in the graph because $f(2)$ is not defined. However, the graph of $g(x)$ in figure 2.2.6 does not have the hole because $g(2) = 4$.

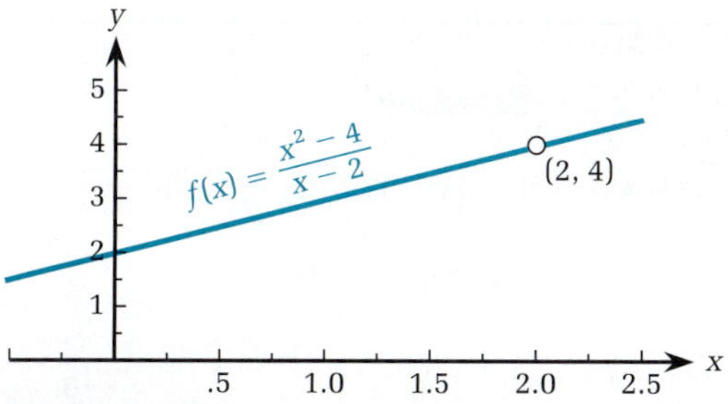

FIGURE 2.2.5

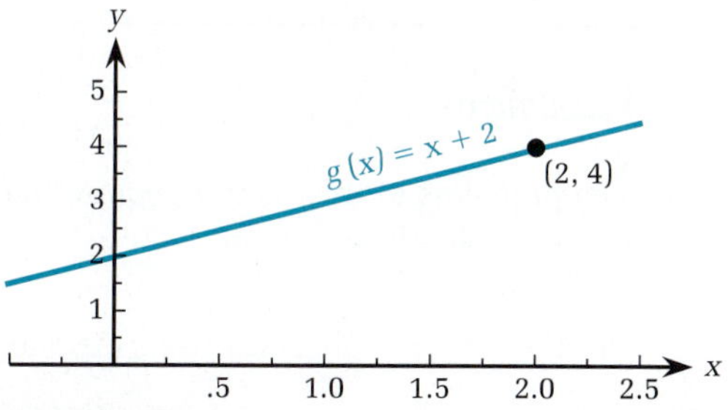

FIGURE 2.2.6

EXAMPLE 9   Using the alternate-function method

Find $\lim\limits_{x \to 1} \dfrac{2x^2 - 5x + 3}{x - 1}$.

SOLUTION

$$\lim_{x \to 1} \frac{2x^2 - 5x + 3}{x - 1} = \lim_{x \to 1} \frac{(x - 1)(2x - 3)}{x - 1}$$

If $x \neq 1$, then $x - 1 \neq 0$, so we can cancel the $(x - 1)$ expressions. Thus, the alternate function is $2x - 3$.

From the fact that $\lim\limits_{x \to 1} (2x - 3) = 2 - 3 = -1$, we have $\lim\limits_{x \to 1} (2x^2 - 5x + 3)/(x - 1) = -1$.

PRACTICE EXERCISE 4

Find $\lim\limits_{x \to 2} \dfrac{x - 2}{x^2 - x - 2}$.

WARNING

When we have a sequence of equations and each of them has "$\lim\limits_{x \to a}$" written as part of the terms, there is a great temptation to quit rewriting that limit symbol. We should resist that temptation for two reasons. First, the rewriting reinforces in our minds that we are doing a limiting process, and second, experience has shown us that many people who drop the "$\lim\limits_{x \to a}$" symbol eventually make mistakes.

2.2 PROBLEMS

Foundations

The problems of this section require the basic skills illustrated by the following:

1. For $f(x) = \dfrac{2x}{x - 2}$, find $f(0), f(3),$ and $f(4)$.

2. Factor $12x^2 - 8x$.   3. Factor $x^2 - 3^2$.

4. Factor $x^2 - x - 12$.

5. Find $(\sqrt{x} + 2) \cdot (\sqrt{x} - 2)$.

Exercises

Find the following limits.

6. $\lim\limits_{x \to 2} (x - 8)$

7. $\lim\limits_{x \to 3} (x + 5)$

8. $\lim\limits_{x \to 3} (2x + 1)$

9. $\lim\limits_{x \to -2} \left(3x - \dfrac{1}{2}\right)$

10. $\lim\limits_{x \to -2} (x^2 + x)$

11. $\lim\limits_{x \to -4} (2x^2 + 3)$

12. $\lim\limits_{x \to -5} (x^3 - 2x)$

13. $\lim\limits_{x \to -2} (3x^3 + 2x^2 - x)$

14. $\lim\limits_{x \to -3} \dfrac{x}{x + 5}$

15. $\lim\limits_{x \to 2} \dfrac{2x}{3x - 1}$

16. $\lim\limits_{x \to 3} \dfrac{x - 3}{x - 3}$

17. $\lim\limits_{x \to 2} \dfrac{2x - 4}{x - 2}$

18. $\lim\limits_{x \to 3} \dfrac{x^2 - 9}{x - 3}$

19. $\lim\limits_{x \to 2} \dfrac{x^2 - 4}{x - 2}$

20. $\lim\limits_{x \to 2} \dfrac{x^2 - x - 2}{x - 2}$

21. $\lim\limits_{x \to -2} \dfrac{x^2 + x - 2}{x + 2}$

22. $\lim\limits_{x \to a} (x^2 - 5x)$

23. $\lim\limits_{x \to a} \dfrac{x^2 + x}{x - 1}$

24. $\lim\limits_{x \to -1} (x^2 + x - 3)$

25. $\lim\limits_{x \to 0} (3x^2 - 2x - 6)$

26. $\lim\limits_{x \to 3} \dfrac{x - 3}{x}$

27. $\lim\limits_{x \to 2} \dfrac{x - 2}{x}$

28. $\lim\limits_{x \to 2} \dfrac{x^2}{x + 2}$

29. $\lim\limits_{x \to 3} \dfrac{x^2 + 1}{x - 1}$

30. $\lim\limits_{x \to 3} \sqrt{2x + 10}$

31. $\lim\limits_{x \to -2} \sqrt{-x + 7}$

32. $\lim\limits_{x \to -2} \dfrac{2x^2 + 3x - 2}{x + 2}$

33. $\lim\limits_{x \to -2} \dfrac{2x^2 + 5x + 2}{x + 2}$

34. $\lim\limits_{x \to -1} \dfrac{x^2 - 1}{x^2 + 3x + 2}$

35. $\lim\limits_{x \to 2} \dfrac{x^2 - x - 2}{x^2 + x - 6}$

36. $\lim\limits_{x \to 2} \sqrt{x^2 - x + 3}$

37. $\lim\limits_{x \to 3} \sqrt{x^2 + 5x - 8}$

38. $\lim\limits_{x \to 2} \dfrac{\sqrt{x + 7} - x^2}{2x + 1}$

39. $\lim\limits_{x \to -1} \dfrac{\sqrt{x + 5} + 2x}{3x - 2}$

40. $\lim\limits_{x \to 4} \dfrac{2 - \sqrt{x}}{x - 4}$ (Hint: $x - 4 = (\sqrt{x} - 2)(\sqrt{x} + 2)$.)

41. $\lim\limits_{x \to 1} \dfrac{1 - \sqrt{x}}{x - 1}$ (Hint: $x - 1 = (\sqrt{x} - 1)(\sqrt{x} + 1)$.)

42. $\lim\limits_{h \to 0} (4x + 4h - 1)$

43. $\lim\limits_{h \to 0} (2x + h + 1)$

44. $\lim\limits_{h \to 0} (3x^2 + 3xh + h^2 + 1)$

45. $\lim\limits_{h \to 0} (4x^2 + 4xh + h^2 + 1)$

![calculator icon] **Graphing Calculator Problems**

For problems 46–48, graph the function over the given interval and use the trace function to estimate the desired limit.

46. $f(x) = \dfrac{x^2 - x - 2}{x - 2}$ for $0 \le x \le 4$, and $\lim\limits_{x \to 2} f(x)$.

47. $g(x) = \dfrac{x^{3/2} - 1}{x - 1}$ for $0 \le x \le 2$, and $\lim\limits_{x \to 1} g(x)$.

48. $k(x) = (1 + 2x)^{1/x}$ for $-1 \le x \le 1$, and $\lim\limits_{x \to 0} k(x)$.

Writing and Discussion Problems

49. Can the alternate-function method always be used to determine a limit?

50. Discuss the following paradox about limits, which was first stated by Zeno, a Greek who lived in Elea twenty-five centuries ago. Suppose Achilles, the fastest runner in all of Greece, is to race a tortoise. If the tortoise is given a head start of 10 yards by Achilles, then the tortoise will never be caught. The reason is that by the time Achilles reaches the starting place of the tortoise, then the tortoise will have moved on to a second position. When Achilles reaches that point, the tortoise will again have moved on. This will continue, with Achilles always having to first reach the position the tortoise had occupied previously. Because this process never ends, Achilles will never catch the tortoise.

Enrichment Problems

Evaluate those limits that exist in problems 51–54.

51. (a) $\lim\limits_{x \to 2^+} \sqrt{x - 2}$ (b) $\lim\limits_{x \to 2^-} \sqrt{x - 2}$
(c) $\lim\limits_{x \to 2} \sqrt{x - 2}$

52. (a) $\lim\limits_{x \to 1^+} \sqrt{x - 1}$ (b) $\lim\limits_{x \to 1^-} \sqrt{x - 1}$
(c) $\lim\limits_{x \to 1} \sqrt{x - 1}$

53. $\lim\limits_{x \to 16} \dfrac{\dfrac{x}{4} - 4}{\sqrt{x} - 4}$ (Hint: multiply numerator and denominator by $\sqrt{x} + 4$.)

54. $\lim\limits_{x \to 9} \dfrac{\dfrac{2}{3}x - 6}{\sqrt{x} - 3}$ (Hint: multiply numerator and denominator by $\sqrt{x} + 3$.)

55. Suppose $f(x) = x^2 - 5x$. Find
$\lim\limits_{h \to 0} \dfrac{f(x + h) - f(x)}{h}$.

56. Suppose $g(x) = \dfrac{2}{x}$. Find $\displaystyle\lim_{h \to 0} \dfrac{g(x + h) - g(x)}{h}$.

57. Suppose $f(x) = 1/(1 + 2^{1/x})$. Using the trace function, or calculating values close to 0, estimate the left- and right-hand limits of $f(x)$, if they exist.

SOLUTIONS TO PRACTICE EXERCISES

1. $\displaystyle\lim_{x \to 2} (3x^2 - 2x) = 3 \cdot 2^2 - 2 \cdot 2$ by L_1

$$= 3 \cdot 4 - 4 = 12 - 4 = 8$$

2. $\displaystyle\lim_{x \to 5} \dfrac{1 + 3x}{2} = \dfrac{\displaystyle\lim_{x \to 5} (1 + 3x)}{\displaystyle\lim_{x \to 5} 2}$ by L_5

$$= \dfrac{1 + 3 \cdot 5}{2} \text{ by } L_1$$

$$= \dfrac{16}{2} = 8$$

3. $\displaystyle\lim_{x \to 3} \dfrac{x^2 - 2x}{\sqrt{x + 1} + x}$

$$= \dfrac{\displaystyle\lim_{x \to 3} (x^2 - 2x)}{\displaystyle\lim_{x \to 3} (\sqrt{x + 1} + x)} \text{ by } L_5$$

$$= \dfrac{3^2 - 2 \cdot 3}{\displaystyle\lim_{x \to 3} \sqrt{x + 1} + \displaystyle\lim_{x \to 3} x} \text{ by } L_1, L_2, \text{ and } L_5$$

$$= \dfrac{9 - 6}{\sqrt{\displaystyle\lim_{x \to 3} (x + 1)} + 3} \begin{array}{l} \text{by } L_1, L_6 \text{ (using } r = 1/2 \\ \text{for the square root)} \end{array}$$

$$= \dfrac{9 - 6}{2 + 3} = \dfrac{3}{5}$$

4. $\displaystyle\lim_{x \to 2} \dfrac{x - 2}{x^2 - x - 2} = \lim_{x \to 2} \dfrac{x - 2}{(x - 2)(x + 1)} =$

$\displaystyle\lim_{x \to 2} \dfrac{1}{x + 1} = 1/3.$

Notice here the use of the alternate function, $1/(x + 1)$, which is equal to the original function except at $x = 2$.

2.3   ## RATES OF CHANGE

As we investigate the relationship between two variables, one of the first questions we ask is: How does a change in one of them affect the other? We often think of changes that occur over time, such as the sales of video games in the United States over the period 1979–1990. Chapter 2 started with this example, and its graph is shown in figure 2.3.1.

Other examples of change with respect to different variables are the effect of soil depth on the concentration of the soil fumigant ethylene dibromide (EDB) under a Hawaiian pineapple field,* shown in figure 2.3.2, and the effect of a particular concentration of a drug on blood pressure, shown in figure 2.3.3.

Average Rate of Change

The definition on page 154 uses x and y as the names for the variables, but the idea of average rate of change can be applied to quantities with any names for the variables.

---

*Loague et al., "Simulation of Organic Chemical Movement in Hawaiian Soils with PRZM: 1. Preliminary Results for Ethylene Dibromide," *Pacific Science* 43:1 (1989), pp. 67–95.

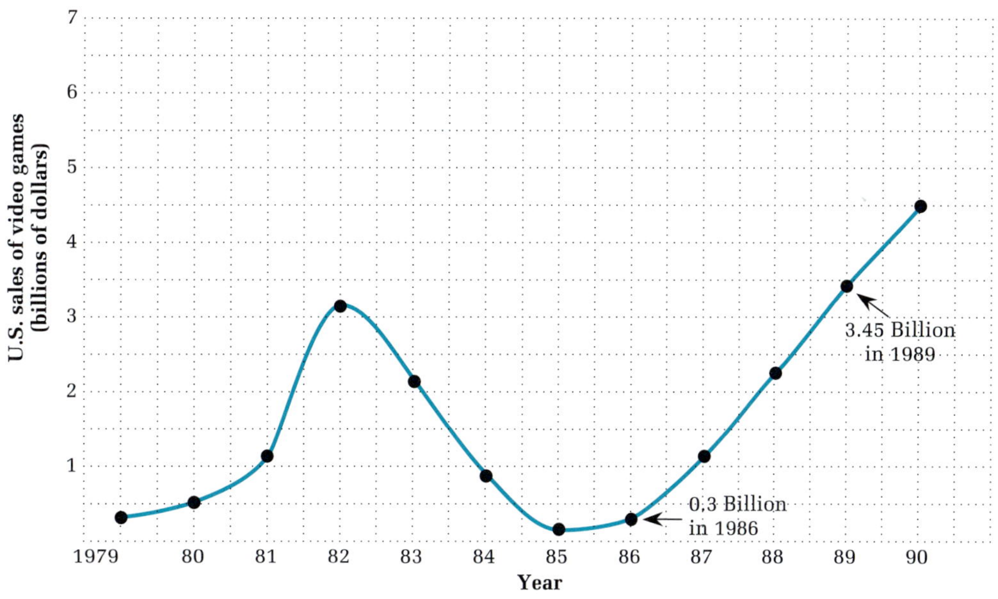

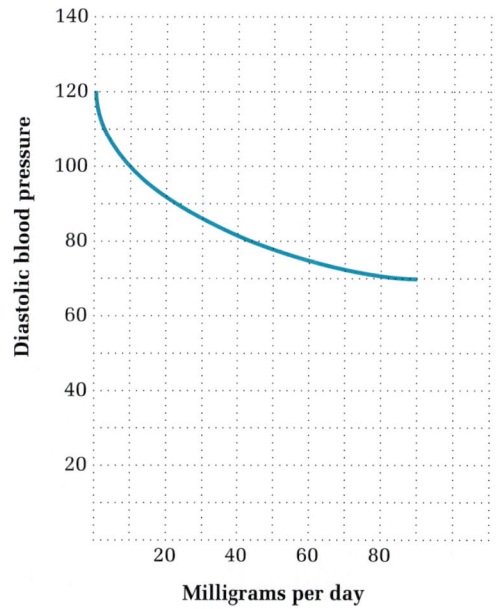

FIGURE 2.3.1

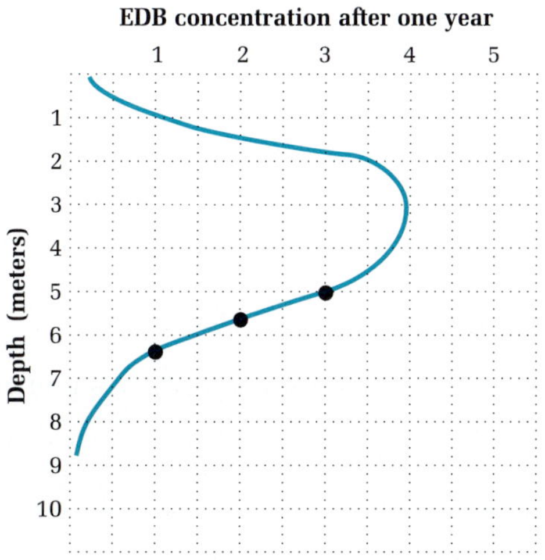

FIGURE 2.3.2

FIGURE 2.3.3

DEFINITION

Average rate of change

Suppose two quantities, x and y, are related. The *average rate of change of y with respect to x* is the ratio of the change in the values of y to the change in the corresponding values of x.

The average rate of change of y with respect to x

$$= \frac{\text{change in values of } y}{\text{change in corresponding values of } x}$$

EXAMPLE 1   Average rate of change

Using the graph in figure 2.3.1, find the average rate of change in sales of video games over the period 1986 to 1989.

SOLUTION

From the graph, we see that sales in 1986 were 0.3 billion dollars and the sales in 1989 were 3.45 billion dollars. Hence, in the period 1986–1989, the average rate of change in sales of video games was:

$$\begin{array}{l}\text{Average rate of change} \\ \text{from 1986 to 1989}\end{array} = \frac{\text{change in sales}}{\text{change in time}}$$

$$= \frac{\text{sales in 1989} - \text{sales in 1986}}{1989 - 1986}$$

$$= \frac{3.45 - 0.3}{1989 - 1986} = \frac{3.15}{3}$$

$$= 1.05 \text{ billion dollars per year}$$

If $y = f(x)$, we can determine an average rate of change for $f(x)$ over the interval $[a, b]$.

DEFINITION

Average rate of change of $f(x)$ over an interval

The *average rate of change of f(x) with respect to x* as x changes from a to b is:

$$\frac{\text{change in values of } f(x)}{\text{corresponding change in values of } x} = \frac{f(b) - f(a)}{b - a}, a \neq b$$

EXAMPLE 2   Average rate of change

Find the average rate of change of $f(x) = x^3/4 - x^2 + 3$ over the interval from $x = 1$ to $x = 4$.

SOLUTION

First evaluate $f(1) = 1^3/4 - 1^2 + 3 = 2.25$ and $f(4) = 4^3/4 - 4^2 + 3 = 16 - 16 + 3 = 3$.

Then the average rate of change is:

$$\frac{f(4) - f(1)}{4 - 1} = \frac{3 - 2.25}{4 - 1} = \frac{0.75}{3} = 0.25.$$

Figure 2.3.4 is a graph of $f(x)$ over the interval [0, 5]. The average rate of change over the interval from 1 to 4 is the slope of the line joining the points $A(1, 2.25)$ and $B(4, 3)$.

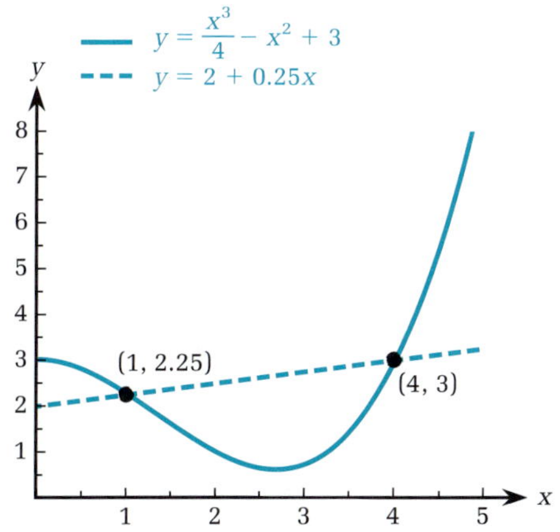

FIGURE 2.3.4

The Average Rate of Change as a Slope

In Example 2 we saw that the average rate of change could be interpreted as the slope of the line joining two points on the graph. This is true for a general function, $f(x)$, as illustrated in figure 2.3.5, in which $A(a, f(a))$ and $B(b, f(b))$ are two arbitrary points on the graph of $y = f(x)$.

Using the Average Rate of Change for Estimation

One use of "average rate of change" is to estimate an unknown value of a variable. In doing so, we are assuming that the actual values are fairly close to values calculated by an approximation that uses a straight line. Example 3 uses an average rate of change to estimate a counter reading on a videocassette recorder (VCR) at a time for which the table in the example does not give the reading.

FIGURE 2.3.5

EXAMPLE 3 The average rate of change, for a function described by a table

TIME (HOURS)	COUNTER READING
1	1560
2	2740
3	3740
4	4640
5	5400
6	6120
7	6760
8	7420

The table at left gives the counter readings for an eight-hour VCR tape, as listed in the user's manual for a particular videocassette recorder.

Let the reading on the counter be represented by a function, $C(t)$, whose graph passes through the particular values. Such a graph is shown in figure 2.3.6. Here t represents hours of recording so that $C(t)$ is a function of the time, t.

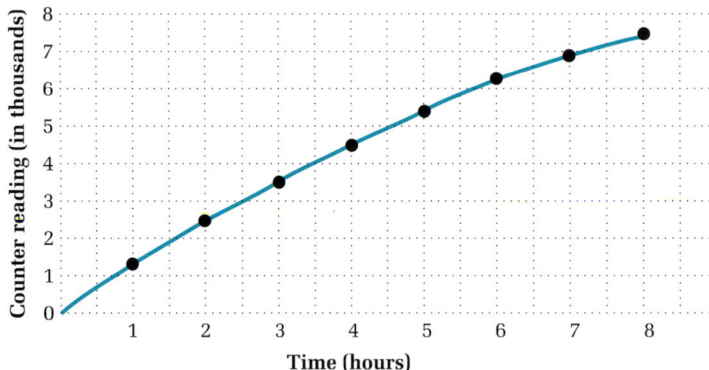

FIGURE 2.3.6

To approximate a value of $C(t)$ for a time at which we do not have a counter reading, we can use an average rate of change. For instance, suppose we want to determine the approximate counter reading for 3 hours and 15 minutes.

SOLUTION

Because 15 minutes corresponds to 0.25 hours, we want to estimate $C(3.25)$. We do this by finding an estimate for a counter reading equivalent to 0.25 hours and adding that amount onto the manual's value for the counter reading at the end of 3 hours.

Over the interval from the end of the third hour to the end of the fourth hour the average rate of change is $(4640 - 3740)/(4 - 3)$ $= 900/1 = 900$ per hour. Hence, a counter reading of $(0.25)900 = 225$ approximately corresponds to 15 minutes. Let us add this value to the reading at 3 hours, which was 3740. Our estimate then is $C(3.25) \simeq 3740 + 225 = 3965$.

PRACTICE EXERCISE 1

Estimate the counter reading at 4 hours, 45 minutes by using the data from the table given in Example 2 for $t = 4$ and $t = 5$. Then use three-fourths of that average, added to the counter reading at $t = 4$.

Average Rates of Change Over Short Intervals

For the VCR example, we are given the counter readings only every hour and so we cannot find average rates of change over intervals of less than an hour. However, for many functions an average rate of change can be computed over an interval of any length whatsoever, no matter how short.

Suppose we know $f(x)$ for all values of x. Of course, for such a function we would not be using an average rate of change to provide an estimate of the value of the function. Nonetheless, the idea of an average rate of change, especially over a very short interval, is fruitful. To see just how this is, let us first illustrate finding average rates of change over progressively smaller intervals.

EXAMPLE 4   Computing the average rate of change over various intervals

Suppose the additional revenue (in millions of dollars) resulting from spending x (thousands of dollars) on advertising is given by $f(x) = 10x - x^2$ for $0 \le x \le 10$.

Find the average rate of change of $f(x)$ between

(a) $x = 1$ and $x = 1.1$ (b) $x = 1$ and $x = 1.01$
(c) $x = 1$ and $x = 1.001$

SOLUTION

(a) $\dfrac{f(1.1) - f(1)}{1.1 - 1} = \dfrac{(11 - 1.21) - (10 - 1)}{0.1} = \dfrac{9.79 - 9}{0.1} = 7.9$

(b) $\dfrac{f(1.01) - f(1)}{1.01 - 1} = \dfrac{(10.1 - 1.0201) - (10 - 1)}{0.01} = \dfrac{0.0799}{0.01} = 7.99$

(c) $\dfrac{f(1.001) - f(1)}{1.001 - 1} = \dfrac{(10.01 - 1.002001) - (10 - 1)}{0.001}$

$= \dfrac{0.007999}{0.01} = 7.999$

Instantaneous Rate of Change

Suppose we consider the limit of the average rate of change as the length of the interval approaches zero.

EXAMPLE 5   The limit of the average rate of change

For the function of Example 4, $f(x) = 10x - x^2$, find the limit of the average rate of change as x approaches 1.

SOLUTION

The average rate of change is $[f(x) - f(1)]/(x - 1)$. Simplifying the numerator and finding the limit as x approaches 1, we have

$\displaystyle\lim_{x \to 1} \dfrac{(10x - x^2) - (10 \cdot 1 - 1^2)}{x - 1}$

$= \displaystyle\lim_{x \to 1} \dfrac{10x - x^2 - 9}{x - 1}$ Rearrange the terms in the numerator to make it easier to factor.

$= \displaystyle\lim_{x \to 1} \dfrac{-x^2 + 10x - 9}{x - 1}$ Factoring the numerator.

$= \displaystyle\lim_{x \to 1} \dfrac{(x - 1)(-x + 9)}{x - 1}$

We now use the alternate-function method for limits from Section 2.2, which allows us to cancel the terms $(x - 1)$ when evaluating the limit as x approaches 1.

$= \displaystyle\lim_{x \to 1} (-x + 9) = -1 + 9 = 8$

The limiting value of the average rate of change is called the *instantaneous rate of change.* An interpretation of the instantaneous rate of change is that at $x = 1$ (corresponding to an expenditure of $1000 on advertising), revenue will increase at a rate of $8,000,000 per additional thousand dollars of advertising.

DEFINITION

Instantaneous rate of change

The *instantaneous rate of change* of $f(x)$ at a is

$$\lim_{x \to a} \frac{f(x) - f(a)}{x - a}, \text{ provided this limit exists.}$$

Notice that in finding an average rate of change, it makes no difference whether $a < x$ or $x < a$ because multiplying both the numerator and the denominator of a quotient by -1 does not change the value of the quotient. Hence

$$\frac{f(x) - f(a)}{x - a} = \frac{-(f(x) - f(a))}{-(x - a)} = \frac{f(a) - f(x)}{a - x}$$

We often find a simpler form of $[f(x) - f(a)]/(x - a)$ by replacing x by $a + h$. Doing so, we get

$$\frac{f(x) - f(a)}{x - a} = \frac{f(a + h) - f(a)}{(a + h) - a}$$

$$= \frac{f(a + h) - f(a)}{h}$$

Using this notation, we can rephrase our definition as follows.

DEFINITION

Instantaneous rate of change

The *instantaneous rate of change* of $f(x)$ at a is

$$\lim_{h \to 0} \frac{f(a + h) - f(a)}{h}, \text{ provided this limit exists.}$$

EXAMPLE 6   Instantaneous rate of change

Let us reconsider Example 5, wherein we found that the average rate of change of $f(x) = 10x - x^2$, at $x = 1$, approached 8 as the length of the interval approached zero. Rework this example, using the notation $a + h$.

SOLUTION

The instantaneous rate of change of $f(x) = 10x - x^2$ at $x = 1$ is

$$\lim_{h \to 0} \frac{f(1 + h) - f(1)}{h} = \lim_{h \to 0} \frac{(10(1 + h) - (1 + h)^2) - (10 \cdot 1 - 1^2)}{h}$$

$$= \lim_{h \to 0} \frac{10 + 10h - 1 - 2h - h^2 - 9}{h}$$

$$= \lim_{h \to 0} \frac{8h - h^2}{h} = \lim_{h \to 0} \frac{h(8 - h)}{h}$$

$$= \lim_{h \to 0} 8 - h = 8$$

Velocity

A familiar situation in which we use the idea of the instantaneous rate of change is that of *speed*. When we are driving a car we speak of an *average speed* and an *instantaneous speed*. The instantaneous speed is often expressed by the function name $v(t)$ and is indicated by the car's speedometer.

DEFINITION

Average and instantaneous velocity

Suppose t measures time and $S(t)$ is the distance traveled by time t along a straight line. Then over the time interval from t to time $t + h$ the *average velocity* is given by

$$\frac{S(t + h) - S(t)}{(t + h) - t} = \frac{S(t + h) - S(t)}{h}$$

and the *instantaneous velocity* at t is given by

$$v(t) = \lim_{h \to 0} \frac{S(t + h) - S(t)}{h}$$

EXAMPLE 7   Average and instantaneous velocity

Suppose the distance that a car travels down a straight highway is given by $S(t) = 3t^2 + 50t$, wherein $S(t)$ is measured in miles and t in hours.

(a) Find the average velocity over the time interval $t = 2$ to $t = 3$.

(b) Find the instantaneous velocity at $t = 2$.

SOLUTION

(a) The average velocity is

$$\frac{\text{change in distance}}{\text{change in time}} = \frac{(\text{location at } t = 3) - (\text{location at } t = 2)}{3 - 2}$$

$$= \frac{S(3) - S(2)}{3 - 2}$$

We need to calculate $S(3)$ and $S(2)$:

$$S(2) = 3 \cdot 2^2 + 50 \cdot 2 = 3 \cdot 4 + 100 = 12 + 100 = 112$$
$$S(3) = 3 \cdot 3^2 + 50 \cdot 3 = 3 \cdot 9 + 150 = 27 + 150 = 177$$

Hence, the average velocity is $(177 - 112)/(3 - 2) = 65$ miles per hour.

(b) The instantaneous velocity requires that we know $S(2)$ and $S(2 + h)$. We found $S(2) = 112$ in part (a). To find $S(2 + h)$, we use the definition of $S(t)$ and find

$$S(2 + h) = 3(2 + h)^2 + 50(2 + h)$$
$$= 3(2^2 + 2 \cdot 2h + h^2) + 50 \cdot 2 + 50h$$
$$= 3(4 + 4h + h^2) + 100 + 50h$$
$$= 12 + 12h + 3h^2 + 100 + 50h = 112 + 62h + 3h^2$$

Thus $\displaystyle \lim_{h \to 0} \frac{S(2 + h) - S(2)}{h} = \lim_{h \to 0} \frac{(112 + 62h + 3h^2) - 112}{h}$

$$= \lim_{h \to 0} \frac{62h + 3h^2}{h} = \lim_{h \to 0} \frac{h(62 + 3h)}{h}$$

$$= \lim_{h \to 0} 62 + 3h = 62 \text{ miles per hour}$$

Many velocity problems have to do with the motion of an object under the effect of gravity.

MOTION RESULTING FROM GRAVITY

Let $S(t)$ measure the number of feet above some reference point (often ground level). If the effect of air resistance is neglected, the formula for vertical motion near the surface of Earth is:

$$S(t) = -16t^2 + v_0 t + s_0$$

The constants v_0 and s_0 are the velocity (in feet per second) and position (in feet), respectively, at time $t = 0$.

The velocity and position at time $t = 0$ are called the *initial velocity* and *initial position,* respectively. The next example illustrates the use of this formula.

EXAMPLE 8   Instantaneous velocity

Suppose Jenn throws a ball straight up into the air and after t seconds the ball's height is given by $S(t) = -16t^2 + 48t$. Find the instantaneous velocity at $t = 1.5$ seconds.

SOLUTION

$$v(1.5) = \lim_{h \to 0} \frac{S(1.5 + h) - S(1.5)}{h}, \text{ so we need to find } S(1.5) \text{ and}$$
$S(1.5 + h)$.

$$S(1.5) = -16(1.5)^2 + 48(1.5)$$
$$= -16(2.25) + 72 = -36 + 72 = 36$$

$$S(1.5 + h) = -16(1.5 + h)^2 + 48(1.5 + h)$$
$$= -16(2.25 + 3h + h^2) + 72 + 48h$$
$$= -36 - 48h - 16h^2 + 72 + 48h$$
$$= 36 - 16h^2$$

Thus $v(1.5) = \lim_{h \to 0} \dfrac{S(1.5 + h) - S(1.5)}{h} = \lim_{h \to 0} \dfrac{36 - 16h^2 - 36}{h}$

$$= \lim_{h \to 0} \frac{-16h^2}{h} = \lim_{h \to 0} -16h = 0$$

When the velocity is zero, the ball is motionless. Thus, $v(1.5) = 0$ means the ball was momentarily stopped, because it had reached the high point in its path. This height we already found to be $S(1.5) = 36$.

Figure 2.3.7 shows the height of the ball over time.

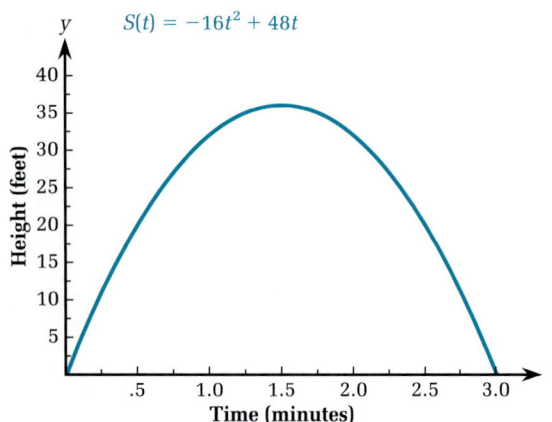

FIGURE 2.3.7

PRACTICE EXERCISE 2

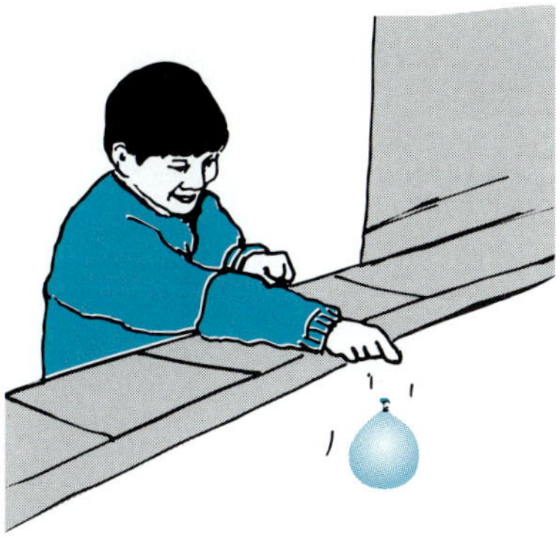

FIGURE 2.3.8

When Frank dropped the water balloon out of his window 50 feet above the courtyard, he knew that its height above ground would be given by $s(t) = 50 - 16t^2$, where the height is measured in feet and t is in seconds.

(a) Find $v(1)$.

(b) Has the balloon hit the ground when t is 1?

Applications

Other examples of instantaneous rates of change include sales, stock prices, advertising budgets, or even the depth of snow, as is illustrated in figure 2.3.9 and discussed in Example 9.

EXAMPLE 9   Instantaneous rate of change of snow depth

Suppose eight inches of new snow falls overnight, but the sun is out and the depth of snow t hours after sunrise is given by $d(t) = 8 - 0.3t^2$ inches (for $0 \leq t \leq 5$). How fast is the snow depth going down four hours after sunrise?

SOLUTION

Because $t = 4$, we need to find $\lim\limits_{h \to 0} \dfrac{d(4 + h) - d(4)}{h}$

$$d(4 + h) = 8 - 0.3(4 + h)^2$$
$$= 8 - 0.3(16 + 8h + h^2)$$
$$= 8 - 4.8 - 2.4h - 0.3h^2$$
$$= 3.2 - 2.4h - 0.3h^2$$

$$d(4) = 8 - 0.3(4)^2$$
$$= 8 - 0.3 \cdot 16 = 8 - 4.8 = 3.2$$

Thus,

$$\lim_{h \to 0} \frac{d(4 + h) - d(4)}{h} = \lim_{h \to 0} \frac{3.2 - 2.4h - 0.3h^2 - 3.2}{h}$$
$$= \lim_{h \to 0} \frac{-2.4h - 0.3h^2}{h}$$
$$= \lim_{h \to 0} -2.4 - 0.3h = -2.4 \text{ inches per hour}$$

A graph showing the snow depth is shown in figure 2.3.9.

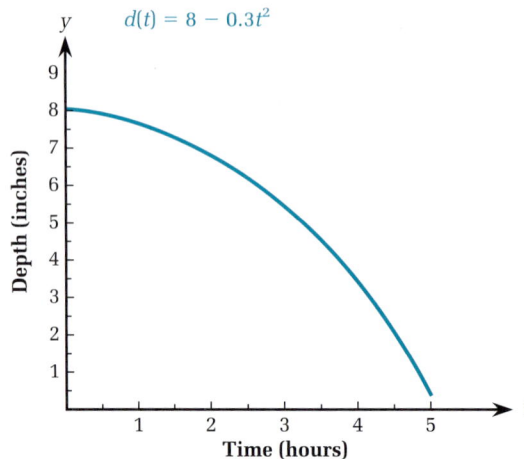

FIGURE 2.3.9

The negative sign in the answer for Example 9 indicates that the snow depth is going down. Generally, we establish our coordinate system so that a decrease in the quantity is indicated by a negative rate of change.

It is quite important to realize that if we find an instantaneous rate of change for a function, then we can apply our results to a variety of applications. For instance, the last example represented the declining depth of snow. However, if the same function represented the number of dollars remaining after t hours of shopping, then after four hours, our money would be decreasing at an instantaneous rate of $2.40 an hour. If the function is the same, then the instantaneous rate is the same. Only the interpretation, that is, the application, is different.

2.3 PROBLEMS

Foundations

The problems of this section require the basic skills illustrated by the following:

1. Find $(3 + h)^2$.

2. Find $h^2 - (3 + h)^2$.

3. For $f(x) = 2x^2 + 1$, find $f(1), f(-1), f(3)$, and $f(x + h)$.

Exercises

By evaluating the functions, find the average rate of change for each function over the given interval.

4. $f(x) = 2x + 1$
   (a) $x = 1$ to $x = 3$ (b) $x = 2$ to $x = 3$

5. $g(x) = 3x - 1$
   (a) $x = 2$ to $x = 5$ (b) $x = 3$ to $x = 4$

6. $g(x) = x^2 + x - 1$
   (a) $x = 1$ to $x = 3$ (b) $x = 2$ to $x = 3$

7. $f(x) = 3x^2 - x$
   (a) $x = 0$ to $x = 3$ (b) $x = 2$ to $x = 3$

8. $f(x) = x^3 - 4x$
   (a) $x = 0$ to $x = 2$ (b) $x = 0$ to $x = 1$

9. $h(x) = 2x^2 - \frac{1}{2}x^4$
   (a) $x = 0$ to $x = 2$ (b) $x = 0$ to $x = 1$

YEAR	VALUE OF STOCK
0	20
1	30
2	45
3	42
4	48
5	50
6	60
7	90
8	120
9	110
10	90

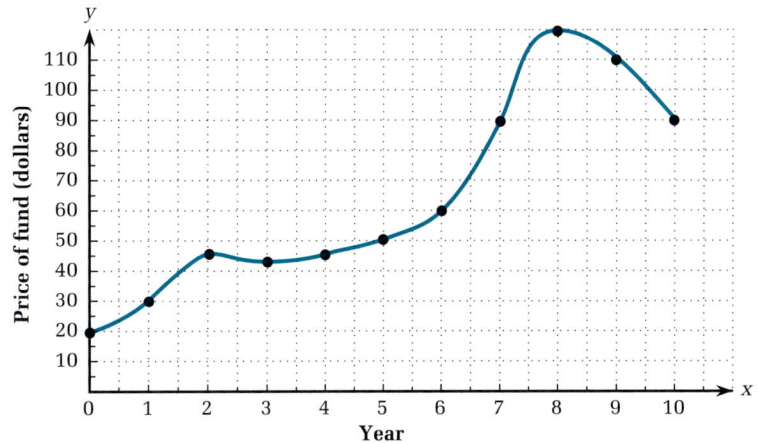

FIGURE 2.3.10

10. $g(x) = \sqrt{x + 2}$
   (a) $x = 0$ to $x = 2$ (b) $x = 0$ to $x = 7$

11. (a) Find the average rate of change of $f(x) = \sqrt{x}$
       over the interval $x = 9$ to $x = 16$.
   (b) Use your answer to part (a) to estimate the
       value of $\sqrt{10}$.

12. (a) Find the average rate of change of $g(x) =$
       $1 + 3x^2$ over the interval $x = 1$ to $x = 2$.
   (b) Use your answer to part (a) to estimate the
       value of $g(1.1)$.

Find the instantaneous rate of change of the functions
in problems 13–16 at the values indicated by using
the formula $\lim\limits_{h \to 0} [f(a + h) - f(a)]/h$.

13. $f(x) = 2x^2 - 1$ at $a = 1$

14. $g(x) = -x^2 + 8$ at $a = 2$

15. $g(x) = x^2 + x - 1$ at $a = 1$

16. $f(x) = 3x^2 - x$ at $a = 0$

17. *(Stock Mutual Fund)*   The price of a major
stock fund had values as shown in the table at
top left. The associated points are shown in figure 2.3.10.

Find the average rate of change over the following intervals:
(a) Last ten years.   (b) Last five years.
(c) Last year.

18. *(Advertising Response)*   The Mirage Lake
Water Co. published a notice in the *Daily Shim-
mer* to increase water rates. The graph in figure
2.3.11 shows the total number of customers who
had seen the ad by day x.

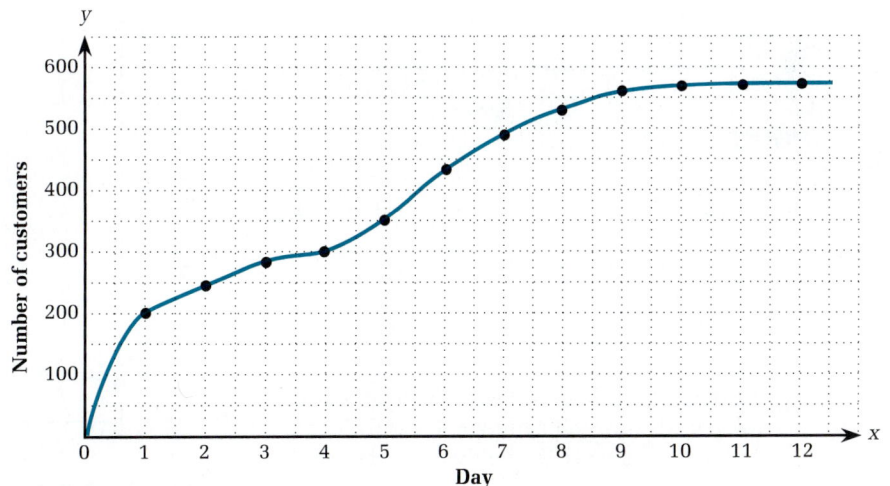

FIGURE 2.3.11

What was the average rate of change of the number of customers who had seen the ad
(a) between $x = 0$ and $x = 10$?
(b) between $x = 0$ and $x = 1$?
(c) between $x = 1$ and $x = 2$?
(d) between $x = 9$ and $x = 10$?

19. **(Cocaine Seizures in U.S.)** The U.S. Drug Enforcement Administration reports that seizures of tons of cocaine were as shown by the accompanying table, graphed in figure 2.3.12.

YEAR	TONS OF COCAINE
1982	5
1983	9
1984	11
1985	24
1986	28
1987	39
1988	53

FIGURE 2.3.12

(a) Find the average rate of increase for the years 1982 to 1984.
(b) Find the average rate of increase for the years 1986 to 1988.
(c) Find the average rate of increase for the years 1987 to 1988.

20. **(Golfing Equipment Sales)** The retail market value of all U.S. golf equipment, in millions of dollars, is shown in the following table,* whose data are graphed in figure 2.3.13.

* Source: U.S. Census of Manufacturers Industry Series, 1958–82.

YEAR	VALUE (MILLIONS OF $)
1958	109.9
1963	186.6
1967	252.3
1972	440
1977	610.3
1982	1028.2
1987	1340.1

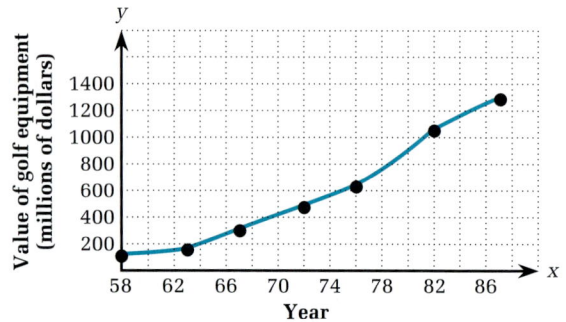

FIGURE 2.3.13

Find the average annual increase in market value over the years
(a) 1958 to 1972.   (b) 1958 to 1987.
(c) 1972 to 1987.

21. **(Paper Money)** The total amount of U.S. paper money in circulation, in billions of dollars,† is shown by the graph in figure 2.3.14.

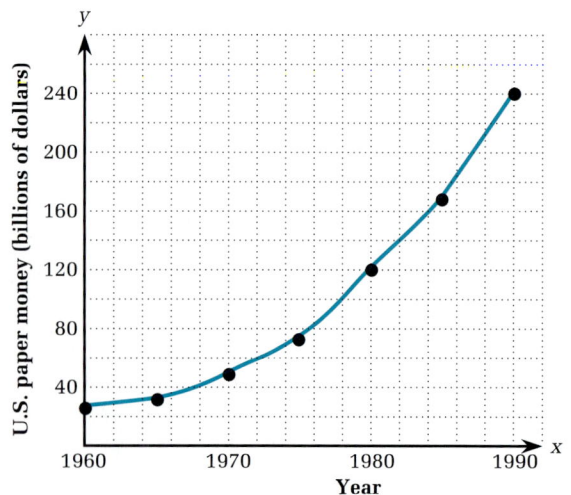

FIGURE 2.3.14

†Source: U.S. Treasury Department, *The New York Times,* reprinted in the *Denver Post,* March 4, 1990.

YEAR	MONEY IN CIRCULATION (BILLIONS OF $)
1960	27
1965	33
1970	49
1975	74
1980	120
1985	172
1990	240

Find the average annual increase in the amount of U.S. paper money in circulation over the years
(a) 1960 to 1970.   (b) 1970 to 1980.
(c) 1960 to 1990.   (d) 1985 to 1990.

22. *(Sales of Facsimile Machines)*   A graph showing the sales of facsimile machines in Europe* is given in figure 2.3.15.

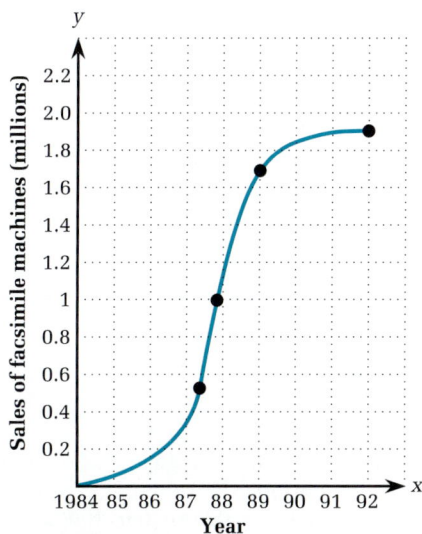

FIGURE 2.3.15

Find the average rate of change of sales during the periods
(a) 1984–1986.   (b) 1986–1989.

23. *(Physician Costs)*   A graph showing the rise in cost of physicians' services in the United States† is shown in figure 2.3.16.

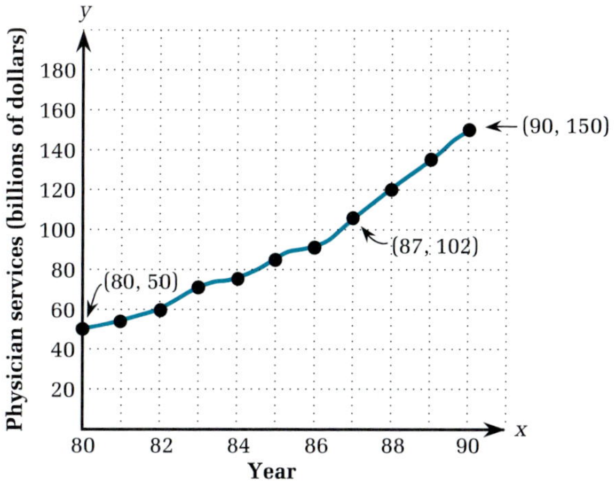

FIGURE 2.3.16

Find the average rate of change over the years
(a) 1980–1981.   (b) 1987–1990.

24. *(Biology, Swimming Rate of Fish)*   The speed at which goldfish (*Carassium auratus*) swim, as a function of the water temperature,‡ is shown in figure 2.3.17. Find the average rate of change of swimming speed when the water temperature changes
(a) from 5°C to 10°C.   (b) from 10°C to 15°C.
(c) from 20°C to 25°C.

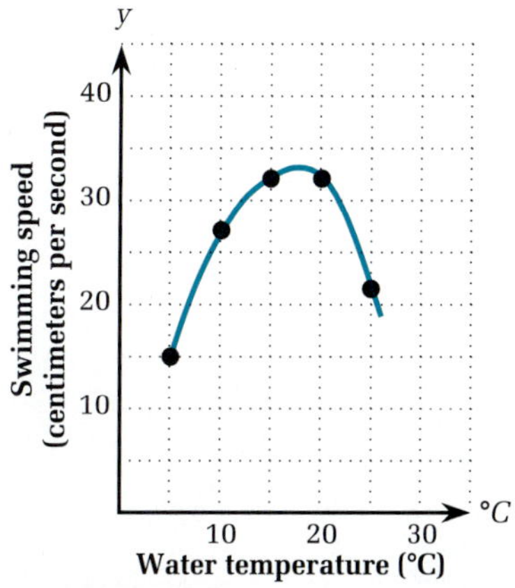

FIGURE 2.3.17

25. *(Biology, Feeding of Shrews)*   The total number of cocoons opened per day by the short-tailed shrew is a function of the density of the cocoons (measured in thousands per acre).§ The graph of this function is shown in figure 2.3.18.

---

* Source: Dataquest, as found in *The Economist*, March 10–15, 1990, p. 27.
† Source: U.S. Department of Commerce.

‡ R. E. Ricklefs, *The Economy of Nature*, 2nd Ed., New York: Clarion Press, 1983, p. 194.
§ Ibid., p. 329.

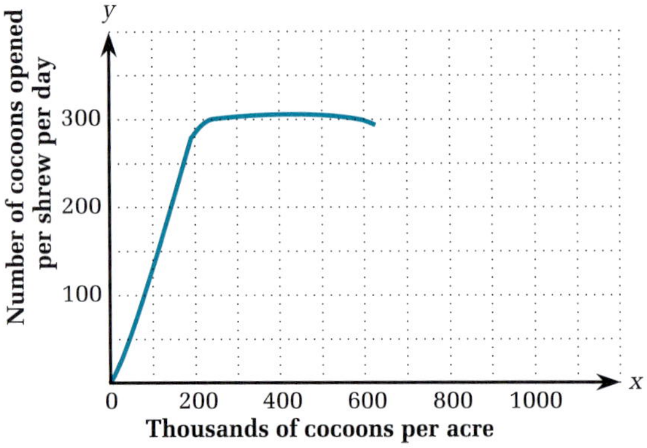

FIGURE 2.3.18

Find the average rate of change of the number of cocoons opened as the density of cocoons (measured in thousands per acre) increases from
(a) 0 to 100. (b) 100 to 200.
(c) 200 to 300.

26. *(Sales of Home Fitness Equipment)* According to the National Sporting Goods Association, in the United States the sales (expressed in millions of dollars) of home fitness equipment over the years 1980 to 1989* were as given in the following table, whose data are graphed in figure 2.3.19.

YEAR	SALES
1980	604
1981	662
1982	723
1983	989
1984	1056
1985	1216
1986	1192
1987	1192
1988	1451
1989	1726

Find the average rate of change over the years
(a) 1980 to 1985. (b) 1985 to 1988.
(c) 1988 to 1989.

* *Denver Post*, November 11, 1990.

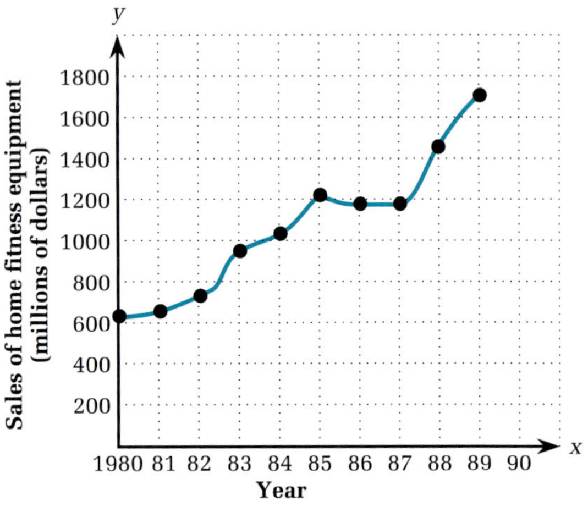

FIGURE 2.3.19

27. *(Biology, Environment, Number of Farms)* The number of farms in the United States over the period 1950–1990† is graphed in figure 2.3.20.

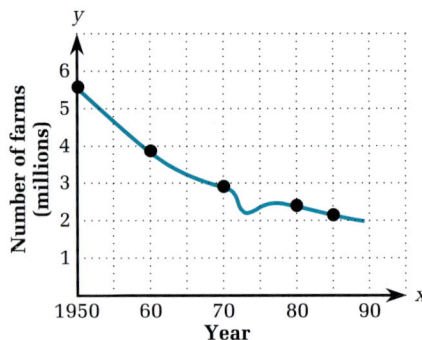

FIGURE 2.3.20

Find the average rate of change in the number of farms in the period
(a) 1950 to 1960. (b) 1950 to 1990.
(c) 1980 to 1990.
(d) If the average rate of change in the number of farms is the same in the period 1990 to 2000 as in the period 1980 to 1990, estimate the number of farms in 2000.

† *The Economist*, March 10–15, 1990.

28. **(Cost of Electricity)** The annual cost of electricity for residential use* is graphed in figure 2.3.21. The dollar amounts given are all in billions of 1989 dollars.

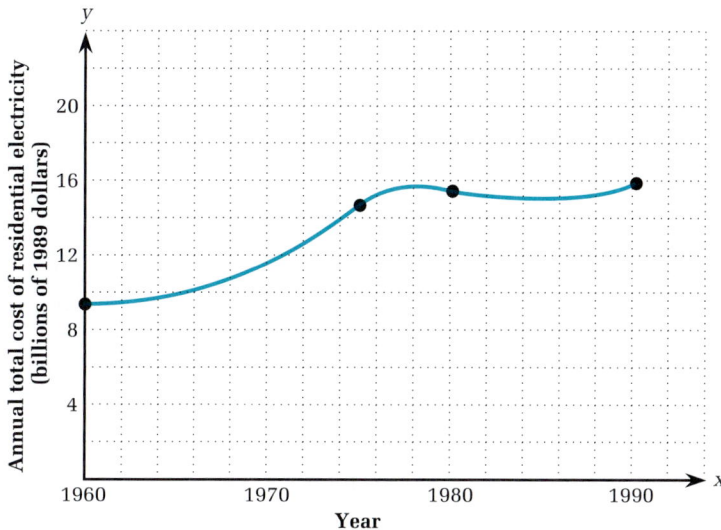

FIGURE 2.3.21

Find the average rate of change over the periods
(a) 1960 to 1970.   (b) 1965 to 1989.
(c) 1975 to 1985.
(d) If the average rate of change over the period 1965 to 1975 had continued to be the rate of change over the period 1975 to 1985, what would be the cost of residential electricity in 1985?

29. **(Franchising)** The concept of franchising trademarks and service to individual stores was begun by the Singer Sewing Machine Co. about 1870. By 1990, more than 7 million people in the United States worked for franchisers in more than 60 industries,† which included such fast-food stores as McDonald's. The graph in figure 2.3.22 shows the total sales made by all companies engaged in franchising during the period 1980 to 1990. The sales are given in billions of dollars.

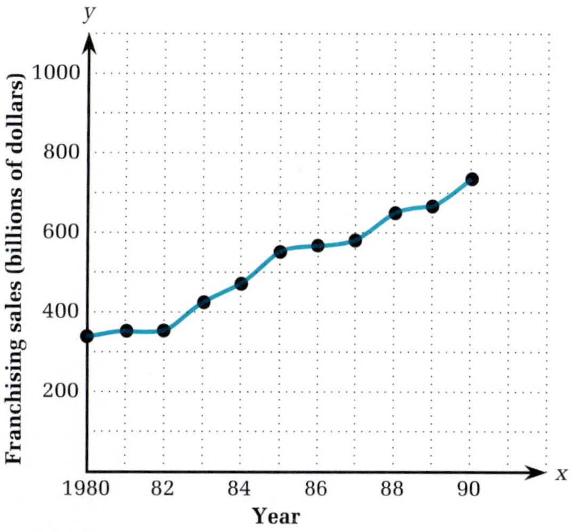

Find the average rate of change in each of the periods
(a) 1980 to 1985.   (b) 1980 to 1990.

FIGURE 2.3.22

(c) 1989 to 1990.
(d) If the average rate of change in the period 1985 to 1990 continues to be the average rate of change for the period 1990 to 2000, find the total sales by all franchisers in 2000.

*R. Berington and A. H. Rosenfield, "Energy for Buildings and Homes," *Scientific American* 263:3 (September 1990), pp. 77–86.
†*Denver Post,* November 11, 1990.

30. *(Velocity)* Suppose the height of a thrown ball is given by $s(t) = -16t^2 + 30t$, where t is in seconds and $s(t)$ is measured in feet. Find $v(3)$.

31. *(Velocity)* Suppose the distance a ski jumper has slid down a ski jump is given by $S(t) = -9t^2$, where t is measured in seconds and $S(t)$ is measured in feet. Find $v(5)$.

Writing and Discussion Problems

32. In Example 3 (about the VCR counter readings), the average rate of change decreased as the number of hours increased.
   (a) Give an explanation of this, with reference to the radius of a circle as compared to its circumference.
   (b) To estimate $C(3.15)$ in Example 3, we used an average rate of change based on readings at hours 3 and 4. Find an estimate of $C(3.15)$ using an average rate of change based on readings at hours 3 and 6. Which would you consider the better estimate? Why?

33. In the definition of instantaneous velocity we did not include the usual warning about "if the limit exists." Is it possible for an object to have its position given by a function $S(t)$ and for there to be a time t_0 such that

$$\lim_{x \to 0} \frac{S(t_0 + h) - S(t_0)}{h}$$

does not exist? Explain your answer.

Enrichment Problems

34. Suppose the height of a ball thrown up in the air is given by $s(t) = -16t^2 + 144t$, where t is in seconds and is measured in feet above the ground. What is the maximum height of the ball? (Hint: Find a time t for which $v(t)$ is zero.)

35. "Shuffle-em Sam" sat down in a poker game with $30 and realized that, surprisingly, the amount of money he had at time t hours after the start of the game could be expressed by the function $A(t) = 30 - 9t^2 + 2t^3$ dollars.
   (a) What amount of money had Sam won (or lost) in the first two hours?
   (b) What was the average rate at which Sam was winning (or losing) money during the first two hours?
   (c) What was the instantaneous rate at which Sam was winning (or losing) money at the end of the second hour?
   (d) If the game lasted five hours, what was the average rate at which Sam was winning (or losing) money over the entire game?
   (e) What was the instantaneous rate at which Sam was winning (or losing) money at the end of the game?
   (f) When the game stopped, why was Sam so eager to keep playing?

SOLUTIONS TO PRACTICE EXERCISES

1. The counter reading goes from 4640 at $t = 4$ to 5400 at $t = 5$. Hence the average is $(5400 - 4640)/(5 - 4) = 760$ per hour. By finding $(0.75)760 = 570$ we have an estimate of $C(4.45) \approx C(4) + 570 = 4640 + 570 = 5210$.

2. (a) To find $v(1)$ for $s(t) = 50 - 16t^2$ we need to determine $\lim_{h \to 0} [s(1 + h) - s(1)]/h$. To do this requires us to find

$$s(1 + h) = 50 - 16(1 + h)^2$$
$$= 50 - 16(1 + 2h + h^2)$$
$$= 50 - 16 - 32h - 16h^2$$
$$= 34 - 32h - 16h^2$$

$$s(1) = 50 - 16(1)^2 = 50 - 16 = 34$$

Then $\lim_{h \to 0} \dfrac{s(1 + h) - s(1)}{h}$

$$= \lim_{h \to 0} \frac{34 - 32h - 16h^2 - 34}{h}$$
$$= \lim_{h \to 0} \frac{-32h - 16h^2}{h}$$
$$= \lim_{h \to 0} -32 - 16h = -32 \text{ feet per second}$$

(b) At $t = 1$, we have $s(1) = 50 - 16(1)^2 = 34$ feet, so the balloon has not hit the ground.

THE DERIVATIVE

In this section, limits are used to extend the idea of "instantaneous rate of change" to a more general setting. We define the derivative of a function and establish the essential rule for finding derivatives. Describing the behavior of a function is much easier to do if we know the derivative of that function.

Geometrically, the derivative allows a generalization from the idea of "a line tangent to a circle" to "a line tangent to the graph of $y = f(x)$."

Tangents

Figure 2.4.1 shows a circle and a line tangent to it at point P.

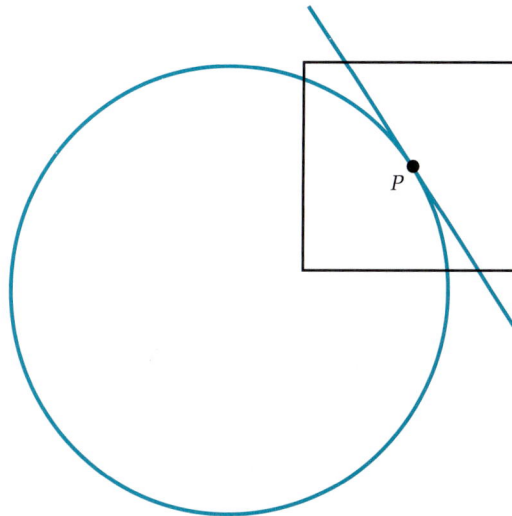

FIGURE 2.4.1

Let us box a portion of the circumference around P in figure 2.4.1 and enlarge the boxed portion as shown in figure 2.4.2. Notice that the exhibited portion of the circumference appears to be more linear than the original circle.

If the boxed portion in figure 2.4.2 is enlarged further, the circumference appears to further straighten out, as shown in figure 2.4.3.

The straightening effect is due to the enlargement and is apparent for the graphs of many functions. Figure 2.4.4 is the graph of a function $y = f(x)$ with P a point on that graph. The box encloses a portion of the graph in the neighborhood of P.

An enlargement of the boxed region of the graph in figure 2.4.4 is shown in figure 2.4.5.

Notice that as we magnify a portion of the curve, that portion appears to become more and more similar to a straight line. There is a "straightening out" effect with each magnification.

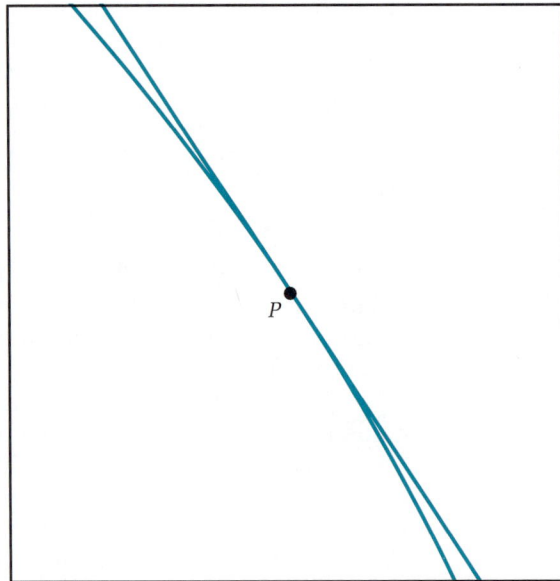

FIGURE 2.4.2

FIGURE 2.4.3

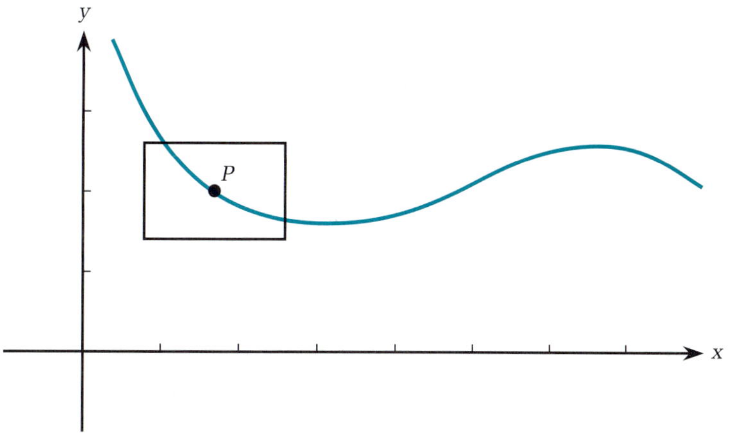

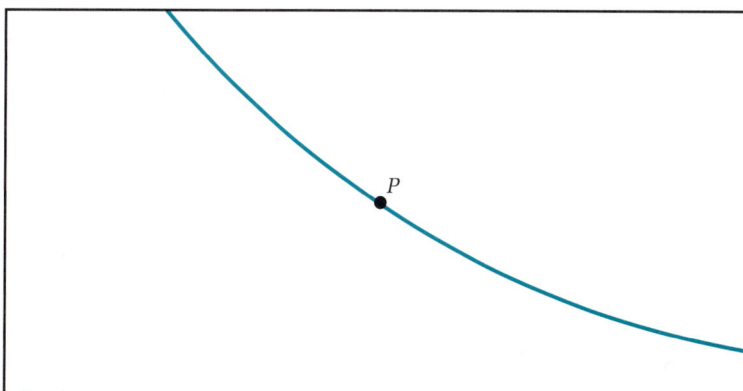

FIGURE 2.4.4

FIGURE 2.4.5

📟 Graphing Calculators

A feature of graphing calculators is their ability to zoom in and enlarge any portion of the graph and redraw it to fill the entire screen.

If this magnifying process were to be repeated over and over, without any stop, all we would be able to see in the portion under consideration would appear to be a straight line, as shown in figure 2.4.6.

Tangent Lines

The straight line that the curve approaches as a result of this magnification process is the *tangent line* to the graph of $f(x)$ at the point P. Among all lines through P, the tangent line is the line that best approximates the curve. Before searching for tangent lines, we should ask ourselves whether every graph actually has a tangent line at every point.

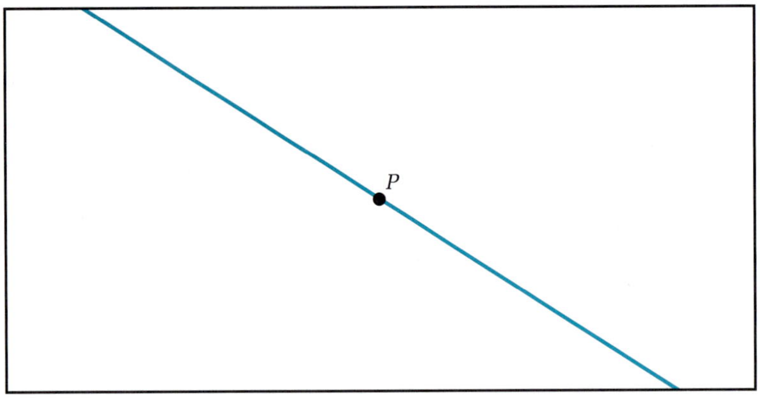

FIGURE 2.4.6

EXAMPLE 1   A graph with a point at which there is no tangent line

Discuss whether a tangent line exists to the graph of $f(x) = |x|$ at $P(0, 0)$.

SOLUTION

The graph of $y = |x|$ is shown in figure 2.4.7.

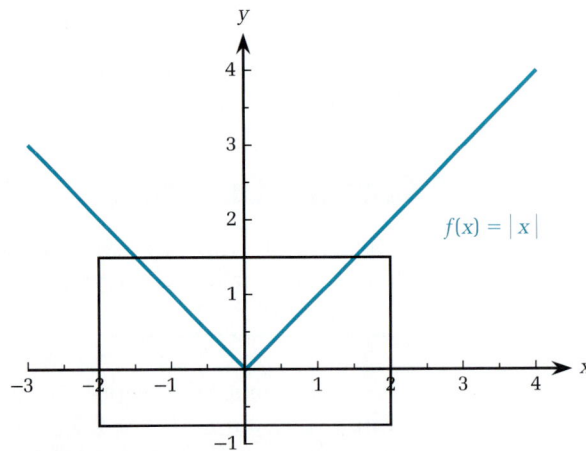

FIGURE 2.4.7

There is no tangent line at the point $P(0, 0)$, which can be seen by looking at a magnification of a neighborhood around $P(0, 0)$, as shown in figure 2.4.8. There is no evidence of a straightening effect at $P(0, 0)$.

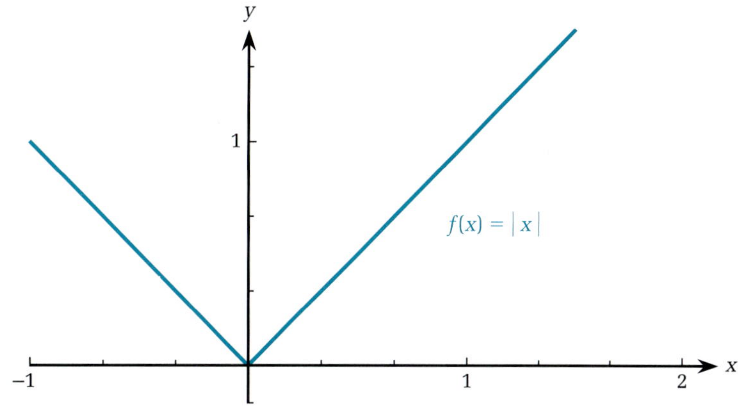

FIGURE 2.4.8

Let us assume we have a function $f(x)$ whose graph does have a tangent line at P; that is, enlargements of the graph in the vicinity of P do show a straightening effect. To find an equation for this tangent line we need to determine its slope. To do this, let us review what a secant line is.

Secant Lines

A line joining two points on a graph is called a *secant line.* Consider the secant line determined by the point $P(x_0, f(x_0))$ and a nearby point $Q(x_0 + h, f(x_0 + h))$, as shown in figure 2.4.9.

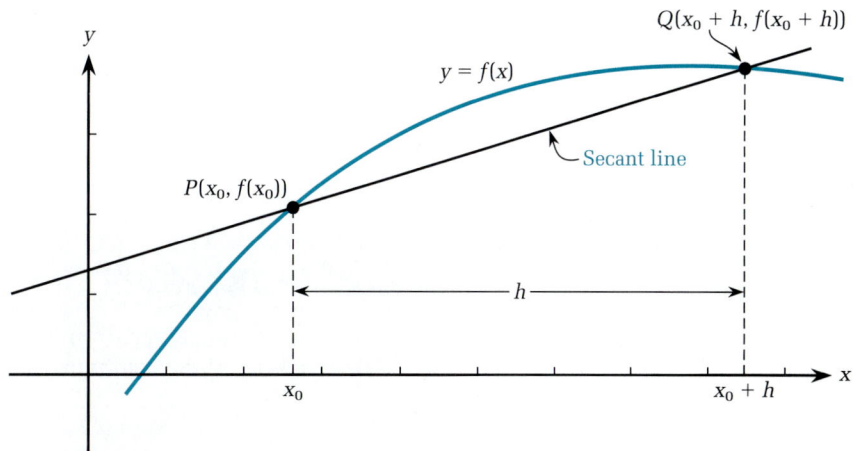

FIGURE 2.4.9

If we consider secant lines determined by various values of h, we see in figure 2.4.10 that as h becomes smaller, and Q becomes closer to P, the secant line approaches the tangent line at the point $P(x_0, f(x_0))$.

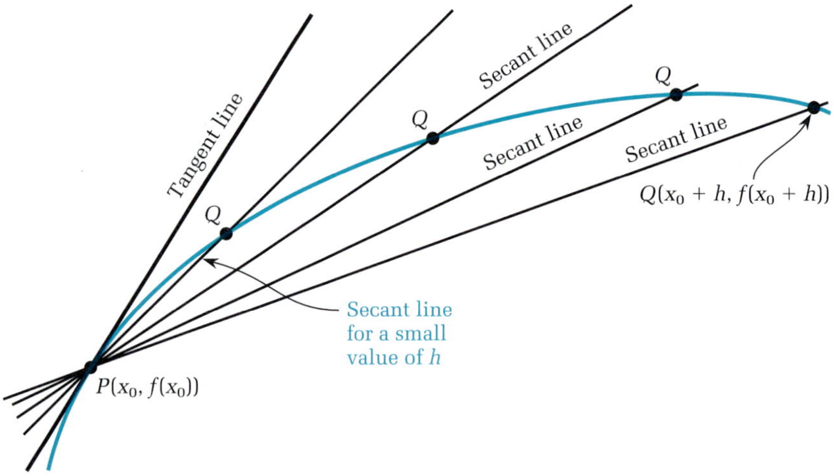

FIGURE 2.4.10

As we take the limit as h approaches zero, the slope of the secant line approaches the slope of the tangent line. If we try to draw a line, using our own straightedge and pencil, we may find it difficult to draw a secant line when h is quite small, just because the points are so close together. Nonetheless, we know any two distinct points do determine a unique line.

Graphing Calculator

The graphing calculator can display each secant line as an enlargement, constantly changing the scale as it zooms in on the point P.

Referring back to figure 2.4.9, let us use symbols to restate our discussion.

SLOPE OF THE SECANT LINE

The slope of the secant line joining $P(x_0, f(x_0))$ and $Q(x_0 + h, f(x + h_0))$ is given by

$$\frac{f(x_0 + h) - f(x_0)}{(x_0 + h) - x_0} = \frac{f(x_0 + h) - f(x_0)}{h}$$

EXAMPLE 2 Finding the slope of a secant line

Suppose $f(x) = x^2 - 3x + 1$. Find the slope of the secant line joining $P(1, f(1))$ and $Q(1 + h, f(1 + h))$.

SOLUTION

To find the coordinates of P and Q we need to find $f(1) = (1)^2 - 3 \cdot 1 + 1 = -1$ and $f(1 + h) = (1 + h)^2 - 3(1 + h) + 1 = (1 + 2h + h^2) - 3 - 3h + 1 = h^2 - h - 1$. Thus,

$$\frac{f(1 + h) - f(1)}{h} = \frac{h^2 - h - 1 - (-1)}{h} = \frac{h^2 - h}{h}$$

$$= \frac{(h - 1) \cdot h}{h} = h - 1 \text{ because } h \neq 0$$

Hence, the slope of the secant line is $h - 1$. For instance, if $h = 2$, then Q has coordinates $(3, 1)$ and the slope of the line PQ is $2 - 1 = 1$, as can be seen on figure 2.4.11.

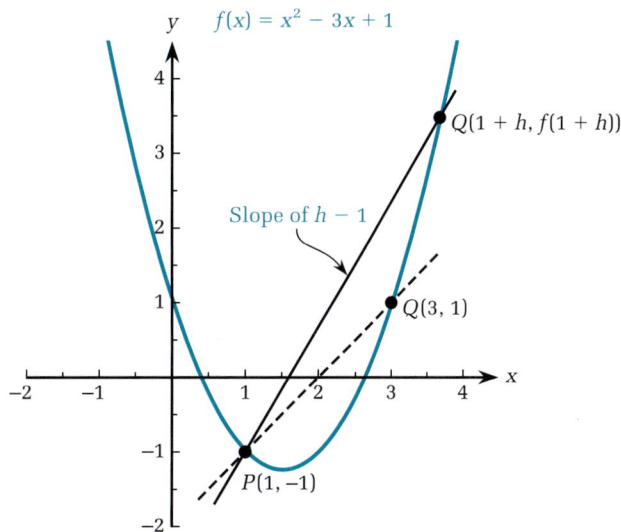

FIGURE 2.4.11

PRACTICE EXERCISE 1

For $f(x) = 3x^2 - x$, find the slope of the secant line joining the points $P(2, f(2))$ and $Q(2 + h, f(2 + h))$.

Tangent Lines and Their Slopes

In our first discussion of tangent lines, we did not have a method for determining whether a tangent line actually existed at the point $P(x_0, f(x_0))$. Our concept of magnification, while quite understandable, lacked a means of implementation. Now we can use the slopes of secant lines joining the points $Q(x_0 + h, f(x_0 + h))$ and $P(x_0, f(x_0))$ to provide a means for determining whether a tangent line exists.

The key idea is that if there is a tangent line to the graph of $y = f(x)$ at $P(x_0, f(x_0))$ then as we magnify the region around the point $P(x_0, f(x_0))$, the secant lines drawn between displayed points on the graph will appear to approach the tangent line. Eventually the visible portion of the secant lines would be indistinguishable from the tangent line because the enlargements force the screen to display such a small region.

In the following we assume tangent lines are not vertical. An example of a vertical tangent line is given in the enrichment problems.

SLOPE OF THE TANGENT LINE TO THE GRAPH OF $y = f(x)$

A tangent line exists at the point $P(x_0, f(x_0))$ provided

$$\lim_{h \to 0} \frac{f(x_0 + h) - f(x_0)}{h}$$

exists. If indeed this limit does exist, its value *defines* the slope of the tangent line at $P(x_0, f(x_0))$.

EXAMPLE 3   Equation of the tangent line

Find the equation of the tangent line to the graph of $f(x) = 2x^2 - x$ at the point $P(1, 1)$.

SOLUTION

The graph of $f(x)$ is shown in figure 2.4.12.

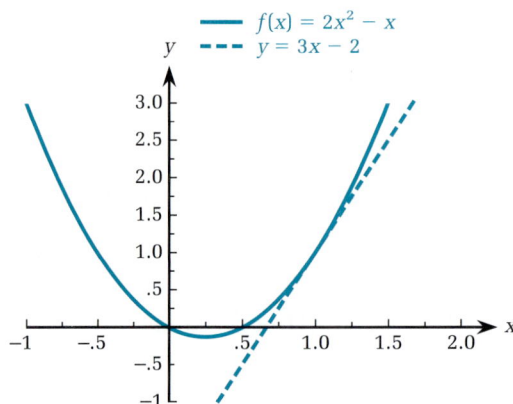

FIGURE 2.4.12

To find the slope of the tangent line, use $x_0 = 1$:

$$\lim_{h \to 0} \frac{[f(x_0 + h)] - [f(x_0)]}{h}$$

$$= \lim_{h \to 0} \frac{[2(1 + h)^2 - (1 + h)] - [2 \cdot 1^2 - 1]}{h}$$

(Here the brackets "[ ]" are used about $f(x_0 + h)$ and $f(x_0)$ to emphasize those terms in the numerator that are equal to each function value.)

$$= \lim_{h \to 0} \frac{[2(1 + 2h + h^2) - 1 - h] - [2 - 1]}{h}$$

$$= \lim_{h \to 0} \frac{[2 + 4h + 2h^2 - 1 - h] - [1]}{h}$$

$$= \lim_{h \to 0} \frac{3h + 2h^2}{h}$$

To evaluate this limit, we use the alternate function method of Section 2.2. To do this, notice that $3 + 2h$ is an alternate function for $(3h + 2h^2)/h$ because $3h + 2h^2 = (3 + 2h)h$.

Hence, $\lim_{h \to 0} [(3h + 2h^2)/h] = \lim_{h \to 0} 3 + 2h = 3$. Thus, the tangent line at $P(1, 1)$ has slope 3.

Recall that if m is the slope of a line, and the point $P(x_0, y_0)$ is on the line, then the *point-slope* form of the line states that the equation of the line is $(y - y_0)/(x - x_0) = m$. In our case, $m = 3$ and $x_0 = y_0 = 1$. Hence, the equation of the tangent line may be written as $(y - 1)/(x - 1) = 3$, which can be rewritten as $y - 1 = 3(x - 1)$. Distributing the multiplication by 3 and then adding 1 to each side, we have $y = 3x - 3 + 1 = 3x - 2$. The form $y = mx + b$ is the *slope-intercept* form of the equation for a line and allows us to easily sketch the graph. ●

PRACTICE EXERCISE 2

Find the equation of the tangent line to the graph of $f(x) = x^2 - 3$ at $Q(2, 1)$.

At this point, let us emphasize that the evaluation of $\lim_{h \to 0} [f(x_0 + h) - f(x_0)]/h$ is performed by using an alternate function and finding the limit of the alternate function as h approaches zero. We *cannot* evaluate the original quotient by setting $h = 0$ because in doing so we would be attempting to divide by zero. Speaking in geometrical terms, when $h = 0$ we are attempting to find a secant line determined by the point $(x_0, f(x_0))$ and itself. That is impossible. To determine a line requires two points.

Tangent Lines Determined by a Graph

Suppose a function is described only by a graph. In such a situation it is still possible to estimate the slope of a tangent line at some point on the graph. We carefully draw the graph and attempt to locate two points on the tangent line whose coordinates are easy to determine. The coordinates of these are then used to determine the slope.

EXAMPLE 4   Slope determined graphically

Figure 2.4.13 is the graph of a function $g(x)$. Estimate the slope of the tangent line at the point $P(1, 3)$.

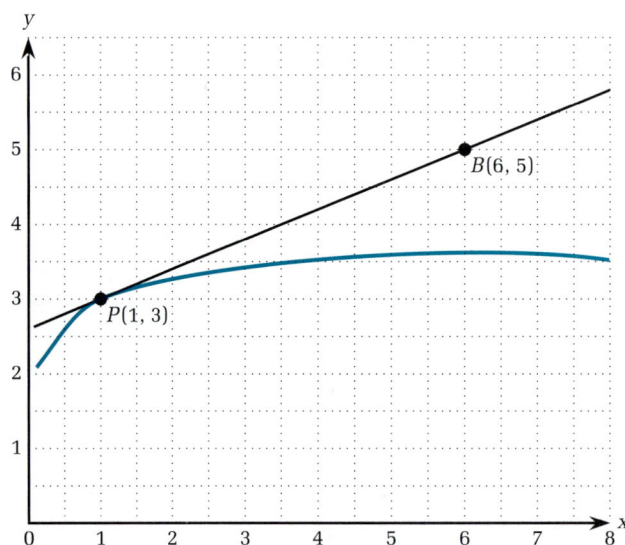

FIGURE 2.4.13

SOLUTION

There are an infinite number of points we could choose as a second point, but it appears the tangent line goes very close to (or even through) the point $B(6, 5)$. If we use that point and the point $P(1, 3)$, the slope is approximately $(5 - 3)/(6 - 1) = 2/5 = 0.4$.

The Difference Quotient

Finding the slope of the tangent line involves evaluating exactly the same limit as we encountered in the last section when we were finding the instantaneous rate of change of $f(x)$ at x_0.

In both situations, we found $\lim_{h \to 0} [f(x_0 + h) - f(x_0)]/h$.

DEFINITION

Difference quotient

Suppose $y = f(x)$. The *difference quotient of $f(x)$* is $\dfrac{f(x + h) - f(x)}{h}$

The Derivative

We want a formula that gives the slope of the tangent line, not just at one point on the graph but at any point. This is analogous to having a formula that gives the instantaneous rate of change for any value of the function. To be successful, we need a function to determine these values. Such a function is called a *derived function* or more simply, *the derivative*. The process of finding such a function is called *differentiation*.

DEFINITION

The derivative

Suppose $y = f(x)$. The *derivative* of $y = f(x)$ is

$$\lim_{h \to 0} \frac{f(x + h) - f(x)}{h}, \text{ provided this limit exists.}$$

The following notations are used for the derivative: $f'(x)$, y', dy/dx, or $\dfrac{d}{dx} f(x)$.

Remarks on Notation.   Regardless of which notation we use, we may always say "the derivative of $f(x)$, with respect to x" or "the derivative of y with respect to x." When there is no doubt of the variable, we may omit the phrase "with respect to."

The individual notations are read as:

$f'(x)$ is "f prime at x" or "f prime of x"

y' is "y prime"

$\dfrac{dy}{dx}$ is "dee-y, dee-x" or "dee-y over dee-x"

$\dfrac{d}{dx} f(x)$ is "dee, dee-x of f" or "dee-f, dee-x"

Each particular notation has its advantages and disadvantages. It is quite easy to write y', but we may have no idea what variable y is a function of. The notation $f'(x)$ is quite convenient if we want to evaluate the derivative at some particular value. The notation dy/dx

or $(d/dx)f(x)$ reminds us that the derivative is the result of operating on the function $f(x)$ to create a new function and that the derivative is being taken with respect to changes in x.

When writing dy/dx or df/dx we also are emphasizing the idea that the derivative is the limit of the quotient of a **d**-ifference in y-values over a corresponding **d**-ifference in x-values. This notation was introduced by Gottfried Wilhelm Leibniz (1646–1716), one of the creators of calculus, and it will be used in many situations as we develop formulas and applications.

Even this variety of notations for derivatives is not an exhaustive list of all the notations in common use. However, they will suffice for us.

Let us return to focus on the concept of a derivative. There is a significant difference, and we emphasize the contrast, between finding the slope of a tangent line at *a single point* on the graph, and using a function to find slopes of the tangent lines *anywhere* on the graph. We shift our perspective from determining values at isolated points to determining functions. The derivative may be evaluated at a particular value, but the derivative *is* a function.

Finding Derivatives

Following is a four-step rule for finding the derivative of a function, using the definition.

FOUR-STEP RULE FOR FINDING $f'(x)$

If the derivative of $f(x)$ does exist, then finding $f'(x)$ consists of the following steps:

1. Find $f(x + h)$.

2. Determine $f(x + h) - f(x)$.

3. Simplify the expression for the difference quotient $[f(x + h) - f(x)]/h$. If possible, create an alternate function by canceling a factor of $h \neq 0$.

4. Finally, determine $\lim\limits_{h \to 0} \dfrac{f(x + h) - f(x)}{h}$. This limit is $f'(x)$.

To find a derivative, in Step 3 we cannot use the rules for limits from Section 2.2 directly on the difference quotient, for we have a quotient whose denominator, h, is approaching zero. For that reason, we search for an alternate function, one without a denominator approaching zero.

WARNING

The derivative is a limit. It is $\lim\limits_{h \to 0} [f(x + h) - f(x)]/h$. If we once write that we are taking a limit, we must remember to keep writing the $\boxed{\lim\limits_{h \to 0}}$ part of the expression for the derivative at each subsequent step until we actually evaluate the limit. In part, the rewriting reminds us of the limit process we want to perform. Additionally, it simply is not true that $x(x + h) = x^2$ unless $h = 0$, and we do not want h to be 0; we want the limit as h *approaches* 0.

It is to avoid this pitfall that when we use our four-step strategy, we write only the difference quotient and algebraic variations of it until we have found a satisfactory alternate function.

EXAMPLE 5   Using the four-step rule for finding the derivative

Find the derivative of $f(x) = x^2$.

SOLUTION

Numbering our steps as in the outline above, we have

1. $f(x + h) = (x + h)^2 = x^2 + 2xh + h^2$
2. $f(x + h) - f(x) = x^2 + 2xh + h^2 - x^2 = 2xh + h^2$
3. $[f(x + h) - f(x)]/h = (2xh + h^2)/h = [(2x + h)h]/h = 2x + h$

We now have an alternate function whose limit we can evaluate as h approaches zero.

4. Hence, we have $\lim\limits_{h \to 0} [f(x + h) - f(x)]/h = \lim\limits_{h \to 0} (2x + h) = 2x$.

Thus, $f'(x) = 2x$.

EXAMPLE 6   Using the four-step rule for finding the derivative

Find the derivative of $f(x) = x^3$.

SOLUTION

Again, we will follow the numbering in the procedure above.

1. $f(x + h) = (x + h)^3 = x^3 + 3x^2h + 3xh^2 + h^3$

2. $f(x + h) - f(x) = x^3 + 3x^2h + 3xh^2 + h^3 - x^3 = 3x^2h + 3xh^2 + h^3$

3. $\dfrac{f(x + h - f(x)}{h} = \dfrac{3x^2h + 3xh^2 + h^3}{h} = \dfrac{(3x^2 + 3xh + h^2)h}{h} = 3x^2 + 3xh + h^2$

This alternate function has a limit as h approaches zero. The term $3x^2$ is unaffected by what happens to h, but the term $3xh$ goes to zero as h goes to zero, and certainly h^2 goes to zero as h does.

4. Hence, $\lim\limits_{h \to 0} \dfrac{f(x + h) - f(x)}{h} = \lim\limits_{h \to 0} (3x^2 + 3xh + h^2) = 3x^2.$

Thus, $f'(x) = 3x^2.$

EXAMPLE 7 Using the four-step rule for finding the derivative

Suppose $f(x) = x^2 + 3x - 1$. Find $\dfrac{dy}{dx}$.

SOLUTION

1. $f(x + h) = (x + h)^2 + 3(x + h) - 1$
$= x^2 + 2xh + h^2 + 3x + 3h - 1$

2. $f(x + h) - f(x) =$
$[x^2 + 2xh + h^2 + 3x + 3h - 1] - [x^2 + 3x - 1].$
The reason for using the "[" and "]" symbols is to help us keep track of which is $f(x + h)$ and which is $f(x)$. Making sure the "$-$" sign is distributed over each term of $f(x)$, we have

$$[x^2 + 2xh + h^2 + 3x + 3h - 1] - x^2 - 3x + 1 =$$
$$2xh + h^2 + 3h = (2x + h + 3)h$$

3. $[f(x + h) - f(x)]/h = [(2x + h + 3)h]/h$, which gives the alternate function $2x + h + 3$

4. Finally, $\lim\limits_{h \to 0} [f(x + h) - f(x)]/h = \lim\limits_{h \to 0} (2x + h + 3) = 2x + 3.$

Thus, $dy/dx = 2x + 3.$

PRACTICE EXERCISE 3

For $f(x) = x^2 - x + 5$, find $f'(x)$.

More Remarks on Notation.   Although we often write our functions as functions of x, nonetheless we may use a different name for the function and its variable. For example, the function $f(x) = x^2 - x$ is the same function as $g(t) = t^2 - t$.

Derivatives also may have different names for the functions and their variables. However, just as the function does not depend on its name or the name of its variable, neither does the derivative. It is *very important* to realize that all of the following statements are equivalent.

If $f(x) = 2x^2 - x + 6$, then $f'(x) = 4x - 1$.

If $f(t) = 2t^2 - t + 6$, then $f'(t) = 4t - 1$.

If $g(z) = 2z^2 - z + 1$, then $g'(z) = 4z - 1$.

When calculating the derivative, the process of finding an alternate function often involves some algebraic manipulation, as we see in Example 8.

EXAMPLE 8   Using the four-step rule for finding the derivative

Find the derivative of $f(x) = \dfrac{3}{x}$.

SOLUTION

Following the four-step process:

1. $f(x + h) = 3/(x + h)$

2. $f(x + h) - f(x) = 3/(x + h) - 3/x$

3. $[f(x + h) - f(x)]/h = \left(\dfrac{3}{x + h} - \dfrac{3}{x} \right) \Big/ h$

We want to create an alternate function, canceling $h \neq 0$ if possible. First, let us use $(x + h)x$ as a common denominator for the numerator to give us

$$\frac{\left(\dfrac{3x - 3(x + h)}{(x + h)x} \right)}{h}$$

Rewriting h as $h/1$, we can invert the denominator and multiply, giving us

$$\left(\frac{3x - 3x - 3h}{(x + h)x} \right) \cdot \left(\frac{1}{h} \right) = \frac{-3h}{(x + h)xh} = \frac{-3}{(x + h)x}$$

Finally, we have an alternate function whose limit as h approaches zero is easy for us to determine.

4. $\lim\limits_{h \to 0} \dfrac{-3}{x(x + h)} = -\dfrac{3}{x^2}$

Hence, if $f(x) = 3/x$, then $f'(x) = -3/x^2$.

Continuity and Alternate Functions

For each of our examples, the alternate function was continuous at zero. Thus, to find its limit as h approached zero we merely had to evaluate the alternate functions at zero. Another way of thinking about those alternate functions is as follows.

Let $D(h) = \dfrac{f(x + h) - f(x)}{h}$ be the difference quotient. Then $D(h)$ is a function of h, but not defined at $h = 0$ because division by zero is not defined. The alternate functions we have found are functions of h that are defined at $h = 0$ and are equal to $D(h)$ at values of h other than zero. Suppose $A(h)$ is the alternate function for $D(h)$. By the Alternate Function Theorem, $\lim\limits_{h \to 0} D(h) = \lim\limits_{h \to 0} A(h)$. Then, because $\lim\limits_{h \to 0} A(h) = A(0)$ we can evaluate $\lim\limits_{h \to 0} D(h)$ by finding $A(0)$.

Graphically, figure 2.4.14 shows $D(h)$ as a function with a limit as h approaches zero, but with a puncture point at $x = 0$; whereas $A(h)$ is equal to $D(h)$ for nonzero values of h and $A(0)$ is defined.

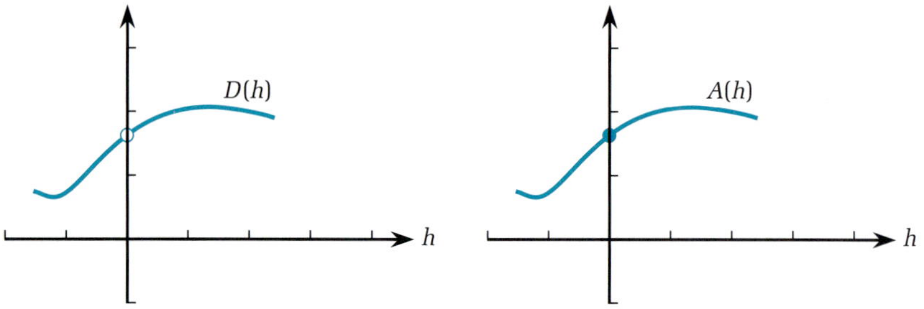

FIGURE 2.4.14

Differentiability and Continuity

For the derivative of $f(x)$ to exist at $x = a$, the graph of the function must be unbroken at the point $P(a, f(a))$ because the existence of a tangent line forces a smoothness of the graph. In other words, the function must be continuous for $x = a$. However, there are functions that are continuous at a point but do not have a derivative there. We saw in Example 1 that $y = |x|$ is such a function at the point $P(0, 0)$. The graph of another such function is shown in Example 9.

EXAMPLE 9   A function that is continuous at a point, but not differentiable there

Explain why the function $f(x) = (x - 1)^{2/3}$, whose graph is shown in figure 2.4.15, is continuous at the point $P(1, 0)$, but there is no derivative for $x = 1$.

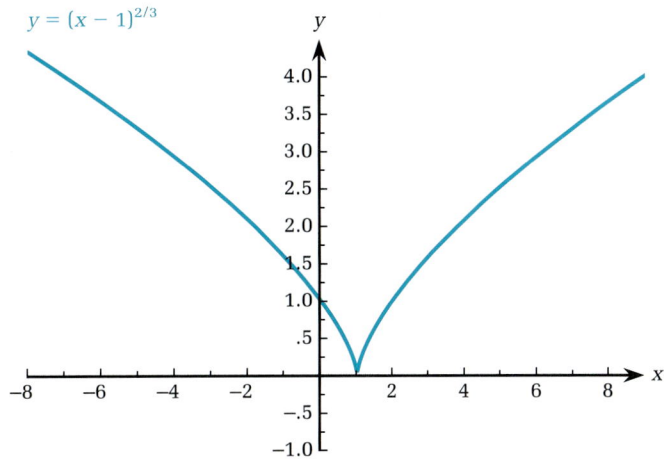

FIGURE 2.4.15

SOLUTION

According to figure 2.4.15, the graph is unbroken in the neighborhood of $P(1, 0)$ and can be drawn without lifting our pencil from the paper. However, in figure 2.4.16 and figure 2.4.17, which show progressively greater enlargements, there is no indication of a straightening out.

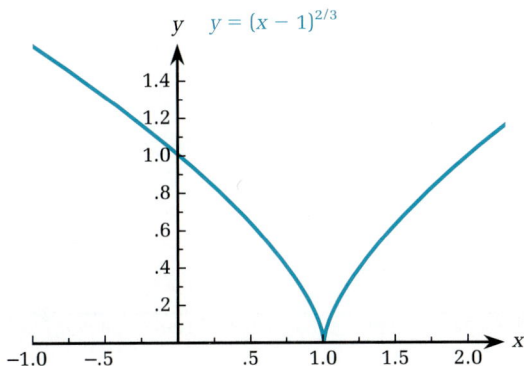

FIGURE 2.4.16

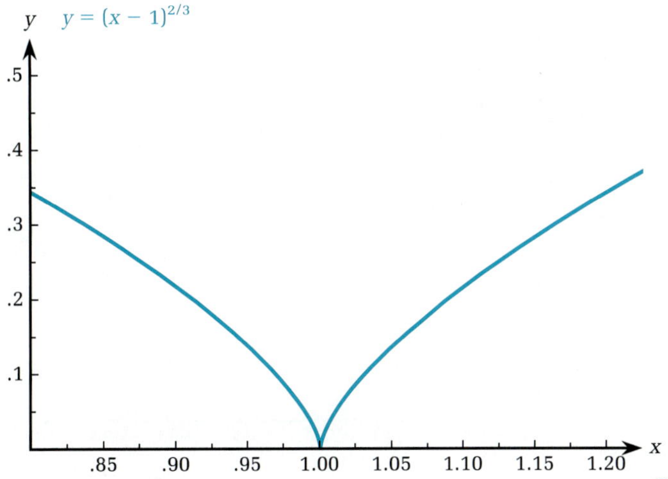

FIGURE 2.4.17

2.4 PROBLEMS

Foundations

The problems of this section require the basic skills illustrated by the following:

1. Let the line l contain the points $A(1, 3)$ and $B(5, 2)$.
   (a) Find the slope of l.
   (b) Find an equation for l.

2. Simplify $2/(x + h) - 2/x$ by finding a common denominator.

Exercises

3. Which of the graphs in figures 2.4.18−2.4.21 have tangents at the indicated points A, B, C, and D?

(a)

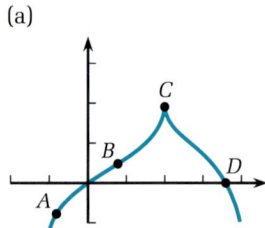

FIGURE 2.4.18

(b)

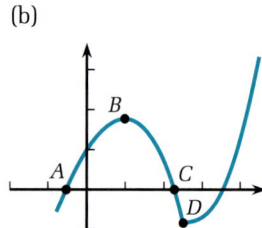

FIGURE 2.4.19

(c)

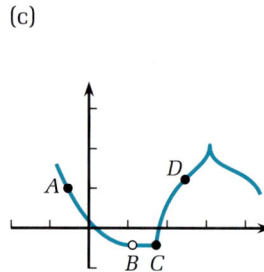

FIGURE 2.4.20

(d)

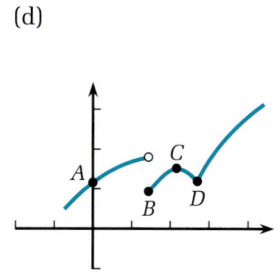

FIGURE 2.4.21

For problems 4−9, estimate graphically the slopes of the tangent lines at the points indicated.

4. *(Space Research Expenditures)*   The annual expenditure by private companies on applied research and development in space research is shown in figure 2.4.22.*

*Source: *The Scientist,* March 4, 1991.

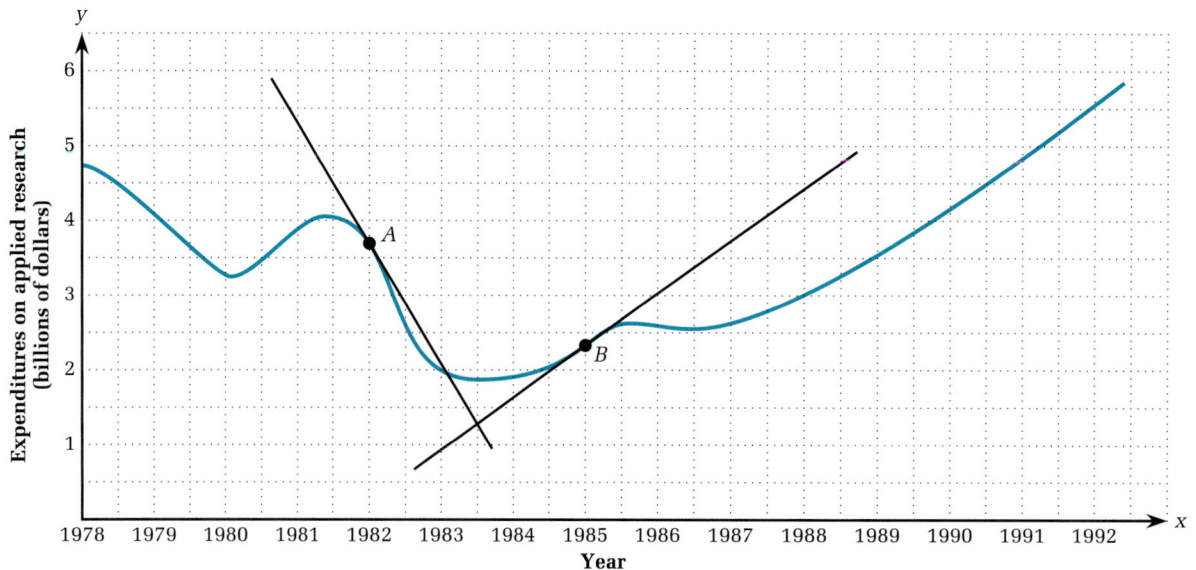

FIGURE 2.4.22

5. **(Health Research Expenditures)**   The U.S. federal government support for basic research in health is shown in figure 2.4.23.*

FIGURE 2.4.23

6. **(Coin Prices)**   Coin collectors purchase rolls of uncirculated coins as an investment, as well as for the pleasure of collecting coins. The graph in figure 2.4.24 shows the average cost of buying an uncirculated roll of 1955 half dollars.†

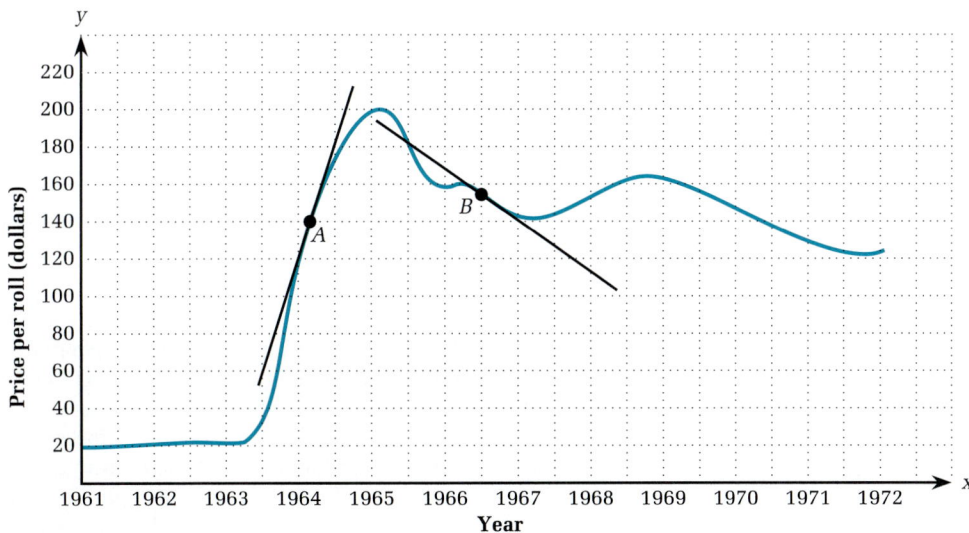

7. **(Energy Conservation, Refrigerators)**   To encourage conservation of electricity, manufacturers of refrigerators have produced more energy-efficient appliances over the years. The graph in figure 2.4.25 shows the average number of kilowatts used by a refrigerator manufactured during the year indicated.‡

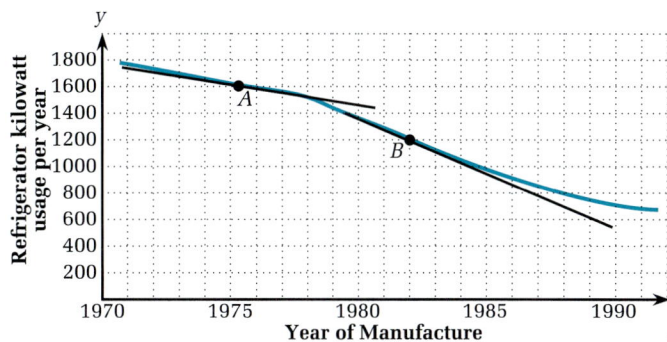

FIGURE 2.4.25

FIGURE 2.4.24

*Source: *The Scientist*, March 4, 1991.
†Source: Montroll and Badger, *Quantitative Aspects of Social Phenomena*, New York: Gordon and Breach, 1974.

‡Source: A. P. Fickett et al., ''Efficient Use of Electricity,'' *Scientific American* 264:3 (Sept. 1990).

8. (Biology, Human Growth) The growth of human beings is shown in figure 2.4.26.*

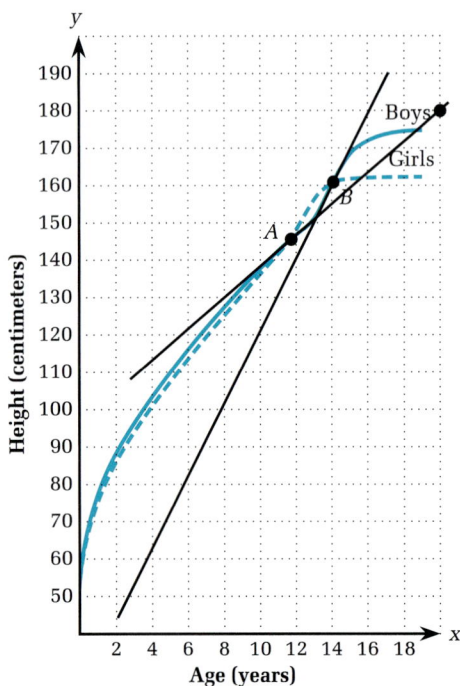

FIGURE 2.4.26

9. (Rollercoaster Prices) The *demand function* that relates the price to the number of tickets sold to the Cascade Thunder Rollercoaster is graphed in figure 2.4.27.

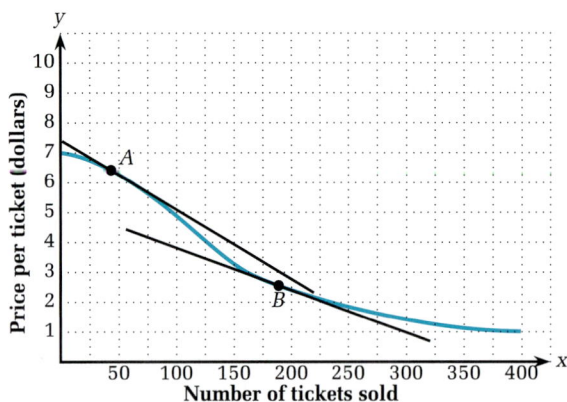

FIGURE 2.4.27

Find the equations for the tangent lines to the graphs of the functions in problems 10–19 at the points indicated.

10. $f(x) = 3x - 2$, $P(2, 4)$

*Source: *Encyclopedia Britannica,* "Growth and Development."

11. $f(x) = 2x - 5$, $P(1, -3)$

12. $f(x) = x^2 + x$, $P(-1, 0)$

13. $g(t) = t^2 + 2t$, $P(0, 0)$

14. $g(x) = 2x^2 + 3$, $Q(2, 11)$

15. $g(x) = 3x^2 + 1$, $Q(-1, 4)$

16. $f(x) = x^2 - 2x + 1$, $Q(0, 1)$

17. $f(x) = x^2 + 2x + 1$, $Q(3, 16)$

18. $f(x) = -1/x$, $P(-1, 1)$

19. $f(x) = 2/x$, $P(1, 2)$

Find the derivatives of the functions in problems 20–29.

20. $g(x) = 5 + 3x$
21. $g(x) = -3 + 2x$
22. $f(x) = x^2 - 2x$
23. $g(t) = t^2 - 5t$
24. $f(t) = -3t^2 + 2$
25. $g(x) = -x^2 + 3$
26. $g(x) = 2x^2 + x - 3$
27. $f(x) = 3x^2 - x + 1$
28. $g(x) = \dfrac{-2}{3x}$
29. $g(x) = \dfrac{-1}{2x}$

Graphing Calculator Problems

30. By means of a graphing calculator, produce a graph of $y = x^{1/n}$ and estimate the slope of the tangent line at $x = 2$ for
(a) $n = 3$. (b) $n = 4$.

31. By means of a graphing calculator, produce a graph of $y = x^2/\sqrt{2x + 3}$ and estimate the slope of the tangent line for
(a) $x = 1$. (b) $x = 2$.

32. A graphing calculator may estimate a derivative by finding a secant line for a very small interval and asserting that the value found is the derivative. This process does well if the function is differentiable at the point, but unfortunately it fails if the function is merely continuous there and not differentiable. If your graphing calculator numerically estimates a derivative, try finding the derivative of $f(x) = |x|$ at $x = 0$. Compare your answer with that of Example 1.

Writing and Discussion Problems

33. Using the function $f(x) = x^3 - 3x + 2$, discuss the following statements, each of which might be a proposed characterization of a tangent line.

(a) A tangent line has only one point in common with the graph of the function.

(b) A tangent line does not cross the graph of the function.

Enrichment Problems

34. Verify that the derivative of $f(x) = \sqrt{x}$ is $f'(x) = 1/(2\sqrt{x})$ by forming the difference quotient and then multiplying numerator and denominator by $\sqrt{x+h} + \sqrt{x}$ before finding the limit as h approaches zero.

35. Verify that the derivatives of $f(x) = \sqrt[3]{x}$ is $f'(x) = (1/3)x^{-2/3}$ by forming the difference quotient and then multiplying numerator and denominator by $(x+h)^{2/3} + (x+h)^{1/3}x^{1/3} + x^{2/3}$ before taking the limit as h approaches zero.

36. The function $f(x) = \sqrt[3]{x}$ is defined for all values of x and is continuous. Problem 35 shows that $f'(x)$ is not defined at $x = 0$. By graphing $f(x) = \sqrt[3]{x}$ show that a graph may have a tangent line even when the derivative of the functions does not exist.

SOLUTIONS TO PRACTICE EXERCISES

1. $f(2) = 3 \cdot 2^2 - 2 = 3 \cdot 4 - 2 = 12 - 2 = 10$ and $f(2 + h) = 3 \cdot (2 + h)^2 - (2 + h) = 3(4 + 4h + h^2) - 2 - h = 12 + 12h + 3h^2 - 2 - h = 10 + 11h + 3h^2$

Thus, $\dfrac{f(2 + h) - f(2)}{h}$

$$= \frac{10 + 11h + 3h^2 - 10}{h}$$

$$= \frac{11h + 3h^2}{h} = \frac{(11 + 3h) \cdot h}{h}$$

Because $h \neq 0$, we can reduce this to $= 11 + 3h$.

2. To find the equation of the tangent line to the graph of $f(x) = x^2 - 3$ at $Q(2, 1)$ we first need to find the slope of the tangent line.

$$\lim_{h \to 0} \frac{f(x_0 + h) - f(x_0)}{h}$$

$$= \lim_{h \to 0} \frac{f(2 + h) - f(2)}{h}$$

$$= \lim_{h \to 0} \frac{[(2 + h)^2 - 3] - [2^2 - 3]}{h}$$

$$= \lim_{h \to 0} \frac{(2^2 + 2 \cdot 2h + h^2 - 3) - 1}{h}$$

$$= \lim_{h \to 0} \frac{4 + 4h + h^2 - 4}{h}$$

$$= \lim_{h \to 0} \frac{4h + h^2}{h} = \lim_{h \to 0} (4 + h) = 4$$

Thus, the slope of the tangent line is 4. Because the tangent line passes through the point $Q(2, 1)$, the point-slope form of an equation of a line gives us $(y - y_0)/(x - x_0) = 4$, and with $y_0 = 1$ and $x_0 = 2$ we have $y - 1 = 4(x - 2)$, which is equivalent to $y = 4x - 8 + 1 = 4x - 7$.

3. Following the four-step rule:
1. $f(x + h) = [(x + h)^2 - (x + h) + 5]$
2. $f(x + h) - f(x)$
   $$= [x^2 + 2xh + h^2 - x - h + 5] - [x^2 - x + 5]$$
   $$= x^2 + 2xh + h^2 - x - h + 5 - x^2 + x - 5$$
   $$= 2xh + h^2 - h$$
3. $\dfrac{f(x + h) - f(x)}{h} = \dfrac{2xh + h^2 - h}{h} =$
   $$\frac{(2x + h - 1)h}{h} = 2x + h - 1$$
4. $\lim\limits_{h \to 0} (2x + h - 1) = 2x - 1$

$f'(x) = 2x - 1$

2.5 DIFFERENTIATION RULES FOR POWERS, SUMS, AND DIFFERENCES

The derivative of $y = f(x)$ may be written as y', or $f'(x)$, or dy/dx, or even as $(d/dx)f$. Whatever the representation, we have found the derivative by evaluating

$$\lim_{h \to 0} \frac{f(x + h) - f(x)}{h}$$

Determining this limit requires a considerable effort on our part. In this section we begin to develop rules that reduce that effort.

Applications in this section include the marginal utility of money (what a dollar is really "worth"), the muscular wing power necessary for a bird to fly, and "handicapping" footraces so as to make competition keener for the runners. In addition, we discuss marginal analysis and apply it to finance.

Derivative of a Constant Function

The simplest functions are those that are constants. For these, no matter what input is provided, the value of the function remains unchanged. That is, there is some constant c such that $f(x) = c$ for all values of x.

The graph of $y = f(x) = c$ is a horizontal line, as shown in figure 2.5.1.

The fact that $f(x) = c$ for any x whatsoever means that $f(x + h) = c$ because $x + h$ is simply another input value. As a result, calculating $f'(x)$ by using its definition as a limit, we have

$$f'(x) = \lim_{h \to 0} \frac{f(x + h) - f(x)}{h}$$

$$= \lim_{h \to 0} \frac{c - c}{h}$$

$$= \lim_{h \to 0} \frac{0}{h} = \lim_{h \to 0} 0 = 0$$

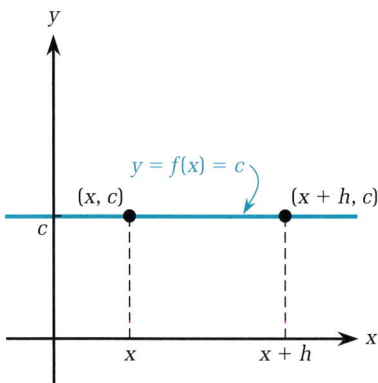

FIGURE 2.5.1

This result is summarized in the following rule for the derivative of a constant function. In the statement of the rule, we express the derivative in several alternate forms and also give the rule in words. The reason for a statement in words is that when we are working with a particular function and we use the rule we do not usually say, "The reason is Rule I," for that would mean nothing to someone unfamiliar with the numbering in this text. Instead, we say "the reason is that the derivative of any constant function is zero."

I. DERIVATIVE OF A CONSTANT FUNCTION

Suppose c is any real number.

If $f(x) = c$, then $f'(x) = 0$, or equivalently,

$$\frac{d}{dx} f(x) = 0$$

Alternate formulations:

$$\frac{d}{dx} c = 0$$

If $y = c$ then $y' = 0$, or equivalently, $dy/dx = 0$.

The derivative of any constant function is zero.

EXAMPLE 1   The derivative of a constant function

(a) Suppose $f(x) = 3$. Then $f'(x) = 0$.

(b) If $f(x) = -2$, then $\dfrac{d}{dx} f(x) = 0$.

(c) $\dfrac{d}{dx} \sqrt{3} = 0$.

(d) If $y = 2\pi$ then $y' = 0$.

Derivative of Powers of x

Table 2.5.1 summarizes some results obtained by us in the previous section. Some of them were examples and some were problems. For the functions that originally did not express x to a power, such an expression is given here.

TABLE 2.5.1

FUNCTION	DERIVATIVE
x	1
x^2	$2x$
x^3	$3x^2$
$\dfrac{1}{x} = x^{-1}$	$\dfrac{-1}{x^2} = -x^{-2}$
$\sqrt{x} = x^{1/2}$	$\dfrac{1}{2\sqrt{x}} = \dfrac{1}{2} x^{-1/2}$

These derivatives are all special cases of the following rule.

II. THE POWER RULE

Suppose n is any real number.

If $f(x) = x^n$, then $f'(x) = nx^{n-1}$, or equivalently,

$$\frac{d}{dx} f(x) = nx^{n-1}$$

Alternate formulations:

$$\frac{d}{dx} x^n = nx^{n-1}$$

If $y = x^n$, then $y' = nx^{n-1}$, or equivalently, $\dfrac{dy}{dx} = nx^{n-1}$.

The derivative of x^n is $n \cdot x^{n-1}$ ("the product of n times x raised to the power n − 1").

A justification of the Power Rule if n is a positive integer is fairly straightforward and is included in an appendix at the end of this section. Notice that if $f(x) = x$, then writing $x = x^1$ we have $n = 1$ and $f'(x) = 1 \cdot x^{1-1} = x^0 = 1$. This result agrees with the formula in table 2.5.1.

EXAMPLE 2   Power Rule

Find the derivative of each of the following:

(a) $f(x) = x^4$

(b) x^{16}

(c) x^{-2}

(d) $y = x^{1.43}$

(e) $y = \sqrt[3]{x}$

SOLUTION

(original power)

(a) If $f(x) = x^4$, then $f'(x) = 4x^3$

(original power reduced by 1)

(original power)

(b) $\dfrac{d}{dx} x^{16} = 16x^{16-1} = 16x^{15}$

(original power reduced by 1)

(c) $\dfrac{d}{dx} x^{-2} = -2x^{-2-1} = -2x^{-3}$

WARNING

Notice here in (c) that because the original power must go down by one, the new power is -3. A common mistake, and one very easy to make, is to have $(d/dx)\, x^{-2}$ incorrectly be $-2x^{-1}$. Be careful!

(d) If $y = x^{1.43}$, then $dy/dx = 1.43x^{1.43-1} = 1.43x^{0.43}$.

(e) The form $y = \sqrt[3]{x}$ is expressing y as a cube root. In order to use the Power Rule to find y' we first must rewrite the cube root as a fractional exponent. For $y = x^{1/3}$, the Power Rule gives $y' = (1/3)\, x^{1/3-1} = (1/3)\, x^{-2/3}$. This can be rewritten, using positive exponents, as $1/(3x^{2/3})$.

PRACTICE EXERCISE 1

Find the derivatives of each of the following:
(a) $f(x) = x^{14}$ (b) x^{-4} (c) $y = x^{3.8}$ (d) $x^{-1/3}$

Derivative of $f(x) = \sqrt{x}$

The Power Rule is so easy to apply that whenever we encounter roots and radicals, our usual procedure is to convert them into powers. For instance, in Example 2 we wrote $\sqrt[3]{x}$ as $x^{1/3}$ and took the derivative in that form. However, the function $f(x) = \sqrt{x}$ occurs so frequently that we may want to learn a special rule for its derivative. The situation is similar to the fact that we have a $\boxed{\sqrt{x}}$ key on our calculator, even though the $\boxed{x^y}$ key can be used to evaluate $\sqrt{x}$ by letting $y = 0.5$.

To determine the derivative of $f(x) = \sqrt{x}$ by the Power Rule we rewrite $\sqrt{x}$ as $x^{0.5}$ and then have $\dfrac{d}{dx} \sqrt{x} = \dfrac{d}{dx} x^{0.5} = 0.5x^{0.5-1} = 0.5x^{-0.5}$

Because $0.5x^{-0.5} = (1/2)(1/x^{0.5}) = 1/(2\sqrt{x})$, we find the derivative of $\sqrt{x}$ is given by the formula

$$\frac{d}{dx} \sqrt{x} = \frac{1}{2\sqrt{x}}$$

It makes no difference whether we write a square root in the decimal exponent form as $x^{0.5}$ or in the radical form as $\sqrt{x}$. The form we use will depend on what we need to do with it.

Constant Multiple Rule

The next rule covers the situation of differentiating a product of a constant number and a function of x.

III. CONSTANT MULTIPLE RULE

Suppose $g(x) = c \cdot f(x)$, where c is any real number, $f(x)$ is a function of x, and $f'(x)$ exists.

Then $g'(x) = cf'(x)$, or equivalently,

$$\frac{d}{dx} cf(x) = c \frac{d}{dx} f(x)$$

Alternate formulations:

$$\frac{d}{dx} [cf(x)] = c \frac{d}{dx} f(x)$$

If $y = cf(x)$ then $y' = cf'(x)$, or equivalently,

$$\frac{dy}{dx} = c \frac{d}{dx} f(x)$$

The derivative of a product of a constant and a function is the product of that constant multiplied by the derivative of the function.

The justification for this rule follows. Let the real number c, the function $f(x)$, and $g(x) = cf(x)$ be as given in the Constant Multiple Rule. The derivative of $g(x)$ is defined by the limit

$$\frac{d}{dx} g(x) = \lim_{h \to 0} \frac{g(x + h) - g(x)}{h}$$

Use the definition of $g(x) = cf(x)$ to get

$$\frac{d}{dx} g(x) = \lim_{h \to 0} \frac{cf(x + h) - cf(x)}{h}$$

$$= \lim_{h \to 0} \frac{c[f(x + h) - f(x)]}{h} \quad \text{by factoring } c \text{ from both terms in the numerator}$$

$$= c \lim_{h \to 0} \frac{f(x + h) - f(x)}{h} \quad \text{by using properties of limits}$$

$$= c \frac{d}{dx} f(x) \quad \text{by the definition of } \frac{d}{dx} f(x)$$

In Example 3, as well as other examples later in the section, a variety of notations for derivatives is used. This is done to help develop a comprehensive familiarity with different notations for derivatives.

EXAMPLE 3   Derivatives using the Constant Multiple Rule

Find derivatives of the following functions.
(a) $f(x) = 3x^4$ (b) $2x^{-1}$ (c) $f(x) = kx$ for any constant k
(d) $y = 5/\sqrt{x}$

SOLUTION

(a) If $f(x) = 3x^4$, then $f'(x) = 3 \cdot (d/dx) x^4 = 3 \cdot 4x^3 = 12x^3$.

(b) $(d/dx)2x^{-1} = 2 \cdot (d/dx) x^{-1} = 2 \cdot (-1 \cdot x^{-2}) = -2x^{-2}$.

(c) If k is any constant and $f(x) = kx$, then $f'(x) = k \cdot (d/dx)x = k \cdot 1 = k$.

(d) First rewrite y so we may use the Power Rule.

$$y = 5\left(\frac{1}{\sqrt{x}}\right) = 5\left(\frac{1}{x^{1/2}}\right) = 5 \cdot x^{-1/2}$$

By the Constant Multiple Rule

$$y' = \frac{d}{dx}(5 \cdot x^{-1/2}) = 5 \cdot \frac{d}{dx} x^{-1/2}$$

$$= 5 \cdot \left(-\frac{1}{2} x^{-3/2}\right) \text{ by the Power Rule}$$

$$= -\frac{5}{2} x^{-3/2}$$

PRACTICE EXERCISE 2

Find derivatives of the following functions.
(a) $f(x) = 7x^3$ (b) $3x^{-2}$ (c) $y = 4.5x^{1.2}$ (d) $5x^{-1/3}$

EXAMPLE 4   Application of the Constant Multiple Rule in forestry

For many coniferous trees, their height, H, is related to their diameter, x (as measured 3 feet above the ground level), by the formula

$H = 63x^{2/3}$.* For example, a spruce tree with a diameter of 2 feet is approximately $63(2)^{2/3} \simeq 100$ feet tall.

(a) Find H'.

(b) Evaluate H' when $x = 2$ and interpret your answer.

SOLUTION

(a)
$$H' = \frac{d}{dx} 63x^{2/3} = 63 \frac{d}{dx} x^{2/3} \text{ by the Constant Multiple Rule}$$

$$= 63 \left(\frac{2}{3} x^{-1/3} \right) \text{ by the Power Rule}$$

$$= 42x^{-1/3}$$

(b) At $x = 2$, we have $H'(2) = 42(2)^{-1/3} \simeq 42(0.7937) \simeq 33$.

This value means that when the diameter of the tree is 2 feet, then the height of the tree is increasing at a rate of about 33 feet per additional foot of diameter.

Derivatives of a Sum or Difference of Functions

The next rule states how to find the derivative of the sum or difference of two functions.

IV. DERIVATIVE OF A SUM OR DIFFERENCE OF FUNCTIONS

Suppose $f(x)$ and $g(x)$ are functions such that $f'(x)$ and $g'(x)$ exist.

$[f(x) \pm g(x)]' = f'(x) \pm g'(x)$, or equivalently,

$$\frac{d}{dx} [f(x) \pm g(x)] = \frac{d}{dx} f(x) \pm \frac{d}{dx} g(x)$$

Alternate formulations:

If $y = f(x) \pm g(x)$, then $\frac{dy}{dx} = y' = f'(x) \pm g'(x)$.

The derivative of the sum (or difference) of functions is the sum (or difference, respectively) of the derivatives of each of the functions.

*T. McMahon, "The Mechanical Design of Trees," *Scientific American* 233:1 (July 1975), pp. 93–103.

The following demonstrates the correctness of the rule for the addition of two functions.

Let $s(x) = f(x) + g(x)$. By the definition of $s'(x)$, we have

$$\frac{d}{dx} s(x) = \lim_{h \to 0} \frac{s(x + h) - s(x)}{h}$$

$$= \lim_{h \to 0} \frac{[f(x + h) + g(x + h)] - [f(x) + g(x)]}{h}$$

$$= \lim_{h \to 0} \frac{[f(x + h) - f(x)] + [g(x + h) - g(x)]}{h} \text{ by rearranging the numerator}$$

$$= \lim_{h \to 0} \frac{f(x + h) - f(x)}{h} + \lim_{h \to 0} \frac{g(x + h) - g(x)}{h}$$

since the limit of a sum is the sum of the limits

$$= \frac{d}{dx} f(x) + \frac{d}{dx} g(x) \text{ by the definition of these derivatives}$$

The corresponding result for the difference of two functions, $f(x) - g(x)$, can be shown similarly.

The rules we have listed so far enable us to find the derivative of any function that is created by raising the variable to powers, by multiplying powers by constants, and by adding or subtracting such expressions.

EXAMPLE 5   A combining of rules for differentiation

Find the derivative of $3x^2 - 2x + 5$.

SOLUTION

$$\frac{d}{dx} [3x^2 - 2x + 5]$$

$$= \frac{d}{dx} 3x^2 - \frac{d}{dx} 2x + \frac{d}{dx} 5 \text{ by Sum and Difference Rules}$$

$$= 3\frac{d}{dx} x^2 - 2\frac{d}{dx} x + \frac{d}{dx} 5 \text{ by Constant Multiple Rule}$$

$$= 3 \cdot 2x - 2 \cdot 1 + 0 \text{ by Power and Constant Rules}$$

$$= 6x - 2$$

We have been stating our rules for functions of the variable x. They are just as correct for functions of any variable because the name of the variable does not affect the function or its behavior. In the next example, we have a function of t, but its derivative is calculated in exactly the same manner as if it were a function of x.

EXAMPLE 6 Finding a derivative with respect to the variable t

Find the derivative with respect to t of $y = 16t^{3.4} + \dfrac{3}{t}$.

SOLUTION

First rewrite $3/t$ as $3t^{-1}$ so that we can later use the Power Rule. Then the derivative is

$$y' = \frac{d}{dt}[16t^{3.4} + 3t^{-1}]$$

$$= 16 \cdot \frac{d}{dt} t^{3.4} + 3 \cdot \frac{d}{dt} t^{-1} \quad \text{by the Sum and the Constant Multiple Rules}$$

$$= 16 \cdot 3.4 \cdot t^{2.4} + 3 \cdot (-1)t^{-2} \quad \text{by the Power Rule}$$

$$= 54.4t^{2.4} - 3t^{-2}$$

PRACTICE EXERCISE 3

Find the derivative with respect to t of $f(t) = 2t^4 - 3t + \dfrac{5}{t}$.

Example 6 and Practice Exercise 3 are the first instances of our using the phrase "the derivative with respect to." Previously, we had been writing the functions as functions of x and so the only variable being considered was x. This was deliberately done so that the statements of the rules for differentiation did not become awkwardly phrased. However, it is very important to realize that whenever any derivative is being taken, it is being taken with respect to some variable. It never hurts to state explicitly what the variable is, and in situations when there are several variables in a formula it is essential to give the variable to which we are taking the derivative.

EXAMPLE 7 Stopping distances for automobiles

Suppose a person is driving a car at a speed of v miles per hour and suddenly is given a signal to stop. The number of feet from the point at which the driver first sees the signal until the car is stopped is approximately $S(v) = 1.1v + 0.054v^2$.* For example, at 55 miles per hour, $S(55) = 1.1(55) + 0.054(55)^2 = 60.5 + 163.35 \approx 224$ ft.

*F. R. Giordano and Maurice D. Weir, *A First Course in Mathematical Modeling,* Monterey, Calif.: Brooks/Cole, 1985, p. 86.

(a) Find $S'(v)$.

(b) Find $S'(55)$ and interpret your answer.

SOLUTION

$$S'(v) = \frac{d}{dv}(1.1v) + \frac{d}{dv}(0.054v^2)$$

$$= 1.1 + 0.054\,\frac{d}{dv}\,v^2 = 1.1 + 0.054(2v)$$

$$= 1.1 + 0.108v$$

$S'(55) = 1.1 + 0.108(55) = 1.1 + 5.94 = 7.04 \approx 7$, which means that at 55 miles per hour, the additional number of feet required to stop is increasing at a rate of about 7 feet for each additional mile per hour of speed.

Remark on Notation.   The notation $f'(x)$ represents the derivative of $f(x)$, and $f'(a)$ represents $f'(x)$ evaluated at $x = a$. For instance, if $f(x) = 3x^2 + x - 8$, then $f'(x) = 6x + 1$ and $f'(2) = 6 \cdot 2 + 1 = 12 + 1 = 13$. If we use the dy/dx notation, we can indicate that the derivative is to be evaluated by drawing a vertical line and giving the evaluating value at the lower end of the line. Using this notation, if $y = f(x)$, then $f'(a)$ is written as $dy/dx\big|_{x=a}$. The vertical bar symbol is read as "the derivative, evaluated at $x = a$."

EXAMPLE 8   Notation for derivatives

For $y = 3x^2 + x - 8$, find $\dfrac{dy}{dx}\bigg|_{x=2}$.

SOLUTION

$\dfrac{d}{dx}[3x^2 + x - 8] = 3 \cdot 2x + 1 = 6x + 1$. Evaluating the derivative at $x = 2$ gives us $(6x + 1)\big|_{x=2} = 6 \cdot 2 + 1 = 13$.

PRACTICE EXERCISE 4

For $y = 2x^3 - 3x + 2$, find $\dfrac{dy}{dx}\bigg|_{x=2}$.

PRACTICE EXERCISE 5

For $f(x) = 3\sqrt{x} + 2x^{2.5}$, find $f'(4)$.

Solving $f'(x) = c$ for x

Given a function, $f(x)$, we may need to find values of x for which $f'(x)$ is a particular value. We will discover later that one of the commonest of these questions is: Determine all values of x such that $f'(x) = 0$.

EXAMPLE 9   Solving $f'(x) = c$

Suppose $f(x) = x^3 - x$.

(a) Find the value(s) of x for which $f'(x) = 2$.

(b) Find the value(s) of x for which $f'(x) = 0$.

SOLUTION

The derivative is $f'(x) = 3x^2 - 1$.

(a) If $3x^2 - 1 = 2$, then $3x^2 = 3$, and so $x^2 = 1$. Hence $x = \pm 1$.

(b) $3x^2 - 1 = 0$ if $x^2 = 1/3$. Solving for x we have $x = \pm \sqrt{1/3}$.

WARNING

When writing the derivative of $f(x)$, we often write $f'(x)$. However, when we have $f(x)$ given as a formula, such as $3x^2 + 1$, we cannot place a "prime" on $3x^2 + 1$ without the very real risk of confusion between $(3x^2 + 1)'$ and $(3x^2 + 1)^1$. We therefore usually write the derivative of $3x^2 + 1$ as $\dfrac{d}{dx}(3x^2 + 1)$.

Marginal Analysis

In Chapter 1 we introduced functions that described revenue, cost, profit, and demand. Using derivatives, we can now analyze the rates of change of these functions. Rates of change are of intense interest to economists and businesspeople. The economist uses the word *marginal* in referring to a rate of change. For instance, the *marginal cost* of one airplane passenger would be the additional expenses associated with that passenger (the additional fuel needed to carry that person's weight, the additional cost of food for that passenger, and so forth). When an airline sells tickets, it uses the techniques of *yield management* to determine how to price seats so as to yield a positive *marginal profit*.

EXAMPLE 10   Marginal analysis of cost

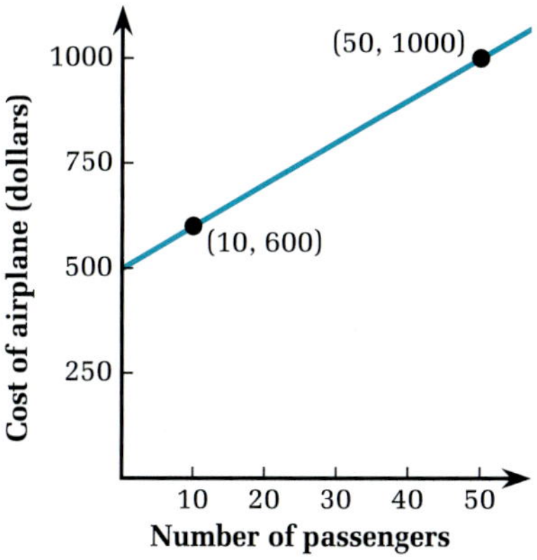

FIGURE 2.5.2

Suppose it costs an airline $500 to fly a plane from Phoenix to Mirage City. This is the expense of the crew, the fuel, depreciation, and so forth. If it costs an additional $10 for each passenger, then the *marginal cost* for each passenger is $10. For x passengers, the cost is given by the function $C(x) = 500 + 10x$. Its graph is shown in figure 2.5.2.

The derivative of $C(x) = 500 + 10x$ is

$$C'(x) = \frac{d}{dx}(500 + 10x)$$

$$= \frac{d}{dx}(500) + \frac{d}{dx}(10x)$$

$$= 0 + 10$$

$$= 10$$

The derivative of the cost function, $C(x)$, is the marginal cost function, $C'(x)$.

The cost function in this example is quite a simple function. It provides the airline with a linear model whereby each additional passenger adds the same amount to the total cost. Generally, the situation is more complicated and the cost function has to be quite a bit more complex.

Whenever and wherever an economist can model costs by a function that can be differentiated, then the *marginal cost function* is the derivative of that function. That is, if $C(x)$ is the cost of x units then $C'(x)$ is the marginal cost function.

The marginal cost function is used in two different ways. First, it is used to estimate the marginal cost of a particular item; second, it is used to determine the rate of change of cost and so analyze costs. In figure 2.5.3 we see an illustration of using the marginal cost function to estimate the marginal cost of a particular item.

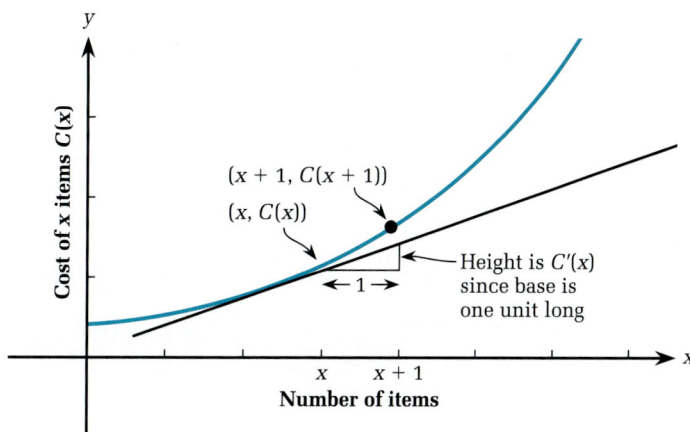

FIGURE 2.5.3

Because $C(x + 1)$ is the cost of $x + 1$ items, the actual marginal cost of the $(x + 1)$-st item is:

$$C(x + 1) - C(x)$$

If we use $C'(x)$, the marginal cost at x, to estimate the marginal cost of the $(x + 1)$-st item, then we are using a straight line to approximate the graph of $C(x)$ in the vicinity of $(x, C(x))$. We find the slope of this straight line by evaluating $C'(x)$ and constructing a triangle with base 1 and whose hypotenuse lies on the line. Notice that the correctness of this estimation depends on how much the graph of $y = C(x)$ has veered from the tangent line between x and $x + 1$.

The remarks we have made about marginal cost can also be applied to *marginal profit, marginal revenue,* and *marginal demand.*

EXAMPLE 11   Marginal analysis of cost

Suppose the cost of x tons of sandstone is $C(x) = 150x + 3x^2$ dollars.

(a) Find the marginal cost at $x = 4$.

(b) Use the marginal cost at $x = 4$ to estimate the marginal cost of the fifth ton.

(c) Find the actual marginal cost of the fifth ton, as measured beyond the fourth ton.

SOLUTION

(a) The marginal cost is the derivative of the cost function, so

$$C'(x) = 150 + 3 \cdot 2x = 150 + 6x$$
$$C'(4) = 150 + 6 \cdot 4 = 150 + 24 = 174$$

(b) The marginal cost of the fifth ton is approximately $C'(4) = 174$

(c) The actual marginal cost is

$$C(5) - C(4) = (150 \cdot 5 + 3 \cdot 5^2) - (150 \cdot 4 + 3 \cdot 4^2)$$
$$= (750 + 75) - (600 + 48)$$
$$= 825 - 648 = 177$$

In Example 12 we use marginals to analyze the behavior of an economic function.

EXAMPLE 12   Marginal analysis of revenue

Suppose a revenue function is given by $R(x) = 6x^2 - x^3$.

(a) Find the marginal revenue at $x = 2$ and $x = 5$ and interpret the values.

(b) When is the marginal revenue equal to 0?

SOLUTION

(a) Marginal revenue is the derivative of revenue, so

$$R'(x) = 6 \cdot 2x - 3x^2 = 12x - 3x^2$$
$$R'(2) = 12 \cdot 2 - 3 \cdot 2^2 = 24 - 3 \cdot 4 = 24 - 12 = 12$$
$$R'(5) = 12 \cdot 5 - 3 \cdot 5^2 = 60 - 3 \cdot 25 = 60 - 75 = -15$$

At $x = 2$, the revenue is increasing as x increases, but by $x = 5$ the revenue is decreasing as x increases.

(b) To determine when the marginal revenue is zero, we must solve $12x - 3x^2 = 0$.

Factor $3x$ out of $12x - 3x^2$ to get $3x(4 - x) = 0$. Because a product is zero only if a factor of the product is zero, the marginal revenue is zero at $x = 0$ and at $x = 4$. A graph of $R(x)$ is shown in figure 2.5.4.

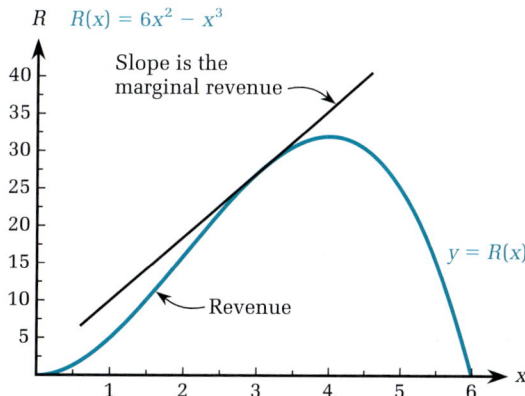

FIGURE 2.5.4

The next example concerns the *marginal utility of money.* Attempts have long been made to quantify the well-known idea that additional money is valued only in relation to the amount of money already possessed. For example, $5 is worth far less to a millionaire than to a college student. The function that describes the worth of money is known as a *utility function,* and one of the first investigations of such functions was published in 1738 by the famous Swiss mathematician Daniel Bernoulli (1700–1782).

EXAMPLE 13   Decreasing marginal utility of money, seriousness of theft

In determining the utility function for thefts of money, T. Sellin and M. E. Wolfgang found the utility function for each potential juror to be of the form $U(x) = k \cdot x^{0.17}$, where x is the number of dollars.* The value of k varies from person to person, but for each person, his or her own valuation of money followed this form. To illustrate the idea, suppose $k = 0.05$ for a particular person. This would mean that person would regard theft of $100 and $1000 as being: $U(100) = 0.05(100)^{0.17} = 0.1094$ and $U(1000) \approx 0.05(1000)^{0.17} \approx 0.1618$, respectively. In other words, it is not ten times as "bad" to steal $1000 as to steal $100. It is not even twice as "bad."

The derivative of $U(x)$ is the *marginal utility of money*. For $U(k) = kx^{0.17}$ we have $U'(x) = k \cdot 0.17x^{0.17-1} = 0.17kx^{-0.83} = 0.17 k/x^{0.83}$. For the individual value of $k = 0.05$ given above, we have $U'(x) = (0.17 \cdot 0.05)/x^{0.83} = 0.0085/x^{0.83}$. Evaluating this at 100 and at 1000, we find $U'(100) = 0.000186$ and $U'(1000) = 0.0000275$. Dividing $U'(100)$ by $U'(1000)$, a criminologist may conclude it is 6.76 times worse to steal an additional dollar at the $100 level than at the $1000 level.

*T. Sellin and M. E. Wolfgang, *The Measure of Delinquency*, New York: Wiley, 1964.

2.5 PROBLEMS

Foundations

The problems of this section require the basic skills illustrated by the following.

Rewrite problems 1–6 using single exponent notation.

1. $\sqrt{4x}$ **2.** $\sqrt[3]{8x}$ **3.** $\dfrac{2}{\sqrt{x}}$

4. $\dfrac{1}{x^3}$ **5.** $\dfrac{x^2}{\sqrt{x}}$ **6.** $\dfrac{3x^2}{x^{-3}}$

Solve problems 7 and 8 for x.

7. $9x^2 - 2 = 34$ **8.** $20 - 2x > 0$

Exercises

Find the derivative of each of the functions in problems 9–41.

9. $f(x) = 7$ **10.** $g(x) = -3.5$

11. $f(x) = \sqrt{5\pi}$ **12.** $f(x) = x^2$

13. $f(x) = x^{10}$ **14.** $g(x) = 5x^{28}$

15. $g(x) = 3x^7$ **16.** $f(x) = -7x^2$

17. $f(x) = -3x^8$ **18.** $y = \sqrt{x}$

19. $y = 2\sqrt{x}$ **20.** $y = x^{1/2}$

21. $y = x^{0.5}$ **22.** $f(x) = -2x^{0.5}$

23. $f(x) = -3x^{1/2}$ **24.** $f(x) = \dfrac{1}{x^2}$

25. $g(x) = \dfrac{3}{x^2}$ **26.** $f(w) = 4w^{3.2}$

27. $g(w) = 2w^{2.4}$ **28.** $g(u) = 5u^{-2.1}$

29. $f(u) = 2u^{-1.5}$

30. $f(x) = x^4 + 3x^2 - 2$

31. $f(x) = x^3 - 5x^2 + x - 5$

32. $g(x) = 5x^3 + x^2 - 3x$

33. $g(x) = 2x^4 + 6x^2 - 3x$

34. $g(x) = \dfrac{2}{\sqrt{x}} + 5x^3$ **35.** $f(x) = \dfrac{3}{\sqrt{x}} - 7x^4$

36. $y = \sqrt[3]{x^2} + \dfrac{2}{x^2}$ **37.** $y = \sqrt[3]{x} - \dfrac{5}{x^2}$

38. $h(u) = 5u^3 - \dfrac{3}{\sqrt{u^5}}$ **39.** $f(u) = \dfrac{2}{\sqrt{u^3}} - 4u^2$

40. $g(x) = 3x^{3.4} + x^{-4} + \dfrac{3}{x^{2.3}}$

41. $f(x) = -5x^{5.2} + \dfrac{4}{x^{-3}}$

Evaluate the derivatives as indicated in problems 42−48.

42. $f(x) = \dfrac{2}{x}, f'(3)$ **43.** $f(x) = 5x^2, f'(2)$

44. $g(t) = 4t^2 + t^{-2}, g'(2)$

45. $f(u) = 3\sqrt{u} + 2u^7, f'(1)$

46. $g(x) = 15x^{3/2}, g'(4)$

47. $y = 3x^3 - 2x, \dfrac{dy}{dx}\bigg|_{x=2}$

48. $y = x^2 + \sqrt[3]{x}, \dfrac{dy}{dx}\bigg|_{x=1}$

49. For $f(x) = 3x^3 - 2x$, find the value(s) of x for which $f'(x) = 34$.

50. For $f(x) = x^4 - 8x^2 + 32x$, find the value(s) of x for which $f'(x) = 32$.

51. For $g(x) = 3x^2 - 42x$, find the value(s) of x for which $g'(x) = 0$.

52. For $h(x) = 5x^2 - 20x$, find the value(s) of x for which $h'(x) = 0$.

53. *(Marginal Analysis of Cost)*   Suppose the cost of x items is $C(x) = 200 + 0.35x^2$.
   (a) Find the marginal cost at x.
   (b) Find the marginal cost at $x = 20$ and use it to estimate the marginal cost of the 21st item.
   (c) Compare your estimate in part (b) with the actual marginal cost of the 21st item.

54. *(Marginal Analysis of Cost)*   Suppose the cost of x items is $C(x) = 21x^2 - x$.
   (a) Find the marginal cost at x.
   (b) Find the marginal cost at $x = 15$ and use it to estimate the marginal cost of the 16th item.
   (c) Compare your estimate in part (b) with the actual marginal cost of the 16th item.

55. *(Marginal Analysis of Revenue)*   Suppose revenue is $R(x) = 5x^2 + 2x - 3$. Find the marginal revenue at x.

56. *(Marginal Analysis of Revenue)*   Suppose revenue is $R(x) = 3x^2 - 4x + 2$. Find the marginal revenue at x.

57. *(Utility of Money)*   Suppose a juror serving on a case involving financial embezzlement has a personal function for the utility of money, $U(x) = 0.05x^{0.17}$, where x represents millions of dollars and $U(x)$ is a "badness" value.

   (a) Find $U(1)$ and $U(1000)$ (Note: Because x represents millions, $x = 1000$ is one billion dollars).
   (b) Find the marginal utility at $x = 100$.

58. *(Total and Marginal Utility)*   The amount of satisfaction received by a consumer from consuming x units of a particular product is known as the *total utility, $U(x)$*. As x increases, $U'(x)$, the rate of change of $U(x)$, is known as the *marginal utility*. Suppose the graph in figure 2.5.5 represents the satisfaction gained from watching x movies a week. The total utility function for this graph is $U(x) = 20x - x^2$.

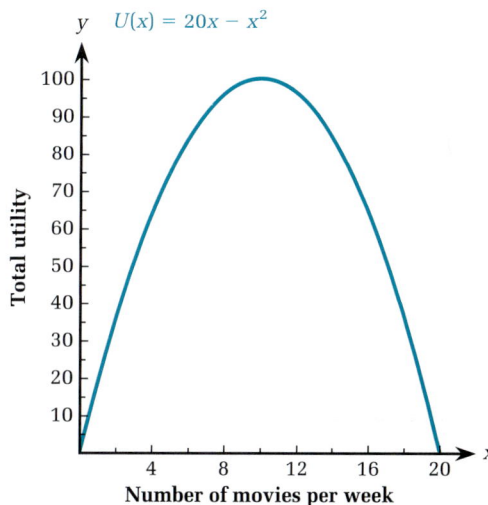

FIGURE 2.5.5

   (a) Find the marginal utility, $U'(x)$.
   (b) Suppose that at present five movies a week are being viewed. Would satisfaction be increased by viewing more or fewer movies? Why?
   (c) Suppose that at present 12 movies a week are being viewed. Would satisfaction be increased by viewing more or fewer movies? Why?
   (d) Why would a marginal utility of zero correspond to maximum satisfaction?
   (e) For what value of x is the marginal utility equal to zero?

59. *(Marginal Analysis of Revenue)*   The Batesville Burger revenue function is $R(x) = -0.00025x^2 + 1.44x$. Find the marginal revenue at $x = 2000$.

60. *(Vanity License Revenue)*   In Chapter 1 we had an example that calculated the revenue from the sale of vanity auto and motorcycle plates to be

$R(x) = 275,000 + 2800x - 40x^2$, where x is the price of each individual plate and $R(x)$ is the total revenue to the state. What is the marginal revenue at $x = 30$?

61. The formula for the area of a (circular) pizza is $A(d) = (\pi/4)\,d^2$, where d is the diameter in inches and $A(d)$ is the area in square inches.
   (a) Find $A'(d)$.
   (b) Evaluate $A'(14)$ and interpret your answer.

62. **(Medicine, Blood Alcohol Level)** Suppose Jackie's blood alcohol level after drinking two cans of beer is given by $L(t) = -0.012t^3 + 0.175t$, where t is in hours and $L(t)$ is in tenths of a percent of alcohol.
   (a) At what rate is the percent of blood alcohol changing after one hour?
   (b) At what point is the rate of change equal to zero?

63. **(Marginal Revenue)** The city would pay Cleanway Dump Co. a total of $200,000 + 5000\sqrt{t}$ dollars by the end of year t over the life of a 20-year contract. What is the marginal revenue for Cleanway at $t = 10$?

64. If an object is thrown upward to a height of h feet, then the time until the object returns to its starting place is approximately $T(h) = 0.5\sqrt{h}$ seconds.
   (a) Find $T'(h)$.
   (b) Evaluate $T'(100)$ and interpret your answer.

65. An object dropped from a height of x feet will be going approximately $v(x) = 5.45\sqrt{x}$ miles per hour when it hits the ground.
   (a) Find $v'(x)$.
   (b) Evaluate $v'(100)$ and interpret your answer.

66. An object thrown straight up in the air with an initial velocity of v_0 will attain a maximum height given by $H(v_0) = v_0^2/64$, with v_0 measured in feet per second and $H(v_0)$ being measured in feet.
   (a) Find $H'(v_0)$.
   (b) Evaluate $H'(88)$ and interpret your answer.

67. **(Froude's Law in Fishery Management and Boatbuilding)** In the study of fish, or in the design of a ship, the maximum velocity, V_{max}, is proportional to the square root of its length, l. In symbols, $V_{max} = k\sqrt{l}$, where k is a constant that depends on the object.
   (a) Find V'_{max}.
   (b) Compare the rate that the maximum veloc-

ity is increasing for a shark 4 feet long with the rate the maximum velocity is increasing for a shark 16 feet long.

68. **(Forestry)** In order to decide when to cut a section of trees, foresters want to know the average diameter of the trees. For coniferous trees (such as spruce) the diameter, measured 3 feet above the ground, is approximately $D(x) = 0.002x^{3/2}$ feet for a tree that is x feet tall.*
   (a) Find $D'(x)$.
   (b) Evaluate $D'(100)$ and interpret your answer.

69. **(Biology, Power of Bird Flight)** Adults of the bird known as the "laughing gull" (*Larus atricilla*) weigh about 0.32 kilograms (about 0.7 pound). The power input of the laughing gull has been experimentally determined to be expressible as $P(v) = 0.090v^2 - 1.29v + 23.9$ watts, where v is measured in meters per second.†
   (a) Find $P'(v)$.
   (b) Find $P'(10)$ and interpret your answer.

70. **(Biology, Power of Bird Flight)** Refer to problem 69. For a laughing gull flying 10.8 meters per second, the power input required has been experimentally determined to be approximately $P(m) = 36.4m^{0.325}$ watts, where m is the mass of the bird, measured in grams.‡
   (a) Find $\dfrac{d}{dm}P$.
   (b) Find $\dfrac{d}{dm}P\big|_{x=320}$ and interpret your answer.

71. **(Biology, Bird Wingspan)** It has been determined experimentally that the wingspan of a bird with a mass of m kilograms is approximately $w = 1.1(m)^{1/3}$ meters. Except for hummingbirds, this formula holds over the range 0.004 kilograms to 10 kilograms.§
   (a) Find dw/dm.
   (b) Evaluate $dw/dm\big|_{x=0.5}$ and interpret your answer.

72. **(Social Sciences, Traveling to Work)** Suppose y is the average distance (one way) a person travels to work. It has been determined experimentally that the average distance for y can be

---

*Thomas A. McMahon, "The Mechanical Design of Trees," *Scientific American,* 233:1 (July 1975), pp. 93–103.
†V. A. Tucker, "Bird Metabolism During Flight: Evaluation of a Theory," *Journal of Experimental Biology* 58 (1973), pp. 689–709.
‡Ibid.
§Ibid.

approximated by the formula $y = 0.85\sqrt{A}$, wherein A is the area of the city in square miles and y is measured in miles.*
(a) Find y'.
(b) Evaluate y' for a city of area 100 square miles and interpret your answer.

73. **(Social Science, Footracing)**   A function (called a "frontier function") that approximates the absolute fastest time for a woman to run five kilometers is $f(x) = 0.0049x^2 - 0.2629x + 18.2708 + 17.1793/x$ minutes.†
(a) Find $f'(x)$.
(b) Evaluate $f'(20)$ and interpret your result.

Graphing Calculator Problems

74. Graph $f(x) = x^n$ over $-1 \le x \le 2$ for $n = 0.5, 1, 2,$ and 3.

75. Graph $f_1(x) = -x^2$, $f_2(x) = 6\sqrt{x}$, and $f_3(x) = -x^2 + 6\sqrt{x}$.
(a) Estimate the slopes of the tangent lines for each of these graphs for $x = 1$.
(b) Estimate a solution of $f_3'(x) = 2$.
(c) Estimate a solution of $f_3'(x) = 1$.

---

*R. J. Smeed, *Journal of Transport and Policy II* 1 (1968).
†J. Camm and T. Grogan, "An Application of Frontier Analysis: Handicapping Running Races," *Interfaces* 18:6 (Nov.–Dec. 1988), pp. 52–60.

Writing and Discussion Problems

76. How is $f'(a)$ different from $(f(a))'$? Illustrate your discussion by using $f(x) = x^2$ and $a = 3$. Show that $f'(3) \ne (f(3))'$.

77. Is it worthwhile to learn a separate rule for the derivative of $f(x) = \sqrt{x}$ rather than always rewriting $\sqrt{x}$ as $x^{0.5}$ and using the Power Rule? How can the $\boxed{1/x}$ key on your calculator be used to solve problems involving division? Are there other keys that may be useful, but are not necessary? Why would they be included?

Enrichment Problems

Find derivatives for the functions in problems 78–83. (Note: Each function may need to be rewritten in another form before we can use our rules.)

78. $h(x) = (2x)^2$ 79. $f(x) = (3x)^2$

80. $f(x) = (2x + 1)^2$ 81. $f(x) = (x - 2)^2$

82. $g(x) = \sqrt{3x}$ 83. $f(x) = \sqrt{2x}$

84. If we know a cost function, $C(x)$, and a revenue function, $R(x)$, what relationship between the marginal cost and the marginal revenue would cause us to recommend an increase in x? Why?

85. Refer to problem 63. What is the number of years during which Cleanway continues to receive at least an additional \$600 each year?

SOLUTIONS TO PRACTICE EXERCISES

---

1. Find the derivatives of each of the following.

(a) $\dfrac{d}{dx} x^{14} = 14x^{13}$ (b) $\dfrac{d}{dx} x^{-4} = -4x^{-5}$

(c) $\dfrac{d}{dx} x^{3.8} = 3.8x^{2.8}$

(d) $\dfrac{d}{dx} x^{-1/3} = -\dfrac{1}{3} x^{-4/3}$

2. Find derivatives of the following.

(a) $\dfrac{d}{dx} 7x^3 = 21x^2$ (b) $\dfrac{d}{dx} 3x^{-2} = -6x^{-3}$

(c) $\dfrac{d}{dx} 4.5x^{1.2} = 5.4x^{0.2}$

(d) $\dfrac{d}{dx} 5x^{-1/3} = 5 \cdot \dfrac{d}{dx} x^{-1/3} =$
$5 \cdot \left(-\dfrac{1}{3} \cdot x^{-4/3}\right) = -\dfrac{5}{3} x^{-4/3}$

3. To find the derivative of $2t^4 - 3t + 5/t$, change $5/t$ to $5t^{-1}$. Hence, $(d/dt)(2t^4 - 3t + 5t^{-1}) = 8t^3 - 3 - 5t^{-2} = 8t^3 - 3 - 5/t^2$. Note that we have to write d/dt rather than d/dx because the derivative is being taken with respect to t.

4. For $y = 2x^3 - 3x + 2$, $dy/dx\big|_{x=2} = (6x^2 - 3)\big|_{x=2} = 21$.

5. For $f(x) = 3\sqrt{x} + 2x^{2.5}$, to find $f'(4)$ first write $\sqrt{x}$ as $x^{0.5}$ so that $f'(x) = 3 \cdot (0.5)x^{-0.5} + 5x^{1.5}$, giving us $f'(4) = 1.5 \cdot (4)^{-0.5} + 5 \cdot (4)^{1.5} = 1.5 \cdot 0.5 + 5 \cdot 8 = 0.75 + 40 = 40.75$.

Appendix: A Partial Verification of the Power Rule

This appendix shows that $\dfrac{d}{dx}x^n = n \cdot x^{n-1}$ for n, a positive integer.

By definition $\dfrac{d}{dx}x^n = \lim\limits_{h \to 0} \dfrac{(x+h)^n - x^n}{h}$. Consider the following powers of $(x + h)$:

$$(x + h)^2 = x^2 + 2xh + h^2$$
$$(x + h)^3 = x^3 + 3x^2h + 3xh^2 + h^3$$

To find $(x + h)^{n+1}$ for any integer power n, multiply $(x + h)^n$ by one more $(x + h)$ term. For instance,

$$(x + h)^4 = (x + h) \cdot (x + h)^3 = (x + h) \cdot (x^3 + 3x^2h + 3xh^2 + h^3)$$
$$= x^4 + 4x^3h + 6x^2h^2 + 4xh^3 + h^4$$

Notice that each time $(x + h)^n$ has x^n as its first term, $n \cdot x^{n-1} \cdot h$ is its second term, and all the other terms have h to a power 2 or greater. Thus, we can factor out h^2 from all the terms from the third term on. For example, $(x + h)^4 = x^4 + 4x^3h + h^2(6x^2 + 4xh + h^2)$.

Hence, when n is any positive integer, $(x + h)^n = x^n + n \cdot x^{n-1} \cdot h + h^2 \cdot P(x)$ for some polynomial, $P(x)$. Using this result in our definition of the derivative, we have,

$$\lim_{h \to 0} \frac{(x+h)^n - x^n}{h}$$
$$= \lim_{h \to 0} \frac{[x^n + n \cdot x^{n-1} \cdot h + h^2 \cdot P(x)] - x^n}{h}$$
$$= \lim_{h \to 0} \frac{n \cdot x^{n-1} \cdot h + h^2 \cdot P(x)}{h} \quad \text{by canceling out } x^n \text{ terms}$$
$$= \lim_{h \to 0} \frac{h(n \cdot x^{n-1} + h \cdot P(x))}{h} \quad \text{by factoring out one } h$$
$$= \lim_{h \to 0} \left(nx^{n-1} + h \cdot P(x)\right) \text{ by canceling } h$$
$$= nx^{n-1}$$

In the last step we have used the fact that for any polynomial $P(x)$, as $h \to 0$, it is true that $h \cdot P(x) \to 0$.

2.6 DIFFERENTIATION RULES FOR PRODUCTS AND QUOTIENTS

The derivative of a sum of functions is simply the sum of the derivatives of each function. That is, $(f(x) + g(x))' = f'(x) + g'(x)$. It would be very convenient for us if the derivative of a product of two functions was similarly always the product of the derivative of each function. However, that is *not* true, as we can show by using $y = x^7$.

For $y = x^7$ we know from the Power Rule that $y' = 7x^6$. Write x^7 as a product. There are many ways to do this, but let us use $x^7 = x^5 x^2$. The derivative of x^5 is $5x^4$ and the derivative of x^2 is $2x$. The product of these derivatives is $5x^4 \cdot 2x = 10x^5$ rather than the correct derivative of x^7, $7x^6$.

Thus, in finding $(f(x) \cdot g(x))'$, in general $(f(x) \cdot g(x))' \neq f'(x) \cdot g'(x)$. Nor is it true that the derivative of a quotient is the quotient of derivatives. Unfortunately, the correct rules are more complicated.

The Derivative of a Product

The correct rule for the derivative of a product is given by the Product Rule, whose justification is included in the appendix to this section.

V. PRODUCT RULE FOR DERIVATIVES

Suppose $f(x)$ and $g(x)$ are functions such that $f'(x)$ and $g'(x)$ exist. Then $[f(x) \cdot g(x)]' = f(x) \cdot g'(x) + g(x) \cdot f'(x)$, or equivalently,

$$\frac{d}{dx}[f(x) \cdot g(x)] = f(x) \cdot \left(\frac{d}{dx}g(x) + g(x) \cdot \left(\frac{d}{dx}f(x)\right)\right)$$

Alternate formulations:

If $y = f(x) \cdot g(x)$, then $y' = f(x) \cdot g'(x) + g(x) \cdot f'(x)$.

The derivative of a product is the sum of the product of the first multiplied by the derivative of the second, added to the product of the second multiplied by the derivative of the first.

If we abbreviate $f(x)$ by f and $g(x)$ by g, we can express the product rule as:

PRODUCT RULE

$$(fg)' = fg' + gf'$$

Let us apply this rule to the example from the opening of the section.

EXAMPLE 1   Product Rule for derivatives

Find the derivative of x^7, written as the product $x^5 \cdot x^2$.

SOLUTION

Let $f(x) = x^5$ and $g(x) = x^2$. Then $f'(x) = 5x^4$ and $g'(x) = 2x$.

By the Product Rule we have $\dfrac{d}{dx}(x^7)$

$$= \frac{d}{dx}(x^5 \cdot x^2) = x^5 \cdot \frac{d}{dx}(x^2) + x^2 \cdot \frac{d}{dx}(x^5)$$

$$= x^5 \cdot 2x + x^2 \cdot 5x^4 = 2x^6 + 5x^6 = 7x^6$$

Before considering another example, let us notice that our discussion about the derivative of $y = x^7$ illustrates a good strategy for remembering (and learning) formulas.

> **If we are not certain we correctly remember a particular "rule," we should try it with a special case for which we are absolutely certain of the answer.**

This strategy is not infallible, for there is always the possibility that wrong methods yield correct answers in unusual situations (see the Writing and Discussion problems at the end of this section for an outlandish example). However, wrong methods usually yield wrong answers.

EXAMPLE 2   Product Rule

Find the derivative of $y = (3x^3 - 20x)(x^2)$.

SOLUTION

$$\frac{d}{dx}\left((3x^3 - 20x)(x^2)\right)$$

$$= (3x^3 - 20x) \cdot \frac{d}{dx}(x^2) + x^2 \cdot \left(\frac{d}{dx}(3x^3 - 20x)\right)$$

$$= (3x^3 - 20x) \cdot 2x + x^2 \cdot (3 \cdot 3x^2 - 20)$$

$$= 6x^4 - 40x^2 + 9x^4 - 20x^2$$

$$= 15x^4 - 60x^2$$

EXAMPLE 3   Product Rule

For $f(x) = 3x^{2/3}(x^{-2} + x^3)$, find $f'(x)$.

SOLUTION

By the Product Rule,

$$f'(x) = (3x^{2/3}) \cdot \frac{d}{dx}(x^{-2} + x^3) + (x^{-2} + x^3) \cdot \frac{d}{dx}(3x^{2/3})$$

(In finding the derivative of x^{-2} remember that if we subtract 1 from -2 we have -3.)

$$= (3x^{2/3}) \cdot (-2x^{-3} + 3x^2) + (x^{-2} + x^3) \cdot \frac{2}{3} \cdot 3x^{-1/3}$$

$$= -6x^{-7/3} + 9x^{8/3} + 2x^{-7/3} + 2x^{8/3}$$

$$= -4x^{-7/3} + 11x^{8/3}$$

PRACTICE EXERCISE 1

Find $\dfrac{d}{dx}((x^3 - x^2)(3x^2 + 2x))$.

Let us check the answer in Example 2 by expressing $(3x^3 - 20x)(x^2)$ as $3x^5 - 20x^3$. Then

$$\frac{d}{dx}(3x^5 - 20x^3) = 3 \cdot 5x^4 - 20 \cdot 3x^2 = 15x^4 - 60x^2$$

This is the same derivative as we obtained by applying the Product Rule. The fact that there are two different paths to the answer gives us an opportunity to check our work. We often find there are alternate approaches to solving a problem. Although no one wants to do needless work, the following advice may help us avoid errors.

> **If possible, try to solve problems by using different methods. If you reach different answers, then reexamine your work.**

An answer often must be simplified before it can be checked or is useful. Unfortunately, there is no general rule about which is the best or more simplified form of an answer. What a "simplified form" is may depend on our future plans for using the answer.

Consider the example we just did. If we want to evaluate the derivative at some value for x, then the form $15x^4 - 60x^2$ may be the preferred ("simplified") form. For instance, if we want to evaluate the derivative at $x = 1$ we may proceed as follows:

$$15x^4 - 60x^2 \Big|_{x=1} = 15(1)^4 - 60(1)^2 = 15 - 60 = -45$$

On the other hand, suppose we want to know when the derivative is equal to zero. To set a sum or a difference equal to zero is not nearly as useful as setting a product equal to zero. This is because the only way a product can be zero is for a factor of that product to be zero. Thus, if we are solving the equation $y' = 0$, the preferred form of the function would be $15x^2(x^2 - 4)$ because in setting that equal to zero, we find that $x^2 = 0$ or $x^2 - 4 = 0$, which yields x-values of 0 and ± 2, respectively. Thus, $y' = 0$ if $x = 0$ or $x = \pm 2$.

Applications

The next example applies the Product Rule for Derivatives to determining revenue, a situation we have encountered before.

EXAMPLE 4   The Product Rule applied to a revenue function

Suppose the demand function for sales of dinner show tickets is $p(x) = -(1/6000)\,x^2 + 20$.

(a) Find the revenue function $R(x)$.

(b) Evaluate $R'(120)$.

(c) Find values of x for which $R'(x) = 0$.

SOLUTION

(a) The revenue function is given by the product of the number of tickets sold times the price (demand function) of each ticket. That is, $R(x) = x \cdot p(x) = x \cdot [-(1/6000)\,x^2 + 20]$. Using the Product Rule on $x \cdot p(x)$ gives

$$R'(x) = x \cdot p'(x) + p(x) \cdot 1$$

$$= x \left(-\frac{2}{6000}\,x \right) + \left(-\frac{1}{6000}\,x^2 + 20 \right) \cdot 1$$

$$= -\frac{2}{6000}\,x^2 - \frac{1}{6000}\,x^2 + 20$$

$$= -\frac{3}{6000}\,x^2 + 20 = -\frac{1}{2000}\,x^2 + 20$$

(b) $R'(120) = -\dfrac{1}{2000}(120)^2 + 20 = -\dfrac{14,400}{2000} + 20$

$\qquad\qquad = -7.2 + 20 = 12.8$

(c) $R'(x) = 0$ if $-(1/2000)\,x^2 + 20 = 0$. This occurs when $(1/2000)\,x^2 = 20$. Hence, $x^2 = 40,000$ and therefore $x = 200$. ●

The Derivative of a Quotient

Suppose y is the quotient of two functions $y = [f(x)]/[g(x)]$ and we want to find y'. To show that the quotient formula is not simply the quotient of the derivatives, suppose we write x^3 as the quotient x^7/x^4. By the Power Rule, $\dfrac{d}{dx}x^3 = 3x^2$. However, $\dfrac{d}{dx}x^7 = 7x^6$, $\dfrac{d}{dx}x^4 = 4x^3$ and their quotient is $(7x^6)/(4x^3) = (7/4)\,x^3 \neq 3x^2$.

Following is the Quotient Rule for Derivatives. One of the problems in the Enrichment Problem section at the end of this section outlines a justification of this rule.

VI. QUOTIENT RULE FOR DERIVATIVES

Suppose $f(x)$ and $g(x)$ are functions such that $f'(x)$ and $g'(x)$ both exist and $g(x) \neq 0$. Then

$$\left(\frac{f(x)}{g(x)}\right)' = \frac{g(x) \cdot f'(x) - f(x) \cdot g'(x)}{(g(x))^2}$$

or, equivalently

$$\frac{d}{dx}\left(\frac{f(x)}{g(x)}\right) = \frac{g(x) \cdot \dfrac{d}{dx}f(x) - f(x) \cdot \left(\dfrac{d}{dx}g(x)\right)}{(g(x))^2}$$

In words:

The derivative of a quotient is a quotient, whose denominator is the square of the original denominator and whose numerator is a difference consisting of the product of the original denominator times the derivative of the original numerator minus a second product consisting of the original numerator times the derivative of the original denominator.

Abbreviating $f(x)$ by f and $g(x)$ by g, we can write the equation for the derivative of a quotient as:

QUOTIENT RULE

$$\left(\frac{f}{g}\right)' = \frac{gf' - fg'}{g^2}$$

The rule for quotients is more complicated than the rule for products. For products, if we make a mistake on which one of the functions to differentiate first, it makes no difference in the final answer because $gf' + fg' = fg' + gf'$. For quotients, however, the subtraction in the numerator causes the order of differentiation to become crucial.

The following example uses the Quotient Rule to find the derivative of x^7/x^4. Of course, if we truly were seeking this derivative in the most straightforward manner, we would write the function as x^3. However, we use a quotient form here only to illustrate the rule.

EXAMPLE 5   Quotient Rule

Find the derivative of $\dfrac{x^7}{x^4}$.

SOLUTION

Let $f(x) = x^7$ and $g(x) = x^4$.

$$\frac{d}{dx}\left(\frac{x^7}{x^4}\right) = \frac{x^4 \cdot 7x^6 - x^7 \cdot 4x^3}{(x^4)^2} = \frac{7x^{10} - 4x^{10}}{x^8} = \frac{3x^{10}}{x^8} = 3x^2,$$

which is truly the correct derivative because $\dfrac{d}{dx}x^3 = 3x^2$.

Let us use the Quotient Rule to find the derivative of a more complicated expression.

EXAMPLE 6   Quotient Rule

Find $\dfrac{d}{dx}\left(\dfrac{1 + x^2}{2x + x^3}\right)$.

SOLUTION

$$\frac{d}{dx}\left(\frac{1+x^2}{2x+x^3}\right) = \frac{(2x+x^3)\cdot 2x - (1+x^2)\cdot(2+3x^2)}{(2x+x^3)^2}$$

$$= \frac{(4x^2+2x^4)-(2+3x^2+2x^2+3x^4)}{(2x+x^3)^2}$$

$$= \frac{-x^4-x^2-2}{(2x+x^3)^2}$$

PRACTICE EXERCISE 2

Find $\dfrac{d}{dx}\left(\dfrac{2x+1}{x^2+3x}\right)$.

Let us reconsider an example we first encountered in Section 1.6.

EXAMPLE 7   Biology, application of the Quotient Rule to "diminishing returns"

Suppose the function $f(x) = (5x+1)/(100-x)$ for $0 < x < 100$ gives the number of minutes a bird searches in order to collect x grams of all the seeds in one section of the field. For example, to collect 80 grams the bird must search for $f(80) = (5\cdot 80 + 1)/(100-80) = (400+1)/(100-80) = 401/20 = 20.05$ minutes.

(a) What is the instantaneous rate of time per gram of the field being totally collected?

(b) Evaluate the instantaneous rate at 80 grams and 95 grams and give an interpretation. See figure 2.6.1 for a graph illustrating the situation.

SOLUTION

(a) The derivative, which gives the instantaneous rate, is

$$f'(x) = \frac{(100-x)\cdot 5 - (5x+1)\cdot(-1)}{(100-x)^2} = \frac{501}{(100-x)^2}$$

(b) $f'(80) = 501/(100-80)^2 = 501/20^2 = 501/400 = 1.2525$, and $f'(95) = 501/5^2 = 501/25 = 20.04$.

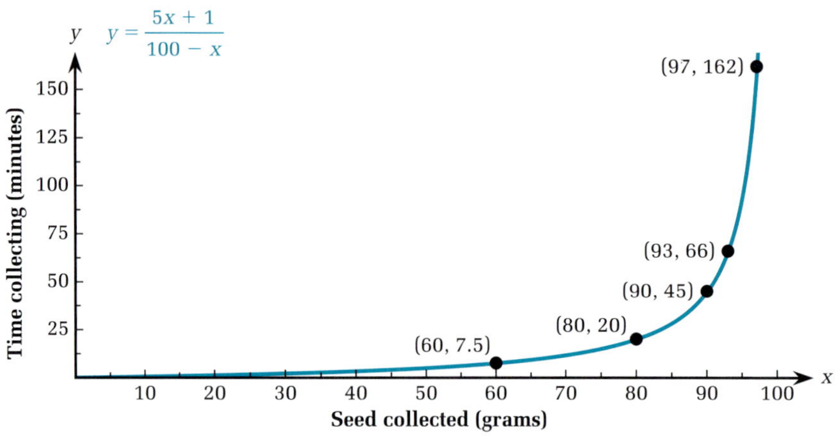

$$y = \frac{5x + 1}{100 - x}$$

FIGURE 2.6.1

Hence, if the bird has collected 80 grams of the seed, then to collect an additional gram will take 1.2525 minutes. If 95 grams have been collected, an additional gram will require 20.04 minutes.

Note that it is unreasonable to give times with such meaningless precision. It is better to use approximate values such as 1.2 minutes at 80 grams and 20 minutes at 95 grams.

Averages

A natural application of quotients occurs when we find averages. An average is defined as some total quantity divided by some measure of the number of units involved. For example, managers need to know the total cost of producing x units, but they also need to know the *average cost per unit*. To find this, they divide the total cost of production by the quantity produced.

As a particular example, if 5000 cassette players are produced at a total cost of \$375,000, the average cost per player is \$375,000/5000 = \$75.

DEFINITION

Average cost, revenue, and profit

If $C(x)$ gives the total cost of producing x units, the *average cost* function is

$$AC(x) = \frac{C(x)}{x}$$

and gives the average cost per unit. *Average revenue* and *average profit* functions are defined similarly.

EXAMPLE 8   Average cost, truck rentals

Suppose a rental truck costs \$59 plus 30 cents per mile. What is the average cost per mile? How is the average cost affected by the number of miles?

SOLUTION

Let x stand for the number of miles and $C(x)$ be the total cost. Since \$0.30x is the cost per mile we have $C(x) = 59 + 0.30x$. Hence $AC(x) = (59 + 0.30x)/x$. The larger x becomes, the smaller is $AC(x)$. For example, for 10 and 200 miles we have $AC(10) = \$6.20$ (per mile) and $AC(200) = \$0.595$ (per mile), respectively. The average cost decreases as the number of miles increases.

We may wish to determine the *rate* of decrease of an average. Because derivatives yield instantaneous rates, and averages are quotients, we will need to find the derivative of a quotient.

Recall that we have used the word *marginal* to refer to the derivative of several business concepts, such as revenue and profit. We now will define *marginal averages.*

DEFINITION   Marginal average cost

The *marginal average cost (MAC)* function is the derivative of the average cost (AC) function.

In symbols,

$$MAC(x) = \frac{d}{dx} AC(x) = \frac{d}{dx} \left(\frac{C(x)}{x} \right).$$

When we use the rule for the derivative of a quotient, we have $MAC(x) = [xC'(x) - C(x)]/x^2$.

This formula for $MAC(x)$ is an excellent example of a formula we should *not* memorize. What we *should* do is learn the concept of "average" and know the rule for the derivative of a quotient.

EXAMPLE 9   Marginal average cost

What is the $MAC(x)$ for $C(x) = 59 + 0.30x$ (the average cost function from Example 8)?

SOLUTION

$$AC(x) = \frac{59 + 0.30x}{x}$$

$$MAC(x) = \frac{d}{dx}\,AC(x) = \frac{d}{dx}\left(\frac{C(x)}{x}\right) = \frac{d}{dx}\left(\frac{59 + 0.30x}{x}\right)$$

$$= \frac{x(0.30) - (59 + 0.30x)(1)}{x^2}$$

$$= \frac{0.3x - 59 - 0.3x}{x^2} = \frac{-59}{x^2}$$

After x miles, the average cost per mile is changing by $-59/x^2$ dollars (or *decreasing* by $59/x^2$ dollars). The larger x is, the smaller is the change in average cost per mile.

Checking our work by using a different method of finding the derivative in Example 9, we could have simplified the average cost function, $AC(x)$, to be $AC(x) = 59x^{-1} + 0.3$. Then $MAC(x) = -59x^{-2}$ by the Power Rule.

In many applications, not all of the items that are manufactured or bought are actually salable or usable, so we need to determine an *adjusted average cost*.

DEFINITION

Adjusted average cost

The *adjusted average cost* is the average cost of those items that are salable or usable.

EXAMPLE 10   Application, adjusted average cost

Suppose the cost of inflating x balloons is $C(x) = 500 + 2x$ but we know that the first three of the balloons are tests and unsalable.

(a) Find the average cost of producing x salable balloons.

(b) Find the marginal average cost of producing x salable balloons.

SOLUTION

(a) If we inflate x balloons, then $x - 3$ are salable. The adjusted average cost is then $AAC(x) = [C(x)]/(x - 3) = (500 + 2x)/(x - 3)$.

(b) The marginal adjusted average cost is the derivative of the adjusted average cost, so

$$MAAC(x) = (AAC(x))' = \left(\frac{500 + 2x}{x - 3} \right)'$$

$$= \frac{(x - 3) \cdot 2 - (500 + 2x) \cdot 1}{(x - 3)^2}$$

$$= \frac{2x - 6 - 500 - 2x}{(x - 3)^2} = \frac{-506}{(x - 3)^2}$$

The next example considers harvesting a forest planting. At this stage in our development of the topic, we will not take into consideration the fact that money payable in the future is less valuable than money payable today. In Chapter 4 we will be able to add that level of complexity and give a more accurate analysis of the following situation.

EXAMPLE 11   Maximize average value, forest harvest

Suppose a forest has been planted and its value after t years is given by $V(t)$, as shown in figure 2.6.2. Find the best time to harvest the forest in order to produce the greatest average revenue.

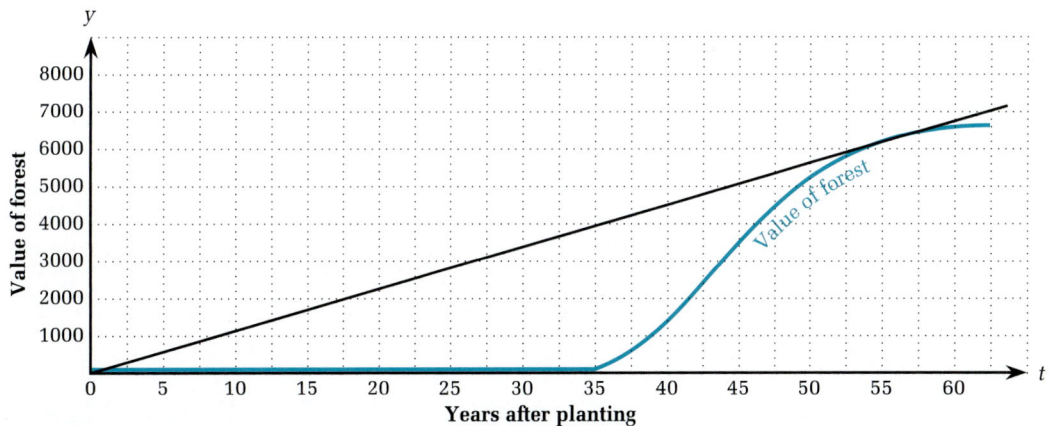

FIGURE 2.6.2

SOLUTION

We are going to show that the best time to harvest the trees is the time when a tangent line to the curve in figure 2.6.2 at that time passes through the origin of the graph.

After t years the value of the forest is $V(t)$, so if we harvested then we would have earned an average return of $F(t) = [V(t)]/t$ per year over that period of time. A graph of $F(t)$ is shown in figure 2.6.3.

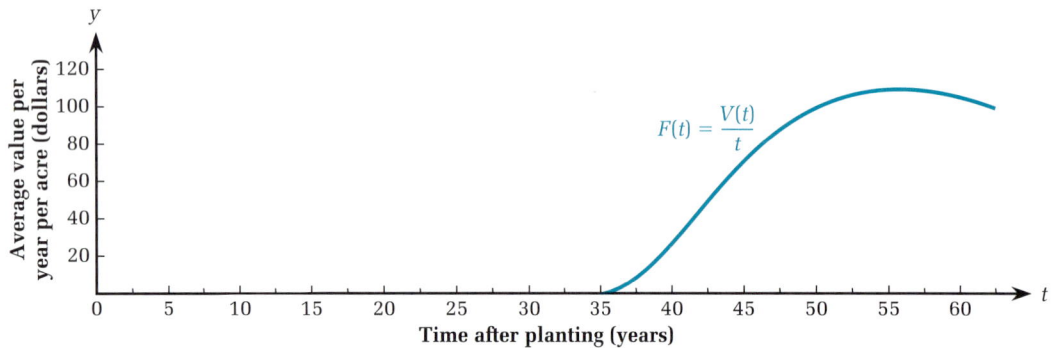

FIGURE 2.6.3

We are interested in the largest possible value for $F(t)$. This would be when the tangent line to the graph of $F(t)$ has a slope of zero.

By the Quotient Rule, $F'(t) = [t \cdot V'(t) - V(t) \cdot 1]/t^2$, and this is zero only when the numerator is zero. Thus, we need to have $t \cdot V'(t) - V(t) = 0$. This is equivalent to solving $V'(t) = [V(t)]/t$.

Notice the slope of the line from the point $P(t, V(t))$ to the origin $O(0, 0)$ is $[V(t) - 0]/(t - 0) = [V(t)]/t$. Thus, when the tangent line to the graph of $V(t)$, whose slope is given by $V'(t)$, passes through the origin, then we have the greatest average revenue from the tree planting.

None of our analysis depends on our actually having an algebraic description of the function $V(t)$, and, in fact, foresters rarely possess such an algebraic formula. Nonetheless, we have shown by calculus that if we are given a value curve, no matter how it was created, then a simple geometric technique can be used to find the optimal time to maximize our average revenue.

Let us highlight the result as follows:

MAXIMIZING AVERAGE VALUE, USING A GRAPHICAL TECHNIQUE

Suppose $V(t)$ is the value of an asset, which over time has an initial increase and then levels off or even decreases. Then the maximum average value occurs at the time t for which the line tangent to the graph of $y = V(t)$ in figure 2.6.4 passes through the origin.

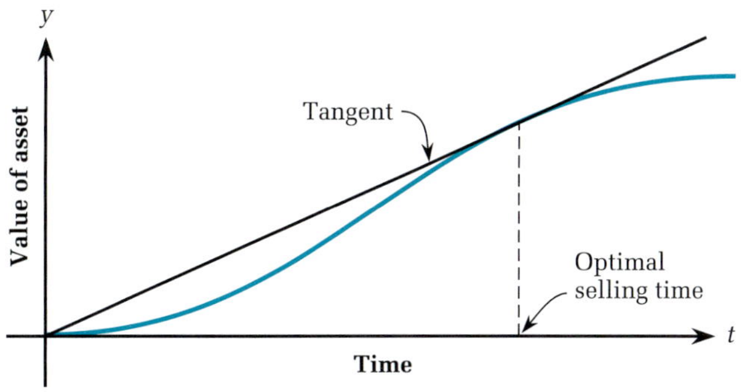

FIGURE 2.6.4

The last application of this section comes from the topic of human health and uses the above geometrical result.

EXAMPLE 12   Maximize average value; biology, hemoglobin levels, "blood packing" by athletes

The blood of mammals contains red blood cells whose hemoglobin transports oxygen to the cells. As the concentration of red blood cells increases, the blood carries more oxygen. Some athletes have experimented with the technique known as "blood packing," whereby extra red blood cells are transfused into their blood in an attempt to improve oxygen transport.

As the proportion of red blood cells increases, the blood becomes more viscous (thick and sticky) and the heart must work harder to pump the blood through the body. When the concentration of red blood cells is low, the increased work imposed on the heart by an increase in the viscosity is minor compared to the gain from the additional available oxygen. However, eventually the viscosity increases to a point where further additional work is not worth the extra oxygen transported.

The relationship between the concentration of hemoglobin and the relative viscosity of the blood is shown in figure 2.6.5.* On the graph, water is given the relative viscosity of 1.

Most mammals normally have an actual concentration of approximately 150 grams of hemoglobin per liter of blood. Notice the line tangent to the curve at this point also passes through the origin. Apparently mammals have evolved an optimal strategy for having the heart pump blood that consists of the "best" viscosity.

*K. Schmidt-Nielsen, *Scaling: Why Is Animal Size So Important?*, Cambridge: Cambridge University Press, 1984, p. 116.

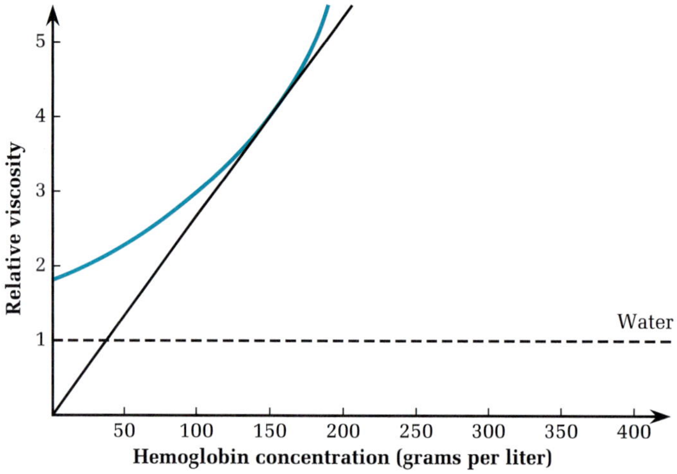

FIGURE 2.6.5

2.6 PROBLEMS

Foundations

The problems of this section require the basic skills illustrated by the following:

1. Rewrite each of the following with a single term x^a for a real number a.

(a) $x^3 x^2$ (b) $x^{-3} x^2$ (c) $\dfrac{x}{x^{-2}}$

(d) $2x^2 \left(5x^{-3} + \dfrac{4}{x^2} \right)$

Exercises

Problems 2–10 are products or quotients that have not been algebraically simplified. Use the Product or Quotient Rule to take their derivatives. *Then, starting* over, first do the algebraic simplification and use rules from Section 2.5 to take the derivative. Check that both methods give the same answer.

2. $x^4 x^2$ 3. $x^3(x^2 - 1)$ 4. $(x^4 + 3)x^3$

5. $x^{-2} \cdot x^{1.5}$ 6. $x^{-1.5} \cdot x^3$

7. $\dfrac{x^2}{x^4}$ 8. $\dfrac{x^3}{x^2}$

9. $x^2(2 - x^3)$ 10. $x^{-3}(x^4 + x)$

Differentiate the functions in problems 11–40.

11. $f(x) = x^2(x^2 - 2)$ 12. $g(x) = x^4(x^3 - 5)$

13. $f(x) = (x^2 + x)(x - 3)$

14. $f(x) = (x^2 - 3x)(2x + 3)$

15. $f(x) = (x^3 + 4x - 1)(2x + 5)$

16. $f(x) = (3x^3 - 2x^2 + x)(5x + 3)$

17. $f(x) = (x^2 + 1)\sqrt{x}$

18. $g(x) = \sqrt{x}(3x^4 - 2x)$

19. $f(x) = x^{4.5}(x^3 - 2x^2)$

20. $g(x) = x^{-2.1}(x^2 - 5x)$

21. $f(x) = x^4(\sqrt{x} - 2x)$

22. $f(x) = (x + 2x^3)(3x - \sqrt{x})$

23. $g(x) = \dfrac{4}{x - 2}$ 24. $f(x) = \dfrac{2}{x + 5}$

25. $f(x) = \dfrac{x^2}{2x + 3}$ 26. $h(x) = \dfrac{3x}{x^2 - 1}$

27. $k(x) = \dfrac{-x}{x^2 + 1}$ 28. $f(x) = \dfrac{-2x^2}{x + 2}$

29. $g(x) = \dfrac{x^2 + 2}{x - 1}$ 30. $g(x) = \dfrac{2x^2 + 3}{x + 2}$

31. $g(x) = \dfrac{x - 4}{x^4}$ 32. $g(x) = \dfrac{x + 3}{x^3}$

33. $f(x) = \dfrac{4x - x^2}{x + 1}$ 34. $f(x) = \dfrac{3x - x^2}{5x + 1}$

35. $g(u) = \dfrac{2u^2 + 5u - 1}{u + 2}$ 36. $g(u) = \dfrac{u^2 - 3u + 2}{3u + 1}$

37. $f(x) = \dfrac{3x^2 - 4}{3\sqrt{x}}$ **38.** $f(x) = \dfrac{2x^3 + 3x}{5\sqrt{x}}$

39. $f(x) = \dfrac{\sqrt{x} - x}{x^2 - 1}$ **40.** $f(x) = \dfrac{2\sqrt{x} + x^3}{x^2 + 3x}$

41. Find an equation for the tangent line to $f(x) = (x - 1)/(x^2 + 1)$ at $P(1, 0)$.

42. Find an equation for the tangent line to $g(x) = (x + 1)/(x^2 - 8)$ at $P(3, 4)$.

43. *(Cost-Benefit)*   The Environmental Protection Agency decides to clean up a radioactive waste site in Utah. The cost of removing x% of the waste is given by $C(x) = 200/(100 - x)$ in millions of dollars.
   (a) Find the instantaneous rate of change of expense as a function of the clean-up percent.
   (b) Evaluate your answer at 50%.
   (c) Evaluate your answer at 90%.
   (d) Interpret your answers at the 50% and 90% levels in terms of dollars per additional percent of clean-up.

44. *(Cost-Benefit)*   The EDF is considering a new coal-fired electric generating plant. It estimates that removing x% of the sulfur dioxide and other pollutants will cost $C(x) = 120/(100 - x)$ million dollars.
   (a) Find the instantaneous rate of change of expense as a function of the removal percent.
   (b) Evaluate your answer at 90%.
   (c) Interpret your answer in terms of dollars per additional percent of clean-up.

45. *(Yield to Maturity, YTM)*   On an 8% bond with 10 years to maturity and having a current market price of $\$x$, the approximate YTM (expressed in percent) is given by: YTM$(x) = (36,000 - 20x)/(1000 + x)$.
   (a) Find the instantaneous rate of change of YTM with respect to market price x.
   (b) Evaluate your result at a market price of $850, and interpret your answer in terms of change of YTM with respect to an additional dollar.

46. *(Marginal Average Cost)*   A video store has a fixed cost of $500 a week and a handling cost of $0.50 for each video it rents that week.
   (a) What is the average cost per video if it rents x videos during the week?
   (b) What is the marginal average cost?
   (c) Evaluate the marginal average cost at $x = 750$ and interpret your answer.

47. *(Marginal Average Profit)*   (See problem 46 for cost.) A video store rents videos for $2 each.
   (a) What is its average profit per video if it rents x videos in the week?
   (b) What is its marginal average profit?
   (c) Evaluate the marginal average profit at $x = 750$ and interpret your answer.

48. *(Marginal Adjusted Average Revenue and Profit)*   Suppose the cost of decorating x cakes is $C(x) = 5 + 2x$ dollars. Furthermore, suppose one cake must be used for a demonstration and so remains unsalable.
   (a) Find the average cost of decorating x salable cakes.
   (b) Find the marginal average cost of decorating x salable cakes.

49. *(Marginal Adjusted Average Revenue and Profit)*   Suppose Lavish Airlines determines the demand function is $p(x) = -(1/100)x^2 + 20$ when they sell x tickets.
   (a) Determine an expression for $R(x)$, the revenue function.
   (b) Find the marginal revenue at $x = 20$.
   (c) Suppose Lavish Airlines always gives away a total of two free tickets on each flight to travel agents to encourage the agents to "Sell Lavish." Find the average revenue per passenger on the *sale* of x tickets. (There are $x + 2$ passengers aboard the flight.)
   (d) Find the marginal average revenue per passenger on the sale of x tickets.

50. *(Memory)*   The longer a list of words, the smaller the percentage of people who can remember them for 10 minutes. Everyone can remember one word but only a very small percentage can remember a list of 10 words for 10 minutes. Suppose the formula for the percentage of the population that can remember x words is $L(x) = 1000/(x^3 - 4x^2 + 13)$. Check that $L(1) = 100$, $L(5) = 26.3$, $L(10) = 1.63$, and $L(15) = 0.4$.
   (a) Find the formula for the instantaneous rate of change at x words.
   (b) Evaluate your answer for a list of five words.
   (c) Interpret your answer to part (b).

51. *(Maximize Average Value, Biology, Forestry)*   Aspen trees (*Populus tremuloides*) grow in groves that have mats of interconnected roots. When the mature trees are cut, a new grove will regrow from the root mat. Figure 2.6.6 is a graph showing the value of cutting an acre of regrowth

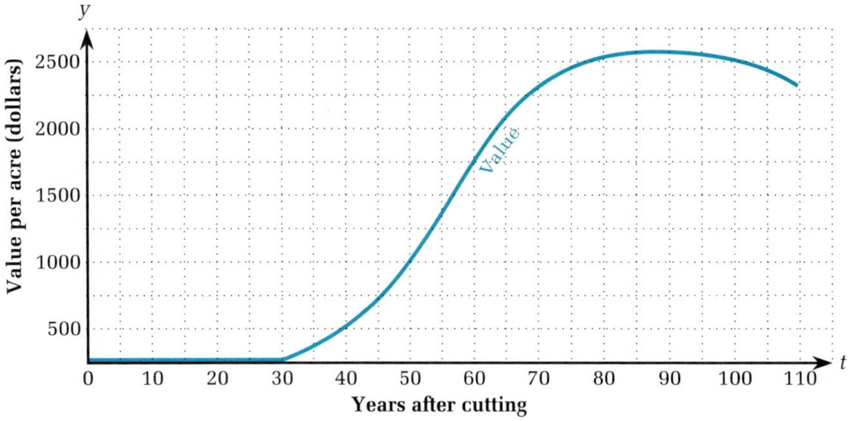

FIGURE 2.6.6

as a function of the number of years since the last cutting. Using the geometric technique illustrated in Example 11, find the time to cut the mature trees that will produce a maximum average revenue per acre.

52. (Maximize Average Value, Cost of Information) The graph shown in figure 2.6.7 represents the amount of information gained as a function of the time spent gathering that information. The information is measured in number of items and the time in hours.
  (a) Use the graphical technique of Example 11 to estimate the time and the amount of information that maximizes the average amount of information.
  (b) Estimate the time beyond which new information is coming in at a rate of less than one new item per hour.

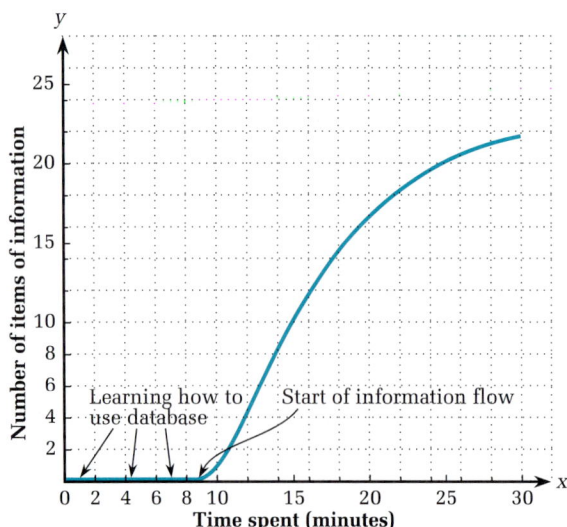

FIGURE 2.6.7

Writing and Discussion Problems

53. Suppose $y = x^a$ for some positive integer a. Show that if $y = x^n x^m$ for any integers n and m, it is always true that $y' \neq (nx^{n-1}) \cdot (mx^{m-1})$. (Thus, in Example 1 it made no difference how we wrote x^7 as a product of other powers of x.)

54. Write a short presentation on "fortuitous cancellation," whereby fractions are reduced by canceling digits without changing the value of the fraction. For example $19/95 = 1/5$, and the 9's can be fortuitously canceled, whereas $23/35 \neq 2/5$, so the 3's cannot be fortuitously canceled. Find all those two-digit numbers whose fractions are examples of fortuitous cancellation.

Enrichment Problems

55. Show that the following steps justify the Quotient Rule for derivatives.
    Let k, f, and g stand for the functions $k(x)$, $f(x)$, and $g(x)$, and assume all of the following derivatives (with respect to x) exist and $g(x) \neq 0$.
    Supply reasons, or explain the algebraic manipulations involved at each step.
    Let the quotient be written as $k = f/g$. Then $kg = f$, giving us

$$kg' + gk' = f'$$
$$gk' = f' - kg'$$
$$k' = \frac{f' - kg'}{g}$$
$$k' = \frac{f' - \frac{f}{g}g'}{g}$$
$$k' = \frac{gf' - fg'}{g^2}$$

SOLUTIONS TO PRACTICE EXERCISES

1.
$$\frac{d}{dx}\overset{\overset{f}{\downarrow}}{(x^3-x^2)}\overset{\overset{g}{\downarrow}}{(3x^2+2x)}$$

$$= \overset{\overset{f}{\downarrow}}{(x^3-x^2)}\overset{\overset{g'}{\downarrow}}{(6x+2)} + \overset{\overset{g}{\downarrow}}{(3x^2+2x)}\overset{\overset{f'}{\downarrow}}{(3x^2-2x)}$$

$$= 6x^4 - 6x^3 + 2x^3 - 2x^2 + 9x^4 - 6x^3 + 6x^3 - 4x^2$$

$$= 15x^4 - 4x^3 - 6x^2$$

2.
$$\frac{d}{dx}\left(\frac{\overset{f}{\downarrow}}{\underset{\underset{g}{\uparrow}}{2x+1}}{x^2+3x}\right)$$

$$= \frac{\overset{\overset{g}{\downarrow}}{(x^2+3x)}\cdot \overset{\overset{f'}{\downarrow}}{2} - \overset{\overset{f}{\downarrow}}{(2x+1)}\overset{\overset{g'}{\downarrow}}{(2x+3)}}{\underset{\underset{g^2}{\uparrow}}{(x^2+3x)^2}}$$

$$= \frac{2x^2 + 6x - (4x^2 + 8x + 3)}{(x^2+3x)^2}$$

$$= \frac{-2x^2 - 2x - 3}{(x^2+3x)^2}$$

Appendix: A Verification of the Product Rule for Derivatives

$$\frac{d}{dx}\left(f(x)\cdot g(x)\right) = f(x)\cdot\left(\frac{d}{dx}g(x)\right) + g(x)\cdot\left(\frac{d}{dx}f(x)\right)$$

We use the definition of the derivative as a limit of a difference quotient.

$$\frac{d}{dx}\left(f(x)\cdot g(x)\right) = \lim_{h\to 0}\frac{f(x+h)\cdot g(x+h) - f(x)\cdot g(x)}{h}$$

We now use a common technique (in mathematics) of choosing a term to subtract from and add back to our expression that will enable us to make further progress. Such a term often seems to appear out of thin air. However, finding such terms results from experience and some serious thought about what will do the job. In some ways, the use of these terms is similar to a picklock in the hands of an expert, who simply says, "Watch this." For our situation, the proper term is $f(x+h)\cdot g(x)$, and we have

$$= \lim_{h\to 0}\frac{f(x+h)\cdot g(x+h) - f(x+h)\cdot g(x) + f(x+h)\cdot g(x) - f(x)\cdot g(x)}{h}$$

Factoring and rewriting this, we get

$$= \lim_{h \to 0} \frac{f(x + h) \cdot [g(x + h) - g(x)] + g(x) \cdot [f(x + h) - f(x)]}{h}$$

$$= \lim_{h \to 0} \left(f(x + h) \cdot \frac{g(x + h) - g(x)}{h} + g(x) \cdot \frac{f(x + h) - f(x)}{h} \right)$$

$$= \lim_{h \to 0} f(x + h) \cdot \lim_{h \to 0} \left(\frac{g(x + h) - g(x)}{h} \right)$$

$$+ \lim_{h \to 0} g(x) \cdot \lim_{h \to 0} \left(\frac{f(x + h) - f(x)}{h} \right)$$

$$(*) = \lim_{h \to 0} f(x + h) \cdot \left(\frac{d}{dx} g(x) \right) + \lim_{h \to 0} g(x) \cdot \left(\frac{d}{dx} f(x) \right)$$

Because we are assuming $f(x)$ is differentiable, $f(x)$ certainly is continuous. Thus $\lim_{h \to 0} f(x + h) = f(x)$. Because $g(x)$ does not contain the variable h, $\lim_{h \to 0} g(x) = g(x)$.

Putting all of these facts together we have $(*)$ equal to

$$f(x) \cdot \left(\frac{d}{dx} g(x) \right) + g(x) \cdot \left(\frac{d}{dx} f(x) \right)$$

2.7 THE CHAIN RULE

Previous sections described how to find derivatives of functions built up by the addition, subtraction, multiplication, and division of other functions. In this section we find the derivative of a function created by yet another technique of combining functions, known as *composition of functions*. Examples of such functions are $(x^2 + 3)^7$ and $\sqrt{x^7 + 2x}$.

Before presenting rules for differentiating functions created by composition, let us briefly review the concept of function composition.

Composition and Decomposition of Functions

Suppose the function $g(x)$ is defined for values in a set A and gives values in a set B. Furthermore, suppose a function $f(x)$ is defined for values in the set B and gives values in the set C. A schematic of such a process is shown in figure 2.7.1.

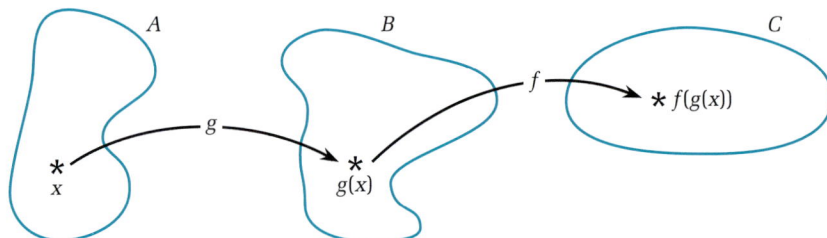

FIGURE 2.7.1

The function $y = f(g(x))$ is defined for values in A and produces values over in C.

DEFINITION

Composite function

Suppose f and g are functions and the range of g lies in the domain of f. The *composite function* is the function h defined by $h(x) = f(g(x))$. The *domain* of h consists of those values in the domain of g for which the value of $g(x)$ lies in the domain of f.

EXAMPLE 1   Composite function

Suppose $f(x) = 12/(x + 1)$ and $g(x) = 5x$. Find $f(g(x))$ and $g(f(x))$ and the domain of each.

SOLUTION

We find $f(g(x))$ by first determining $g(x)$ and then substituting $g(x)$ into f. Doing so we have $f(g(x)) = f(5x) = 12/(5x + 1)$.

To find the domain of $f(g(x))$, we notice the domain of $g(x)$ consists of all the real numbers. However, $f(g(x))$ has a denominator of $5x + 1$, and division by zero is not defined. Hence, the domain of $f(g(x))$ consists of all the real numbers except for x such that $5x + 1 = 0$. Solving this for x, we find the domain of $f(g(x))$ to be all real numbers except $x = -1/5$.

We find $g(f(x))$ by beginning with $f(x)$ and substituting $f(x)$ into g. We get $g(f(x)) = g\left(\dfrac{12}{x + 1}\right) = 5\left(\dfrac{12}{x + 1}\right) = \dfrac{60}{x + 1}$. The domain of $g(f(x))$ thus consists of all real numbers except $x = -1$.   ●

WARNING

It is usually the case that $f(g(x)) \neq g(f(x))$ and we must be quite careful of the order in which we compose functions.

In calculus, the true usefulness of the composite function concept comes from our ability to *decompose* functions into functions that are simpler and with which we are more familiar. In particular, we want to use functions whose derivatives we already know. Decomposition is not a unique process because the same function can be decomposed several ways.

EXAMPLE 2   Decomposition of a function

Express the function $y = (x^2 + 4)^3$ as a composite function.

SOLUTION

Let $f(x) = x^3$ and $g(x) = x^2 + 4$. Then $y = (x^2 + 3)^3 = f(x^2 + 4) = f(g(x))$.

A second decomposition would be to let $r(x) = (x + 4)^3$ and $s(x) = x^2$. Then $r(s(x)) = r(x^2) = (x^2 + 4)^3$.

Both of these decompositions are valid, but the first seems to be more "natural." ●

Consider the effect of a change in the value of x. Any such change will change the value of $g(x)$, and consequently the value of $y = f(g(x))$ is also changed.

EXAMPLE 3   Change in a composite function

Suppose $f(x) = x^3$ and $g(x) = x^2 + 4$. Find the change in $y = f(g(x))$ as x changes from 1 to 3.

SOLUTION

If x changes from 1 to 3, then $g(x)$ changes from $g(1) = 1^2 + 4 = 5$ to $g(3) = 3^2 + 4 = 13$.

By using these values in $f(x) = x^3$, we find $f(5) = 5^3 = 125$ and $f(13) = (13)^3 = 2197$. A schematic of this process is shown in figure 2.7.2.

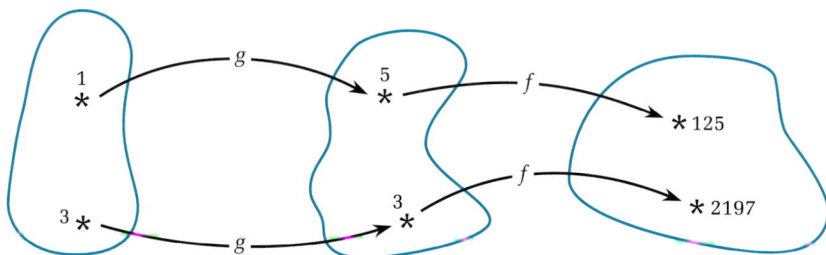

FIGURE 2.7.2

As x changes from 1 to 3, then $y = f(g(x)) = (x^2 + 4)^3$ changes from 125 to 2197. ●

Chain Rule

The derivative of $y = f(g(x))$ with respect to x is the *instantaneous rate of change.* To find y', we use the Chain Rule, explained below. A justification of the Chain Rule is found in the appendix to this section.

VII. THE CHAIN RULE

The derivative of $y = f(g(x))$ with respect to x is given by

$$\frac{d}{dx}(f(g(x))) = \frac{d}{dg}f \cdot \frac{d}{dx}g$$

Alternate formulations:

Equivalently, $y' = f'(g(x)) \cdot g'(x)$

The derivative of $f(g(x))$ is the product of the derivative of f with respect to g times the derivative of g with respect to x.

If we keep in mind that f is a function of $g(x)$ and g is a function of x, then a very compact form of the Chain Rule is

CHAIN RULE

$$\frac{df}{dx} = \frac{df}{dg}\frac{dg}{dx}$$

Written in the form $df/dx = (df/dg)(dg/dx)$, the Chain Rule appears to be a simple algebraic cancellation of the dg terms. Although we are not considering df/dg and dg/dx as fractions, nonetheless this form does suggest a means of remembering the rule.

The Chain Rule multiplies rates. Notice that df/dg is the *rate of change of f with respect to g* and dg/dx is the *rate of change of g with respect to x*. We multiply these rates together to find *the rate of change of f with respect to x*.

As an example, think of a worker's income per week. If she works for $7 per hour for 8 hours per day, then she earns 7 (dollars per hour) · 8 (hours per day) = 56 (dollars per day).

EXAMPLE 4 The derivative of a composed function by the Chain Rule

Find the derivative with respect to x of $(x^2 + 4)^3$.

SOLUTION

Let $f(g) = g^3$ and $g(x) = x^2 + 4$. The derivatives of these functions can be found by the Power Rule and the Sum Rule. We do need to be careful that we take the derivative with respect to the correct variables. We need the derivatives of f with respect to g and we need the derivative of g with respect to x.

$$\frac{d}{dg}f(g) = 3g^2 \text{ and } \frac{d}{dx}g = 2x$$

Writing $(x^2 + 4)^3$ as $f(g(x))$, we have

$$\frac{d}{dx}(x^2 + 4)^3 = \frac{d}{dx}f(g(x))$$

$$= \frac{d}{dg}f \cdot \frac{d}{dx}g \text{ by the Chain Rule}$$

$$= 3g^2 \cdot 2x$$

Substituting for g, we have

$$\frac{d}{dx}(x^2 + 4)^3 = 3(x^2 + 4)^2 \cdot 2x = 6x\,(x^2 + 4)^2$$

We could stop at this point, for we have found the derivative, but let us have a cross-check on our work by using an alternate means for finding the derivative.

For $y = (x^2 + 4)^3$ we can multiply this expression out as a polynomial. Doing so gives

$$y = x^6 + 12x^4 + 48x^2 + 64$$

Differentiating y by using the Power Rule from the previous section, we have $y' = 6x^5 + 12 \cdot 4x^3 + 48 \cdot 2x = 6x^5 + 48x^3 + 96x$. This is equivalent to $6x(x^2 + 4)^2$, so both methods arrive at the same result.

In Example 4 we found two methods of finding the derivative of $(x^2 + 4)^3$. When the powers involved are small, such as a cube, we may be tempted to use the alternate approach of multiplying terms out rather than using the Chain Rule. However, were we to attempt this for Example 5, we would find ourselves doing an incredible amount of algebra.

EXAMPLE 5   Chain Rule, compound interest

Suppose $2000 is deposited in an account paying r percent interest, compounded monthly. The value of the account after 10 years is given by $A(r) = 2000[1 + (r/1200)]^{120}$.

(a) Find the derivative with respect to r of $A(r)$.

(b) Find and interpret $A'(7)$.

SOLUTION

(a) Think of $A(r)$ as being composed of two functions, $f(x) = 2000x^{120}$ and $g(x) = 1 + (x/1200)$. Then

$$f(g(r)) = f\left(1 + \frac{r}{1200}\right) = 2000\left(1 + \frac{r}{1200}\right)^{120}$$

$$= A(r).$$

To find $A'(r)$ we need to find $f'(g(r))$ and $g'(r)$.

$$f'(x) = 2000 \cdot 120x^{120-1} = 240{,}000x^{119}$$

$$g'(x) = 0 + \frac{1}{1200} = \frac{1}{1200}$$

By the Chain Rule,

$$A'(r) = f'(g(r)) \cdot g'(r) = 240{,}000(g(r))^{119} \cdot \frac{1}{1200} = 200(g(r))^{119}$$

Substituting $g(r) = 1 + (r/1200)$, we have $A'(r) = 200[1 + (r/1200)]^{119}$.

(b) An interpretation of this derivative is to think of $A'(r)$ as giving the instantaneous change in the 10-year value of the account as r changes.

To do the particular example of an interest rate of 7 percent we have

$$A'(7) = 200\left(1 + \frac{7}{1200}\right)^{119}$$

$$= 200\,(1 + 0.0058333333)^{119}$$

$$= 200\,(1.0058333333)^{119}$$

$$= 200\,(1.9980063) = 399.60$$

(The reason for the many places used for the decimal expression of 7/1200 is the very smallness of the number and the fact that numbers will be taken to a high power.)

In other words, at 7 percent the value of the account is rising at an instantaneous rate of $399.60 per each unit of increase in the percentage of interest.

For the sake of comparison, we can evaluate $A(7) = 4019.32$ and $A(8) = 4439.29$. The actual increase in total value between earning 7 percent and 8 percent over 10 years is thus $419.96.

If $y = f(g(x))$, we may think of f as being the "outside function" and g as being the "inside function." Using such terminology, y' is the product of the derivative of the outside function, taken with respect to the inside function, multiplied by the derivative of the inside function. Quite frequently the outside function is a power function. In fact, that is such a common situation that a special case of the Chain Rule is referred to as the Generalized Power Rule.

Generalized Power Rule

VIII. GENERALIZED POWER RULE

For any real number n and any function $f(x)$ for which $f'(x)$ exists,

$$\frac{d}{dx}[f(x)]^n = n[f(x)]^{n-1} \cdot \frac{d}{dx}f(x)$$

Alternate formulations:

If $y = [f(x)]^n$, then $y' = n \cdot (f(x))^{n-1} \cdot f'(x)$.

The derivative of a power of $f(x)$ is given by the product of that power multiplied together with $f(x)$ raised to a power one less than the original power and that then times the derivative of $f(x)$.

Thinking of $f(x)$ as the "inside function" and x^n as the "outside function," we see that indeed we have taken the derivative of x^n, which is $n(x^{n-1})$, and evaluated this at $f(x)$. Then we multiplied this by $f'(x)$, which is the derivative of the "inside function."

EXAMPLE 6   Generalized Power Rule

Check that finding $\dfrac{d}{dx}(x^2)^3$ by the Generalized Power Rule yields the same derivative as finding $\dfrac{d}{dx}x^6$ directly.

SOLUTION

Let $f(x) = x^2$, so that $f'(x) = 2x$. The Generalized Chain Rule states that $\dfrac{d}{dx}(f(x))^3 = 3 \cdot (f(x))^{3-1} \cdot f'(x) = 3 \cdot (f(x))^2 \cdot f'(x)$. Replacing $f(x)$ by x^2, we have $\dfrac{d}{dx}(x^2)^3 = 3 \cdot (x^2)^{3-1} \cdot 2x = 3(x^2)^2 \cdot 2x$, which by algebra gives us $3x^4 \cdot 2x = 6x^5$.

EXAMPLE 7   Generalized Power Rule

Find $\dfrac{d}{dx}(5x^2 + 3)^7$.

SOLUTION

Letting $f(x) = 5x^2 + 3$ and $n = 7$, we have

$$\left\{\frac{d}{dx}(f(x))^7 = 7 \cdot (f(x))^6 \cdot \frac{d}{dx}f(x)\right\}$$

$$\frac{d}{dx}(5x^2 + 3)^7 = 7(5x^2 + 3)^6 \cdot \frac{d}{dx}(5x^2 + 3)$$

$$= 7(5x^2 + 3)^6 \cdot (5 \cdot 2x) = 70x(5x^2 + 3)^6$$

This last is the preferred form of the answer. The reason is that usually we will be interested in knowing when the derivative is zero, and so we prefer a factored form of the derivative.

PRACTICE EXERCISE 1

Find $\dfrac{d}{dx}(2x^3 + 5)^2$.

The next example finds a derivative involving a square root by rewriting the square root as a fractional power and using the Generalized Power Rule.

EXAMPLE 8   Generalized Power Rule used with a square root

Find $\dfrac{d}{dx}\sqrt{x^7 + 3}$.

SOLUTION

Rewrite the square root as a 1/2 power, so we have

$$\frac{d}{dx}(x^7 + 3)^{1/2} = \frac{1}{2}(x^7 + 3)^{(1/2)-1} \cdot \frac{d}{dx}(x^7 + 3)$$

$$= \frac{1}{2}(x^7 + 3)^{-1/2} \cdot 7x^6$$

$$= \frac{7}{2} \cdot x^6(x^7 + 3)^{-1/2}$$

We may leave the answer in this form or rewrite the negative fractional exponent in terms of a positive fractional exponent and then as a square root. Doing these steps gives us

$$\frac{d}{dx}(x^7 + 3)^{1/2} = \frac{7}{2}x^6(x^7 + 3)^{-1/2} = \frac{7x^6}{2(x^7 + 3)^{1/2}} = \frac{7x^6}{2\sqrt{x^7 + 3}}$$

We have seen that $\frac{d}{dx}\sqrt{x} = 1/(2\sqrt{x})$. When dealing with square roots, we often do not bother to change to fractional powers of $1/2$ and then use the Generalized Power Rule. Instead we use the Chain Rule. Either approach to finding the derivative yields the same answer.

EXAMPLE 9   Chain Rule used with a square root

Find $\frac{d}{dx}\sqrt{2x^2 + 1}$.

SOLUTION

To emphasize our procedure, let $f(x) = 2x^2 + 1$. Using $\frac{d}{dx}\sqrt{x} = 1/(2\sqrt{x})$ and substituting $f(x)$ for x we have, by the Chain Rule, $\frac{d}{dx}\sqrt{f(x)} = 1/(2\sqrt{f(x)}) \cdot \frac{d}{dx}f(x)$.

Using the fact that $f'(x) = 4x$, and substituting for $f(x)$ and $\frac{d}{dx}f(x)$, we have $\frac{d}{dx}\sqrt{2x^2 + 1} = 1/(2\sqrt{2x^2 + 1}) \cdot 4x = 2x/\sqrt{2x^2 + 1}$.

PRACTICE EXERCISE 2

Find the derivative of $\sqrt{x^7 + 3}$ by both the Chain Rule and the Generalized Power Rule.

As we find the derivatives of other functions, we will be using the Chain Rule many times in future sections. We can use the Chain Rule or Generalized Power Rule in conjunction with any of our previous formulas. In Example 10 we find the derivative of a product.

EXAMPLE 10   Generalized Power Rule

Find $\frac{d}{dx}[(3x + 2)^4(x^2 - 5)^3]$.

SOLUTION

Using the Product Rule and the Generalized Power Rule,

$$\frac{d}{dx}[(3x + 2)^4(x^2 - 5)^3]$$

$$= (3x + 2)^4 \cdot \left[\frac{d}{dx}(x^2 - 5)^3\right] + (x^2 - 5)^3 \cdot \left[\frac{d}{dx}(3x + 2)^4\right]$$

(∗) $= (3x + 2)^4 \cdot [3(x^2 - 5)^2 \cdot 2x] + (x^2 - 5)^3 \cdot [4(3x + 2)^3 \cdot 3]$

To simplify the expression in (∗), we notice the second term contains $(3x + 2)^3$ and the first term contains $(3x + 2)^4$. Hence we can factor out $(3x + 2)^3$ from both terms. Similarly, we can factor out $(x^2 - 5)^2$ from both terms. Lastly, because $4 \cdot 3 = 12$ and $3 \cdot 2 = 6$, we can factor 6 out of each term.

With a common factor of $6 \cdot (3x + 2)^3 \cdot (x^2 - 5)^2$, we can rewrite (∗) as

(∗∗) $= 6(3x + 2)^3(x^2 - 5)^2[2(x^2 - 5) + x(3x + 2)]$

             $= 6(3x + 2)^3(x^2 - 5)^2(2x^2 - 10 + 3x^2 + 2x)$

(∗∗∗) $= 6(3x + 2)^3(x^2 - 5)^2(5x^2 + 2x - 10)$

Expressing the derivative as this product of terms is the preferred form of the answer.

PRACTICE EXERCISE 3

Find $\dfrac{d}{dx}\left(\dfrac{(2x + 1)^2}{1 - x^2}\right).$

EXAMPLE 11   Application to biology, diminishing returns

Let us return to the example in Section 2.6 in which a bird collects x grams of seeds in one section of the field by collecting for $f(x) = (5x + 1)/(100 - x)$ minutes. For this function we showed $f'(x) = 501/(100 - x)^2$. This derivative gave us a rate, the instantaneous rate of time spent to find an additional gram of all the seeds present.

Suppose this bird expends energy at a rate of 0.05 Calories per minute while collecting seeds. If the bird spends t minutes collecting, then $g(t) = 0.05t$ is the number of Calories spent on collecting. $f(x)$ is the number of minutes spent, so if we let $t = f(x)$, then $g(f(x))$ is the number of Calories expended by the bird to collect x grams of the seed.

To determine the rate at which Calories are used to collect additional grams of seeds, we need to determine $\frac{d}{dx} g\, (f\,(x))$, where x is the number of grams of seeds.

By the Chain Rule,

$$\frac{dg}{dx} = \frac{dg}{df} \cdot \frac{df}{dx} = (0.05) \cdot \frac{501}{(100 - x)^2}.$$

This is measured in Calories spent on collecting an additional gram of seed. To find some sample values, calculate dg/dx for $x = 95$ and $x = 98$.

SOLUTION

If 95 grams have been collected, the bird will spend

$$0.05 \cdot \frac{501}{(100 - 95)^2} = 0.05 \cdot \frac{501}{5^2} = 0.05 \cdot \frac{501}{25} \simeq 1 \text{ Calorie}$$

per additional gram of seed. If 98 grams have been collected, the bird will spend

$$0.05 \cdot \frac{501}{(100 - 98)^2} \simeq 6 \text{ Calories}$$

per additional gram of seed.

2.7 PROBLEMS

Foundations

The problems of this section require the basic skills illustrated by the following:

Simplify the functions in problems 1–3.

1. $3x(2 + 4x)$ **2.** $\sqrt{x}\,(x^{0.5} - 2x)$

3. $(2x)^3(x^2 - 5)$

4. Write $\sqrt[4]{x^2 - 2x}$ in exponential form.

5. Solve $\sqrt{2x - 1} = 3$.

6. Write $3(x - 4)^2(2x + 3)^2 + 2(x - 4)^3(2x + 3)$ as a product of factors.

Exercises

7. For $f(t) = t^2 + 1$ and $g(x) = 3x + 2$, find:
(a) $f(g(x))$ (b) the domain of $f(g(x))$
(c) $\frac{d}{dx} f(g(x))$

8. For $f(t) = 2t^{-4} + 3$ and $g(x) = 2x - 1$, find:
(a) $f(g(x))$ (b) the domain of $f(g(x))$
(c) $\frac{d}{dx} f(g(x))$

9. For $f(t) = 3/(t - 2)$ and $g(x) = x^2 + x$, find:
(a) $f(g(x))$ (b) the domain of $f(g(x))$
(c) $\frac{d}{dx} f(g(x))$

10. For $f(t) = (t - 5)/2$ and $g(x) = 1/x$, find:
(a) $f(g(x))$ (b) $\frac{d}{dx} f(g(x))$

Find the derivative of the functions in problems 11–30 by the Chain Rule or the Generalized Power Rule.

11. $y = (1 + 3x)^8$ **12.** $y = (1 + 4x)^8$

13. $y = (2 + x^2)^5$ **14.** $y = (x^2 - 3)^5$

15. $y = (2x + x^2)^8$ **16.** $y = (3x - x^3)^4$

17. $y = \sqrt{3x + 4}$ **18.** $y = \sqrt{4x + 1}$

19. $f(x) = \sqrt{2x^2 - 3}$ **20.** $f(x) = \sqrt{2x^3 - 5}$

21. $g(x) = \sqrt[3]{x^3 + 1}$ **22.** $g(x) = \sqrt[4]{x^2 - 3}$

23. $f(x) = (3x - 1)^{1.2}$ **24.** $f(x) = (7x + 1)^{3.1}$

25. $g(x) = (x^2 + 2x)^{0.25}$

26. $g(x) = (3x^2 - 4x)^{1.25}$

27. $f(x) = (x^2 + 2)^{-10}$ **28.** $f(x) = (x^3 + 1)^{-12}$

29. $g(x) = (2x^{-2} + 1)^{-4}$ **30.** $g(x) = (3x^{-3} - 2)^{-4}$

Find the derivative of problems 31–38 by using the Product or Quotient Rule together with the Chain Rule or Generalized Power Rule as necessary.

31. $g(x) = x^2 \cdot (x^4 - x)^3$ **32.** $g(x) = 2x \cdot (x^3 + 2x)^4$

33. $f(x) = x^{-3}(x^2 + 2x)^5$ **34.** $f(x) = x^{-5}(x^3 - 4x)^2$

35. $f(x) = (x^2 + 1)^2/(3x^2 - 2)$

36. $g(x) = (2x^4 - x)^2/(2x^3 + 5)$

37. $g(x) = (3x^2 - 1)/(2x - 3)^2$

38. $f(x) = (x^4 - x)/(3x^2 - 1)^3$

39. Suppose $G(x) = (x - 4)^3(2x + 3)^2$.
   (a) Find $G'(x)$.
   (b) Solve for x such that $G'(x) = 0$.

40. Let $f(x) = (2x - 1)^{3/2} - 9x$.
   (a) Find $f'(x)$.
   (b) Solve for x such that $f'(x) = 0$.

41. Find an equation for the line tangent to $f(x) = \sqrt{x^3 - 2}$ at $P(3, 5)$.

42. Find an equation for the line tangent to $g(x) = \sqrt[3]{x^2 - 12}$ at $P(2, -2)$.

43. Given the problem of finding the derivative of $(x^2 - 3)/(x^3 + 1)$, some people prefer to avoid using the rule for the derivative of a quotient, so they rewrite the quotient as $(x^2 - 3) \cdot (x^3 + 1)^{-1}$ and then use the Product and General Power rules. Show that the two methods give the same derivative.

44. Given the problem of finding the derivative of $(x^2 + 1)/(2x^2 + 3)$, some people prefer to avoid using the rule for the derivative of a quotient, so they rewrite the quotient as $(x^2 + 1) \cdot (2x^2 + 3)^{-1}$ and then use the Product and General Power rules. Show that the two methods give the same derivative.

45. *(Airplane Engine Expense, Learning Experience Curves)*   The cost of maintaining airplane engines or other machinery could be approximated by a cost function curve such as $C(x) =$

$(0.7x^2 + 2.75x + 0.28)/(x + 0.53)^2$, where $C(x)$ is in thousands of dollars and x is in months.
   (a) Find $C'(x)$.
   (b) Evaluate $C'(0.1)$ and $C'(1)$.
   (c) Interpret your results.

46. *(Annuity)*   Suppose $100 is deposited at the end of each year into an account on which the interest rate is i (expressed decimally), compounded annually. The total accumulated amount in the account at the end of 30 years is given by $V(i) = 100 [(1 + i)^{30} - 1)/i]$.
   (a) Find $V(0.05)$.
   (b) Find $V'(i)$.
   (c) Find $V'(0.05)$ and interpret your answer.

47. *(Waste Removal)*   In Section 1.5 we found Mirage Lake Water Co. could remove $P(t) = (101t - 3)/(2 + t)$ percent of the impurities if they invested t thousand dollars. Suppose they are investing money in removal at the rate of $2000 a day.
   (a) Find the rate at which the percentage of impurities is being reduced, as a function of the number of days.
   (b) Evaluate this rate for days 2, 10, and 20.

48. *(Marginal Revenue Product)*   Economists define the *marginal revenue product* to be the rate at which revenue is changing as a function of the number of workers. Let w be the number of workers. Thus, if $R(x)$ is the revenue from the production and sale of x items, and if x is a function of the number of workers, then dR/dw is the marginal revenue product. Suppose $R(x) = 50x - x^2$ and $x(w) = 30 - w$. Find the marginal revenue product at $w = 10$ and interpret your answer.

49. *(Marginal Revenue Product)*   Refer to problem 48. Suppose a demand function is $p(x) = 100 - x^2$, where x is the number of items produced, and w workers can produce $\sqrt{w + 2}$ items a day.
   (a) Find the revenue function.
   (b) Find the marginal revenue product at $w = 5$ and interpret your answer.

50. *(Marginal Revenue Product)*   Refer to problem 48. The Bakery knows the daily demand function for wedding cakes is $p(x) = -(1/5)x + 25$ dollars, where x is the number of cakes. When w cake decorators are hired, they can each decorate four cakes a day.
   (a) Find the revenue function.
   (b) Find the marginal revenue product at $x = 2$ and interpret your answer.

Writing and Discussion Problems

51. Suppose $f(x) = 2x^2$ and $g(x) = \sqrt{x}$. Explain why the domain of $f(g(x))$ is not the set of all real numbers, even though $2(\sqrt{x})^2 = 2x$ and the domain of $r(x) = 2x$ *is* the set of all real numbers.

52. Why is decomposition of functions important?

53. Why is it important to state what the derivative is being taken with respect to?

Enrichment Problems

54. Suppose $r(x) = x^3 + 3x^2 + 2x + 1$ and $t(x) = x + 1$. Find a function $s(x)$ such that $r(x) = s(t(x))$.

55. State a Chain Rule formula for dy/dx if $y = f(g(h(x)))$.

SOLUTIONS TO PRACTICE EXERCISES

1. $\dfrac{d}{dx}(2x^3 + 5)^2 = 2(2x^3 + 5)^1 \cdot 2 \cdot 3x^2 =$
$12x^2(2x^3 + 5)$

2. By the Chain Rule, $\dfrac{d}{dx}\sqrt{x^7 + 3} =$

$\dfrac{1}{2\sqrt{x^7 + 3}} \cdot \dfrac{d}{dx}(x^7 + 3) = \dfrac{7x^6}{2\sqrt{x^7 + 3}}$. To use the Generalized Power Rule, we write $\sqrt{x^7 + 3}$ as

$(x^7 + 3)^{1/2}$ and find $\dfrac{d}{dx}(x^7 + 3)^{1/2} =$

$(1/2)(x^7 + 3)^{1/2-1} \cdot 7x^6 =$
$(7/2)x^6(x^7 + 3)^{-1/2}$, an equivalent answer.

3. From the Quotient and Generalized Power rules, we have

$\dfrac{d}{dx}\left(\dfrac{(2x + 1)^2}{1 - x^2}\right) =$

$$\dfrac{(1 - x^2) \cdot 2(2x + 1)^1 \cdot 2 - (2x + 1)^2 \cdot (-2x)}{(1 - x^2)^2}$$

In the numerator, we can factor out $(2(2x + 1))$, giving

$$\dfrac{(2(2x + 1)) \cdot [2 - 2x^2 + (2x + 1) \cdot x]}{(1 - x^2)^2}$$

$$= \dfrac{2(2x + 1)(2 + x)}{(1 - x^2)^2}$$

Appendix: The Chain Rule

An informal justification of the Chain Rule proceeds as follows. Consider the definition of the derivative

$$\dfrac{d}{dx}f(g(x)) = \lim_{h \to 0} \dfrac{f(g(x + h)) - f(g(x))}{h}$$

Provided $g(x + h) - g(x) \neq 0$, the difference quotient may be rewritten to give us

(#) $\dfrac{d}{dx}f(g(x)) = \lim_{h \to 0} \left(\dfrac{f(g(x + h)) - f(g(x))}{h}\right)\left(\dfrac{g(x + h) - g(x)}{g(x + h) - g(x)}\right)$

$= \lim_{h \to 0} \dfrac{f(g(x + h)) - f(g(x))}{g(x + h) - g(x)} \cdot \lim_{h \to 0} \dfrac{g(x + h) - g(x)}{h}$

Because $g(x)$ is differentiable, it is continuous. Thus, $\lim\limits_{h \to 0} g(x + h) = g(x)$, and we have

$$\lim_{h \to 0} \dfrac{f(g(x + h)) - f(g(x))}{g(x + h) - g(x)} = \lim_{g(x+h) \to g(x)} \dfrac{f(g(x + h)) - f(g(x))}{g(x + h) - g(x)}$$

which is simply $\dfrac{d}{dg}f$, although it is written in a way we usually do not use.

Substituting this back into (#) above, we find $\dfrac{d}{dx}f(g(x)) =$ $\dfrac{d}{dg}f \cdot \dfrac{d}{dx}g$, which is the Chain Rule.

There is a difficulty with this as a "proof," namely, what if $g(x + h) - g(x) = 0$? This flaw can be overcome, but to do so here would be beyond the mathematical level of this book.

2.8   IMPLICIT DIFFERENTIATION

Consider the equation $x^3 + y^3 - 6xy = 0$. The graph of this equation is an example of a type of graph known as the *Folium of Descartes,* and it is shown in figure 2.8.1.

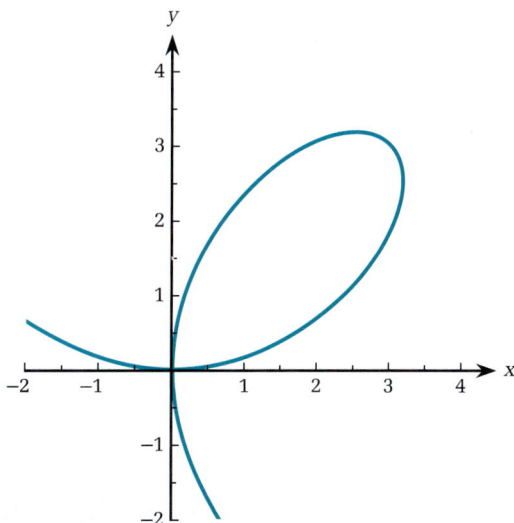

FIGURE 2.8.1

There is no function of x such that its graph is the same as that in figure 2.8.1 because vertical lines would intersect the graph in several places. Even so, the Chain Rule may be used to determine the slope of the tangent line to the graph at most points. For example, we will show that at the point $P(3, 3)$ the slope of the tangent line is -1.

In general, given an equation involving the variables x and y, we will use the Chain Rule to determine an expression for dy/dx in terms of x and y.

Implicit and Explicit Functions

Suppose the variables x and y are related by an equation. If y is expressed as $y = f(x)$ for some function of x, then we say y is *explic-*

itly represented as a function of x. For example, both $y = 2x - 1$ and $y = \sqrt{3x + 1}$ are explicit functions of x.

If the relationship between values of x and y is not originally given by an explicit representation, it may still be possible to find one. For instance, the equation $y - 2x = 6$ has the same graph as $y = 2x + 6$. That is, the values of x and y that satisfy $y - 2x = 6$ are exactly the same values that satisfy $y = 2x + 6$. As one example, in either form of the equation, if $x = -1$, then $y = 4$.

DEFINITION

Implicit function

The variable y is an *implicit function of* x if there is an equation relating x and y and any vertical line intersects the graph of the equation in (at most) one point.

The equation $y = 2x + 6$ expresses y as an explicit function of x. However, the equation $y - 2x = 6$ defines y *implicitly* as a function of x.

Beginning with an implicit relationship, we often try to find an explicit one. For instance, consider the formula $2y = x + 6$. As this stands, the value of y is defined implicitly in terms of x. If we solve the equation for y in terms of x, we find $y = (x + 6)/2$, or equivalently, $y = (1/2)x + 3$. In these forms y is expressed *explicitly* in terms of x.

EXAMPLE 1   The graph of a circle and implicitly defined functions

In Section 1.2 we found the equation for a circle. The equation $x^2 + y^2 = 25$ has the graph shown in figure 2.8.2. This graph does not define a function because vertical lines may intersect it in more than one point. For instance, both $P(3, 4)$ and $Q(3, -4)$ are on the graph. However, the graph can be represented if we combine the graphs of the two functions, $f_1(x) = \sqrt{25 - x^2}$ and $f_2(x) = -\sqrt{25 - x^2}$.   ●

It is possible that near a point on the graph of an equation there is a function whose graph is the same as the graph of the equation for nearby points. For instance, consider the point $P(3, 4)$ on the graph of the equation $x^2 + y^2 = 25$. All of the points on that graph that are near $P(3, 4)$ are also points on the graph of $f_1(x) = \sqrt{25 - x^2}$, as is shown in figure 2.8.3 and the enlargement of the boxed portion of that graph shown in figure 2.8.4.

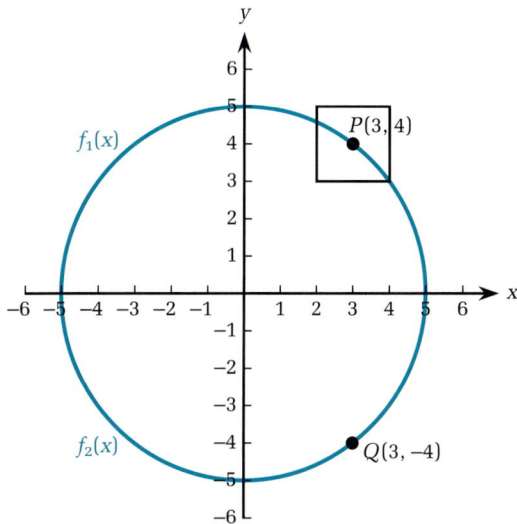

FIGURE 2.8.2

FIGURE 2.8.3

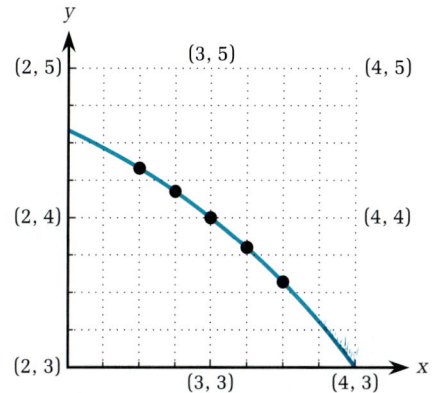

FIGURE 2.8.4

For the general situation, suppose we choose a point on the graph of an equation at which the graph is not intersecting itself and the tangent line is not vertical. Then in some sufficiently great enlargement of a box centered on that point any vertical line will intersect the graph of the equation in only one point. This means that a function exists whose graph duplicates that of the equation at least in the vicinity of the chosen point. This is shown in figure 2.8.5.

Given an equation relating the variables x and y and a point on the graph $P(x_0, y_0)$, suppose $f(x)$ is a function whose graph is the same as the graph of the equation in the vicinity of $P(x_0, y_0)$. If we assume that $f'(x)$ exists, then finding the value of the derivative is possible by using the original equation. The process of doing so is termed *implicit differentiation*.

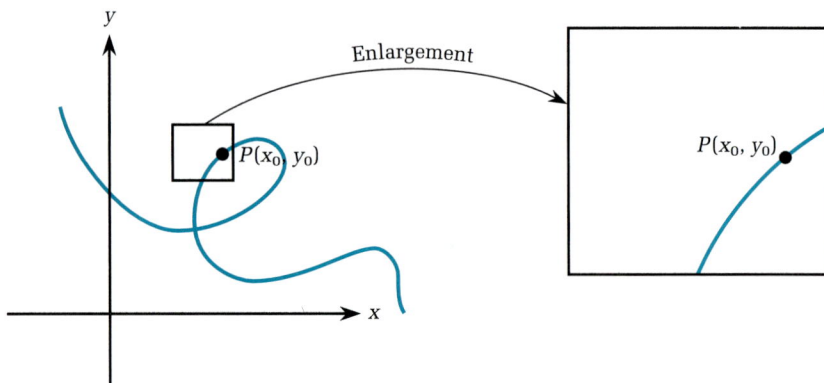

FIGURE 2.8.5

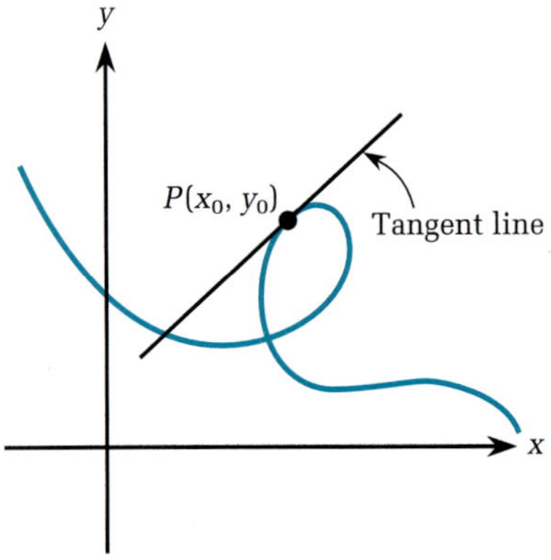

FIGURE 2.8.6

Figure 2.8.6 shows the graph of an equation and the tangent line at $P(x_0, y_0)$.

Implicit Differentiation

Before we attempt a complicated expression, let us illustrate the method of implicit differentiation on a simple problem.

EXAMPLE 2   Implicit differentiation

Suppose $y^2 = x$. Find an expression for dy/dx and evaluate it at $P(4, 2)$.

SOLUTION

FIGURE 2.8.7

In figure 2.8.7 we see the graph of $y^2 = x$ is such that y cannot be described as one function of x. Vertical lines intersect the graph more than once. Before doing implicit differentiation, let us answer the question by using explicit functions. Doing so will give us a check on the method of implicit differentiation.

There are explicit functions whose combined graphs represent the graph of $y^2 = x$. The upper branch is given by $y = \sqrt{x}$ and the lower branch by $y = -\sqrt{x}$. Because the point $P(4, 2)$ is on the upper branch, we find the derivative $\dfrac{d}{dx}\sqrt{x} = 1/(2\sqrt{x})$ and evaluate this at $x = 4$ to give us $1/(2\sqrt{4}) = 1/(2 \cdot 2) = 1/4$.

To illustrate implicit differentiation, suppose we leave the equation as $y^2 = x$ and *assume* that in the neighborhood of the $P(4, 2)$ there is some function whose graph is the same as $y^2 = x$. Because

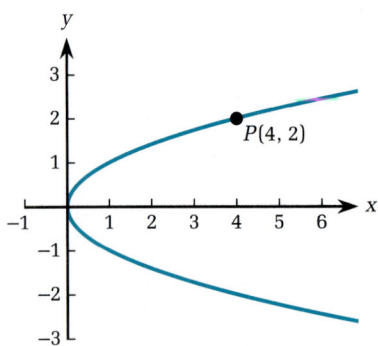

the derivatives of each side of an equation would also be equal, we have

$$(*) \qquad \frac{d}{dx} y^2 = \frac{d}{dx} x$$

On the left side, $\frac{d}{dx} y^2 = 2y \cdot \frac{d}{dx} y$, by the Chain Rule. Here the value of $\frac{d}{dx} y$ is unknown, but we can regard this expression as a variable and simply represent it as dy/dx. On the right side of equation $(*)$ we have $\frac{d}{dx} x = 1$. Hence,

$$(**) \qquad 2y \frac{dy}{dx} = 1$$

Solving this for dy/dx gives

$$(***) \qquad \frac{dy}{dx} = \frac{1}{2y}$$

Notice that we do not have an explicit representation for dy/dx in terms of x. This is to be expected because we do not have y itself explicitly represented as a function of x. Nonetheless, we can determine the value of dy/dx at $P(4, 2)$ because we can substitute $x = 4$, $y = 2$ into equation $(***)$ to get

$$\frac{dy}{dx} = \frac{1}{2 \cdot 2} = \frac{1}{4}$$

This is the same value we obtained by the explicit approach. ●

Let us try a more complicated example, one in which we lack any explicit function whose graph in the neighborhood of the point is the same as the graph of the equation. We continue to assume such a function exists and that it has a derivative.

EXAMPLE 3   Implicit differentiation

Suppose $y = f(x)$ has the same graph as $x^2y + y^3 = 3y + 4$ near $P(1, 2)$. Find dy/dx at $P(1, 2)$.

SOLUTION

Take the derivative with respect to x on each side of the equation.

$$\frac{d}{dx}(x^2y + y^3) = \frac{d}{dx}(3y + 4)$$

The derivative of a sum of functions is the sum of their derivatives, so

$$\frac{d}{dx}(x^2y) + \frac{d}{dx}y^3 = \frac{d}{dx}3y + \frac{d}{dx}4$$

Using the formulas for derivatives of products and powers and the Chain Rule, we find

(*) $$\left\{x^2 \cdot \frac{d}{dx}y + y \cdot \left(\frac{d}{dx}x^2\right)\right\} + 3y^2 \cdot \frac{d}{dx}y = 3\frac{d}{dx}y + 0$$

Notice in equation (*) that whenever the "inside" function is y and we need to find $\frac{d}{dx}$ of that function, we are unable to do more than simply put $\frac{d}{dx}y$ as a "placeholder" for the derivative. Write dy/dx for $\frac{d}{dx}y$. Evaluate $\frac{d}{dx}x^2 = 2x$, and rewrite equation (*) as

(**) $$x^2\frac{dy}{dx} + y \cdot 2x + 3y^2\frac{dy}{dx} = 3\frac{dy}{dx}$$

Collect onto one side of the equality all of those terms containing dy/dx and place all the other terms on the other side, to get

(***) $$x^2\frac{dy}{dx} + 3y^2\frac{dy}{dx} - 3\frac{dy}{dx} = -2xy$$

Factor dy/dx from the terms on the left.

$$(x^2 + 3y^2 - 3)\frac{dy}{dx} = -2xy$$

Solving for dy/dx, we have

(****) $$\frac{dy}{dx} = \frac{-2xy}{x^2 + 3y^2 - 3}$$

This provides us with an expression for dy/dx. The expression contains both x and y as variables, but it is usable because in order to find the derivative at some particular point on the graph, all we need to do is substitute the x- and y-values at that point into equation (****). Doing so at $P(1, 2)$, we have

$$\frac{dy}{dx} = \frac{-2 \cdot 1 \cdot 2}{1^2 + 3 \cdot 2^2 - 3} = \frac{-4}{1 + 12 - 3} = \frac{-4}{10} = -0.4.$$

●

To summarize the process:

> ## FINDING A DERIVATIVE WITH RESPECT TO *x* BY IMPLICIT DIFFERENTIATION
>
> 1. Go through the defining equation, taking the derivative of each term with respect to *x*. Doing so, there will be terms involving $\frac{d}{dx}y$. Because *y* as a function of *x* is not known, at such a stage write dy/dx or y' as a placeholder for the derivative.
>
> 2. Collect all terms that have a factor of dy/dx on one side of the equality sign and factor dy/dx out of those terms.
>
> 3. Divide both sides by the other factor of the dy/dx side of the equation.
>
> 4. Evaluate dy/dx for the values of *x* and *y* at the point given.

One application of implicit differentiation is to determine in what direction the curve is going at some known point on a curve.

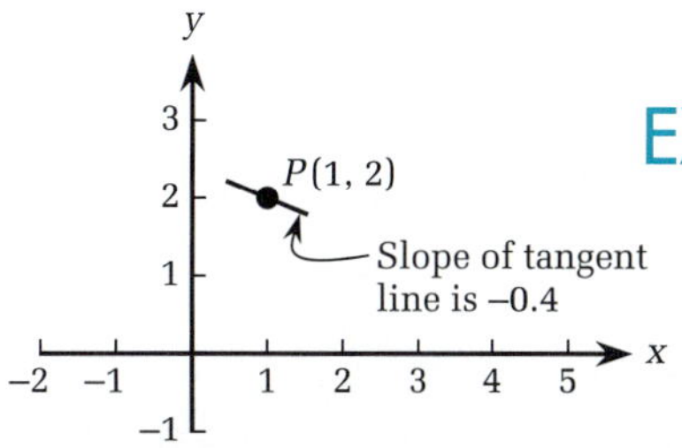

FIGURE 2.8.8

EXAMPLE 4   Drawing a tangent line

In Example 3, the point $P(1, 2)$ is on the graph of $x^2y + y^3 = 3y + 4$ and the tangent line has slope -0.4 at that point. Hence, the curve is going down as *x* increases, as shown in figure 2.8.8. If we were attempting to find other points on the graph near $P(1, 2)$, we would first consider examining points near that tangent line.

EXAMPLE 5   Using implicit differentiation to find the slope of the tangent line

Find the slope of the tangent to the graph of $x^3 + y^3 - 6xy = 0$ at the point $P(3, 3)$.

SOLUTION

This is shown in figure 2.8.9.

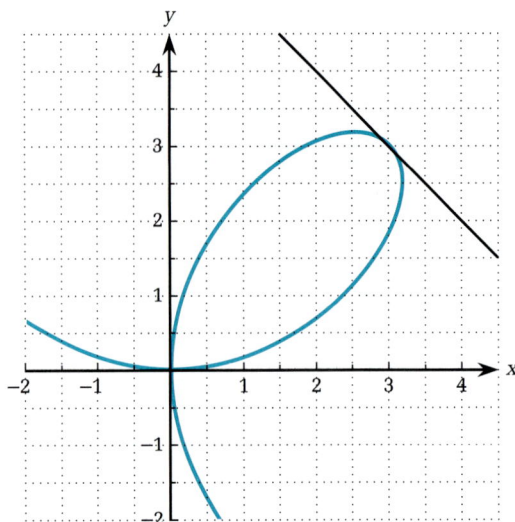

FIGURE 2.8.9

We assume there is a function $y = f(x)$ whose graph in the neighborhood of the point is the same as the given graph. By taking the derivative of both sides of $x^3 + y^3 - 6xy = 0$, we have

$$\frac{d}{dx}(x^3 + y^3 - 6xy) = \frac{d}{dx}0.$$

Using y' to represent dy/dx, we have

$$3x^2 + 3y^2(y') - 6(x \cdot y' + y \cdot 1) = 0.$$

Simplifying this expression and collecting all the terms involving y' together, we get

$$3x^2 + 3y^2y' - 6xy' - 6y = 0$$
$$3x^2 - 6y = -3y^2y' + 6xy' = (-3y^2 + 6x)y'$$

Dividing both sides by $-3y^2 + 6x$ yields

$$y' = \frac{3x^2 - 6y}{-3y^2 + 6x}$$

To find the slope of the tangent line at $P(3, 3)$, we need to evaluate y' at $x = 3$, $y = 3$:

$$y' = \frac{3(3)^2 - 6(3)}{-3(3)^2 + 6(3)} = \frac{27 - 18}{-27 + 18} = \frac{9}{-9} = -1$$

PRACTICE EXERCISE 1

Suppose $x^2y - y^2 = -2$. Find dy/dx in terms of x and y and evaluate it at the point $P(1, 2)$.

2.8 PROBLEMS

Foundations

The problems of this section require the basic skills illustrated by the following:

1. Evaluate $xy - 2x + y$ at $P(1, 1)$.
2. Evaluate $(y - x^2)y^2 - x$ at $P(-1, 1)$.
3. Find the derivative of $(x^2 + 2x)^2\sqrt{2x - 1}$.
4. Find an equation for the tangent line to the graph of $y = 7/(3x - 1)$ for $x = 5$.

Exercises

For problems 5–15, assume each of the given equations determines a function $y = f(x)$ in the neighborhood of the indicated point.
(a) Use implicit differentiation to find dy/dx in terms of x and y.
(b) Evaluate dy/dx at the point given.

5. $xy - 2x + y = 0$ at $P(1, 1)$
6. $2x + 3xy - 4y = 1$ at $P(1, 1)$
7. $x^2 + 2x - y^2 = -1$ at $P(1, 2)$
8. $x^2 - 2xy + y^2 = 1$ at $P(2, 1)$
9. $2x^2 - 5x^2y + y^3 = 2$ at $P(1, 0)$
10. $-2x + x^2y - y^3 = -1$ at $P(0, 1)$
11. $(x^2 + y)x^3 = 3x + y$ at $P(-1, 1)$
12. $(y - x^2)y^2 = x + y$ at $P(-1, 1)$
13. $y\sqrt{x} + x\sqrt{y} - 3x = 3$ at $P(9, 4)$
14. $x^{1/3}y^2 - \sqrt{xy} = 4$ at $P(8, 2)$
15. $(x^2 + 2y)^2\sqrt{2x - 1} = 7y + 1$ at $P(1, 0)$

For the following equations, assume a function exists whose graph in the neighborhood of the indicated point is the same as the graph determined by the equation. Find the equation of the line tangent to the graph at the point indicated.

16. $x^3 - xy = y^3 + 8$ at $P(2, 0)$
17. $x^3 = 3xy - y^3 - y$ at $P(1, 1)$
18. $x^2 + xy + y^3 = 3y + 4$ at $P(2, 1)$
19. $x^2 - xy - y^2 = 4x - 7$ at $P(2, 1)$

Graphing Calculator Problems

When y is not a function of x, then in order to estimate dy/dx from an equation in x and y, a graphing calculator has to be given values for x and y in parametric or polar forms. In such a form, the BOX function can be used to enlarge a screen until there appears to be a graph of a function of x. Then by using the scale shown, or by finding other points on the graph and using graph paper to plot those points, we can estimate dy/dx geometrically.

20. For the range $-4 \le x \le 4$, $-4 \le y \le 4$, and $-2 \le t \le 2$, graph in parametric mode $x(t) = t^3 - 2t$, $y(t) = t^3 + t$. The screen will show that y is not a function of x. Use the BOX or other zoom feature near $t = -0.5$ to include only a part of the graph for which there would be a function with the same graph. Zoom in on the region near the value $t = -0.5$ and estimate the slope of the tangent line to the graph there.

21. For the range $-10 \le x \le 10$, $-10 \le y \le 10$, and $0 \le \theta \le 20$, graph in polar form $r = 0.5\theta$. The screen will show that y is not a function of x. Use the BOX or other zoom feature near $\theta = 4$ to include only a part of the graph for which there would be a function with the same graph. Zoom in on the region near the value $\theta = 4$ and estimate the slope of the tangent line to the graph there.

22. For the range $-2 \le x \le 4$, $-2 \le y \le 4$, and $0 \le \theta \le 2\pi$, graph in polar form $r = (6 \cdot \cos\theta \cdot \sin\theta)/[(\cos\theta)^3 + (\sin\theta)^3]$. Compare this graph with the graph in figure 2.8.1. Use the BOX or other zoom feature near $\theta = \pi/4$ to include only a part of the graph for which there would be a function with the same graph. Zoom in on the region near the value $\theta = \pi/4$ and estimate the slope of the tangent line to the graph there.

Writing and Discussion Problems

23. Give examples that show the difference between an implicit function and an explicit function.

24. If the graph of an equation has a vertical tangent line at a point P, discuss the possibilities for enlargements of a box containing P.

Enrichment Problems

25. Is there an explicit function $y = f(x)$ whose graph is the same as the graph of $x = 2y + 1$?

26. Is there an explicit function $y = f(x)$ whose graph is the same as the graph of $x = y^3 + y$?

SOLUTION TO PRACTICE EXERCISE

1. From $\dfrac{d}{dx}(x^2 y - y^2) = \dfrac{d}{dx}(-2)$ we have

$[x^2 (dy/dx) + y \cdot 2x] - 2y (dy/dx) = 0$.

    Collecting together terms with dy/dx and factoring, we get $(x^2 - 2y)(dy/dx) = -2xy$, or

$dy/dx = -2xy/(x^2 - 2y)$, which at $P(1, 2)$ is equal to $(-2 \cdot 1 \cdot 2)/(1^2 - 2 \cdot 2) = -4/(-3) = 4/3$.

2.9

RELATED RATES

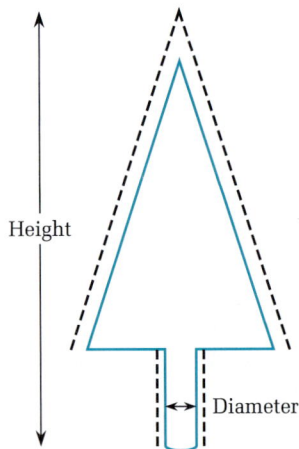

FIGURE 2.9.1

If variables are related, then so are their rates of change. This section covers relationships between different rates of change. Generally, these interconnections are discovered by using the Chain Rule and implicit differentiation.

    Consider some examples of related rates. In the world of business, the amount of money spent on advertising is related to sales. Hence, the rate of change of spending on advertising is related to the rate of change of sales. In the field of medicine, the blood flow in an artery is related to the diameter of that blood vessel, and so the change in the blood flow is related to a constricting or relaxing change in the diameter of the artery. In the study of biology, the height of a tree is related to the diameter of the tree trunk. Therefore, the rate of change of the height of a tree is related to the rate of change of its diameter. The type of question we often encounter is: If a tree is growing taller at a rate of five inches a year, how is that rate related to the rate of increase of the tree's diameter? See figure 2.9.1.

Solving Problems Involving Related Rates

Problems involving related rates are usually phrased in words, leaving it up to us to transform descriptions into formulas and interpret the results of differentiation. The following example illustrates a four-step approach to solving related rates problems. The steps are numbered in the example so we can refer to them later.

EXAMPLE 1   Rates of change of the area and radius of a circle

Suppose the area of a circle is increasing by five square feet each hour. Find the rate of change of the radius when the area is 100 square feet.

SOLUTION

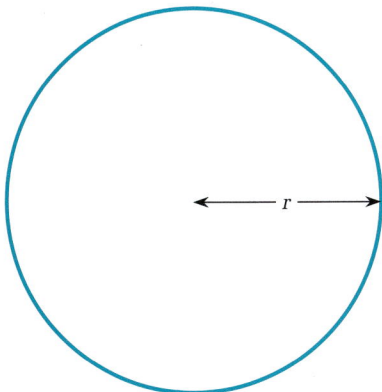

FIGURE 2.9.2

1. Figure 2.9.2 shows a circle. Let r represent the radius of the circle in feet and A stand for the area in square feet. Let t represent time, measured in hours.

2. The area and the radius of a circle are related by the formula

$$A = \pi r^2$$

   The rate of change of area with respect to time is the derivative of A with respect to t, dA/dt, measured in square feet per hour. The rate of change of radius with respect to time is the derivative of r with respect to t, dr/dt, measured in feet per hour.

3. By the Chain Rule and implicit differentiation with respect to t, we can differentiate $A = \pi r^2$ to obtain $dA/dt = \pi \cdot 2r \cdot dr/dt$.

4. Knowing that $A = 100$ and $A = \pi r^2$, we find the radius at that time by solving this equation for r. This gives $100 = \pi r^2$, which implies $r = \sqrt{\dfrac{100}{\pi}}$. Substituting for r, π, and $\dfrac{dA}{dt} = 5$ in the formula for $\dfrac{dA}{dt}$, we have $5 = 2\pi \sqrt{\dfrac{100}{\pi}} \cdot \dfrac{dr}{dt}$. Hence $dr/dt \approx 0.141$ feet per hour.

   The four steps in the example can be followed whenever we need to solve problems involving related rates.

THE FOUR-STEP APPROACH TO SOLVING RELATED RATES PROBLEMS

1. Choose symbols for all the variables, both those known and those to be determined. Be careful that the units of measurement are consistent. If possible, label a sketch of the situation.

2. Write equations that relate the variables to each other. These may be geometric formulas or other known information.

3. Use the Chain Rule and implicit differentiation to take derivatives with respect to the desired variable.

4. Substitute the known information and solve for the desired rate of change. Make certain to give the appropriate units of measure.

Many related rates problems are phrased in geometrical terms, and we will return to another one shortly, but first let us consider several business applications.

Applications in Business

EXAMPLE 2   Related rates in marketing

The president of PZ Co. is interested in marketing floppy discs. The company's Market Research Department estimates that profits from the sale of x cases of discs will be $P(x) = \sqrt{x}$ thousand dollars, and furthermore, if PZ buys t hours of advertising airtime, then the sales will be $x(t) = 10t/(t + 10)$ cases of discs. Find the rate of increase of profit with respect to hours of airtime when $t = 90$.

SOLUTION

1 & 2. The first two steps, determining symbols for the variables and stating equations involving them, has already been done by the statement of the problem.

3. To determine dP/dt use the Chain Rule in the form $dP/dt = dP/dx \cdot dx/dt$. Then we get

$$\frac{dP}{dt} = \frac{d}{dx}\sqrt{x} \cdot \frac{d}{dt}\left(\frac{10t}{t + 10}\right)$$

$$= \frac{1}{2\sqrt{x}} \cdot \left(\frac{(t + 10) \cdot 10 - (10t) \cdot 1}{(t + 10)^2}\right)$$

$$= \frac{1}{2\sqrt{x}} \cdot \left(\frac{100}{(t + 10)^2}\right)$$

4. When $t = 90$, then $x = (10 \cdot 90)/(90 + 10) = 900/100 = 9$, which gives us

$$\frac{dP}{dt} = \frac{1}{2\sqrt{9}} \cdot \left(\frac{100}{(90 + 10)^2}\right) = \frac{100}{6 \cdot 100^2} = \frac{1}{600}$$

This is in thousands of dollars per hour of airtime, so our interpretation is that profits would be increasing at the rate of $(1/600)\$1000 = \$1.66$ per additional hour of advertising airtime.

An alternate approach to solving the last example would be to substitute the value for x directly into $P(x) = \sqrt{x}$ to obtain $P(t) = \sqrt{10t/(t + 10)}$ and then find dP/dt directly. We can check that the answer will be the same, but we wanted to illustrate here the *related rate* concept that: "The rate of change of profit with respect to advertising airtime" equals "the rate of change of profits with respect to sales" times "the rate of change of sales with respect to advertising airtime."

PRACTICE EXERCISE 1

In Example 2, find the increase in the rate of profit with respect to advertising airtime when $t = 10$.

Related rate problems often involve changes with respect to time. However, that is not the only sort of related rate we want to solve. The next application discusses the connection between "the rate of change of price with respect to quantity" and "the rate of change of quantity with respect to price."

Related Rates with Marginal Demand

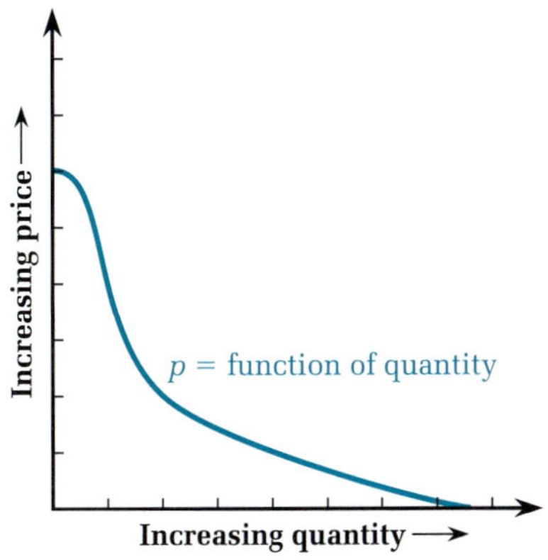

Suppose $p = p(x)$ is a demand function, for which p is price and x is the quantity. One such function is graphed in figure 2.9.3 in the standard manner, with quantity along the horizontal axis and price along the vertical axis. In this way, price is given as a function of quantity. The graph in figure 2.9.4 shows the same relationships between price and quantity, but with quantity graphed as a function of price.

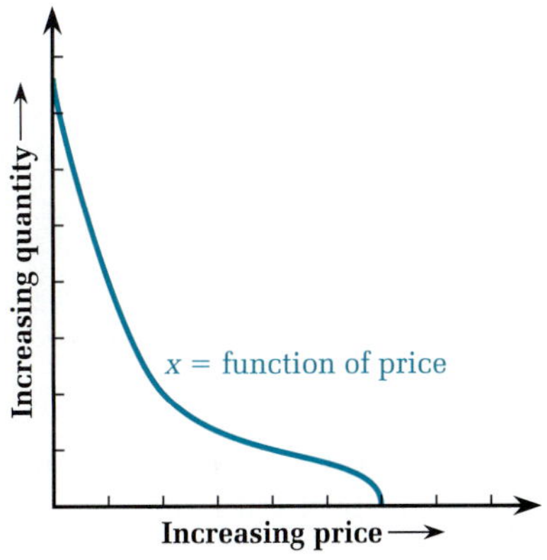

We can determine the relation between the rate that sales decrease as a function of price, and the rate that price decreases as a function of quantity. These are related rates.

Of the four steps, both 1 and 2 have been taken care of by the statement of the situation.

Let us suppose the equation $p = p(x)$ implicitly defines a function for x in terms of p. Because $p = p(x)$ is a demand function, dp/dx is the rate of change of price with respect to quantity *(marginal demand)* and dx/dp is the rate of change of quantity with respect to price.

3. By the Chain Rule, if we begin with $p = p(x)$ and use implicit differentiation with respect to p we have $\dfrac{dp}{dp} = \dfrac{dp}{dx} \cdot \dfrac{dx}{dp}$.
   Because $dp/dp = 1$, we have $dp/dx \cdot dx/dp = 1$, which shows $\dfrac{dx}{dp} = \dfrac{1}{\dfrac{dp}{dx}}$.

4. Hence, the rate of change of quantity with respect to price is equal to the reciprocal of the marginal demand (the rate of change of price with respect to quantity).

FIGURE 2.9.3

Increasing price → (vertical axis)
Increasing quantity → (horizontal axis)
$p = $ function of quantity

FIGURE 2.9.4

Increasing quantity → (vertical axis)
Increasing price → (horizontal axis)
$x = $ function of price

To summarize:

RELATED MARGINAL DEMAND RATES

Suppose $p = p(x)$ is a demand function, whereby price, p, is a function of quantity, x. Then

$$\frac{dp}{dx} \cdot \frac{dx}{dp} = 1$$

EXAMPLE 3   Economics, marginal demand

Consider the demand curve $p = 1000/(x^2 + 1)$, where p is in dollars and x is units of some commodity. A graph of this relationship is shown in figure 2.9.5.

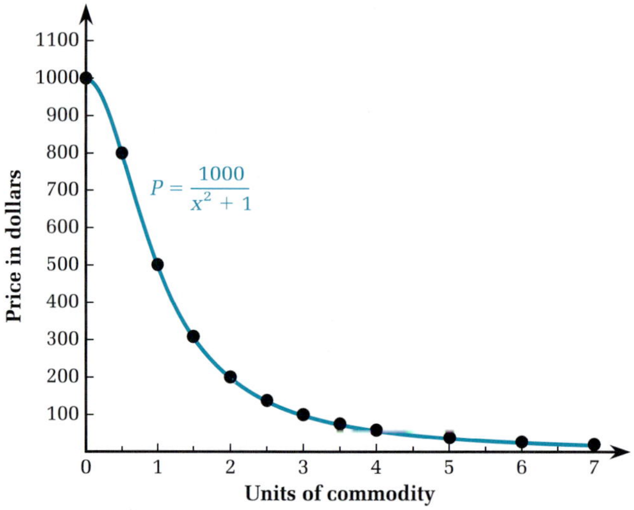

FIGURE 2.9.5

Find the marginal demand at $x = 6$ and find the rate of change of quantity with respect to price at $x = 6$. Interpret your results.

SOLUTION

To simplify taking the derivative, rewrite the relationship as $p = 1000(x^2 + 1)^{-1}$. The marginal demand is $dp/dx = -1000(x^2 + 1)^{-2} \cdot 2x = -2000x/(x^2 + 1)^2$.

At $x = 6$ the marginal demand is $(-2000 \cdot 6)/(6^2 + 1)^2 = -12,000/37^2 \approx -8.7655$. This means that if the quantity is increasing at $x = 6$, the price has to be falling at a rate of 8.7655 dollars per unit increase.

The rate of change of quantity with respect to price is $dx/dp = 1/(dp/dx) = 1/(-8.7655) \simeq -0.1141$. This means that if the price is increasing when the demand is $x = 6$, then the demand will be decreasing at the rate of -0.1141 units per dollar increase in price.

Related Rates in Implicit Form

We now return to problems of related rates with respect to time. The next example illustrates the situation in which we do not find an explicit formula for one variable in terms of another, but simply leave the equation in implicit form.

EXAMPLE 4   Related rates, motion pictures

A movie camera is mounted on an elevator on the outside of Awesome Towers Hotel. As the elevator rises at the rate of 20 ft/sec, the camera will be aimed at a fight scene occurring in the back seat of a convertible racing away from the hotel at a speed of 88 ft/sec. Focusing of the camera will be done by a computer, so it needs to know at every instant how fast the distance from the camera lens to the back seat is changing. Find the rate of change of distance from the camera to the car at 10 seconds. Assume that both the elevator and the car were at the same point at $t = 0$.

SOLUTION

Use the four-step approach.

1. See the schematic in figure 2.9.6.

   Let h be the height of the elevator, c be the distance of the car, and D be the distance between the two. We are being asked to find dD/dt when $t = 10$.

2. By the Pythagorean Theorem for right triangles, $D^2 = h^2 + c^2$.

3. Taking derivatives on both sides with respect to t, $\dfrac{d}{dt}D^2 = \dfrac{d}{dt}(h^2 + c^2)$, which gives

   (##) $2D \cdot \dfrac{dD}{dt} = 2h \cdot \dfrac{dh}{dt} + 2c + \dfrac{dc}{dt}$

4. We were given that $dh/dt = 20$ and $dc/dt = 88$. Furthermore, at $t = 10$ seconds, $h = 20 \cdot 10 = 200$ ft and $c = 88 \cdot 10 = 880$. We can use these to find D at $t = 10$ by again using

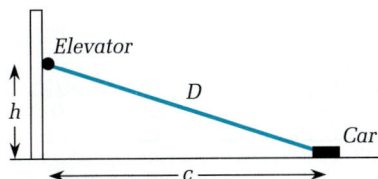

FIGURE 2.9.6

$$D^2 = h^2 + c^2 = (200)^2 + (880)^2 = 40,000 + 774,400 = 814,400, \text{ or } D = \sqrt{814,400} = 902.4 \text{ ft.}$$

Substituting these values into equation (##) gives

$$2 \cdot 902.4 \cdot \frac{dD}{dt} = 2 \cdot 200 \cdot 20 + 2 \cdot 880 \cdot 88$$

$$1804.8 \cdot \frac{dD}{dt} = 8000 + 154,880 = 162,880$$

$$\frac{dD}{dt} = \frac{162,880}{1804.8} = 90.25 \text{ ft/sec}$$

This value for dD/dt is reasonable because if the elevator and car were going in exactly opposite directions, their combined velocity would be 108 ft/sec, and if they were going directly toward each other, their head-on velocity would be 68 ft/sec.

WARNING

In solving problems involving related rates, sometimes numerical values are substituted too early. Keep in mind that derivatives represent rates of changes, and we must not fix the variables as constants prior to taking the derivative. We must allow the variables to change values.

To illustrate this warning, in Example 4, at $t = 10$ we found $h = 200$ and $c = 880$. If we had used these values in our equation for D^2 before we took the derivative, then D^2 would have been a number, namely 814,400. The derivative of this number would have been zero. We need to wait until we are ready to evaluate the derivatives before we substitute numeric values.

In other words, to create the related rates equation involving derivatives we first must differentiate. Only then may we substitute specific values.

2.9 PROBLEMS

Foundations

The problems of this section require the basic skills illustrated by the following:

1. If revenue is given by $R(x) = 250x - x^2$ and the cost is $C(x) = 5x + 0.1x^2$, then find the equation for profit.

2. Find $\dfrac{d}{dt}\sqrt{3t + 10}$.

3. Find $\dfrac{d}{dh}0.002h^{3/2}$.

Exercises

For the demand functions given in problems 4–11, find the following *related demand rates*:
(a) dp/dx
(b) dx/dp (the rate of change of quantity with respect to price)

4. $p = -x^2 + 15$

5. $p = -x^2 - 2x + 19$

6. $xp^3 + p = 5$

7. $x^2p + 2x = 7$

8. $px + p^2 = -20x + 400$ at $x = 15, p = 5$

9. $120x + p^2 + px = 2400$ at $x = 10, p = 30$

10. $xp^3 + 2xp + p = 10$ at $x = 3, p = 1$

11. $x^2 p^2 + x + 2p = 8$ at $x = 2, p = 1$

12. *(State Lotteries)*   In 1989, Pennsylvania had a lotto game with a jackpot of more than 100 million dollars. Such state lotteries are operated by private companies, who typically take about 15% of all the money wagered for their expenses and profit. Suppose that S. K. Y. Lotto Co. is operating a lotto game and collecting 15% for their expenses and profit. They notice that sales for the weekly game have totaled approximately $S(t) = 250 + 10t^3$ thousand dollars by day t. What is their rate of expenses and profit on the seventh day?

13. *(Related Rates, Cable TV Shopping)*   When a discount of x percent is offered on an item, ONAIR Products can sell $s(x) = 101 - 50/(x + 1)$ percent of that item ($0 \leq x \leq 49$). Over a half hour, the discount offered is $x = \sqrt{3t} + 10$ (t in minutes, $0 \leq t \leq 30$). Find ds/dt at $t = 10$ and interpret your answer.

14. *(Biology, Forestry)*   A spruce tree has a conical shape in which the height is approximately three times the diameter of the branches at the base. The volume of a cone is $(1/3)\pi r^2 h$, where r is the radius of the base and h is the height (see figure 2.9.7.

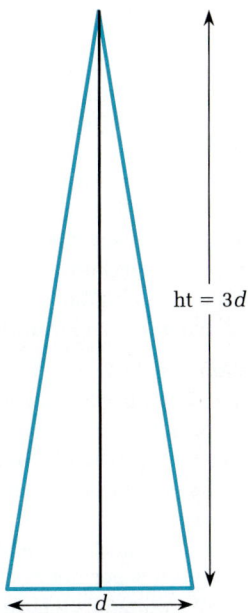

ht = 3d

FIGURE 2.9.7

(a) Suppose a tree is growing at a rate of six inches a year. If the tree is 30 feet tall, what is the rate of change of volume of the tree?

(b) Suppose a tree has a height given by $h(t) = \sqrt{t}$ after growing for t years. What is the rate of change of the volume of the tree when the tree is 25 years old?

15. *(Forestry)*   The diameter of the trunk of a western coniferous tree is related to the height of the tree by the formula $d = (0.002)h^{3/2}$.* Suppose the height in year t is approximated by $h(t) = 100t^2/(1 + t^2)$ feet.
   (a) Find the rate of change of the diameter when $t = 10$.
   (b) Find the rate of change of the diameter when $t = 20$.

16. *(Related Rates of Revenue, Cost, and Profit)*   Suppose a revenue function is $R(x) = 250x - x^2$ and a cost function is $C(x) = 5x + 0.1x^2$. Furthermore, if $x = 10$, then $dx/dt = 5$. For $x = 10$, find:
   (a) dR/dt
   (b) dC/dt
   (c) dP/dt (the rate of change of profit)

17. *(Related Rates of Revenue, Cost, and Profit)*   Suppose the revenue function is $R(x) = 50x - 2x^2$ and the cost function is $C(x) = 3x + 0.05x^2$. Furthermore, if $x = 5$ then $dx/dt = 7$. For $x = 5$, find:
   (a) dR/dt
   (b) dC/dt
   (c) dP/dt (the rate of change of profit)

   Problem 18 comes from the work of Jean Léonard Marie Poiseuille (1797–1869), who lived in France and studied the circulation system of blood in mammals. He was the first person to measure the expansion of an artery during a contraction of the heart (the artery expanded by 1/23 of its diameter). He revolutionized the measurement of blood pressure.

   Blood pressure had been measured by inserting a brass tube directly into a major artery and letting the blood flow out and up through a connected glass tube as high as the heart could pump it. That was the way Stephen Hales (1677–1761), the first investigator of blood pressure, had learned a horse's heart could pump blood eight feet up a tall tube. Because the blood coagulated outside the body, the glass tube had to have a fairly large diameter. A test consumed a lot of blood.

   Poiseuille greatly improved the process by using a mercury-filled manometer, which allowed a shorter tube, and by using calcium carbonate to pre-

*T. McMahon, "The Mechanical Design of Trees," *Scientific American* 233: 1 (July 1975), pp. 93–103.

vent clotting, which allowed a narrower tube. Even so, a blood-pressure check was a very different experience in Paris in 1840 than what we casually undergo today.

18. **(Biomedical, Poiseuille's Law for Arterial Blood Flow)**   Blood has greater velocity in the center of an artery than it does near the edge. For an artery with radius R, the velocity at a point r from the center of the artery is given by $V = c(R^2 - r^2)$, where c is a constant that has to be determined for the particular artery. See figure 2.9.8.

FIGURE 2.9.8

Suppose an artery has a radius R of 0.08 mm (millimeters), and because of a drug that constricts arteries (such as ergonovine), the radius is decreasing at the rate of 0.00005 mm per minute (that is, $dR/dt = -0.00005$). At a location 0.02 mm from the center of the artery (and leaving c unevaluated), what is dV/dt?

19. **(Related Rates, Biomedical—Tumor Size)**   A spherical tumor is being treated by a chemical that works on the surface of the tumor. It reduces the radius of the tumor by 0.1 inches per day. What is the rate of reduction of the volume of the tumor when the tumor has radius 2 inches? (A sphere of radius r has volume $V = (4/3)\pi r^3$.)

20. Suppose a spill of honey has occurred, resulting in a pool with radius 20 inches. Around the entire circumference of the pool an army of ants is eating as quickly as possible. They are shrinking the radius of the pool at the rate of 0.2 inches per minute. What is the rate of reduction of the area of the pool of honey when the pool has a radius of 16 inches?

21. **(Movies, Automatic Focusing)**   A camera is positioned 1200 feet away from the base of a space

shuttle launch. Computer controls are used to keep the shuttle in focus at all times. Thus, the computer must know the rate at which the distance from the camera to the shuttle is changing. The camera operator has estimated the shuttle's velocity is going to be 110 feet per second when the shuttle has risen 500 feet into the air. At that instant, how fast is the distance from the shuttle to the camera changing? See figure 2.9.9.

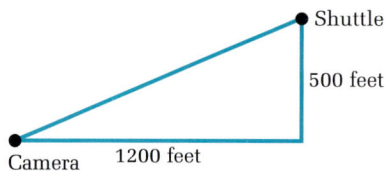

FIGURE 2.9.9

22. Redo problem 21 with the camera 100 yards from the base of the launch.

23. **(Oil Spills at Sea)**   There have been oil spills as great as 25 million gallons from oil tankers. Consider the following situation for a more limited spill of 10 million gallons.

When the oil is first dumped on the water it makes a very thick pool, which thins out as it spreads. Assume that the shape it takes is always cylindrical.

There are about eight gallons to the cubic foot, so 10 million gallons would be approximately 1.25 million cubic feet. The formula for the volume of a cylinder is $\pi r^2 h$, where r is the radius and h is the height. If we assume the pool was a cylinder 10 ft deep, it would have a radius of approximately 200 feet.

Actually, the rate at which the radius grows depends on a variety of things, but let us assume that $r(t) = 100t$ feet, where t is in minutes. With that assumption, and knowing that the total volume of oil is constant (so $dV/dt = 0$), find dh/dt when $t = 240$ minutes.

24. **(Froude's Law, Sharks)**   In Section 2.5 we introduced Froude's Law: The maximum velocity of a fish of length l feet is approximated by $V_{max}(l) = k\sqrt{l}$ for some constant k. The Blacktip reef shark (*Carcharhinus melanopterus*), a member of the Requiem class of sharks, often grows to a length of nine feet. Such sharks have been clocked swimming at 18 miles per hour.

Suppose a shark of t years has length $l(t) = 10t/(2 + t)$ feet. Find $\dfrac{d}{dt} V_{max}$ at $t = 2$ and interpret your result.

Enrichment Problems

25. **(Ecology, Nonreturnable Wine Jars)**   In Roman times vast quantities of inexpensive wine were shipped from Spain to Rome in non-returnable clay jars, which were then smashed and piled behind the docks. By the second century A.D. the mound, known as Monte Testaccio, contained an estimated 40 million jars. These formed a mound 140 feet high with a diameter of 1200 feet.*

*Montroll and Badger, *Introduction to Quantitative Aspects of Social Behavior,* New York: Gordon and Breach Science Publishers, 1974, p. 2.

Use the formula for the volume of a cone, $V = (1/3)\pi r^2 h$ (r radius of base, h height), and assume the ratio of height to diameter remained the same as Monte Testaccio grew. If $dV/dt = 250{,}000$ cubic feet per year, then what were the rates dh/dt and dr/dt at the time the mound was 140 feet high?

SOLUTION TO PRACTICE EXERCISE

1. When $t = 10$, then $x(10) = (10 \cdot 10)/(10 + 10) = 100/20 = 5$. Hence,

$$\frac{dP}{dt} = \frac{1}{2\sqrt{5}} \cdot \left(\frac{100}{(10 + 10)^2}\right) = \frac{1}{2\sqrt{5}}\left(\frac{100}{20^2}\right)$$

$$= \frac{100}{800\sqrt{5}} \approx 0.0559$$

CHAPTER 2 REVIEW

Discuss or define:

1. Graphing conventions
2. Limits (one-sided, two-sided)
3. Alternate functions for rational functions
4. Average rate of change
5. Secant line
6. Instantaneous rate of change
7. Difference quotient
8. Derivative
9. Alternate notations for the derivative
10. Tangent line
11. Marginals (cost, revenue, profit)
12. Average and instantaneous velocity
13. Instantaneous rate of change
14. Average cost, revenue, and profit
15. Marginal averages
16. Chain Rule
17. Generalized Power Rule
18. Implicit derivatives
19. Related rates

Formulas for Derivatives

The derivative of $f(x)$ is

$$f'(x) = \lim_{h \to 0}\frac{f(x + h) - f(x)}{h}$$

If c and n are constants and if derivatives exist for $f(x)$ and $g(x)$, then

I. $\dfrac{d}{dx}c = 0$

II. *(Power Rule)* $\dfrac{d}{dx}x^n = n \cdot x^{n-1}$

III. *(Constant Multiple Rule)* $\dfrac{d}{dx}(cf) = c \cdot \dfrac{d}{dx}f$

IV. *(Sum and Difference Rules)*
$$\dfrac{d}{dx}(f \pm g) = \dfrac{d}{dx}f \pm \dfrac{d}{dx}g$$

V. *(Product Rule)* $\dfrac{d}{dx}(f \cdot g) = f \cdot \dfrac{d}{dx}g + g \cdot \dfrac{d}{dx}f$

VI. *(Quotient Rule)* $\dfrac{d}{dx}\left(\dfrac{f}{g}\right) = \dfrac{g \cdot \dfrac{d}{dx}f - f \cdot \dfrac{d}{dx}g}{g^2}$,
provided $g(x) \neq 0$

VII. *(Chain Rule)* $\dfrac{d}{dx}(f(g(x))) = \left(\dfrac{d}{dg}(f)\right) \cdot \dfrac{d}{dx}g$

VIII. *(Generalized Power Rule)*
$$\dfrac{d}{dx}(f)^a = a[f]^{a-1} \cdot \dfrac{d}{dx}f$$

REVIEW PROBLEMS FOR CHAPTER 2

1. Use the graph of $f(x)$ shown in figure 2.10.1 to answer the parts of this question.

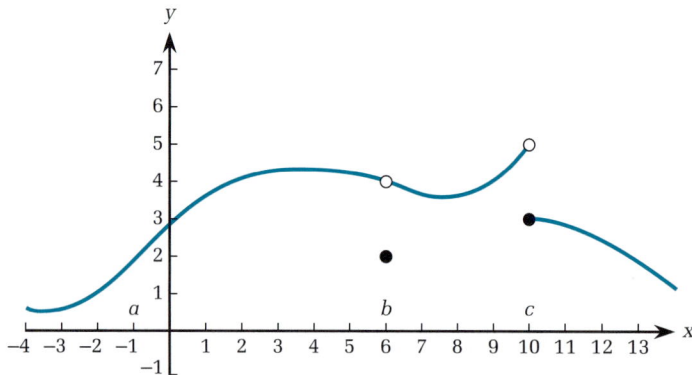

FIGURE 2.10.1

Find those limits and values of $f(x)$ that exist.
(a) $\lim\limits_{x \to a} f(x)$ (b) $f(a)$ (c) $\lim\limits_{x \to b^-} f(x)$
(d) $\lim\limits_{x \to b^+} f(x)$ (e) $\lim\limits_{x \to b} f(x)$ (f) $f(b)$
(g) $\lim\limits_{x \to c^-} f(x)$ (h) $\lim\limits_{x \to c} f(x)$ (i) $f(c)$

For Problems 2–9, find those limits that exist.

2. (a) $\lim\limits_{x \to 3} \dfrac{x + 3}{2x}$ (b) $\lim\limits_{x \to 3} \dfrac{x + 2}{x - 3}$
(c) $\lim\limits_{x \to 3} \dfrac{x^2 - x - 6}{x - 3}$

3. (a) $\lim\limits_{x \to 2} \dfrac{x - 5}{x + 2}$ (b) $\lim\limits_{x \to 2} \dfrac{x}{x - 2}$
(c) $\lim\limits_{x \to 2} \dfrac{x^2 + x - 6}{x - 2}$

4. (a) $\lim\limits_{x \to 3^-} \dfrac{x - 3}{|x - 3|}$ (b) $\lim\limits_{x \to 3^+} \dfrac{x - 3}{|x - 3|}$
(c) $\lim\limits_{x \to 3} \dfrac{x - 3}{|x - 3|}$

5. (a) $\lim\limits_{x \to 3^-} \sqrt{3 - x}$ (b) $\lim\limits_{x \to 3^+} \sqrt{3 - x}$

6. $\lim\limits_{h \to 0} (3h^2 - 2h + 3)$ **7.** $\lim\limits_{h \to a} (3h + 2)$

8. $\lim\limits_{h \to 0} \dfrac{3h^2 + 2h}{h}$ **9.** $\lim\limits_{h \to 0} \dfrac{(a + h)^3 - a^3}{h}$

10. Suppose ground beef is \$2.40 a pound.
(a) What range of prices appears on packages that weigh within two ounces of one pound?
(b) What range of weights would appear on packages that cost within 20 cents of \$2.00?

11. Find the slope of the line from $(2, f(2))$ to $(3, f(3))$ for $f(x) = 1/(2x)$.

12. Find the slope of the line from $(2, f(2))$ to $(4, f(4))$ for $f(x) = 2/x$.

13. Find the average rate of change of $f(x) = 3x - \sqrt{2x}$ over the interval $x = 0$ to $x = 8$.

14. Find the average rate of change of $g(x) = 24x - 3x^2$ over the intervals:
   (a) $x = 0$ to $x = 4$ (b) $x = 0$ to $x = 8$

15. Find the instantaneous rate of change for $f(x) = 10x - 2x^2$ at $x = 2$.

16. Find the instantaneous rate of change for $g(x) = x^2 + 3$ at $x = 1$.

17. The graph in figure 2.10.2 gives ticket prices for the Denver Broncos football games from the founding of the team in 1963 until 1989.

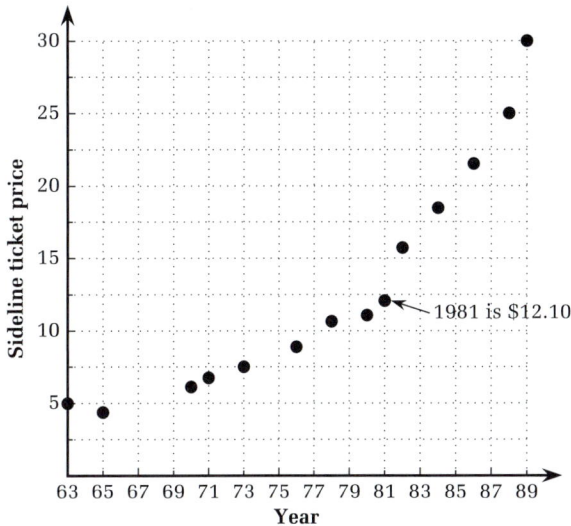

FIGURE 2.10.2

   (a) Find the average rate of increase in ticket prices from 1963 to 1989.
   (b) Find the average rate of increase in ticket prices from 1981 to 1989.
   (c) How many years did it take for ticket prices to increase 20%, starting in 1963?
   (d) What percent did ticket prices increase from 1988 to 1989?

18. The graph in figure 2.10.3 summarizes a study done on the relationship between the density (number of cars per mile) and the flow (number of cars per hour) of cars going through the Lincoln Tunnel.*
   (a) If the density increases from 20 to 40 cars per mile, estimate the average change in flow.

*Montroll and Badger, *Quantitative Aspects of Social Behavior*, New York: Gordon and Breach Science Publishers, 1974, p. 248.

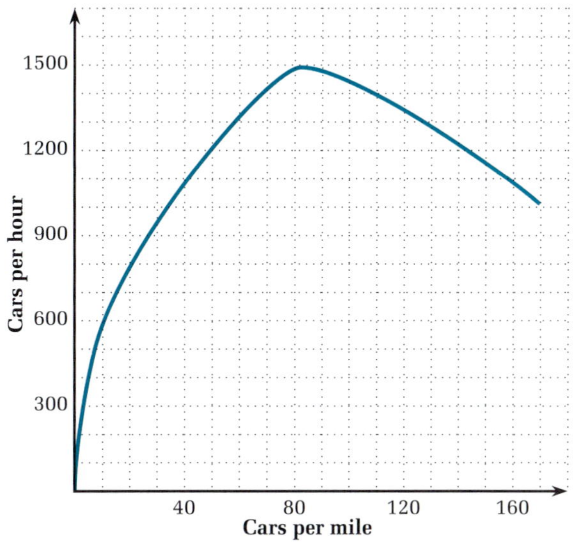

FIGURE 2.10.3

   (b) If the density increases from 40 to 140 cars per mile, estimate the average change in flow.
   (c) If the density increases from 80 to 160 cars per mile, estimate the average change in flow.
   (d) Estimate the slope of the tangent line at a density of 80 cars per mile.

19. Find the difference quotient for $f(x) = x^2 + 3x$.

20. Find the difference quotient for $f(x) = 2/x^2$.

21. Find the difference quotient for $f(x) = 2\sqrt{3x}$.

22. Use the definition of the derivative as $f'(x) = \lim_{h \to 0} \dfrac{f(x + h) - f(x)}{h}$ to find the derivatives of the following functions.
   (a) $f(x) = 2x^2 - 3x$ (b) $f(x) = \dfrac{1}{x + 2}$
   (c) $f(x) = 2x - \dfrac{3}{x}$

23. Find an equation for the tangent line to $f(x) = \sqrt{x + 2}$ when $x = 7$.

24. Find an equation for the tangent line to $g(x) = 3x/(x^2 + 1)$ when $x = 3$.

Find derivatives of the functions in problems 25–38.

25. $f(x) = 2x^3 - x + 5$ **26.** $g(x) = 3x^{-2} - x^5$

27. $f(x) = 3x^2 - \dfrac{1}{x}$ **28.** $h(x) = x^2 + 3x + \dfrac{1}{x^2}$

29. $f(t) = \sqrt{5t}$

30. $g(x) = (3x - 1)(x^2 + 8x)$

31. $f(x) = \left(x + \dfrac{1}{2}\right)(-x^2 + 2x)$

32. $f(x) = (x^3 - 2)(3\sqrt{x} + 4)$

33. $g(x) = (2x^3 + x^{-2})^5$ **34.** $f(t) = (t^2 + t)^{-3}$

35. $g(x) = \dfrac{3x - 5}{x^2 + 3}$ **36.** $g(x) = \dfrac{x^2 - x}{16 - x^2}$

37. $f(x) = (3x^2 - x + 2)\sqrt{x + 2}$

38. $f(x) = \dfrac{\sqrt{x^2 - x}}{x^2}$

For problems 39–42, evaluate the derivative at the indicated value.

39. For $f(x) = \sqrt{2x^3 + 2x + 5}$, find $f'(2)$.

40. For $f(x) = (x^2 + 1)^{-5}$, find $f'(1)$.

41. For $y = (x^2 + 1)^4$, find $\dfrac{dy}{dx}$ at $x = 1$.

42. For $y = \dfrac{2x}{x^2 + 1}$, find $\dfrac{dy}{dx}$ at $x = 2$.

For problems 43 and 44, use implicit differentiation to find the slope of the line tangent to the curve determined by the given equation at the indicated point.

43. $x^3 - 4x^2y + y^3 + 7 = 0$ at $P(2, 1)$

44. $x^3 - 3x^2y + y^2 + \dfrac{2}{x} = 1$ at $P(1, 2)$

45. For the demand function $p(x) = -x^3 + 5x + 100$, find:

(a) $\dfrac{dp}{dx}$ (b) $\dfrac{dx}{dp}$

46. Suppose that a revenue function is $R(x) = 100x/(x + 1)$. Find the marginal revenue when $x = 1/2$.

47. The cost of x items is $C(x) = 100 + 25x^2$ in dollars.
(a) Find the average cost per item if 15 items are bought.
(b) Find the marginal average cost at $x = 15$.

48. Suppose that white cotton T-shirts cost $3 each and that after being silk-screened they are sold at a price determined by a demand curve of $p(x) = -0.01x^2 + 7$. Find the marginal profit at $x = 5$ and interpret your answer.

49. The Petroglyph Art Shop always wastes three stones when beginning a new job of stone carving. Suppose the cost of carving x stones is $C(x) = 3x + 25/x$.
(a) Find the adjusted average cost of the $(x - 3)$ salable stones.
(b) Find the marginal adjusted average cost at $x = 5$.

50. The demand function for Mom's cinnamon rolls is $p(x) = 0.09 - 0.003x$.
(a) Find $R(x)$.
(b) Find the marginal revenue when $x = 10$.

51. *(Handicapping Footraces)*   The frontier function that describes the fastest a man can run five kilometers is $f(x) = 11.1674 - 0.0625x + 0.0023x^2 + 64.9426/x$ minutes, where x is the age of the man.* Find $f'(20)$ and interpret your value.

52. Suppose that the utility of money is given by $U(x) = 0.2x^{0.2}$, wherein x is measured in millions of dollars. Find the marginal utility at $x = 2$.

53. The rate of oxygen consumption by mammals is approximated by $O(m) = 11.6(m)^{0.76}$, for which m is the mammal's weight in kilograms and $O(m)$ is measured in milliliters of oxygen per minute.†
(a) Find $O'(m)$.
(b) Evaluate $O'(90)$ and interpret your answer.

*J. D. Camm and T. J. Grogan, "An Application of Frontier Analysis: Handicapping Running Races," *Interfaces*, 18:6 (Nov.–Dec. 1988), pp. 52–60.
†Knut Schmidt-Nielsen, *Scaling: Why Is Animal Size So Important?* Cambridge: Cambridge U. Press, 1984, p. 105.

54. The power input required by the bird known as the budgerigar has been determined experimentally to be approximately $P(v) = 0.180(v - 9.7)^2 + 3.25$. In this expression, v is the velocity measured in meters per second and $P(v)$ is measured in watts.*
(a) Find $P'(v)$.
(b) Evaluate $P'(6)$ and interpret your answer.

55. Suppose a ball is thrown up and its height t seconds later is $H(t) = -16t^2 + 40t + 6$ feet above ground.
(a) When is the ball at its highest point?
(b) How high is its highest point?

56. The Recycle Club estimates there are 100 pounds of aluminum cans left after the picnic. The number of minutes it takes to collect x pounds of those cans is $t(x) = 10 + 200/\sqrt{100 - x}$.
(a) Find $t(19)$ and $t(75)$.
(b) Find $t'(19)$ and $t'(75)$.
(c) Give interpretations of $t'(19)$ and $t'(75)$.
(d) Suppose the members of the club doing the collecting are expending Calories at the rate of 60 Calories per minute. Find the rate of expenditure of Calories per pound of cans collected when 75 pounds have been collected.

*Ibid., p. 90.

57. A balloon is being blown up at the rate of two cubic feet per minute. What is the instantaneous rate of change of the radius with respect to time when $r = 1$ foot? (The volume of a sphere is $V = (4/3)\pi r^3$, where the radius is r.)

58. A circular skin wound is healing and its radius is decreasing at the rate of two millimeters per day. (Remember, in this situation dr/dt is negative.) How fast is the area of the wound decreasing when the wound has a radius of 14 millimeters?

GRAPHING AND OTHER APPLICATIONS OF DERIVATIVES

OUR UNDERSTANDING OF A FUNCTION IS GREATLY AIDED BY A DISPLAY OF its graph, or at least a portion of the graph. These displays are called *windows* or *screens* of the graph, whether they are truly electronic screens or produced freehand by connecting together some points whose coordinates have somehow been calculated. Until now we have not discussed how to choose screens. In this chapter we learn how to find and display those screens that show such interesting sections of the graph as high or low points, asymptotes, or breaks in the graph.

Computers or graphing calculators easily calculate many values of the function. However, if we calculate millions of values but none of those values are near "interesting" points on the graph, their display on a screen does not help us understand the function.

For instance, suppose $f(x) = 2x^3 - 9x^2 + 12x - 2$. Figures 3.0.1–3.0.3 show three screens, each with indicated ranges.

$y = 2x^3 - 9x^2 + 12x - 2$

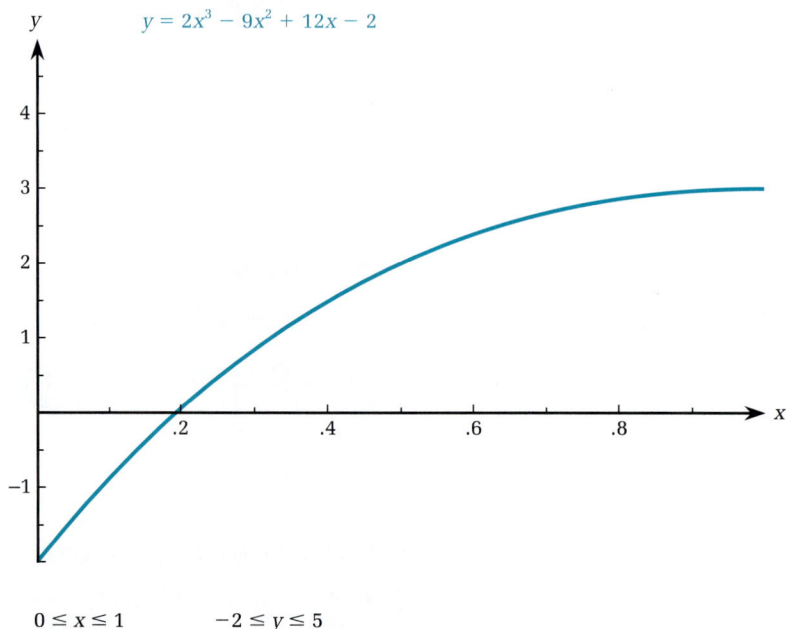

$0 \le x \le 1$ $-2 \le y \le 5$

FIGURE 3.0.1

$y = 2x^3 - 9x^2 + 12x - 2$

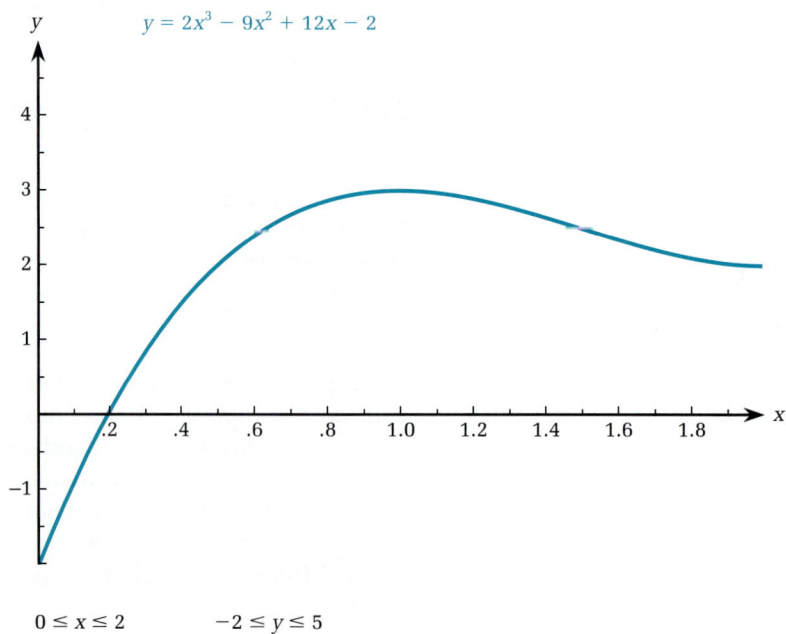

$0 \le x \le 2$ $-2 \le y \le 5$

FIGURE 3.0.2

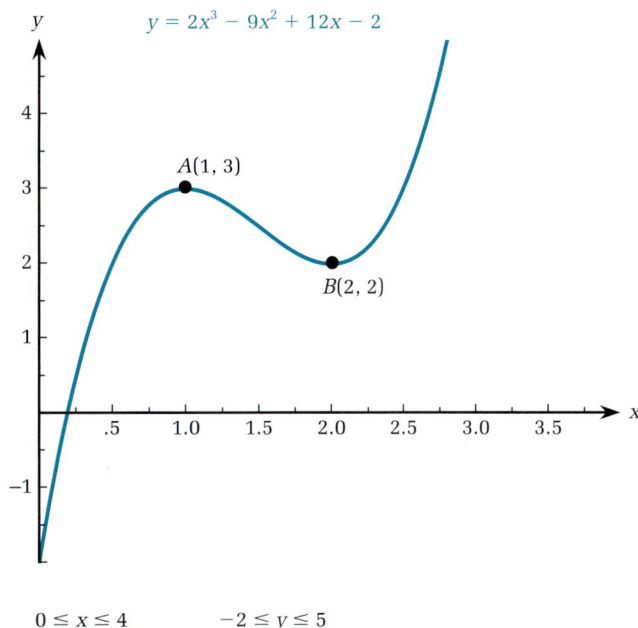

$$y = 2x^3 - 9x^2 + 12x - 2$$

$A(1, 3)$

$B(2, 2)$

$0 \leq x \leq 4$ $-2 \leq y \leq 5$

FIGURE 3.0.3

If we look at only figure 3.0.1, we have no idea whether the graph has any high or low points. If we look at figure 3.0.2, we can see a high point, but we do not know what happens elsewhere. In figure 3.0.3 we see both a high and a low point, but still we are uncertain whether there are other high or low points.

In addition, the coordinates of the high and low points in figure 3.0.3 are not known with certainty. It appears that the coordinates of the high point at A are $(1, 3)$ and the coordinates of the low point at B are $(2, 2)$. Although we can check that $(1, 3)$ and $(2, 2)$ are on the graph, there conceivably could be locations near these points that have greater and smaller values of $f(x)$.

Using the ⌷Trace⌷ function on a graphing calculator does not provide a complete answer because determining high and low points cannot be done precisely by using it. The reason is that as we continue tapping the ⌷▷⌷ key, the value of x may never be the exact value that yields the maximum (or minimum) value of $f(x)$. Even if the maximum were reached, we would not know it, for we would not be certain that somewhere in the vicinity there was not a value of x that yielded a greater or lesser value.

Although approximations are often satisfactory, nonetheless there are situations where we cannot, or do not want to, find an approximation. We want an exact answer. Therefore, we use a method that guarantees we have found the high or low points. Such points are often related to the "optimal" solution to a situation.

Optimization means finding the "best" solution to a problem. Examples include the "best" size of order, the "best" price of a bus ticket, or the "best" number of workers. The reason for putting quotation marks around the word "best" is that what is best is related to goals, and may differ for various decision makers. For example, customers would choose as a "best" number of grocery-store clerks a number that insures no customer ever has to wait in a line, whereas to the store manager the "best" number is such that there never is an idle clerk waiting for a customer to arrive.

A vital optimization question for all people is *maximum sustained yield.* How much of a renewable resource can be taken without the supply declining? To consider one example, the United States National Park Service has determined that the springs in Death Valley are fed by water that fell as rain approximately 5,000 to 10,000 years ago. Any nearby drilling of wells and pumping of wellwater must be carefully planned so that these springs continue to flow.

The hunting of whales or the catching of certain species of fish has led to their near extinction. Some species of birds, such as auks and passenger pigeons, are now gone forever. When is the harvest too great?

3.1   FUNDAMENTALS OF GRAPHING

In Section 2.1 we defined the function $f(x)$ to be continuous at $x = a$ if the function had a limit as x approached a and if that limit was $f(a)$. In algebraic notation, $f(a) = \lim_{x \to a} f(x)$. This can be interpreted geometrically by saying that a continuous function is one whose graph can be drawn through the point $(a, f(a))$ without lifting our pencil from the paper. An example of such a function and value a is shown in figure 3.1.1.

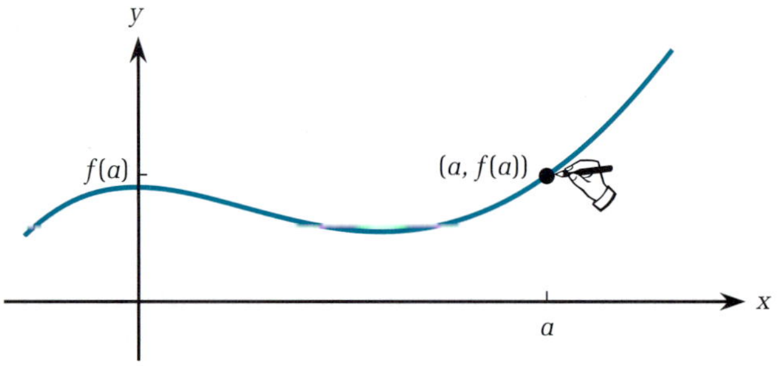

FIGURE 3.1.1

In this section we first discuss intervals and then define *continuity over an interval.* After investigating some consequences of continuity, we study *intervals of increase* and *intervals of decrease.*

Notation for Intervals

The *open interval from a to b* consists of all the real numbers between a and b, not including a and b. This is the set $\{x : a < x < b\}$, which in *interval notation* is (a, b). The *closed interval from a to b* consists of all the real numbers from *a* to *b*, including *a* and *b*. This is the set $\{x : a \leq x \leq b\}$, which in *interval notation* is [*a, b*]. If we wish to include only one endpoint and not the other one, we can write [*a, b*) or (*a, b*], depending on whether only the *a* is included or only the *b* is included. The *interior points* of an interval are the values between the endpoints.

EXAMPLE 1   Descriptions of intervals

Describe in words and in set notation each of the following intervals: (2, 5), [2, 4), (1, 5], [2.5, 5.5].

SOLUTION

(2, 5) is the set of real numbers from 2 to 5, not including 2 or 5, $\{x : 2 < x < 5\}$.

[2, 4) is the set of real numbers from 2 to 4, including 2 but not 4, $\{x : 2 \leq x < 4\}$.

(1, 5] is the set of real numbers from 1 to 5, including 5, but not 1, $\{x : 1 < x \leq 5\}$.

[2.5, 5.5] is the set of real numbers from 2.5 to 5.5, including both 2.5 and 5.5, $\{x : 2.5 \leq x \leq 5.5\}$. ●

PRACTICE EXERCISE 1

Describe in words and in set notation each of the following intervals: (a) [0, 3], (b) [2, 4), (c) [−0.5, 5], (d) (3.1 , 4].

To display an interval on the real number line, use a closed circle (●) to show the endpoint is included and an open circle (○) to show the endpoint is not included.

EXAMPLE 2   Displays of intervals

Several intervals are shown in figure 3.1.2.

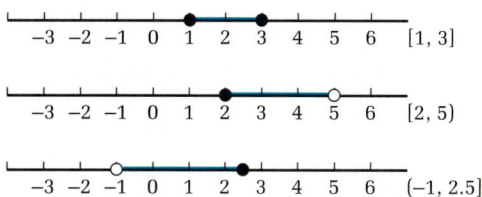

FIGURE 3.1.2

PRACTICE EXERCISE 2

On the real number line, display the intervals of Practice Exercise 1.

The interval (a, ∞) is the set of all the real numbers greater than a; in other words, the set $\{x : a < x\}$. Similarly, $(-\infty, b)$ is the set of all real numbers less than b. If the interval is to include the value of a or b, we write $[a, \infty)$ or $(-\infty, b]$, as appropriate.

EXAMPLE 3   Unbounded intervals

Describe the intervals $[5, \infty)$ and $(-\infty, 3)$.

SOLUTION

The interval $[5, \infty)$ consists of those real numbers that are 5 or any number larger than 5, without any upper limit on their value. The interval $(-\infty, 3)$ consists of those real numbers less than 3, without any lower limit on their value.

WARNING

Remember that the symbol "∞" is not a real number, but rather a shorthand way of expressing an idea of unboundedness. Because "∞" is not a real number, it makes no sense to write an expression such as $[5, \infty]$ because ∞ cannot be included in the interval.

Continuity over an Interval

To define continuity over an interval, we might be tempted to insist that the function must be continuous at every point of the interval. However, doing so would cause us difficulty at the endpoints of closed intervals. The reason is that the definition of $\lim\limits_{x \to a} f(x)$ requires that there be values on *both* sides of $x = a$ that can be used as inputs for the function. As a consequence, if a value of x is an endpoint, then insistence on continuity at each point would mean no function would be continuous over a closed interval. However, we do want the possibility to exist that a function is said to be continuous over a closed interval.

DEFINITION   Continuous over an interval

The function $f(x)$ is *continuous over an interval* if

(i) $f(x)$ is continuous at every interior point of the interval, and

(ii) if a is an endpoint of the interval and included in the interval, then $f(a)$ is equal to the appropriate one-sided limit.

EXAMPLE 4   Continuity over an interval

Assume that $f(x) = \sqrt{x}$ is continuous at every interior point of the interval $[0, 9]$. Show that $f(x)$ is continuous over the entire interval $[0, 9]$.

SOLUTION

The interval $[0, 9]$ includes both endpoints. $\sqrt{0} = 0 = \lim\limits_{x \to 0^+} \sqrt{x}$ and $\sqrt{9} = 3 = \lim\limits_{x \to 9^-} \sqrt{x}$, so we have that $f(x)$ is continuous over $[0, 9]$.

In the last example we assumed the function was continuous at every interior point. Which functions are continuous over intervals? Without becoming overly formal, and realizing there will be additional functions that are continuous over intervals, the answer is shown next.

FUNCTIONS THAT ARE CONTINUOUS OVER INTERVALS

Polynomial functions

Rational functions (for values for which the denominators are nonzero)

Power functions (for values for which they are defined)

Functions created from other continuous functions by the algebra of functions, including composition (provided they are defined)

EXAMPLE 5 Intervals of continuity

Discuss the continuity of $f(x) = \dfrac{2x}{x^2 - 4}$ over $[0, \infty)$.

SOLUTION

$f(x)$ is a rational function, so it is continuous unless the denominator is zero. $x^2 - 4 = 0$ if $x = \pm 2$. Thus, $f(x)$ is continuous over the intervals $[0, 2)$ and $(2, \infty)$. A graph of this function is shown in figure 3.1.3.

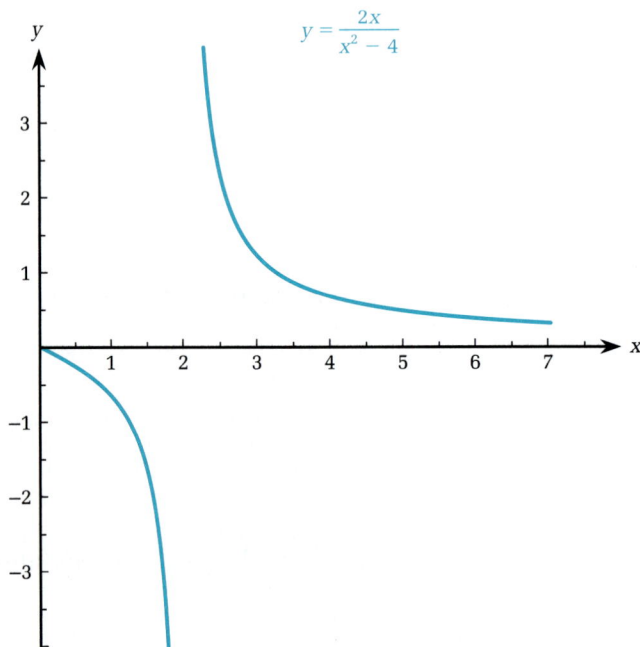

$$y = \frac{2x}{x^2 - 4}$$

FIGURE 3.1.3

EXAMPLE 6 Intervals of continuity

Discuss the continuity of $g(x) = \dfrac{\sqrt{x + 1}}{x - 2}$.

SOLUTION

First, consider $\sqrt{x + 1}$. The function $f(x) = x + 1$ is continuous because it is a polynomial. The function $h(x) = x^{1/2}$ is a power function, so it is continuous whenever defined. By function composition, $\sqrt{x + 1} = h(f(x))$ is continuous wherever it is defined. The denominator, $x - 2$, is continuous because it is a polynomial, and $g(x)$ is

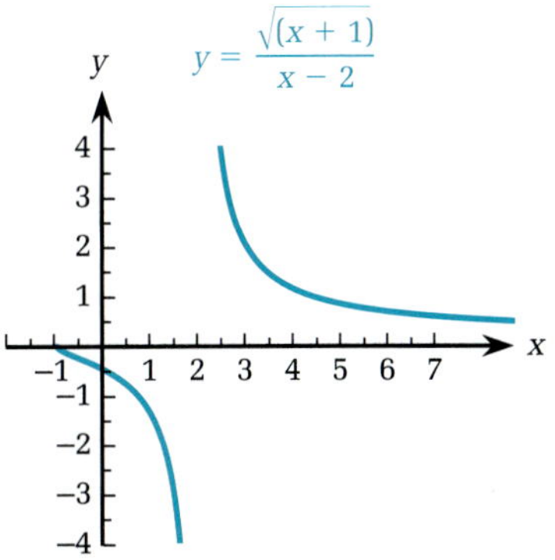

FIGURE 3.1.4

thus the quotient of two continuous functions. Hence, $g(x)$ is continuous for all values of x for which $g(x)$ is defined.

For $\sqrt{x + 1}$ to be defined, we need $x + 1 \geq 0$. This requires $x \geq -1$. Additionally, we must make certain that in evaluating $(\sqrt{x + 1})/(x - 2)$ the denominator is not zero. Hence, $x \neq 2$. Combining these facts, we have $g(x)$ is continuous over the intervals $[-1, 2)$ and $(2, \infty)$.

A graph of $g(x)$ is shown in figure 3.1.4.

PRACTICE EXERCISE 3

Discuss the continuity of $f(x) = \dfrac{\sqrt{x + 2}}{x}$.

Intermediate Value Property

All continuous functions have unbroken graphs; that is, there are no skips or missing points. Hence, if we know two values of the function, then any value between those must also be taken by the function.

FIGURE 3.1.5

INTERMEDIATE VALUE PROPERTY FOR CONTINUOUS FUNCTIONS

If $f(x)$ is a continuous function over the interval $[a, b]$ and M is any value between the values of $f(a)$ and $f(b)$, then $f(c) = M$ for some c between a and b.

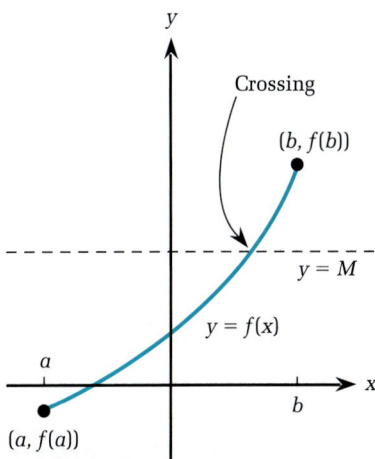

To see a geometric interpretation of the Intermediate Value Property, let us suppose $f(a) < f(b)$ and M is such that $f(a) < M < f(b)$.

If $f(x)$ is continuous over $[a, b]$, then the horizontal line $y = M$ must intersect the graph of $y = f(x)$ somewhere between $x = a$ and $x = b$, as shown in figure 3.1.5. The graph may cross the line $y = M$ several times, as the graph of $g(x)$ does in figure 3.1.6, but there certainly is at least one crossing.

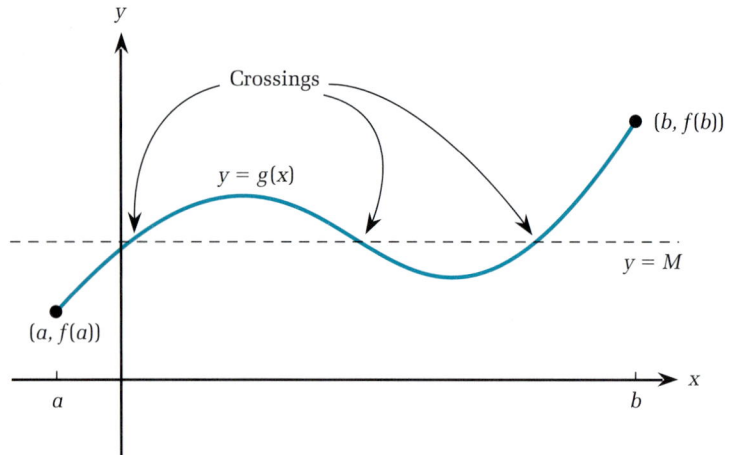

FIGURE 3.1.6

A similar situation results from assuming $f(a) > f(b)$.

EXAMPLE 7   Intermediate value property, application to U.S. automobile market share

The following table indicates the U.S. market share held by the Big Three automakers (Ford, General Motors, and Chrysler) at the start of each year for the period 1978–1988.

1978	1979	1980	1981	1982	1983	1984	1985	1986	1987	1988
84	78.5	74	73	73	74.8	75.2	74	72	67	69.4

A continuous function that passes through these points is graphed in figure 3.1.7.

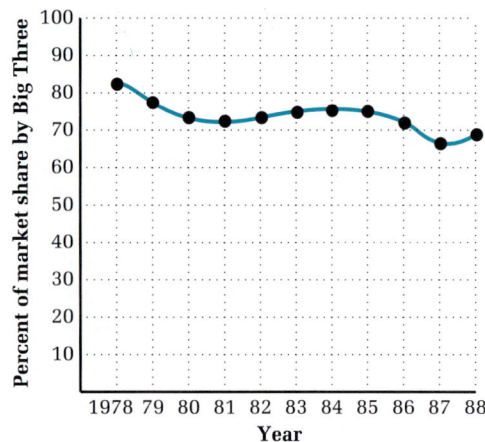

FIGURE 3.1.7

The Big Three had 72 percent of the market at the start of 1986 and only 67 percent at the end of that year, so there must have been some time during the year when they had exactly 70 percent. ●

An important and computationally useful consequence of the Intermediate Value Property is that a continuous function that changes sign between $x = a$ and $x = b$ must be zero somewhere in the interval $[a, b]$. By repeated applications of the property, we can focus in on such a zero.

EXAMPLE 8   Intermediate value property, locating zeroes of a function

Suppose $f(x) = x^3 + 3x - 2$. Show that $f(c) = 0$ for some c such that $0 < c < 1$, and develop a process for approximating c.

SOLUTION

$f(0) = -2$ and $f(1) = 1^3 + 3 \cdot 1 - 2 = 2$. Because $-2 < 0 < 2$, by the Intermediate Value Property there must exist some value c, between 0 and 1, such that $f(c) = 0$. See figure 3.1.8.

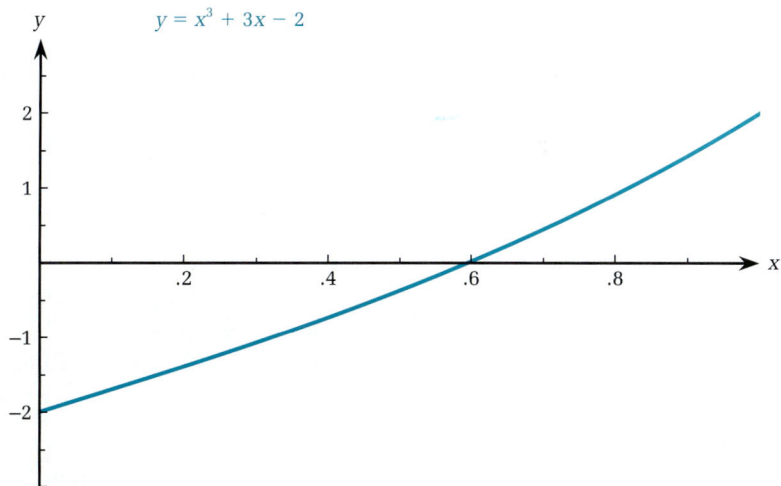

FIGURE 3.1.8

So far, this example has shown us there is a zero of $f(x) = x^3 + 3x - 2$ between $x = 0$ and $x = 1$. We may locate c with greater precision by the following procedure: Let a be a value of x between 0 and 1. Calculate $f(a)$. If $f(a) = 0$, we are done. If $f(a) \neq 0$, then we can apply the Intermediate Value Property a second time.

A common technique used to find better approximations is *bisection.* This method considers the value of $f(x)$ at the midpoint of the interval.

For example, we know $f(0) = -2$ and $f(1) = 2$. The midpoint of $[0, 1]$ is $1/2$. Calculate $f(1/2) = (1/2)^3 + 3 \cdot (1/2) - 2 = 1/8 + 3/2 - 2 = -3/8$, a negative number. Because $f(x)$ changes sign from negative to positive in the interval $[1/2, 1]$, there must be a zero in that interval. (If $f(1/2)$ had been positive, then $f(x)$ would have changed sign in the interval $[0, 1/2]$ and we would have looked for a zero there.)

By repeating this process we can locate the x-coordinate of the zero with any degree of precision we desire.   ●

Graphing Calculators

We can locate zeroes by using the ⬚Trace⬚ function on a graphing calculator and then tapping the ⬚▷⬚ key until we see the value of y change from negative to positive.

PRACTICE EXERCISE 4

Suppose $h(x) = (7 - x^2)/(x^2 + 1)$. Show that $h(c) = 0$ for some c such that $0 < c < 3$.

EXAMPLE 9   Application—Intermediate Value Property

Using her new graphing calculator, Angela wants a display of $f(x) = 50/x + x^2/10 - 13$, and she wants to show a part of the graph at which $f(x) = 0$. Find an interval using consecutive integers that she can use.

SOLUTION

By substitution, Angela finds $f(1) = 37.1$, $f(2) = 12.4$, $f(3) = 4.57$, $f(4) = 1.1$, and $f(5) = -0.5$. Thus, she sets the calculator to display $f(x)$ over the interval $[4, 5]$.   ●

Interval of Constant Sign

One aspect of a function is where it is positive or negative.

DEFINITION   Interval of constant sign

Suppose $f(x)$ has the same sign, either positive or negative, throughout an interval. Then that interval is an *interval of constant sign.*

EXAMPLE 10   Intervals of constant sign

Find the intervals of constant sign for $f(x) = 3x + 2$.

SOLUTION

Solving $3x + 2 > 0$ we have $3x > -2$, or equivalently $x > -2/3$. Hence, $(-2/3, \infty)$ is an interval of constant sign, over which $3x + 2$ is always positive. Similarly, $(\infty, -2/3)$ is an interval of constant sign, over which $3x + 2$ is always negative.

Consider the situation in which $x = a$ and $x = b$ are immediately adjacent zeroes of a continuous function. That is, $f(a) = 0$ and $f(b) = 0$, but $f(x) \neq 0$ for any x such that $a < x < b$. By the Intermediate Value Property we know $f(x)$ does not change sign in the interval (a, b). Hence, if we want to learn whether $f(x)$ is positive or negative over that interval, we can find $f(x)$ for any convenient value of x in (a, b) and we are then certain $f(x)$ has the same sign over the entire interval.

The value used to test the sign of $f(x)$ is called a *test value*. Often test values are integers, but any convenient number may be used. Test values may be used from any intervals over which we are certain $f(x)$ is never zero, including intervals of the forms (a, ∞) and $(-\infty, a)$.

USING TEST VALUES TO DETERMINE INTERVALS OF CONSTANT SIGN FOR CONTINUOUS FUNCTIONS

Suppose $f(x)$ is a continuous function.

1. Determine all values of x for which $f(x) = 0$.

2. Use the values of step (1) to divide the domain of $f(x)$ into intervals.

3. For each interval, use a test value to determine the sign of $f(x)$ over that interval.

EXAMPLE 11   Test values for intervals of constant sign

Let $f(x) = (x + 1)(x - 1)(x - 2)$. Find those intervals over which $f(x)$ is positive and those intervals over which $f(x)$ is negative.

SOLUTION

If $f(x) = 0$, then $x + 1 = 0$, $x - 1 = 0$, or $x - 2 = 0$.

Thus, the intervals over which $f(x)$ is not zero are $(-\infty, -1)$, $(-1, 1)$, $(1, 2)$, and $(2, \infty)$. Test values from these intervals could be -2, 0, 1.5, and 3.

By direct calculation, $f(-2) = (-2 + 1)(-2 - 1)(-2 - 2) = -12$. Similarly, $f(0) = 2$, $f(1.5) = -0.625$, and $f(3) = 8$. Hence, $f(x)$ is positive over both $(-1, 1)$ and $(2, \infty)$ and negative over $(-\infty, -1)$ and $(1, 2)$. A convenient way to display this information is to use a sign line, as shown in figure 3.1.9.

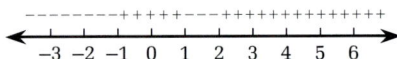

FIGURE 3.1.9

Intervals of Increase or Decrease

Suppose we consider the values of $f(x)$ over a particular interval. Those values may increase, decrease, or remain constant. In the following definition, the values of a and b must be arbitrary values of the interval. It is not enough that $f(a)$ is less than $f(b)$ for just some values of a that are less than some values of b in the interval.

DEFINITION   Interval of increase or decrease

Suppose $f(x)$ is a function defined over an interval and a and b are any two arbitrary values of that interval such that $a < b$.

$f(x)$ is *increasing over the interval* if $f(a) < f(b)$ for every possible choice of a and b.

Similarly, $f(x)$ is *decreasing over the interval* if $f(a) > f(b)$ for every possible choice of a and b.

Notice that the description of a function as *increasing* or *decreasing* is given relative to some particular interval. A function may be increasing (or decreasing) over part of some specified interval but not over all of the interval.

When we graph functions in the usual manner of "left to right," any increasing function rises and any decreasing function falls. When graphing a function, if we come to the edge of the part of the graph shown, we draw

(i)   a ● if that point is included

(ii)   a ○ if the point is omitted

(iii)   no special sign if the graph is meant to continue beyond the part shown.

EXAMPLE 12   Graphing intervals of increase and decrease

The graph of $g(x) = -x^2 - 6x + 9$ is shown in figure 3.1.10. The function increases over $(-\infty, -3)$ and decreases over $(-3, \infty)$.

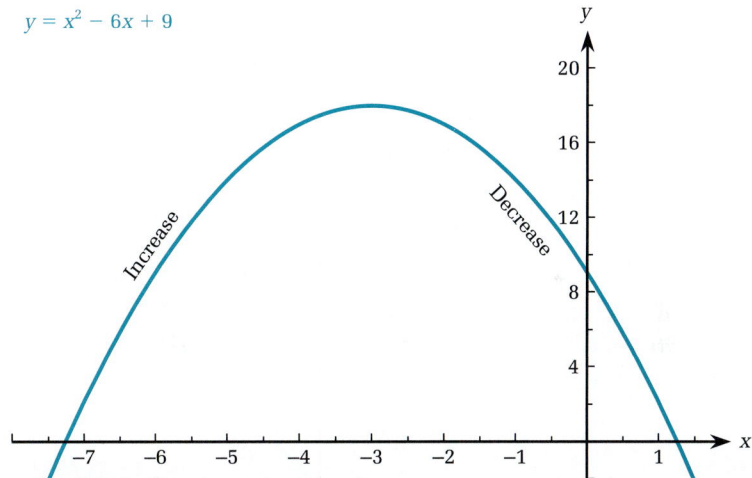

$y = x^2 - 6x + 9$

Increase

Decrease

FIGURE 3.1.10

DEFINITION

Increasing or decreasing at a point

A function is *increasing at a point* if the function is increasing over some open interval containing that point. The definition of *decreasing at a point* is similar.

WARNING

A function is *never* said to be increasing or decreasing at the endpoint of any interval.

To decide whether $f(x)$ is increasing or decreasing at $x = a$ we often consider the value of $f'(a)$, provided it exists and is not zero. The following theorem is justified in the appendix at the end of this section.

> ## THEOREM 3.1   USING THE DERIVATIVE TO DETERMINE INCREASE OR DECREASE
>
> Suppose $f'(a)$ exists.
>
> If $f'(a) > 0$, then $f(x)$ is increasing at a.
>
> If $f'(a) < 0$, then $f(x)$ is decreasing at a.

Graphs showing functions that are increasing and decreasing are shown in figure 3.1.11.

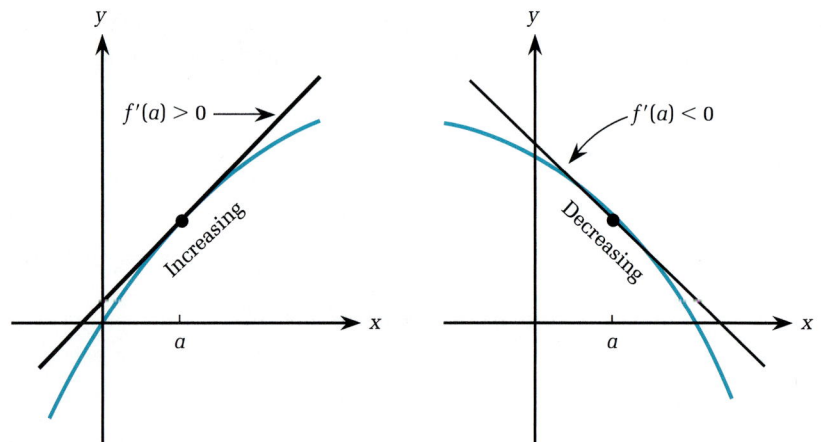

FIGURE 3.1.11

EXAMPLE 13   Using $f'(x)$ to determine an interval of increase

Suppose $f(x) = -x^2 - 6x + 9$. Show that if $a < -3$, then $f'(a) > 0$. Then use Theorem 3.1 to show that $f(x)$ is increasing over $(-\infty, -3)$.

SOLUTION

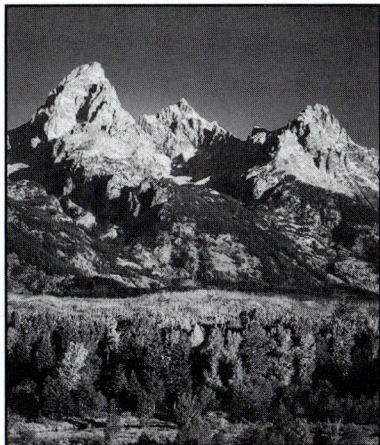

We may realize that we have already graphed $f(x)$ in figure 3.1.10, so we do know that $(-\infty, -3)$ is an interval of increase of $-x^2 - 6x + 9$. However, the point of this example is to illustrate Theorem 3.1 in a situation for which the answer is already apparent to us.

The derivative is $f'(x) = -2x - 6$. We want to show that if $a < -3$, then $-2a - 6 > 0$.

We would be able to say $-2a - 6 > 0$ if we could say $-2a > 6$. Hence we need to have $\dfrac{-2a}{-2} < \dfrac{6}{-2} = -3$. (Recall that dividing both sides of an inequality by a negative number reverses the sign of the inequality.)

Thus, in order to claim $f'(a) > 0$ we need to have $a < -3$, which is exactly what we can assume about a. Using Theorem 3.1, $f(x)$ is increasing at a for all values of $(-\infty, -3)$. ●

PRACTICE EXERCISE 5

For $f(x) = x^2 - x - 6$, show that if $a > 0.5$, then $f'(a) > 0$. Use Theorem 3.1 to show $f(x)$ is increasing over $(0.5, \infty)$.

If we have a product of factors, we may find it convenient to produce a *sign line* that shows which values of x produce a positive value and which produce a negative value. The use of a sign line is shown in Example 14.

EXAMPLE 14 Using sign lines for intervals of increase and decrease

Find the largest interval(s) over which $g(x) = x^3 - 3x^2 + 6$ is increasing. Do the same for $g(x)$ decreasing.

SOLUTION

The derivative is $g'(x) = 3x^2 - 6x$, which should be factored to give $g'(x) = 3x(x - 2)$.

To have $g'(x) > 0$, we need both x and $(x - 2)$ to have the same sign. The sign line of each, as well as the sign line of their product, is shown in figure 3.1.12.

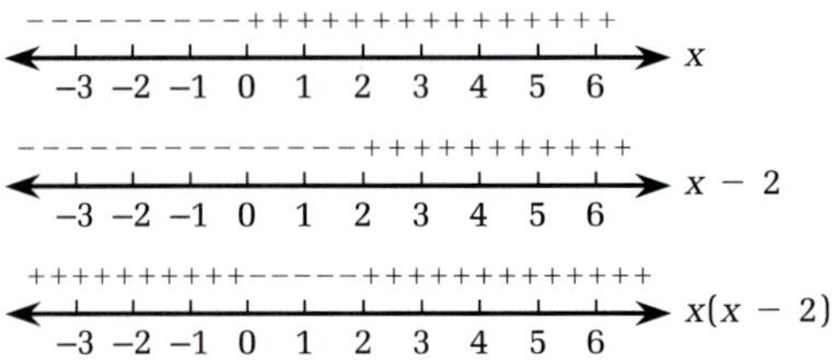

FIGURE 3.1.12

Hence, $3x(x-2) > 0$ if $x < 0$ or $x > 2$.

The graph of $g(x)$ is shown in figure 3.1.13.

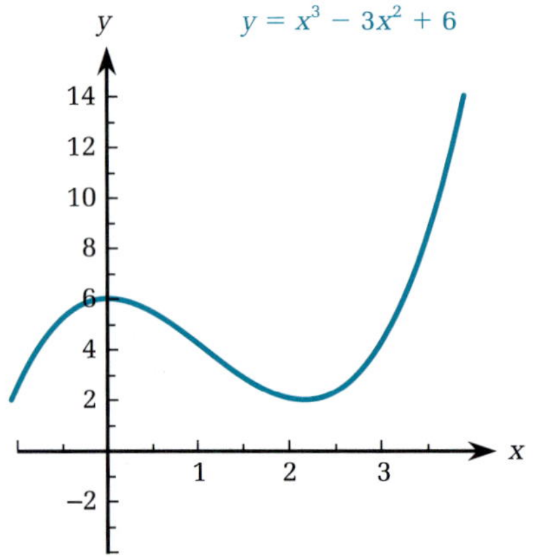

FIGURE 3.1.13

3.1 PROBLEMS

Foundations

The problems of this section require the basic skills illustrated by the following:

Use interval notation to describe the sets in problems 1–8.

1. $\{x: -2 < x \le 0\}$

2. $\{x: -1 \le x \le 5\}$

3. $\{x: 3 \le x \le 7.5\}$

4. $\{x: 1 < x \le 1.3\}$

5. $\{x: 1 < x < 3\}$

6. $\{x: 2 < x < 5\}$

7. $\{x: 1.5 < x\}$

8. $\{x: 4 < x\}$

Use set notation to describe the intervals in problems 9–16.

9. $(-2, 3.5)$

10. $(-1, 3)$

11. $[1, 5]$

12. $[2, 8]$

13. $(1, \infty)$

14. $[5, \infty)$

15. $[1, 2)$

16. $(2, 4]$

17. Solve for x if $3x - 4 = 0$.

18. Solve for x if $x^2 - 3 = 0$.

19. Find those values of t such that $3t - 1 > 0$.

20. Find those values of x such that $x^3 - 4x > 0$.

Exercises

Discuss the continuity of each of the functions in problems 21–32.

21. $f(x) = 3x - 4$

22. $g(x) = x + 2$

23. $f(x) = \dfrac{|x + 1|}{x + 1}$

24. $f(x) = \dfrac{x - 2}{|x - 2|}$

25. $g(x) = \sqrt{2x + 3}$

26. $f(t) = \sqrt{3t - 1}$

27. $g(x) = \dfrac{4x^2 - 9}{2x - 3}$

28. $r(x) = \dfrac{x}{x^2 - 1}$

29. $f(x) = \dfrac{2x}{x^2 - 3}$

30. $w(x) = \dfrac{\sqrt{x + 1}}{x}$

31. $r(t) = \dfrac{\sqrt{2t + 1}}{t}$

32. $h(x) = \begin{cases} 2x + 1 & \text{if } x \le 0 \\ x & \text{if } x > 0 \end{cases}$

33. Let $f(x) = x - \lfloor x \rfloor$, where $\lfloor x \rfloor$ is the greatest integer in x (also called the floor integer of x). (For example, $\lfloor 2.9 \rfloor = 2$.) Graph $f(x)$ and determine where it is continuous.

34. Let $g(x) = \lceil x \rceil - x$, where $\lceil x \rceil$ is the ceiling integer function of x. (For example, $\lceil 2.75 \rceil = 3$ and $\lceil 4.1 \rceil = 5$.) Graph $g(x)$ and determine where it is continuous.

35. The Parking Spot charges $10 per car for parking up to four hours. Beyond four hours, it charges $1 per half-hour or any fraction thereof. Let $C(x)$ be the cost of parking x hours.
    (a) Express $C(x)$ using the ceiling function $\lceil x \rceil$ (see problem 34) and defining $C(x)$ in pieces if necessary.
    (b) For what values of x is $C(x)$ continuous?
    (c) Sketch a graph of $C(x)$ for $0 \le x \le 8$.

36. The Airport Courier charges $10 for any parcel weighing up to one pound. For heavier parcels they charge another $1 for each additional pound or fraction thereof. Let $C(x)$ be the cost of shipping a parcel weighing x pounds.
    (a) Express $C(x)$ using the ceiling function $\lceil x \rceil$ (see problem 34) and defining $C(x)$ in pieces if necessary.
    (b) For what values of x is $C(x)$ continuous?
    (c) Sketch a graph of $C(x)$ for $0 \le x \le 6$.

37. Use the Intermediate Value Property to show that for $f(x) = (x - 1)/(x - 3)$, there is a value, c, in the interval $(0, 2)$ such that $f(c) = 0$.

38. Use the Intermediate Value Property to show that for $f(x) = x^3 + 100/x - 60$, there is a value, c, in the interval $(1, 2)$ such that $f(c) = 0$.

39. Suppose $f(x) = x^3 - x - 1$.
    (a) Use the Intermediate Value Property to show that there is a value, c, in the interval $(0, 2)$ such that $f(c) = 0$.
    (b) Find $f(1)$ and decide whether or not there is a value, c, in $(0, 1)$ or in $(1, 2)$ such that $f(c) = 0$.

40. Suppose $f(x) = 1/x$.
    (a) Show $f(5) > 0$ and $f(-2) < 0$.
    (b) Can we use the Intermediate Value Property to assert there is some c in $[-2, 5]$ such that $f(c) = 0$? Why or why not?

41. Suppose $f(x) = \begin{cases} cx + 1 & \text{if } x \le 5 \\ x^2 & \text{if } x > 5 \end{cases}$.
    Find a value for c such that $f(x)$ is continuous at $x = 5$.

For problems 42–49 find those intervals over which the function is positive and those intervals over which the function is negative.

42. $f(x) = x^2(x - 1)$

43. $f(x) = (x - 2)^2(x + 1)$

44. $g(x) = (x^2 - x - 2)(x - 1)$

45. $g(x) = (x^2 - 2x - 3)(x - 2)$

46. $f(x) = x^2 - x + 1$ 47. $f(x) = 2x^2 + x + 1$

48. $g(x) = x^2 - 3x$ 49. $g(x) = x^3 - 4x$

50. Suppose $f(x) = 3x^2 - 12x - 1$. Show that $f'(a) > 0$ when $a > 2$. Use that result to decide whether $f(x)$ is increasing or decreasing over $(2, \infty)$.

51. Suppose $f(x) = -x^2 + 6x - 2$. Show that $f'(a) > 0$ when $a < 3$. Use that result to decide whether $f(x)$ is increasing or decreasing over $(-\infty, 3)$.

52. Show that for $f(x) = 2/(x + 2)$, we have $f'(x) < 0$ over $(0, \infty)$.

53. Show that for $f(x) = x/(x + 1)$, we have $f'(x) > 0$ over $(0, \infty)$.

54. For $f(x) = 2x^2 - x + 5$, find those values of x for which $f'(x) > 0$.

55. For $f(x) = x^2 + 2x - 2$, find those values of x for which $f'(x) < 0$.

Graphing Calculator Problems

56. For $f(x) = x^3 + 100/x - 60$, approximate to within 0.01 the value, c, in the interval $(1, 2)$ such that $f(c) = 0$.

57. For $f(x) = x^3 - x - 1$, approximate to within 0.01 the value, c, in the interval $(0, 2)$ such that $f(c) = 0$.

Problems for Writing and Discussion

58. The concept of "increasing or decreasing at a point" does not apply to endpoints of intervals. Explain the reason for this by considering the graph of $f(x) = x^2$. By our definitions, $f(x)$ is decreasing over the interval $(-\infty, 0)$ and increasing over the interval $(0, \infty)$. However, what would we say about the increase or decrease at the point $x = 0$?

59. In several problems we have used the *ceiling integer function,* whose value is the integer immediately above (or equal to) the value of x. This is denoted by $f(x) = \lceil x \rceil$.

It is possible to write this function using the *floor integer function,* often called the *greatest integer function.* The notation for this function is $\lfloor x \rfloor$, or often we use $[x]$.

For instance, suppose the cost of parking a car at a parking lot is \$2 for the first hour and 50¢ for each additional hour or fraction thereof. If we represent the cost by $C(x)$, an algebraic description that uses the ceiling integer function is

$$C(x) = \begin{cases} 2 & \text{if } x \le 1 \\ 2 + 0.50 \cdot \lceil x - 1 \rceil & \text{if } x > 0 \end{cases}$$

The second part of this definition for $C(x)$ can also be expressed as $2 - 0.50 \cdot [1 - x]$ if $x > 1$.

Show that indeed both expressions are the same function and discuss the advantages and disadvantages of using either notation.

Enrichment Problems

Find the largest intervals over which the functions in problems 60–64 are increasing or decreasing.

60. $x^3 - 6x^2 + 3$ **61.** $x^3 + 3x^2 - 9x + 1$

62. $3x^4 + 8x^3$ **63.** $4x^5 - 5x^4$

64. $3x^4 + 8x^3 - 6x^2 - 24x$

65. Let $g(x) = 2x - \lfloor x \rfloor$. Graph $g(x)$ and discuss the continuity of $g(x)$ over $[0, 3]$.

66. Let $h(x) = 2x - \lceil x \rceil$. Graph $h(x)$ and discuss the continuity of $h(x)$ over $[0, 3]$.

67. Let $f(x) = x - \lceil 2x \rceil$. Graph $f(x)$ and discuss the continuity of $f(x)$ over $[0, 3]$.

68. Suppose $f(x) = \begin{cases} c^2 x & \text{if } x \ge 1 \\ cx + 2 & \text{if } x < 1 \end{cases}$. Find a value for c such that $f(x)$ is continuous at $x = 1$.

69. Suppose $g(x) = \begin{cases} x^2 - cx & \text{if } x \le 2 \\ x + 1 & \text{if } x > 2 \end{cases}$. Find a value for c such that $g(x)$ is continuous at $x = 2$.

70. Suppose $f(0) = 1$, $f(1) = 0$, and $f(x)$ is continuous over $[0, 1]$. Show there exists a value, c, such that $f(c) = c$ and $0 < c < 1$.

71. Find the largest interval(s) over which $6x^{5/2} - 10x^{3/2}$ is increasing or decreasing.

72. We have seen that if $f'(a) > 0$, then $f(x)$ is increasing at a. Considering $f(x) = x^3$, show the derivative at a can be zero and yet the function can be increasing at a.

SOLUTIONS TO PRACTICE EXERCISES

1. (a) $[0, 3] = \{x : 0 \le x \le 3\}$
  (b) $[2, 4) = \{x : 2 \le x < 4\}$
  (c) $[-0.5, 5] = \{x : -0.5 \le x \le 5\}$
  (d) $(3.1, 4] = \{x : 3.1 < x \le 4\}$

2. On the number line, the intervals of Practice Exercise 1 are shown in figure 3.1.14.

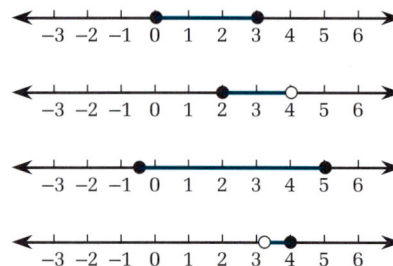

FIGURE 3.1.14

3. For $f(x) = (\sqrt{x + 2})/x$ we need the numerator defined, which means we need $x + 2 \geq 0$, and we must have $x \neq 0$ so that we do not attempt a division by zero. Thus, $x \geq -2$, with $x \neq 0$ is the domain of $f(x)$, and over that domain the function is continuous. Using interval notation, $f(x)$ is continuous over $[-2, 0)$ and $(0, \infty)$.

4. $h(0) = 7$ and $h(3) = (7 - 9)/10 = -0.2$. Because $7 > 0 > -0.2$, and $(7 - x^2)/(x^2 + 1)$ is continu-ous over the interval $[0, 3]$, the Intermediate Value Property assures the existence of a value c such that $0 < c < 3$ and $h(c) = 0$.

5. For $f(x) = x^2 - x - 6$, $f'(x) = 2x - 1$. To have $f'(a) > 0$, we need $2a - 1 > 0$, which will be true for $2a > 1$, or equivalently, for $a > 1/2 = 0.5$. By Theorem 3.1, $f(x)$ is increasing for $(0.5, \infty)$.

Appendix: Theorem 3.1

$f'(a) > 0$ implies $f(x)$ is increasing at a. An informal proof follows.

In order for $f'(x) = \lim\limits_{h \to 0} \dfrac{f(a + h) - f(a)}{h}$ to be positive, $[f(a + h) - f(a)]/h$ itself must be positive, at least for sufficiently small values of h. When h is positive, then $a + h$ is to the right of a and we have a positive quotient with a positive denominator. Thus the numerator must also be positive and we have $f(a + h) - f(a) > 0$, so $f(a + h) > f(a)$.

When h is negative, then $a + h$ is to the left of a and we have a positive quotient with a negative denominator. Hence the numerator must also be negative and we have $f(a + h) - f(a) < 0$, which gives us $f(a + h) < f(a)$. Together, these give us that $f(x)$ is increasing at a.

3.2

RELATIVE AND ABSOLUTE EXTREME VALUES: USING THE FIRST DERIVATIVE

This is the first of three sections on graphing. Our goal is to be able to find screens, or portions of graphs, that show the regions of any high or low points on a graph. Our method consists of two steps.

The first step is to find all possible values of x for which the point $P(x, f(x))$ could be a high or low point. The process we use to do this does not guarantee that each of the values found in this first step will yield a high or low point, but we can guarantee that any high or low point is associated with one of these values.

The second step is to sort through these values in order to determine which of them actually are extreme values. Several tests are used for this sorting. The test given in this section is called the First Derivative Test.

Extreme Points and Values

Suppose the graph of a function $y = f(x)$ is as shown in figure 3.2.1. Let us define different types of high and low points, first doing so in words and then using inequalities.

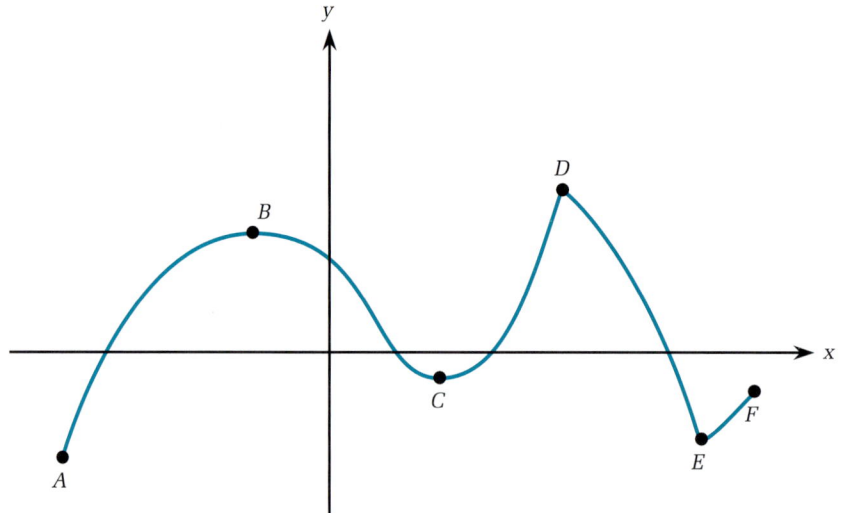

FIGURE 3.2.1

The points labeled B and D in figure 3.2.1 are higher than the points near them. We say these are *relative maximum points* of the graph and the values of $f(x)$ at those points are *relative maximum values* of $f(x)$. Similarly, the points C and E are lower than the points near them, so we say these are *relative minimum points* of the graph and the values of $f(x)$ at those points are *relative minimum values* of $f(x)$.

We apply the word "relative" to maximum or minimum points only if there are points of the graph on each side of the point. The points A and F are *not* considered to be *relative* minimum or maximum points.

DEFINITION

Relative maximum and minimum values of $f(x)$

The value $f(a)$ is a *relative maximum value* if $f(a) \geq f(x)$ for all x in some interval about a. *Relative minimum value* is defined similarly, with $f(a) \leq f(x)$.

Referring to the graph in figure 3.2.1, we see that the point A is an *absolute minimum point,* lower than any other point on the graph. The point D is an *absolute maximum point,* higher than any other point on the graph. The values of $f(x)$ at A and D are *absolute minimum* and *absolute maximum values,* respectively.

DEFINITION

Absolute maximum and minimum values of $f(x)$

The value $f(a)$ is an *absolute maximum value* if $f(a) \geq f(x)$ for all x in the domain of $f(x)$. *Absolute minimum value* is defined similarly, with $f(a) \leq f(x)$.

The term *"extreme values"* (sometimes called *"extrema,"* which is the plural form of the word *"extremum"*) refers to any and all of the maximum or minimum values of the function. Similarly, *"extreme points"* refer to the maximum or minimum points on the graph.

WARNING

(1) The word "maximum" usually means "maximum value," not "maximum point." If we wish to refer to a maximum point on a graph, we will need to be explicit. Similarly for "minimum."

(2) We use the phrase "extreme value of $f(x)$" to emphasize we are finding a value of $f(x)$, but as we become more familiar with what we are doing, the term "value" may be omitted and we may speak of an "extreme of $f(x)$."

(3) Points at the ends of the graph may be absolute extreme points, but they are never relative extreme points because "relative extreme" values occur only at interior points.

The graphs used in Examples 1 and 2 illustrate the terms for extreme values.

EXAMPLE 1   Relative and absolute extreme values

Suppose $f(x) = 2x^3 - 3x^2 + 3$. Using the graphs of figures 3.2.2 and 3.2.3, estimate the extreme values for $y = f(x)$ over (a) $[-1, \infty)$, and (b) $[-0.25, \infty)$.

FIGURE 3.2.2

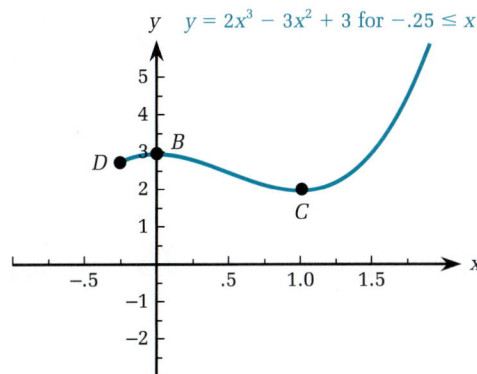

FIGURE 3.2.3

SOLUTION

(a) Over $[-1, \infty)$ there appears to be an absolute minimum value of -2 at $A(-1, -2)$, a relative maximum value of 3 at $B(0, 3)$, and a relative minimum value of 2 at $C(1, 2)$. There is no absolute maximum value at all on the interval $[-1, \infty)$.

(b) Over $[-0.25, \infty)$ the value of 2 at $C(1, 2)$ now becomes an absolute minimum value. This is because the value of the function at the point D is greater than the value of the function at $C(1, 2)$.

In Example 1 we actually displayed only a part of the graph because the interval $[-1, \infty)$ is unbounded and any actual display is bounded by the page or the screen that we use. Thus, the claim that no absolute maximum value existed must be based on some evidence other than any displayed portion of the graph. As we work through these several sections we will become able to find such evidence.

EXAMPLE 2   Relative and absolute extreme values

The graph of $g(x) = x^{2/3}$ over the interval $[-1, 8]$ is shown in figure 3.2.4. Discuss any extreme values.

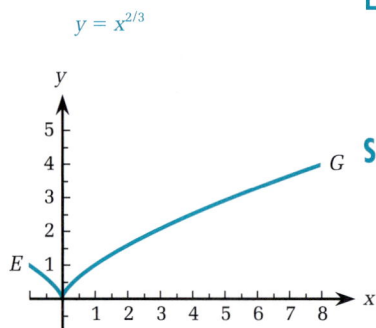

$y = x^{2/3}$

FIGURE 3.2.4

SOLUTION

The function has an absolute (and also relative) minimum value of 0 at $F(0, 0)$ and an absolute maximum value of 4 at $G(8, 4)$. The point $E(-1, 1)$ does not provide a relative maximum value because it is at an endpoint and no endpoint can provide any kind of relative maximum or minimum value. Endpoints are either absolute extreme values or they are not any kind of extreme value.

Critical Numbers and Critical Points

Consider the graphs for Examples 1 and 2. Notice that for those x-values at which we had relative extreme values, the derivatives were zero or did not exist. Such values are identified in the following definition.

DEFINITION

Critical number and critical point

A *critical number* for $f(x)$ is any x-value in the domain of $f(x)$ for which $f'(x) = 0$ or $f'(x)$ does not exist.

A point on the graph of $f(x)$ is a *critical point* if the x-value is a critical number.

The critical numbers are well named because it is with them that we start our search for extreme values of a function. Let us make certain of the critical numbers for Examples 1 and 2.

EXAMPLE 3   Critical numbers

Find the critical numbers for $f(x) = 2x^3 - 3x^2 + 3$.

SOLUTION

First, we determine $f'(x)$ so that we may find values of x for which the derivative is zero or not defined.

$f'(x) = 6x^2 - 6x$, which can be factored into $6x(x - 1)$. Solving $f'(x) = 0$ gives $6x(x - 1) = 0$. Because a product is zero only when a factor is zero, we have $x = 0$ or $x - 1 = 0$. Hence we find 0 and 1 are critical numbers.

The derivative of the function is always defined, so there are no critical numbers arising from the non-existence of $f'(x)$. Hence, the critical numbers are 0 and 1.

EXAMPLE 4   Critical numbers

Find the critical numbers for $f(x) = x^{2/3}$ over $[-1, 8]$.

SOLUTION

$f'(x) = (2/3)x^{-1/3} = 2/(3x)^{1/3}$, which is never zero, but which is undefined if $x = 0$. Hence the only critical number is 0.

The critical numbers are only those values in the domain for which $f'(x) = 0$, or $f'(x)$ does not exist. If we solve $f'(x) = 0$ and find a solution outside of the domain, then that value is not a critical number. Because of this, in the following Practice Exercise there is only one critical number.

PRACTICE EXERCISE 1

Find the critical number(s) for $f(x) = x^3 - 3x + 1$ over $[0, \infty)$.

Extreme Values of Continuous Functions

The search for extreme values of a continuous function begins with the critical numbers and any endpoints of the domain. We consider these values because of the following two facts.

RELATIVE EXTREME VALUES AND DERIVATIVES

If $f(x)$ has a relative extreme value at $x = a$, then a is a critical number for $f(x)$. That is, $f'(a) = 0$, or $f'(x)$ does not exist at $x = a$.

EXTREME VALUES OF CONTINUOUS FUNCTIONS OVER CLOSED INTERVALS

A continuous function, considered over a closed interval, has absolute maximum and minimum values there.

The first of these says that if we determine all of the critical numbers for a function, $f(x)$, then we are certain that all of the relative extreme values are located among those numbers. There may be critical numbers that do not lead to extreme values, but each relative extreme value is associated with a critical number.

The second of these says that for continuous functions over closed intervals, we are certain that there really are absolute extreme values. Because any absolute extreme must be a relative extreme or occur at an endpoint of the interval over which $f(x)$ is defined, by checking all of these values we must find the absolute maximum and minimum values of the functions.

To illustrate some of the possibilities, consider the graph of a continuous function, $f(x)$, in figure 3.2.5.

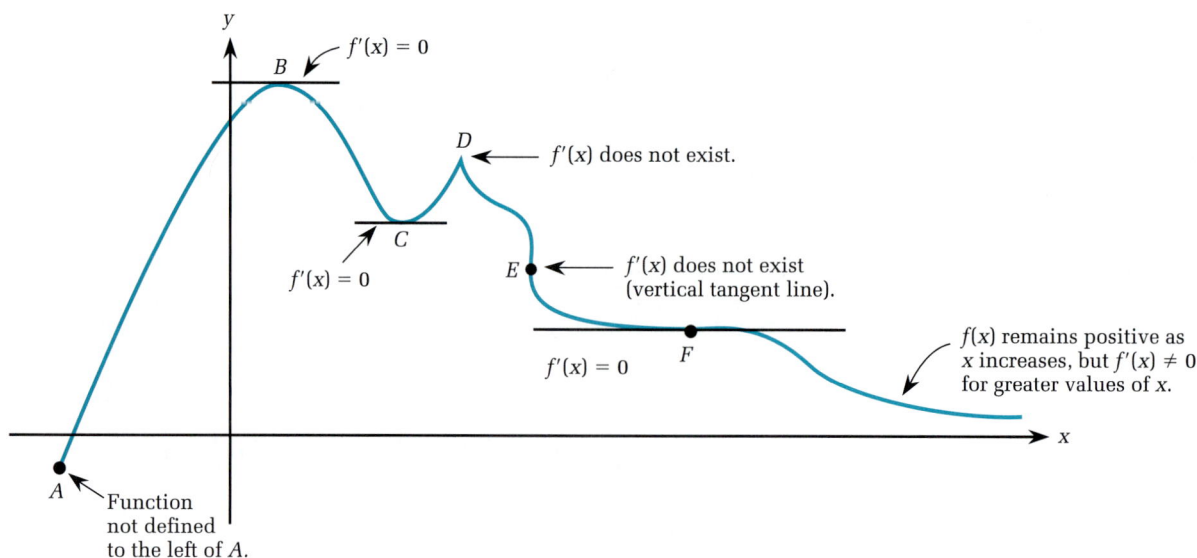

FIGURE 3.2.5

The point A is the absolute minimum point; B is a relative maximum point, as well as the absolute maximum point; C is a relative minimum point; and D is a relative maximum point. Notice that at each of the points A, B, C, and D the derivative does not exist or the derivative is zero.

The derivative at E does not exist because the tangent line is vertical at that point. Still, E is not an extreme point of any sort. The derivative at F is also zero, but F is not an extreme point because after the curve levels off, it then continues its fall to the right of F, although never falling below the value at A.

The points E and F illustrate that we may find points at which the derivative is zero or non-existent that are not extreme points of any sort. However, at *every* extreme point the derivative is zero, or does not exist, or we are at an endpoint on the graph.

Hence, we start our search for extreme points and extreme values by considering the behavior of $f(x)$ at every critical number and any domain endpoints. For continuous functions defined over closed intervals, we are certain to find absolute extreme values and we may find relative extreme values.

Making a list of the critical numbers and any endpoints of the domain is done in the same spirit as casting out a fishing net. When we pull the net in, we may find it contains more than we wanted, perhaps some floating weeds or bottles. However, we are absolutely certain that every possible maximum and minimum is also trapped in our net. We examine what we have caught by studying the function for each value we have netted, then sorting out and discarding what we do not want. The remainder of this section, as well as Section 3.3, provides various methods for this sorting.

The First Derivative Test

In Section 3.1, we found that if $f'(a) > 0$, then $f(x)$ was increasing at $x = a$, and if $f'(a) < 0$, then $f(x)$ was decreasing at $x = a$. Using this information we can state a test that evaluates potential extreme values.

THE FIRST DERIVATIVE TEST

Suppose $f(x)$ has a critical number $x = c$. If $f'(x)$ exists for values of x near c, then $f(c)$ is

1. *A relative maximum* if $f'(x) > 0$ just to the left of c and $f'(x) < 0$ just to the right of c.

2. *A relative minimum* if $f'(x) < 0$ just to the left of c and $f'(x) > 0$ just to the right of c.

3. *Neither a relative maximum nor minimum* if $f'(x)$ has the same sign on both sides of c.

The phrase "just to the left" means values for x that are close enough to the c value that they are not to the left of another critical number; a similar definition holds for "just to the right." In most cases, the derivative is itself a continuous function, and because such functions do not change sign without being zero, we can select any point just to the right or left of c and do the evaluation of the derivative at that point to determine the sign of the derivative on that side of c. Of course, we try to select values for which the actual computation is easy.

The graphs in figure 3.2.6 show the possibilities for a function with $x = c$ being a critical number. A through D all have $f'(c) = 0$; E through H all have $f'(c)$ not existing.

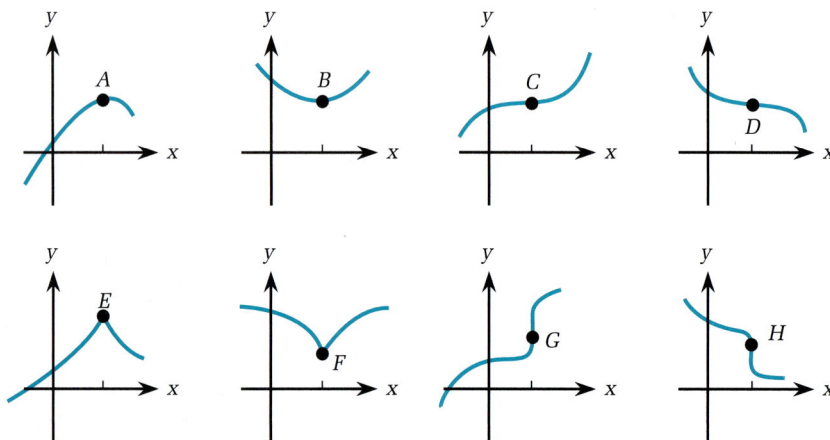

FIGURE 3.2.6

In A and E the function has a relative maximum value for $x = c$, and in B and F the function has a relative minimum value for $x = c$. The other four graphs illustrate functions for which $x = c$ is a critical number but no extreme value exists for $x = c$.

The First Derivative Test does not test for absolute maximum or minimum values, only for relative maximum or minimum values. Nonetheless, it is important that a graph show any relative extreme values, even if they are not absolute extreme values.

Determining and Graphing Extreme Values

EXAMPLE 5   Using the First Derivative Test to determine and graph extreme values

Find the critical numbers and any extreme values for $f(x) = 2x^3 - 15x^2 + 36x$ over [0,5]. Graph the function.

SOLUTION

Because $f'(x)$ is always defined, the only critical numbers are values of x for which the derivative is zero. The derivative is $f'(x) = 6x^2 - 30x + 36 = 6(x^2 - 5x + 6) = 6(x - 2)(x - 3)$. Hence $f'(x) = 0$ for $x = 2$ and 3. These x-values are the critical numbers.

The behavior of $f(x)$ over $[0, 5]$ must be considered over the sub-intervals created by using endpoints of the original interval and any critical numbers. We determine whether $f'(x)$ is positive or negative in each of these subintervals. Then we know whether $f(x)$ is increasing or decreasing in each subinterval. One way to determine the sign of $f'(x)$ is to actually evaluate $f'(x)$ at some point in each subinterval. Because our particular $f'(x)$ is a continuous function, it will not change sign within an interval.

From the subinterval $[0, 2]$ we may choose any point to evaluate $f'(x)$. For $x = 1$, $f'(1) = 6 \cdot (1 - 2)(1 - 3) = 6 \cdot (-1)(-2) = 12$. Because $12 > 0$, we know $f(x)$ is increasing between $x = 0$ and $x = 2$.

Next, evaluate $f'(x)$ on $[2, 3]$. It is convenient to do evaluations at integers when we can, but in this interval we cannot do so. Using $x = 2.5$ gives us $f'(2.5) = 6 \cdot (2.5 - 2)(2.5 - 3) = 6(0.5)(-0.5) = -1.5$. Because $-1.5 < 0$, $f(x)$ is decreasing between $x = 2$ and $x = 3$.

Finally, evaluate $f'(x)$ between 3 and 5. For $x = 4$ we find $f'(4) = 6 \cdot (4 - 2)(4 - 3) = 6 \cdot 2 \cdot 1 = 12 > 0$. Thus $f(x)$ is increasing between $x = 3$ and $x = 5$.

The following summarizes the procedure. Beginning at the left endpoint of the interval, $f'(x)$ is positive from 0 to 2, with $f'(2) = 0$. Then $f'(x)$ is negative on the right side of $x = 2$, on up to $x = 3$, where it is zero again. On the right side of $x = 3$, $f'(x)$ is positive to the endpoint of $x = 5$.

A sign line for $f'(x)$ is shown in figure 3.2.7.

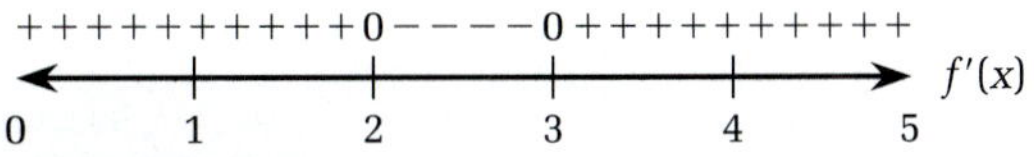

FIGURE 3.2.7

By the First Derivative Test, there is a relative maximum when $x = 2$ and a relative minimum when $x = 3$.

We are not done, for we do not know whether the values at the endpoints are greater or less than the values of the relative maximum and minimum. The only way to determine this is to find all those values from $f(x) = 2x^3 - 15x^2 + 36x$, the original function.

$$f(0) = 0$$
$$f(2) = 2 \cdot 2^3 - 15 \cdot 2^2 + 36 \cdot 2 = 16 - 60 + 72 = 28$$
$$f(3) = 2 \cdot 3^3 - 15 \cdot 3^2 + 36 \cdot 3 = 54 - 135 + 108 = 27$$
$$f(5) = 2 \cdot 5^3 - 15 \cdot 5^2 + 36 \cdot 5 = 250 - 375 + 180 = 55$$

We conclude that 0 is an absolute minimum value, occurring at $A(0, 0)$; that 28 is a relative maximum value, occurring at $B(2, 28)$; that 27 is a relative minimum value, occurring at $C(3, 27)$; and that 55 is an absolute maximum value, occurring at $D(5, 55)$.

The graph of this is shown in figure 3.2.8.

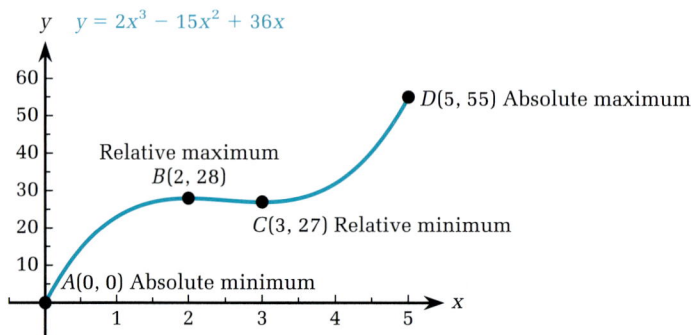

FIGURE 3.2.8

We can summarize our procedure for graphing a function that is continuous over the closed interval [a, b] in the following sequence of steps.

FINDING EXTREME VALUES OF A CONTINUOUS FUNCTION, $f(x)$, OVER THE INTERVAL [a, b]

1. Determine the critical numbers for $f(x)$ over $[a, b]$. These are values of x in $[a, b]$ such that $f'(x) = 0$ or $f'(x)$ is undefined.

2. Subdivide the interval, using the critical numbers.

3. Determine whether $f'(x)$ is positive or negative in each of these subintervals.

4. When possible, use the First Derivative Test to determine which (if any) critical numbers yield relative extreme values.

5. Compare all relative extreme function values with the function values at the endpoints of the interval. The largest of all of these is the absolute maximum and the smallest of these is the absolute minimum.

PRACTICE EXERCISE 2

Find the critical numbers and any extreme values for $f(x) = 4x^3 - 3x^2 - 6x + 7$ over $[-2, 2]$ and graph the function there. (Hint: $f'(x)$ factors.)

A function may have no relative extreme values in some interval under consideration. In fact, it may have no relative extreme values for any interval. The next example is a function whose only extreme values occur at endpoints.

EXAMPLE 6   Graphing a function with extreme values only at endpoints

Find the critical numbers and any extreme values for $f(x) = x^3 - 3x^2 + 3x - 1$ over $[0, 3]$. Graph the function there.

SOLUTION

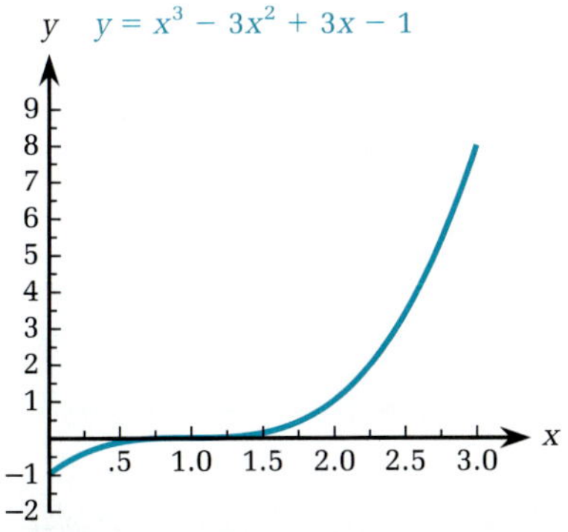

$f'(x) = 3x^2 - 6x + 3 = 3(x^2 - 2x + 1) = 3(x - 1)^2$. Hence $x = 1$ is the only critical number resulting from solving $f'(x) = 0$. The endpoints of the interval are 0 and 3.

Any nonzero number squared is positive, and $f'(x) = 3(x - 1)^2$. Hence, for $x \neq 1$, we have $f'(x) > 0$ and so $f(x)$ is always increasing. Therefore, $f(1)$ cannot be a relative maximum or relative minimum value and so any maximum and/or minimum values for $f(x)$ must occur at the endpoints.

Because $f(x)$ is always increasing, the absolute minimum must occur at the left endpoint and the absolute maximum at the right endpoint. Thus, $f(0) = -1$ is the absolute minimum value and $f(3) = 8$ is the absolute maximum value. The graph is shown in figure 3.2.9.

FIGURE 3.2.9

⌨ Graphing Calculators

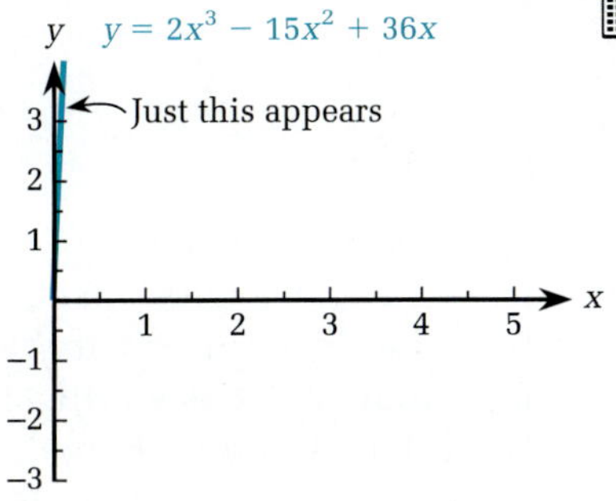

Suppose we must decide on the range values for a screen. For instance, consider the function of Example 5, which was $f(x) = 2x^3 - 15x^2 + 36x$ over $[0, 5]$. Suppose we had no idea of the nature of the graph.

First, we set the x-range to be $0 \leq x \leq 5$, because that is the interval. For the y-range, we may try leaving the default y-values in place. Suppose these are $-3.1 \leq y \leq 3.1$. Entering $f(x)$ and then pressing $\boxed{\text{EXE}}$ yields the display in figure 3.2.10.

This is not helpful. We need a range for y that provides more of a view.

FIGURE 3.2.10

ESTABLISHING RANGES FOR A GRAPHING CALCULATOR

To find the x- and y-ranges for a graph of $f(x)$ over $[a, b]$

1. Use $a \leq x \leq b$.

2. For y use the minimum and maximum values of $f(x)$, evaluated at the critical numbers and both a and b.

Using our work in Example 5 we know the critical numbers for $f(x) = 2x^3 - 15x^2 + 36x$ are 2 and 3. By evaluation, we find $f(2) = 28$ and $f(3) = 27$. Considering $f(x)$ at the endpoints of $[0, 5]$ we have $f(0) = 0$ and $f(5) = 55$. Using the minimum and maximum of these, we set the range of y to be $0 \leq y \leq 55$. The graph we get is the one that was shown in figure 3.2.8.

Using the ⎡Trace⎤, we may obtain a good approximation to the relative maximum or minimum values.

PRACTICE EXERCISE 3

Determine ranges for x and y so that the screen shows the graph of $y = g(x) = -x^3 + 5x^2 + x - 5$ over $[0, 5]$.

EXAMPLE 7   Finding and graphing extreme values

Find the critical numbers and any extreme values for $f(x) = \dfrac{x}{x^2 + 1}$ over $[0, 3]$.

SOLUTION

First determine

$$f'(x) = \frac{(x^2 + 1) \cdot 1 - x \cdot 2x}{(x^2 + 1)^2} = \frac{x^2 + 1 - 2x^2}{(x^2 + 1)^2} = \frac{1 - x^2}{(x^2 + 1)^2}$$

Since $x^2 + 1 \neq 0$ for any x, we know $f'(x)$ always exists. Solving $f'(x) = 0$ involves solving $1 - x^2 = 0$ because the only way a quotient can be zero is for the numerator to be zero and the denominator to be nonzero. Thus, $f'(x) = 0$ when $x = \pm 1$. The value $x = 1$ is a critical number, but the value $x = -1$ is not in the interval and so is not a critical number. Using the critical number and the original endpoints, we subdivide the interval into $[0, 1]$ and $[1, 3]$.

The denominator of the derivative is always positive, hence the sign of the numerator will determine the sign of the quotient.

Let us consider numbers near 1. At $x = 1/2$, we have $1 - x^2 = 1 - 1/4 > 0$. At $x = 2$, we have $1 - x^2 = 1 - 4 < 0$. Thus, by the First Derivative Test there is a relative maximum for the critical number $x = 1$.

Because $f(0) = 0/(0 + 1) = 0$, $f(1) = 1/(1^2 + 1) = 1/2 = 0.5$, and $f(3) = 3/(3^2 + 1) = 3/10 = 0.3$, we have that $A(0, 0)$ is an absolute minimum point and $B(1, 0.5)$ is a relative and absolute maximum point.

In terms of extreme values, we say 0 is an absolute minimum value and 0.5 is both a relative maximum value and an absolute maximum value. The graph is shown in figure 3.2.11.

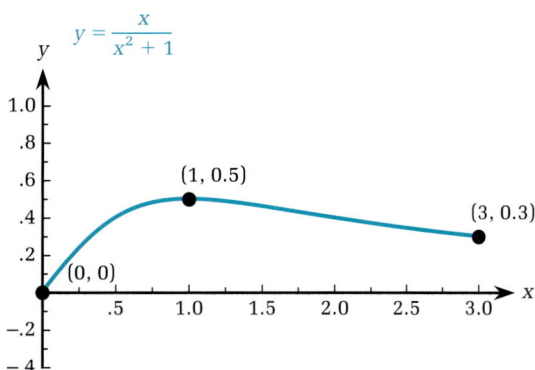

FIGURE 3.2.11

WARNING

Quite often the point at which the maximum (or minimum) value occurs is itself mistakenly called the maximum (or minimum) value. Thus, in our last example someone may say that the maximum value is $B(1, 0.5)$ instead of saying that the maximum value is 0.5. We should avoid confusing a maximum point with a maximum value. The y-coordinate of a point is not the point itself.

An even more serious error is to give the critical number as itself being the maximum or minimum value.

PRACTICE EXERCISE 4

Find the critical numbers and any extreme values for $f(x) = x^3 + 48/x - 30$ over $[1, 3]$.

In Example 8 we have a function for which we can be certain that solving $f'(x) = 0$ will yield an absolute minimum x-value without even applying the First Derivative Test. Notice the interval is unbounded.

EXAMPLE 8   Determining extreme values

Find any relative maximum or minimum values for $f(x) = 4x + 1/x$ over $(0, \infty)$.

SOLUTION

$f'(x) = 4 - 1/x^2$. This is zero when $4 = 1/x^2$, which implies $x = \pm 1/2$. Only $1/2$ lies in $(0, \infty)$, so $x = -1/2$ would not be a critical number, but $x = 1/2$ would be.

Because $x > 0$, we have $f(x) > 0$. Furthermore, for x close to zero, the term $1/x$ is quite large. For a large value of x we see that $4x$ is also large. Hence, the function must have some minimum value, and because we know there is only one critical number, then the minimum must be determined by that number.

The value $f(0.5) = 4(0.5) + 1/0.5 = 2 + 2 = 4$ is a relative minimum value.

The value 4 is also an absolute minimum value, as can be seen from the graph of $f(x)$ in figure 3.2.12.

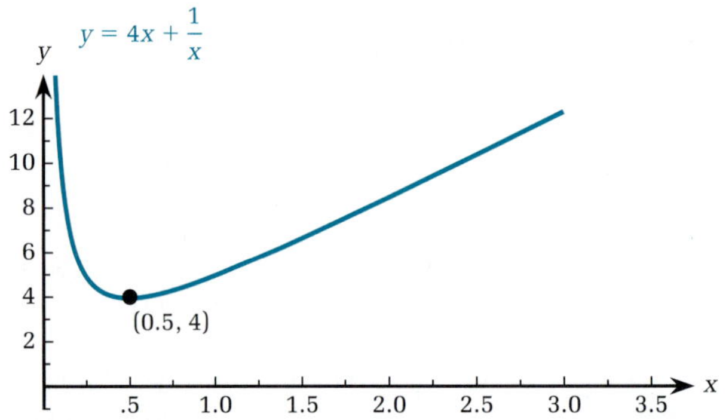

FIGURE 3.2.12

In this section we have seen that if $f(x)$ is continuous over a closed interval, then there must be extreme values. Example 8 showed that there may even be some extreme values if the interval is unbounded. Example 9 shows that if the interval is unbounded, there may be no extreme values of any sort, even though the function is continuous.

EXAMPLE 9   Continuous function with no extreme values

Show that function $g(x) = 1/x$ has no extreme values over $(0, \infty)$.

SOLUTION

The derivative is $g'(x) = -1/x^2$, which is never zero. It is possible for $-1/x^2$ to be undefined, namely for $x = 0$, but that value of x is

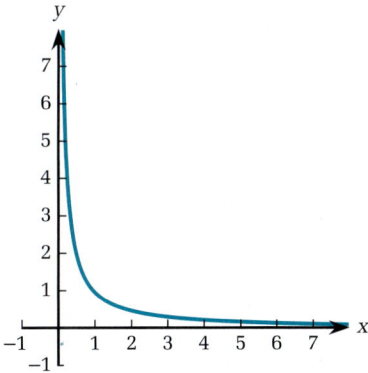

FIGURE 3.2.13

not in the interval $(0, \infty)$, so it is not a critical number. Hence, there are no critical numbers. Furthermore, $g'(x) < 0$ for all $x > 0$, so that the function is decreasing throughout its domain. There are no extreme values. The graph of $g(x)$ is unbounded, but shown in figure 3.2.13 is a portion of the graph, showing that it approaches, but does not reach, the x-axis.

3.2 PROBLEMS

Foundations

The problems of this section require the basic skills illustrated by the following:

1. For $f(x) = -2x^2 - 8x$, find (a) $f(2)$, (b) $f(-3)$, and (c) $f'(x)$, and (d) solve $f'(x) = 0$.

2. Find the derivative of $(2x - 1)^3$.

3. Write $x^4 - 6x^2 + 9$ in factored form.

Exercises

On the graphs of problems 4–8, indicate which points are (a) relative extreme points and/or (b) absolute extreme points. For each point, indicate whether it is a maximum or minimum extreme point.

4.

FIGURE 3.3.14

5.

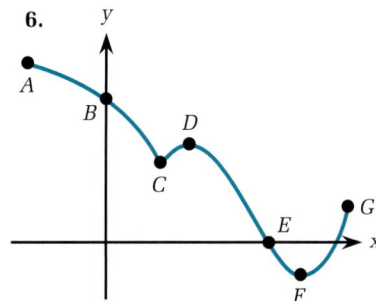

FIGURE 3.2.15

6.

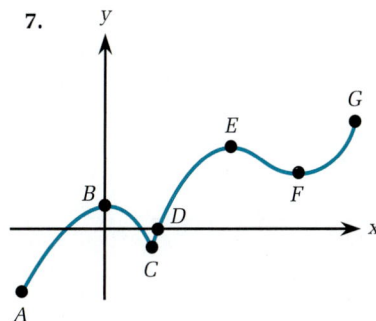

FIGURE 3.2.16

7.

FIGURE 3.2.17

On the graphs of problems 8–11, indicate which points are critical because:
(a) $f'(x) = 0$ at the point.
(b) the derivative does not exist at the point.

8.

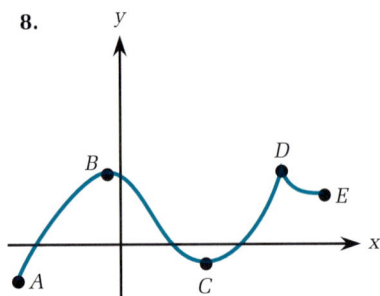

FIGURE 3.2.18

9.

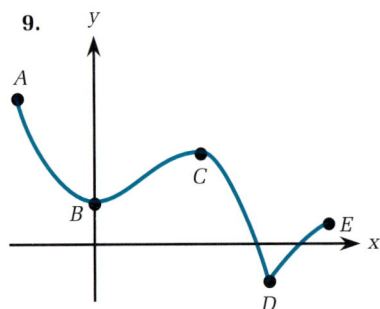

FIGURE 3.2.19

10.

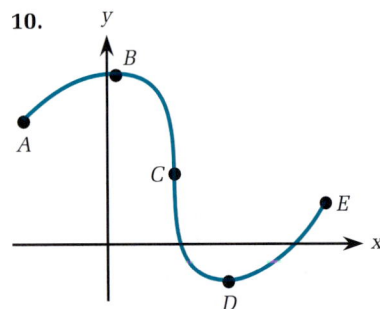

FIGURE 3.2.20

11.

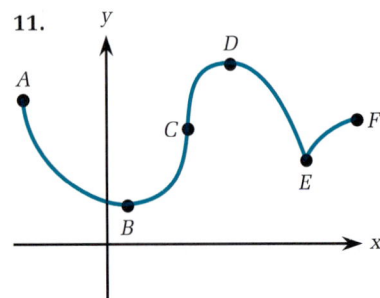

FIGURE 3.2.21

Find the critical numbers for each of the functions in problems 12–31.

12. $f(x) = x^2 + 1$

13. $f(x) = 5x^2 + x$

14. $g(x) = 2x + 8$

15. $f(x) = -3x - 7$

16. $f(x) = x^2 - x + 1$

17. $h(x) = 3x^2 + 2x - 1$

18. $f(t) = t^3 - t^2 - 6$

19. $g(t) = t^3 - 2t^2 + 5$

20. $f(x) = x^2 - x$ over $[0, 2]$

21. $f(x) = 2x^2 - x$ over $[0, 5]$

22. $g(x) = x^3 - x + 5$ over $[0, 5]$

23. $f(x) = x^3 - 2x^2 + 1$ over $[1, 5]$

24. $f(x) = \dfrac{5}{x - 1}$ over $[2, 5]$

25. $g(x) = \dfrac{2}{x + 1}$ over $[0, 3]$

26. $g(x) = x^{1/3}$ over $[-1, 2]$

27. $f(t) = (t - 1)^{1/3}$ over $[0, 3]$

28. $f(t) = t^{2/3} + 1$ over $[-1, 1]$

29. $g(x) = \dfrac{x}{\sqrt{x + 1}}$ over $[0, \infty)$

30. $f(x) = \dfrac{x^2}{x - 3}$ over $[4, \infty)$

31. $h(x) = x + \dfrac{2}{x}$ over $[1, \infty)$

For each function in problems 32–49, find: (a) the critical numbers, (b) any relative maximum or minimum values, and (c) the absolute minimum and maximum values. (d) Also sketch the graph.

32. $f(x) = 3x^2 - 18x + 3$ over $[0, 4]$

33. $f(x) = -2x^2 + 4x - 1$ over $[-2, 1]$

34. $f(x) = (2x - 1)^3$ over $[-1, 1]$

35. $f(x) = (3x - 2)^3$ over $[0, 1]$

36. $f(x) = 2x^3 - 15x^2 + 36x$ over $[-2, 2]$

37. $f(x) = x^3 + 3x^2 - 9x + 1$ over $[-2, 2]$

38. $f(x) = x^3 - 3x^2 + 6$ over $[-1, 3]$

39. $f(x) = x^4 - 2x^2 + 2$ over $[-2, 2]$

40. $f(x) = x^4 - 8x^2 + 20$ over $[0, 3]$

41. $f(x) = (x^2 - 1)^3$ over $[0, 2]$

42. $f(x) = (x^2 - 4)^3$ over $[0, 3]$

43. $f(x) = \dfrac{1}{x^2 + 1}$ over $[-1, 2]$

44. $f(x) = \dfrac{x}{x^2 + 4}$ over $[-1, 3]$

45. $f(x) = \dfrac{\sqrt{x}}{x^2 + 1}$ over $[0, 1]$

46. $f(x) = \sqrt{x + 1}$ over $[-1, 2]$

47. $f(x) = 2\sqrt{x}$ over $[0, 2]$

48. $f(x) = 3x^{2/3}$ over $[-1, 2]$

49. $f(x) = x^{2/3} - 1$ over $[-2, 1]$

Graphing Calculator Problems

By finding y-values at critical numbers and endpoints, give x- and y-ranges for screens for problems 50 and 51.

50. $f(x) = x^3 - 6x^2 + 3$ over $[-1, 5]$

51. $g(x) = x^4 - 6x^2 + 9$ over $[-1, 3]$

52. Suppose $y = f(x) = x^4 - 6x^2 + 4x$ is to be graphed over $[-2.5, 2]$. To determine a range for y you want to solve $f'(x) = 0$ to find critical numbers for $f(x)$, which you can then use to estimate the range. Use $\boxed{\text{Trace}}$ or some other method to estimate solutions to $f'(x) = 0$ and use those estimates to find an approximate range for y.

Problems for Writing and Discussion

53. Discuss whether or not the following is true. If $f(x)$ is increasing over an interval, then $f'(x) > 0$ for all x in that interval. (Compare this with Theorem 3.1 and Example 5 of this section.)

54. In Chapter 1 we stated that if we had a quadratic $y = ax^2 + bx + c$, then the x-coordinate of the high or low point of the graph was $-b/(2a)$. Show by derivatives that this result is correct.

55. Throughout much of this section we insisted that the function under consideration be continuous and defined over a closed interval. The reason for these conditions is that they guarantee $f(x)$ will have both an absolute maximum and absolute minimum for x-values somewhere in, or at the endpoints of, the interval.
   (a) Show that even though the function $f(x) = 1/x$ is continuous over the interval $(0, 1)$, nonetheless $f(x)$ has no maximum value there.
   (b) Show that the function $g(x) = x - \lfloor x \rfloor$ is defined over $[0, 2]$, but has no maximum value there.

Enrichment Problems

For the functions in problems 56–63, determine (a) any critical numbers, (b) any relative maximum or minimum values that exist, and (c) any absolute maximum or minimum values (if any exist).

56. $g(x) = \dfrac{x + 2}{x - 2}$

57. $f(x) = \dfrac{x}{x - 1}$

58. $f(x) = x + 4/x^2$

59. $g(x) = x + 1/x$

60. $f(x) = \dfrac{x^2 - x}{16 - x^2}$

61. $h(x) = \dfrac{x^2 - 2x}{4 - x^2}$

62. $f(x) = |x^2 - 3|$ over $[0, 3]$

63. $f(x) = x(x - 2)^4$ over $[0, 2]$

64. Find an equation for a quadratic $f(x) = ax^2 + bx + c$ that has a minimum at $P(2, 1)$ and goes through $Q(0, 2)$. (Hint: $f'(2) = 0$. Why?)

65. Show that $f(x) = x^3 + 6x$ has no relative maximum or minimum values.

SOLUTIONS TO PRACTICE EXERCISES

1. $f'(x) = 3x^2 - 3 = 3(x^2 - 1) = 3(x + 1)(x - 1)$. Solving $f'(x) = 0$, we have $x = -1$ and $x = 1$. The number -1 is *not* a critical number because -1 is not in the interval $[0, \infty)$. Hence $x = 1$ is the only critical number arising from $f'(x) = 0$. Because $f'(x)$ always exists, no critical numbers arise from $f'(x)$ not existing.

2. For $f(x) = 4x^3 - 3x^2 - 6x + 7$ over $[-2, 2]$, the derivative always exists. $f'(x) = 12x^2 - 6x - 6 = 6(2x^2 - x - 1) = 6(2x + 1)(x - 1)$. Thus, $f'(x) = 0$ when $x = -0.5$ or $x = 1$. The critical numbers are -0.5 and 1.

   Let -1 and 0 be the numbers near -0.5 to use in the First Derivative Test and let 0

and 1.5 be the numbers near 1 for the same test.

$$f'(-1) = 6(-2 + 1)(-1 - 1) = 6(-1)(-2)$$
$$= 12 > 0$$
$$f'(0) = 6(0 + 1)(0 - 1) = 6(1)(-1)$$
$$= -6 < 0$$
$$f'(1.5) = 6(3 + 1)(1.5 - 1) = 6(4)(0.5)$$
$$= 12 > 0$$

Hence there must be a relative maximum for $x = -0.5$ and a relative minimum for $x = 1$. Evaluating $f(x)$ at all of the critical numbers and endpoints gives us

$$f(-2) = -32 - 12 + 12 + 7 = -25$$
$$f(-0.5) = -0.5 - 0.75 + 3 + 7 = 8.75$$
$$f(1) = 4 - 3 - 6 + 7 = 2$$
$$f(2) = 32 - 12 - 12 + 7 = 15$$

Thus, we have an absolute minimum at $A(-2, -25)$, a relative maximum at $B(-0.5, 8.75)$, a relative minimum at $C(1, 2)$, and an absolute maximum at $D(2, 15)$. The graph is shown in figure 3.2.22.

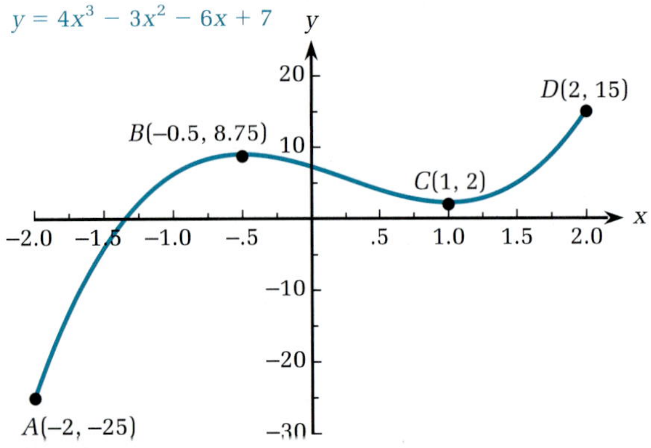

FIGURE 3.2.22

3. Evaluating $g(x)$ at the endpoints gives $g(0) = -5$ and $g(5) = 0$. For $g(x) = -x^3 + 5x^2 + x - 5$, the derivative is $g'(x) = -3x^2 + 10x + 1$. By the quadratic formula, $g'(x) = 0$ for

$$x = \frac{-10 \pm \sqrt{10^2 - 4 \cdot (-3) \cdot 1}}{2 \cdot (-3)}$$
$$= \frac{-10 \pm \sqrt{100 + 12}}{-6} = \frac{-10 \pm \sqrt{112}}{-6}$$

We are establishing a range, so we use a fairly rough approximation for these values of x. If we use $\sqrt{112} \approx 10.6$ we have $x \approx$
$$\frac{-10 - 10.6}{-6} \approx 3.4 \text{ and } x \approx \frac{-10 + 10.6}{-6} \approx -0.5.$$
Considering these two values, only 3.4 is in

the interval $[0, 5]$, and for that value, $g(3.4) = -(3.4)^3 + 5(3.4)^2 + (3.4) - 5 \approx -39 + 57.8 + 3.4 - 5 = 17.2$.

The screen using an x-range of $0 \le x \le 5$ and a y-range of $-5 \le y \le 18$ is shown in figure 3.2.23.

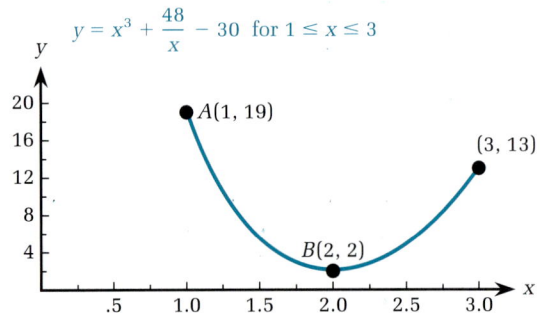

FIGURE 3.2.23

We could now enlarge the part of this graph near the maximum point to estimate the maximum value.

4. For $f(x) = x^3 + 48/x - 30$, we have

$$f'(x) = 3x^2 - \frac{48}{x^2} = \frac{3x^4 - 48}{x^2}$$

Over the interval $[1, 3]$, the derivative is always defined. For $f'(x) = 0$, we have $3x^4 - 48 = 0$, which gives $x^4 = 16$. Write $x^4 - 16$ in factored form as $(x^2 + 4)(x - 2)(x + 2)$. Hence we have $x = \pm 2$. Only $+2$ is in the interval $[1, 3]$. Thus, the only critical number is 2.

Using the trial values of $x = 1.5$ (between 1 and 2) and $x = 2.5$ (between 2 and 3), we have

$$f'(1.5) = \frac{3(1.5)^4 - 48}{(1.5)^2} \approx 14.5033$$

and

$$f'(2.5) = \frac{3(2.5)^4 - 48}{(2.5)^2} \approx 11.07$$

By the First Derivative Test, $f(2) = 2^3 + 48/2 - 30 = 8 + 24 - 30 = 2$ is a relative minimum value.

Checking the endpoints, we have $f(1) = 19$ and $f(3) = 13$. Hence, the absolute maximum value of 19 occurs at $A(1, 19)$ and the absolute (and also relative) minimum value of 2 occurs at $B(2, 2)$.

The graph is shown in figure 3.2.24.

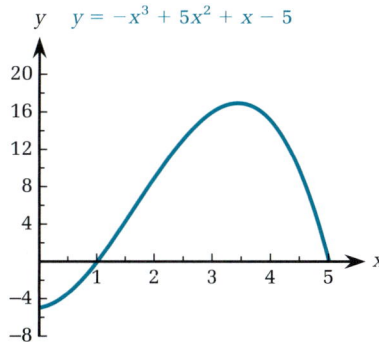

$$y \quad y = -x^3 + 5x^2 + x - 5$$

FIGURE 3.2.24

CONCAVITY AND THE SECOND DERIVATIVE

So far in this chapter we have been determining intervals over which functions increase, decrease, or remain constant, with our attention focused on the search for extreme values of the functions. In this section we determine the rates of change of those increases and decreases.

One example of a rate of increase that is commonly reported by the news media concerns the Consumer Price Index (CPI), which is a measure of the cost of a particular assortment of goods and services. The Consumer Price Index is revised monthly to reflect the change in prices due to inflation or deflation. Many funding programs use the CPI. For example, Social Security benefits are related to the CPI.

The CPI was established by the U.S. Department of Commerce with 1967 as the base year, using 100 as the value of the CPI for that year. By 1985, the CPI had risen to about 320, meaning that the same selection of goods and services that had cost $100 in 1967 cost about $320 in 1985. The federal government, embarrassed by the ever higher CPI numbers, decided to reindex the CPI. This reindexing meant the 1967 CPI value dropped from 100 to 33.4, but, of course, this reindexing provided only a cosmetic approach to the reporting of inflation.

Regardless of how the CPI is indexed, the annual percentage change in the CPI remains unaffected because it is simply the amount of the change divided by the CPI at the start of the year. The graphs in figures 3.3.1 and 3.3.2 show the revised CPI and the annual rate of change in the CPI, respectively, over the period 1960–1990.*

Inflation occurs in each of the years shown here, but during some years even the rates of inflation were going up, so that inflation was getting worse.

Let $CPI(t)$ represent the Consumer Price Index as a function of time. The graph $CPI(t)$ goes through the points in figure 3.3.1. The derivative of $CPI(t)$ is the rate of change of $CPI(t)$, so its graph will pass through the points in figure 3.3.2. What if we find the derivative

*Source: *Statistical Abstracts*. Washington, D.C.: Department of Commerce, 1992.

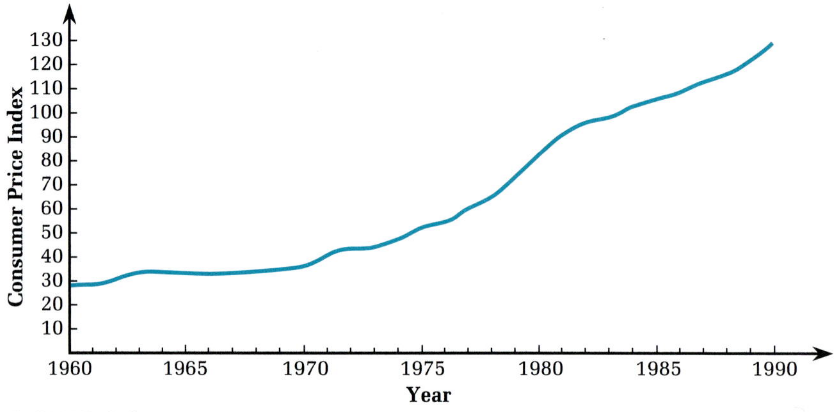

FIGURE 3.3.1

FIGURE 3.3.2

of the derivative of $CPI(t)$? How are its values related to the shape of the graph of $CPI(t)$ and the graph of its derivative?

Concavity

Consider the graph shown in figure 3.3.3.

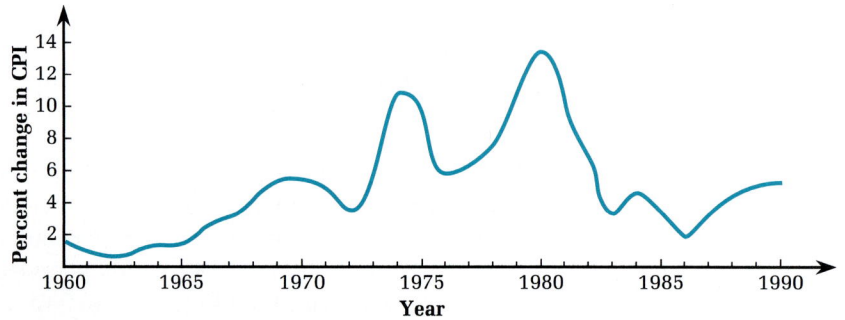

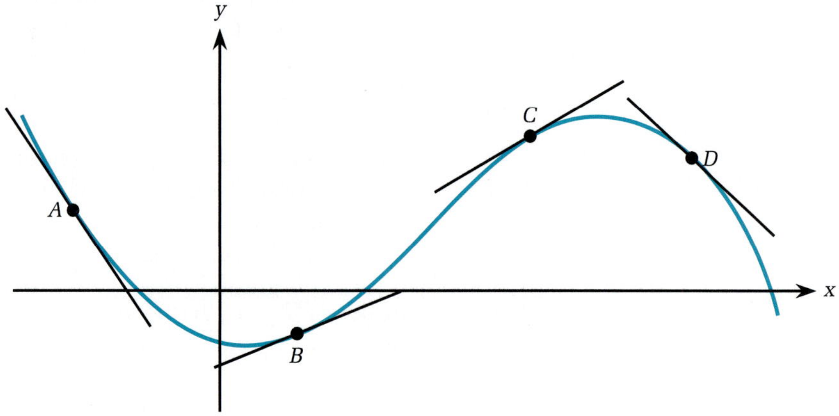

FIGURE 3.3.3

At point A the function is decreasing, but the rate of decrease is slowing. At point B the function is increasing, and the rate of increase is quickening. At both A and B the graph is curving upward and a tangent line would lie below the graph.

At point C the function is still increasing, but the rate of increase is slowing. At point D the function is decreasing and the rate of decrease is accelerating. At C and D a tangent line would lie above the graph.

DEFINITION

Concave at a point and concave over an interval

The graph of $y = f(x)$ is *concave up* at a point $P(a, f(a))$ if the graph lies above the tangent line for each value in some open interval containing a. The graph is *concave down* at $P(a, f(a))$ if the graph lies below the tangent line for each value in some open interval containing a.

A graph is *concave up over an interval* if it is concave up at every point of that interval. A graph is *concave down over an interval* if it is concave down at every point of that interval.

Using this terminology, we can say that the graph of the Consumer Price Index is concave up for the 1977–1980 period and concave down for the period 1982–1985.

Concavity is determined by the rate of change of the rate of change, and the rates of changes are related to derivatives, so therefore what we need to investigate concavity is the derivative of the derivative. This function is called the *second derivative*.

The Second Derivative

Taking a derivative produces a function. If the process is repeated on the derived function, we find the *second derivative*.

DEFINITION

Second derivative

The *second derivative* of $y = f(x)$ is the derivative of $y' = f'(x)$. It may be expressed by any of the following:

$$y'', \ f''(x), \ \frac{d^2y}{dx^2}, \ \text{or} \ \frac{d^2}{dx^2} f(x).$$

The notation y'' is quite common, but its use requires that we are certain with respect to what variable the derivative is being taken.

The notation $\dfrac{d^2}{dx^2} f(x)$ is meant to emphasize that we are finding $\dfrac{d}{dx}\left(\dfrac{d}{dx} f(x)\right)$.

The little 2's are *not* exponents, although they may look like them.

WARNING

EXAMPLE 1   Finding a second derivative

For $f(x) = x^3 - 6x^2 + 9x + 1$, find $f''(x)$.

SOLUTION

We begin by finding the first derivative: $f'(x) = 3x^2 - 6 \cdot 2x + 9 = 3x^2 - 12x + 9$.

To take the derivative of $f'(x)$ we proceed as we have done earlier in taking derivatives. The fact that $f'(x)$ is itself a derivative does not change the fact that it is a function and our processes for finding derivatives can be applied to it. Doing so, we have $f''(x) = 3 \cdot 2x - 12 = 6x - 12$. ●

EXAMPLE 2   Finding a second derivative

Find y'' for $y = \sqrt{3x + 1}$.

SOLUTION

First, rewrite the square root as a fractional power: $y = (3x + 1)^{1/2}$. Then take the derivative by the Generalized Power Rule: $y' = (1/2)(3x + 1)^{(1/2)-1} \cdot 3 = (3/2)(3x + 1)^{-1/2}$.

The derivative of this is found by again applying the Generalized Power Rule. Hence, $y'' = 3/2 \cdot (-1/2)(3x + 1)^{-(1/2)-1} \cdot 3$, which can be simplified as $y'' = (-9/4)(3x + 1)^{-3/2}$. ●

PRACTICE EXERCISE 1

Find y'' for $y = x^{3/2} + x^4$.

By using the second derivative, we can determine whether a graph is concave up or concave down. If $f''(a) > 0$, then $f'(x)$ is increasing at a, with increasing values for $f'(x)$ meaning the tangent lines are becoming steeper. Thus we have shown the first part of the following result. The second part is shown similarly.

THEOREM 3.2

If $f''(a) > 0$, then $f(x)$ is concave up at $P(a, f(a))$.

If $f''(a) < 0$, then $f(x)$ is concave down at $P(a, f(a))$.

The term *concavity* is used to refer to whether the curve is concave up or down.

EXAMPLE 3   Determining concavity

Find intervals for which the graph of $f(x) = x^3 + x^2 - 5x + 1$ is concave up and intervals for which the graph is concave down.

SOLUTION

$f'(x) = 3x^2 + 2x - 5$ and $f''(x) = 6x + 2$.

By Theorem 3.2, if $f''(x) > 0$ then the graph is concave up. To find such values of x we solve $6x + 2 > 0$. If $6x > -2$, then $x > -1/3$. Hence, the curve is concave up in the interval $(-1/3, \infty)$.

Similarly, if $6x + 2 < 0$, then $x < -1/3$, so the graph is concave down in the interval $(-\infty, -1/3)$. ●

EXAMPLE 4   Determining concavity

Find intervals for which the graph of $g(x) = (2x - 1)^{2/3}$ is concave up and intervals for which the graph is concave down.

SOLUTION

$$g'(x) = \frac{2}{3}(2x - 1)^{-1/3} \cdot 2 = \frac{4}{3}(2x - 1)^{-1/3} \text{ and}$$

$$g''(x) = \frac{4}{3} \cdot \left(-\frac{1}{3}\right)(2x - 1)^{-4/3} \cdot 2 = -\frac{8}{9}(2x - 1)^{-4/3}.$$

To determine the sign of $g'(x)$, it is helpful to rewrite this last expression using only positive exponents: $g''(x) = -\frac{8}{9}\left(\dfrac{1}{(2x - 1)^{4/3}}\right)$.

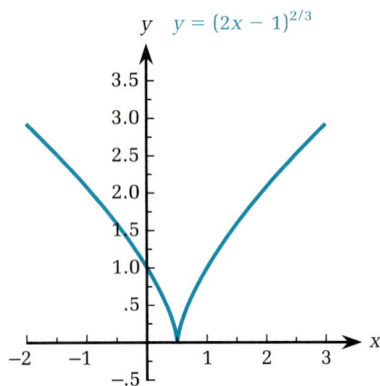

FIGURE 3.3.4

The 4/3 power of any expression is always positive. Hence, $g''(x) < 0$ for all x for which it is defined, which is all x except for $2x - 1 = 0$. As a result, $g''(x) < 0$ except for $x = 1/2$.

The graph is concave down over $(-\infty, 1/2)$ and $(1/2, \infty)$, as shown in figure 3.3.4. Do *not* say the function is always concave down. The reason is that at the point $P(1/2, 0)$, the function has no concavity because it has no tangent line.

Second Derivative Test

Section 3.2 presented the First Derivative Test, which can be used to find relative extreme values. We now present a test based on the second derivative, which also can be used for that purpose. This test is useful when the second derivative is easily calculated and evaluated for a critical number.

SECOND DERIVATIVE TEST

Given $f(x)$, suppose that $f'(a) = 0$ and $f''(a)$ exists. The value $f(a)$ is

(i) a relative maximum if $f''(a) < 0$.

(ii) a relative minimum if $f''(a) > 0$.

(iii) of uncertain status if $f''(a) = 0$. (Use some other means of evaluation.)

The first two cases are illustrated in figure 3.3.5.

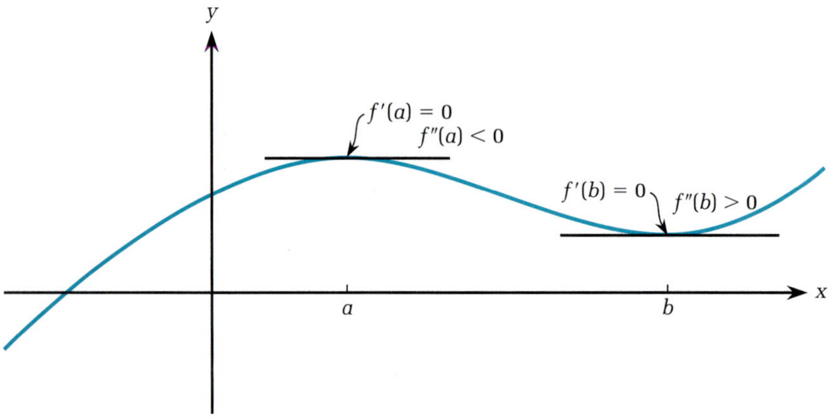

FIGURE 3.3.5

To show the third case, that is, $f''(a) = 0$ is not helpful, consider the functions x^2, $-x^2$, and x^3. As shown in figure 3.3.6, x^2 has a relative minimum at $P(0, 0)$, $-x^2$ has a relative maximum there, and x^3 has neither a relative maximum nor a relative minimum at $P(0, 0)$. Yet, for each of these functions, $f''(0) = 0$. These examples show the need to use the First Derivative Test, or to evaluate the function for values near $x = a$.

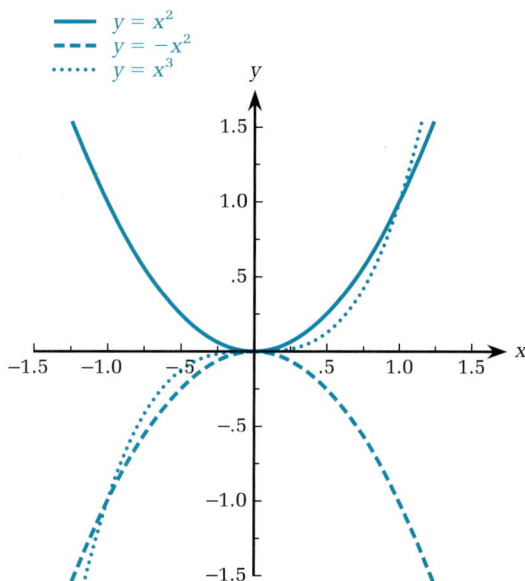

FIGURE 3.3.6

EXAMPLE 5   Finding extreme values using the Second Derivative Test

Find any relative maximum and minimum values for $f(x) = x^3 - 2x^2 + x + 1$. Sketch a graph of $f(x)$.

SOLUTION

To begin, find the first and second derivatives, $f'(x) = 3x^2 - 4x + 1 = (3x - 1)(x - 1)$ and $f''(x) = 6x - 4$.

Solving $f'(x) = 0$, we find $x = 1/3$ and $x = 1$ are critical numbers. The second derivative must be evaluated at each of these numbers.

$f''(1/3) = 6(1/3) - 4 = 2 - 4 = -2 < 0$. Hence, by the Second Derivative Test, the x-value $1/3$ must be a relative maximum. The value is found by evaluating $f(x)$ at $1/3$. $f(1/3) = (1/3)^3 - 2(1/3)^2 + 1/3 + 1 = 31/27$.

At $x = 1$, we find $f''(1) = 6 \cdot 1 - 4 = 2 > 0$. By the Second Derivative Test, the x-value 1 must be a relative minimum. Its value is $f(1) = 1^3 - 2 \cdot 1^2 + 1 + 1 = 1$.

In summary, the relative maximum point is $A(1/3, 31/27)$ and the relative minimum point is $B(1, 1)$. The graph is shown in figure 3.3.7.

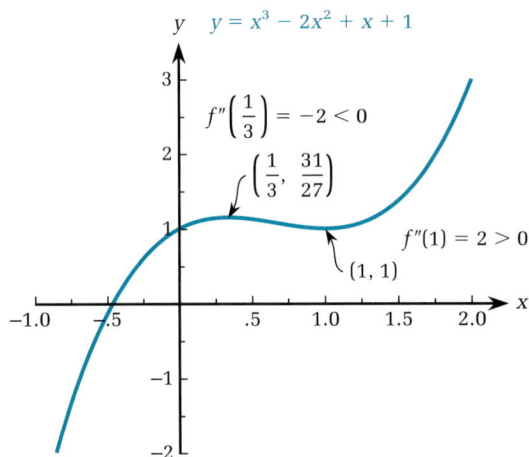

$y = x^3 - 2x^2 + x + 1$

$f''\left(\frac{1}{3}\right) = -2 < 0$

$\left(\frac{1}{3}, \frac{31}{27}\right)$

$f''(1) = 2 > 0$

$(1, 1)$

FIGURE 3.3.7

Imaginary bowls, as shown in figure 3.3.8, are one way of remembering whether a positive second derivative goes with a relative minimum or maximum. A bowl that can hold something is "positive"; the bowl sides are sloping up. A bowl that cannot hold anything is "negative," and has sides that slope down.

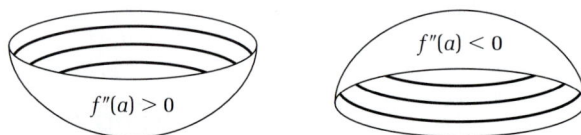

$f''(a) > 0$

$f''(a) < 0$

FIGURE 3.3.8

PRACTICE EXERCISE 2

Find any relative maximum or minimum points for $f(x) = 10x^3 - 21x^2 + 12x + 5$ and sketch the graph. (Hint: $f'(x)$ factors.)

Inflection Points

The point on a curve where the rate of change stops increasing and starts decreasing, or the other way around, is of interest. In a business application, this point is the *point of diminishing returns,* an example of which is shown in figure 3.3.9. In epidemiology, this point is the *maximum rate of infection,* as illustrated in figure 3.3.10.*

*Source: C.-E. A. Winslow, *Man and Epidemics,* Princeton, N.J.: Princeton University Press, 1952.

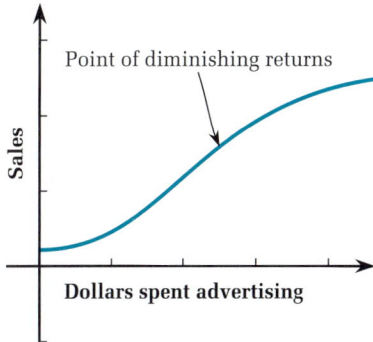

FIGURE 3.3.9

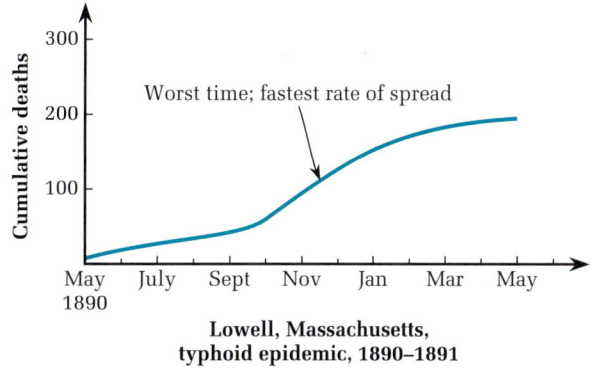

FIGURE 3.3.10

The rate of change is given by the derivative. At the point where that rate stops increasing and starts decreasing, the curve is steepest. There the derivative must have reached its maximum value. Hence, the second derivative either does not exist or it must be zero, as we saw by our earlier work on the relative extreme values of a function. See figure 3.3.11.

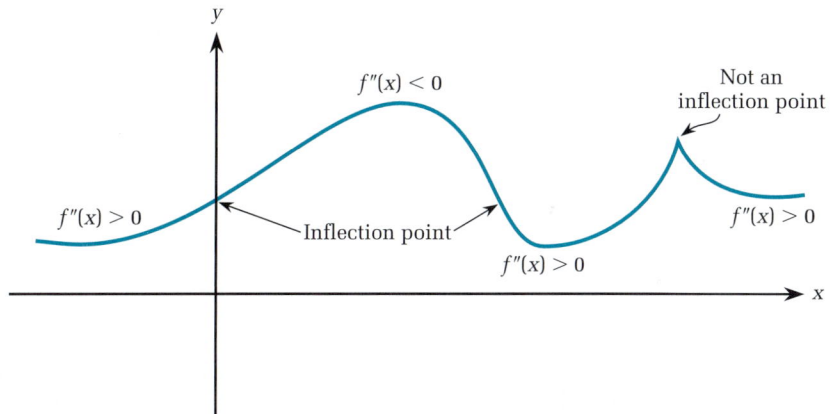

FIGURE 3.3.11

DEFINITION

Inflection point

An *inflection point* is any point on the graph of $y = f(x)$ where $f''(x)$ changes sign.

Before showing applications of inflection points, let us find these points for several functions.

EXAMPLE 6   Inflection points

Find any inflection points for $f(x) = x^3 - 6x^2 + 18$.

SOLUTION

$f'(x) = 3x^2 - 12x$ and $f''(x) = 6x - 12 = 6(x - 2)$.

Solving $x - 2 = 0$ we have $x = 2$. Evaluating $f''(x)$ for convenient numbers near 2, we find $f''(1) = -6$ and $f''(3) = 6$. Hence $f''(x)$ does go from negative to positive at 2. Because $f''(x)$ changes sign at 2, that is the x-coordinate of the inflection point. $f(2) = 8 - 24 + 18 = 2$, so the point $P(2, 2)$ is an inflection point. A graph of $f(x)$ is shown in figure 3.3.12.

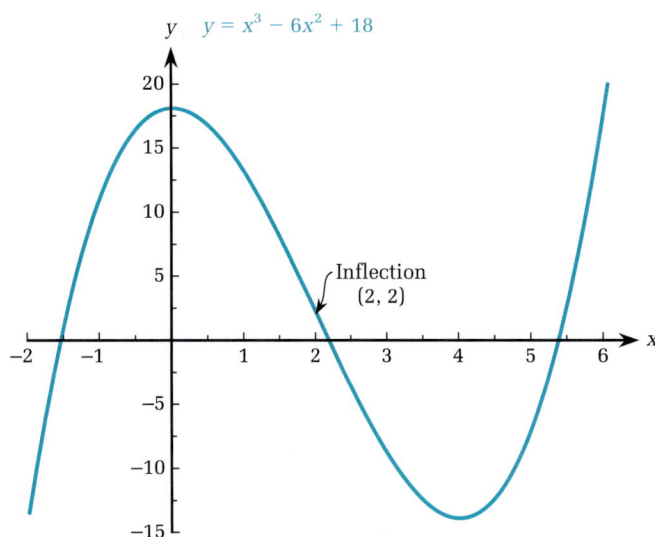

FIGURE 3.3.12

PRACTICE EXERCISE 3

Find any inflection points for $f(x) = x^3 - 9x^2 + 3x$.

From Example 6, it may seem that it is enough simply to solve $f''(x) = 0$ for x and claim that value is the x-value of the inflection point. However, the functions in the next two examples show that, first, $f''(a)$ may be zero and yet the point $P(a, f(a))$ is not an inflection point, and second, an inflection point may exist at a point where the second derivative does not exist at all.

EXAMPLE 7   A function $f(x)$ such that $f''(a) = 0$, but $(a, f(a))$ is not an inflection point

Show that there are no inflection points on the graph of $f(x) = x^4$.

SOLUTION

For $f(x) = x^4$, we find $f'(x) = 4x^3$ and $f''(x) = 12x^2$, which never changes sign, so there is no inflection point. Nonetheless, $f''(0) = 0$. A graph of $f(x)$ is shown in figure 3.3.13.

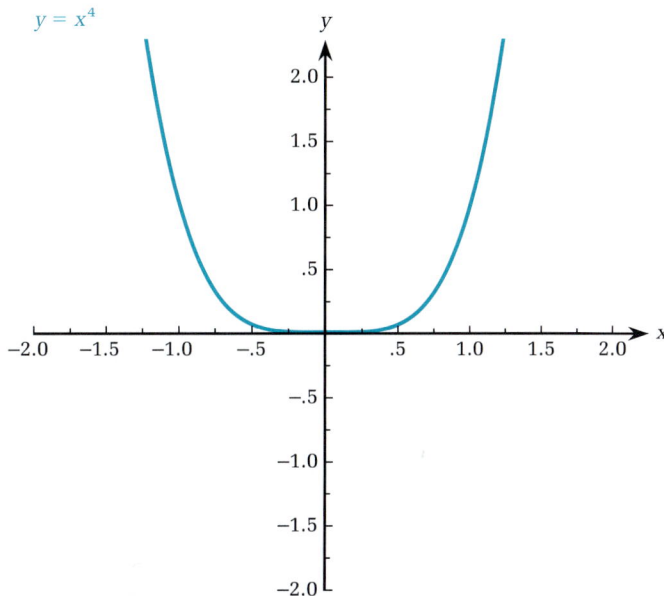

$y = x^4$

FIGURE 3.3.13

EXAMPLE 8   A function $g(x)$ such that $(a, g(a))$ is an inflection point, but $g''(a)$ does not exist

Show that the graph of $g(x) = x^{5/3}$ has an inflection point at $Q(0, 0)$.

SOLUTION

The first and second derivatives are $g'(x) = (5/3)x^{2/3}$ and $g''(x) = 5/3 \cdot (2/3)x^{-1/3}$.

$g''(x)$ does not exist at $x = 0$ and yet there is an inflection point at $Q(0,0)$ because $g''(x) = (10/9)x^{-1/3}$, which is positive for positive values of x and negative for negative values of x. The graph of $g(x)$ is shown in figure 3.3.14.

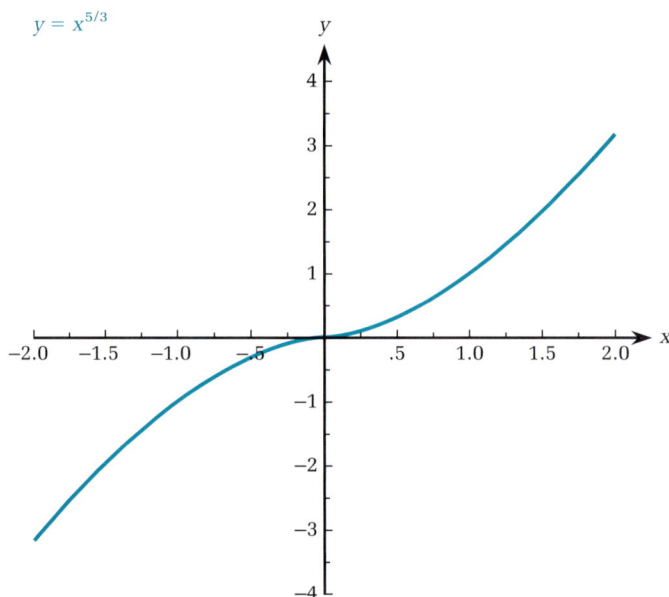

$y = x^{5/3}$

FIGURE 3.3.14

TO FIND INFLECTION POINTS

1. Determine those values a for which $f''(a)$ does not exist or is zero.

2. Decide whether or not $f''(x)$ changes sign at those values. If it does, the point $(a, f(a))$ is an inflection point.

Graphing Calculators

A graphing calculator may help to find inflection points. If we graph a function, we do not know if the chosen screen includes all the inflection points. The screen may suggest the location of inflection points, but the number of pixels is very small. It is best to find the second derivative and then check by using [Trace] to estimate when it changes sign.

EXAMPLE 9

Approximate the inflection points of $f(x) = 6x^5 - 15x^4 - 10x^3 + 30x^2$. Use as a first screen $-4 \leq x \leq 4$ and $-20 \leq y \leq 20$.

SOLUTION

Figure 3.3.15 shows the first screen.

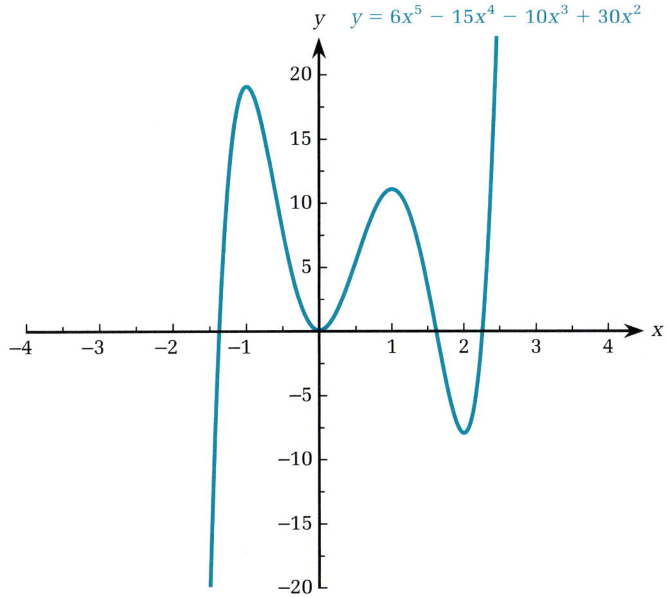

FIGURE 3.3.15

It appears there are three inflection points. The derivative of $f(x)$ is $f'(x) = 30x^4 - 60x^3 - 30x^2 + 60x = 30(x^4 - 2x^3 - x^2 + 2x)$. The second derivative is $f''(x) = 30(4x^3 - 6x^2 - 2x + 2)$.

We are estimating values for which $f''(x)$ changes sign, so let us graph $4x^3 - 6x^2 - 2x + 2$. The range for x we know to be $-2 \le x \le 2$. The range for y can be left as $-20 \le y \le 20$. In fact, it makes no difference what range we use for y as long as it contains both some negative and positive values. The reason is that we intend to $\boxed{\text{Trace}}$ over the curve and find those x-values when the y-values change sign.

Using $\boxed{\text{Trace}}$, we can estimate the following crossings of the x-axis: $x \approx -0.64, \approx 0.47$, and ≈ 1.66. Each of these estimates for the x-coordinate of the inflection points can be improved by zooming in on the crossing of the x-axis by the graph of $y = 4x^3 - 6x^2 - 2x + 2$.

The y-coordinates of each inflection point can be estimated by $\boxed{\text{Trace}}$ along the graph of the original function. A better estimate can be found by evaluation with the original function. For instance, $f(0.47) = 6(0.47)^5 - 15(0.47)^4 - 10(0.47)^3 + 30(0.47)^2$.   ●

The next example involves an inflection point and a "point of diminishing returns."

EXAMPLE 10   The inflection point as the "point of diminishing returns"

Suppose the New Products Committee is thinking about buying the franchise for a new puzzle being marketed by the Puzzle-It-Out Co.

After looking over the sales for the first few months, they are excited by the prospects. Not only are sales increasing, but even the monthly increase of sales is increasing. Their data are displayed in table 3.3.1.

TABLE 3.3.1

MONTH	SALES	INCREASE OVER PREVIOUS MONTHS	
1	29	29	
2	112	83	$(112 - 29 = 83)$
3	243	131	$(243 - 112 = 131)$
4	416	173	$(416 - 243 = 173)$
5	625	209	$(625 - 416 = 209)$

Full of enthusiasm, they call a meeting with the Vice President for Finance. While waiting for the meeting to begin, they look over their figures again and discuss what the curve for future sales might be.

The graph of $f(x) = 30x^2 - x^3$ fits the data exactly. Graphing that function indicates future sales would increase, and even the rate of increase would increase, but only for a time.

They find $f'(x) = 60x - 3x^2$ and $f''(x) = 60 - 6x$. To find a maximum, they set $f'(x) = 0$. Hence $60x - 3x^2 = 3x(20 - x) = 0$ at $x = 0$ and $x = 20$.

The function does continue to increase until $x = 20$, and $f(20) = 4000$. However, the rate of increase will slow long before $x = 20$. In fact, as x increases, $f''(x) > 0$ and the curve is concave up until $x = 10$, where $f''(10) = 0$. But for $x > 10$ the curve is concave down and $f''(x) < 0$. All this is a way of saying that $P(10, 2000)$ is an inflection point on the curve, as shown in figure 3.3.16.

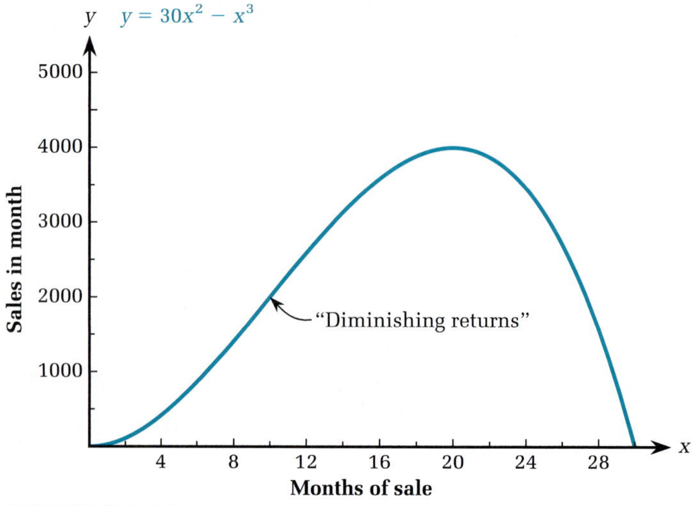

FIGURE 3.3.16

Thinking about this, the chair of the New Products Committee tells the secretary of the Vice President for Finance, "We need to post-pone this meeting—indefinitely." ●

The next example gives a physical interpretation to the idea of an inflection point.

EXAMPLE 11   Bicycle riding—a physical interpretation of an inflection point

Imagine that an immense version of the graph in figure 3.3.16 is laid out on a large, flat, open parking lot and a stripe of paint is put on the graph. Imagine we ride a bicycle along the stripe from left to right. As we start out from $A(0, 0)$ we have to lean to our left. By the time we get to the point $B(10, 2000)$ we are sitting up straight, and past there we have to lean to our right. Any point on our trip where we change which side we lean toward is an *inflection point*. ●

Higher Order Derivatives

The process of taking derivatives need not stop with taking a second derivative. There can be a third derivative, a fourth derivative, and so on.

DEFINITION   *n*th derivatives and the order of a derivative

Suppose $f(x)$ and $g(x)$ are such that $g(x)$ is the result of taking the derivative of $f(x)$ and its derived functions a total of n times. Then $g(x)$ is the nth derivative of $f(x)$, and it is called a derivative of *order n*.

The derivatives of $f(x)$ are written as $f'(x)$ for the first deriva-tive, $f''(x)$ for the second derivative, and $f'''(x)$ for the third derivative. Subsequent derivatives are written $f^{(4)}(x)$, $f^{(5)}(x)$, $f^{(6)}(x)$, . . . , $f^{(n)}(x)$, and so on.

EXAMPLE 12   Finding higher order derivatives

Suppose $f(x) = 3x^5 - 5x^3 + 4x - 1$. Find $f'''(x)$, $f^{(4)}(x)$, and a form for the nth order derivative.

SOLUTION

We need to find the derivative of each order, starting with the first derivative.

$$f'(x) = 3 \cdot 5x^4 - 5 \cdot 3x^2 + 4 = 15x^4 - 15x^2 + 4$$
$$f''(x) = 15 \cdot 4x^3 - 15 \cdot 2x = 60x^3 - 30x$$
$$f'''(x) = 60 \cdot 3x^2 - 30 = 180x^2 - 30$$
$$f^{(4)}(x) = 180 \cdot 2x = 360x$$
$$f^{(5)}(x) = 360$$
$$f^{(6)}(x) = 0$$
$$f^{(n)}(x) = 0 \text{ for all } n \geq 6$$

3.3 PROBLEMS

Foundations

The problems of this section require the basic skills illustrated by the following:

Find the derivatives of problems 1–3 as indicated.

1. $\dfrac{d}{dx}(x^{-3} + 2x^2)$ **2.** $\dfrac{d}{dx}\left(2x^{3.1} - \dfrac{2}{3x}\right)$

3. $\dfrac{d}{dx}\left(\dfrac{4x}{x^2 + 1}\right)$

4. Find the critical numbers of $f(x) = 2x^3 - 24x$.

5. Solve $3x^2 - 7x + 2 = 0$.

6. Solve $x^4 - 3x^2 = 0$.

7. Solve $-x^2 - 2x + 15 > 0$.

Exercises

Find y'' for each of the problems 8–17.

8. $y = 5x - 2$ **9.** $y = 5x + 7$

10. $y = x^{-3} + 2x^2$ **11.** $y = x^{-2} - x^3$

12. $y = 2x^{3.4} + \dfrac{1}{x}$ **13.** $y = 3x^{3.1} - \dfrac{2}{3x}$

14. $y = \sqrt{x + 4}$ **15.** $y = \sqrt{3x - 2}$

16. $y = (2x + 3)^7$ **17.** $y = (2x^3 + 1)^6$

18. Find $\dfrac{d^2}{dx^2}\left(2x^3 + \dfrac{3}{x^2}\right)$.

19. Find $\dfrac{d^2}{dx^2}(6x^2 - \sqrt{x + 2})$.

20. For $y = 2x^3 - 5x^2$, find $\dfrac{d^2y}{dx^2}\bigg|_{x=1}$.

21. For $y = x^3 + x$, find $\dfrac{d^2y}{dx^2}\bigg|_{x=1}$.

22. **(Consumer Price Index)**   Using the graph given at the start of this section, estimate the second derivative of the Consumer Price Index function in the years (a) 1975 and (b) 1985.

The graphs in problems 23–26 represent functions.
(a) For each, estimate intervals over which the graph is concave up and intervals for which the graph is concave down.
(b) Estimate the x-coordinate on any inflection points.

23.

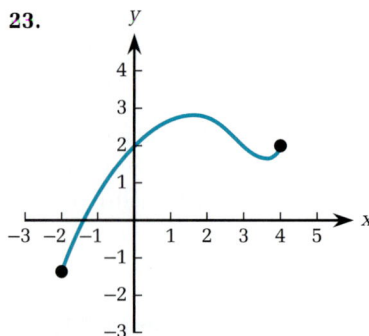

FIGURE 3.3.17

24.

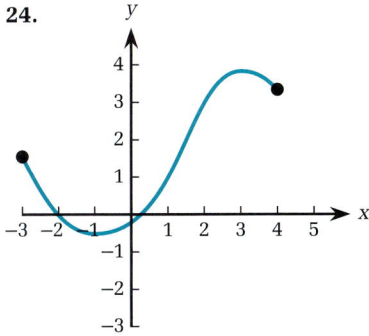

FIGURE 3.3.18

25.

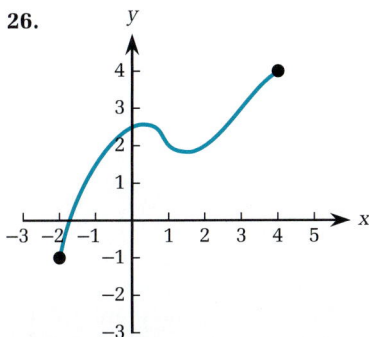

FIGURE 3.3.19

26.

FIGURE 3.3.20

For each function in problems 27–40, find the open intervals over which the graph is concave up and those for which the graph is concave down.

27. $f(x) = x^2 - 3x + 3$ **28.** $f(x) = 3 + x - x^2$

29. $g(x) = x^3 - x$ **30.** $g(x) = x^3 - 3x$

31. $f(x) = 2x^3 - 5x^2 + 4x - 1$

32. $f(x) = 2x^3 - 7x^2 + 4x + 3$

33. $f(x) = x^4 - 2x^2$

34. $f(x) = 3x^4 - 4x^3 + 5$

35. $g(x) = 3x^5 - 2x^3$ **36.** $g(x) = x^5 - 5x^3$

37. $f(x) = \dfrac{1}{x^2}$ **38.** $g(x) = \dfrac{3}{(x-1)^2}$

39. $f(x) = \dfrac{x+4}{\sqrt{x}}$ **40.** $g(x) = \dfrac{x+1}{\sqrt{x}}$

For the functions in problems 41–48:
(a) Find the critical numbers.
(b) Find any relative and any absolute maximum and minimum values.
(c) Find any inflection points.
(d) Sketch the graph.

41. $f(x) = 2x^3 + x^2 - 4x + 7$ over $[0, 1]$

42. $f(x) = x^3 + 3x^2 + 3x + 2$ over $[-2, 2]$

43. $f(x) = 4x^3 - 3x^2 - 6x + 1$ over $[-1, 2]$

44. $g(x) = x^3 + x^2 - x + 2$ over $[-1/2, 2]$

45. $f(x) = \dfrac{4}{x} + 9x$ over $[0.5, 2]$

46. $f(x) = \dfrac{4x}{x^2 + 1}$ over $[0, 4]$

47. $g(x) = \dfrac{x^2 - 2}{x^2 + 4}$ over $[-1, 3]$

48. $f(x) = x^{4/5}$ over $[-1, 2]$

Find the third derivatives of the functions in problems 49–52.

49. $f(x) = 3x^5 + x^3$ **50.** $g(x) = x^5 - 3x^4$

51. $f(x) = \dfrac{1}{1+x}$ **52.** $h(x) = \dfrac{2}{x+3}$

53. Suppose that the sales on day x are given by $s(x) = 100 + 30x^2 - x^3$ over $[1, 15]$.
(a) What day had the greatest sales?
(b) On what day was the rate of sales increase the greatest?

54. A new electric jogging machine was put in at the health club. The percentage of members using the machine on day x was given by $p(x) = 6x^2 - x^3$ $(0 \leq x \leq 6)$.
(a) What day had the greatest rate of increase by new users?
(b) What was the maximum percentage of member users?

55. Suppose the percentage of aluminum cans that are not recycled by week t is given by $g(t) = 27\left(\dfrac{100 + t^2}{27 + t^2}\right)$.

(a) Find the week during which the rate of cans being recycled is greatest.

(b) What is the greatest rate (per week) that cans are being recycled?

56. **(Inflection Points on a Graph of Videocassette Recorder Sales)** Figure 3.3.21 shows percentages of American households with VCRs.* Estimate any inflection point(s) on the curve.

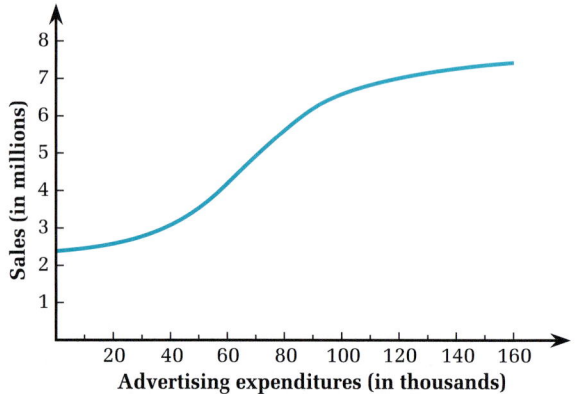

FIGURE 3.3.21

57. The following table gives the percentages of American households with telephones.† Plot the data and estimate the inflection point in the period 1935–1980.

YEAR	PERCENTAGE
1920	35
1925	38.7
1930	40.9
1935	31.8
1940	36.9
1945	46.2
1950	61.8
1955	71.5
1960	78.3
1965	84.6
1970	90.5
1980	93

*Source (1978–1988): Television Bureau of Advertising; source (1992): Motion Picture Association of America.
†Source: *Historical Abstracts of U.S.* Washington, D.C.: U.S. Department of Commerce, 1958.

58. **(Sales as a Function of Advertising Dollars— "Bangs for the Bucks").** In 1990, Adolph Coors Co. shipped 11.8 million barrels of their Coors Light beer and spent 71.7 million dollars advertising it.‡ Advertising expenditures affect sales, and how much to spend is a difficult decision. Figure 3.3.22 shows a hypothetical graph of sales as a function of dollars spent on advertising some product. Find the inflection point and interpret its meaning.

FIGURE 3.3.22

Graphing Calculator Problems

59. Consider the function $f(x) = x^6 - 2x^5 + 3x$. Display this graph over $[-2, 2]$ and estimate by ⬚Trace⬚ the value for x at the inflection point. You will need to work out the best range values for y. *After* you have tried using ⬚Trace⬚ on the graph of $f(x)$, then use ⬚Trace⬚ on the graphs of $f'(x)$ and $f''(x)$. On which graph do you think your use of ⬚Trace⬚ gave you a more accurate location of the inflection point? What are the limitations of this method of finding inflection points? What are the advantages?

60. Estimate the inflection points of the graph of $y = 7x^5 + 18x^4 + 3x^3 - 14x^2$.

Problems for Writing and Discussion

61. What are the advantages and disadvantages of using the Second Derivative Test rather than the First Derivative Test?

‡Source: *Denver Post,* April 16, 1991.

62. Does the second derivative of $f(x)$ have to exist at a point in order for that point to be an inflection point?

63. Does there have to be an inflection point between every two relative extreme points?

64. Many newspapers publish hourly temperatures for a city. Using that, or some other source, graph the hourly temperatures over a 24-hour period for some location. Illustrate any inflection points. Do there always have to be two? Can there be more than two? Can there be an odd number of inflection points over one 24-hour period?

65. Investors in stocks and bonds are often advised, "Buy low. Sell high." Why might better advice be, "Don't buy low. Buy when the rate of increase is high"? Sketch a stock price and indicate places to buy and sell.

Enrichment Problems

66. Show that a quadratic function of the form $f(x) = ax^2 + bx + c$ does not have any inflection point.

67. Show that the graph of a cubic polynomial of the form $f(x) = ax^3 + bx^2 + cx + d$ has exactly one inflection point, and give the x-coordinate of the inflection point.

68. Suppose a production cost function, $C(x)$, is such that $C''(x) > 0$ (that is, the graph is concave upward as in figure 3.3.23). Show that the least average production cost occurs when the average cost $AC(x) = (C(x))/x$ is equal to the marginal cost, $MC(x) = C'(x)$. (Hint: Find a critical number for $AC(x)$ and use the Second Derivative Test.)

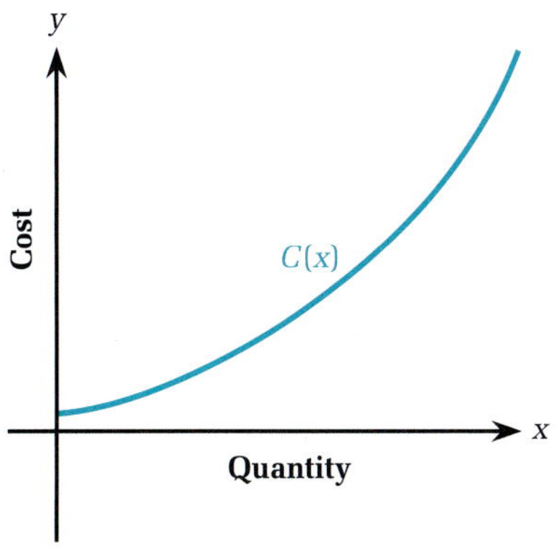

FIGURE 3.3.23

69. If two functions are equal, then every derivative of them is equal. Use this to find values for a, b, and c such that $x^3 - 9x^2 + 32x - 35 = a(x - 3)^3 + b(x - 3)^2 + c(x - 3) + 7$.

SOLUTIONS TO PRACTICE EXERCISES

1. For $y = x^{3/2} + x^4$, $y' = (3/2)x^{1/2} + 4x^3$, $y'' = (3/2)(1/2)x^{-1/2} + 12x^2$.

2. For $f(x) = 10x^3 - 21x^2 + 12x + 5$, $f'(x) = 30x^2 - 42x + 12 = 6(5x^2 - 7x + 2) = 6(5x - 2)(x - 1)$. Thus, the critical numbers are 2/5 and 1.

$f''(x) = 60x - 42$, so $f''(2/5) = 24 - 42 < 0$ and $f''(1) = 60 - 42 > 0$. By the Second Derivative Test, $A(2/5, f(2/5))$ is a relative maximum point and $B(1, f(1))$ is a relative minimum point.

When we evaluate these we get $A(0.4, 7.08)$ for the relative maximum point and $B(1, 6)$ for the relative minimum point. A graph of this function is shown in figure 3.3.24.

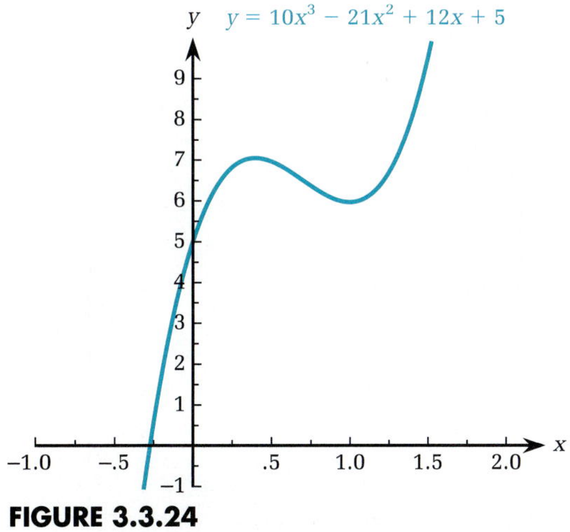

FIGURE 3.3.24

3. For $f(x) = x^3 - 9x^2 + 3x$, $f'(x) = 3x^2 - 18x + 3$ and then $f''(x) = 6x - 18$.

$f''(x) = 0$ when $6x - 18 = 0$, which is true for $x = 3$. Choose values of 2 and 4 as x-values on either side of 3. By evaluating $f''(2) = 6 \cdot 2 - 18 = -6$ and $f''(4) = 6 \cdot 4 - 18 = 6$, we find the second derivative changes sign at $x = 3$. Hence $P(3, f(3)) = P(3, -45)$ is the inflection point.

3.4 GRAPHS INVOLVING INFINITE LIMITS AND ASYMPTOTES

In this section, we conclude our discussion of graphing by investigating horizontal and vertical asymptotes and then summarizing our techniques for graphing.

Let us start by considering the recycling of aluminum cans, 92 billion of which were manufactured in the United States in 1992. Shown in figures 3.4.1–3.4.4 is a sequence of graphs depicting a possible function of the percentage of aluminum cans that have *not* been recycled by week t. The actual function is unknown, but it has this general character. The horizontal scales become compressed over this sequence to give us the sense of what is happening over progressively longer intervals of time.

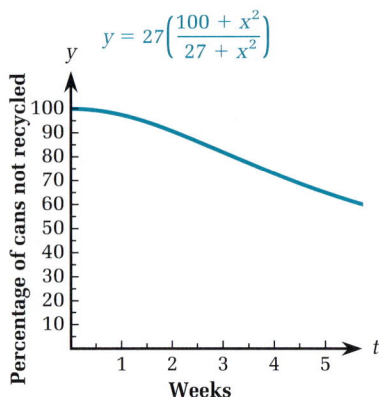

$$y = 27\left(\frac{100 + x^2}{27 + x^2}\right)$$

FIGURE 3.4.1

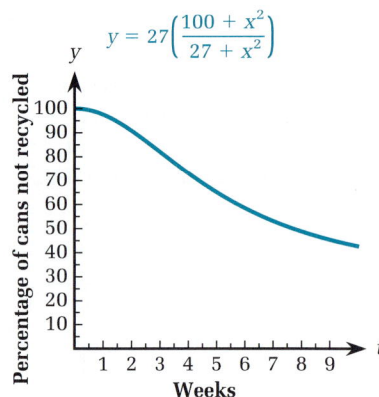

$$y = 27\left(\frac{100 + x^2}{27 + x^2}\right)$$

FIGURE 3.4.2

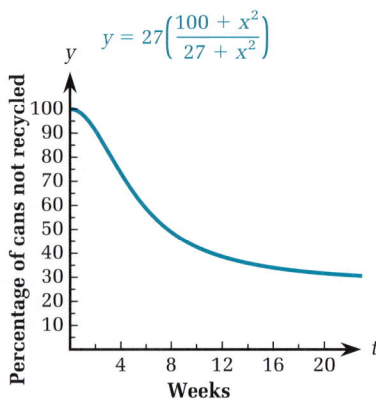

$$y = 27\left(\frac{100 + x^2}{27 + x^2}\right)$$

FIGURE 3.4.3

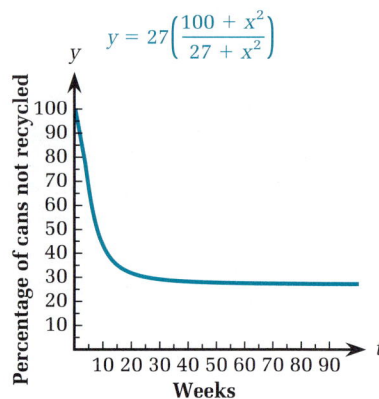

$$y = 27\left(\frac{100 + x^2}{27 + x^2}\right)$$

FIGURE 3.4.4

From these graphs we can estimate that approximately 27% of all aluminum cans are never recycled, but are lost in landfills or are otherwise unavailable for reuse.

The graphs in figures 3.4.1–3.4.4 were modeled by using the function $g(t) = 27\left(\dfrac{100 + t^2}{27 + t^2}\right)$, where $g(t)$ gives the percentage of aluminum cans that have not been recycled by week t. Asking, "What happens over the long time?" is asking for the limit of $g(t)$ as t increases without bound. The phrase used to describe this is "the limit as t goes to infinity." In symbols, $\lim\limits_{t\to\infty} g(t) = 27$.

Infinity and Limits

Suppose we want to graph $y = f(x)$. In doing so, imagine that we readjust the horizontal scales several times, just as we did above, so that the screen of the graph shows more and more of the x-axis. If it happens that the graph appears to flatten out and approach a horizontal line, then the y-value of that line is the limit of the function as x increases without bound. We use the phrase "the limit as x goes to infinity." Do not think of "infinity" as being some immense real number, whose symbol is ∞ and which x is getting close to, however. Instead, think of x continuing out, beyond any bound.

The general situation is described by the following:

DEFINITION

Limit at infinity

The *limit as x goes to infinity of f(x) is L*, written $\lim\limits_{x\to\infty} f(x) = L$, means that $f(x)$ is arbitrarily close to L when x is sufficiently large. The *limit as x goes to negative infinity of f(x) is K*, written $\lim\limits_{x\to-\infty} f(x) = K$, is defined in a similar manner.

In order to simplify our discussion, we will deal only with increasing values of x, or $\lim\limits_{x\to\infty} f(x)$. Often there are similar results for $\lim\limits_{x\to-\infty} f(x)$.

In Section 2.2 we listed six properties of limits (L_1 to L_6), each stated for x approaching a real number. In the following table are given properties for limits at infinity.

VALUES OF LIMITS AT INFINITY

$L_{\infty 1}$ $\lim\limits_{x\to\infty} x^n = \infty$ and $\lim\limits_{x\to\infty} \dfrac{1}{cx^n} = 0$
for any constant c and any positive number n

$L_{\infty 2}$ $\lim\limits_{x\to\infty}(f(x) + g(x)) = \lim\limits_{x\to\infty} f(x) + \lim\limits_{x\to\infty} g(x)$

$L_{\infty 3}$ $\lim\limits_{x\to\infty}(f(x) \cdot g(x)) = \lim\limits_{x\to\infty} f(x) \cdot \lim\limits_{x\to\infty} g(x)$

$L_{\infty 4}$ $\lim\limits_{x\to\infty}\left(\dfrac{g(x)}{f(x)}\right) = \dfrac{\lim\limits_{x\to\infty} g(x)}{\lim\limits_{x\to\infty} f(x)}$ (provided $\lim\limits_{x\to\infty} f(x) \neq 0$)

Using these rules we can show the limit at infinity of rational functions.

THE LIMIT AT INFINITY OF RATIONAL FUNCTIONS

Suppose $p(x)$ is a polynomial of degree n and the coefficient of its higher power is P. Suppose $q(x)$ is a polynomial of degree m and the coefficient of its highest power is Q.

$L_{\infty 5}$ $\lim\limits_{x\to\infty}\left(\dfrac{p(x)}{q(x)}\right) = \begin{cases} \text{Does not exist if } n > m. \\ 0 \text{ if } n < m \\ \dfrac{P}{Q} \text{ if } n = m \end{cases}$

EXAMPLE 1   Limits at infinity

Find the following limits, if they exist.

(a) $\lim\limits_{x\to\infty} \dfrac{3}{2x + 1}$ (b) $\lim\limits_{x\to\infty} \dfrac{x}{2x + 1}$ (c) $\lim\limits_{x\to\infty} \dfrac{x^2}{2x + 1}$

(d) $\lim\limits_{x\to\infty} 2x + 1$

SOLUTION

(a) Use $L_{\infty 5}$. For $p(x) = 3$, we have $n = 0$ (because $3x^0 = 3$) and $P = 3$. For $q(x) = 2x + 1$, we have $m = 1$ and $Q = 2$. Because $n < m$, $\lim\limits_{x\to\infty} 3/(2x + 1) = 0$.

(b) Use L$_{\infty 5}$. For $p(x) = x$, we have $n = 1$ and $P = 1$. For $q(x) = 2x + 1$, we have $m = 1$ and $Q = 2$. Because $m = n$,
$\lim\limits_{x \to \infty} x/(2x + 1) = P/Q = 1/2$.

(c) Use L$_{\infty 5}$. For $p(x) = x^2$, we have $n = 2$ and $P = 1$. For $q(x) = 2x + 1$, we have $m = 1$ and $Q = 2$. Because $n > m$, the limit does not exist.

(d) $\lim\limits_{x \to \infty} 2x + 1$ does not exist because $2x + 1$ becomes unboundedly large as x does.

PRACTICE EXERCISE 1

Find the following limits, if they exist.

(a) $\lim\limits_{x \to \infty} \dfrac{x - 1}{x^2 + 1}$ **(b)** $\lim\limits_{x \to \infty} \dfrac{1}{x - 1}$ **(c)** $\lim\limits_{x \to \infty} \dfrac{3x}{2 - x}$

(d) $\lim\limits_{x \to \infty} \dfrac{x^2}{x - 1}$

Horizontal Asymptotes

Recall that the definition of an *asymptote* is: If the graph of a function appears to be straightening out and approaching a line, then that line is called an *asymptote* of the graph. Asymptotes are indicated on a graph by a dashed line.

> ### HORIZONTAL ASYMPTOTES AND LIMITS AT INFINITY
>
> The line $y = M$ is a *horizontal asymptote* of $y = f(x)$ if $\lim\limits_{x \to \infty} f(x) = M$.

EXAMPLE 2   Horizontal asymptote

Find the horizontal asymptote of $f(x) = 1 + \dfrac{2}{x + 1}$.

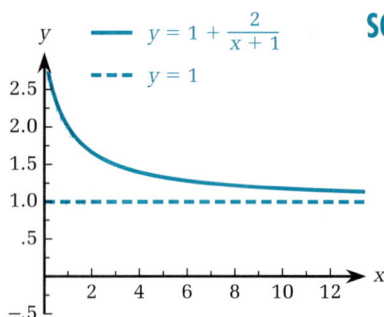

FIGURE 3.4.5

SOLUTION

By $L_{\infty 5}$ we have $\lim\limits_{x \to \infty} \dfrac{2}{x + 1} = 0$. Using $L_{\infty 2}$, we have $\lim\limits_{x \to \infty} \left(\dfrac{1 + 2}{x + 1} \right) = 1 + 0 = 1$. Thus, the horizontal asymptote is $y = 1$, as shown in figure 3.4.5.

In order for the graph of $f(x)$ to have a horizontal asymptote, $\lim\limits_{x \to \infty} f(x)$ must exist. In the next example, this limit does not exist.

EXAMPLE 3   A function with no horizontal asymptote

Show that the graph of $f(x) = \dfrac{2x + 3}{5}$ has no horizontal asymptote.

SOLUTION

By $L_{\infty 5}$, we know $\lim\limits_{x \to \infty} (2x + 3)/5$ does not exist and, correspondingly, there is no horizontal asymptote.

EXAMPLE 4   Horizontal asymptote

Show that $f(x) = \dfrac{1 + 3x}{x}$ has a horizontal asymptote of $y = 3$.

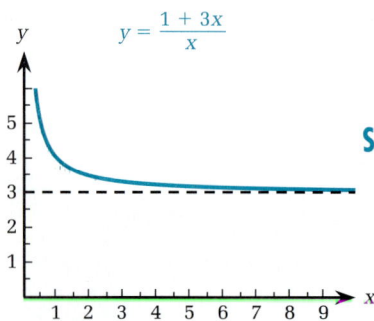

FIGURE 3.4.6

SOLUTION

Use $L_{\infty 5}$. To evaluate $\lim\limits_{x \to \infty} (1 + 3x)/x$, let $p(x) = 1 + 3x$ and $q(x) = x$. Then $n = m = 1$ and we have $P = 3$ and $Q = 1$. Hence, $\lim\limits_{x \to \infty} (1 + 3x)/x = 3$.

A graph of $f(x)$ is shown in figure 3.4.6.

EXAMPLE 5   Horizontal asymptote

Find the horizontal asymptote for $g(x) = \dfrac{x}{x^2 + 1}$.

SOLUTION

Use $L_{\infty 5}$. For $p(x) = x$ and $q(x) = x^2 + 1$, we have $n = 1$, $m = 2$, $P = 1$, and $Q = 1$. Because $n < m$, $\lim\limits_{x \to \infty} \dfrac{x}{x^2 + 1} = 0$ is the horizontal asymptote.

PRACTICE EXERCISE 2

Find the horizontal asymptotes for:

(a) $f(x) = \dfrac{2x - 3}{3x}$ (b) $g(x) = \dfrac{3x}{x^2 - 1}$

An application of limits at infinity and asymptotes is shown in Example 6.

EXAMPLE 6   Horizontal asymptotes, strawberry production

A strawberry is harvestable when it is ripe and can be shipped to market. Suppose an approximation for the percentage (expressed decimally) of the total number of strawberries in a field that are harvestable by day t is $R(t) = \dfrac{9t^2 - 10t + 9}{10t^2 + 10}$, where t is measured in days and $t = 1$ is the first day the field has plants ready to be picked.

Find $R(1)$, $R(2)$, $R(3)$, $R(4)$, and $R(5)$, and estimate the eventual total percentage of harvestable strawberries in the field.

SOLUTION

$$R(1) = \frac{9 \cdot 1^2 - 10 \cdot 1 + 9}{10 \cdot 1^2 + 10} = \frac{9 - 10 + 9}{10 + 10} = \frac{8}{20} = 0.4, \text{ which means}$$

40% are harvestable on the first day.

On the second day, $R(2) = 0.5$, which means 50% are harvestable by that day, an increase over the first day. Continuing to evaluate $R(t)$ at $t = 3$, 4, and 5 gives $R(3) = 0.6$, $R(4) = 0.66$, and $R(5) = 0.71$. If you wait until day 5 to do any picking, on that day a total of 71% of the strawberries are ripe.

To find the eventual total percentage, we need to find $\lim_{x \to \infty} R(t) = \lim_{x \to \infty} (9t^2 - 10t + 9)/(10t^2 + 10)$.

Because both numerator and denominator are polynomials, we let $p(t) = 9t^2 - 10t + 9$ and $q(t) = 10t^2 + 10$. With both $n = 2$ and $m = 2$ we use $L_{\infty 5}$ and find the limit to be $9/10 = 0.9$. This means that the limit is 90%.

In actuality, pickers begin picking on day one because the berries do not remain harvestable for more than a few days.   ●

Vertical Asymptotes

Vertical asymptotes of the graph of a rational function were mentioned in Chapter 1. Such asymptotes commonly are caused by

x-values that produce zero in a denominator while producing a non-zero value in the numerator. Example 7 shows such a situation.

EXAMPLE 7   Vertical asymptotes

Find any vertical asymptotes for $f(x) = \dfrac{3x}{x^2 - 1}$.

SOLUTION

The denominator $x^2 - 1 = 0$ when $x = \pm 1$, and for those x-values, the numerator is not zero, so not only do $\lim_{x \to 1} f(x)$ and $\lim_{x \to -1} f(x)$ fail to exist, the function actually "blows up," as shown in figure 3.4.7.

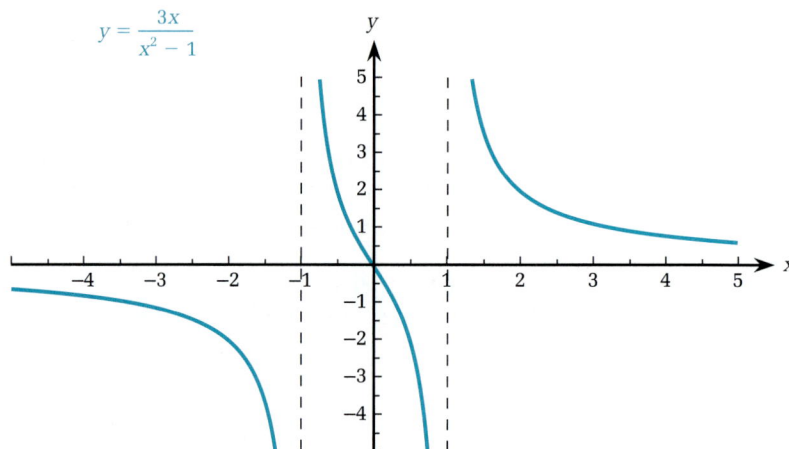

FIGURE 3.4.7

PRACTICE EXERCISE 3

Find any vertical asymptotes of $f(x) = \dfrac{5}{x^2 - x - 6}$.

Summary of Graphing Techniques

Collecting together all the derivative tests and other information, we can summarize the techniques of graphing.

SUMMARY OF GRAPHING TECHNIQUES

1. Determine the domain.

2. Use dashed lines to represent vertical asymptotes (usually at x-values for which the function "blows up" by having a zero denominator but nonzero numerator) and horizontal asymptotes ($\lim\limits_{x \to \infty} f(x)$).

3. Find $f'(x)$.

4. Find the critical numbers for x by listing (a) solutions of $f'(x) = 0$, and (b) values of x for which $f'(x)$ does not exist.

5. Evaluate $f(x)$ at each critical number and any endpoints of the domain.

6a. If it is easy to evaluate $f''(x)$ at the solutions found in (4a), do so and use the Second Derivative Test.

   b. If it is difficult to evaluate $f''(x)$ at a (4a) solution, or if $f''(x) = 0$ there, use the First Derivative Test or check values of $f(x)$ on either side of your solutions to (4a).

7. For the values of (4a), use the First Derivative Test or check values of $f(x)$ on either side.

8. Compare all the relative maximum and minimum values you have found with the values of $f(x)$ at any endpoints.

9. (This step is often optional.) Locate inflection points by using $f''(x)$ and going though the process of searching for maximum and minimum values of $f'(x)$.

10. Using the information you have found in steps (1)–(9), sketch a graph, indicating asymptotes and the points of all types of extreme values.

Example 8 draws upon many of the skills we have acquired over these several sections on graphing. It uses a function that appears quite simple, but whose graph is fairly difficult to describe.

EXAMPLE 8

Sketch the graph of $f(x) = \dfrac{2}{x^2 - x - 2}$.

SOLUTION

1. *Domain.* The denominator factors as $x^2 - x - 2 = (x - 2)(x + 1)$. Because $f(x)$ is not defined if the denominator is zero, the domain consists of all real numbers except $x = 2$ and -1.

2. *Vertical Asymptotes.* There are vertical asymptotes at $x = 2$ and $x = -1$ because the denominator is zero but the numerator is not zero at those values.
   Horizontal Asymptotes. By finding $\lim_{x \to \infty} f(x) = 0$, we see $y = 0$ is the horizontal asymptote.

3. *Find $f'(x)$.* Use the quotient rule to find the derivative

$$f'(x) = \frac{(x^2 - x - 2) \cdot 0 - 2 \cdot (2x - 1)}{(x^2 - x - 2)^2} = \frac{-4x + 2}{(x^2 - x - 2)^2}$$

4. *Find Critical Numbers.*
   (a) The critical numbers arising from the derivative being zero are found by setting the numerator of $f'(x)$ equal to zero. If $-4x + 2 = 0$, then $x = 1/2$.
   (b) There are no critical numbers arising from values of x for which $f'(x)$ is not defined. $f'(x)$ is not defined at 2 and -1, but those are not critical numbers because the function itself was not defined at those x-values.

5. *Evaluate $f(x)$ at Each Critical Number.*

$$f(0.5) = \frac{2}{(0.5)^2 - 0.5 - 2} = \frac{2}{-2.25} = -\frac{8}{9} \simeq -0.8889$$

6. *Decide What to Do About Finding and Using the Second Derivative.* Taking the second derivative is not really difficult, but it appears to be tedious. So we use the First Derivative Test to check for a relative maximum or minimum. For $f'(x) = (-4x + 2)/(x^2 - x - 2)^2$, the numerator goes from being positive to being negative at $x = 0.5$ (because $-4x + 2 > 0$ only when $2 > 4x$, which requires $0.5 > x$) and the denominator is always positive (because it is a squared term).

   Thus, $f'(x) > 0$ for $x < 0.5$ and $f'(x) < 0$ for $x > 0.5$. Of course, $f'(x)$ does not exist at -1 and 2. Let us see what we can say about the function so far.

   Because the numerator of $f(x)$ is 2, $f(x)$ is never zero, for the only way a quotient is zero is if the numerator is 0. Thus, the function never crosses the x-axis. In addition, the function is trapped between, or to the sides of, the two vertical asymptotes of $x = 2$ and -1. Lastly, we know $f'(0.5) = 0$, forcing the tangent to be horizontal at the point on the curve $P(0.5, -8/9)$. That point is a relative maximum by the First Derivative Test.

Table 3.4.1 helps us to visualize the graph. We choose a value for x that is to the left of $x = -1$, several values near $x = 0.5$, and a value to the right of $x = 2$.

TABLE 3.4.1

x	$f(x) = \dfrac{2}{x^2 - x - 2}$
-2	0.5
0	-1
0.5	$-8/9$
1	-1
3	0.5

The corresponding points are plotted in figure 3.4.8, and a curve is drawn through them, with vertical and horizontal asymptotes indicated.

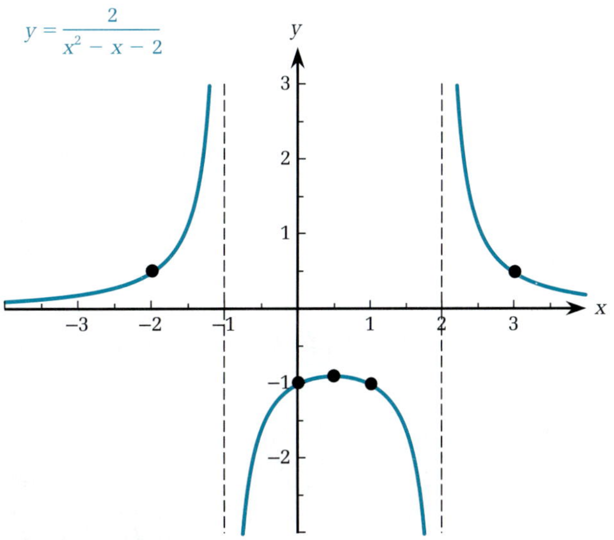

FIGURE 3.4.8

In Example 8, we did not consider a continuous function over a closed interval, but had we done so we would have had to evaluate the function at those interval endpoints and compare the values there with any relative extreme values we had found. Also, there were no inflection points to plot.

Graphing Calculator

If we have a graphing calculator, we can easily enter the function, but we still need to determine the range of values that produce the "interesting" points of extreme values and asymptotes.

SUMMARY OF GRAPHING TECHNIQUES FOR USE WITH A GRAPHING CALCULATOR

(G1) Determine domain and vertical asymptotes.

(G2) Find the critical numbers by determining $f'(x)$ and then finding x-values for which $f'(x) = 0$ or does not exist.

(G3) Evaluate $f(x)$ at every critical and endpoint number.

(G4) Set the range for x to be *from* slightly to the left of the smallest critical number, endpoint, or x-value of any vertical asymptote *to* slightly to the right of the same set of numbers.

(G5) Set the range for y to be *from* slightly less than each value of $f(x)$ at critical numbers or endpoints *to* slightly more than each of those numbers.

(G6) To check for horizontal asymptotes, use the $\boxed{\text{Zoom}}$ feature several times and see if a straightening appears. If the scale becomes lost on the y-axis, then manually set the range of y to be unchanged each time and only increase the interval for the values of x.

(G7) Locate inflection points by using $f''(x)$ and going though the process of searching for maximum and minimum values of $f'(x)$.

EXAMPLE 9   Determining ranges for an initial screen for graphing calculators

Suppose $f(x) = \dfrac{2}{x^2 - x - 2}$. Determine ranges for an initial screen.

SOLUTION

Use the computations of the preceding example. We found vertical asymptotes to occur at $x = 2$ and $x = -1$. There is a critical number at $x = 1/2$, and $f(1/2) = -8/9$. We may choose an initial screen to be $-2 \leq x \leq 3$ and $-2 \leq y \leq 1$. Many other choices would also be possible.   ●

The particular screen of Example 9 will show nothing about the horizontal asymptote, and we did not expect it to do so. To find the graph, we will need to [Zoom] back. Doing so does reveal the x-axis to be a horizontal asymptote.

There is a lot more that can be done with graphing, but we have done quite enough for now.

3.4 PROBLEMS

Foundations

The problems of this section require the basic skills illustrated by the following:

1. Determine the degrees and the leading coefficients of both numerators and denominators:
   (a) $\dfrac{3x^2 - x + 2}{2x^2 - x}$ (b) $\dfrac{3 - 2x}{x + 1}$

2. Find the critical numbers for $f(x) = x + 4/x^2$ over $[0.5, 3]$.

Exercises

Find those limits that exist in problems 3–12.

3. $\displaystyle\lim_{x\to\infty} \dfrac{2x}{x^2 - 1}$

4. $\displaystyle\lim_{x\to\infty} \dfrac{x}{x^2 + 1}$

5. $\displaystyle\lim_{x\to\infty} \dfrac{2x + 1}{x}$

6. $\displaystyle\lim_{x\to\infty} \dfrac{1 - x}{x}$

7. $\displaystyle\lim_{x\to\infty} (3x - 1)$

8. $\displaystyle\lim_{x\to\infty} (2x + 3)$

9. $\displaystyle\lim_{x\to-\infty} \dfrac{x - 1}{x + 1}$

10. $\displaystyle\lim_{x\to-\infty} \dfrac{3x + 1}{1 - 2x}$

11. $\displaystyle\lim_{x\to\infty} \dfrac{1 + x - 2x^2}{3x^2 + 1}$

12. $\displaystyle\lim_{x\to\infty} \dfrac{-2x^2 - x + 3}{3x + x^2}$

Determine the horizontal and vertical asymptotes of each function in problems 11–34.

13. $f(x) = \dfrac{1}{x - 2}$

14. $f(x) = \dfrac{2}{x - 3}$

15. $g(x) = \dfrac{x + 3}{2 - x}$

16. $g(x) = \dfrac{2x - 3}{1 - x}$

17. $g(x) = \dfrac{3 - 2x}{x + 1}$

18. $g(x) = \dfrac{3x - 2}{2 + x}$

19. $f(x) = \dfrac{1}{x^2 - 4}$

20. $f(x) = \dfrac{3}{x^2 - 9}$

21. $f(x) = \dfrac{x}{x^2 - 1}$

22. $f(x) = \dfrac{2x}{x^2 - 3}$

23. $f(x) = \dfrac{2x + 1}{x^2 - x}$

24. $f(x) = \dfrac{x - 2}{x^2 - 3x}$

25. $f(x) = \dfrac{1 + 5x}{x^2 + 2x}$

26. $f(x) = \dfrac{2 - 3x}{x^2 - 3x}$

27. $f(x) = \dfrac{x^2 - x}{3x^2}$

28. $f(x) = \dfrac{x^2 + 2x}{2x^2}$

29. $f(x) = \dfrac{x^3 - 2x^2}{2x + 1}$

30. $f(x) = \dfrac{x^3 + 2x^2}{x - 1}$

31. $f(x) = \dfrac{3x^2 - x + 2}{2x^2 - x}$

32. $f(x) = \dfrac{2x^3 + x}{x^3 - x}$

33. $g(x) = \dfrac{x^2 - x - 12}{x + 3}$

34. $g(x) = \dfrac{x^2 + 5x + 6}{x + 2}$

35. Suppose the percentage of aluminum cans that are not recycled by day t is $g(t) = (1000 + 10t^2)/(10 + t^2)$. What percentage of cans do not get recycled in the long term?

36. Suppose the E. T. Pizza Co. puts coupon ads on cars at the mall. After t days, the percent redeemed is $r(t) = 100 - (100 + 95t^2)/(1 + t^2)$. What percentage of coupons are redeemed in the long term?

Sketch the graphs of the functions in problems 37–60, locating and labeling any horizontal or vertical asymptotes and any relative or absolute maximum or minimum points.

37. $f(x) = \dfrac{3}{x - 2}$

38. $f(x) = \dfrac{2}{x + 1}$

39. $f(x) = 4x + \dfrac{1}{x}$

40. $f(x) = x + \dfrac{4}{x}$

41. $f(x) = \dfrac{2x}{x-1}$ **42.** $f(x) = \dfrac{x}{2x-1}$

43. $f(x) = 4x + \dfrac{1}{\sqrt{x}}$ **44.** $f(x) = x + \dfrac{2}{\sqrt{x}}$

45. $f(x) = x + \dfrac{4}{x^2}$ over $[0.5, 3]$

46. $f(x) = \dfrac{x}{4} + \dfrac{1}{x^2}$ over $[1, 4]$

47. $f(x) = \dfrac{5}{x^2}$ over $[1, 5]$ **48.** $f(x) = \dfrac{2}{x^2}$ over $[1, 3]$

49. $f(x) = \dfrac{3}{x^2 - 2x}$ over $(0, \infty)$

50. $f(x) = \dfrac{5}{2x^2 - 5x}$ over $(0, \infty)$

51. $f(x) = \dfrac{1}{x^2 + 2}$ over $[-1, 4]$

52. $f(x) = \dfrac{3}{x^2 + 1}$ over $[-1, 2]$

53. $f(x) = \dfrac{1}{x^2 - 4x + 5}$ over $[1, 3]$

54. $f(x) = \dfrac{1}{x^2 - 2x + 2}$ over $[0.5, 2]$

55. $f(x) = 3x^{2/3} - x$ over $[0, 10]$

56. $f(x) = 3x^{2/3} - x$ over $[0, 4]$

57. $f(x) = \dfrac{x}{2x^2 + 1}$ over $[0, 2]$

58. $f(x) = \dfrac{x}{x^2 + 1}$ over $[0, 2]$

59. $g(x) = 3x^{1/3} - x$ over $[-1, 2]$

60. $g(x) = 6x^{2/3} - 4x$ over $[-1, 2]$

▨ Graphing Calculator Problems

Find the ranges for an initial screen for problems 61 and 62.

61. $f(x) = \dfrac{x-2}{x^2 - x - 5}$

62. $g(x) = \dfrac{x - 103}{x^2 + 10x - 200}$

63. Sketch the graph of $(x^5 - 5x)/(x - 2)$, finding any relative maximum or minimum values.

Problems for Writing and Discussion

64. Suppose $y = M$ is a horizontal asymptote of $y = f(x)$. Can there be a value $x = a$ such that $f(a) = M$? [Hint: Consider $f(x) = x^3/(2x^3 + x - 1)$.]

$L_{5\infty}$ was stated for a rational function. Describe how it should be modified so that we can find the limit as $x \to \infty$ of the functions in problems 65 and 66.

65. $f(x) = \dfrac{\sqrt{3x}}{x + 2}$ **66.** $f(x) = \dfrac{\sqrt{2x} + 1}{3x^2}$

67. Why would $x = 2$ *not* be a vertical asymptote for $f(x) = \dfrac{\sqrt{x - 5}}{x - 2}$?

68. Why would $x = 0$ *not* be a vertical asymptote for $f(x) = \dfrac{\sqrt{x - 1}}{x}$?

Enrichment Problems

69. The graph of $g(x) = (x + 1 - \sqrt{x^2 + 1})/2x$ has two different horizontal asymptotes. Find each.

Any line that the graph of a function approaches but does not intersect is an asymptote. We have considered only horizontal and vertical asymptotes, but an asymptote can be *oblique,* that is, neither vertical nor horizontal but at a slant. For problems 70–73, find the oblique asymptote and sketch the graph of each function.

70. $f(x) = x + \dfrac{1}{x}$ **71.** $g(x) = 2x - \dfrac{1}{x}$

72. $g(x) = \dfrac{x^2 + 1}{x - 2}$ **73.** $f(x) = \dfrac{x^2 - 3}{x + 2}$

74. By $L_{\infty 5}$, $\lim\limits_{x \to \infty} (1 + x)/x = 1$. How large does x have to be to guarantee $(1 + x)/x$ is within 0.01 of 1?

SOLUTIONS TO PRACTICE EXERCISES

1. (a) To find this limit, use L$_{\infty}$5. $p(x)$ has degree 1 and $q(x)$ has degree 2. By L$_{\infty}$5 we have $\lim_{x \to \infty} (x - 1)/(x^2 + 1) = 0$.

   (b) $\lim_{x \to \infty} 1/(x - 1) = 0$.

   (c) $\lim_{x \to \infty} 3x/(2 - x) = -3/1 = -3$. Use L$_{\infty}$5. The degrees of $p(x)$ and $q(x)$ are both one, $P = 3$ and $Q = -1$. So the limit is $P/Q = -3$.

   (d) Again, using L$_{\infty}$5 we find $x^2/(x - 1)$ does not exist.

2. (a) The horizontal asymptote for $f(x) = (2x - 3)/3x$ is found from $\lim_{x \to \infty} (2x - 3)/3x = 2/3$ by L$_{\infty}$5.

   (b) The horizontal asymptote for $g(x) = 3x/(x^2 - 1)$ is found from $\lim_{x \to \infty} 3x/(x^2 - 1) = 0$ by L$_{\infty}$5.

3. The vertical asymptotes of $f(x) = 5/(x^2 - x - 6)$ occur when $x^2 - x - 6 = 0$. Because $x^2 - x - 6 = (x - 3)(x + 2)$, so the asymptotes are at $x = 3$ and $x = -2$.

3.5 OPTIMIZATION

Success in achieving our goals depends on sustained and effective effort and the wise use of our resources. One resource is calculus. This section, and the two sections following, demonstrate some of the "payoff" power of calculus. This is done by applying calculus to situations in which we want to optimize something. Think of these examples as somewhat of a "calculus buffet," where we are expected to learn what is available, how to take what we want and need, and how to finish what we start. However, we should not be intimidated by the fact there are so many possible applications. After all, we are not expected to heap everything on our plate at once.

Word Problems

A significant hurdle often appears when we solve optimization problems. It is that the problems are not presented in the form of equations with variables. Instead, we begin with only a description of a situation. It is up to us to create a mathematical model and establish the equations whose manipulation and solution can be translated back into action. In brief, we need to solve "word problems."

For many people, the need to solve a word problem creates a high level of anxiety, and they try quite hard to avoid such problems. That is understandable, but it also is unproductive. In an effort to provide some help in solving word problems, the following outline is offered as a suggested agenda for action. It does not guarantee success, but it may help.

SOLVING WORD PROBLEMS

1. Read the problem over once to learn the general situation. Attempt to explain the situation to someone else. Talking, even if you actually do it silently and only to yourself, is quite effective in focusing your thoughts.

2. Read the problem a second time, determining just what is known, what is unknown, and what would be an acceptable solution and/or an ideal solution.

3. Choose names and symbols for all the quantities. There is great power in good notation.

4. Write the choices down on paper, together with a sketch of the physical situation when appropriate.

5. Think about examples and formulas that relate to the problem and may help you get from your starting point to the solution.

6. Express the quantity that is to be maximized or minimized as a function of just one variable. This may require algebraic manipulation of several formulas.

7. Do the necessary mathematical computations—carefully.

8. Upon arriving at a solution, ask yourself if the solution makes sense. Think how it can be cross-checked, other than the obvious check from the answers in the back of the text.

Word problems are difficult. We should not expect ourselves to be able to write down an answer immediately. Sometimes an answer comes easily, but usually we must think for a few minutes. Do not simply sit, staring glassy-eyed at the book. It is vital that *something gets started.* Get the mind working on something, even if it is not clear just what it should be working on eventually. *Do not worry about false starts.* Just as "warming up" is desirable before strenuous physical activity, "getting something down on paper" is almost a necessity for successful mental activity.

There is great variety among optimization problems. We consider several common types and then go on to some miscellaneous situations.

Enclosure Problems

We may be asked to enclose or divide up a region or volume in some manner. Often we are given a fixed perimeter and asked to find the maximum area, or a fixed area and asked to find the smallest perimeter. Following are examples of both types.

EXAMPLE 1   Maximizing enclosed area with fixed perimeter

Juan wants to fence his garden. He has 1000 feet of fencing and wants a rectangular garden. What dimensions will give him the largest garden? How large will that garden be?

SOLUTION

Juan could simply lay out a rectangle and hope for the best. Three sample rectangles are shown in figure 3.5.1. Can calculus help? He decides that if he were to label one side x and the other side y, then the enclosed area would be $A = x \cdot y$. Because he has two sides of length x and two sides of length y, he would use $L = 2x + 2y$ feet of fence.

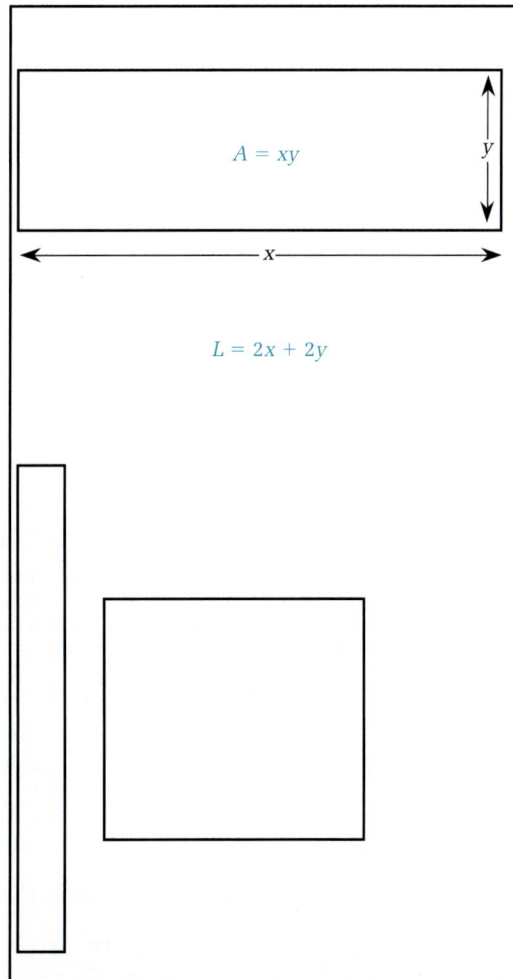

FIGURE 3.5.1

As Juan imagines different configurations, he thinks of the sides as being movable. If he makes x quite big, then y would have to be quite small. In the extreme case, the entire 1000 feet of fence would be used up by 500 feet down one side and 500 feet coming back, with nothing enclosed. The same thing would happen if x got too small. So the values of x go from 0 to 500, and the corresponding y-values go from 500 to 0.

Juan sets up a maximization problem. The area is expressed as being dependent on two variables, x and y, and not just one variable. However, he realizes that no matter what x and y he chooses, the total length of fence used, L, is $2x + 2y$. He only has 1000 feet of fence, so $2x + 2y = 1000$. From this it follows that $x + y = 500$. Hence, $y = 500 - x$.

The area formula is $A = xy$. Substituting $y = 500 - x$ in the area formula, he gets $A = x(500 - x) = 500x - x^2$, which is a function of the x variable alone.

Taking the derivative, $A'(x) = 500 - 2x$. Setting $A'(x) = 0$ and solving for x, a critical number is $x = 250$. (Juan knows that maximum values for functions can occur at endpoints, in this case $x = 0$ and $x = 500$, but he also knows the endpoints give him no area at all and so surely do not yield maximum values for his area.)

If $x = 250$, then $y = 250$ also and the area is $250 \cdot 250 = 62,500$ square feet.

This value really is a maximum, as Juan can tell by testing values near $x = 250$, or by using the First Derivative Test, or the Second Derivative Test, which for this problem is the easiest test of all to apply because $A''(x) = -2 < 0$.

To review Juan's strategy for solving the problem, using the numbered steps:

1 and 2. Juan had 1000 feet of fencing and wanted to create a rectangular area of maximum size.

3 and 4. He chose the variables x and y as names for the lengths of sides, and A and L for area and length, respectively. He labeled a diagram with these choices.

5. He used the formulas for area and length: $A = xy$ and $L = 2x + 2y$, respectively.

6. He found a formula for area that used only one variable. He did this by solving $1000 = 2x + 2y$ for y in terms of x and substituting into $A = xy$. Then $A = 500x - x^2$.

7. He maximized A by finding the derivative of A and checking that the critical number did provide a maximum.

8. He checked that he had found a maximum by using the Second Derivative Test. ●

Enclosure problems often concern subdividing areas, or different costs for parts of the perimeter, or some other complication.

EXAMPLE 2   Maximizing enclosed subdivided area with fixed perimeter

Ellen has 1000 feet of fence and she is raising two varieties of chickens. She wants to enclose as much area as she can in a rectangular manner, but she needs to subdivide the area into two smaller regions of equal area. How should she place her fencing so as to maximize the total area?

SOLUTION

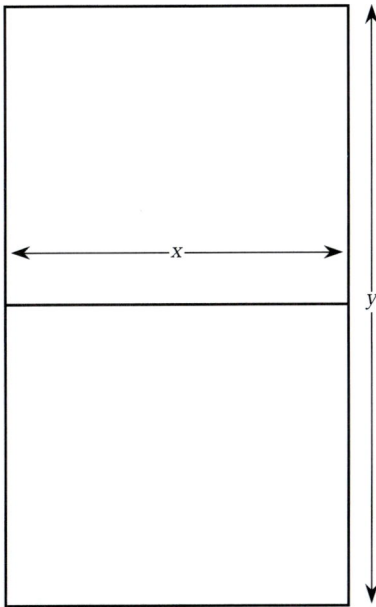

FIGURE 3.5.2

Drawing a sketch, Ellen labels the long sides as y and the shorter sides as x (see figure 3.5.2). There are three x lengths and two y lengths. She is not really sure whether the three pieces are short side or long side, but she is guessing that she would not want to be increasing three lengths if she could be increasing only two lengths.

Let A be the area and L the length of fencing. $A = x \cdot y$ as before, but now $L = 3x + 2y = 1000$. Solving the second equation for y, Ellen finds $y = (1000 - 3x)/2 = 500 - 3x/2$. Substituting for y in $A = x \cdot y$ gives

$$A = x\left(500 - \frac{3x}{2}\right) = 500x - \left(\frac{3}{2}\right)x^2, \text{ with } 0 \le x \le \frac{1000}{3}$$

Finding critical numbers from $A' = 500 - (3/2) \cdot 2x = 500 - 3x = 0$ gives $x = 500/3 \approx 166.7$. This value for x may be substituted directly into $A = 500x - (3/2)x^2$, or Ellen can find y and use $A = xy$. Because she needs to know y anyhow, she does the latter and uses $x = 500/3$ in the formula for y to give

$$y = 500 - \frac{3x}{2} = 500 - \frac{3(500/3)}{2} = 500 - \frac{500}{2} = 250$$

Hence, the area enclosed is $A = (500/3) \cdot 250 \approx 166.7 \cdot 250 \approx 41,667$ square feet.

This is a maximum value of A, as can be seen from the Second Derivative Test, $A'' = -3 < 0$. ●

In solving this problem, the length of fencing was rounded to the nearest tenth of a foot. The decision to use "tenths of a foot," rather than following our usual practice of rounding to four decimal places, was based on the actual interpretation of the problem. There is no definite rule that determines how many places of accuracy to use. For each problem we can only offer an answer that seems practical and usable.

PRACTICE EXERCISE 1

Jean is constructing a rectangular holding area for wild horses in Idaho. She has 900 feet of fence. She notices a long, straight cliff whose base she can use for one side of the holding area.

(a) What dimensions produce the largest holding area?

(b) What is that area?

To this point, we have been given a fixed perimeter and have been attempting to maximize the area. In Example 3, the area is fixed and the costs of the perimeter are to be minimized.

EXAMPLE 3   Minimizing perimeter costs of a fixed area

Find dimensions that minimize costs for a grocery store that is to be built in the shape of a rectangle and must contain 18,000 square foot. The front of the store costs $300 a foot and the sides and back of the store each cost $175 a foot.

SOLUTION

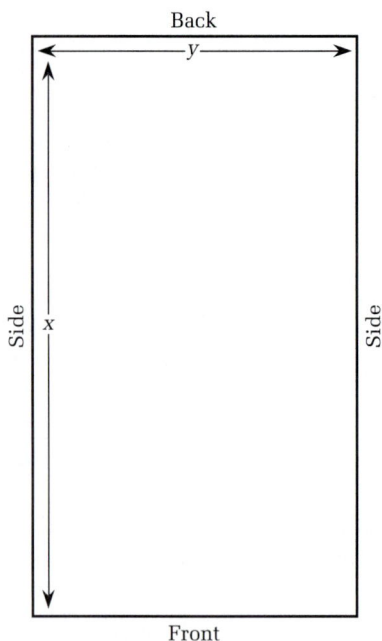

FIGURE 3.5.3

Let the sides be labeled x and the front and back be labeled y, as in figure 3.5.3. Then $A = x \cdot y = 18,000$.

The cost is C = front + two sides + back = $300y + 2 \cdot 175x + 175y$ = $475y + 350x$.

Solving for y in the area equation, $xy = 18,000$, gives $y = 18,000/x$. Substituting for y into C, we have $C = 475(18,000/x) + 350x = 8,550,000/x + 350x$. In this form, the cost, C, is a function of just one variable, x. We may write C as $C(x)$ to emphasize the fact that C is a function of x.

Taking the derivative to find critical numbers, $C'(x) = -8,550,000/x^2 + 350$.

Setting $C'(x) = 0$ gives $x^2 = 8,550,000/350 \approx 24,428.6$, so that $x \approx 156.3$. Substituting this value in the equation for y gives $y \approx 18,000/156 \approx 115.2$.

Using these values for x and y gives a store of area equal to $x \cdot y \approx 156.3 \cdot 115.2 = 18,005.76$ square feet (slightly over the 18,000 because of round-off errors), which costs about $C(156) = 8,550,000/156.3 + 350 \cdot 156.3 \approx \$109,408$.

This truly is a minimum cost, as can be seen from realizing C'' is always positive.

Maximize Revenue and Production

The revenue received by selling a quantity of items at a particular unit price is given by the product of that quantity times the price. Let x represent the quantity sold, $p(x)$ the demand function, which gives the price received for each x, and $R(x)$ the total revenue received. Expressed as an equation, $R(x) = x \cdot p(x)$.

Let us reconsider some examples from Section 1.5.

EXAMPLE 4 Maximize revenue, with given demand function

Based on actual market research, the demand curve for Minute Rice was $p(x) = -0.01x + 1.69$, where $p(x)$ was in dollars per box and x was the number of boxes sold.*

(a) For what quantity is total revenue maximized?

(b) What price should be charged in order to gain that revenue?

(c) What is the maximum revenue?

SOLUTION

(a) $R(x) = x(-0.01x + 1.69) = -0.01x^2 + 1.69x$, which is a form that is easier to differentiate.

The derivative is $R'(x) = -0.02x + 1.69$. Setting $R'(x) = 0$ and solving for x yields $x = 84.5$ (which the grocer might round to 85). Using the Second Derivative Test, the revenue will be a maximum because $R''(x) = -0.02 < 0$.

(b) The price is $p(85) = -0.01(85) + 1.69 = -0.85 + 1.69 = \$0.84$.

(c) The maximum revenue is $R(85) = 85 \cdot 0.84 = \$71.40$.

This is the same answer we got in Section 1.5. The reason we were able to do this problem previously without calculus was the very special nature of the demand curve—it was linear. This made the revenue function a quadratic and even without calculus we could find the vertex of the graph of a quadratic equation. However, we need calculus to maximize revenue with nonlinear demand curves, such as in Example 5.

EXAMPLE 5 Maximize revenue for a product with a nonlinear demand curve

Jetways Earplugs are sold in boxes with 100 pairs per box. The demand function is $p(x) = -2x^2 + 2400$, where x is the number of boxes and $p(x)$ is the price in dollars per box. The demand function is shown in figure 3.5.4.

*Lodish and Reibstein, "New Goldmines and Minefields in MR," *Harvard Business Review,* January 1986, p. 169.

(a) What number of boxes produces the maximum revenue?

(b) What is the price at that level of sales?

(c) What is the maximum revenue?

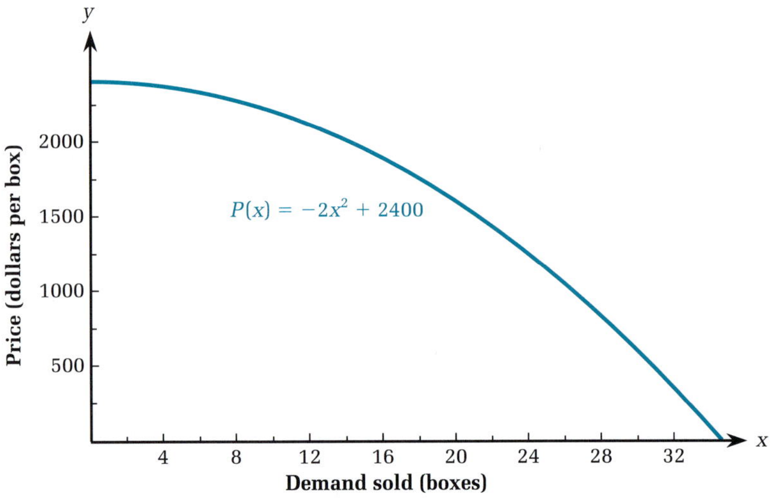

FIGURE 3.5.4

SOLUTION

(a) Let $R(x)$ be the revenue on the sale of x boxes. Regardless of whether the demand curve is linear or not, it is still true that revenue is the product of quantity times price:
$$R(x) = x \cdot p(x) = x(-2x^2 + 2400) = -2x^3 + 2400x.$$
To maximize $R(x)$, we find value(s) of x for which $R'(x) = 0$. The derivative is $R'(x) = -6x^2 + 2400$. If $R'(x) = 0$, then $6x^2 = 2400$. Solving for x gives $x^2 = 400$, and so $x = \pm 20$. Because we must allow only positive values for x, we check only $x = 20$.

Use the Second Derivative Test. $R''(x) = -12x$, so that $R''(20) = -12(20) = -240 < 0$. Hence $R(20)$ is a maximum. The graph of $R(x)$ is shown in figure 3.5.5.

(b) The price received from the sale of 20 boxes is $p(20) = -2(20)^2 + 2400 = -800 + 2400 = 1600$ dollars per box (of 100).

(c) The revenue on that sale is $R(20) = x \cdot p(x) = 20 \cdot 1600 = \$32,000$.

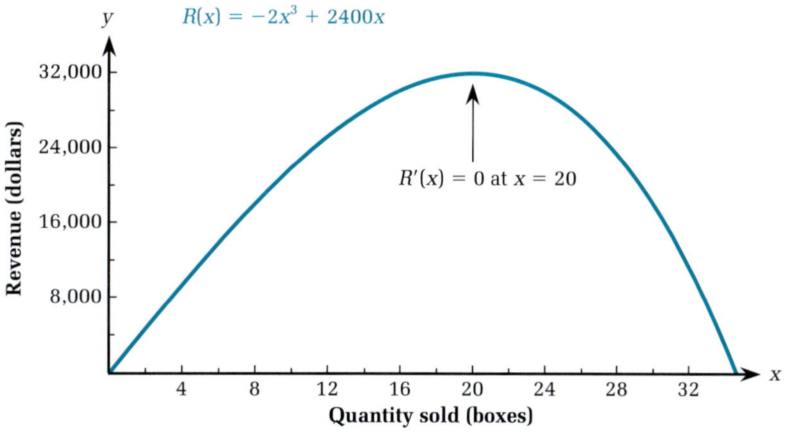

$$R(x) = -2x^3 + 2400x$$

$R'(x) = 0$ at $x = 20$

FIGURE 3.5.5

The demand function is usually not initially provided in algebraic form. More typically, some data are provided and we then have to create the demand function. Consider another problem from Section 1.5.

EXAMPLE 6 Finding an equation for demand and maximizing revenue

When the Little Dinner Theatre priced their tickets at $25, they sold 300. When they raised their price to $30, they sold only 250. Assume the demand function is linear.

(a) Find the demand function.

(b) Find the number of tickets they should sell to maximize their revenue.

(c) Find the price of those tickets.

(d) Find the maximum revenue.

SOLUTION

(a) The assumption that the demand function is linear means $p(x) = ax + b$ for some values of a and b. We can find the slope, a, from knowing $p(300) = 25$ and $p(250) = 30$. These give us

$$a = \frac{\text{change in price}}{\text{change in quantity sold}} = \frac{30 - 25}{250 - 300} = \frac{5}{-50} = -0.1$$

Hence the line has the form $p(x) = -0.1x + b$ for some value of b. To evaluate b, we use the fact that $p(300) = 25$. Using 300 for x, we have $p(300) = -0.1(300) + b = 25$.

Solving $-30 + b = 25$ gives $b = 55$. Thus the linear demand function is $p(x) = -0.1x + 55$.

(b) The revenue function is $R(x) = x \cdot p(x) = x(-0.1x + 55)$. Because we plan to take a derivative, we rewrite $R(x)$ as a polynomial, $R(x) = -0.1x^2 + 55x$.

The derivative is $R'(x) = -0.2x + 55$. To find critical numbers, solve $R'(x) = 0$. Doing so gives $x = 55/0.2 = 275$.

This value produces a maximum revenue because $R'' = -0.2 < 0$.

(c) The price of the 275 tickets is found to be $p(275) = -0.1(275) + 55 = -\$27.50 + 55 = \$27.50$.

(d) The maximum revenue is $275 \cdot 27.50 = \$7,562.50$.

PRACTICE EXERCISE 2

Batesville Burgers sells 150 burgers when they are priced at $0.75 each and 200 burgers when they are priced at $0.50 each. Assume a linear demand curve.

(a) Find the demand function.

(b) Find the revenue function for Batesville Burgers.

(c) What price should be chosen in order to maximize revenue?

(d) How many burgers will sell at that price?

(e) What will be the maximum revenue?

Example 7 models an attempt to maximize total production.

EXAMPLE 7   Maximize orchard production

Apple orchards are commonly planted with 100 trees per acre. J APPLE Co. finds that when it plants apple trees at a density of 100 per acre, the yield per tree is six bushels. If it plants trees more densely, the yield of each and every tree declines. In fact, if it plants $100 + z$ trees per acre, then the yield per tree is $y(z) = 6 - 0.05z$ bushels.

(a) How many trees should be planted per acre in order to maximize total production?

(b) What is the maximum total production?

SOLUTION

(a) Let z be the number of trees over 100 that are planted per acre. So each acre is planted with $100 + z$ trees. Let $T(z)$ be the function of z that gives the total yield per acre.

Total yield per acre is the product of the number of trees per acre times the yield per tree. $T(z) = (100 + z) \cdot (6 - 0.05z) = 600 + 6z - 5z - 0.05z^2$. Combining similar terms, we have $T(z) = 600 + z - 0.05z^2$.

In order to find critical numbers, set the derivative equal to zero. From $T'(z) = 1 - 0.1z = 0$, we have $1 = 0.1z$. To solve this for z we divide each side by 0.1, giving us $z = 10$.

Because $T''(z) = -0.1$ for all z, there is a maximum at $z = 10$ by the Second Derivative Test. Because z is the number of trees over the 100 number, the best yield per acre is given by planting 110 trees per acre.

(b) Each tree will produce $6 - (0.05 \cdot 10) = 5.5$ bushels and each acre will have a total yield of $110 \cdot 5.5 = 605$ bushels.

EXAMPLE 8   Maximizing rental income

FJ Realty manages an 80-unit apartment house. Each unit rents for $400 a month and every unit is currently rented. Thus, their total revenue is $80 \cdot 400 = \$32,000$. They decide to raise the rent by $10 per unit. One of the tenants decides that $410 a month is too much and moves out. Still, the total income from the apartment house is $79 \cdot 410 = \$32,390$, which is $390 more than it was. Another $10 raise in the rent results in yet another tenant leaving. Total income is then $78 \cdot 420 = \$32,760$, which is $370 more than previously.

If they lose only one tenant each time they raise the rent by $10, what rent produces the maximum income and what is that income?

SOLUTION

Suppose we let z stand for the number of $10 raises in rent. Then, after z raises, the rent is $400 + 10z$. They lose one tenant each time the rent goes up $10, so if there are z $10 raises, there are $80 - z$ tenants left.

Total income, $T(z)$, is a function of z. Thus,

$$T(z) = \text{(dollars of rent per unit)} \cdot \text{(number of units rented)}$$
$$= (400 + 10z)(80 - z)$$
$$= 32,000 + 400z - 10z^2$$

Taking the derivative, $T'(z) = 400 - 20z$. Solving for $T'(z) = 0$, we find $z = 20$. $T''(20) = -20$, so there is a maximum at $z = 20$.

The most money for FJ Realty is achieved if the rent is $\$(400 + 10 \cdot 20) = \$600$ and there are $80 - 20 = 60$ units rented. In that situation, the total income is $\$600 \cdot 60 = \$36,000$.

These last two examples, although originating in different situations, have the same method of attack. The variable z is assumed to be the number of increases (or "bumps") in some action. In both of the examples, the "yield per unit" function was assumed to be linear, even though we know full well that only some serious study can determine that function. It probably is not linear in either case. But whatever it is, if we can take the derivative we have a tool to help us to determine the maximum yield.

Example 9 involves a production function that employs a non-linear yield per unit.

EXAMPLE 9   Maximizing production using a nonlinear yield per unit function

A sandwich company has 10 employees, and each one can make 150 sandwiches per hour. Adding new employees will lower "sandwich per employee" productivity because of crowding in the kitchen. When z new employees are added, the number of sandwiches will be $(150 - 2z^2)$ per employee. Find the maximum number of sandwiches per hour the company can produce, and determine both the number of employees involved and the production per employee that yields that maximum.

SOLUTION

The total production can be expressed as a function of the number of new employees, z, as $P(z)$ = (number of employees) · (number of sandwiches per employee) = $(10 + z) \cdot (150 - 2z^2)$, which, when multiplied out, is $P(z) = 1500 + 150z - 20z^2 - 2z^3$.

Taking the derivative, $P'(z) = 150 - 40z - 6z^2$. We want to find z such that $P'(z) = 0$. The quadratic equation does not factor, but the Quadratic Formula yields the solutions:

$$z = \frac{-(-40) \pm \sqrt{(-40)^2 - 4(-6)150}}{2(-6)}$$

$$= \frac{40 \pm \sqrt{5200}}{-12} \simeq \frac{40 \pm 72.1110}{-12}$$

The value for z must be positive, so we use the solution $z \simeq (40 - 72.1110)/(-12) \simeq 2.6759$.

To have the number of additional employees be an integer, we need to check $P(2)$ and $P(3)$. $P(2) = (10 + 2) \cdot (150 - 2 \cdot 2^2) = 12 \cdot 142 = 1704$ and $P(3) = (10 + 3) \cdot (150 - 2 \cdot 3^2) = 13 \cdot 132 = 1716$.

Hence, the maximum production possible per hour is 1716 sandwiches, produced by 13 workers, each producing 132 sandwiches.

Further Applications

A quite different problem of optimization is shown by Example 10. It is phrased in terms of a lifeguard rescuing a drowning swimmer, but exactly the same mathematics is used in minimizing the cost of laying pipe from an offshore drilling platform or minimizing the cost of placing fiber-optic cable.

EXAMPLE 10   Minimizing costs of two-cost linear events

Jack is swimming 600 feet offshore and 500 feet down the beach from Marie when suddenly she sees he is in trouble. She can run 8 feet per second and swim 4 feet per second. She wants to reach him as quickly as possible. How far down the beach should she run first before swimming out to him in a straight line? A diagram is shown in figure 3.5.6.

SOLUTION

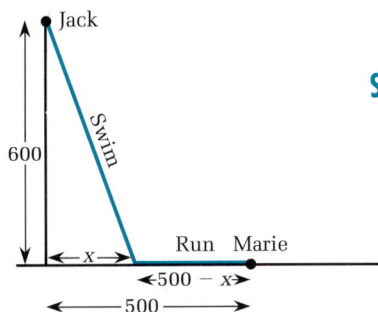

FIGURE 3.5.6

Let x be the distance on the shoreline from a point directly toward shore from Jack to the point at which Marie will start swimming. On the diagram, Marie will run the distance labeled $500 - x$ and then swim along the hypotenuse of the right triangle. The length of that hypotenuse is $\sqrt{(600)^2 + x^2} = \sqrt{360,000 + x^2}$ by the Pythagorean Theorem.

Marie's time is determined by dividing each distance in her effort by the appropriate rate. On shore it is $500 - x$ feet divided by her running speed of 8 feet per second. Her time in the water is $\sqrt{360,000 + x^2}$ feet divided by her swimming speed of 4 feet per second. Thus, the total time will be $T(x) = (500 - x)/8 + (\sqrt{360,000 + x^2})/4$ seconds.

To minimize this time, take the derivative and set it equal to zero.

$$T'(x) = -\frac{1}{8} + \left(\frac{1}{4}\right)\left(\frac{1}{2\sqrt{360,000 + x^2}} \cdot (2x)\right)$$

$$= -\frac{1}{8} + \frac{x}{4\sqrt{360,000 + x^2}}$$

Finding a common denominator gives

$$T'(x) = \frac{-\sqrt{360,000 + x^2} + 2x}{8\sqrt{360,000 + x^2}}$$

For $T'(x)$ to be zero, the numerator must be zero. To solve $-\sqrt{360,000 + x^2} + 2x = 0$, we put the $2x$ on the right-hand side and square both sides to get $360,000 + x^2 = 4x^2$. Thus, $360,000 = 3x^2$, which gives $x \simeq 346.4$.

The best time is achieved by running $(500 - 346) = 154$ feet (taking 19.2 seconds) and swimming $\sqrt{360,000 + (346)^2} = 693$ feet (taking 173.2 seconds). Altogether Marie takes about 192.4 seconds to get to Jack.

We went to a great deal of effort to obtain this answer. How does it compare with Marie simply running 500 feet down the beach, and then swimming 600 feet out to Jack? Her time would be $500/8 = 62.5$ seconds running and $600/4 = 150$ seconds swimming for a total of 212.5 seconds, about 20 seconds slower. (What could possibly happen to Jack in only 20 seconds out at sea?)

What about a direct swim to Jack? Marie would swim a distance of $\sqrt{500^2 + 600^2} \simeq 781$ feet at 4 feet per second. This would take $781/4 \simeq 195.3$, only 3 seconds slower than the fastest approach. ●

In reality, no one would expect Marie to solve this problem before taking action. It certainly would take her more than 20 seconds to solve it. However, a problem having a similar solution, and of great interest in business and industry, follows.

The cost of laying pipeline or fiber-optic cable is greater at sea than on land. Thus, companies try to angle toward shore in the manner of the swimmer in Example 10, rather than making a direct line to shore and then going along the shore.

As another example, it was long noticed that birds migrating from Europe to Africa went around the Mediterranean if possible. Although this may be partly due to food availability, it may also be due to the lack of updrafts over water. As storks pass Istanbul, they wait until midday to resume their flight. Thus, if a bird is leaving an island and heading for a point down along the shore, it may also take this angled approach.

3.5 PROBLEMS

Fundamentals

The problems of this section require the basic skills illustrated by the following:

1. (a) Find the total length of fencing needed for the rectangular field and all its subdivisions, as shown in figure 3.5.7.
   P(b) Find the total area enclosed.

2. Suppose in problem 1 that the cost of fencing on the outside is $8 per foot and the cost on the inside partitions is $5 per foot. Find an expression for the total cost of fencing in terms of x and y.

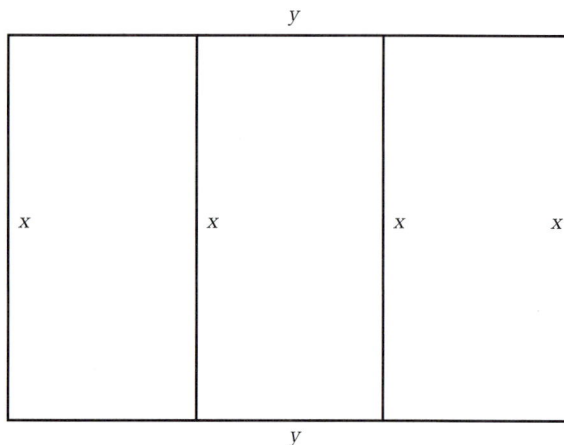

FIGURE 3.5.7

Exercises

Enclosure Problems

3. Suppose an enclosure is made next to the base of a very tall cliff, which can serve as one side. The enclosure is to be rectangular (see figure 3.5.8). There is 600 feet of fencing available.
   (a) What are the dimensions of the largest possible enclosure?
   (b) What is the area of that enclosure?

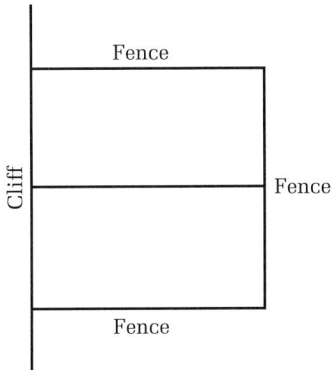

FIGURE 3.5.8

4. Suppose 1200 feet of fencing is to be used to make a rectangular enclosure, which is to be divided into three sections (see figure 3.5.9).
   (a) What are the dimensions of the largest possible enclosure?
   (b) What is the area of that enclosure?

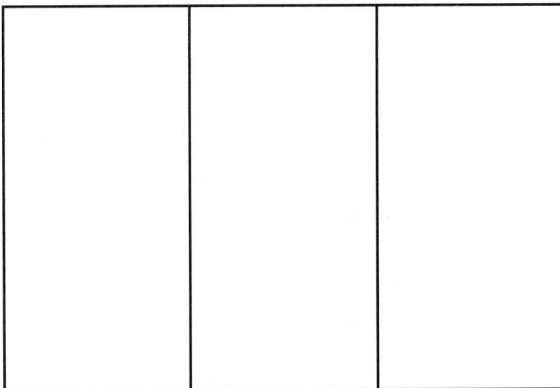

FIGURE 3.5.9

5. Suppose the planners for a mud football contest need to lay out two identical playing fields. They have 1200 feet of fencing and the fields can be side-by-side as shown in figure 3.5.10.

Find the dimensions and the maximum area of each field.

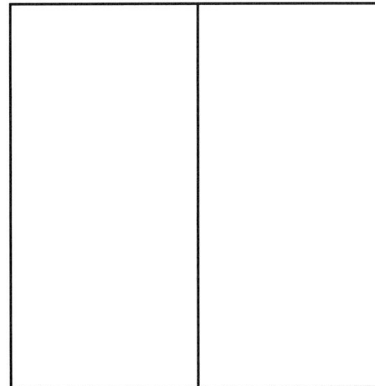

FIGURE 3.5.10

6. WR Amusement Co. needs to construct a new rectangular storage yard adjacent to a highway. It will be enclosed with a fence that costs $25 per foot along the highway side and $5 per foot along the other three sides (see figure 3.5.11). They have $5000 to spend on the fence.
   (a) What dimensions create the largest storage yard?
   (b) What is the area of the largest storage yard?

Highway

FIGURE 3.5.11

7. Sue has $1800 to construct a rectangular enclosure that consists of a high surrounding fence and a lower inside fence that divides the enclosure in half (see figure 3.5.12). The high fence costs $6 a foot and the low fence costs $3 a foot.
   (a) What are the dimensions of the largest possible enclosure?
   (b) What is the area of that enclosure?

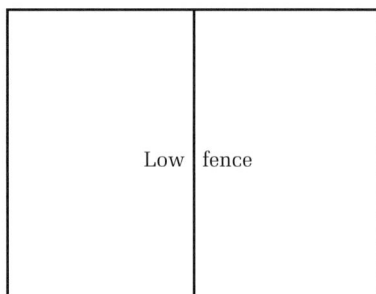

FIGURE 3.5.12

8. Vandanna intends to build a roofless canvas booth to take to county fairs and flea markets. It needs to be rectangular and have 500 square feet of floor space. The front wall (facing the midway) is to be 3 feet high and the sidewalls and back wall are to be 8 feet high (see figure 3.5.13).
   (a) What dimensions will use the least number of square feet of canvas?
   (b) What is the least amount of canvas?
   (c) What would be the effect on the answers to (a) and (b) if the booth had a flat roof of canvas?

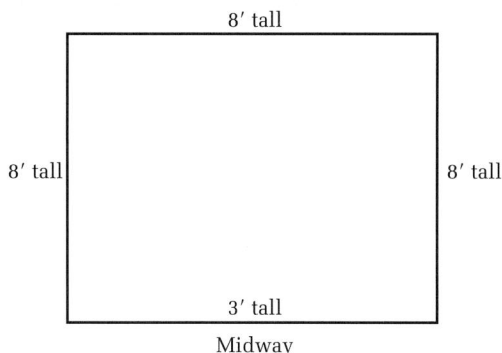

FIGURE 3.5.13

9. Apple Crisp Fruit is building a new rectangular roadside stand facing a highway. They have $1400 available. Alongside the highway, the cost is $25 per foot. On the sides and the back, the cost is $10 per foot.
   (a) What dimensions yield the largest possible area?
   (b) What is the largest possible area?

Maximizing Revenue

10. The Old Movie Pie Company sells frozen pies, packed eight to a box. The demand curve is $p(x) = 102 - 3x$, where x is in boxes and $p(x)$ is in dollars per box.
    (a) Determine the revenue function.
    (b) Find the number of boxes that will maximize revenue.
    (c) Find the associated price per box.
    (d) Find the amount of the maximum revenue.

11. Suppose the demand function for peaches is $p(x) = 15 - 0.1x$, where x is in bushels and $p(x)$ is in dollars per bushel.
    (a) Find the revenue function.
    (b) Find the number of bushels that will maximize revenue.
    (c) Find the associated price per bushel.
    (d) Find the amount of the maximum revenue.

12. Suppose the demand function for apples is $p(x) = 12 - 0.1x$, where x is in bushels and $p(x)$ is in dollars per bushel.
    (a) Find the revenue function.
    (b) Find the number of bushels that will maximize revenue.
    (c) Find the associated price per bushel.
    (d) Find the amount of the maximum revenue.

13. The chess club made a deal with Stalemate Tours to charter a 35-seat bus. The contract called for $25 from each club member. However, Stalemate put a clause in the contract that stated, "For each empty seat on the bus, the price per passenger shall increase by $1." What is the maximum that Stalemate Tours can collect from the busload?

14. The Tombstone Ghosts football team can sell 10,000 tickets at $10 each. For every increase in price of 25 cents, the demand decreases by 200. For example, at $11, the sales would be 9200.
    (a) Find the demand function.
    (b) Find the revenue function.
    (c) Determine the price that will maximize total revenue.
    (d) Find the number of tickets sold to generate that total revenue.
    (e) Find the total revenue.

15. Cheesecakes? Cheesecakes! finds that at $15 each, they sell 150 cakes a day. If they raise their prices by $1 per cheesecake, they will sell 12 fewer cakes. That happens each time they raise their prices. For example, at $16 they will sell 138 cakes and at $17 they will sell 126 cakes.
    (a) Find the demand function for cheesecakes.
    (b) How should they price their cheesecakes so as to maximize weekly revenue?
    (c) What is that revenue?
    (d) How many cheesecakes will they be selling at that price?

16. Cheesecakes? Cheesecakes! currently has 20 stores, each grossing $125,000 a month. Adding a new outlet will attract some new customers, but it will take some customers away from the other stores. Each new store is estimated to cause the gross sales to drop by $2000 at each store. For example, if they decide to open three new stores, each of their 23 stores will gross $119,000 a month.
    (a) How many new outlets should they open in order to maximize total monthly gross?
    (b) What will be that gross?

17. A tree farm sells seedling trees in units of 1000 trees. If x is the number of units and $p(x)$ is the price per unit, the demand function is $p(x) = -x^2 + 300$.
    (a) Find the revenue function.
    (b) Find the pricing that will maximize total revenue.
    (c) Find the maximum total revenue.
    (d) Find the number of units that are shipped to achieve that revenue.

18. A nut farm has a demand curve of $p(x) = 10,000/(x + 1)^2$ dollars for x tons of nuts.
    (a) Find the revenue function.
    (b) Find the pricing that will maximize total revenue.
    (c) How many tons of nuts are shipped to achieve that maximum?
    (d) What is the maximum revenue?

19. Suppose a demand curve is given by $p(x) = 24 - \sqrt{x}$.
    (a) Find the equation representing revenue.
    (b) Find the marginal revenue at $x = 9$.
    (c) Find the value of x that produces the maximum revenue.
    (d) Find the price that produces the maximum revenue.
    (e) Find the maximum revenue.

20. Carol has a river rafting business. Charging $8 for a two-hour trip, she sells 500 tickets a week. Every time the price increases by 25 cents, there are 10 fewer tickets sold. For example, at $8.75 she sells 470 tickets.
    (a) Find the demand function.
    (b) Determine the pricing that maximizes Carol's total revenue.
    (c) How many tickets are sold to produce that maximum total revenue?
    (d) What is the maximum total revenue?

21. Refer to problem 20. Suppose Carol has to spend $5 on lunch for each ticket holder (for example, an $8 ticket makes her only $3 now).
    (a) What effect does this have on her pricing structure if she still wants to maximize revenue?
    (b) Find the number of tickets she must sell to maximize total revenue.
    (c) Find that maximum revenue.

22. *(Maximizing Profit)*   At the time of the study in Example 4, the wholesale price of Minute Rice was $0.69 a box. Using the demand curve for x boxes of $p(x) = -0.01x + 1.69$, find (a) the retail price that will produce the maximum profit and (b) the maximum profit.

23. *(Maximizing Profit)*   Use the demand function of problem 22, but suppose that the wholesale price is raised to $0.79. Should the 10-cent increase be passed on to the customer, or is the optimal strategy different? (That is, find the maximum profit and the retail price that will produce that profit.)

24. *(Maximizing Profit)*   Suppose a demand function for a Chocolate Fantasia charity event is $p(x) = -1/6000(x^2) + 20$, for x tickets sold and with $p(x)$ the price in dollars. Suppose the average cost to the sponsor of the event is $15 per person.
    (a) Find the profit function.
    (b) Find the number of tickets that produce a maximum profit.
    (c) Find the price per ticket that yields that maximum.
    (d) Find the maximum profit.

25. *(Maximizing Production)*   Lazy Snake Oil Co. currently has 20 wells drilled, and each is producing 200 barrels of oil per day. New wells may increase total production, but because of decreasing pressure, each and every well will lose some capacity. In fact, for each new well brought into production, each well will produce five barrels per day less. For example, when two new wells go in, then each of the 22 wells would be producing 190 barrels of oil per day.
    (a) How many new wells should be drilled in order to maximize total daily production?
    (b) What will be that total production?

26. *(Maximizing Processing of Student Schedules)*   At registration there are six data-entry clerks doing scheduling. Each can process 50 student schedules per hour. If x new clerks are hired, the

productivity per clerk will drop to $50 - x^2$ schedules per hour due to machine-time delays.
(a) Find the total maximum number of schedules that can be processed per hour.
(b) Find the number of clerks involved.
(c) Find the number of schedules each clerk is processing.

27. Refer to problem 26. Suppose a new software program allows each clerk to enter $100 - x^2$ schedules per hour.
(a) Find the total maximum number of schedules that can be processed per hour.
(b) Find the number of clerks involved.
(c) Find the number of schedules each clerk is processing.

28. A fiber-optic line is to be laid from an island three miles offshore to a point that is 10 miles away from a point directly landward of the island (see figure 3.5.14). The cost of the line is twice as great in water as on land. At what point on land should they make landfall with the line in order to minimize total costs?

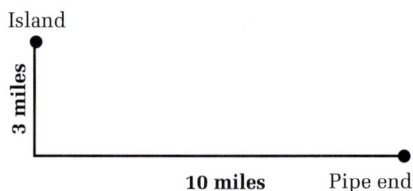

Island

3 miles

10 miles   Pipe end

FIGURE 3.5.14

29. **(Golf Drives)** Based on work by Brian Bolt, when a golf ball is smashed by a club moving V meters per second and of clubhead weight M grams, then the golf ball is propelled to a velocity of U meters per second according to the formula $U = 1.7MV/(M + 46)$.* (The 46 is the weight of the golf ball in grams, and the 1.7 has to do with the elasticity of the ball. Typical clubhead weights range from 100 to 300 grams.)
   Although Evan Williams, twice winner of America's long-driving championship, had his clubhead speed measured at an amazing 130 mph (which is about 58 meters per second), the typical golfer has a velocity, V, that is slower with a heavier club. In fact, a reasonable approximation is $V = 40 - 0.02M$ meters per second for a clubhead of M grams. Using $V = 40 - 0.02M$, find the clubhead weight that maximizes U.

*B. Bolt, "Tennis, Golf and Loose Gravel; Insight from Easy Mathematical Models," *UMAP Journal* 4:1 (1983), pp. 5–18.

30. **(Waterpower)** The power developed by an undershot waterwheel (where the water passes under the wheel) can be expressed as $P = k(R^2v - 2Rv^2 + v^3)$, where R is the rate water is moving into the waterwheel and v is the velocity of the wheel.† Find the value of v, expressed in terms of R, which maximizes P. (In fact, v is controlled by how much we "load" the wheel by having it perform a task.)

31. Divide 12 into two numbers, a and b, so that their product $a \cdot b$ is a maximum. (For example, 12 could be divided into 2 and 10 and the product would be 20. Is that a maximum?)

32. Suppose x and y are such that $x - y = 6$. What is the minimum for $x \cdot y$? (For example, $10 - 4 = 6$, but is $10 \cdot 4 = 40$ the smallest possible product?)

33. Find the maximum of x^2y, provided $x + y = 9$.

34. Find the maximum of xy, provided $y + x^2 = 15$.

35. Suppose that square corner cuts, each r long, are going to be made in an 8-by-12 sheet of cardboard and then the sides will be folded up and taped together to make a box (see figure 3.5.15).
(a) What should be the value for r that maximizes the volume of the box?
(b) What is that volume?

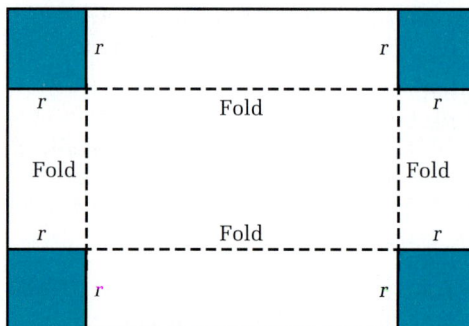

r Fold r

r

Fold   Fold

r Fold

r r

FIGURE 3.5.15

Writing and Discussion Problem

36. Suppose Emilio and Jack each have 60 feet of fencing and six fenceposts. Discuss the possible fields they can enclose for each of the following situations.
(a) Each has a separate rectangular field.
(b) The fields are rectangular, but can adjoin.
(c) The fields do not have to be rectangular.
(d) They can share a field.

†B. Bolt, "Waterpower," *UMAP Journal* 5:3 (1984), pp. 257–265.

Enrichment Problems

37. A cylindrical can of radius r and height h is to contain 25 cubic inches. The volume of the can is $V = \pi r^2 h$ and the total surface area is $S = 2\pi rh + 2\pi r^2$. Find the values for r and h that minimize the total surface area.

38. Modern frozen juice packaging often uses paper for the sides and metal for the top and bottom of the container. Suppose the paper costs only 1/3 as much as the metal and that the can contains 60 cubic inches of frozen juice concentrate. Find the height and radius of the most economical can. (Hint: The actual costs of the metal and paper do not make any difference, as long as the metal costs three times what the paper costs. Let

paper cost k cents per square inch and metal cost $3k$ cents per square inch.)

39. A racetrack is to be laid out in the shape of an oval, the ends of which are semicircles and the sides of which are parallel lines (see figure 3.5.16). The track is 1760 feet long.
  (a) What dimensions will produce the maximum area for the rectangular center of the oval?
  (b) What is that maximum rectangular area?

FIGURE 3.5.16

SOLUTIONS TO PRACTICE EXERCISES

1. See figure 3.5.17. If x stands for one side and y for the other side, the total length of fence used is $2x + y = 900$. The area enclosed is $A = x \cdot y$. To eliminate one variable, solve $2x + y = 900$ for y and have $y = 900 - 2x$. Then $A = x \cdot (900 - 2x) = 900x - 2x^2$. With A thereby expressed as a function of one variable, we can write $A(x) = 900x - 2x^2$ and take the derivative, $A'(x) = 900 - 4x$. So $A'(x) = 0$ when $x = 900/4 = 225$. $A''(x) = -4 < 0$, giving us that at $x = 225$ and $y = 900 - 2(225) = 900 - 450 = 450$, Jean has a maximum area of $225 \cdot 450 = 101{,}250$ square feet.

2. (a) The slope of the demand curve is $(0.5 - 0.75)/(200 - 150) = -0.005$, and the equation of the demand function is $p(x) = -0.005x + 1.5$, which is obtained by taking $p(x) = -0.005x + b$ and using the fact that $f(200) = 0.50$.
  (b) $R(x) = x \cdot p(x) = -0.005x^2 + 1.5x$. $R'(x) = -0.01x + 1.5$. $R'(x) = 0$ when $x = 150$. $R''(x) = -0.01$ for any x.
  (c) $p(150) = \$0.75$.
  (d) 150 burgers.
  (e) Selling 150 burgers at $0.75 each produces a maximum revenue of $112.50.

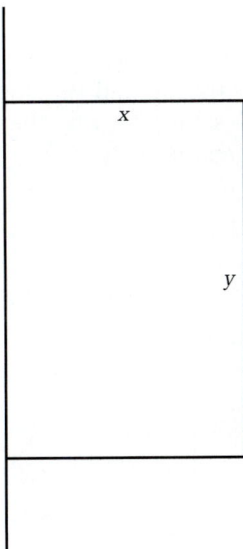

FIGURE 3.5.17

APPLICATIONS TO BUSINESS AND ECONOMICS

This section covers several topics in optimization, each of which may be studied independently. First is the determination of the optimal quantity of an item to be ordered or produced when the item must be ordered or produced in batches. For instance, if a company uses computer ribbons throughout the year, how many should they order each time they place an order?

Next we discuss the scheduling of routine maintenance. Our example is the cleaning of a Boeing 747.

The final part of the section concerns harvesting in batch sizes that will maintain a sustained yield. For instance, how many pounds of shrimp can be collected each year without overfishing the shrimp-beds and causing a decline in total production?

Economic Order Quantity (EOQ)

A garden center selling bales of peat moss, a factory making cans from sheets of aluminum, an office using paper clips, a mint making gold coins—these are quite different enterprises, yet they all share the common problem of ordering and carrying items in inventory.

Each time they place an order, there is a fixed cost of ordering a shipment, no matter whether they order a few items or millions. On an annual basis, they can reduce this cost by ordering infrequently. On the other hand, following a policy of infrequent orders means that each individual shipment must be large. The annual cost of storing the inventory will soar as the size of the shipment increases.

As a historical note, in the 1960s American automobile manufacturers were accustomed to having enough parts on hand to last several days. By the 1980s, manufacturers tried to have only enough for a few hours. A totally efficient, vertically integrated assembly line may have almost no inventory on hand, relying on the arrival of each part at just the moment it is needed.

The graph in figure 3.6.1 shows an ordering pattern in which demand is continuous and uniformly linear throughout the year. An order is instantly filled, each shipment is the same size, and the number of units goes to zero before restocking occurs.

FIGURE 3.6.1

These are all physical assumptions. In addition, we make an assumption about accounting. It is standard accounting practice to calculate the inventory carrying cost (physical storage space, cost of borrowed money, etc.) on just one-half of the total order size. Hence, rather than worrying about how much of the inventory is present at any particular time, we simply assume that one-half of the order size is in inventory all of the time.

This problem covers demand over some interval of time. The interval could be a year, a month, or some other period. There is no loss of generality if we assume the time interval is one year. Then by using "annual" or "yearly," we avoid the awkward phrase "interval of time."

Let us set up the problem of ordering and storing some commodity over a year. First, choose names for the variables involved:

D = demand for the commodity over the year, measured in number of units.

P = cost of placing an order for one shipment of the commodity.

S = cost of storing one unit for one year.

x = number of units in one shipment. This is the order size.

$T(x)$ = total cost for one year of ordering and storing the commodity.

To determine $T(x)$ we need to know and add together the "costs of placing orders for the year" plus the "storage costs for a year."

The "cost of placing orders for the year" is the product of "the cost for placing one order" times "how often over the year orders are placed." The product of the number of units in one shipment times the number of orders over the year must equal D, the total demand. Hence, if x units are ordered at a time and D is the annual demand, then D/x orders are placed in a year. In consequence, the cost of placing orders for the year is $P \cdot (D/x)$.

The "storage cost for the year" equals the product of "the number of units in storage" times "cost of storing one unit for one year." The formula is $(1/2)x \cdot S$. (Recall that we assume one-half the order is present all the time.)

Combining these two costs, we find

$$T(x) = P \cdot \frac{D}{x} + \frac{1}{2}x \cdot S$$

A typical $T(x)$ is graphed in figure 3.6.2. Notice total cost is great for small orders (due to frequent ordering) and total cost is also high for large orders (due to stockpiling inventory).

Finding the minimum value of $T(x)$ is called the *Economic Order Quantity (EOQ)* problem. In general, to find the minimum for $T(x)$ we solve $T'(x) = 0$ and check that the critical number yields a minimum value.

$$T(x) = \frac{1000}{x} + 2.5x$$

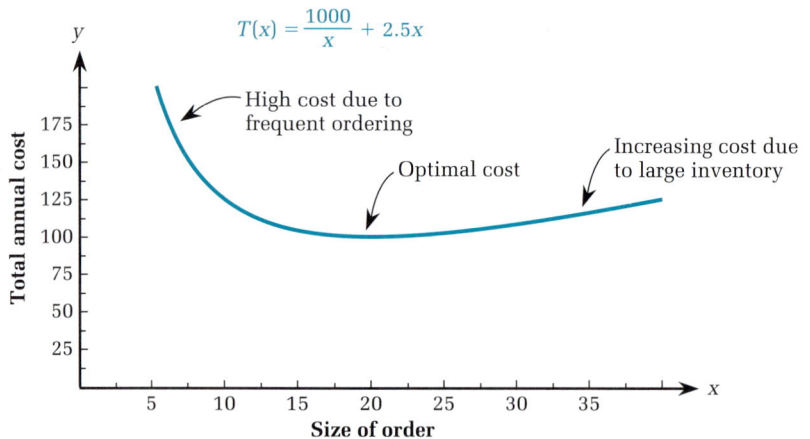

FIGURE 3.6.2

EXAMPLE 1   Economic Order Quantity (EOQ)

The SMP Co. uses 100 cases of computer printer ribbons per year. Placing an order costs $10 and the cost of storing one case for the entire year is $5. The company wishes to minimize total annual ordering and storing costs for computer ribbons.

(a) Find the optimal order size.

(b) How often will an order be placed?

SOLUTION

$D = 100$, $P = \$10$, and $S = \$5$. The total annual cost is therefore $T(x) = 10 \cdot 100/x + (1/2)x \cdot 5 = 1000/x + 5x/2$, whose graph is shown in figure 3.6.2. To minimize $T(x)$, find the derivative and set it equal to zero. $T'(x) = -1000/x^2 + 5/2 = 0$ gives us $1000/x^2 = 5/2$. Solving this for x^2, we have $(1000 \cdot 2)/5 = x^2$. Finally, from $x^2 = 400$ we find $x = 20$. Thus, each order should be for a lot of 20 cases of computer printer ribbons.

One means of determining that this value truly gives a minimum is to look at the second derivative, which is $2000/x^3$ and so is always positive for $x > 0$.

At $x = 20$, the total cost $T(20) = 1000/20 + (5 \cdot 20)/2 = 50 + 50 = 100$.

When solving an EOQ, we can determine how often the order must be placed by dividing the annual demand by the size of the optimal order. For our example, with $D = 100$ and $x = 20$, we find that we place an order $100/20 = 5$ times per year.   ●

Perhaps surprisingly, the cost of the item itself does not affect the solution to the question of the optimal EOQ. The truth of this statement is demonstrated in an enrichment problem at the end of this section.

PRACTICE EXERCISE 1

The gymnasium at a small university orders 2000 towels per year. It costs $40 to place an order and $1 to store one towel for one year.

(a) What is the most economical order quantity to minimize the total costs of ordering and storing towels for a year?

(b) What is that minimal annual total cost?

There are many variations to the EOQ problem. One of them is to assume storage space must be provided for all of the inventory that is purchased in each order. An example of this is when the space provided for a shipment of frozen food is not available for apples as the supply of frozen food is being used up. We consider this variation, termed "The Alternate Storage Cost for EOQ," in the enrichment problems at the end of this section.

Production Runs

A situation that is very similar to the EOQ concerns using a machine that can make a variety of items, but only one type of item at a time. Thus, the machine is set up to make a batch of one thing, and when enough of those are produced to satisfy demand for a while, the machine is reprogrammed or reconfigured to do a different job. The cost of resetting the machine corresponds to the cost of placing the order in the EOQ problem. The cost of inventory is treated exactly as in EOQ. Just as there, the cost of materials has no effect on how many items are in one batch or how often a batch is made.

EXAMPLE 2   Production runs

The Smelter Mountain Nail Company sells 5000 tons of roofing nails on a continuous basis throughout the year. It costs them $25 to set up the machine to make roofing nails and it costs $100 to store one ton of nails for an entire year. They want to minimize annual total production costs of roofing nails. When they set up the machine to make nails, how many should they make at one time? How many times per year will they make roofing nails? What is their minimal total production costs of roofing nails?

SOLUTION

Let x stand for number of tons in one production run. The number of times per year they have to set up will be $5000/x$, and they will be charged for storing one-half of their total run for the entire year. Thus, the annual total production cost will be $C(x) = (5000/x) \cdot 25 + (x/2) \cdot 100$ dollars per year.

This is exactly the same format as the EOQ problems. We will continue our solution by taking the derivative and setting it equal to zero. Simplifying our equation for $C(x)$, we get $C(x) = 125,000/x + 50x$. Taking the derivative gives $C'(x) = -125,000/x^2 + 50$, which is equal to zero when $125,000/x^2 = 50$. Multiply each side by x^2 and divide by 50. Then $x^2 = 125,000/50 = 2500$. Solving for x, we have that $x = 50$ tons.

They will make roofing nails $5000/50 = 100$ times over the year. The total cost will be $C(50) = 125,000/50 + 50 \cdot 50 = 2500 + 2500 = \$5000$.

In both Examples 1 and 2, notice that when the minimum annual total cost is achieved, exactly the same amount of money is being spent on ordering (or setting up) as is being spent on storing. In our last example, $2500 is spent on setting up and $2500 is spent on storing.

This even splitting of costs applies only to the situation in which the demand is uniformly linear and one-half the order (or production run) is assumed to be present all of the time. Nonetheless, when these conditions do apply, this equality of costs provides us a very handy check on our computations.

PRACTICE EXERCISE 2

The Lovable Pet Co. sells 50,000 boxes per year of Hearty II dog biscuits in a uniformly linear manner. It costs $5 to set the machine and $0.125 to store one box of the Hearty II biscuits for a year. They wish to minimize total annual production costs of Hearty II biscuits.

(a) Each time they set the machine, how many should they make?

(b) How many times per year will they be making Hearty II?

(c) What will be the minimal total annual costs of producing Hearty II biscuits?

In the examples we have considered, the answers were integers. If an answer is not an integer but the problem requires an integer solution, then for the best feasible solution we evaluate the cost function for the integers on each side of the optimal (but unattainable) order or production batch size.

Scheduled Maintenance

Example 3 demonstrates how to determine the optimal time to perform one kind of routine maintenance. The values used are suggested by an article reporting on work by W. S. Weir of Seattle University.*

*W. S. Weir, *Chronicle of Higher Education,* March 28, 1990, p. A6.

EXAMPLE 3 How to schedule the cleaning of a Boeing 747

In 1990 the United States forbade smoking on domestic air flights of under six hours in duration. Before that date, tar and other smoke-related by-products would solidify on the inside skin of the aircraft due to the subzero temperature of the skin at cruising altitudes. The rate of such deposits was about 2/3 pound per hour of flight time.

The Boeing 747 consumes about five pounds of fuel per 100 pounds of weight for each hour of flight time. We will show in Chapter 5 that after k hours of flight time, the plane would have burned $k^2/60$ pounds of fuel lifting solidified tar.

Let us assume a useful life of 25,000 hours of total flight time. If we assume the company will go through the process of cleaning after k hours of flight time, then in the 25,000 hours there would be $25,000/k$ cycles of burning $k^2/60$ pounds of fuel and $25,000/k$ cleanings of the accumulated tar.

If we assume eight pounds of fuel is one gallon and each gallon costs $1.50, then in 25,000 hours, the total spent on flying the tar around is

$$\left(\frac{\frac{25,000}{k} \cdot \frac{k^2}{60}}{8} \right) \cdot 1.50 = 78.125k$$

If we assume the cost of one cleaning is $2000, then again in the 25,000 hours, the total cost of cleaning is $25,000/k \cdot 2000 = 50,000,000/k$. Hence, if the tar is cleaned after k hours of flight time, the total cost over the 25,000 hours would be

$$T(k) = 78.125k + \frac{50,000,000}{k}$$

This function has the same form as the EOQ equation. To minimize $T(k)$ we take the derivative and set $T'(k)$ equal to zero. $T'(k) = 78.125 - 50,000,000/k^2 = 0$ implies $k^2 = 50,000,000/78.125 = 640,000$, so that $k = 800$ hours. ●

A situation similar to Example 3 arises in deciding when to scrape barnacles off of the hull of a ship or any other decision involving a fixed cleaning cost and a rising cost of continuing not to clean.

Sustainable Yields

We now investigate *sustainable yields.* These are the amounts that can be harvested indefinitely without reducing the capacity for future harvesting. Critical to this analysis is what biologists call a *reproduction curve.*

DEFINITION

Reproduction curve

If P is the population at the beginning of one cycle of reproduction and $f(P)$ is the population at the end of one cycle, then the graph of $y = f(P)$ is a *reproduction curve*.

For example, suppose every fall we count the number of elk in a park. Based on data from many years we create the curve shown in figure 3.6.3.

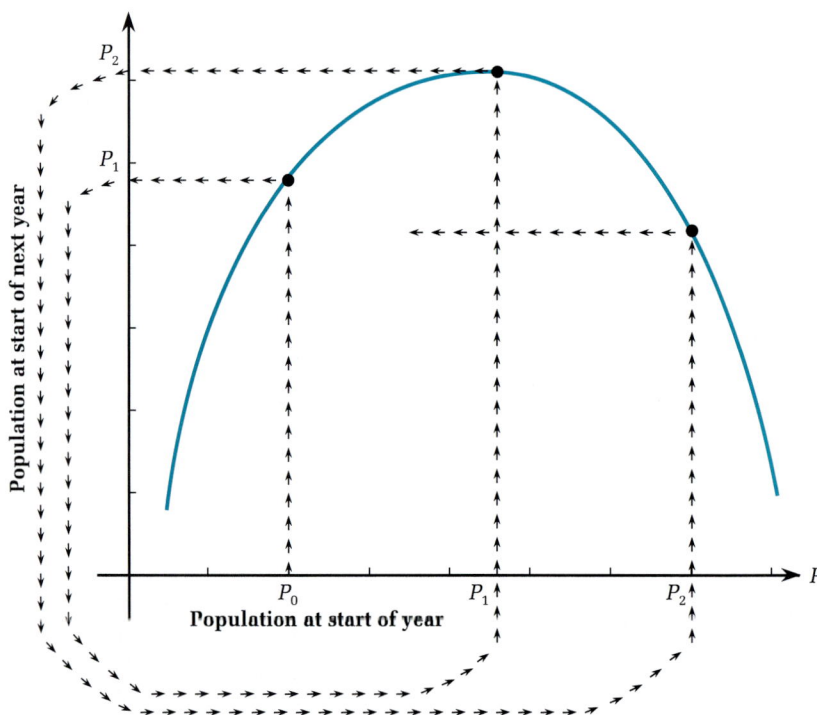

FIGURE 3.6.3

To use such a reproduction curve, assume that at the start of some year the population is P_0. Locate P_0 along the x-axis, read vertically up to the curve, and then go horizontally over to the y-axis to find the population, $f(P_0)$, at the end of the year.

If P_0 is the population at the start, let $P_1 = f(P_0)$ be the population at the end of the first year. The P_1 value is used for the start of the second year, and $P_2 = f(P_1)$ is the population at the end of the second year. This notation continues along the years. For example, P_7 is the population at the end of the seventh year, and is a function of P_6.

The reproduction curve for elk may vary from year to year, depending on the amount of winter snow cover, available grazing, freezing temperatures, forest fires, and many other variables. However, a wildlife biologist can offer a general guess as to the shape of the curve. For a fish farmer who can control the environment of the fish, the reproduction curve is quite stable.

On the left side of a typical curve, that is, for years when the population at the beginning of the year is small, under favorable conditions the increase from year to year may be extremely great. For example, when rabbits were first introduced to Australia, one pair produced 50 to 100 rabbits a year because offspring born in the spring could themselves breed within a short time.

In the central part of the curve we have a maximum possible population. On the right side we find the population curve declining, perhaps because competition for resources results in fewer animals at the end of the year than were present at the beginning of the year.

The actual history of any population depends on many things, but certainly the reproduction curve can predict the stability of a population. Consider two scenarios. In the first, shown in figure 3.6.4, the reproduction curve is only slightly above the *maintenance curve* of exact replacement of population. In the second, shown in figure 3.6.5, the reproduction curve soars above the maintenance curve.

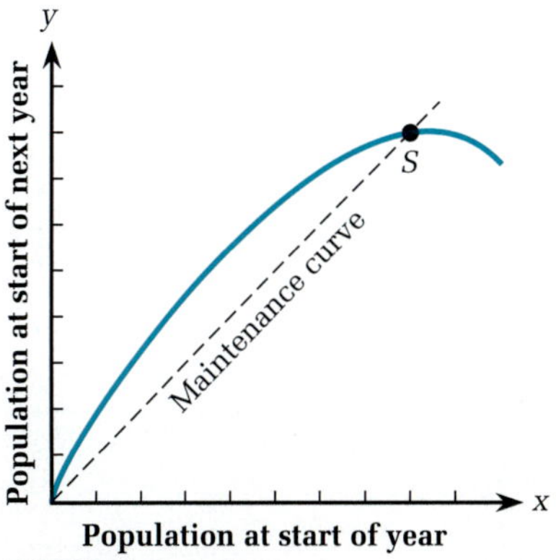

FIGURE 3.6.4

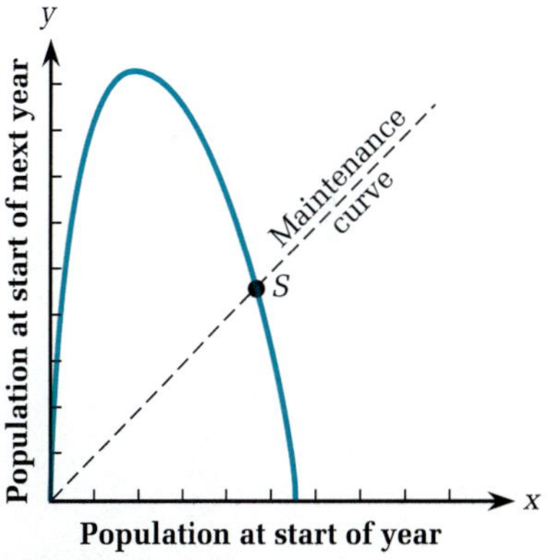

FIGURE 3.6.5

The points marked *S* are *stability points* for the reproduction curve. That is, if we start with *S*, we will end with *S* after one generation.

If a population behaves as in the first scenario, then unexpected changes in the population (a catastrophe perhaps) will not be a disaster because the population tends to stabilize. However, in the second scenario, wild swings in population occur quite often and may even lead to extinction.

For each of these scenarios, assume we start with a population of $P_0 = 10$ and track populations over six years by following the arrows up to the reproduction curve, finding the new population, and then carrying the new value around to the bottom to use as an input value for another year. Figures 3.6.6 and 3.6.7 show this on the reproduction curves.

Figures 3.6.8 and 3.6.9 show graphs of the annual populations under each scenario.

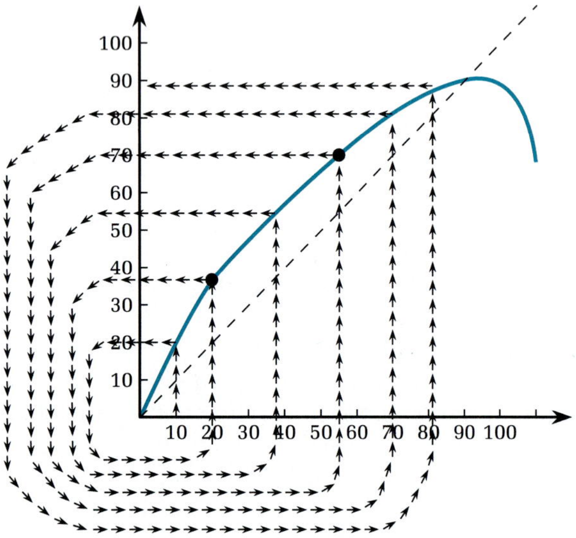

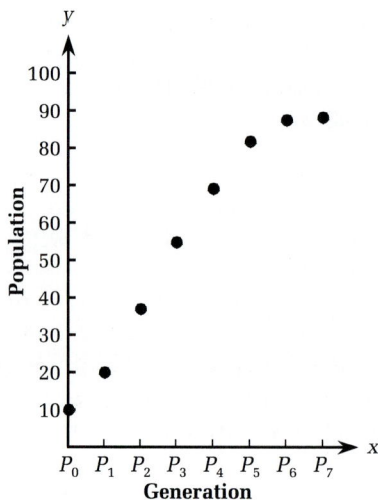

Stabilizes at 90

$P_0 = 10$ $P_4 = 70$
$P_1 = 20$ $P_5 = 81$
$P_2 = 38$ $P_6 = 88$
$P_3 = 55$

FIGURE 3.6.6

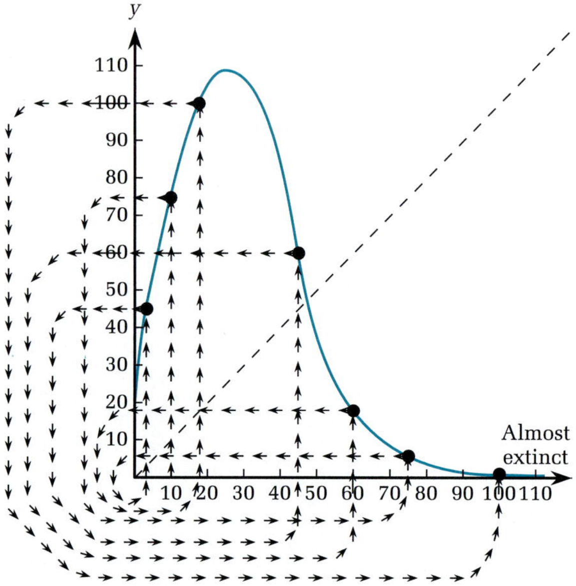

$P_0 = 10$ $P_4 = 60$
$P_1 = 75$ $P_5 = 18$
$P_2 = 6$ $P_6 = 100$
$P_3 = 45$ $P_7 = $ almost extinct

FIGURE 3.6.7

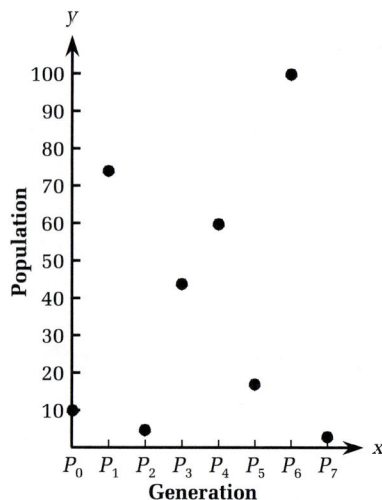

FIGURE 3.6.8

FIGURE 3.6.9

Suppose we consider the difference $f(P) - P$. If exactly that number of individuals were removed from the population, then next year's starting population would be just what this year's had been. For example, if we have 100 dandelion plants at the start of this season and at the end we have 1000, by removing 900 we will enter next season with the same number as we did this season. Thus, 900 is a harvest that could be sustained year after year.

DEFINITION

Sustainable annual harvest

If $f(P)$ is a reproduction function, then the *sustainable annual harvest*, $H(P)$, is the amount of a population that can be removed annually without changing the total population.

$$H(P) = f(P) - P$$

On the graph in figure 3.6.10, $H(P)$ is the difference between $y = f(P)$ and $y = P$.

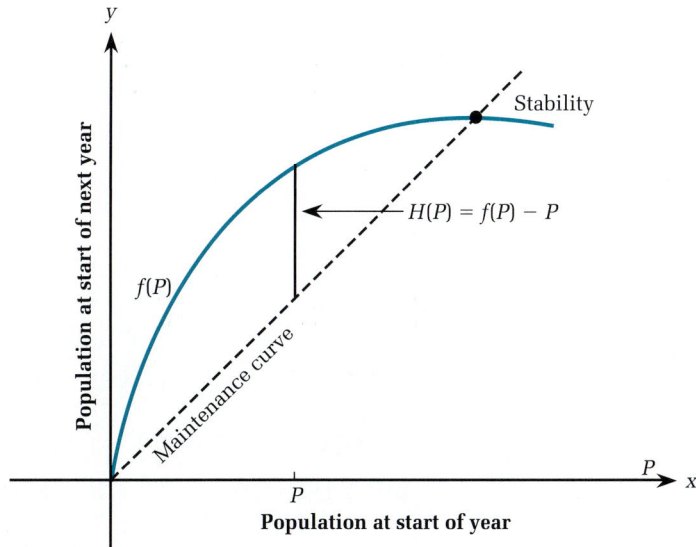

FIGURE 3.6.10

We are interested in the maximum sustainable annual harvest. This problem is approached as any other maximization problem: We take the derivative and set it equal to zero.

Because $H'(P) = (f(P) - P)' = f'(x) - 1$, we have the following result.

If $H'(P) = f'(P) - 1 = 0$ has solution $P = B$, and if $H(B)$ is a maximum, then $H(B)$ is the *maximum sustainable annual harvest*. This result is summarized below.

> ## MAXIMUM SUSTAINABLE HARVEST
>
> If $f(P)$ is an annual reproduction curve, then the *maximum sustainable harvest* occurs at B such that $\dfrac{d}{dP} f(P) = 1$. The amount of the harvest is given by $H(B) = f(B) - B$.

Interest in the maximum sustainable yield comes from many situations. For example, most states have hunting seasons for deer and other animals. Officials want to have the largest possible number of big game taken year after year. Ocean and lake fisheries are regulated by federal and state laws intended to maintain viable populations. Collecting plants is subject to regulation so that overcollecting does not occur.

EXAMPLE 4   Maximum sustainable harvest

Suppose the reproduction curve for dandelions is $f(P) = 1000P - P^2$.

(a) What is the sustainable harvest from a population of 100?

(b) What is the maximum sustainable annual harvest?

SOLUTION

$$H(P) = f(P) - P = (1000P - P^2) - P = 999P - P^2$$

(a) $H(100) = 999(100) - (100)^2 = 99{,}900 - 10{,}000 = 89{,}900.$

(b) $H'(P) = 999 - 2P$. When $H'(P) = 0$, then $P = 499.5$ is the B value. Of course, we have to use 499 or 500 and check each.

$$H(499) = 999 \cdot 499 - (499)^2 = 498{,}501 - 249{,}001 = 249{,}500$$
$$H(500) = 999 \cdot 500 - (500)^2 = 499{,}500 - 250{,}000 = 249{,}500$$

Thus, starting with 500 dandelions, we can harvest almost a quarter of a million plants at the end of the year and still start next year with 500 plants.

We can determine the maximum sustainable yield even if the reproduction curve is not described by a formula. This is possible because solving the equation $\dfrac{d}{dP} f(P) = 1$ is equivalent to finding the point on the graph where the slope of the tangent line is 1. Our method then is to graph the reproduction curve with identical horizontal and vertical scales, then adjust a straightedge until it appears there is a slope of 1 for the tangent line. The value of P there is the value that yields the maximum sustainable harvest, which is calculated as being $f(P) - P$.

EXAMPLE 5   Using a tangent line of slope 1 to determine the maximum sustainable harvest

Find the maximum sustainable yield for the species whose reproduction curve is graphed in figure 3.6.11.

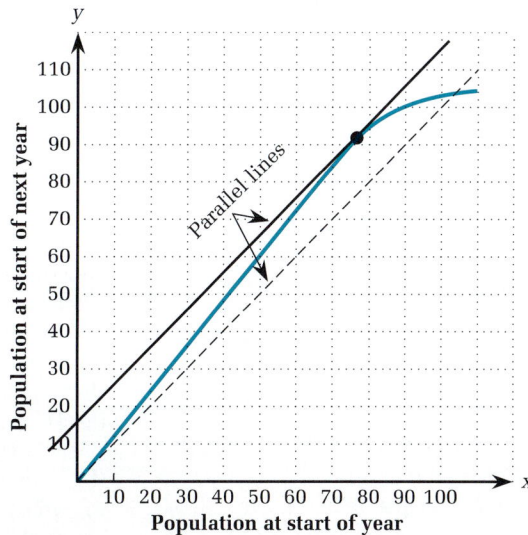

FIGURE 3.6.11

SOLUTION

First, place the straightedge by lining it up along the diagonal corners of the grid. This gives it a slope of 1. Then carefully slide the straightedge (without changing its tilt) until it just touches the curve. By doing this, it appears the tangent line has slope 1 at the point $(75, 90)$. The maximum sustainable yield is thus $90 - 75 = 15$ units per generation.

PRACTICE EXERCISE 3

Suppose a reproduction curve is $f(P) = 10\sqrt{P}$.

(a) Find the sustainable harvest at $P = 4$.

(b) Find the maximum sustainable annual harvest.

3.6 PROBLEMS

Foundations

The problems of this section require the basic skills illustrated by the following:

Find the derivative with respect to x of the functions in problems 1 and 2.

1. $\dfrac{96{,}000}{x}$ **2.** $\dfrac{3}{2}x$

3. Solve $-\dfrac{12{,}500}{x^2} + 5 = 0$.

Exercises

For the Economic Order Quantity and production run problems, always assume that the supply is used up in a uniform, linear manner over the entire interval of time given. Furthermore, use the standard accounting practice of considering one-half the order size to be on hand at all times.

Economic Order Quantity (EOQ)

4. Barb's Gym orders 800 towels per year. It costs Barb $12 to place an order and it costs her $3 to store a towel for an entire year. She wants to minimize her total annual costs of ordering and storing towels.
  (a) How many should she order at a time?
  (b) How often should she order?
  (c) What will be her minimum total annual cost?

5. Tip-o-pen Co. uses 500 barrels of ink per year. It costs $25 to place an order and $10 to store one barrel for an entire year. The company would like to minimize the total annual costs of ordering and storing ink.

  (a) How many barrels should it order at a time?
  (b) How often should it order?
  (c) What will be the minimum total annual cost?

6. The office coffee club uses 75 one-pound cans of coffee per year. It costs $3 to place an order and $4.50 to store one can for the entire year. The club would like to minimize the total annual costs of ordering and storing coffee.
  (a) How many cans should it order at a time?
  (b) How often should it place an order?
  (c) What will be the minimum total annual cost?

7. Swisshole Cheesecake Co. uses 200,000 lb of chocolate per year. It costs $1000 to place and stock an order. It costs $4 to store one pound of chocolate a year. The company would like to minimize the total annual costs of ordering and storing chocolate.
  (a) How many pounds should be ordered at one time?
  (b) How often should it place an order?
  (c) What will be the minimum total annual cost?

8. Swisshole Cheesecake Co. uses 5000 lb of cheese per year. It costs $25 to place an order and $4 to store one pound of cheese for the year. The company would like to minimize the total annual costs of ordering and storing cheese.
  (a) How many pounds should be ordered at one time?
  (b) How often should it place an order?
  (c) What will be the minimum total annual cost?

9. Rework problem 8 using the same values except for changing the storage cost to $1 per pound for the year.

10. Earth Life Force (ELF) Cookies uses 12,100 pounds of walnuts per year. It costs $16 to place an order and $2 to store one pound of walnuts for one year. The company wishes to minimize total annual costs of ordering and storing walnuts.
    (a) How many pounds should ELF Cookies order at one time?
    (b) What is the minimum total annual cost of ordering and storing walnuts?

Production Runs

11. Jake's Brakes produces 375 brake shoes per year. It costs Jake $5 to set up the machine and $6 to store a brake shoe for the year.
    (a) How large a production run should Jake have so as to incur a minimum annual cost of both setting up to make brake shoes and storing brake shoes?
    (b) What is that minimum annual cost?

12. JR Fabrics makes backpacks. It sells 128 of model G during the year. It costs $8 to set up the pattern to make G and $8 to store one model G for the entire year.
    (a) How large a production run should JR Fabrics have to incur a minimum total annual cost of both setting up the pattern and storing model G backpacks?
    (b) What is that minimum annual cost?

13. JR Fabrics makes "Where The Heck Is . . ." T-shirts with a customized city name. They sell 126 per month with "Smalltown, USA" on them. It costs $7 to set up a production run and $0.25 to store one T-shirt for a month.
    (a) How large a production run should JR Fabrics have to incur a minimum total monthly cost of both setting up a production run and storing these T-shirts?
    (b) What is that minimum monthly cost?

14. Rework problem 13 using 64 T-shirts per month, a cost of $8 to set up a production run, and the same $0.25 per month to store one T-shirt.

15. **(Scheduled Maintenance)** Refer to Example 3. Suppose the Boeing 747 has a life of 100,000 hours and it costs $10,000 to clean the tar. Find the interval of flight time, k, that minimizes the total expense of flying and cleaning tar.

Maximum Sustainable Harvest

16. It is believed that the reproduction curve for the snowshoe hare is $f(P) = -0.025P^2 + 4P$.
    (a) Find the maximum sustainable harvest for this curve.

    (b) Find the initial population necessary for that harvest.

17. Suppose the reproduction curve for Australian rabbits is $f(P) = -P^3 + 49P$, where P is in millions.
    (a) Find the maximum sustainable harvest.
    (b) Find the initial population necessary for that harvest.

18. Suppose a reproduction curve is $f(P) = 20\sqrt{P}$.
    (a) Find the maximum sustainable harvest.
    (b) Find the initial population necessary for that harvest.

19. Suppose a reproduction curve is $f(P) = 25\sqrt{P}$.
    (a) Find the maximum sustainable harvest.
    (b) Find the initial population necessary for that harvest.

20. Suppose a reproduction curve is given by the graph in figure 3.6.12. Find the maximum sustainable yield.

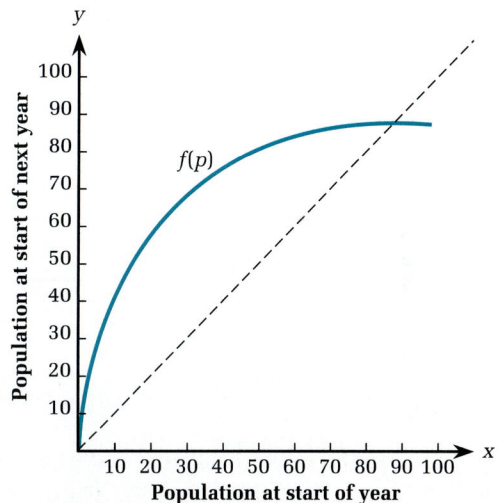

FIGURE 3.6.12

Problems for Writing and Discussion

21. There is the possibility that a theoretical reproduction curve is such that $y = f(P)$ never crosses the curve $y = P$.
    (a) Suppose a reproduction curve is $y = 1.1P$. What is the population after 10 generations if the initial population is 100?
    (b) What happens to a population if its reproduction curve does not cross the line $y = P$?

(c) If the reproduction curve is given by $y = 1.1P$, why might you suspect the domain cannot be "for all values of P"?

Enrichment Problems

22. Suppose that placing an order costs P dollars, the annual demand is D items, and the cost of storing one unit for one year is S dollars. Starting with $T(x) = (D/x) \cdot P + (x/2) \cdot S$, show that the optimal economic order quantity is $\sqrt{2DP/S}$ and that the order should be placed $\sqrt{(DS)/(2P)}$ times a year.

23. To demonstrate that the cost of an item does not affect the solution to the EOQ, suppose the item costs \$$R$ per unit. The cost of each order is $P + R \cdot x$ and in the year we order D/x times. Thus, the dollar cost of all the orders equals "dollar cost per order" times "number of orders per year," which is given by $(P + R \cdot x) \cdot D/x = P \cdot D/x + R \cdot D$. The cost of storing, as before, equals $(x/2) \cdot S$. Added together, these equal $P \cdot D/x + R \cdot D + (x/2) \cdot S$. Suppose we call this function $f(x)$. Show the minimum of $f(x)$ occurs at the same critical number that solves the EOQ problem.

24. **(Routine Maintenance)**   Refer to Example 3. Suppose the cost of cleaning is \$$C$ dollars. Find the interval of time k that minimizes the total cost of flying and cleaning tar when the total life of the plane is 100,000 hours.

Alternate Storage Cost for EOQ

Suppose that we must provide storage and handling for all the inventory ordered at one time. For example, space set aside for frozen food cannot be used for the storage of tomatoes. Using the notation of the last problem, the total annual cost would be given by $T(x) = (D/x) \cdot P + x \cdot S$ because the entire order of size x is being stored.

25. Rework problem 7 using this alternate storage cost approach. Compare your answer to the answer for problem 7. Why do you think the optimal lot size becomes smaller?

26. Rework problem 8 using this alternate storage cost approach. Compare your answer to the answer for problem 8. Why do you think the optimal lot size becomes smaller?

SOLUTIONS TO PRACTICE EXERCISES

1. (a) Total annual cost of x units is:

$$T(x) = \left(\frac{\text{annual demand}}{\text{lot size}} \cdot \text{ordering cost}\right)$$
$$+ \left(\frac{\text{lot size}}{2} \cdot \text{storage cost of one unit}\right)$$
$$= \frac{2000}{x} \cdot 40 + \frac{x}{2} \cdot 1$$
$$= \frac{80,000}{x} + \frac{x}{2}$$
$$T'(x) = -\frac{80,000}{x^2} + \frac{1}{2}$$

Setting $T'(x) = 0$, we have $x^2 = 160,000$, so that $x = 400$. This is an optimal lot size because $T''(x) > 0$ for all $x > 0$.

(b) $T(400) = 80,000/400 + 400/2 = 200 + 200 = 400$.

2. (a) To have a production run of size x, the annual production cost is:

$$C(x) = \left(\frac{\text{total production}}{\text{run size}} \cdot \text{cost of set-up}\right)$$
$$+ \left(\frac{\text{run size}}{2} \cdot \text{annual storage cost per unit}\right)$$
$$= \frac{50,000}{x} \cdot 5 + \frac{x}{2} \cdot 0.125$$
$$= \frac{250,000}{x} + 0.0625x$$

$C'(x) = -250,000/x^2 + 0.0625$, and setting $C'(x) = 0$ we have $x^2 = 250,000/0.0625 = 4,000,000$, giving us $x = 2000$.

(b) The number of runs a year is (total production/run size) $= 50,000/2000 = 25$.

(c) The minimal annual costs will be $250,000/2000 + 0.0625 \cdot 2000 = 125 + 125 = 250$.

3. (a) $H(P) = f(P) - P = 10\sqrt{P} - P$
$H(4) = 10\sqrt{4} - 4 = 10 \cdot 2 - 4 = 20 - 4 = 16$

(b) $H'(P) = 10 \cdot 1/(2\sqrt{P}) - 1$. So when $H'(P) = 0$, then $5/\sqrt{P} = 1$, giving $5 = \sqrt{P}$. The maximum sustainable harvest is when $P = 25$ and then $H(25) = 10\sqrt{25} - 25 = 10 \cdot 5 - 25 = 50 - 25 = 25$.

3.7 DIFFERENTIALS

Before the availability and widespread use of calculators, computations that are quite simple today were quite difficult. Even finding a decimal value for $\sqrt{2}$ was tedious and somewhat complicated. Tables were cumbersome to use, and no matter how detailed the table, the exact values we were seeking always seemed to be omitted.

Differentials, the topic of this section, were then, and remain today, an aid when finding approximations. In Chapter 5 we will find that differentials have another, very important application. However, that application will require that we learn many new concepts. Hence, at this point we restrict ourselves to the use of differentials as approximations. Examples of applications follow.

In 1892, Mikimoto Kōkichi (1858–1954) discovered how to grow cultured pearls. He and his son-in-law founded the Japanese cultured pearl industry. A "seed" of mussel is inserted into an oyster, and over a period of up to three years the oyster deposits nacre to create a pearl. The layer of nacre must be at least 0.5 millimeter thick for the pearl to be genuine. We can use differentials to estimate the volume of nacre deposited on a "seed" of diameter 0.5 centimeter in order to create a genuine pearl.

Estimating the cost of gold-plating a basketball for the NCAA champion team can also be done with differentials.

Tangent Line Approximations

We used tangent lines in Section 2.4 to introduce the idea of a derivative. Now, let us use the tangent line to approximate the value of a function.

EXAMPLE 1   Using a tangent line approximation

Find approximations for $\sqrt{2}$ and $\sqrt{17}$.

SOLUTION

Let us first approximate $\sqrt{2}$. Given this problem, we would certainly use a calculator and find that $\sqrt{2} \approx 1.4142136$. However, we want to illustrate the use of a tangent line approximation, so let us pretend our calculator broke or picked an inconvenient time to run out of battery power. We can still find an approximation by using calculus.

Let $f(x) = \sqrt{x}$. Knowing values for integers squared, we do know a few precise points on the graph of $y = \sqrt{x}$. For instance, $\sqrt{0} = 0$,

$\sqrt{1} = 1$, $\sqrt{4} = 2$, $\sqrt{9} = 3$, and so on, because these x-values are themselves squares. A sketch of the graph of $f(x)$ appears in figure 3.7.1.

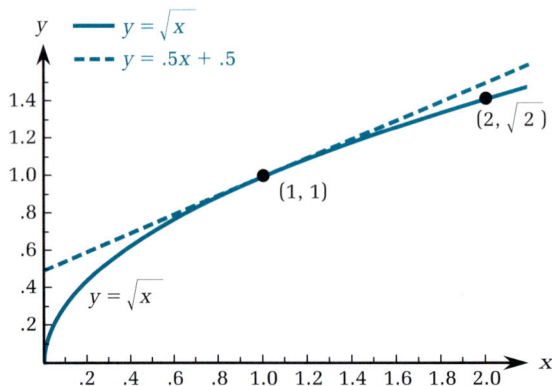

FIGURE 3.7.1

We could use $\sqrt{2} \simeq \sqrt{1} = 1$, but that is a poor approximation. Can we do better?

We could try numbers larger than 1, square them and compare the square to 2. For example, $(1.2)^2 = 1.44$, which means 1.2 is a better approximation than 1. Or we could try $(1.4)^2 = 1.96$ to find that 1.4 is an even better approximation. This method may yield better and better approximations, but the repeated squaring involved will take considerable time and effort.

A better plan for an approximation consists of leaving the $y = \sqrt{x}$ curve at the point $P(1, 1)$ and following the tangent line from that point, as shown in figure 3.7.2. If we do not go too far, the tangent line will remain close to the curve, and so for a given value of x the associated y-value on the tangent line will not be too different than the exact y-value on the curve.

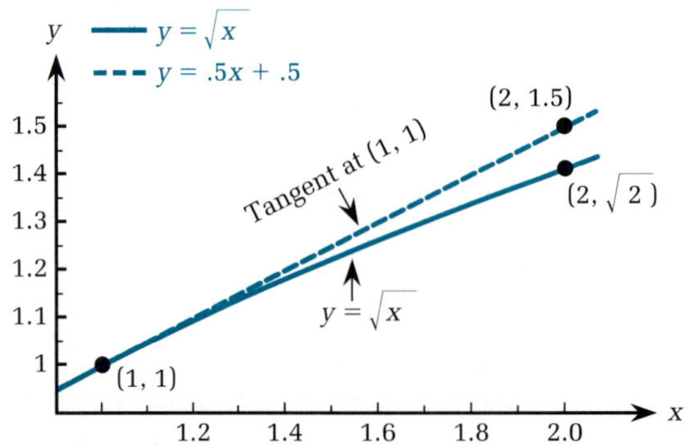

FIGURE 3.7.2

Without worrying about what "too far" means at the moment, let us go ahead and try this idea for an approximation.

The slope of the tangent line to $y = \sqrt{x}$ at $P(1, 1)$ is found by evaluating the derivative, $\dfrac{d}{dx}\sqrt{x} = 1/(2\sqrt{x})$, at $x = 1$.

$$\left.\frac{1}{2\sqrt{x}}\right|_{x=1} = \frac{1}{2\sqrt{1}} = \frac{1}{2 \cdot 1} = \frac{1}{2} = 0.5$$

The tangent line goes through $P(1, 1)$, so with a slope of 0.5, it has the point-slope equation of $(y - 1)/(x - 1) = 0.5$. This is equivalent to $y - 1 = 0.5(x - 1)$, which simplifies to $y = 0.5x + 0.5$. On the tangent line, substituting $x = 2$ gives $y = 0.5 \cdot 2 + 0.5 = 1 + 0.5 = 1.5$, which provides an approximation of $\sqrt{2} \approx 1.5$.

We can check $(1.5)^2 = 2.25$. This is not a very good approximation and the reason is that we left the curve quite far away from the x-value we were interested in using. However, we did so in this example so that the figures we used would have some space between the tangent line and the graph of $y = \sqrt{x}$. To illustrate the process working much better numerically, let us turn to estimating $\sqrt{17}$.

If we leave the $\sqrt{x}$ curve at 16 and proceed on the tangent curve until $x = 17$, we find that the slope of the tangent line is

$$\left.\frac{1}{2\sqrt{x}}\right|_{x=16} = \frac{1}{2\sqrt{16}} = \frac{1}{2 \cdot 4} = \frac{1}{8} = 0.125$$

The equation of the tangent line is $y = 0.125x + 2$.

Substituting $x = 17$, we have $y = 4.125$, which is then our approximation for $\sqrt{17}$. By squaring, we find $(4.125)^2 = 17.015625$, so that we have found a fairly good approximation.

TANGENT LINE APPROXIMATION

If $f'(a)$ exists, then $f(x) \simeq f(a) + (f'(a))(x - a)$.

PRACTICE EXERCISE 1

Use the tangent line of Example 1 to approximate $\sqrt{16.5}$.

Definitions of Differentials

Consider the graph of a general function $y = f(x)$ in figure 3.7.3.

Suppose $P(x_0, f(x_0))$ and $Q(x, f(x))$ are two points on the graph. Let the actual difference in the y-values be represented by Δy and the actual difference in the x-values by Δx. Then $\Delta x = x - x_0$ and $\Delta y = f(x) - f(x_0)$.

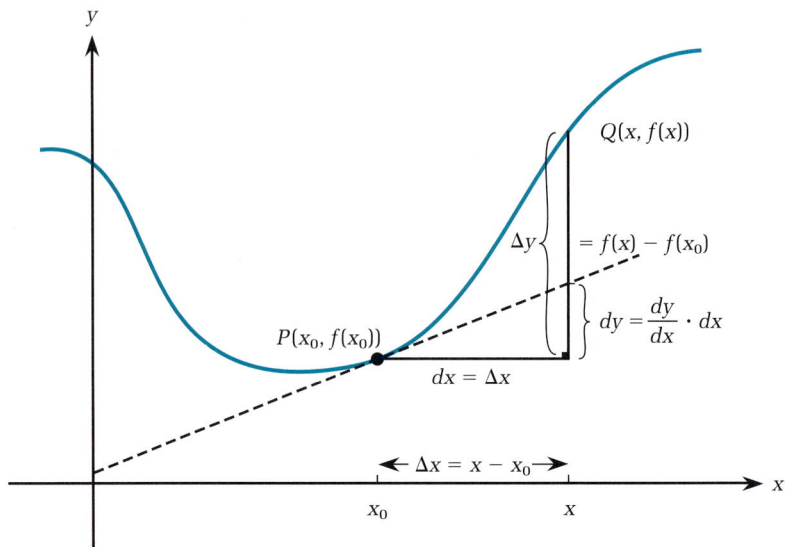

FIGURE 3.7.3

By the definition of dy/dx, we have $dy/dx = \lim_{\Delta x \to 0} \Delta y/\Delta x$. Hence, $\Delta y/\Delta x \simeq dy/dx$, which implies $\Delta y \simeq dy/dx \cdot \Delta x$.

We want to assign a name to the expression $dy/dx \cdot \Delta x$ because it is a good approximation to Δy. This is the *differential*.

DEFINITION

Differential

The *differential of x*, written as dx, is defined to be $dx = \Delta x$.
The *differential of y*, written as dy, is defined to be $dy = dy/dx \cdot \Delta x$.

Here we see an advantage of the notation dy/dx for the derivative. We can write the differential of y, which is dy, as equal to the product of the derivative of y with respect to x, written as dy/dx, multiplied by the differential of x, which is dx. Now, think of "dy" or "dx" as being actual, concrete numbers. The ratio of dy over dx is equal to dy/dx. The ratio of the differentials is then equal to the derivative, revealing to us a compelling reason for writing the derivative as dy/dx.

EXAMPLE 2   Finding differentials

For $y = f(x) = \dfrac{1}{x + 1}$, $x = 2$, and $\Delta x = 0.1$, find dx, Δy, and dy.

SOLUTION

The value dx is always Δx, so $dx = 0.1$. The value of Δy is the actual change in y as x changes from 2 to $2 + 0.1$. To find Δy we need to evaluate $\Delta y = f(2 + 0.1) - f(2)$.

Doing so, $\Delta y = f(2.1) - f(2) = 1/(2.1 + 1) - 1/(2 + 1) = \dfrac{1}{3.1} - \dfrac{1}{3} = \dfrac{3 \cdot 1 - 1 \cdot 3.1}{(3.1)(3)} = -0.1/9.3$.

The definition of dy is

$(*)$
$$dy = \frac{dy}{dx}\bigg|_{x=2} \cdot dx$$

The derivative is $dy/dx = (d/dx)[1/(x + 1)] = -1/(x + 1)^2$.

Evaluated at $x = 2$, the derivative is $-1/(2 + 1)^2 = -1/3^2 = -1/9$.

Because $dx = 0.1$, substituting for both terms in equation $(*)$ gives $dy = -1/3^2 \cdot 0.1 = (-0.1)/9$. Using a calculator to compare these values, we have $\Delta y \simeq -0.010752688$, $dy \simeq -0.011111111$. ●

PRACTICE EXERCISE 2

Let $y = \sqrt{x}$.

(a) Find dy when $dx = 2$, and $x = 16$.

(b) Use $4 + dy$ to estimate $\sqrt{18}$.

The differential dx can be positive or negative as needed. For instance, we can approximate $\sqrt{15}$ by using $x = 16$ and $dx = -1$ and so obtain

$$dy = \frac{dy}{dx}\bigg|_{x=16} \cdot dx = \frac{1}{2\sqrt{16}} \cdot (-1) = -\frac{1}{8}$$

Hence, $\sqrt{15} \simeq \sqrt{16} + (-1/8) = 4 - (1/8) = 3.875$.

Differentials with Implicitly Defined Functions

Implicit differentiation was introduced in Section 2.8. Therein we encountered situations in which we could not explicitly represent y as a function of x, but nevertheless we could find dy/dx. In such a situation we also may find differentials.

EXAMPLE 3 Finding the differentials of an implicitly defined function

Assume the equation $x^2y + y^2 - 5x = 1$ determines y as a function of x. Find dy when $x = 1$, $y = 2$, and $dx = 0.1$.

SOLUTION

Using implicit differentiation $2x \cdot y + x^2 \cdot dy/dx + 2y \cdot dy/dx - 5 = 0$. Substituting $x = 1$, $y = 2$, and $dx = 0.1$ gives us

$$2 \cdot 1 \cdot 2 + 1^2 \cdot \frac{dy}{dx} + 2 \cdot 2 \cdot \frac{dy}{dx} - 5 = 0$$

Solving for dy/dx, we have $5(dy/dx) - 1 = 0$, so $dy/dx = 1/5$. Finally, $dy = dy/dx \cdot dx = 1/5 \cdot 0.1 = 0.02$.

Applications

EXAMPLE 4 Approximating the cost of gold plating

An official NCAA basketball has a 30-inch circumference, which means the radius is approximately 4.775 inches. Suppose we wish to award a gold-plated basketball to the NCAA championship team. We want to estimate the cost of the gold plating.

For jewelers, gold electroplating is legally at least seven-millionths of an inch thick, and it must be at least 10-karat gold. (Pure gold is 24-karat, and 10-karat is 10/24 pure gold.) Each cubic inch of 10-karat gold has approximately 4.64 ounces of pure gold. Assume pure gold is $400 per ounce. Use differentials to estimate the cost of plating a coating seven-millionths of an inch thick of 10-karat gold.

SOLUTION

The volume of a sphere is $V(r) = (4/3)\pi r^3$. We want to estimate ΔV, the actual change in volume, by dV, the differential.

$$dV = \frac{4}{3}(\pi \cdot 3r^2)dr = 4\pi r^2 dr$$

Using $r = 4.775$ and $dr = 0.000007$, we have $dV = 4\pi(4.775)^2(0.000007) \approx 0.002006$ cubic inches.

To find the price, P, we multiply the number of cubic inches by both 4.64 (the number of ounces of pure gold in a cubic inch) and $400 (the cost per ounce). We then get $P \approx 0.002006 \cdot 4.64 \cdot 400 = 3.72.

If we have a profit, cost, or revenue function, the *marginal* profit, cost, or revenue has been defined as the derivative of the appropriate function. Differentials give us the estimated profit of the $(n + 1)$-st unit because we can use $dx = 1$ to estimate the change in profit from the nth item to the $(n + 1)$-st item.

We are using $dx = 1$, so $dy = dy/dx \cdot 1 = dy/dx$.

Another way of saying this is that the marginal profit is the differential of the profit function. Similarly, the differentials of the cost and revenue functions are the marginals of their respective functions.

Approximate Range of Values

Another use of differentials is illustrated by Example 5.

EXAMPLE 5   Approximate range of values, cookie fillings

An automatic cookie machine is set to dispense creme filling between two chocolate wafer cookies. When the machine is set on position x, it squirts exactly $10x^2 + x$ micro-ounces of filling. At the beginning of each shift the machine setting is rechecked, but during the day the setting may drift.

Suppose the machine is set on 100 and it does not vary more than 5 on either side. What is the possible variation in the filling?

SOLUTION

Let $F(x)$ represent the number of ounces of filling.

$F(x) = 10x^2 + x$, so $F'(x) = 20x + 1$. For $dx = 5$ and $x = 100$, we have dy given by $dy = (20(100) + 1) \cdot 5 = 10{,}005$ micro-ounces on either side of $F(100) = 10 \cdot 100^2 + 100 = 100{,}100$. (As an aside to those filling cookies, 100,000 micro-ounces is equivalent to 1/10 of an ounce.)

To the question asked in Example 5, a reasonable response is, "Why not simply calculate $F(105)$ and $F(95)$?" That certainly could be done. However, usually the bakers ask the question "going the other way," as in Example 6.

EXAMPLE 6   Approximate range of values

Suppose the machine of Example 5 needs to be set so that the amount of filling does not vary more than 5000 micro-ounces from the ideal filling of 100,100 micro-ounces created by having $x = 100$. What are the possible x-values?

SOLUTION

We could solve the quadratics $10x^2 + x = 95,100$ and $10x^2 + x = 105,100$ by the quadratic formula, but the numbers involved are quite awkward. It is much easier to take $dy = (dy/dx)dx$ and let $dy = 5000$.

The derivative $dy/dx = 20x + 1$ is evaluated at $x = 100$ to be $dy/dx = 2001$.

Substituting in the equation $dy = (dy/dx)dx$, we get $5000 = 2001 \cdot dx$.

Solving for dx, we have $dx = 5000/2001 \approx 2.5$. We are safe from cookie overfill or underfill if we keep x in the interval 97.5 to 102.5.

PRACTICE EXERCISE 3

Use differentials to approximate how close x must be to 10 to make certain that $f(x) = 2x^2 - x$ is within 1 of $f(10) = 190$.

Notice that in Example 6 and in Practice Exercise 3 we have reversed the differentials. Rather than having dx be exact and dy be the approximation, we have used dy as exact and dx as an approximation. However, because the Chain Rule states that $\dfrac{dy}{dx} \cdot \dfrac{dx}{dy} = 1$, we have $dx = \dfrac{dx}{dy} \cdot dy = \dfrac{1}{\dfrac{dy}{dx}} \cdot dy$

As a result, beginning with $y = f(x)$ and assuming that this function implicitly gives us x as a function of y, we can find the dx directly in terms of dy.

3.7 PROBLEMS

Foundations

The problems of this section require the basic skills illustrated by the following:

Find the derivatives of the functions in problems 1 and 2.

1. $x^2 - 2x$ **2.** $\sqrt{3x + 1}$

3. Use implicit differentiation to find an expression for dp/dx if $px + p^2 = -20x + 400$.

Exercises

For problems 4–11, find (a) Δy and (b) dy for the given function and values of x and Δx.

4. $f(x) = x^2 - 2x$, $x = 3$, $\Delta x = 0.01$

5. $f(x) = x^2 + x$, $x = 2$, $\Delta x = 0.1$

6. $f(x) = 3/x$, $x = 3$, $\Delta x = 0.1$

7. $f(x) = -2/x$, $x = 1$, $\Delta x = 0.2$

8. $f(x) = 3x - 5$, $x = 2$, $\Delta x = 0.01$

9. $f(x) = 2x + 1$, $x = 3$, $\Delta x = 0.02$

10. $f(x) = \sqrt{3x + 1}$, $x = 8$, $\Delta x = -0.1$

11. $f(x) = \sqrt{x}$, $x = 9$, $\Delta x = -1$

Differentials with Implicitly Defined Functions

For problems 12 and 13, use implicit differentiation to find dy for the given values of x, y, and dx.

12. $x^3 - xy = y^3 + 8$; $x = 2$, $y = 0$, $dx = 0.05$

13. $-2x + x^2y - y^3 = -1$; $x = 0$, $y = 1$, $dx = 0.2$

The equations in problems 14–17 implicitly define demand functions. For each, use implicit differentiation to find dp/dx and then find dp for the given values of x, p, and dx.

14. $px + p^2 = -20x + 400$; $x = 15$, $p = 5$, $dx = 0.5$

15. $2px + p^2 = -100x + 2500$; $x = 10$, $p = 30$, $dx = 2$

16. $x^2p^2 + x + 2p = 8$; $x = 2$, $p = 1$, $dx = 0.2$

17. $xp^3 + xp + x^2 + 3p = 4$; $x = 1$, $p = 3$, $dx = 0.2$

18. *(Approximations, Spheres in Space)*   Suppose a sphere of radius 50 feet is to be built for a space satellite.
   (a) Estimate the volume of a seven-millionth-inch thick coating on the sphere.
   (b) Find the cost of the coating if we use pure silver. Assume pure silver costs $4 an ounce and there are 6.1 ounces per cubic inch.

19. *(Approximations, Cultured Pearls)*   For background on this problem, see the beginning of this section. To create a cultured pearl, a seed of mussel is inserted into an oyster. After the oyster has deposited a layer of nacre at least 0.5 millimeter thick, then the pearl is considered genuine. Use differentials to estimate the volume of nacre deposited on a "seed" of diameter 1 centimeter in order to create a genuine pearl. (Hint: A diameter of 1 centimeter means a radius of 0.5 centimeter.)

20. *(Marginal Revenue)*   (a) Find the marginal revenue at $x = 100$ if $R(x) = x^3 - 2x$.
   (b) Use it to estimate the revenue at $x = 101$.

21. *(Marginal Revenue)*   (a) Find the marginal revenue at $x = 25$ if $R(x) = 2x^2 + 3x$.
   (b) Use it to estimate the revenue at $x = 26$.

22. *(Marginal Profit)*   (a) Find the marginal profit at $x = 120$ if $P(x) = \sqrt{x^2 + 2x}$.
   (b) Use it to estimate the profit at $x = 119$.

23. *(Marginal Profit)*   (a) Find the marginal profit at $x = 15$ if $P(x) = 100 - 1/x^2$.
   (b) Use it to estimate the profit at $x = 14$.

24. *(Approximate Range of Values)*   Suppose $f(x) = 3x^2 + 2$. Estimate how close $f(x)$ is to $f(9)$ if x is within 0.1 of 9.

25. *(Approximate Range of Values)*   Suppose $f(x) = 2x^2 - 3$. Estimate how close $f(x)$ is to $f(5)$ if x is within 0.1 of 5.

26. *(Approximate Range of Values)*   For $f(x) = 1/(2x) + x$, estimate how close x must be to $x = 5$ to make the values of $f(x)$ be within approximately 0.1 of $f(5) = 5.1$.

27. *(Approximate Range of Values)*   For $f(x) = 1/\sqrt{x} + x$, estimate how close x must be to $x = 16$ to make the values of $f(x)$ within approximately 0.1 of 16.25.

28. *(Approximate Range of Values, Chocolate Coating for Malted-Milk Balls)*   The Chocolate PropWorks is making immense malted-milk balls, about the size of basketballs, for a TV ad. The radius of the ball is about 5 inches. How close does the radius have to be to 5 so that the volume is within 25 cubic inches of the 523.6 cubic inches they would get at exactly a 5-inch radius? (Hint: The formula for the volume of a sphere of radius r is $V = (4/3)\pi r^3$.)

29. *(Approximate Range of Values, Blast Furnace Temperatures)*   The temperature, in degrees Fahrenheit, of a blast furnace is given by $T(x) = 3000x/(x + 1)$, where x is the setting on a gas-line value. For example, at a setting of $x = 4$, the temperature is $T(4) = (3000 \cdot 4)/(4 + 1) = 12{,}000/5 = 2400°F$.
   (a) Use differentials to estimate how close the temperature is to $2400°F$ if $x = 4$ and $dx = 0.1$.
   (b) How close does the setting have to be to $x = 4$ if the temperature is $2400°F$ and $dy = 50°F$?

Problems for Writing and Discussion

30. *(Mean Value Theorem)*   Suppose $f(x)$ is continuous over the interval $[a, b]$ and $f'(x)$ exists over the interval (a, b). The *Mean Value Theorem* states that under these conditions there is a value c in (a, b) such that

$$f'(c) = \frac{f(b) - f(a)}{b - a}$$

A geometric interpretation of this result is that there is some point on the graph of $y = f(x)$ at which the tangent line is parallel to the secant line joining the points $P(a, f(a))$ and $Q(b, f(b))$, as is shown in figure 3.7.4.

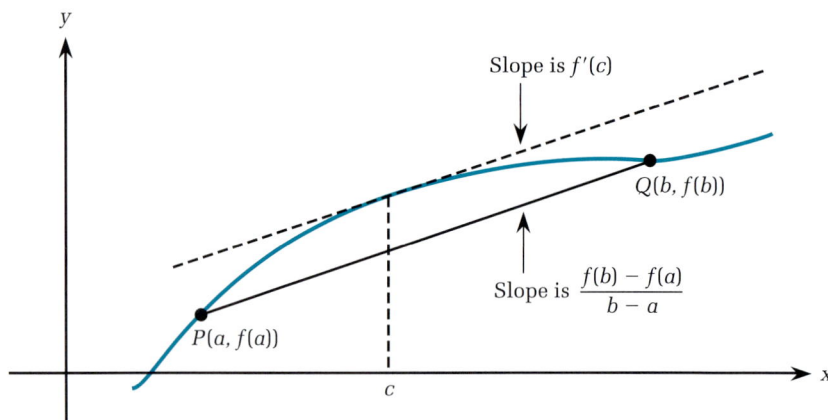

FIGURE 3.7.4

(a) Find c for $f(x) = x^2$ over [2, 3].
(b) Find c for $f(x) = \sqrt{x}$ over [1, 4].
(c) Use the Mean Value Theorem to demonstrate that a runner who covers five miles in 28 minutes must have been running exactly 10 miles per hour at some point along the route.

Enrichment Problems

31. **(Approximation, Waste Removal Coatings)**
    Suppose a storage drum is a cylinder of radius one foot and height three feet.
    (a) Find the surface area.
    (b) To store chemical wastes, its entire inside is to be given a zinc coating, which is to be 0.1 inch thick. Assume zinc costs $0.50 per pound and weighs 445 pounds per cubic foot. Estimate the cost of coating each storage drum by coating the area found in part (a).

32. When $f(x)$ is approximated by using a tangent line at the point $P(a, f(a))$, we are using differentials and the formula is

    (A) $\qquad f(x) \simeq f(a) + f'(a)(x - a)$

    A better approximation results from using

    (B) $f(x) \simeq f(a) + f'(a)(x - a) + \dfrac{f''(a)}{2}(x - a)^2$

    (a) Show that equation (B) gives a better approximation to $\sqrt{2}$ than we found in Example 1.
    (b) Show that equation (B) gives a better approximation to $\sqrt{18}$ than we found in Practice Exercise 1.

SOLUTIONS TO PRACTICE EXERCISES

1. $\sqrt{16.5} \simeq 4 + 0.125 \cdot 0.5 = 4 + 0.0625 = 4.0625$.
   Checking, $(4.0625)^2 = 16.5039$.

2. (a) $dy/dx = 1/(2\sqrt{x})$, so that at 16, $dy/dx = 1/8 = 0.125$. Because $dx = 2$, we have $dy = 0.125 \cdot 2 = 0.25$.

   (b) $\sqrt{18} \simeq 4 + dy = 4 + 0.25 = 4.25$.

3. $f'(x) = 4x - 1$ so $f'(10) = 39$ and $dy = 1$. $dy = dy/dx \cdot dx$ gives us $1 = 39 \cdot dx$. Solving, we have $dx = 1/39 \simeq 0.02564$, which thus is our estimation of how close x must be to 10.

CHAPTER 3 REVIEW

Define or discuss:

1. Open intervals, closed intervals

2. Continuous over an interval

3. Intermediate Value Property of continuous functions

4. A function increasing (or decreasing) over an interval

5. A function increasing (or decreasing) at a point

6. Screen of display

7. Relative and absolute extreme points

8. Relative and absolute extreme values

9. Critical point

10. Critical number

11. First Derivative Test

12. Inflection point

13. Concavity

14. Second Derivative Test

15. Limits at infinity

16. Horizontal and vertical asymptotes

17. How to solve word problems

18. Optimization

19. Economic Order Quantity (EOQ)

20. Sustainable annual yield

21. Differentials

REVIEW PROBLEMS FOR CHAPTER 3

1. Find the largest intervals over which $f(x) = x^3 - 6x$ is (a) increasing, (b) decreasing.

2. Find the largest intervals over which $g(x) = (-3 + 4x)/(x^2 + 1)$ is (a) increasing, (b) decreasing.

On the graphs in problems 3–6 indicate any (a) relative extreme points, (b) absolute extreme points, and (c) inflection points.

3.

![graph]

FIGURE 3.8.1

4.

![graph]

FIGURE 3.8.2

5.

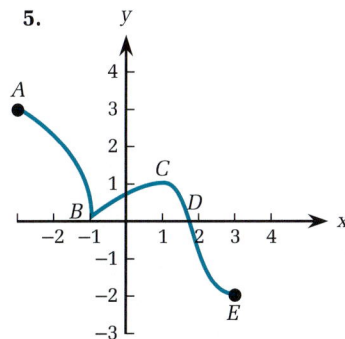

FIGURE 3.8.3

6.

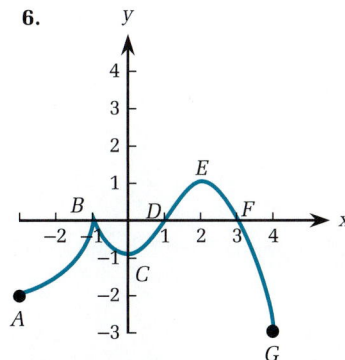

FIGURE 3.8.4

For each of the functions in problems 7–26, find (a) the critical numbers, (b) any relative maximum or minimum values, and (c) any absolute maximum or minimum values. Where no domain is specified, use the set of all real numbers as the domain. (d) Sketch a graph.

7. $f(x) = x^2 + x - 3$ $[-1, 2]$

8. $f(x) = 2x^2 - x + 1$ $[0, 2]$

9. $f(x) = x^3 + 6x^2 + 9x - 5$ $[-2, 2]$

10. $f(x) = 2x^3 - x^2 - 6$ $[-1, 1]$

11. $g(x) = 2x^3 - 9x^2 + 6$ $[-1, 5]$

12. $h(x) = 5\sqrt{x} - x$ $[0, \infty)$

13. $f(x) = x^{2/3} - 2x + 2$ $[0, \infty)$

14. $g(x) = \sqrt{x} - x + 2$ $[0, 3]$

15. $f(x) = \dfrac{x}{x^2 + 4}$ $[0, \infty)$

16. $f(x) = \dfrac{2x}{x^2 + 1}$ $[0, \infty)$

17. $f(x) = \dfrac{3x + 1}{x^2 + 11}$ $[-\dfrac{1}{3}, \infty)$

18. $f(x) = x + \dfrac{2}{x}$

19. $f(x) = x^2 + \dfrac{16}{x}$

20. $g(x) = 27x^2 + \dfrac{16}{x}$ $(0, \infty)$

21. $g(x) = |4x^2 - 1|$ $[0, 1]$

22. $g(x) = |x - 1|$ $[-1, 2]$

23. $f(x) = \dfrac{x + 1}{\sqrt{x - 2}}$ $[3, \infty)$

24. $f(x) = \dfrac{x^2}{\sqrt{x - 2}}$ $(2, \infty)$

25. $h(x) = 12x^{1/3} - x$

26. $h(x) = 3x^{2/3} - x$

Find any inflection points for the functions in problems 27–30.

27. $f(x) = x^3 + 6x^2 + 9x - 5$

28. $f(x) = 2x^3 - x^2 - 6$

29. $f(x) = \dfrac{x}{x^2 + 4}$

30. $f(x) = \dfrac{2x}{x^2 + 1}$

For the functions in problems 31–36, find any horizontal asymptotes and vertical asymptotes.

31. $h(x) = \dfrac{2x - 1}{x - 1}$

32. $g(x) = \dfrac{2x}{x^2 + 1}$

33. $f(x) = \dfrac{x + 1}{x^2 + 3}$

34. $k(x) = \dfrac{3x^2 - 1}{2x}$

35. $r(x) = \dfrac{x^2 - 4}{1 - 4x^2}$

36. $f(x) = \dfrac{3x^2 + 6x}{x^2 - 4}$

37. Suppose that $P(x) = -x^3 + 15x^2$ $(0 \le x \le 12)$ is the profit in thousands of dollars from selling x thousand cases of grapes.
    (a) Graph the function.
    (b) Find the maximum profit.
    (c) Find the number of cases that produce that profit.
    (d) Find the inflection point.
    (e) Give an economic interpretation of the inflection point.

38. Suppose the percentage of one lot of glass bottles not recycled by day t is given by $R(t) = (1000 + 5t^2)/(10 + t^2)$.
    (a) Find the day of greatest change in the percentage of bottles being recycled.
    (b) What is the limit of the percentage of the lot being recycled as t approaches infinity?

39. Suppose Sam has $10,000 to construct a rectangular enclosure surrounded by a high outside fence and divided in half by a lower inside fence. The high fence costs $10 per foot and the low fence costs $5 per foot.
    (a) What are the dimensions of the largest possible enclosure?
    (b) What is the area of that enclosure?

40. Suppose $y = x^{3/2} + x^7/42$. Find (a) y''' and (b) $y^{(4)}$.

41. If $f(x) = x^5 - 3x^2$, find $\dfrac{d^3}{dx^3} f(x)$.

42. A rectangular lot is to be created and fenced in alongside a street. On the street side the fence is $5 per foot and on the other sides it is $3 per foot.
    (a) For $10,000, what is the maximum area that can be enclosed?
    (b) What are the dimensions of that area?
    (c) Suppose the lot must contain 19,200 square feet. Which dimensions create a lot of least fencing cost?

43. A rectangular garden is to consist of a rectangle subdivided into six equal-sized plots as shown in figure 3.8.5.

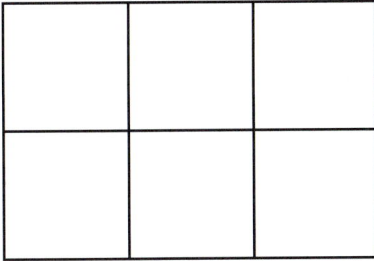

FIGURE 3.8.5

(a) Find the maximum area that can be created by using 1200 feet of fencing.
(b) What are the dimensions of each plot?
(c) What is the maximum area of each plot?

44. A 60-unit apartment house is renting at $300 per unit. If the rent is raised by $10 \cdot k$ dollars, then there will be only $(60 - k)$ apartments rented.
(a) Find the rent per unit.
(b) Find the number of units rented that produces a maximum total revenue.
(c) What is the maximum possible total revenue?

45. Erasers have a demand function of $p(x) = 100/(x + 1)$ and a cost function given by $C(x) = 5 + 0.25 \cdot x$.
(a) Find a formula for the revenue, $R(x)$.
(b) Find the maximum of the profit function, $P(x) = R(x) - C(x)$.
(c) What value for x produces the maximum profit?

46. Autumn Mesa Farms has 50 employees harvesting Golden Delicious apples. Each employee can pick 10 bushels of apples per hour. If x new employees are added, then because of crowding the total number of bushels of apples picked per hour by each employee will be $10 - 0.1x$.
(a) Find the number of new employees to add so as to maximize the total number of bushels picked per hour.
(b) Suppose the revenue per bushel is $2 and the pickers are paid $4 per hour. How many (if any) additional pickers should be hired in order to maximize the total revenue per hour?

47. C & A Hardware sells 60 table saws during a year. It costs them $30 to make an order and pay for its freight. It costs them $25 to store one saw for one year. They use the standard accounting practice of assuming one-half the inventory is present all the time.
(a) How many table saws should they order at

a time in order to minimize the annual total costs of ordering and shipping?
(b) How often should they order?
(c) What is the minimum annual total cost of ordering and storing table saws?

48. Bayshore Communications needs to lay a telemetry cable to connect a well three miles directly off the coast with a reading station located five miles down the coast, as shown in figure 3.8.6. The cost of laying the cable is $500 per mile on land and $1000 per mile underwater. The cable will be laid in a line to a point on shore and then along the shore to the reading station.

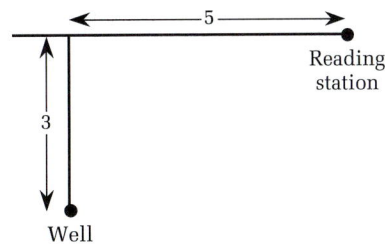

FIGURE 3.8.6

(a) Where should the cable be laid to minimize the total cost?
(b) Find the minimum total cost.

49. Suppose the annual reproduction rate for edible mushrooms is $f(P) = 10P - 0.15P^2$. Find the maximum sustainable yield.

50. Suppose a reproduction curve is given by figure 3.8.7. Find the maximum sustainable yield.

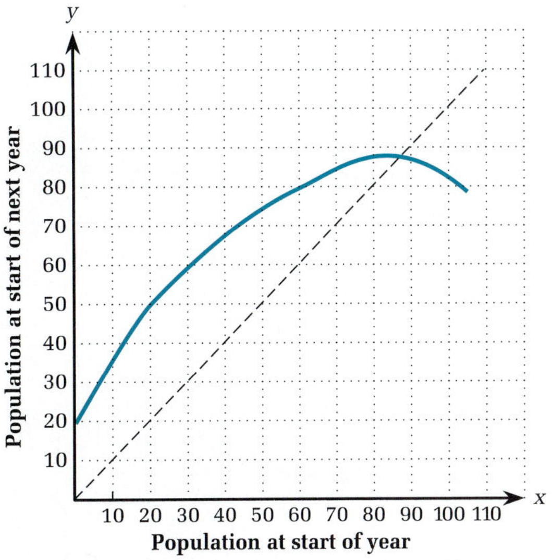

FIGURE 3.8.7

51. Approximate $\sqrt{26}$ by using differentials.

52. Suppose $f(x) = (\sqrt{x} + 1)^2$. Use differentials to estimate how close x has to be to 9 to make sure that $f(x)$ is within 0.1 of 16.

Enrichment Problems

53. Refer to problem 47 and discuss how their ordering pattern would change if the store incurred storage fees on the entire inventory of table saws for the entire year, rather than on just one-half of the inventory.

54. Suppose a ferryboat has an operating cost per hour that is proportional to the cube of the boat's speed. Going upriver a distance d miles and then returning, and with the river's current being c miles per hour, what operating speed will minimize the total cost of the ferryboat's trip?

CHAPTER 4

EXPONENTIAL AND LOGARITHMIC FUNCTIONS

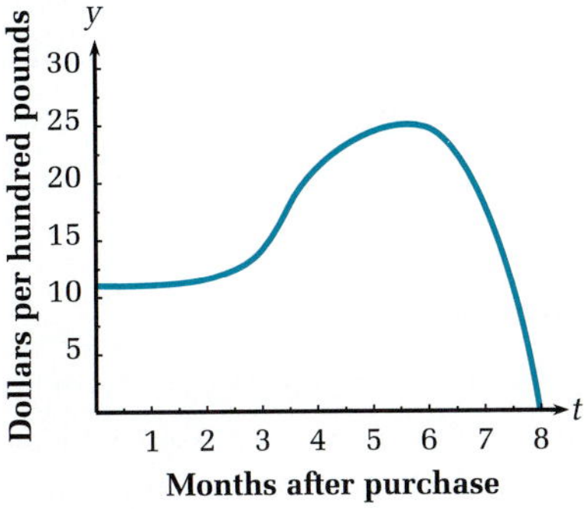

FIGURE 4.0.1

THIS CHAPTER COVERS EXPONENTIAL FUNCTIONS AND THEIR INVERSE functions, the logarithmic functions. These functions and their derivatives have numerous applications. Let us just mention one here.

Suppose a commodities broker purchases an apple crop. Over the fall and winter months the apples increase in value, due to increasing scarcity. Eventually the quality of the stored apples starts to decline and thus so does their value. Let $V(t)$ represent their value over the eight months after purchase. The graph of $y = V(t)$ is shown in figure 4.0.1.

The maximum value of $V(t)$ is 25, which occurs when $t = 6$. However, the best time to sell may not be at six months because values need to be adjusted to take into account the effect of interest. The key idea is that one dollar payable to us in six months is not as valuable as a dollar payable today. The reason is that the dollar paid to us today would earn interest over the next six months. To find the best time to sell, we will apply derivatives to the concept of *present value*.

4.1 EXPONENTIAL FUNCTIONS

For a real number b and a positive integer n, the notation b^n represents the product

$$b^n = \overbrace{b \cdot b \cdot \ \cdots \ \cdot b}^{n \text{ factors}}$$

In the following definition, y may be any real number, not just an integer.

DEFINITION

Base and exponent power

In the expression x^y, the *base* is x and *exponent power* is y.

Exponent powers were introduced in Section 1.7, and some elementary properties were presented there. Throughout that section, the base was a variable and the exponent power was a constant. Such functions are known as *power functions,* and examples of them include x^3, x^{-2}, or $7\sqrt{x}$ (written as $7x^{0.5}$).

In this section we study functions for which the roles of constant and variable are reversed. Here the base is to be the constant and the exponent power is a variable or a function of a variable. Examples of such functions include 2^x, 5^{x-1} and $(2.7)^{-x^2}$.

Review of Properties of Exponents

The following chart summarizes the rules for exponents, as given in Section 1.7. The *base* is assumed to be a positive real number so that no matter whether the exponent is positive or negative, the expression b^x is defined.

PROPERTIES OF EXPONENTS

For positive real numbers b and c and any real numbers x and z:

$E1 \quad b^x \cdot b^z = b^{x+z}$ $\qquad$ $E2 \quad b^0 = 1$

$E3 \quad b^{-x} = \dfrac{1}{b^x}$ $\qquad$ $E4 \quad (b \cdot c)^z = b^z \cdot c^z$

$E5 \quad \dfrac{b^x}{b^z} = b^{x-z}$ $\qquad$ $E6 \quad (b^z)^x = b^{z \cdot x}$

$E7 \quad \left(\dfrac{b}{c}\right)^x = \dfrac{b^x}{c^x}$

The following example illustrates these rules by asking us to perform computations and compare results.

EXAMPLE 1 Exponent notation

Verify each of the following.
(a) $2^3 \cdot 2^5 = 2^8$ $\qquad$ (b) $3^4/3 = 3^3$ $\qquad$ (c) $10^3 = 2^3 \cdot 5^3$

SOLUTION

(a) By $E1$, we have $2^3 \cdot 2^5 = 2^{3+5} = 2^8$. We check that $2^3 = 8$, $2^5 = 32$, $2^8 = 256$, and indeed $8 \cdot 32 = 256$.

(b) By $E5$, we know $3^4/3 = 3^{4-1} = 3^3$. By direct computation, $3^4 = 81$ and $81/3 = 27$, which is also 3^3.

(c) $10^3 = 1000$ and $10 = 2 \cdot 5$. By $E4$, $(2 \cdot 5)^3 = 2^3 \cdot 5^3$. Direct computation gives $2^3 = 8$, $5^3 = 125$, and $8 \cdot 125 = 1000$. $\quad$ •

PRACTICE EXERCISE 1

Evaluate each of the following by using the rules $E1-E7$ and comparing the results obtained through direct computation when possible.
(i) 3^0 $\qquad$ (ii) $2^{-3} \, 2^4$ $\qquad$ (iii) $2^3 5^3$ $\qquad$ (iv) $(3 \cdot 5^2)^3$ $\qquad$ (v) $(3a^2)^3$
(vi) $(4 \cdot 2^{-3})^2$

If an expression involves the nth root of a number, it often is best to express the *radical* or *root form* of the number, such as $\sqrt[n]{b}$, in its *exponential form,* as $b^{1/n}$.

EXAMPLE 2 Expressing a root in exponential form

Rewrite $\sqrt{b}$, $\sqrt[5]{b}$ and $\sqrt[x]{b^3}$ in exponential form.

SOLUTION

$\sqrt{b} = b^{1/2}$ and $\sqrt[5]{b} = b^{1/5}$. Finally, combining roots and exponents, $\sqrt[x]{b^3} = (b^3)^{1/x} = b^{3/x}$.

Interest: Simple, Compound, and Continuous

To illustrate exponential expressions and motivate their use in functions, let us calculate interest paid on deposits of money.

Interest paid at the end of an interval of time is said to be *simple interest.* An example would be any bond or account that pays interest in one lump sum at the end of the time interval of the deposit. This interest is not redeposited into the account. An account pays *compound interest* when the earned interest is redeposited into the account and then itself earns interest at the rate of the original investment.

EXAMPLE 3 Simple interest

Suppose we bought a 12% bond for $1000. What is the annual payment if the interest is annual simple interest?

SOLUTION

The *simple interest* payment is $120 each year because 12% of $1000 is $0.12 \cdot 1000 = 120$.

We now consider compound interest over uniform intervals of time. These can be years, months, or any other interval. The increase in value is illustrated in figure 4.1.1, wherein the shaded regions indicate the interest added into the value that existed at the beginning of that interval. As we have more and more intervals of time, we find the amount of money continually growing.

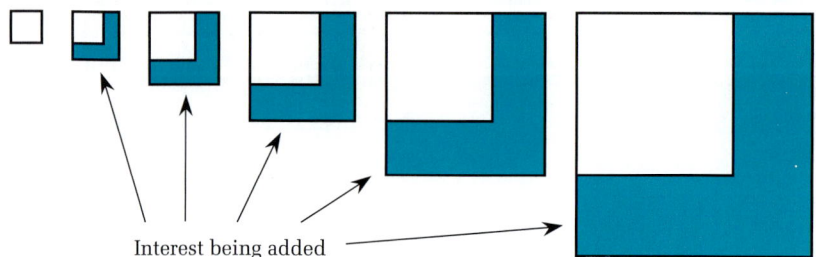

Interest being added

FIGURE 4.1.1

Suppose we begin with P dollars and are paid compound interest at the annual rate of r percent for x years. If each year is divided into n intervals, then there are a total of nx intervals.

First, write the annual interest rate in decimal form as i. For example, if $r = 12\%$, use $i = 0.12$. Because the interest is being paid at n regular intervals during the year, at the end of each interval, interest is paid at the rate of i/n.

The interest on P dollars at the end of an interval is $(i/n)P$. Hence, beginning an interval with P dollars, at the end of the interval there are $P + (i/n)P$ dollars. Factoring a P, this amount can be written as $P[1 + (i/n)]$. Over a second interval, we begin with $\{P[1 + (i/n)]\}$ dollars and we add $(i/n)\{P[1 + (i/n)]\}$ dollars to give us a total of

$$(*) \qquad \left\{ P\left(1 + \frac{i}{n}\right) \right\} + \left(\frac{i}{n}\right)\left\{ P\left(1 + \frac{i}{n}\right) \right\}$$

The expression $(*)$ can be factored and rewritten as

$$\left\{ P\left(1 + \frac{i}{n}\right) \right\}\left(1 + \frac{i}{n}\right) = P\left(1 + \frac{i}{n}\right)^2$$

Because for each interval the amount at the end of the interval is the product of the amount at the start of the interval multiplied by $[1 + (i/n)]$, we have the following formula.

INTERVAL COMPOUNDING OF INTEREST

If P dollars are invested at an annual interest rate of r percent and the interest is compounded n times a year, then the amount after x years is

$$(4.1.1) \qquad V_n(x) = P\left(1 + \frac{i}{n}\right)^{nx}, \qquad \text{where } i = \frac{r}{100}.$$

EXAMPLE 4   Interval compounding of interest

Suppose $1000 is placed in a retirement account that pays 12% compound interest. Find the value of the account at the end of 25 years if the interval of compounding is (a) annual, (b) semiannual, or (c) monthly.

SOLUTION

Use $V_n(x) = P(1 + i/n)^{nx}$ with $i = 0.12$ and $x = 25$.

(a) For $n = 1$, $V_1(25) = 1000(1 + 0.12/1)^{1 \cdot 25} = 1000(1.12)^{25} = 17,000.06$.

(b) For $n = 2$, $V_2(25) = 1000(1 + 0.12/2)^{2 \cdot 25} = 1000(1 + 0.06)^{50} = 18{,}420.15$.

(c) For $n = 12$, $V_{12}(25) = 1000(1 + 0.12/12)^{12 \cdot 25} = 1000(1 + 0.01)^{300} = 19{,}788.47$.

PRACTICE EXERCISE 2

Find the value of an account that compounds monthly. Assume $1000 is invested, the annual rate is 9%, and the money is left for 10 years.

As the intervals of time become shorter, there is a gain in the final amount due to the more frequent compounding. Using daily compounding for the account in Example 4, $V_{365}(25) = \$20{,}075.63$. Is there any limit to the growth of the account? In symbols, is there a limit to $1000[1 + (0.12/n)]^{25n}$? The general question is, does $\lim_{n\to\infty} P(1 + i/n)^{nx}$ exist? The answer is yes.

It can be shown that $\lim_{n\to\infty} (1 + i/n)^n = e^i$, where the symbol "$e$" stands for a particular constant number, just as "π" stands for the constant ratio of the circumference of a circle to the diameter of that circle. The evaluation of e^x is usually done with the $\boxed{e^x}$ key on our calculator. The first few digits in the decimal expansion of e are 2.7182.

Using the fact that $\lim_{n\to\infty} (1 + i/n)^n = e^i$ and the properties of limits, we have

$$\lim_{n\to\infty} P\left(1 + \frac{i}{n}\right)^{nx} = \lim_{n\to\infty} P\left(\left(1 + \frac{i}{n}\right)^n\right)^x$$

$$= P\left(\lim_{n\to\infty}\left(1 + \frac{i}{n}\right)^n\right)^x = P(e^i)^x = Pe^{ix}$$

In summary, we have $\lim_{n\to\infty} V_n(x) = Pe^{ix}$.

The result of letting the number of time intervals for compounding increase beyond any bound (and so the corresponding length of each time interval decreases toward zero) is said to be *continuous compounding*.

CONTINUOUS COMPOUNDING OF INTEREST

If P dollars are invested and interest is compounded continuously, then after x years the value is

(4.1.2) $V(x) = P \cdot e^{ix}$,   where i is the interest rate, expressed decimally.

WARNING

We need to be very careful when we encounter or use any formula involving interest rates, so that we understand which form is being used. We use i to stand for the decimal form of interest and r to stand for the percentage form. Thus, $i = r/100$.

EXAMPLE 5   Continuous compounding

Find the value after 25 years of $1000, invested at 12%, with continuous compounding.

SOLUTION

In $V(x) = P \cdot e^{ix}$ we have $P = 1000$, $i = 0.12$, and $t = 25$.

$V(25) = 1000 \cdot e^{(0.12)25} = 1000 \cdot e^3 = 1000 \cdot 20.08554 = \$20,085.54.$

Comparing $V_{365}(25) = \$20,075.63$, the result of daily compounding, with $V(25) = \$20,085.54$, we see there is only $9.91 difference.

The graphs of $V_1(x)$, $V_2(x)$, $V_4(x)$, and $V(x)$ are shown in figure 4.1.2. In addition, the straight line toward the bottom of the figure shows what little money simple interest would provide over the same period. To emphasize the difference between these functions, figure 4.1.3 shows an enlargement of the region near the end of the time period.

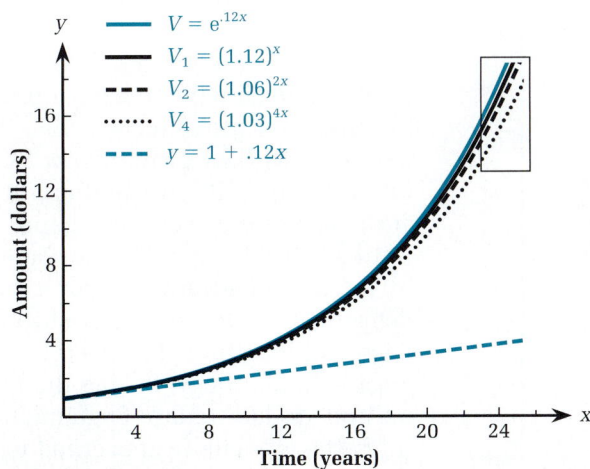

FIGURE 4.1.2

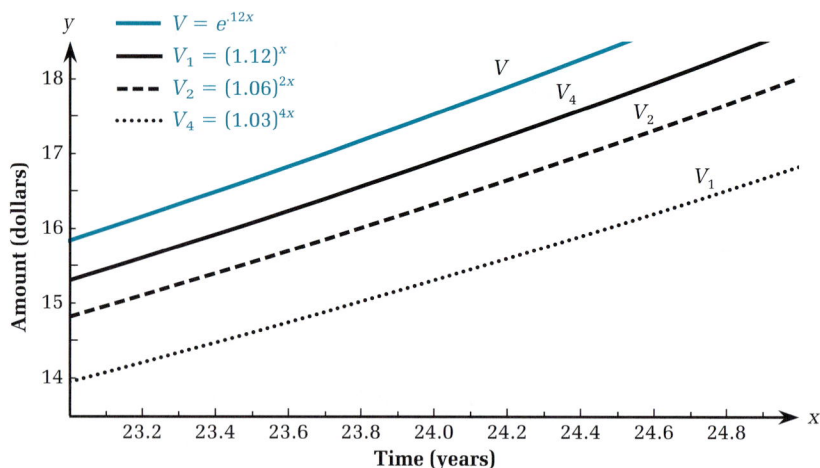

FIGURE 4.1.3

PRACTICE EXERCISE 3

Suppose we invest $2000 in an Individual Retirement Account (IRA) advertising "continuous compounding" at 8%. What will our account be worth after five years?

MATHEMATICAL INNOVATORS
Leonhard Euler (1707–1783)

The first use of the symbol "e" was by Leonhard Euler, one of the greatest and certainly the most prolific of all mathematicians. Euler (the name is pronounced "oiler") was born near Basel, Switzerland, but worked primarily in St. Petersburg and Berlin. He was interested in both pure and applied mathematics, winning a prize at the age of nineteen for a paper on the arrangement of masts on sailing ships. Later in life, he shared in an award of £20,000 offered by the British government for a method enabling a ship captain to determine the location of a ship at sea to within one-half of one degree of longitude. In the eighteenth century, £20,000 was a very great sum of money, having the buying power of more than a half million dollars today. England, like any country with shipping interests, was vitally concerned with the location of sea routes and the navies of both friendly and competing countries.

Exponential Functions and Their Graphs

The graphs in figure 4.1.3 were examples of exponential functions.

DEFINITION

Exponential function

An *exponential function* is of the form $f(x) = a \cdot b^x$ for $b > 0$, $b \neq 1$, and $a \neq 0$.

The value of b is the *base.* The domain of $f(x)$ is the set of all real numbers.

In the definition of exponential function we insist that $b > 0$ so that b^x exists for any value of x, positive or negative. Furthermore, we require $b \neq 1$ because otherwise $1^x = 1$ for all x-values, and we would have $f(x) = a$, a function we call a constant function.

An exponential function, such as $f(x) = 2^x$, is defined for all real numbers x. However, if we plot by hand usually we evaluate 2^x for only a few values of x, plot the resulting points and then connect those points in a smooth, continuous curve. Generally, integer values are used for x. This is done only because 2^x is most easily evaluated at integers, particularly if we do not use a calculator. Nonetheless, if x is equal to a non-integer number, for example, $\sqrt{3}$, the value $2^{\sqrt{3}}$ does exist and is an exact real number. Using a calculator, an approximation is $2^{\sqrt{3}} \simeq 3.322$.

If we have located some points on the graph and think too great a "gap" appears between those points, then we may locate additional points to provide a clearer graph. There is no general rule as to how many points are sufficient.

EXAMPLE 6   The graph of an exponential function

Graph $y = f(x) = 2^x$ for $-3 \leq x \leq 3$.

SOLUTION

x	-3	-2	-1	0	1	2	3
2^x	$\frac{1}{8}$	$\frac{1}{4}$	$\frac{1}{2}$	1	2	4	8

The table at left gives $y = 2^x$ for some values of x.

We want to connect these points by a smooth, continuous curve. However, we have quite a gap between $(2, 4)$ and $(3, 8)$. To locate a point in this region, find $f(2.5) = 2^{2.5} \simeq 5.66$. Plotting the point $(2.5, 5.66)$ helps us correctly sketch the graph, shown in figure 4.1.4.

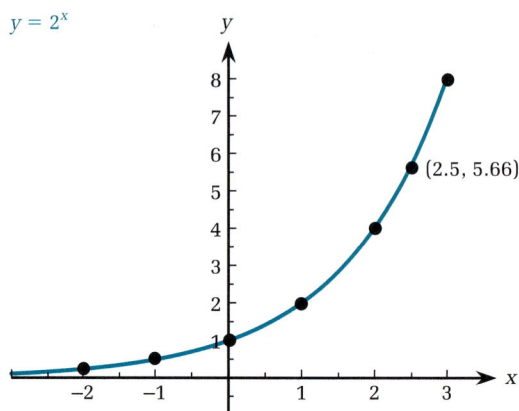

$y = 2^x$

FIGURE 4.1.4

Notice in Example 6, the approximation of 5.66 is used for $2^{2.5}$, rather than using our usual four-place approximation of $2^{2.5} \simeq 5.6569$. This is done because when plotting a point, it is unlikely we can place the point with four-place accuracy. As always, the context in which a problem has arisen will determine the necessary precision of the graph.

PRACTICE EXERCISE 4

Sketch a graph of $f(x) = 3^x$.

Both Example 6 and Practice Exercise 4 used bases that were greater than one. Those functions increase as do their x-values. The next example shows what happens when the base is less than one.

EXAMPLE 7   The graph of an exponential function with base less than one

Sketch a graph of $f(x) = (2/3)^x$ using two-place accuracy and $-3 \leq x \leq 4$.

SOLUTION

A table of values for x and $(2/3)^x$ is given below. The value of -2.5 for x is chosen so as to avoid having a wide gap in the graph.

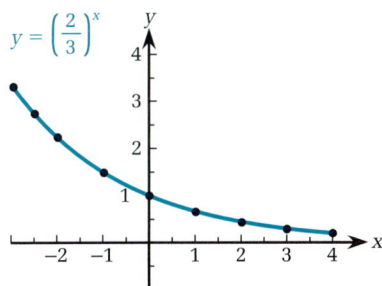

$y = \left(\frac{2}{3}\right)^x$

FIGURE 4.1.5

x	-3	-2.5	-2	-1	0	1	2	3	4
$\left(\frac{2}{3}\right)^x$	$\frac{27}{8} \approx 3.37$	2.76	2.25	1.5	1	0.67	0.44	0.30	0.20

A sketch of the graph of $f(x)$ is shown in figure 4.1.5.

PRACTICE EXERCISE 5

Sketch a graph of the curve $f(x) = (1/2)^x$ for $-3 \le x \le 3$.

The graphs thus far have been merely screens, showing only finite intervals for values of x and y. However, the exponential function $f(x) = b^x$ has as its domain the set of all real numbers. As a result, the graph of $y = b^x$ is unbounded, for x goes beyond any positive or negative bound. What happens to y and how the graph appears is described next.

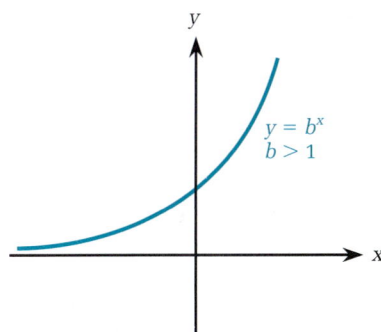

$y = b^x$
$b > 1$

FIGURE 4.1.6

GRAPHS OF EXPONENTIAL FUNCTIONS $f(x) = b^x$

For an exponential function $f(x) = b^x$, $(b > 0, b \ne 1)$,

$$f(0) = b^0 = 1 \text{ and } b^x > 0 \text{ for all } x.$$

I. If $b > 1$, then $\lim\limits_{x \to -\infty} b^x = 0$. The function is always increasing as x increases, and the function becomes unboundedly large for extremely large values of x.

II. If $0 < b < 1$, then the function is unboundedly large as x goes to negative infinity. The function is always decreasing as x increases, and $\lim\limits_{x \to \infty} b^x = 0$.

The graphs of such functions are shown in figures 4.1.6 and 4.1.7.

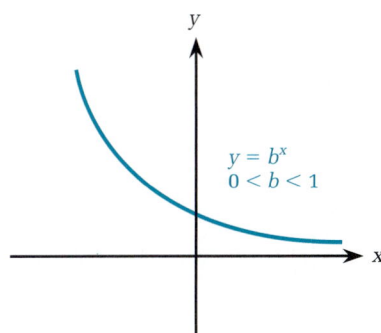

$y = b^x$
$0 < b < 1$

FIGURE 4.1.7

COMPARING THE GRAPHS OF $y = b^x$ FOR DIFFERENT VALUES OF b

Suppose b_1 and b_2 are bases of exponential functions and $b_1 < b_2$.

If $x < 0$, then $b_1{}^x > b_2{}^x$.

If $x > 0$, then $b_1{}^x < b_2{}^x$.

The graphs of $y = 2^x$, $y = 3^x$, and $y = 4^x$ are shown in figure 4.1.8. The graph of 2^x is above the graph of 3^x for $x < 0$. They are equal at $x = 0$. For $x > 0$, the graph of 2^x is below the graph of 3^x.

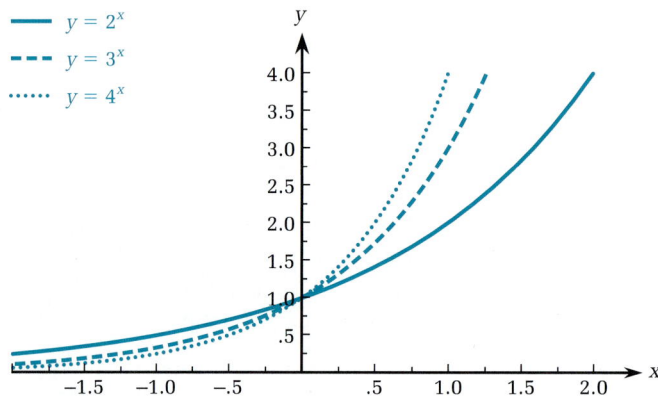

FIGURE 4.1.8

▨ Graphing Calculators

To compare the graphs of $y = b^x$ for various values of b, either do not clear the screen after each graph or use the process to graph several functions at once. Figure 4.1.9 shows the graphs for $y = (1/2)^x$, $y = (1/3)^x$, and $y = (1/4)^x$.

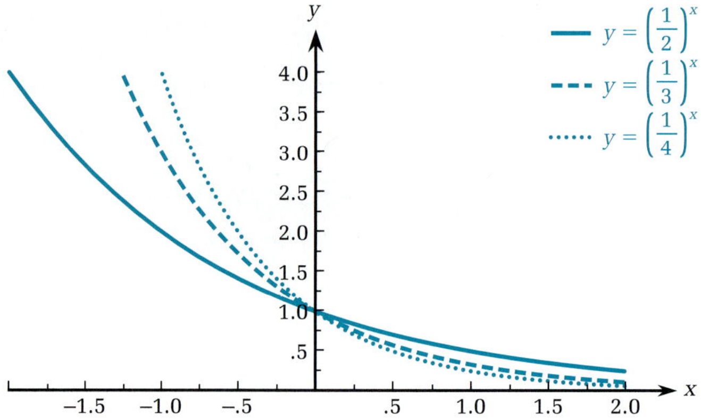

FIGURE 4.1.9

Suppose the graphs are produced consecutively. We then can use $\boxed{\text{Trace}}$ and $\boxed{\triangleright}$ to be on one of the graphs. From there we can jump vertically from graph to graph by using the $\boxed{\triangle}$ and $\boxed{\triangledown}$ keys. If we have three graphs shown, such as in figure 4.1.9, the sequence of graphs visited may depend on whether $x > 0$ or $x < 0$.

A function may be created by the algebra of functions from an exponential function. Such functions are illustrated by the next example.

EXAMPLE 8   Graphs of functions created with exponential functions

On one set of axes, graph the functions $y = 2^x$, $f(x) = 3(2^x)$, and $r(x) = (-1/3)(2^x)$.

SOLUTION

To graph these, we use our knowledge of the more basic exponential component of the graph. Then we plot points that are computationally easy to evaluate and attempt to connect these together. The graph of $y = 2^x$ is shown above as part of figure 4.1.8. Multiplying each y-value by 3 gives the graph for $f(x) = 3(2^x)$, shown in figure 4.1.10. Also shown in that figure is $r(x)$, obtained from the graph of $y = 2^x$ by multiplying by $-(1/3)$.

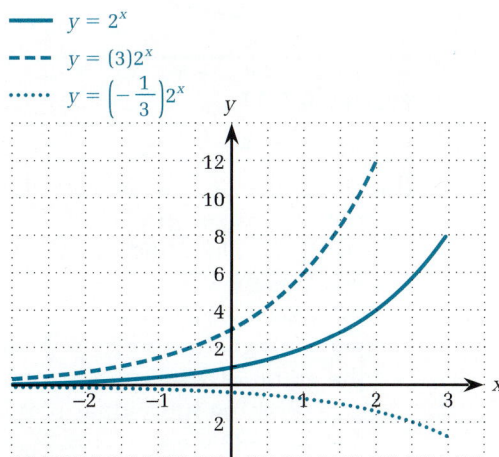

FIGURE 4.1.10

EXAMPLE 9   Graphing

Sketch graphs of (a) $s(x) = 2^{x+1}$ and (b) $g(x) = 3^{2x-1}$.

SOLUTION

(a) For $s(x) = 2^{x+1}$, there are two approaches that we can use. First, we can use Laws of Exponents to write $2^{x+1} = 2^x \cdot 2^1 = 2(2^x)$. Then we could multiply the y-values of 2^x by 2.

Second, we may begin by plotting some points. For $x = 2$, we have $s(2) = 2^{2+1} = 2^3 = 8$. For $x = -2$, we have $s(-2) = 2^{-2+1} = 2^{-1} = 0.5$.

We might go on plotting many points, but we may notice that the graph of 2^{x+1} has the same shape as the graph of 2^x, except it is shifted to the left. That is, when $x + 1 = t$ for some value t, then $x = t - 1$. Thus, the value of x that produces a particular power of 2, such as 2^4, can be found by letting $x + 1$ be that power and solving for x. In the case of 2^4, this would give $x = 3$.

The graph is shown in figure 4.1.11.

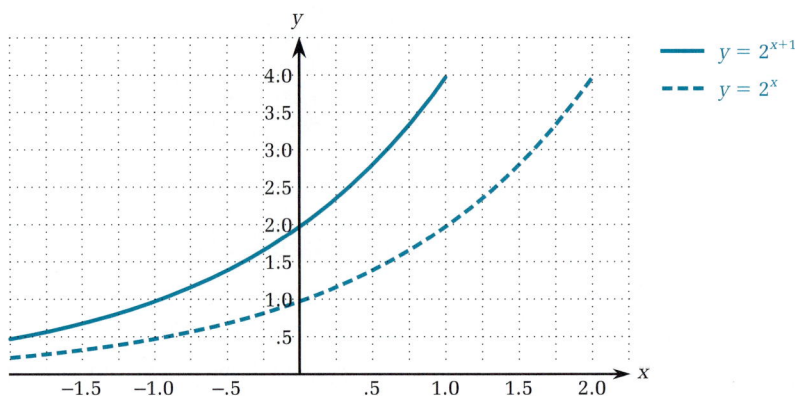

FIGURE 4.1.11

(b) For $g(x) = 3^{2x-1}$, we again have two approaches. First, we may rewrite $g(x)$ as $g(x) = 3^{2x-1} = 3^{2x}3^{-1} = (3^2)^x \cdot (1/3) = (1/3)(9^x)$. Then find the graph of $y = 9^x$ and multiply the y-values by 1/3.

Second, again we can solve for those values of x that give particular values of $2x - 1$ and by comparing these values to the graph of $y = 3^x$ we can sketch the graph of $g(x)$.

Plotting individual points would give $g(0) = 3^{2 \cdot 0 - 1} = 3^{-1} \approx 0.3333$ and $g(1) = 3^{2 \cdot 1 - 1} = 3^1 = 3$. If we want to find that x for which $3^{2x-1} = 3^0$, then $2x - 1 = 0$. Solving for x gives $x = 1/2$. Hence, $g(1/2) = 3^{2 \cdot 1/2 - 1} = 3^{1-1} = 3^0 = 1$. See figure 4.1.12.

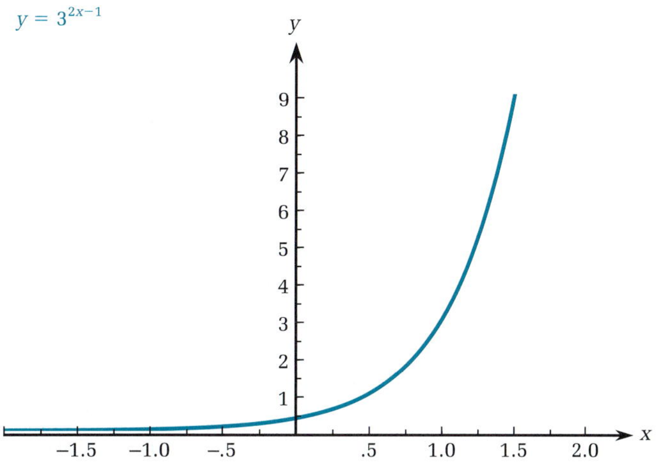

$y = 3^{2x-1}$

FIGURE 4.1.12

The base e is widely used as a base for exponential functions and is called the natural base.

EXAMPLE 10   The graph of $y = e^x$, the natural base

The graph of $y = e^x$ is shown in figure 4.1.13. The values were calculated using the key labeled $\boxed{e^x}$ on a calculator and rounding to four places.

x	-1.5	-1	-0.5	0	0.5	1	1.5
e^x	0.2231	0.3679	0.6065	1	1.6487	2.7183	4.4817

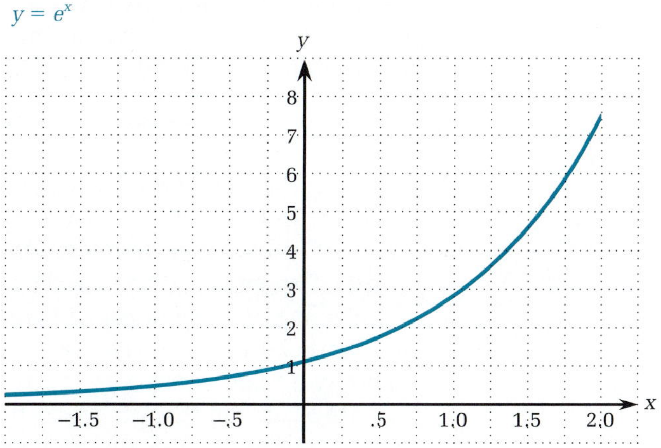

$y = e^x$

FIGURE 4.1.13

Frankly, at this stage if we are asked to graph any but the simplest of exponential functions, we should use a graphing calculator. To determine the ranges for the screens and to find precise values of the functions, we use calculus, which is presented in Section 4.2.

Applications

Human and other populations of regions are often modeled by considering them to be growing (or decreasing) in a continuous manner and doing so at a rate of increase (or decrease) that is constant. With these assumptions, the size of the population is estimated by using the same formula (4.1.2) developed for continuous compounding of interest. The initial population is used in place of an initial amount of money and the rate of population increase replaces the rate of interest.

EXAMPLE 11   Population growth

The population of France in 1990 was about 56 million. The annual growth rate was 0.4%. Assuming that growth rate continues, estimate the population of France in 2025.

SOLUTION

First, $P = 56$, the initial population. To find i, we convert 0.4% to 0.004 for the decimalized rate of growth. The time period is from 1990 to 2025, giving $t = 35$ years of compounding growth.

An estimate for the population given by (4.1.2) is $56 \cdot e^{0.004 \cdot 35} = 56 \cdot e^{0.14} = 56 \cdot 1.1503 = 64.4$ million.   ●

PRACTICE EXERCISE 6

The population of Kenya was about 25 million in 1990. The annual growth rate was 4.2%. Assuming that growth rate continues, estimate the population of Kenya in 2025.

Populations and other changing values may decrease as well as increase. In the case of a decrease, we use a negative rate of growth.

Let us consider depreciation in the financial sense of finding the worth of an object that is wearing out or becoming less valuable in some sense over a period of time. Accountants use several formulas for depreciation. We saw "straight line" depreciation in Chapter 1, whereby a fixed amount is subtracted during each time interval. Here we introduce another standard form.

DEFINITION

Double declining balance depreciation

Suppose the initial value of an item is C and it is to be depreciated by *double declining balance* over a period of N intervals of time. During each interval of time, the fraction of its value subtracted is $2/N$. After x intervals of time, the value of the item is

$$V_{ddb} = C[1 - (2/N)]^x$$

EXAMPLE 12   Double declining balance vs. straight line depreciation

Suppose Bill's Bakery buys a new oven for $5000 and wants to depreciate it over 10 years. Using annual time intervals, find the equations for both double declining balance and straight line depreciation. Graph both equations and estimate the time at which the graphs cross.

SOLUTION

For $N = 10$ and $C = 5000$, by using straight line we subtract 5000/10 dollars a year. Hence $f(x) = 5000 - 500x$ is the equation for straight line depreciation. The formula for double declining balance is $V_{ddb} = 5000 [1- (2/10)]^x = 5000(0.8)^x$. These functions are graphed in figure 4.1.14. It appears the graphs cross at about 8.5 years.

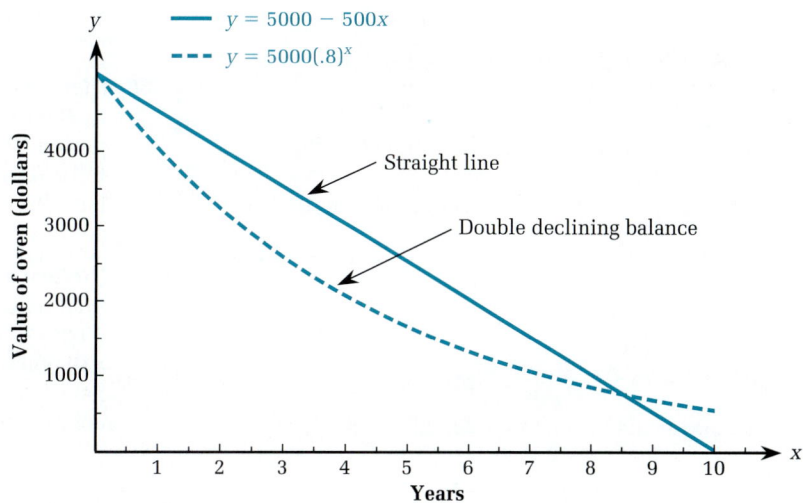

FIGURE 4.1.14

4.1 PROBLEMS

Foundations

The problems of this section require the basic skills illustrated by the following:

Evaluate the expressions in problems 1–5.

1. 2^5 **2.** 2^{-3} **3.** $(1/3)^2$

4. $(-2/3)^3$ **5.** $(2)^{1.3}$

6. Which of the bases in problems 1–5 would not be used as the base of an exponential function? Why?

By using the $\boxed{e^x}$ key on a calculator, evaluate problems 7–10.

7. $e^{1.43}$ **8.** $(2e)^{1.43}$

9. e^{-3} **10.** $(1.5e)^{-3}$

11. What is the decimal form of the following interest rates?
(a) 6%   (b) 0.5%   (c) 10%   (d) -3%

Exercises

In problems 12–16, find the value of an account that was opened with A dollars, paid annual interest at the rate of r percent compounded n times a year, and held for t years.

12. $4000 at 6%, semiannually $(n = 2)$, for 5 years

13. $2500 at 8.2%, monthly $(n = 12)$, for 12 years

14. $7500 at 3.2%, quarterly $(n = 4)$ for 10 years

15. $25,000 at 6.5%, daily $(n = 365)$ for 2 years

Sketch graphs of the functions in problems 16–33.

16. $f(x) = 4^x$ **17.** $f(x) = 5^x$

18. $f(x) = \left(\dfrac{1}{5}\right)^x$ **19.** $f(x) = \left(\dfrac{1}{4}\right)^x$

20. $f(x) = (0.2)^x$ **21.** $f(x) = (0.3)^x$

22. $g(x) = 5(4^x)$ **23.** $g(x) = 4(5^x)$

24. $f(x) = (-2)\left(\dfrac{1}{5}\right)^x$ **25.** $f(x) = (-3)\left(\dfrac{1}{4}\right)^x$

26. $f(x) = 5^{-x}$ **27.** $f(x) = \left(\dfrac{1}{2}\right)^{-x}$

28. $f(x) = e^{-x}$ **29.** $f(x) = \left(\dfrac{1}{e}\right)^x$

30. $g(x) = 2e^x$ **31.** $g(x) = -e^x$

32. $f(x) = 2^{x-1}$ **33.** $g(x) = 3^{x+1}$

34. *(Continuous Compounding)* Suppose $2000 is invested at 10%, compounded continuously. Find the value of the account at the end of five years.

35. *(Continuous Compounding)* Suppose $5000 is invested at 8%, compounded continuously. Find the value of the account at the end of 10 years.

36. *(Continuous Compounding)* The conventional tale is told that the Manhattan Indians traded the island to the Dutch in 1626 for $24 worth of kettles, axes, and cloth.
(a) Show that $24, invested at 5% and compounded continuously from 1626 to 1996, would increase to about 2.6 billion dollars.
(b) Find the value in 1996 if the same $24 had been invested at 5.5% and compounded continuously.

37. *(Return on Common Stock)* The mean annual return on common stock in the period 1926–1984 was 11.7%. Find what a portfolio of $10,000 worth of stock in 1926 would have been worth in 1984 using continuous compounding.

38. *(Return on Corporate Bonds)* The mean annual return on long-term corporate bonds in the period 1926–1984 was 4.7%. Find what a portfolio of $10,000 worth of bonds in 1926 would have been worth in 1984 using continuous compounding.

39. *(Return on U.S. Treasury Bills)* The mean annual return on U.S. Treasury bills in the period 1926–1984 was 3.4%. Find what a portfolio of $10,000 worth of Treasury bills in 1926 would have been worth in 1984 using continuous compounding.

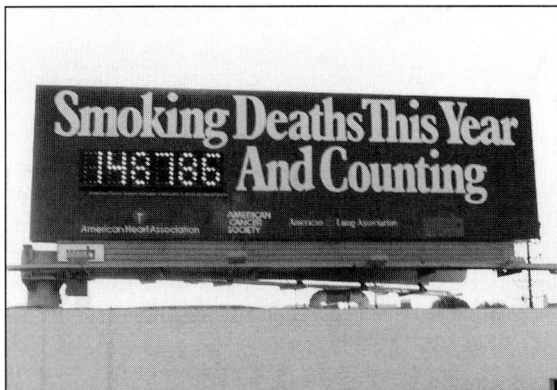

40. (Population) The population of the United States was about 250 million in 1990. The annual growth rate was about 0.86%.
(a) Assuming that the growth rate continues at 0.86%, estimate the population of the United States in 2025.
(b) Assuming a growth rate of 2.4%, estimate the population of the United States in 2025.

41. (Population) The population of Mexico was about 90 million in 1990. The annual growth rate was about 2.4%.
(a) Assuming that the growth rate continues at 2.4%, estimate the population of Mexico in 2025.
(b) Assuming a growth rate of 0.86%, estimate the population of Mexico in 2025.

42. (Biology, Exponential Growth) Common duckweed (*Lemna gibba*) has a daily growth rate of 30% per day when the population of a pond begins to grow. If a pond starts with two pounds of fronds, how many pounds of fronds will be there after seven days?

43. (Biochemical Manufacturing, Exponential Growth) Genetic and other biochemical manufacturers often use a chemostat. In this, a nutrient solution is constantly being added to a vat in which an organism is growing, and the overflow mixture is being harvested in some manner. Using optical density or another means, an optimal growth/harvest rate is maintained. A common organism used is *Escherichia coli* because it is easy to work with and grows very rapidly.

For the sake of simplicity, let us assume that there is enough nutrient present to sustain exponential growth for one hour, even without adding any new solution. *Escherichia coli* has a growth rate of about 3.5% per minute. Starting with one kilogram, how much would be present after one hour?

44. (Biology, Tuberculosis Research) Tuberculosis is one of the world's leading causes of death from infectious disease. However, medical researchers have not worked extensively with the tuberculosis bacterium. One reason is that this bacterium is slow growing. Bacterial geneticists usually work with bacterial types for which a single cell grows into 10 million cells in eight hours. However, the tuberculosis bacterium requires several weeks for such growth.
(a) Assuming each bacterium splits into two at the end of each generation, how many generations occur in growing one cell into 10 million?
(b) How long is one generation if one cell becomes 10 million by the end of three weeks?

45. (Double Declining Balance Depreciation) Suppose an $8000 car is depreciated over 10 years using double declining balance. Find $V_{ddb}(7)$.

46. (Double Declining Balance Depreciation) Cursor Computers wants to use double declining balance depreciation for a $4000 computer over five years. What is the computer's value at the end of four years?

Graphing Calculator Problems

47. Use $\boxed{\text{Trace}}$ to estimate the solution(s) to $3^x = 3 + x$.

48. Graph $E(x) = [1 + (1/x)]^x$ for each of the following ranges.
(a) $0 \le x \le 10$, $0 \le y \le 4$
(b) $0 \le x \le 100$, $0 \le y \le 4$

49. For $E(x)$ of problem 48, use $\boxed{\text{Trace}}$ to estimate the limit as x goes to infinity.

50. Graph $F(x) = (1 + x)^{1/x}$ for the range $0 \le x \le 1$, $0 \le y \le 4$. Use $\boxed{\text{Box}}$ and $\boxed{\text{Trace}}$ to estimate $\lim_{x \to 0} F(x)$.

Writing and Discussion Problems

51. Why is 1 not used as a base for exponential functions?

52. What is the difference between double declining balance and straight line depreciation? What advantages and disadvantages does each have?

53. Each person has two direct ancestors in each generation. We each have one biological mother and father. Thus, if we go back n generations, we have 2^n direct ancestors in that generation (four grandparents, $2^2 = 4$; eight great-grandparents, $2^3 = 8$; and so on). Assuming all your ancestors are distinct, find the number of your ancestors 10 generations ago (about the time of the founding of the United States) and the number of your ancestors 90 generations ago (about the time Caesar and Cleopatra were sailing on the Nile). Discuss the likelihood that 90 generations ago your direct ancestors were all distinct.

Enrichment Problems

Graph the functions in problems 54 and 55.

54. $f(x) = -2 + e^x$ **55.** $f(x) = 3 + e^{-x}$

56. Discuss the relationship between $E(x) = (1 + 1/x)^x$ and $F(t) = (1 + t)^{1/t}$. Which is easier to find using a calculator, $\lim_{t \to 0} (1 + t)^{1/t}$ or $\lim_{x \to \infty} (1 + 1/x)^x$?

Effective Annual Rate of Interest

Problems 57 and 58 require the following background information.

In 1968 Congress passed the "Truth in Lending Act," which required each financial institution to state the "effective annual rate of interest." This is the interest percentage that would produce returns generated by the nominal interest, compounded or not in whatever manner the institution desires, but done so in a single annual payment ("simple annual interest").

For example, 6% compounded continuously for one year will turn a deposit of $1 into the amount of $1 \cdot e^{0.06} = 1.061836547$, so the "effective annual rate of interest" would be 6.184%. If the 6% were compounded quarterly, then the one dollar would be worth $1 \cdot (1 + 0.015)^4 = 1.061363551$, so the "effective annual rate of interest" would be 6.136%.

57. *(Effective Annual Rate of Interest)*   Find the effective annual rate of interest if interest is 8% compounded as follows.
(a) annually   (b) quarterly   (c) monthly
(d) continuously

58. *(Effective Annual Rate of Interest)*   Find the effective annual rate of interest if interest is 5% compounded as follows.
(a) semiannually   (b) monthly
(c) daily (use 365-day year)
(d) continuously

SOLUTIONS TO PRACTICE EXERCISES

1. (i) $3^0 = 1$
   (ii) $2^{-3} 2^4 = 2^{-3+4} = 2^{+1} = 2$, and $2^{-3} = 1/8$, $2^4 = 16$ and $(1/8) \cdot 16 = 2$.
   (iii) $2^3 5^3 = (2 \cdot 5)^3 = 10^3 = 1000$, and indeed $2^3 = 8$, $5^3 = 125$, and $8 \cdot 125 = 1000$.
   (iv) $(3 \cdot 5^2)^3 = 3^3 \cdot (5^2)^3 = 3^3 \cdot 5^{2 \cdot 3} = 3^3 5^6 = 27 \cdot 15{,}625 = 421{,}875$, and $3 \cdot 5^2 = 3 \cdot 25 = 75$, with $(75)^3 = 421{,}875$.
   (v) $(3a^2)^3 = 3^3(a^2)^3 = 3^3 \cdot a^{2 \cdot 3} = 3^3 a^6 = 27a^6$
   (vi) $(4 \cdot 2^{-3})^2 = 4^2 \cdot (2^{-3})^2 = (2^2)^2 \cdot 2^{-6} = 2^4 \cdot 2^{-6} = 2^{4-6} = 2^{-2} = 1/2^2 = 1/4$, which can also be found as $(4 \cdot 2^{-3})^2 = (2^2 \cdot 2^{-3})^2 = 2^{-1 \cdot 2} = 2^{-2} = 1/4$.

2. Because the 9% has to be divided up among 12 months, the amount of interest per month is 9%/12 = 0.75%, which is 0.0075 when expressed decimally.

By the end of 10 years, which is 120 months, she would have $1000 \cdot (1 + 0.0075)^{120} = 1000 \cdot (1.0075)^{120} = 1000 \cdot 2.45136 = \$2{,}451.36$.

3. After five years, our account will be worth $V(5) = 2000 \cdot e^{0.08 \cdot 5} = 2000 \cdot e^{0.4} = 2000 \cdot 1.491824698 = \$2{,}983.65$.

4. $f(x) = 3^x$. See figure 4.1.15.

x	-3	-2	-1	0	1	2	3
3^x	$\frac{1}{27}$	$\frac{1}{9}$	$\frac{1}{3}$	1	3	9	27

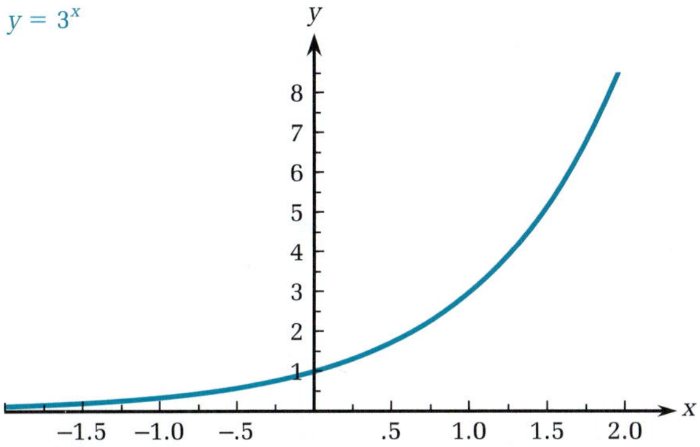

$y = 3^x$

FIGURE 4.1.15

5. $f(x) = (1/2)^x$. See figure 4.1.16.

x	-3	-2	-1	0	1	2	3
$(1/2)^x$	8	4	2	1	$\frac{1}{2}$	$\frac{1}{4}$	$\frac{1}{8}$

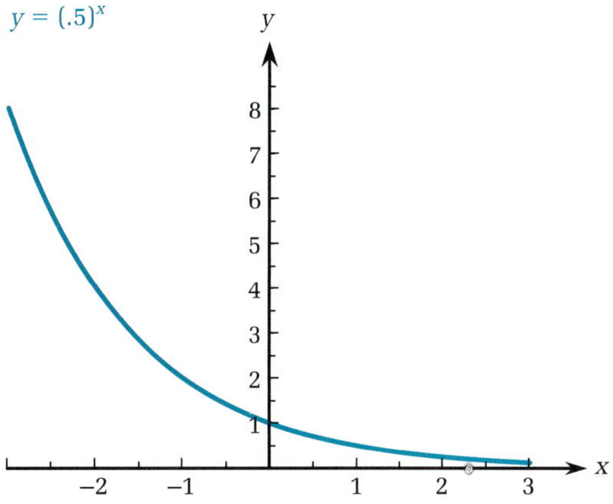

$y = (.5)^x$

FIGURE 4.1.16

6. Starting with a population of 25 million in 1990 and with an annual growth rate of 4.2%, assuming that growth rate continues, an estimation for the population of Kenya in 2025 is $25 \cdot e^{0.042 \cdot 35} = 25 \cdot e^{1.47} = 25 \cdot 4.349 = 108.7$ million.

4.2 DERIVATIVES OF EXPONENTIAL FUNCTIONS

Digital display billboards are fascinating. Flashing numbers continually change. At one Florida theme park, visitors watch the Department of Commerce's ongoing estimation of the U.S. population. Several cities have billboards flashing the number of deaths due to lung cancer. A long-distance company shows their minute-by-minute estimation of savings by customers using its system. Watching such a display we may ask ourselves, "At what rate are these numbers changing?" For instance, "At what rate is the population increasing?"

Discussions about rates of change concern derivatives. In this section we find the derivative of e^x, and more generally, we find the derivative of $e^{f(x)}$ whenever $f(x)$ is itself a differentiable function.

The Derivative of $f(x) = b^x$ as a Limit

We seek the derivative of $f(x) = b^x$ for $b > 0$. Recall that for any function, $f(x)$, the derivative of $f(x)$ is defined as the limit of the difference quotient

$$\frac{d}{dx}f(x) = \lim_{h \to 0} \frac{f(x + h) - f(x)}{h}$$

For $f(x) = b^x$, we have $\dfrac{d}{dx}b^x = \lim\limits_{h \to 0} \dfrac{b^{x+h} - b^x}{h}$. The situation is graphically represented by figure 4.2.1.

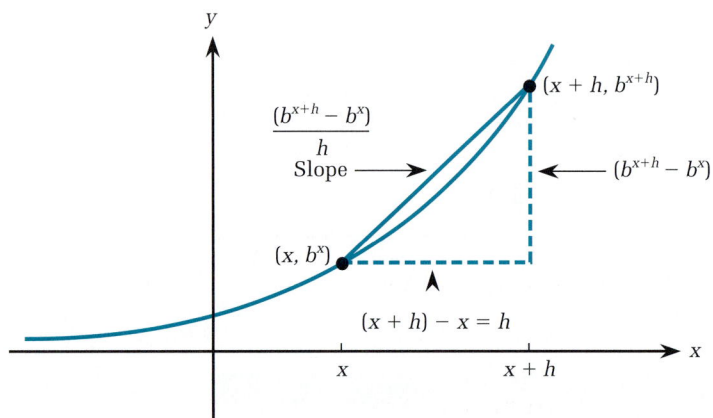

FIGURE 4.2.1

We begin the evaluation of this limit as follows:

$$\lim_{h \to 0} \frac{b^{x+h} - b^x}{h} = \lim_{h \to 0} \frac{b^x b^h - b^x}{h} \text{ by a rule of exponents } (E1)$$

$$= \lim_{h \to 0} \frac{b^x(b^h - 1)}{h} \text{ by factoring out the } b^x$$

$$= b^x \lim_{h \to 0} \frac{(b^h - 1)}{h}$$

Because b^x does not depend on h, it behaves as a constant relative to h.

This is as far as we can go using simple techniques. Now we come to a difficult problem. What is the value of $\lim\limits_{h \to 0} (b^h - 1)/h$? We certainly cannot simply substitute 0 for h, for division by zero is not defined. Also, we have seen that expressions of quotient forms in which both the numerator and denominator go to zero are quotients that do not always approach one in value.

The value of $\lim\limits_{h \to 0} (b^h - 1)/h$ depends very much on the choice of b. Table 4.2.1 shows the result of substituting particular values of h when $b = 2$ and $b = 3$. These two choices for b are not made at random, as will become apparent shortly. Values are rounded to five places.

The graphs of $y = 2^x$ and $y = 3^x$ and their tangent lines at $(0, 1)$ are displayed in figure 4.2.2.

TABLE 4.2.1

h	$\dfrac{b^h - 1}{h}$	
	$b = 2; \dfrac{2^h - 1}{h}$	$b = 3; \dfrac{3^h - 1}{h}$
0.05	0.70530	1.12935
0.01	0.69555	1.10467
0.001	0.69339	1.09922
0.0001	0.69317	1.09867
−0.0001	0.69321	1.09855
−0.001	0.69291	1.09801

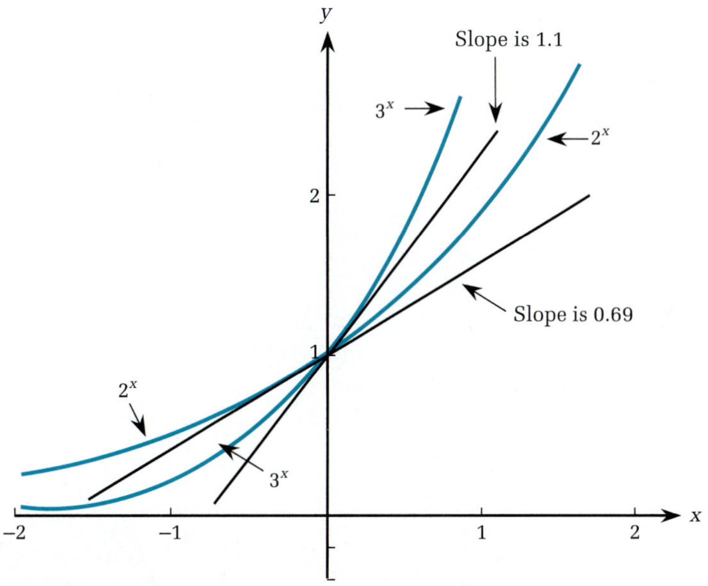

FIGURE 4.2.2

It appears that $\lim_{h \to 0} (2^h - 1)/h$ is about 0.69 and $\lim_{h \to 0} (3^h - 1)/h$ is about 1.10. These values are the slopes of the tangent lines. If these values are the correct limits, then we have $\dfrac{d}{dx} 2^x \simeq 2^x \cdot (0.69)$ and $\dfrac{d}{dx} 3^x \simeq 3^x \cdot (1.1)$.

In Section 4.1 we saw that b^x increases more rapidly with larger values for b. Hence as we consider values of b from 2 to 3, the slopes of the tangent lines to b^x at $(0, 1)$ increase from 0.69 to 1.1. It seems reasonable that for some particular value, the slope of the tangent line is 1. Let this value be represented by e. That is, we are now saying that e is the particular value of b for which

$$\lim_{h \to 0} \frac{e^h - 1}{h} = 1$$

The value of e is approximately 2.718281828. This is the same constant, e, that we encountered in Section 4.1. There we found it equal to $\lim_{x \to \infty} (1 + 1/x)^x$.

Derivative of e^x

Because $\lim_{h \to 0} (e^h - 1)/h = 1$, we have for the derivative of e^x the following:

DERIVATIVE OF e^x

$$\frac{d}{dx} e^x = e^x$$

In graphing $f(x) = e^x$, the slope at the point (x, e^x) is given by e^x. The graph is shown in figure 4.2.3.

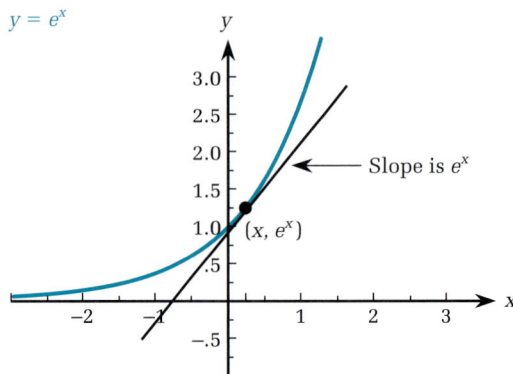

FIGURE 4.2.3

WARNING

The power function, x^n, has as its derivative nx^{n-1}. The exponential function, e^x, has as its derivative e^x. Do not make the mistake of using the Power Rule on exponential functions. It is *not* the case that $\dfrac{d}{dx}e^x$ is xe^{x-1}.

Examples 1–4 combine the result for $\dfrac{d}{dx}e^x$ with other properties of derivatives we had established earlier.

EXAMPLE 1   The derivative of the product of a constant and e^x

$$\frac{d}{dx}5e^x = 5 \cdot \frac{d}{dx}e^x = 5e^x$$

EXAMPLE 2   The Derivative of the Product of a Function of x and e^x

Find $\dfrac{d}{dx}(x \cdot e^x)$.

SOLUTION

To find $\dfrac{d}{dx}(x \cdot e^x)$, we need to realize $(x \cdot e^x)$ is a product of two functions. We should apply the Product Rule, which states that if u and v are differentiable functions, $(u \cdot v)' = u \cdot v' + v \cdot u'$.

With $u = x$ and $v = e^x$, we have

$$\frac{d}{dx}(x \cdot e^x) = x \cdot \left(\frac{d}{dx}e^x\right) + e^x \cdot \left(\frac{d}{dx}x\right)$$

$$= x \cdot e^x + e^x \cdot 1, \text{ where } \frac{d}{dx}x = 1 \text{ and } \frac{d}{dx}e^x = e^x.$$

Now we may factor e^x from each term, giving

$$\frac{d}{dx}(x \cdot e^x) = (x + 1) \cdot e^x$$

PRACTICE EXERCISE 1

Find $\dfrac{d}{dx}(3x \cdot e^x)$.

EXAMPLE 3   The derivative of the product of $\sqrt{x}$ and e^x

Find $\dfrac{d}{dx}(\sqrt{x} \cdot e^x)$.

SOLUTION

$$\frac{d}{dx}(\sqrt{x} \cdot e^x) = \sqrt{x} \cdot \left(\frac{d}{dx}e^x\right) + e^x \cdot \left(\frac{d}{dx}\sqrt{x}\right) \text{ by the Product Rule}$$

$$= \sqrt{x} \cdot e^x + e^x \frac{1}{2\sqrt{x}} \text{ by taking the derivatives.}$$

Finding a common denominator of $2\sqrt{x}$ we have

$$\frac{d}{dx}(\sqrt{x} \cdot e^x) = \frac{(2\sqrt{x})\sqrt{x} \cdot e^x + e^x}{2\sqrt{x}}, \text{ which can be rewritten as}$$

$$= \left(\frac{2x + 1}{2\sqrt{x}}\right) \cdot e^x$$

EXAMPLE 4   The Quotient Rule involving e^x

Find $\dfrac{d}{dx}\left(\dfrac{e^x}{x^2}\right)$.

SOLUTION

This derivative involves the quotient of two functions. The Quotient Rule states that if f and g are differentiable functions and $g^2 \neq 0$, then $(f/g)' = (g \cdot f' - f \cdot g')/g^2$. Therefore,

$$\frac{d}{dx}\left(\frac{e^x}{x^2}\right) = \frac{x^2\left(\dfrac{d}{dx}e^x\right) - e^x\left(\dfrac{d}{dx}(x^2)\right)}{(x^2)^2} \text{ for } x \neq 0,$$

$$= \frac{x^2 \cdot e^x - e^x \cdot 2x}{x^4} \text{ by taking the derivatives.}$$

We may factor xe^x from both terms in the numerator and then cancel x from the numerator and the denominator, to get

$$\frac{d}{dx}\left(\frac{e^x}{x^2}\right) = \frac{x \cdot e^x (x-2)}{x^4} = e^x \left(\frac{x-2}{x^3}\right)$$

PRACTICE EXERCISE 2

If $f(x) = \dfrac{e^x}{3x}$, find $f(x)$.

Derivative of $e^{f(x)}$

To find the derivatives of functions such as e^{3x} or e^{-x^2} we use the Chain Rule because these functions are generated by composition. The Chain Rule states that

$$\frac{d}{dx}g(f(x)) = \left(\frac{d}{df}g(f(x))\right) \cdot \frac{d}{dx}f(x)$$

If $g(x) = e^x$, then $g(f(x)) = e^{f(x)}$, and applying the Chain Rule gives us the following:

THE CHAIN RULE FOR EXPONENTIAL FUNCTIONS

If $f(x)$ is differentiable, then $\dfrac{d}{dx}e^{f(x)} = e^{f(x)} \cdot \dfrac{d}{dx}f(x)$.

Alternatively, we can write $(e^{f(x)})' = e^{f(x)} \cdot f'(x)$.

In the formula for the Chain Rule, the variable x is merely a "dummy" variable. If our function is a function of t, then the rule would be stated as $\dfrac{d}{dt}e^{f(t)} = e^{f(t)} \cdot \dfrac{d}{dt}f(t)$. Similar formulas apply for other variables.

EXAMPLE 5   The Chain Rule for exponential functions

Find $\dfrac{d}{dx}e^{3x}$.

SOLUTION

$$\frac{d}{dx}e^{3x} = e^{3x} \cdot \frac{d}{dx}3x = e^{3x} \cdot 3. \text{ This is usually written as } 3e^{3x}.$$

EXAMPLE 6   The Chain Rule for exponential functions

Find $\dfrac{d}{dx} e^{-x^2}$.

SOLUTION

$$\frac{d}{dx} e^{-x^2} = e^{-x^2} \cdot \frac{d}{dx}(-x^2) = e^{-x^2} \cdot (-2x) = -2x \cdot e^{-x^2}.$$

PRACTICE EXERCISE 3

Find $\dfrac{d}{dx} e^{2x+5}$.

EXAMPLE 7   The Chain Rule for exponential functions

Find $\dfrac{d}{dx} e^{\sqrt{x}+x^2}$.

SOLUTION

$$\frac{d}{dx} e^{\sqrt{x}+x^2} = e^{\sqrt{x}+x^2} \cdot \frac{d}{dx}(\sqrt{x} + x^2) = e^{\sqrt{x}+x^2} \cdot \left(\frac{1}{2\sqrt{x}} + 2x\right).$$

Graphing Exponential Functions

Our ability to find derivatives of exponential functions enables us to find critical numbers and sketch the graphs of such functions in the same manner we did in Chapter 3.

EXAMPLE 8   Finding critical numbers and graphing

Sketch the graph of $f(x) = e^x/x$ over $[0.5, 2]$.

SOLUTION

First we find the critical numbers, those numbers for which the derivative is zero or not defined.

To find the derivative, use the Quotient Rule:

$$f'(x) = \frac{x \cdot \dfrac{d}{dx} e^x - e^x \left(\dfrac{d}{dx} x\right)}{x^2} = \frac{x \cdot e^x - e^x \cdot 1}{x^2} = \frac{(x-1) \cdot e^x}{x^2}$$

For a quotient to be zero, the numerator must be zero. Thus, the derivative is zero only if $(x - 1) \cdot e^x = 0$. A product is zero only if a factor is zero. Because $e^x \neq 0$ for all x, we find $x - 1 = 0$, or $x = 1$.

The derivative is not defined if $x = 0$, but that number is not in our interval, so it is not a critical number. Thus, only 1 is a critical number. We could calculate values of $f(x)$ near $x = 1$, or try to apply the First Derivative Test or the Second Derivative Test to determine whether $x = 1$ is a relative maximum or minimum.

Consider the derivative and notice that $x^2 \geq 0$ for all x and also $e^x \geq 0$ for all x. Hence the sign of the first derivative is completely determined by the sign of $(x - 1)$. Thus, the changing in sign of the $(x - 1)$ term at $x = 1$ permits an easy application of the First Derivative Test.

If $x < 1$, then $x - 1 < 0$, giving $f'(x) < 0$. If $x > 1$, then $x - 1 > 0$, so $f'(x) > 0$. Thus, as x increases past 1, the derivative changes from being negative, to being zero, to being positive. The First Derivative Test indicates that a relative minimum occurs at $x = 1$.

Since the first derivative is negative from 0 to 1 and is positive past 1, we know the absolute maximum occurs at one of the endpoints and is the greater of $f(0.5)$ and $f(2)$. Furthermore, the absolute minimum is the relative minimum.

Finding the values of $f(x)$ at the critical number and the endpoints, we have

$$f(0.5) = e^{0.5}/0.5 \approx 1.64872/0.5 = 3.2974$$
$$f(1) \ \ = e^1/1 \approx 2.7183$$
$$f(2) \ \ = e^2/2 \approx 3.6945$$

Thus, $e^2/2$ is the absolute maximum and e is the absolute minimum over $[0.5, 2]$. A graph of this function is shown in figure 4.2.4.

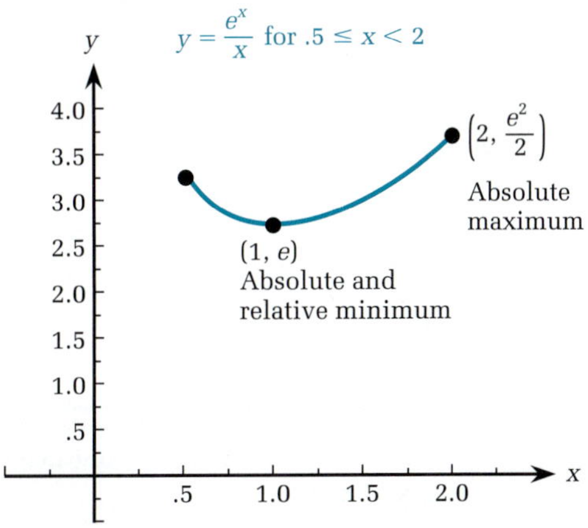

FIGURE 4.2.4

Applications: Revenue, Population, Memory

Any and all of our earlier applications may now be used with exponential functions. For instance, when $p(x)$ is a demand function, then $R(x) = x \cdot p(x)$ is the revenue function, just as in prior sections.

EXAMPLE 9   Maximizing revenue

Suppose the demand for watches is given by $p(x) = 25e^{-0.1x}$, where x is the number of watches and $p(x)$ is the price per watch. The demand function is indicated by the graph in figure 4.2.5 and the revenue function by the graph in figure 4.2.6.

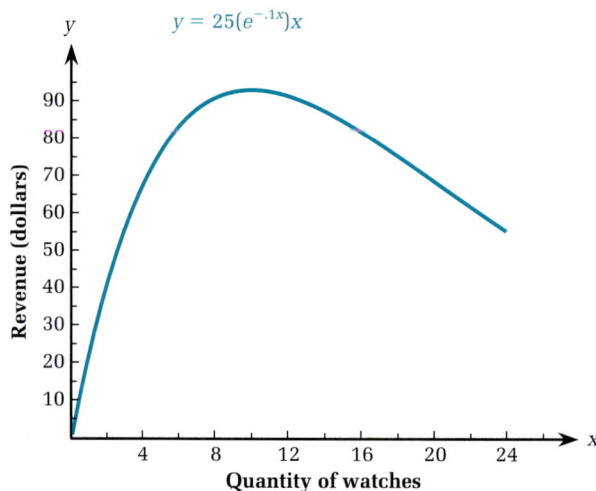

FIGURE 4.2.5

FIGURE 4.2.6

(a) Find $R(x)$ and $R'(x)$.

(b) Find the value of x and the price that will maximize revenue.

(c) Find the maximum revenue.

SOLUTION

(a) The revenue is $R(x) = x \cdot p(x)$. Substituting for $p(x)$, we have $R(x) = x \cdot 25e^{-0.1x} = 25x \cdot e^{-0.1x}$. The derivative is $R'(x) = 25 \cdot e^{-0.1x} + 25x \cdot e^{-0.1x} \cdot (-0.1)$ by the Product and Chain Rules, and $R'(x) = (25 - 2.5x) \cdot e^{-0.1x}$ by factoring $e^{-0.1x}$.

Critical numbers occur if $R'(x) = 0$ or is not defined. The base, e, raised to any power is never zero and it is always defined. Thus, the only critical number occurs if $-2.5x + 25 = 0$. Solving for x we have $25 = 2.5x$, so that $x = 10$.

(b) At $x = 10$ we have $R'(10) = 0$ and $R'(x)$ is changing sign from positive to negative as x increases past 10. Thus, we have a maximum revenue at this quantity. The price is determined by substituting 10 into the formula for demand, $p(x) = 25e^{-0.1x}$, $p(10) = 25e^{-0.1 \cdot 10} = 25e^{-1} = 25/e \approx 9.197$ is the price.

(c) The maximum revenue is given by $R(10) = 10 \cdot 9.197 = 91.97$.

Returning to the example with which we opened this section—

EXAMPLE 10   Exponential growth of population

The population of the United States was about 250 million in 1990 and had an annual growth rate of about 0.86%. If this growth rate remains the same, what is the rate at which people would be added to the population in the year 2000?

SOLUTION

Using the population formula from Section 4.1, and letting t represent the number of years after 1990, we have $P(t) = 250 \cdot e^{0.0086t}$ millions. Notice that we convert 0.86% into the decimal rate of growth, which is 0.0086.

To find the population increase in the year 2000, we need to find $P'(t)$ and evaluate this for $t = 10$, which is the number of years between 1990 and 2000.

The derivative is $P'(t) = 250 \cdot (e^{0.0086t} \cdot (0.0086))$ by the Chain Rule for exponential functions, or $P'(t) = 2.15 \cdot e^{0.0086t}$. Then $P'(10) = 2.15 \cdot e^{0.0086(10)} = 2.15 \cdot e^{0.086} = 2.15 \cdot 1.0898 = 2.3431$.

The population would be growing at the rate of 2.3431 million people a year.

A standard model in psychology is the *Ebbinghaus memory model*. This model is based on the work of Hermann Ebbinghaus (1850–1909), who determined that if a person has learned something (perhaps a list of names or anything else), then a good approximation for the percentage of that material remembered after t units of time is given by

$$P(t) = (100 - a) \cdot e^{-bt} + a \qquad \text{(for some constants } b > 0 \text{ and } 0 < a < 100\text{)}$$

The constants a and b depend on the material and the person. Notice that the value of $P(t)$ is in percent, rather than in a decimalized form of percentage.

EXAMPLE 11   Ebbinghaus memory model

Suppose a memory function is given by $P(t) = 25 \cdot e^{-0.3t} + 75$, t being in days.

(a) Find $P'(t)$.

(b) Evaluate and give an interpretation to both $P(4)$ and $P'(4)$.

SOLUTION

(a) $P'(t) = 25 \cdot (-0.3)e^{-0.3t} = -7.5e^{-0.3t}$

(b) $P(4) = 25 \cdot e^{-0.3(4)} + 75 = 25 \cdot e^{-1.2} + 75 = 25 \cdot 0.3012 + 75$
   $= 7.53 + 75 \approx 83$

   $P'(4) = -7.5 \cdot e^{-0.3(4)} = -7.5 \cdot 0.3012 = -2.259 \approx -2$
   The interpretation of $P(4)$ and $P'(4)$ is that on day 4 the person remembers about 83% of the information and is forgetting it at the rate of about 2% a day.

4.2 PROBLEMS

Foundations

The problems of this section require the basic skills illustrated by the following:

1. State the Product, Quotient, and Chain Rules.

2. Solve $e^x(2x) - 2e^x = 0$.

3. Solve $2xe^x - x^2e^x = 0$.

4. Find $\dfrac{d}{dx}(x^2 + 2x)$.

5. Find $\dfrac{d}{dx}(x^2 - 5x)$.

6. Find $\dfrac{d}{dx}\sqrt{x^2 + 1}$.

7. If a demand function is given by $p(x)$, express the revenue in terms of x and $p(x)$.

Exercises

8. Fill in the following table.

h	$\dfrac{(2.71828)^h - 1}{h}$
1	
0.1	
0.01	
0.001	
0.0001	
−0.0001	

Because $e \simeq 2.71828$ and

$$\frac{d}{dx}(2.71828)^x = \lim_{h \to 0} \frac{(2.71828)^{x+h} - (2.71828)^x}{h}$$

$$= (2.71828)^x \cdot \lim_{h \to 0} \frac{(2.71828)^h - 1}{h},$$

the values in the table strongly support our claim that $\dfrac{d}{dx}e^x = e^x$.

9. Fill in the following table.

h	$\dfrac{\left(\dfrac{1}{2.71828}\right)^h - 1}{h}$
1	
0.1	
0.01	
0.001	
0.0001	
−0.0001	

Because $e \simeq 2.71828$ and

$$\frac{d}{dx}(2.71828)^{-x} = \lim_{h \to 0} \frac{(2.71828)^{-(x+h)} - (2.71828)^{-x}}{h}$$

$$= (2.71828)^{-x} \cdot \lim_{h \to 0} \frac{(2.71828)^{-h} - 1}{h},$$

the values in the table strongly support our claim that $\dfrac{d}{dx}e^{-x} = -e^{-x}$.

Find derivatives of the functions in problems 10–27.

10. $f(x) = e^{2x}$ **11.** $f(x) = e^{0.5x}$

12. $f(x) = 3e^{-2x}$ **13.** $f(x) = 2e^{-3x}$

14. $f(x) = 3x - e^{2x}$ **15.** $f(x) = 2x - e^{3x}$

16. $f(x) = x + e^{x^2}$ **17.** $f(x) = -x^2 + e^{2x}$

18. $f(x) = 2x \cdot e^x$ **19.** $f(x) = 3x \cdot e^x$

20. $f(x) = x^2 \cdot e^{2x}$ **21.** $f(x) = x^3 \cdot e^{-2x}$

22. $f(x) = \sqrt{2x} \cdot e^x$ **23.** $f(x) = \sqrt{x} \cdot e^{2x}$

24. $f(x) = \dfrac{x^2}{e^x}$ **25.** $f(x) = \dfrac{3x}{e^x}$

26. $f(x) = e^{x^2} + 2x$ **27.** $f(x) = e^{x^2 - 5x}$

Graph the functions in problems 28–33, being careful to show any relative or absolute maximum or minimum points. Find the function values at these points.

28. $f(x) = 1 + e^{2x}$ over $[-2, 2]$

29. $f(x) = 3 - e^{-x}$ over $[-2, 2]$

30. $f(x) = \dfrac{e^x}{2x}$ over $[0.5, 2]$

31. $f(x) = \dfrac{e^{2x}}{x}$ over $[0.3, 2]$

32. $f(x) = e^{-2x^2}$

33. $f(x) = e^{-x^2}$

34. Suppose a demand curve is $p(x) = 10e^{-0.2x}$.
(a) Find the value of x that will maximize total revenue.
(b) Find the associated price.
(c) Find the maximum total revenue.

35. Suppose a demand curve is $p(x) = 100e^{-0.3x}$.
(a) Find the value of x that will maximize total revenue.
(b) Find the associated price.
(c) Find the maximum total revenue.

36. The population of Mexico was about 90 million in 1990. The annual growth rate was about 2.4%. Assuming the growth rate remains the same, estimate the population in 2010 and find the rate at which people will be added to the population in that year.

37. The population of New Zealand was about 3.4 million in 1990. The annual growth rate was about 0.7%.
(a) Assuming the growth rate remains the same, estimate the population in 2010 and the rate at which people will be added to the population in that year.
(b) Assume the growth rate from 1990 to 2010 for New Zealand is that of India in 1990 (about 1.7%). Find the population in 2010 and the rate at which people will be added to the population in that year.

38. The population of India was about 810 million in 1990. The annual growth rate was about 1.7%.
(a) Assuming the growth rate remains the same, estimate the population in 2010 and the rate at which people will be added to the population in that year.
(b) Assume the growth rate from 1990 to 2010 for India is that of New Zealand in 1990

(about 0.7%). Find the population in 2010 and the rate at which people will be added to the population in that year.

39. The graph in figure 4.2.7 gives the actual amount of paper money in circulation in the United States.* This graph can be closely approximated by the graph of $A(t) = 30e^{0.07t}$, where t is number of years after 1960.
    (a) Find $A'(t)$.
    (b) Evaluate $A'(30)$.
    (c) Give an interpretation to $A'(30)$.

40. **(Psychology, Ebbinghaus Memory Model)** Suppose a memory function is given by $P(t) = 70 \cdot e^{-0.2t} + 30$, where $P(t)$ is the percentage of the original information still remembered on day t.
    (a) Find $P(t)$.
    (b) Evaluate and give an interpretation to both $P(6)$ and $P'(6)$.

41. **(Psychology, Ebbinghaus Memory Model)** Suppose a memory function is given by $P(t) = 90 \cdot e^{-0.3t} + 10$, where $P(t)$ is the percentage of the original information still remembered by hour t.
    (a) Find $P'(t)$.
    (b) Evaluate and give an interpretation to both $P(5)$ and $P'(5)$.

*Source: *Denver Post,* March 4, 1990, crediting the *NY Times* and U.S. Treasury Dept.

42. **(Learning)** Rob can remember $N(t) = 28 - 27e^{-0.1t}$ names of flavors of ice cream after he has read the list t times.
    (a) Find $N'(t)$.
    (b) Evaluate $N'(7)$.
    (c) Give an interpretation to $N'(7)$.

43. **(Learning)** After t weeks of running a check-stand, a checker might have learned the codes for $C(t) = 100 - 99e^{-0.2t}$ items.
    (a) Find $C'(t)$.
    (b) Evaluate $C'(3)$.
    (c) Give an interpretation to $C'(3)$.

Graphing Calculator Problems

44. Let $0 \le x \le 2$.
    (a) Find a range so that the screen for $f(x) = e^x$ fills most of the screen.
    (b) On the same screen graph e^x; $[1 + x/2]^2$; $[1 + x/10]^{10}$; $[1 + x/100]^{100}$.
    (c) What can you conclude about $[1 + (x/n)]^n$ for large values of n?

45. (a) On the same screen graph e^{-x} and $[1 - x/n]^n$ for $n = 2, 5, 50$.
    (b) What can you conclude about $[1 - x/n]^n$ for large values of n?

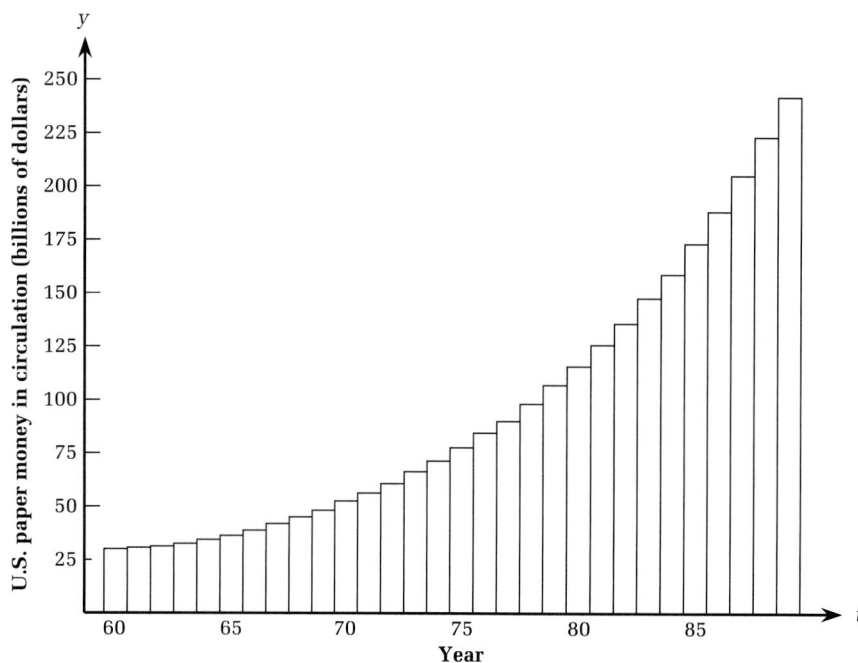

FIGURE 4.2.7

46. In choosing an initial range for a screen to approximate solutions to $e^{f(x)} = g(x)$, what about values of x for which $g(x) \leq 0$? Explain your answer.

47. Approximate both of the solutions of $x + 2 = e^x$.

48. Approximate the solution(s) of $e^{x^2-x} = 1/x + x$.

49. The graph of $f(x) = 10(e^x + e^{-x})$ is called a *catenary curve* and describes the shape of a cable suspended between two posts. Graph both $f(x)$ and $g(x) = 6.8x + 6.8$. Suppose the graph $g(x)$ describes a hillside and imagine there are two posts, each 30 units high and erected on the graph of $g(x)$ at $x = -1$ and $x = 1.5$. Notice these posts will just touch the graph of $f(x)$. Estimate the lowest height the cable is above the ground.

Writing and Discussion Problems

50. Find a copy of *Statistical Abstracts* for the current year and look up the table for populations and the growth of populations. Make projections for three countries, assuming current growth rates as well as some alternate rates. Discuss whether these projections are valid.

Enrichment Problems

Find the derivatives of the functions in problems 51–54.

51. $f(x) = e^{\sqrt{x^2+1}}$

52. $f(x) = e^{\sqrt{x^2-2}}$

53. $f(x) = \sqrt{e^{2x} + x}$

54. $f(x) = \sqrt{e^{-x} - 2x}$

55. Graph $f(x) = x^2 + e^{3x}$ and estimate the relative minimum.

56. Graph $g(x) = e^x/x^2$ over $[1, 3]$, indicating any extreme values.

57. (*Statistics and Probability*)   The function $f(x) = (1/\sqrt{2\pi})\, e^{-x^2/2}$ is a *standard normal density function*. Its graph is a bell-shaped curve, shown in figure 4.2.8. Find the inflection points of this function.

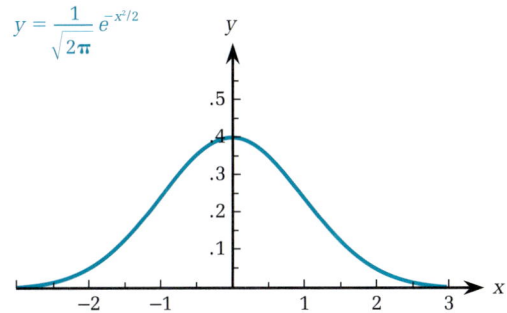

$y = \dfrac{1}{\sqrt{2\pi}}\, e^{-x^2/2}$

FIGURE 4.2.8

58. (*Statistics and Probability*)   A *normal probability density function* is given by $f(x) = (1/(\sigma\sqrt{2\pi}))\, e^{-(x^2/(2\sigma^2))}$, where σ is the standard deviation. Show that the graph of any normal probability density function has inflection points at $x = \pm\sigma$.

59. The function $y = e^x$ has the property that $dy/dx = y$. Find a linear function, $y = ax + b$, that also has this property.

SOLUTIONS TO PRACTICE EXERCISES

1. $\dfrac{d}{dx}\, 3x \cdot e^x = 3x \cdot \dfrac{d}{dx}\, e^x + e^x \dfrac{d}{dx}\, 3x$

$\qquad = 3x \cdot e^x + e^x \cdot 3 = 3e^x(x + 1)$

2. $\dfrac{d}{dx}\!\left(\dfrac{e^x}{3x}\right) = \dfrac{3x \cdot \dfrac{d}{dx}\, e^x - e^x \cdot \left(\dfrac{d}{dx}\, 3x\right)}{(3x)^2}$

$\qquad = \dfrac{3x \cdot e^x - 3 \cdot e^x}{9x^2} = e^x \cdot \left(\dfrac{x - 1}{3x^2}\right)$

3. $\dfrac{d}{dx}\, (e^{2x+5}) = e^{2x+5} \cdot \dfrac{d}{dx}\, (2x + 5) = e^{2x + 5} \cdot 2$

$\qquad = 2e^{2x + 5}$

4.3 LOGARITHMIC FUNCTIONS

The first section of this chapter introduced exponential growth. Therein we found the value of $f(x) = a \cdot e^{g(x)}$ if we knew the value of x. For example, the function $f(x) = 10{,}000e^{0.07x}$ gives the value after x years of an initial investment of \$10,000 at an annual interest rate of 7% compounded continuously. A graph of $f(x) = 10{,}000e^{0.07x}$ is shown in figure 4.3.1. If we want to know its value after eight years, we substitute $x = 8$ and find the value to be \$10,000 $\cdot e^{0.07 \cdot 8} =$ \$17,506.73.

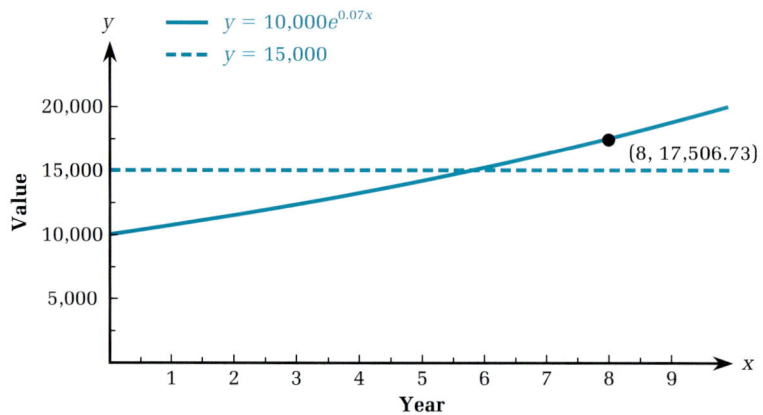

FIGURE 4.3.1

In this section, we consider the question of determining when an exponential function achieves a particular value. On investment growth, we ask, "How long does it take?" For instance, how long does it take \$10,000, invested at an annual rate of 7% compounded continuously, to grow to \$15,000? On the graph in figure 4.3.1, this question is equivalent to asking, "If we draw a horizontal line across at $y = \$15{,}000$, for what value of x does it intersect the graph?"

Another application involves pizza that is delivered. Hot from the oven, the freshly baked pizza starts to cool. Suppose we have a formula for its temperature as a function of time. Can we determine how long it takes for the pizza to reach a particular temperature? This question is of great interest because if pizza is delivered "cold," the customer is unhappy.

The answers to these questions bring us to the topics of logarithms and logarithmic functions.

The first book on logarithms was published in 1614 by John Napier (1550–1617) of Scotland. He created and used logarithms to solve complicated numerical calculations. From his time until the recent wide use of inexpensive pocket calculators, logarithms were taught and used for computations involving large numbers. In fact, Pierre de LaPlace (1749–1827) wrote that the invention of logarithms "by shortening the labors, doubled the life of the astronomer."

Today, the needs of calculus are the primary reason to study exponential and logarithmic functions. However, calculus was not created until 50 years after Napier's death, and he could not have foreseen the essential role logarithms were to play in calculus. Hence, the concept of the logarithm provides an excellent example of a mathematical idea created to solve one problem and then applied to solve quite different problems.

Logarithms

The definition of a logarithm is given in terms of an exponential.

DEFINITION

Logarithm

For $a > 0$, $b > 0$, $b \neq 1$, $\log_b a = c$ provided $b^c = a$.

The expression "$\log_b a = c$" is read *"the logarithm of a to the base b is c,"* or more briefly as *"the log of a to the base b is c."*

This definition can be thought of as a one-line dictionary, which relates an expression involving an exponential with an equivalent expression involving a logarithm. We require a to be positive because the value of b^c is positive for any value of c. The restrictions on b occur because we need a positive base for exponents and we cannot use 1 as a base.

Using the definition of a logarithm, we immediately can transform any exponential equation into a corresponding logarithmic equation and vice versa. Knowing $2^5 = 32$, we can immediately say that $\log_2 32 = 5$. Knowing $3^{-2} = 1/9$, we can write $\log_3 (1/9) = -2$.

EXAMPLE 1   Equivalent exponential and logarithmic equations

Below are shown equivalent exponential and logarithmic equations.

$$3^2 = 9 \qquad\qquad \log_3 9 = 2$$

$$\left(\frac{1}{2}\right)^2 = \frac{1}{4} \qquad\qquad \log_{1/2}\left(\frac{1}{4}\right) = 2$$

$$2^{-1} = \frac{1}{2} \qquad\qquad \log_2\left(\frac{1}{2}\right) = -1$$

$$10^3 = 1000 \qquad\qquad \log_{10} 1000 = 3$$

$$10^{-3} = 0.001 \qquad\qquad \log_{10} 0.001 = -3$$

PRACTICE EXERCISE 1

For each of the following, write an equivalent equation using logarithms.
(a) $2^2 = 4$ (b) $3^{-1} = 1/3$ (c) $(1/3)^2 = 1/9$

To find $\log_b c$, we need to know the power of b that produces c. Usually c is not already written as a power of b, so we must find the power of b or we need to use a calculator or table.

EXAMPLE 2   Determining $\log_b c$ by writing c as a power of b

Find $\log_2 8$ and $\log_4 32$.

SOLUTION

We need to find that power of 2 that produces 8. That is, we need to solve $2^c = 8$. From our experience in finding powers, we know $2^3 = 8$, so $\log_2 8 = 3$.

To solve $4^c = 32$, we write both 4 and 32 as powers of 2. Since $4 = 2^2$, then $4^c = (2^2)^c = 2^{2c}$. Express 32 as 2^5. Equating these we have $2^{2c} = 2^5$, so $2c = 5$. Solving for c, we have $c = 5/2$.

As soon as we know $4^{5/2} = 32$, then we can write $\log_4 32 = 5/2$.

If we cannot easily find a power of b that produces c, then we use a calculator or a table to determine $\log_b c$. There is such a power of b for each positive number c, as can be seen from the graph of $y = b^x$ in figure 4.3.2. There, if we draw the horizontal line $y = c$, that line will cross the graph at some point, as indicated. The x-value of that point is $\log_b c$.

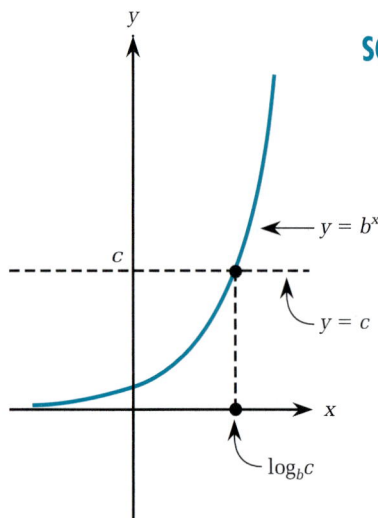

$y = b^x$

$y = c$

$\log_b c$

FIGURE 4.3.2

Common and Natural Logarithms

The definition for logarithms allows us to use any value, b, for the base as long as $b > 0$ and $b \neq 1$. However, ordinarily only the numbers 10 and e (the number whose value is about 2.718281828) are used as base numbers. Because of this, special terms and symbols are used for logarithms using those bases.

On our calculator the common log key is labeled $\boxed{\log}$ and the natural log key is labeled $\boxed{\ln}$. If the base is not 10 or e, then the value of the base must be written down as a subscript following the word "log." We will discover shortly how the calculator can be used to find logarithms to other bases.

DEFINITION   Common and natural logarithms

$\log_{10} x$ is called the *common* log of x (and usually written as $\log x$).

$\log_e x$ is called the *natural* log of x (and usually written as $\ln x$).

EXAMPLE 3   Common logarithms

Find $\log 100$, $\log 10$, and $\log 0.01$.

SOLUTION

We can use our calculators or use the definition of logarithms in terms of exponents. In order to use the definition, we need to know $100 = 10^2$, $10 = 10^1$, and $0.01 = 10^{-2}$.

From those equations we have $\log 100 = 2$, $\log 10 = 1$, and $\log 0.01 = -2$. We should check that using our calculators we get the same values. Make sure to use the $\boxed{\log}$ key.

In general, the definition of $\ln x$, in terms of powers of e, does us little good because the powers of e, such as $\sqrt{e}$ and e^3, occur so rarely in ordinary life. Hence, to find $\ln x$ we are forced to use a calculator or natural log tables.

EXAMPLE 4   Natural logarithms

Using the $\boxed{\ln}$ key, we find $\ln 2 = 0.69314718$ and $\ln 5 = 1.609437912$.

PRACTICE EXERCISE 2

Find $\log 2$, $\log 5$, $\log 0.01$, $\ln 10$, $\ln 5$, and $\ln 1$.

Properties of Logarithms

The following properties are stated both for logarithms using a general base, b, as well as for logarithms to the base e, the natural logarithms. Proofs of these properties are contained in an appendix to this section.

PROPERTIES OF LOGARITHMS

For positive real numbers, r and s, and for a real number $b > 0$, $b \neq 1$:

L1. $\log_b (r \cdot s) = \log_b r + \log_b s$ $\ln (r \cdot s) = \ln r + \ln s$

L2. $\log_b \dfrac{r}{s} = \log_b r - \log_b s$ $\ln \dfrac{r}{s} = \ln r - \ln s$

L3. $\log_b r^t = t \cdot \log_b r$ $\ln r^t = t \cdot \ln r$

L4. $\log_b b = 1$ $\ln e = 1$

L5. $\log_b 1 = 0$ $\ln 1 = 0$

L6. $\log_b b^r = r$ $\ln e^r = r$

If we do not want to calculate using a particular base, we can change the base by the following conversion formula.

L7. $\log_b r = \dfrac{\log_a r}{\log_a b}$ for any $a > 0$, $a \neq 1$

EXAMPLE 5   Properties of logarithms

For illustrations of L1, L2, L3 and L7, consider

(i) Using the product of $2 \cdot 3 = 6$ to find $\ln 6$.
By L1, $\ln (2 \cdot 3) = \ln 2 + \ln 3 = 0.6931 + 1.0986 = 1.7917$.
From our calculator, $\ln 6 = 1.7918$.

(ii) Using the quotient $(3/2) = 1.5$ to find $\ln (3/2)$.
By L2, $\ln (3/2) = \ln 3 - \ln 2 = 1.0986 - 0.6931 = 0.4055$.
From our calculator, $\ln 1.5 = 0.4055$.

(iii) Using the fact that $2^3 = 8$ to find $\ln 8$.
By L3, $\ln 2^3 = 3 \cdot \ln 2 = 3 \cdot 0.6931 = 2.0793$.
From our calculator, $\ln 8 = 2.0794$.

(iv) If we want to go from base 2 to base e, we can use L7 to evaluate $\log_2 8$.
$\log_2 8 = (\ln 8/\ln 2) = (2.0794/0.6931) = 3.0001$.
Because $2^3 = 8$, we know $\log_2 8 = 3$.

Notice that in three out of four of our calculations in Example 5, rounding off made a difference in the last digit. The reason for this discrepancy was our use of four-place approximations for intermediate calculations. We should have used one more place in the intermediate calculations and then rounded down to the number of significant digits for a particular problem. Thus, had we wanted four places of accuracy for the above logarithms, we should have used five-place accuracy in the intermediate calculations. This procedure does not guarantee success, but it suffices for most situations.

As an illustration, had we used ln 2 = 0.69314, then $3 \cdot \ln 2$ would be 2.07942, and we would round that to 2.0794, which we found to be a four-place approximation for ln 8.

Logarithmic Functions

By using the values of a logarithm to a base, we can define a function.

DEFINITION **Logarithmic function**

Suppose $b > 0$, $b \neq 1$. The *logarithmic function of base b* is: $f(x) = \log_b x$ defined for $x > 0$.

The range of a logarithm function includes negative numbers because b^c is defined for c being any real number whatsoever. The reason the domain of $\log_b x$ does not include negative numbers can be seen by considering what would happen if we try to find $\log_3(-2)$. We would be seeking an x such that $3^x = -2$. No such x can exist since the output of the exponential function is always positive.

Graphs of Logarithmic Functions

The graphs of logarithmic functions can be generated by several methods, including the use of a graphing calculator or computer. In the next section of this chapter we introduce the derivative of the logarithmic function and use it as an aid in graphing. For now, let us merely plot a few points to gain a rough idea of the graph of a fairly simple logarithmic function.

EXAMPLE 6 **The graph of a logarithmic function**

Graph $y = f(x) = \log_2 x$.

SOLUTION

Even without using a calculator we can find values of y for certain selected values of x. In particular, by changing $y = \log_2 x$ into $2^y = x$, we see that for values of x that are powers of 2, we can determine $\log_2 x$. Several integer powers of 2, both positive and negative, are shown in the following table.

x	$0.25 = 2^{-2}$	$0.5 = 2^{-1}$	1	2	$4 = 2^2$	$8 = 2^3$
$\log_2 x$	-2	-1	0	1	2	3

Having found these values, we smoothly connect their associated points to produce the graph shown in figure 4.3.3.

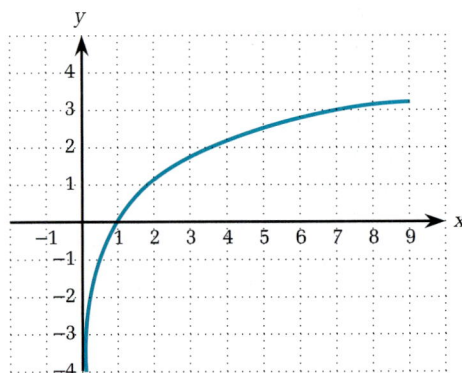

FIGURE 4.3.3

PRACTICE EXERCISE 3

Graph $y = f(x) = \log_3 x$.

To graph $y = f(x) = \ln x$ or $y = \log x$, it is easiest to determine several values by using our calculator and then smoothly connect the associated points. In the following table the values of $\ln x$ are given only to two places because in our graph we lack the necessary precision to plot points determined to four-place accuracy.

x	0.05	0.1	0.5	1	4	7	15
$\ln x$	-3.00	-2.30	-0.69	0	1.39	1.95	2.71

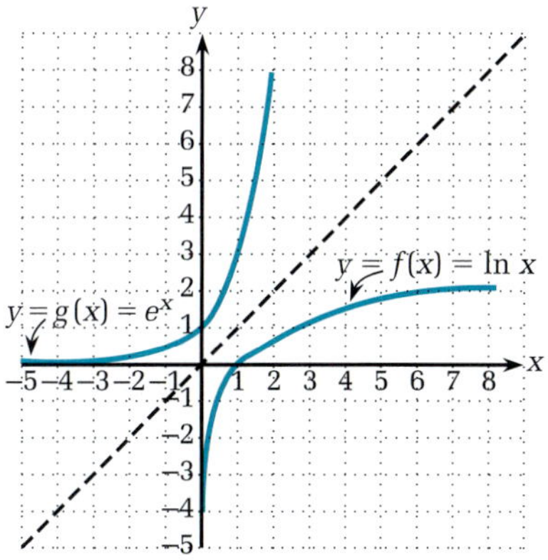

FIGURE 4.3.4

The graph of $f(x) = \ln x$ and the graph of $g(x) = e^x$ are displayed in figure 4.3.4.

Notice that the graph of $f(x) = \ln x$ and the graph of $g(x) = e^x$ are mirror images of each other, reflected in a mirror placed on the line $y = x$. This property is true of every exponential function and its related logarithmic function. That is, for any base b, the graphs of $y = b^x$ and $y = \log_b x$ are mirror image reflections about the graph of the line $y = x$.

We can summarize some of our conclusions in the following properties of logarithmic functions.

PROPERTIES OF LOGARITHMIC FUNCTIONS

The following are true of any logarithmic function of the form $f(x) = \log_b x$:

1. The domain is $(0, \infty)$.

2. The range is $(-\infty, \infty)$.

3. $\log_b 1 = 0$.

4. $\lim\limits_{x \to \infty} \log_b x = \infty$ and $\lim\limits_{x \to 0^+} \log_b x = -\infty$.

Exponential and Logarithmic Equations

Equations involving logarithms or exponents are quite common. Let us consider some examples, the first one being the question with which we started this section.

EXAMPLE 7   Exponential equation involving value of an investment

Suppose \$10,000 is invested at 7% annual interest with continuous compounding. How long does it take to grow to \$15,000?

SOLUTION

We must be careful to convert from the percentage rate, r, to the decimal rate, $i = r/100$, before using the formula that states: Under continuous compounding the amount after t years is given by $P(t) = 10{,}000 \cdot e^{0.07t}$.

To solve $15,000 = 10,000 \cdot e^{0.07t}$ for t, we divide by 10,000 to obtain $e^{0.07t} = 15,000/10,000 = 1.5$. This equation can be converted into an equation with natural logarithms. $e^{0.07t} = 1.5$ is equivalent to $0.07t = \ln 1.5$.

Because $\ln 1.5 = 0.40547$, we have $0.07t = 0.40547$ (notice that we use a five-place form for $\ln 1.5$ because we are not at the final round-off stage). Dividing by 0.07 gives us $t = 0.40547/0.07 = 5.79$ years.

The final answer is in two-place form because we divide up years into only two-place accuracy.

PRACTICE EXERCISE 4

Helen invested $5000 at 8% annual interest, compounded continuously. How long will it be until her account reaches $12,000?

EXAMPLE 8   Doubling times for investments

Suppose A dollars are invested at r percent annual interest compounded continuously. When will the amount double?

SOLUTION

Convert the interest to decimal form, $i = r/100$, in order to use $P(t) = A \cdot e^{it}$. For $P(t)$ to reach $2A$ we need to solve $2A = A \cdot e^{it}$. Dividing each side by A, we have $2 = e^{it}$. Converting this into natural logs, we get $\ln 2 = it$. Because $\ln 2 = 0.693147$, we have $0.693147 = it$, which, when solved for t, gives $t = 0.693147/i$.

Thus, if we divide the decimal interest rate into 0.693147 we will have the doubling time.

Rule of 70, Rule of 72

In solving Example 8, we found the time required for an amount to double under continuous compounding was $t = 0.693147/i$. Because of the general avoidance of decimals when possible, this result gave rise to the rules known in finance as the Rule of 70 and the Rule of 72. In these rules, the interest rates are stated as percentages and the numerator used is 70 or 72, depending on which rule is being used (rather than the more accurate value of 69.3147). Although the Rule of 70 is more accurate in the case of continuous compounding, the Rule of 72 is often used to allow for the slightly longer doubling times required when interest is not being compounded continuously.

EXAMPLE 9   The Rules of 70 and 72

Use the Rules of 70 and 72 to estimate the length of time required for $12,000 to double under continuous compounding with an interest rate of 8.75%.

SOLUTION

$70/8.75 = 8$ years. $72/8.75 = 8.23$ years.

PRACTICE EXERCISE 5

Use the Rule of 72, rather than 70, to estimate the length of time required for $4000 to double under continuous compounding with an interest rate of 4.5%.

Further Exponential/Logarithmic Equations

The following are other examples of equations that are solved by switching back and forth between expressions using logarithms and expressions using exponents.

EXAMPLE 10   Logarithmic equation

Solve $\ln 3x = 1.34$.

SOLUTION

Change to the exponential form, $3x = e^{1.34}$. Use a calculator to evaluate $e^{1.34} = 3.81904$ (again, notice the five-place intermediate form). Solving for x in $3x = 3.81904$, we have $x = 3.81904/3 = 1.2730$.

PRACTICE EXERCISE 6

Solve $\ln 5x = 3$.

EXAMPLE 11   Exponential equation

Solve $e^{x^2} = 5$.

SOLUTION

Change to the natural logarithmic form to get $x^2 = \ln 5$. Use a calculator to find $\ln 5 = 1.60944$. From $x^2 = 1.60944$, we have $x = \pm\sqrt{1.60944} = \pm 1.2686$. ●

The next example can be solved in two different ways. In the first, we take the natural logarithm of each side of an equation. This is a very common technique for solving equations involving exponents. In fact, it is such a common approach that it deserves highlighting.

HINT FOR SOLVING AN EQUATION INVOLVING EXPONENTS

If an equation involves exponents, then taking the natural logarithm of each side of the equation and using properties of logarithms may help in the solution of the equation.

EXAMPLE 12   Solving an equation by taking natural logarithms

Solve $5^x = 8$.

SOLUTION

Take the natural logarithm of each side of the equation to get $\ln 5^x = \ln 8$. So $\ln 5^x = x \cdot \ln 5$ by property L3 of logarithms. From $x \cdot \ln 5 = \ln 8$, we find $x = (\ln 8/\ln 5) = 2.07944/1.60944 = 1.2920$. ●

Alternatively, in Example 12, we could have gone from $5^x = 8$ to the logarithmic form of $\log_5 8 = x$. Then we could use the change of base formula (L7) to go from base 5 to base e, finally obtaining the same answer. However, taking the natural logarithm of each side involves less remembering of formulas, and, furthermore, we will see it is a technique that is essential later.

The next example arises from a function that models the blood serum level after a patient takes medicine. The model is discussed more fully in the exercises.

EXAMPLE 13   Using logarithms to solve an exponential equation

Suppose $f(x) = 5(e^{-x} - e^{-1.2x})$. Solve $f'(x) = 0$.

SOLUTION

Using the derivative formulas of Section 4.2, $f'(x) = 5(-e^{-x} - (-1.2)e^{-1.2x})$. Thus, we must solve $5(-e^{-x} + 1.2e^{-1.2x}) = 0$.

Dividing by 5 and putting $-e^{-x}$ on the right, we get $1.2 \cdot e^{-1.2x} = e^{-x}$.

To collect all the terms involving x on the same side of the equality sign, divide each side by $e^{-1.2x}$ and then simplify by the rules of exponents. $1.2 = (e^{-x}/e^{-1.2x}) = e^{-x} \cdot e^{1.2x} = e^{-x + 1.2x} = e^{0.2x}$.

If we convert this into a logarithmic equation, we have $\ln 1.2 = 0.2x$, which, when we divide by 0.2, gives us $x = 0.18232/0.2 \approx 0.91$.

4.3 PROBLEMS

Foundations

The problems of this section require the basic skills illustrated by the following:

Express problems 1–4 as a power of 2.

1. 4   **2.** 16   **3.** $\dfrac{1}{8}$ **4.** $\dfrac{1}{2}$

Express problems 5–8 as a power of 4.

5. 4   **6.** 16   **7.** $\dfrac{1}{8}$ **8.** $\dfrac{1}{2}$

Exercises

For problems 9–22, write an equivalent equation using exponents.

9. $\log_3 9 = 2$

10. $\log_2 16 = 4$

11. $\log_2 32 = 5$

12. $\log_3 27 = 3$

13. $\log_2 \left(\dfrac{1}{8} \right) = -3$

14. $\log_3 \left(\dfrac{1}{27} \right) = -3$

15. $\log_4 8 = \dfrac{3}{2}$

16. $\log_4 \left(\dfrac{1}{2} \right) = -\dfrac{1}{2}$

17. $\log_{10} 0.001 = -3$

18. $\log_{10} 10{,}000 = 4$

19. $\log_{100} 1000 = 1.5$

20. $\log_{1000} 100 = \dfrac{2}{3}$

21. $\ln 6 = 1.7918$

22. $\ln 7 = 1.9459$

For problems 23–28, write an equivalent equation using logarithms.

23. $4^2 = 16$

24. $2^5 = 32$

25. $3^{-2} = \dfrac{1}{9}$

26. $2^{-4} = \dfrac{1}{16}$

27. $\left(\dfrac{1}{5} \right)^{-2} = 25$

28. $\left(\dfrac{1}{3} \right)^{-1} = 3$

29. Graph $y = \log_{10} x$.

30. Graph $y = \log_5 x$.

31. Graph $y = \log_8 x$.

32. Graph $y = \log_4 x$.

33. Graph $y = \log_{1/3} x$.

34. Graph $y = \log_{1/5} x$.

Evaluate problems 35–42, rounding answers to four places.

35. $\log 6$

36. $\log 8$

37. $\log 25$

38. $\log 50$

39. $\ln 2.3$

40. $\ln 5.1$

41. $\ln 234.1$

42. $\ln 83.1$

For problems 43–46, use L7, the change of base formula, to find the values.

43. $\log_2 5$

44. $\log_3 5$

45. $\log_3 2$

46. $\log_2 3$

Solve each of the equations in problems 47–64 for x.

47. $\log_2 x = 3$

48. $\log_3 x = 2$

49. $\log_3 x = -2$

50. $\log_2 x = -4$

51. $\ln 2x = 5$

52. $\ln 5x = 3$

53. $\ln (x + 1) = 2$

54. $\ln (x - 2) = 1$

55. $e^{2x} = 10$

56. $e^{3x} = 5$

57. $e^{x-2} = 5$

58. $e^{x+2} = 3$

59. $e^{-0.003x} = 20$

60. $e^{-0.025x} = 10$

61. $3^x = 4$

62. $2^x = 5$

63. $\left(\dfrac{1}{2}\right)^x = 3$

64. $\left(\dfrac{1}{3}\right)^x = 4$

65. *(Rule of 70, Rule of 72)* Estimate the time required for a deposit earning 5% annual interest to double in value using continuous compounding
(a) using the Rule of 70,
(b) using the Rule of 72.

66. An accident results in a $50,000 judgment on behalf of a newborn baby. The money is put into an account paying 8% annual interest compounded continuously.
(a) What will be the value of the account when the child is 18?
(b) How old is the child when the account is worth $250,000?

67. An endowment is created for the benefit of the Alphabet Revision Foundation. Initially $10,000 is deposited in an account paying 6% annual interest compounded continuously.
(a) What is the value of the endowment in 10 years?
(b) When will the endowment be worth $25,000?

Problems 68–71 all depend on the following information. In 1989, the Coldwell Banker division of Sears Financial Network did a survey of home prices for homes with the following characteristics: a floor area of approximately 2000 square feet, three bedrooms, two baths, a family room, a two-car garage, in a neighborhood that would reflect the norm for a mid-level executive transferred within a national company. Some of these prices, and the change from the preceding year, are shown in table 4.3.1.

TABLE 4.3.1

MARKET	1988 PRICE	PERCENTAGE CHANGE FROM 1987 PRICE
Beverly Hills	$841,667	10.1
Greenwich, Conn.	671,667	3.4
San Francisco	556,667	24.4
Denver	94,500	-6.2
San Antonio	77,300	-7.9
Corpus Christi	78,000	3.6

Assume those prices continue compounding continuously, at the given rates, indefinitely into the future.

68. How many years would it take for the Beverly Hills home to be worth $1,000,000?

69. How many years would it take for the Greenwich, Conn., home to be worth $750,000?

70. How many years would it take for the Denver home to be worth $75,000?

71. How many years would it take for the San Antonio home to be worth $50,000?

Problems 72–74 depend on Newton's Law of Cooling, which characterizes the rates at which hot things cool off and cold things warm up. More specifically, the rate at which an object changes temperature is proportional to the difference between its temperature and the "room" (the "surrounding," or "ambient") temperature.

Expressed mathematically, if B is the temperature of an object and A is the room temperature, then there is a constant k (which depends on the object that is changing temperature) such that at time t, the temperature T of the object is

$$T = A + (B - A) \cdot 10^{-kt}$$

72. *(Newton's Law of Cooling, Pizza Delivery)* At Tony's Slice of Life, a pizza comes out of the oven, is sliced, and is slipped onto a cardboard sheet. It has already started cooling down and by that time it is 375°F (degrees Fahrenheit). The room temperature is 75°F. Tony notices that the temperature of the pizza is given by $T(t) = 75 + 300 \cdot 10^{-0.14t}$, where t is in minutes. How long until the pizza is 135°F? (Hint: When using base 10, make sure to use the $\boxed{\log}$ key, not the $\boxed{\ln}$ key.)

73. ***(Newton's Law of Cooling, Pizza Delivery)*** Refer to problem 72. Tony found his pizza was getting cold too fast, so he bought some insulated delivery boxes. By using them, the temperature, in degrees Fahrenheit, is given by $T(t) = 75 + 300 \cdot 10^{-0.04t}$, where t is in minutes. How long until the pizza is 135°F?

74. ***(Newton's Law of Cooling, Warming Up Soft Drinks)*** Suppose a case of soft drinks is at 32°F and is taken to a picnic. Their temperature after t minutes is $T(t) = 92 - 60 \cdot 10^{-0.01t}$. Find the length of time until they are at 60°F.

Problems 75–77 concern graphing a function that models the presence of a drug in the bloodstream. A mathematical model for such a condition is given by the function

$$f(t) = \frac{k}{a-b}(e^{-bt} - e^{-at}), \qquad \text{where } a > b > 0$$

The graph of $f(t)$ has the general shape of the graph in figure 4.3.5.

75. Suppose $f(t) = e^{-t} - e^{-2t}$. Determine the maximum value of $f(t)$.

76. Suppose $f(t) = 0.5(e^{-t} - e^{-3t})$. Determine the maximum value of $f(t)$.

77. Suppose $f(t) = 20(e^{-t} - e^{-1.05t})$. Determine the maximum value of $f(t)$.

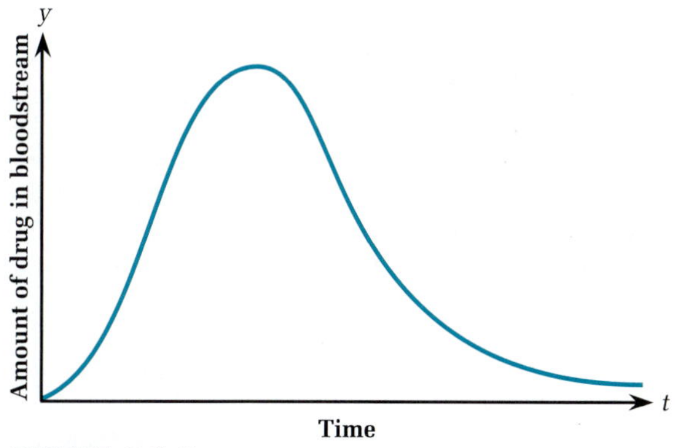

FIGURE 4.3.5

Graphing Calculator Problem

78. Graph the functions in problems 75–77. What effect does changing the values of a and b have on the graph of $f(t) = (k/(a-b))(e^{-bt} - e^{-at})$? If you know the time at which you want the maximum to occur, can you find values for a and b to achieve that?

Writing and Discussion Problem

79. Logarithms did make computations easier for the seventeenth-century scientist, but perhaps Laplace, who was quoted earlier, was incorrect when he said that logarithms doubled the lives of astronomers by halving their time spent computing. What do you think about our use of calculators and computers today? Do they free us to do other things or does the power to do more computations faster, easier, and cheaper simply allow us—maybe even entice us—to carry out more difficult computations, strive for ever faster techniques, and work on problems that we would not have attempted twenty years ago?

Enrichment Problems

80. ***(Pizza Delivery)*** Refer to problem 72. Find the value of k so that $T(t) = 75 + 300 \cdot 10^{-kt}$ will have the pizza reach 135°F in 30 minutes.

81. ***(Home Prices)*** Refer to Table 4.3.1. How many years until the Greenwich home costs the same as the San Francisco home?

82. ***(Home Prices)*** Refer to Table 4.3.1. How many years until the Denver home costs the same as the Corpus Christi home?

83. An endowment of \$10,000 is made. It earns 6% interest compounded continuously, which is allowed to accumulate. How large must the endowment grow for it to support an annual award of \$1000 in perpetuity (that is, without diminishing the capital of the endowment)? (Hint: While the endowment is undergoing accumulation, its value is $10{,}000 \cdot e^{0.06t}$. We need to find a k so that $10{,}000 \cdot e^{0.06k}$ is so large that the interest for one year on that amount is \$1000 more than the amount itself.)

SOLUTIONS TO PRACTICE EXERCISES

1. (a) $2^2 = 4$ is equivalent to $\log_2 4 = 2$.
   (b) $3^{-1} = 1/3$ is equivalent to $\log_3 (1/3) = -1$.
   (c) $(1/3)^2 = 1/9$ is equivalent to $\log_{1/3} (1/9) = 2$.

2. $\log 2 = 0.3010$, $\log 5 = 0.6990$, $\log 0.01 = -2$,
   $\ln 10 = 2.3026$, $\ln 5 = 1.6094$, $\ln 1 = 0$.

3. Graph $y = f(x) = \log_3 x$. The following table contains values of x and $\log_3 x$. Figure 4.3.6 is a graph of those values connected into a smooth curve.

x	$\dfrac{1}{9}$	$\dfrac{1}{3}$	1	3	9
$\log_3 x$	-2	-1	0	1	2

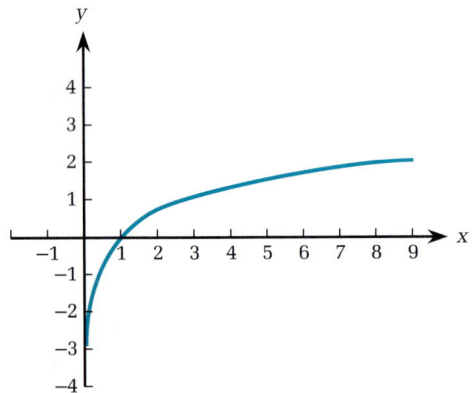

FIGURE 4.3.6

4. If Helen invested $5000 at 8% annual interest compounded continuously, her account would have $12,000 when $12{,}000 = 5000e^{0.08t}$. Dividing by 5000 gives $12/5 = e^{0.08t}$. Written in natural logarithmic form, this is $\ln (12/5) = 0.08t$. $\ln (12/5) = \ln 2.4 = 0.87547$, so we have $t = 0.87547/0.08 = 10.94$ years.

5. Using the Rule of 72, we divide 72 by 4.5 to find it takes 16 years for the $4000 to double.

6. $\ln 5x = 3$ is equivalent to $e^3 = 5x$. Using a calculator, we find $e^3 = 20.08554$, so $x = 20.08554/5 = 4.0171$.

Appendix: Proofs of the Logarithm Properties

L1. $\log_b r \cdot s = \log_b r + \log_b s$

Let $A = \log_b r \cdot s$, $B = \log_b r$, and $C = \log_b s$. Each of these, rewritten in exponential form, would be $b^A = r \cdot s$, $b^B = r$, and $b^C = s$. Then

$$b^A = r \cdot s \qquad \text{by definition of } A$$
$$= b^B \cdot b^C \qquad \text{by the definitions of } B \text{ and } C$$
$$= b^{B+C} \qquad \text{by the E1 property of exponents.}$$

Because $e^x = e^w$ only if $x = w$, we have $A = B + C$, as claimed.

L2. $\log_b r/s = \log_b r - \log_b s$

For this equation, let $t = r/s$. Then $r = ts$ and by L1 we have $\log_b r = \log_b ts = \log_b t + \log_b s$. Replacing the t by r/s, we have $\log_b r = \log_b (r/s) + \log_b s$. By subtracting $\log_b s$ from each side, we have L2.

L3. $\log_b r^t = t \cdot \log_b r$

Let $A = \log_b r^t$ and $B = \log_b r$. Rewriting these in exponential form gives us $b^A = r^t$ and $b^B = r$. Substituting r into $b^A = r^t$ we have $b^A = (b^B)^t = b^{Bt}$ by a property of exponents (E6). Because $e^x = e^w$ only if $x = w$, we have $A = t \cdot B$, which demonstrates L3.

L7. $\log_b x = (\log_a x/\log_a b)$ for any $a > 0$, $a \neq 1$

Let $A = \log_b x$, $B = \log_a x$, and $C = \log_a b$, and rewrite each in exponential form: $b^A = x$, $a^B = x$, and $a^C = b$. Substituting the value for b from the third into the first of these, we have $(a^C)^A = x$. From the second we know x is also a^B. Because $(a^C)^A = a^{CA}$, we have $CA = B$. Dividing by C, we have $A = B/C$, which is L7.

4.4 DERIVATIVES OF LOGARITHMIC FUNCTIONS AND LOGARITHMIC DIFFERENTIATION

In this section we determine the derivatives of logarithmic functions to various bases. The natural logarithms, using base e, are the most important because logarithms using other bases can be translated into natural logarithms. Furthermore, the simplest formulas use e as the base, which is one reason it is called the "natural base."

The material that follows the discussion on relative rate of change may be regarded as optional, and its omission will not cause a difficulty in later sections. The exercises at the end of the section that are based on that material are marked.

Let us begin our discussion by considering several uses of (and concerns about) iodine. A region of the American Midwest was once known as the "goiter belt" because enlargements of the thyroid gland were so common. This medical condition was often caused by a lack of iodine in the diet, and once the problem was identified it was easily remedied by adding iodine to common table salt. Hence we have the sale of "iodized salt" in grocery stores. The thyroid gland needs and concentrates iodine. It does so regardless of the form of the iodine.

A radioactive form of iodine, known as iodine-131 (whose symbol is ^{131}I), is given to treat various thyroid conditions because it becomes concentrated in and thereby irradiates the thyroid without irradiating the rest of the body. On the other hand, too much ^{131}I is harmful. In 1986, the power plant at Chernobyl released massive amounts of ^{131}I into the atmosphere. Milk produced by dairy cattle in the region was unfit for human consumption for a long time afterward.

In Section 4.6 we will show that if we start with P_0 grams of ^{131}I, then the number of days by which time P grams have become harmless is given by

$$T(P) = -11.5 \ln \left(1 - \frac{P}{P_0}\right)$$

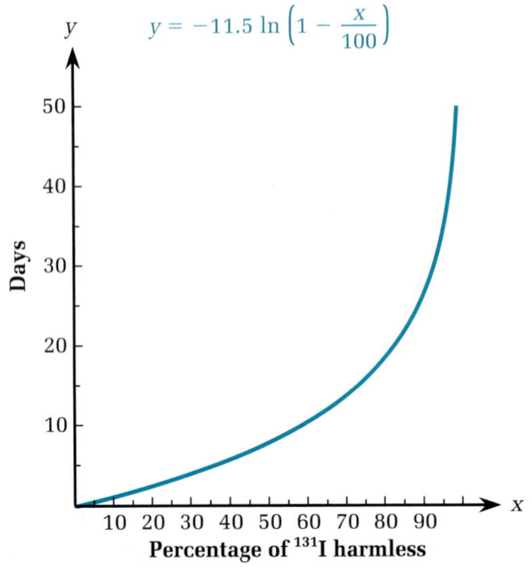

FIGURE 4.4.1

A graph of $T(P)$ is shown in figure 4.4.1.

In order to find the rate at which time must pass in order to transform some quantity of ^{131}I, we need to be able to find the derivative of a logarithmic function. To plan whether or not to destroy their herds, the dairy operators in the vicinity of Chernobyl needed to know how long they could expect that they would have to hold their milk off the market.

Derivative of ln x

To find the derivative of $f(x) = \ln x$, we use the process of implicit differentiation. It may seem strange to do this because we already have an explicit form for $f(x)$ and the derivative is always given by $\lim_{h \to 0} [f(x + h) - f(x)]/h$ if that limit exists. However, the evaluation of the limit is quite difficult, and using the method of implicit differentiation is quite simple.

Implicit differentiation, which was discussed in Section 2.8, does not prove that the derivative exists. Rather, it finds the derivative, provided the derivative actually does exist. As a result, when we use implicit differentiation we begin by asserting that there is some derivative for $f(x) = \ln x$. This is a claim that we justify in this case by considering the graph of $y = \ln x$ in figure 4.4.2. For this graph, we feel confident that tangent lines do exist at every point on the graph. Based on our assumption about the existence of tangent lines, we assume the derivative of $f(x) = \ln x$ does exist.

We first transform $f(x) = \ln x$ into the equivalent exponential equation, $e^{f(x)} = x$. Taking the derivative of both sides gives

$$\frac{d}{dx} e^{f(x)} = \frac{d}{dx} x$$

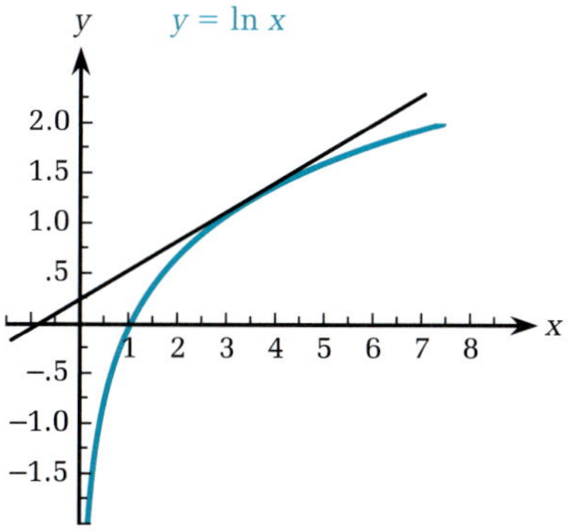

$y = \ln x$

FIGURE 4.4.2

On the left side we use the Chain Rule to get $\dfrac{d}{dx} e^{f(x)} =$ $e^{f(x)} \dfrac{d}{dx} f(x)$. On the right side we have $\dfrac{d}{dx} x = 1$. Combining these gives $e^{f(x)} \cdot \dfrac{d}{dx} f(x) = 1$. Because $e^{f(x)} = x$, we may substitute x for $e^{f(x)}$. This results in $x \cdot \dfrac{d}{dx} f(x) = 1$. Next, divide both sides by x to get $\dfrac{d}{dx} f(x) = 1/x$. With $f(x)$ replaced by $\ln x$, we have shown the correctness of the following equation.

DERIVATIVE OF ln *x*

$$\frac{d}{dx} \ln x = \frac{1}{x}$$

From the graph of $y = \ln x$ in figure 4.4.2 or from the formula $\dfrac{d}{dx} \ln x = 1/x$, it is apparent that for small positive values of x, the tangent line is quite steep, but as x increases, the slope of the tangent line decreases toward zero, although actually always remaining positive. The curve is concave downward, as can be verified from the second derivative, which is $-(1/x^2)$ and therefore always negative.

EXAMPLE 1   The derivative of a natural logarithmic function

Find derivatives of the following:

(a) $2 \cdot \ln x$ (b) $x^2 \cdot \ln x$ (c) $(\ln x)/\sqrt{x}$

SOLUTION

(a) $\dfrac{d}{dx} 2 \cdot \ln x = 2 \cdot \dfrac{d}{dx} \ln x = 2 \cdot \dfrac{1}{x} = \dfrac{2}{x}$

(b) $\dfrac{d}{dx} (x^2 \cdot \ln x) = x^2 \cdot \left(\dfrac{d}{dx} \ln x \right) + (\ln x) \cdot \left(\dfrac{d}{dx} x^2 \right)$ by the Product Rule

$= x^2 \cdot \dfrac{1}{x} + (\ln x) \cdot 2x$ by the Power Rule and the derivative of $\ln x$

$= x + 2x \cdot \ln x$

(c) $\dfrac{d}{dx}\left(\dfrac{\ln x}{\sqrt{x}}\right) = \dfrac{\sqrt{x}\cdot\left(\dfrac{d}{dx}\ln x\right) - \ln x\left(\dfrac{d}{dx}\sqrt{x}\right)}{(\sqrt{x})^2}$ by the Quotient Rule

$$= \dfrac{\sqrt{x}\cdot\dfrac{1}{x} - \ln x\cdot\left(\dfrac{1}{2\sqrt{x}}\right)}{(\sqrt{x})^2}$$

$$= \dfrac{\dfrac{1}{\sqrt{x}} - \dfrac{\ln x}{2\sqrt{x}}}{x}$$

Now, we use $2\sqrt{x}$ as a common denominator in the numerator, to get

$$\dfrac{d}{dx}\left(\dfrac{\ln x}{\sqrt{x}}\right) = \dfrac{\dfrac{2}{2\sqrt{x}} - \dfrac{\ln x}{2\sqrt{x}}}{x}$$

$$= \dfrac{\dfrac{2-\ln x}{2\sqrt{x}}}{x} = \left(\dfrac{2-\ln x}{2\sqrt{x}}\right)\cdot\dfrac{1}{x}$$

$$= \dfrac{2-\ln x}{2\cdot x^{3/2}}$$

PRACTICE EXERCISE 1

Find derivatives of the following:
(a) $5\ln x$ (b) $\sqrt{x}\cdot\ln x$ (c) $(\ln x)/x^3$

Derivative of $\ln|f(x)|$

Recall the Chain Rule from Chapter 3, $\dfrac{d}{dx}f(g(x)) =$

$\dfrac{d}{dg}f(g(x))\cdot\dfrac{d}{dx}g(x)$. We are going to use the Chain Rule with the "outside" function being the natural logarithmic function.

First, consider the special case of $g(x) = |x|$. If $x > 0$ then $|x| = x$, and so $\ln|x| = \ln x$. Then $\dfrac{d}{dx}\ln|x| = \dfrac{d}{dx}\ln x = 1/x$.

If $x < 0$ then $|x| = -x$, and we have $\ln|x| = \ln(-x)$. Taking the derivative of each side and applying the Chain Rule, we have $\dfrac{d}{dx}\ln|x| = \dfrac{d}{dx}\ln(-x) = \dfrac{1}{-x}\cdot\dfrac{d}{dx}(-x)$.

By finding $\dfrac{d}{dx}(-x) = -1$ we continue and obtain $\dfrac{d}{dx} \ln |x| = 1/x$.

Hence, regardless of whether $x < 0$ or $x > 0$, we have $\dfrac{d}{dx} \ln |x| = 1/x$.

Notice that $x \neq 0$ because $\ln 0$ is not defined. In the following result, we do not have to explicitly say the formula is true only if $f(x) \neq 0$, because $\ln |f(x)|$ is not defined for $f(x) = 0$.

The use of the Chain Rule with a general function $f(x)$ gives the following rule.

THE DERIVATIVE OF $\ln |f(x)|$

If $f'(x)$ exists, then

$$\frac{d}{dx} \ln |f(x)| = \frac{1}{f(x)} \cdot \frac{d}{dx} f(x)$$

which is equivalent to writing $(f'(x))/(f(x))$

EXAMPLE 2 Finding the derivative of $\ln |f(x)|$

Find the derivatives of the following:
(a) $\ln (x^2 + 1)$ (b) $x \cdot \ln (\ln x)$ (c) $\sqrt{\ln x}$

SOLUTION

(a) $\dfrac{d}{dx} \ln (x^2 + 1) = \dfrac{1}{x^2 + 1} \cdot \dfrac{d}{dx}(x^2 + 1)$ by the Chain Rule

$$= \frac{1}{x^2 + 1} \cdot 2x = \frac{2x}{x^2 + 1}$$

(b) $\dfrac{d}{dx} x \cdot \ln (\ln x) = x \cdot \dfrac{d}{dx} \ln (\ln x) + (\ln (\ln x)) \cdot \left(\dfrac{d}{dx} x\right)$ by the Product Rule

$$= x \cdot \left(\frac{1}{\ln x} \cdot \frac{d}{dx} \ln x\right) + (\ln (\ln x)) \cdot 1 \text{ by the Chain Rule}$$

$$= x \cdot \frac{1}{\ln x} \cdot \frac{1}{x} + \ln (\ln x) = \frac{1}{\ln x} + \ln (\ln x)$$

(c) $\dfrac{d}{dx} \sqrt{\ln x} = \dfrac{1}{2\sqrt{\ln x}} \cdot \dfrac{d}{dx} \ln x = \dfrac{1}{2\sqrt{\ln x}} \cdot \dfrac{1}{x} = \dfrac{1}{2x\sqrt{\ln x}}$

PRACTICE EXERCISE 2

Find the following:

(a) $\dfrac{d}{dx} \ln (2x + 3)$ (b) $\dfrac{d}{dx} \ln (x^{-2} + x^2)$

As an application of derivatives, consider Example 3, in which we seek to maximize an average profit.

EXAMPLE 3   Graphing a logarithmic function and seeking maximum average profit

Suppose the profit from the sales of x units is given by

$$P(x) = \begin{cases} 0 & \text{if } 0 \le x \le 1/3 \\ \ln 3x & \text{if } x > 1/3 \end{cases}$$

(a) Sketch a graph of $P(x)$ over $(0, 2]$.

(b) Let $f(x)$ be the average profit, and graph $f(x)$ over $(0, 2]$, finding the maximum of $f(x)$ on this interval.

SOLUTION

(a) To determine the graph of $P(x)$, we evaluate the function for some values of x. On the interval $(0, 1/3]$, the graph is horizontal, indicating no profit. Starting at $x = 1/3$, some values for $P(x)$ have been evaluated by using a calculator. The results are displayed in the chart below.

x	1/3	2/3	1	1.5	2
$P(x)$	0	0.6931	1.0986	1.5041	1.7918

For $x > 1/3$, the derivative of $P(x)$ is given by $P'(x) = (1/3x) \cdot 3 = 1/x$, which is positive for all such x-values. Thus $P(x)$ is increasing all the time over the interval $[1/3, 2]$. The graph of $P(x)$ is shown in figure 4.4.3.

(b) The average profit is

$$f(x) = P(x)/x = \begin{cases} 0 & \text{if } 0 < x \le 1/3 \\ (\ln 3x)/x & \text{if } x > 1/3 \end{cases}$$

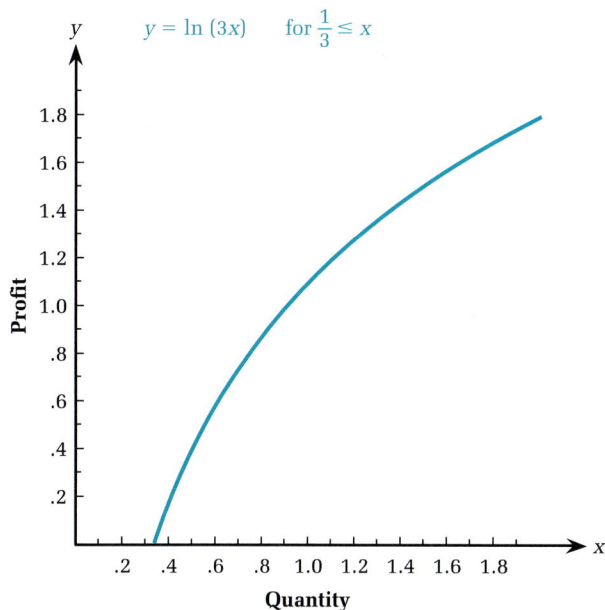

FIGURE 4.4.3

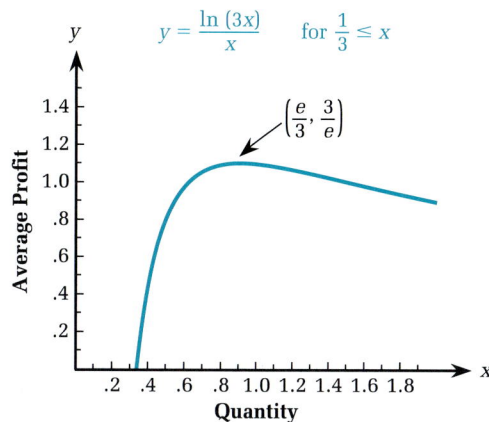

FIGURE 4.4.4

$(\ln 3x)/x$ is positive for $x > 1/3$, so we are certain the maximum must occur somewhere over the interval $[1/3, 2]$. It may occur at an endpoint, and we need to check those. If it does not occur at an endpoint, then it occurs at a point where $f'(x) = 0$.

We find the derivative, using the Quotient Rule, to be

$$f'(x) = \frac{x\left(\dfrac{1}{3x} \cdot 3\right) - (\ln 3x) \cdot 1}{x^2} = \frac{1 - \ln 3x}{x^2}$$

This is zero when the numerator is zero. Thus, we need to solve $1 - \ln 3x = 0$, which gives us $1 = \ln 3x$. By changing to an exponential form of the equation, we have $e^1 = 3x$. Solving for x gives $x = e/3 \simeq 0.9061$. This is a critical number in the interval $[1/3, 2]$.

We need to evaluate $f(x)$ at $\dfrac{e}{3}$ and at the right-hand endpoint, $x = 2$.

$$f\left(\frac{e}{3}\right) = \frac{\ln\left(3\left(\dfrac{e}{3}\right)\right)}{\dfrac{e}{3}} = \frac{\ln e}{\dfrac{e}{3}} = \frac{1}{\dfrac{e}{3}} = \frac{3}{e} \simeq 1.1036$$

$$f(2) = \frac{\ln (3 \cdot 2)}{2} = \frac{\ln 6}{2} \simeq 0.8959$$

Because $f(2) \le f(e/3)$, the maximum for the average profit is 1.1036. A graph of $f(x)$ is shown in figure 4.4.4.

Let us now return to the example in the introduction to this section.

EXAMPLE 4   Iodine removal

If we start with 40 grams of ^{131}I, then the time (measured in days) at which P grams have been removed is given by $T(P) = -(1/0.087)\ln(1 - P/40)$. Find the rate (in terms of days per gram) when $P = 30$.

SOLUTION

$$T'(P) = -\frac{1}{0.087}\left(\frac{1}{\left(1 - \dfrac{P}{40}\right)}\right)\left(-\frac{1}{40}\right) = \frac{1}{(0.087)(40 - P)}.$$

At $P = 30$ we have

$$T'(30) = \frac{1}{(0.087)(40 - 30)} = \frac{1}{0.087(10)} = \frac{1}{0.87} = 1.15 \text{ days per gram.}$$

The relative rate of change involves a descriptive use of logarithms.

Relative Rate of Change

We often have interpreted the derivative as a rate of change. This is especially the case when the function is presented as a function of time. In many situations, rates of change are compared. For example, is the cost of a loaf of bread rising as fast as the cost of a box of oranges? How do they compare with the increases in the prices of cars, houses, or pizzas, or the increase in the minimum wage? Is the amount of federal support for education rising as fast as federal support for health care?

If the cost of a box of oranges is increasing at a rate of $0.30 per year and the cost of FXIX turbo-automobiles is increasing at a rate of $200 per year, which is increasing at a faster rate of dollars? Of course, it is the turbo-autos. However, does that make as much difference as which is increasing at a faster rate *relative* to its current price?

If oranges cost $8.00 a box and turbo-autos cost $20,000 apiece, then the relative rates of change are, for a box of oranges, 0.30/8.00 = 0.0375, and for turbo-autos, 200/20,000 = 0.01. Thus, it is the price of a box of oranges that has a greater relative rate of change.

DEFINITION   Relative rate of change

The *relative rate of change* of $C(t)$ is $C'(t)/C(t)$.

By the Chain Rule, $\dfrac{d}{dt} \ln C(t) = \dfrac{1}{C(t)} \cdot C'(t) = \dfrac{C'(t)}{C(t)}$. Therefore, the derivative of the natural logarithm of the cost function is the relative rate of change of the cost function.

EXAMPLE 5   Relative rate of change

Suppose the cost function is $C(t) = t^2 + e^t$. Find the relative rate of change at $t = 3$.

SOLUTION

$\dfrac{d}{dt} \ln (t^2 + e^t) = (2t + e^t)/(t^2 + e^t)$. At $t = 3$, this is $(2 \cdot 3 + e^3)/(3^2 + e^3) \approx 0.8968$.

PRACTICE EXERCISE 3

Suppose a cost function is $C(t) = e^{2t}$. Find the relative rate of change at $t = 3$.

Notice: The remaining material of this section is not necessary for the understanding of the following chapters of the book. Hence it may be regarded as optional.

Derivative of $\log_b |f(x)|$

In order to find the derivative of a logarithmic function with an arbitrary base, we proceed as follows. Transform the function $\log_b |x|$ by using the change of base formula (L7 in Section 4.3).

$$\log_b |x| = \frac{\ln |x|}{\ln b}$$

Because b is a constant, $\ln b$ is also a constant, as is $1/\ln b$.

Rewriting $(\ln |x|)/(\ln b)$ as $(1/\ln b) \cdot \ln |x|$, we find

$$\frac{d}{dx} \log_b |x| = \frac{d}{dx} \left(\frac{\ln |x|}{\ln b} \right) = \frac{d}{dx} \left(\frac{1}{\ln b} \cdot \ln |x| \right)$$

$$= \frac{1}{\ln b} \cdot \frac{d}{dx} \ln |x| = \frac{1}{\ln b} \cdot \frac{1}{x}$$

DERIVATIVE OF $\log_b |x|$

$$\frac{d}{dx} \log_b |x| = \frac{1}{\ln b} \cdot \frac{1}{x}$$

For $\log_b |f(x)|$, the Chain Rule gives us $\dfrac{d}{dx} \log_b |f(x)| = \dfrac{1}{\ln b} \cdot \dfrac{d}{dx} \ln (|f(x)|)$, and because $\dfrac{d}{dx} \ln (|f(x)|) = \dfrac{1}{f(x)} \cdot \dfrac{d}{dx} f(x)$, we have the following expression.

DERIVATIVE OF $\log_b |f(x)|$

$$\frac{d}{dx} \log_b |f(x)| = \frac{1}{\ln b} \cdot \frac{1}{f(x)} \cdot \frac{d}{dx} f(x)$$

EXAMPLE 6   The derivative of $\log_b |f(x)|$

Find $\dfrac{d}{dx} \log (x^2 + 3x)$.

SOLUTION

Recall that if "log" is used without a subscript, then base 10 is meant. Thus, $b = 10$ and $f(x) = x^2 + 3x$.

$$\frac{d}{dx} \log (x^2 + 3x) = \frac{d}{dx} \log_{10} (x^2 + 3x)$$

$$= \frac{1}{\ln 10} \cdot \frac{1}{x^2 + 3x} \cdot (2x + 3)$$

Logarithmic Differentiation

Although exponential functions grow quite rapidly, the function $f(x) = x^x$ grows even quicker than b^x for any base b whatsoever. To find out just how fast it does grow, we find its derivative. We cannot use the formula for b^x or the formula for x^n because neither the base nor the exponent is a constant. To find the derivative we employ the technique of *logarithmic differentiation.* This consists of taking the natural logarithm of each side of an equation and then using implicit differentiation.

EXAMPLE 9   Logarithmic differentiation

Find $\dfrac{d}{dx} x^x$.

SOLUTION

Let $y = x^x$ and take the natural logarithm of each side. $\ln y = \ln x^x = x \cdot \ln x$ by a property of logarithms (L3).

Use implicit differentiation on each side to get

$$\frac{1}{y} \cdot y' = 1 \cdot \ln x + x \cdot \frac{1}{x}$$

Multiply each side by y and replace y by x^x to obtain

$$y' = x^x(\ln x + 1)$$

The process of *logarithmic differentiation* can be used on quite complicated functions. Let us consider another example.

EXAMPLE 10   Logarithmic differentiation

Find $\dfrac{d}{dx} (3x)^{x^2+1}$.

SOLUTION

Let $y = (3x)^{x^2+1}$ and take the natural logarithm of each side to get $\ln y = \ln (3x)^{x^2+1} = (x^2 + 1) \cdot \ln (3x)$ by a property of logarithms (L3).

By implicit differentiation, we have $(1/y) \cdot y' = (2x) \cdot \ln (3x) + (x^2 + 1) \cdot (1/(3x)) \cdot 3$ (remember the Chain Rule here).

Simplify this by bringing the y over to the right and replacing it by $(3x)^{x^2+1}$. Thus, $y' = (3x)^{x^2+1} \cdot (2x \cdot \ln (3x) + (x^2 + 1) \cdot 1/x)$.

4.4 PROBLEMS

Foundations

The problems of this section require the basic skills illustrated by the following:

Rewrite problems 1–4 so that they have x expressed to only a single power.

1. $\dfrac{x^2}{3x}$ **2.** $(x^2)\sqrt{x}$ **3.** $\dfrac{1/x}{2x^3}$ **4.** $\dfrac{2x^2}{\sqrt{x}}$

For problems 5–7, find an expression that is a quotient rather than a sum or difference.

5. $\dfrac{1}{2x} + x^2$ **6.** $\dfrac{1 + 2x}{x} + 3x$ **7.** $\dfrac{1}{\sqrt{x}} - 2x$

Exercises

Problems 8–45 cover only the material from that part of the section using the formula for $\dfrac{d}{dx}\ln f(x)$. Problems 45–55 cover the optional material.

Find the derivatives of the functions in problems 8–31.

8. $f(x) = 3\ln x$ **9.** $g(x) = (1/2)\ln x$

10. $f(x) = \ln(2x - 4)$ **11.** $f(x) = \ln(3x + 2)$

12. $f(x) = x^2 \ln 3x$ **13.** $f(x) = x\ln 2x$

14. $f(x) = \dfrac{\ln(x + 1)}{2x}$ **15.** $f(x) = \dfrac{\ln(2x + 1)}{x}$

16. $g(x) = 3(\ln x)^2$ **17.** $f(x) = (\ln \tfrac{1}{2}x)^3$

18. $f(x) = \ln(x^3 - x)$ **19.** $g(x) = \ln(x^2 + 2x)$

20. $f(x) = \ln(e^x + x)$ **21.** $f(x) = \ln(e^{2x} - 1)$

22. $f(x) = \ln\left(\dfrac{3x}{x + 1}\right)$ **23.** $f(x) = \ln\left(\dfrac{2x}{x^2 - 1}\right)$

24. $f(x) = x\ln x^{-3}$ **25.** $f(x) = x^2 \ln x^2$

26. $f(x) = 2x^2 \cdot \ln\sqrt{x}$ **27.** $f(x) = x^{-3}\ln x^2$

28. $f(x) = \dfrac{\ln x}{x^3}$ **29.** $f(x) = \dfrac{1 + x^3}{\ln x}$

30. $f(x) = x \cdot \ln x - x$ **31.** $f(x) = \ln|2x - 1|$

Find the second derivatives of each function in problems 32–34.

32. $f(x) = 3\ln x$ **33.** $g(x) = \ln 3x^2$

34. $k(x) = x \cdot \ln x - x$

35. Suppose the average cost per unit is given by $AC(x) = 4/x + \ln x^2$, $(x > 1)$. Find the minimum average cost per unit.

36. Suppose the average cost per unit is given by $AC(x) = 16/x + \ln x$, $(x > 1)$. Find the minimum average cost per unit.

37. Find the equation of the line tangent to the graph of $y = \ln x$ at the point $P(e, 1)$.

38. Find the equation of the line tangent to the graph of $y = \ln 3x$ at the point $P(e/3, 1)$.

Find the relative rates of change for the cost functions in problems 39 and 40 at the values indicated.

39. $C(t) = \dfrac{2t + 1}{t}$ at $t = 5$

40. $C(t) = 5 - e^{-t}$ at $t = 5$

Sketch graphs of the functions in problems 41 and 42, being sure to indicate any maximum or minimum values of any type (absolute or relative).

41. $f(x) = x^2 \ln x$ over $(0, 1]$

42. $f(x) = \dfrac{\ln x}{x}$ over $[1, 3]$

43. Find the minimum of $g(x) = \dfrac{x}{\ln x} - \dfrac{x}{4}$, $x > 1$.

44. If we start with 100 grams of radioactive isotope strontium-90 (symbol ^{90}Sr), then the time (measured in years) at which P grams have been removed is given by

$$T(P) = -41 \ln\left(1 - \dfrac{P}{100}\right)$$

Find the rate (measured in years per gram) at which ^{90}Sr is being transformed when P is 80.

45. *(Traffic Flow)* The graph in figure 4.4.5 shows the relationship between the number of cars passing a point in the Holland Tunnel (measured in cars per hour) as a function of the vehicle concentration (measured in cars per mile). As the tunnel becomes more congested, the number of cars passing a particular point actually decreases. Hence, the traffic engineers want to limit access in order to maximize traffic flow. The same problem occurs on freeways and is often controlled by traffic lights located at on-ramps.

The flow of vehicles is modeled by the function $q(x) = \lambda x \ln(m/x)$ where m is the maximum number of cars that can be packed (bumper to bumper) into a mile, x is the actual density of cars per mile, and λ is a constant that depends on the road and is measured in miles per hour. (Narrow tunnels and wide-open freeways have different psychological effects on drivers. For the Holland Tunnel, λ is about 19.)*

(a) Find the maximum for $q(x)$.
(b) The velocity of individual cars can be modeled by the function $v(x) = \lambda \ln(m/x)$. Find the velocity that allows the maximum flow to occur.

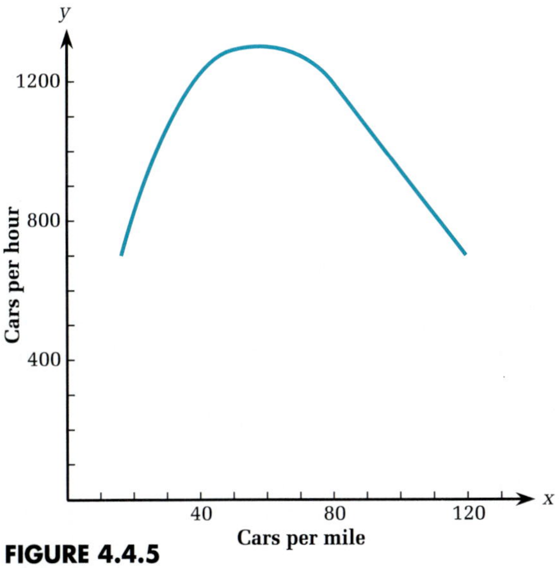

FIGURE 4.4.5

The following problems cover the optional parts of this section.

Find the derivatives of the functions in problems 46–55.

46. $g(x) = \log_3(2x + 4)$
47. $g(x) = \log_2(3x - 1)$
48. $f(x) = \log(x^2 - 1)$
49. $f(x) = \log(x^2 + 2x)$
50. $f(x) = 5^{2x}$
51. $f(x) = 2^{3x}$
52. $f(x) = 3^{x^2+4}$
53. $f(x) = 2^{x^2+1}$
54. $f(x) = x \cdot 3^x$
55. $f(x) = x \cdot 2^{-2x}$

Graphing Calculator Problems

56. Estimate any relative or absolute extreme values of $f(x) = (\ln(x+1))/2x + x^2$.
57. Estimate any relative or absolute extreme values of $f(x) = \ln(e^x - 2x)$.
58. Find the domain of $f(x) = \ln[2x/(x^2 - 1)]$ and sketch its graph.
59. Find the domain of $f(x) = \ln(e^x + x)$ and sketch its graph.

Enrichment Problems

Use logarithmic differentiation to find derivatives of the functions in problems 60–65.

60. $f(x) = x^{2x}$
61. $f(x) = x^{-x}$
62. $f(x) = (2x + 5)^x$
63. $f(x) = (3x + 1)^x$
64. $f(x) = x^{x^2}$
65. $f(x) = x^{\sqrt{x}}$

66. One of the useful keys on our calculator is the factorial key, whose symbol is $\boxed{x!}$. When a positive integer is displayed and the $\boxed{x!}$ key is pressed, the value calculated consists of the product of all positive integers that are less than or equal to the original integer. For example, if $\boxed{5}$ is displayed and $\boxed{x!}$ is pressed, then $5! = 5 \cdot 4 \cdot 3 \cdot 2 \cdot 1 = 120$, and the display is $\boxed{120}$.

As the insurance business developed in London in the eighteenth century, the estimation of certain risks required evaluation of $x!$ for large integer values of x. *Stirling's formula* (in honor of James Stirling) was published in 1738, and gave the approximation of

$$x! \simeq x^x e^{-x} \sqrt{2\pi x}$$

Although the function $x!$ is defined only for integers, the function $f(x) = x^x e^{-x} \sqrt{2\pi x}$ is differentiable by logarithmic differentiation. Find $f'(x)$.

*Montroll and Badger, *Introduction to Quantitative Aspects of Social Phenomena*, New York: Gordon and Breach, 1974, p. 247.

SOLUTIONS TO PRACTICE EXERCISES

1. (a) $\dfrac{d}{dx} 5 \ln x = 5 \cdot \dfrac{1}{x}$

(b) $\dfrac{d}{dx} \sqrt{x} \cdot \ln x = \sqrt{x} \cdot \dfrac{1}{x} + (\ln x) \dfrac{1}{2\sqrt{x}}$

(c) $\dfrac{d}{dx} \dfrac{\ln x}{x^3} = \dfrac{x^3 \cdot \dfrac{1}{x} - (\ln x) \cdot 3x^2}{(x^3)^2}$

$\qquad = \dfrac{x^2 - 3x^2 \cdot \ln x}{x^6} = \dfrac{1 - 3 \ln x}{x^4}$

2. (a) $\dfrac{d}{dx} \ln (2x + 3) = \dfrac{1}{2x + 3} \cdot 2 = \dfrac{2}{2x + 3}$

(b) $\dfrac{d}{dx} \ln (x^{-2} + x^2) = \dfrac{1}{x^{-2} + x^2} \cdot (-2x^{-3} + 2x)$

3. For $C(t) = e^{2t}$, the relative rate of change at $t = 3$ is $C'(t)/C(t) = 2e^{2t}/e^{2t} = 2$. (In fact, for this $C(t)$ the relative rate of change does not depend on t.)

4. $\dfrac{d}{dx} \log_2 (x^3 + x^2) = \dfrac{1}{x^3 + x^2} \cdot \dfrac{1}{\ln 2} \cdot (3x^2 + 2x)$

5. $\dfrac{d}{dx} 3^x = 3^x \cdot \ln 3$

6. $\dfrac{d}{dx} 3^{x^2-x} = 3^{x^2-x} \cdot (2x - 1) \cdot \ln 3$

4.5   MODELS OF GROWTH AND DECAY

The mathematical modeling of processes has a long and rich history. In the time of the famous painter Rembrandt van Rijn (1606−1669), Dutch towns first experimented with the practice of selling bonds to raise money. For a fixed purchase price, the city promised to pay each bondholder a lifetime of annual payments. Unfortunately for city treasuries, the cities did not consider the age of the purchaser. Thus, young and old alike received equal payments until their deaths. It quickly became apparent that youthful purchasers would bankrupt the cities, which promptly quit selling such bonds.

We should realize that in seventeenth-century Holland there were no models of lifespans. Such models were first established by the insurance underwriters in eighteenth-century London. They had been writing ship insurance and were beginning to write life insurance. They needed a basis for setting premiums.

Today, decisions on entitlement programs, such as Medicare, should depend on detailed models of potential costs. Unfortunately for all of us, far too often the models presented depend as much (or more) on political considerations as on fiscal, physical, and mathematical realities.

In addition to models from business, we discuss here topics from medicine and other subjects wherein exponential functions are widely used for model-building.

Exponential Growth

Before discussing bounded growth, let us review unbounded growth. We encountered exponential curves and exponential growth in Section 4.1. In that section, the growth models of money and populations were of *uninhibited* growth—expansion continued with no bound to the growth.

INNOVATORS IN APPLICATIONS OF MATHEMATICS
Thomas Malthus (1766–1834) and Pierre-François Verhulst (1804–1849)

Verhulst

Thomas Malthus achieved fame for his writings on what were called "gluts" in his time and are called "business depressions" today. In his famous book published in 1803, *An Essay on the Principle of Population,* Malthus, using the phrase "a slight acquaintance with numbers will show," argued that anything growing at a compound rate must eventually surpass anything growing at a simple rate. Thus, the growing population would exceed its capacity to support itself adequately. Malthus was concerned with the crowding, loss of quality of life, and human misery that would occur in a teeming society that had "standing room only."

The king of Holland, alarmed by Malthus's essay and concerned for the future of his own country, commissioned Pierre-François Verhulst to investigate the matter of population. In his report of 1844, Verhulst reported there were many factors that affected populations, among them food, shelter, and sanitation. His conclusion was that rather than the population continuing to increase exponentially, there would be a leveling off, as shown in figure 4.5.1. Verhulst named this model for population the *logistic model.* A Belgian himself, Verhulst offered as his "bottom line" that the population of Belgium would approach, but not exceed, a total of 9,400,000.

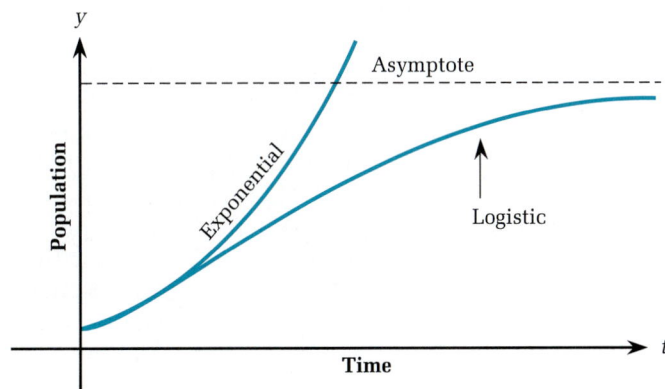

FIGURE 4.5.1

THE MODEL OF EXPONENTIAL GROWTH

Let $P(t)$ be the population at time t. Exponential growth occurs if the rate of growth, k, is directly proportional to the population. In symbols, $dP/dt = kP$, writing k in decimal form. If P_0 is the population at $t = 0$, then the equation for exponential growth is:

(4.5.1)
$$P(t) = P_0 e^{kt}$$

In Section 4.1 the growth rates were given. In general, that is not the case; rather, we have to determine the growth rate, as we do in Example 1.

EXAMPLE 1   Determining the rate of exponential growth

Suppose the city of San Diego is growing exponentially. In 1980 the population was 875,800 and in 1990 the population was 1,110,500.

(a) Find the equation of exponential growth.

(b) Project the population in 2010.

SOLUTION

(a) Let $t = 0$ correspond to 1980. Then $P_0 = 875,800$ and formula (4.5.1) is of the form $P(t) = 875,800e^{kt}$.

To determine k, substitute $t = 10$ (1990 is 10 years after 1980) and $P(10) = 1,110,500$, to get $1,110,500 = 875,800e^{10k}$.

This can be solved for k by using logarithms. First divide each side by 875,800 and then transform into an equivalent equation involving logarithms. $1,110,500/875,800 = 1.268 = e^{10k}$. Hence, $\ln 1.268 = 10k$. Solving for k gives $k = (\ln 1.268)/10 = 0.0237$. Thus, the equation of exponential growth is $P(t) = 875,800e^{0.0237t}$.

(b) The projected population for 2010 is found by letting $t = 30$ (the number of years after 1980).

$$P(30) = 875,800e^{0.0237 \cdot 30} = 875,800e^{0.711}$$
$$\approx 875,800 \cdot 2.036 \approx 1,783,000$$

PRACTICE EXERCISE 1

The population of Columbus, Ohio, was 565,000 in 1980 and 632,910 in 1990. Suppose the growth is exponential. Find the equation of exponential growth.

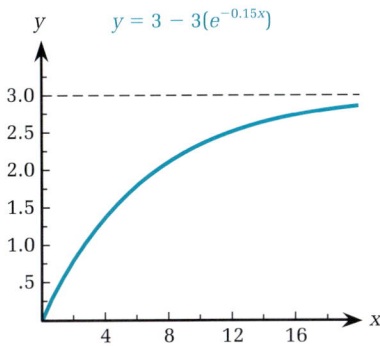

$y = 3 - 3(e^{-0.15x})$

FIGURE 4.5.2

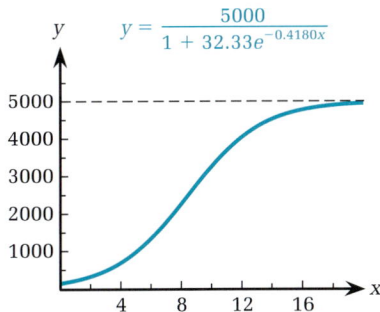

$y = \dfrac{5000}{1 + 32.33e^{-0.4180x}}$

FIGURE 4.5.3

Bounded Growth

There are almost always bounds to growth. However, knowing limits exist does not tell us what those limits are or what will happen along the way. We do not even know whether some steady-state equilibrium condition is possible. Crystal balls are notoriously hazy, frequently opaque, and always difficult to read. One economist said that the most important thing about making economic forecasts was to make sure that they were made frequently.

We investigate only two types of bounded growth. The first is bounded exponential growth, illustrated by the graph in figure 4.5.2, and the second is logistic growth, illustrated by the graph in figure 4.5.3.

Bounded Exponential Change

In some situations, growth is characterized by a rapid burst of change at the very beginning, followed by a gradual leveling off, and finally the graph of growth approaches a horizontal asymptote. One type of function whose graph exhibits this sort of behavior is a function whose rate of change is proportional to the difference between its current value and its limiting potential. It can be shown that the following formula is correct.

BOUNDED EXPONENTIAL GROWTH

Suppose $f(x)$ is a function such that $\dfrac{d}{dx} f(x) = k(M - f(x))$ and $f(0) = y_0$. Then

(4.5.2) $$f(x) = M - (M - y_0)e^{-kx}$$

We can show that formula (4.5.2) is a solution to the differential equation by taking the derivative of $f(x)$. Doing so, we have

$$\frac{d}{dx} f(x) = \frac{d}{dx} (M - (M - y_0)e^{-kx})$$

$$= 0 - (M - y_0)e^{-kx}(-k)$$

$$= k(M - y_0)e^{-kx}$$

$$= k(M - f(x))$$

In addition, it is true that $f(0) = y_0$.

EXAMPLE 2   Bounded exponential growth

Let the number of visitors to a new national park be given by $V(t)$. Suppose the rate of change of $V(t)$ is 15% of the difference between $V(t)$ and the estimated maximum capacity of the park, which is three million visitors a year.

(a) Find the number of visitors in the fifth year.

(b) Find the rate of change in the number of visitors that year.

SOLUTION

We want to find $V(t)$ and $V'(t)$, both evaluated at $t = 5$. Remembering to convert percentages into decimal values, we have $V'(t) = 0.15(3 - V(t))$. Thus, formula (4.5.2) assures us that $V(t) = M - (M - y_0)e^{-kx}$. The park is new, so $y_0 = 0$. The maximum capacity is 3 million, so $M = 3$. Using $k = 0.15$, we have $V(t) = 3 - 3e^{-0.15t}$.

(a) $V(5) = 3 - 3e^{-0.15(5)} = 3 - 3e^{-0.75} \approx 3 - 3 \cdot 0.4724 = 1.5829$ million visitors.

(b) $V'(5) \approx 0.15(3 - 1.5829) = 0.15 \cdot 1.4171 = 0.2126$ new visitors that year.

The graph for this example is shown in figure 4.5.2.

In Example 2, we knew the designed maximum capacity of the park. Usually, in modeling real situations we do not know that value. The general situation is that we have some data and we want to estimate the maximum.

For an advertisement, a charity solicitation by mail, or an offer to subscribe to a magazine, the function describing the *eventual total response* is actually of the form $R(t) = M(1 - e^{-kt})$. This is formula (4.5.2) with $y_0 = 0$, which is reasonable because no one responds before the offer is made. It can be shown that if r people respond by time $t = 1$ and s respond by time $t = 2$, then $1 + e^{-k} = s/r$. This can be solved for k by converting to natural logarithms. Then M can be found because $M = r/(1 - e^{-k})$.

EXAMPLE 3   Eventual total response

Suppose Alfredo's Sweats posts a sign offering all members a free month of swimming at their new pool. By the end of the first week 40% have responded, and by the end of the second week a total of 50% have responded. Assume the response function is a bounded exponential function.

(a) Find the response function.

(b) Find the eventual total response.

(c) Sketch a graph of the response function.

SOLUTION

(a) $s = 50$ and $r = 40$, so we have $1 + e^{-k} = s/r = 50/40 = 1.25$. Then $e^{-k} = 1.25 - 1 = 0.25$.

Converting to logarithms, $-k = \ln 0.25 = -1.3863$, so $k = 1.3863$, and $M = 40/(1 - e^{-1.3863}) = 40/(1 - 0.25) = 40/0.75 = 53.3$. Hence, the total response function is $R(t) = 53.3(1 - e^{-1.3863t})$ percent.

(b) The eventual total response will be approximately 53.3%.

(c) The graph of $R(t)$ is shown in figure 4.5.4.

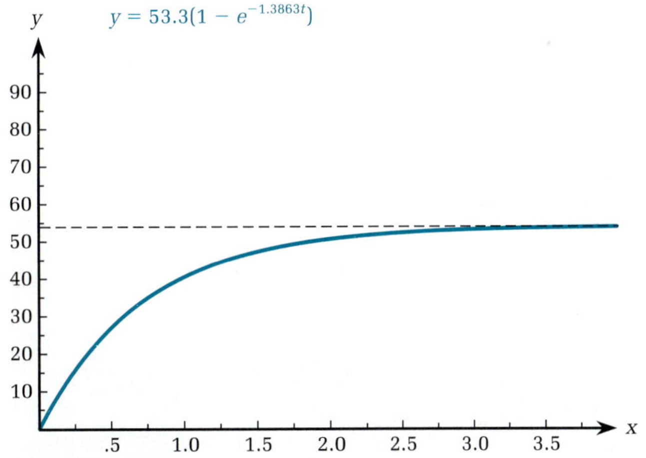

FIGURE 4.5.4

In some situations, we know we have a response function, but we do not know the day the offer is first made. Surprisingly, there still is a method for calculating the eventual total response.

Let A, B, and C be three observations of the cumulative total response at three equally spaced times. The eventual total response, M, can be found from

$$M = \frac{A \cdot C - B^2}{A + C - 2B}$$

It makes no difference which days are chosen as long as the time between the first and second observations is exactly as long as the time between the second and third. The demonstration of this result is left as an enrichment problem at the end of this section.

EXAMPLE 4   Eventual total response

Suppose an advertisement for Jim's Investment Gems is inserted in each copy of a magazine. Such inserts are known as "blow-ins." By some day there had been 2814 responses. Two days later, there had been 3754 responses, and two days after that there had been a total of 4524 responses. Estimate the eventual total response.

SOLUTION

From the formula, with $A = 2814$, $B = 3754$, and $C = 4524$, we have

$$K = \frac{(2814)(4524) - (3754)^2}{2814 + 4524 - 2(3754)} = \frac{12{,}730{,}536 - 14{,}092{,}516}{-170}$$

$$= \frac{1{,}361{,}980}{170} \simeq 8012$$

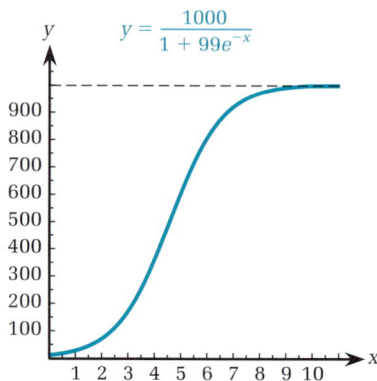

$y = \dfrac{1000}{1 + 99e^{-x}}$

FIGURE 4.5.5

Logistic Functions and Sigmoid Curves

The graphs of *logistic* functions are S-shaped curves. These are called *sigmoid curves,* named for the letter "sigma" in the Greek alphabet, which corresponds to the letter "S" in English. These functions approach limits and may do so in the manner indicated by the graph in figure 4.5.5.

Such a function may model the spread of a rumor, a disease, or a new technology. It may even model the size of a leaf growing on a plant.

THE MODEL OF LOGISTIC GROWTH

Logistic growth occurs if there is an absolute limit to the possible population and if the rate of growth is directly proportional to both the population and the remaining capacity for growth.

If we let $P = P(t)$ be the population at time t and M be the absolute limit, then $dP/dt = sP(M - P)$ for some constant s.

If we let P_0 be the population at $t = 0$, then the equation for logistic growth is

$$(4.5.3) \qquad P(t) = \frac{M}{1 + \left(\dfrac{M - P_0}{P_0}\right) \cdot e^{-sMt}}$$

The constant s can be thought of as some measure of "sociability." In order to make the formula somewhat less intimidating, let us rewrite it using as constants $S = (M - P_0)/P_0$ and $c = sM$.

THE GENERAL LOGISTIC FUNCTION

If $f(t)$ is a logistic function with an absolute limit to the population of M, then

$$(4.5.4) \qquad f(t) = \frac{M}{1 + S \cdot e^{-ct}} \qquad \text{for positive constants } S \text{ and } c.$$

The next example illustrates a logistic model.

EXAMPLE 5   Spreading a rumor, a logistic model

Suppose you heard from a friend, "Someone told me free pizzas are going to be handed out at 4 P.M. Friday."

Let $P(t)$ represent the number of people who have heard the rumor by time t and let M stand for the total number of people who eventually will hear this rumor. Assume the rate that new people hear the rumor is proportional to both the number of people who have already heard it (and are spreading it) and the number who have not heard it (a number that is decreasing as more and more hear the rumor). This is just the situation for which we can use the logistic model.

Let $P_0 = 10$, $M = 1000$, and $s = 0.001$. Sketch a graph of $P(t)$.

SOLUTION

Substituting in (4.5.3), we have $P(t) = 1000/(1 + 99 \cdot e^{-t})$.

Typical values of t in hours and $P(t)$ are shown in the following table, and these values are those graphed in figure 4.5.5.

t	0	1	2	3	4	5	6	7	8	9	10	11
$P(t)$	10	27	69	169	355	600	803	917	968	988	996	998

Maximum Rate of Increase of Logistic Functions (Inflection Points)

In Example 5, the rate at which the rumor spreads appears to be greatest when about half of the people have heard the rumor. That could be checked by finding $P'(t)$ in terms of t and maximizing it.

Finding the maximum of P' may be difficult to do by determining $P''(t)$ because P itself is quite a complicated function and its derivative is even more so. However, such an effort is unnecessary because we had defined $P(t)$ by the differential equation $P'(t) = sP(M - P)$, which equals $sMP - sP^2$, actually a quadratic polynomial. We know from graphing quadratic polynomials that the maximum value of this one is at $M/2$. Hence, the maximum of $P'(t)$, which is the maximum rate of increase for $P(t)$, occurs at $P(t) = M/2$. Looking at the graph in figure 4.5.6, we see this is the inflection point on the graph. Hence, to find the maximum rate of increase, solve $P(t) = M/2$ for t and substitute this value in $P'(t)$.

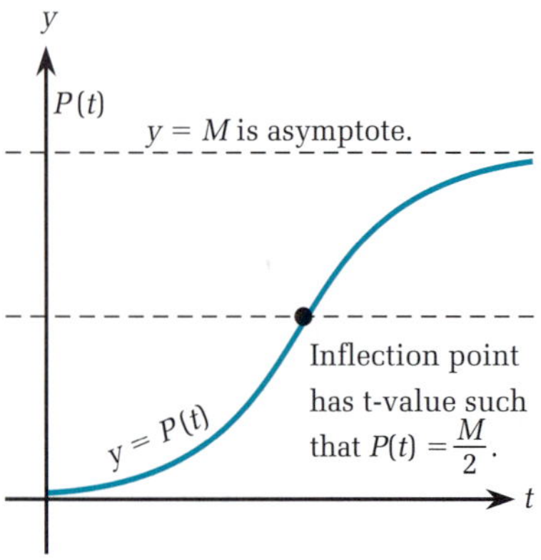

Let $S = (M - P_0)/P_0$ and $c = sM$, so we can work with the form $f(t) = M/(1 + S \cdot e^{-ct})$. To find the time t at the inflection point, we solve $(1/2)M = M/(1 + S \cdot e^{-ct})$ for t.

Dividing by M gives $1/2 = 1/(1 + S \cdot e^{-ct})$. Hence, $1 + S \cdot e^{-ct} = 2$, so that $S \cdot e^{-ct} = 1$. Dividing by S we have $e^{-ct} = 1/S$, so $S = e^{ct}$. Then, using logarithms, $\ln S = ct$. Solving for t we have $t = (\ln S)/c$.

This result is summarized in the following rule.

y = M is asymptote.

Inflection point has t-value such that $P(t) = \dfrac{M}{2}$.

$y = P(t)$

FIGURE 4.5.6

TO FIND THE MAXIMUM RATE OF INCREASE FOR A LOGISTIC FUNCTION (THE INFLECTION POINT)

If a logistic function is given by $f(t) = M/(1 + S \cdot e^{-ct})$, then the maximum rate of increase occurs at the time at which $f(t) = (1/2)M$. This is given by

(4.5.5)
$$t = \frac{\ln S}{c}.$$

The maximum rate of increase is $f'\left(\dfrac{\ln S}{c}\right)$.

An enrichment problem is to show $f''(t) = 0$ when $t = (\ln S)/c$. From biology comes another example of logistic growth.

EXAMPLE 6   Modeling the size of leaves and plants

The size of a plant leaf grows in a logistic manner with area, $A(t)$, given by $A(t) = K/[1 + (K - 1)e^{-rt}]$, where K and r are positive constants and t is the number of 12-hour periods.*

For $K = 140$ and $r = 1.2$, graph $A(t)$ for $t = 0, 1, 2, 3, 4, 5, 6, 7, 8, 9$, and 10. Find the time at which the leaf is growing most rapidly.

SOLUTION

Substituting the given constants, we have

$$A(t) = \frac{140}{1 + (139) \cdot e^{-1.2t}}$$

Values of $A(t)$ are shown in the following table, and the graph of $A(t)$ is shown in figure 4.5.7.

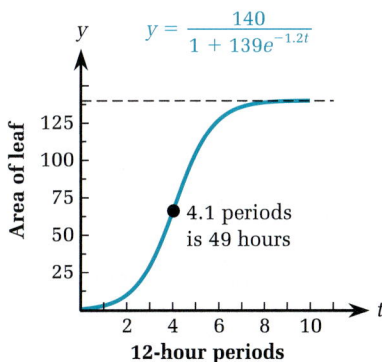

$y = \dfrac{140}{1 + 139e^{-1.2t}}$

4.1 periods is 49 hours

FIGURE 4.5.7

t	0	1	2	3	4	5	6	7	8	9	10
$A(t)$	1	3.3	10.3	29.2	65.3	104.1	126.8	135.8	138.7	139.6	139.9

To find the time of most rapid growth, we solve $t = (\ln S)/c = (\ln 139)/1.2 \approx 4.1$.

Because t is measuring 12-hour periods, $t = 4.1$ means about 49 hours.

TO FIND A LOGISTIC MODEL KNOWING THE MAXIMUM POPULATION AND TWO POINTS ON THE CURVE

If $f(t)$ is a logistic function with a maximum limit M to the population, then we know

$$f(t) = \frac{M}{1 + S \cdot e^{-ct}}$$

for positive constants S and c. (This is equation 4.5.4.) Substitute the two known points on the graph of $y = f(t)$ and evaluate S and c.

This is done in the following logistic model of the spread of a disease such as the flu or the common cold.

*R. Svoboda, "The Calculus of Leaves: A Modeling Project," *The UMAP Journal* 4:1 (1983), pp. 19–27.

EXAMPLE 7   Determining the logistic model from known points: epidemics

The graph showing the spread of the common cold through a susceptible population approximates a logistic curve. Let $P(t)$ be the total number of people who by day t have or have had a particular strain of a rhinovirus (and who are now immune from catching another cold from that particular strain).

Suppose $P(0) = 150$ cases, $P(5) = 1000$, and the total susceptible population is 5000. Find $P(t)$ and evaluate $P(12)$.

SOLUTION

$M = 5000$, so $P(t) = 5000/(1 + S \cdot e^{-ct})$. Now we use our knowledge of $P(0) = 150$ and $P(5) = 1000$ to evaluate S and c.

$P(0) = 5000/(1 + S) = 150$ gives $1 + S = 5000/150 = 33.33$. Thus $S = 33.33 - 1 = 32.33$. Substituting for S gives $P(t) = 5000/(1 + 32.33 \cdot e^{-ct})$.

Substituting $t = 5$ and $P(t) = 1000$ gives $1000 = 5000/(1 + 32.33 \cdot e^{-c \cdot 5})$.

To solve for c, $1 + 32.33 \cdot e^{-5c} = 5000/1000 = 5$, so $32.33 \cdot e^{-5c} = 5 - 1 = 4$. Therefore, $e^{-5c} = 4/32.33 = 0.1237$.

Converting to natural logarithms gives $-5c = \ln 0.1237 = -2.0899$, and finally, $c = 2.0899/5 = 0.4180$. Thus, the equation is $P(t) = 5000/(1 + 32.33 \cdot e^{-0.4180t})$, whose graph is shown in figure 4.5.3.

$$P(12) = \frac{5000}{1 + 32.33 \cdot e^{-0.4180 \cdot 12}} = \frac{5000}{1 + 32.33 \cdot e^{-5.016}}$$

$$\simeq \frac{5000}{1 + 0.2144} = \frac{5000}{1.2144} \simeq 4117$$

Graphs of logistic functions are fairly symmetric about the line $y = (1/2)M$, as shown in figure 4.5.8. When attempting to describe and graph a particular function, we can exploit this (almost) symmetry of the logistic curve and use the fact that at $t = (\ln S)/c$ we have $y = (1/2)M$. This allows us to make the following estimate.

THE SIGNIFICANT DURATION OF A LOGISTIC EVENT

If $f(t) = M/(1 + S \cdot e^{-ct})$ is a logistic model, then the time by which essentially all that is going to happen has happened is given by

(4.5.6) $t = \dfrac{2 \ln S}{c}$ (significant duration)

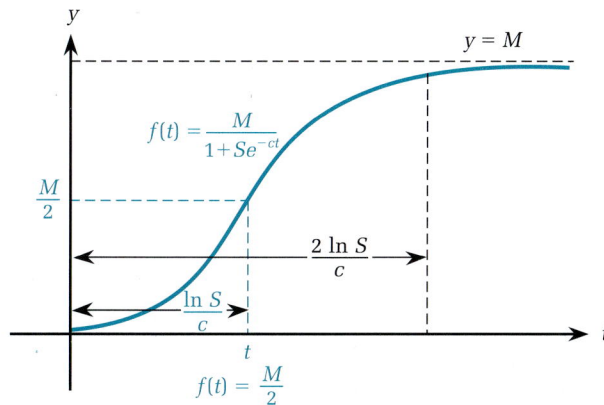

FIGURE 4.5.8

The significant duration is the time by which essentially all the susceptible people have caught the cold, or heard the rumor, or whatever.

EXAMPLE 8

Use the function of Example 7, $P(t) = 5000/(1 + 32.33 \cdot e^{-0.4180t})$, to answer the following.

(a) Find the day by which approximately one-half of the susceptible population had (or have had) the particular strain.

(b) Estimate the day by which essentially all have had it.

(c) Find the value of $P(t)$ for the day in part (b).

SOLUTION

(a) Using the formula above, approximately one-half will have had it by day $t = (\ln S)/c$. Substituting for S and c gives $t = (\ln 32.33)/0.4180 = 3.4760/0.4180 = 8.3$ days.

(b) $2 \cdot 8.3 = 16.6$. Thus, an estimation for the date when essentially everyone has had it would be 17 days.

(c) Actually computing $P(17)$, we have

$$P(17) = \frac{5000}{1 + 32.33 \cdot e^{-0.4180(17)}} = \frac{5000}{1 + 32.33 \cdot e^{-7.106}}$$

$$= \frac{5000}{1 + 0.02652} \cong 4871.$$

The value of 4871 is fairly close to 5000, the total susceptible population.

▓ Graphing Calculators: Estimating Screen Ranges

One of the most important uses of finding the "significant duration" value of formula (4.5.6) is in setting the ranges when graphing the logistic function. From our work in Examples 6 and 7, we want to graph that function over the interval [0, 20] or some similar interval that ends somewhat beyond the value of 17. If we use too long an interval, such as [0, 100], as shown in figure 4.5.9, the left-hand side becomes too crowded to show us a meaningful estimate of the inflection point. On the other hand, an interval that is too short, say perhaps [0, 5], as shown in figure 4.5.10, does not show us enough of the graph.

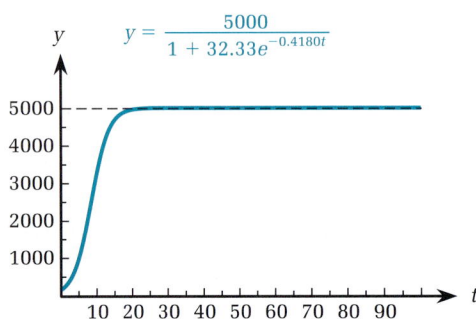

$$y = \frac{5000}{1 + 32.33e^{-0.4180t}}$$

FIGURE 4.5.9

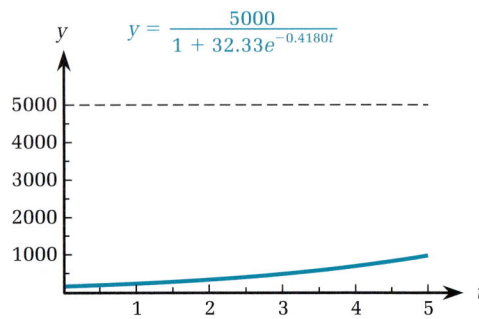

$$y = \frac{5000}{1 + 32.33e^{-0.4180t}}$$

FIGURE 4.5.10

PRACTICE EXERCISE 2

Suppose $f(t) = 3000/(1 + Se^{-ct})$, $f(0) = 10$, and $f(2) = 25$.

(a) Find S and c.

(b) Find t such that $f(t) = 1500$.

(c) Find t such that $f(t)$ is "essentially" 3000.

Applications to Radioactive Decay and Blood Plasma Drug Levels

In Section 4.4 we discussed a formula that would give the time necessary for a certain amount of the radioactive isotope of iodine, ^{131}I, to have changed into something else. Suppose we want to find a formula that gives the amount of ^{131}I that remains after time t. That is, rather than letting time be a function of how much of the isotope has become harmless, we want the amount of the remaining ^{131}I to be a function of time.

The rate of radioactive decay is always proportional to the amount present. That is, if we let $A(t)$ be the amount at time t, then $dA/dt = kA$ for some negative constant k. The asymptote is 0, and if we let A_0 represent the initial amount, then $A(t) = A_0 \cdot e^{kt}$.

Associated with every isotope is its *half-life,* the time it takes for one-half of the original substance to radioactively decay into other elements. The half-life for radioactive iodine, ^{131}I, is eight days. For example, if we start with four pounds, then after eight days we have $(1/2)(4) = 2$ pounds, after another eight days we have $(1/2)(2) = 1$ pound, after another eight days we have $(1/2)(1) = 1/2$ pound, and so on. Hence, after a total of 24 days we have 1/2 pound of our original two pounds left because we have gone through three half-life periods.

Use $A(t) = A_0 \cdot e^{kt}$ and let h be the half-life. Then $A(h) = (1/2) A_0 = A_0 \cdot e^{kh}$. We can solve for h by canceling A_0 and converting to logarithms. We have $1/2 = e^{kh,}$ so $\ln 0.5 = kh$, which gives $h = (\ln 0.5)/k$.

THE QUANTITY OF A RADIOACTIVE ISOTOPE

Suppose h is the half-life of a radioactive isotope. Starting with an amount A_0 of this isotope, the amount remaining at time t is

(4.5.7) $A(t) = A_0 \cdot e^{kt},$ where $k = \dfrac{\ln 0.5}{h}$

Conversely, if $A(t) = A_0 \cdot e^{kt}$, the half-life of the isotope is $h = (\ln 0.5)/k$.

Notice that in the formula for $A(t)$ the value of k is always negative because h is positive and $\ln 0.5 = -0.6931471$.

EXAMPLE 9   Quantity of a radioactive isotope

Tritium, a radioactive isotope of hydrogen, has a half-life of 12 years. Find the formula for the amount of tritium remaining after t years when we start with two pounds.

(a) When will we have 0.2 pounds?

(b) When $t = 5$, at what rate is the tritium decaying?

SOLUTION

$k = (\ln 0.5)/12 = -0.0577623$ and $A_0 = 2$, so by substituting in formula (4.5.7) we have $A(t) = 2 \cdot e^{-0.0577623t}$.

(a) We have 0.2 pounds when $0.2 = 2 \cdot e^{-0.0577623t}$. Dividing by 2 gives $0.2/2 = e^{-0.0577623t}$. Converting to natural logarithms gives $\ln 0.1 = -0.0577623t$. Solving for t, $t = (\ln 0.1)/(-0.0577623) = -2.302585/(-0.0577623) = 39.86$ years.

(b) To answer this, we need to find the derivative of $A(t)$. $A'(t) = (2 \cdot e^{-0.0577623t}) \cdot (-0.0577623)$, and this, evaluated at $t = 5$, is

$$A'(5) = (2 \cdot e^{-0.0577623(5)}) \cdot (-0.0577623)$$
$$= -2 \cdot 0.7491534 \cdot 0.0577623$$
$$= -0.08655 \text{ pounds per year.}$$

PRACTICE EXERCISE 3

Refer to Example 9. When will there be 1.2 pounds of tritium, and at that time at what rate will the tritium be decaying?

At this point let us consider an application to levels of drugs in blood plasma. Drugs introduced into the body are gradually eliminated. We are going to introduce two models of this process. The first model assumes the drug is already in the blood and is going to be eliminated. Furthermore, it assumes the amount of the drug remaining in the blood plasma has a half-life time, just as do radioactive isotopes. Thus, the equations for radioactive decay can be applied.

Nursing and other medical handbooks list the half-life values for drugs. These are listed using the symbol $\boxed{t_{\frac{1}{2}}}$. Table 4.5.1 lists $\boxed{t_{\frac{1}{2}}}$ times for several common drugs.

TABLE 4.5.1

DRUG	$t_{\frac{1}{2}}$
Aspirin	15 min
Cimetidine (for ulcers)	2 hr
Digoxin (a heart stimulant)	35 hr
Nadolol (a beta blocker)	20–24 hr
Penicillin V	30 min

EXAMPLE 10 Determining blood plasma drug levels

Suppose someone takes two aspirin tablets of 325 mg each. The level in the blood plasma reaches a peak and then in the following hours the amount of medicine in the bloodstream decreases. If it is assumed that the peak blood plasma level was 600 mg (not the full 650 because some was eliminated on the way to the peak level), then find an approximation for the amount of aspirin remaining t hours after the peak. How much is remaining after four hours?

SOLUTION

The value for the exponent in the drug level is calculated just like the exponent for radioactive decay. If a drug has a half-life of h hours, and there was D_0 at the peak, then the amount at time t is given by $D(t) = D_0 \cdot e^{[(\ln 0.5)/h] \cdot t}$.

For aspirin $h = 0.25$ (15 minutes is 1/4 of an hour), so we have $(\ln 0.5)/0.25 = -2.7726$, and $D(t) = 600 \cdot e^{-2.7726t}$ with t in hours.

A graph of $D(t)$ is shown in figure 4.5.11.

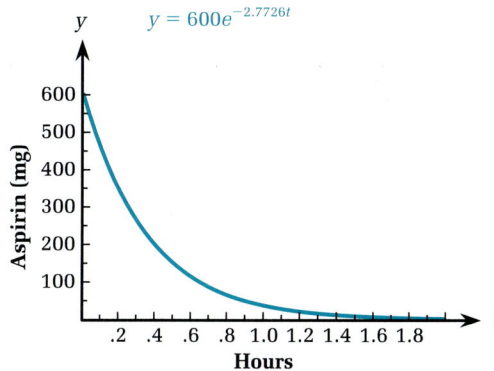

$y = 600e^{-2.7726t}$

FIGURE 4.5.11

After four hours, $D(4) = 600 \cdot e^{-2.77 \cdot 4} = 600 \cdot e^{-11.08} = 600 \cdot 0.0000154 = 0.009$ mg remains.

Consider a second model for drugs in blood plasma, one that takes into account the increasing concentration when the drug is first appearing in the blood. Such a model may be expressed in the form

$$D(t) = k_1(e^{-k_2 t} - e^{-k_3 t}) \text{ for constants } k_1, k_2, \text{ and } k_3 \ (k_3 > k_2).$$

The graph of $D(t) = 10(e^{-0.1t} - e^{-t})$ for $0 \le t \le 15$ is shown in figure 4.5.12.

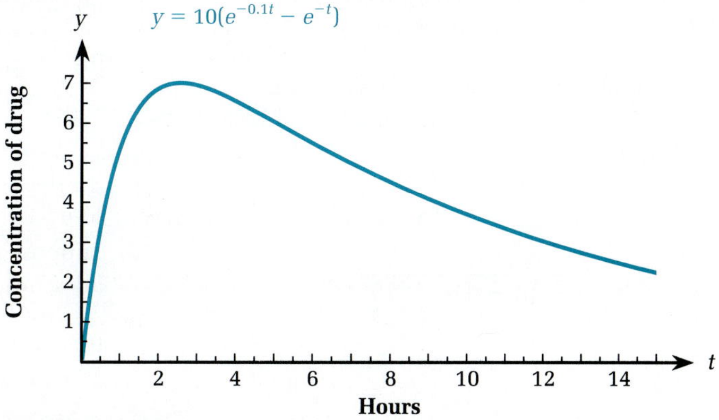

$y = 10(e^{-0.1t} - e^{-t})$

FIGURE 4.5.12

4.5 PROBLEMS

Foundations

The problems of this section require the basic skills illustrated by the following:

1. Evaluate e^{2x} for $x = 3$ and for $x = 3.7$.

2. Write $e^{-0.1x} = 2y + 1$ in the form of a logarithmic equation for x in terms of y.

3. Solve $1 + e^{-2x} = 6$ for x.

Exercises

4. Suppose the city of Boston is growing exponentially. In 1980 the population was 563,000 and in 1990 the population was 574,280.
   (a) Find the equation of exponential growth.
   (b) Project the population in 2010.

5. Suppose the city of Jacksonville is growing exponentially. In 1980 the population was 541,000 and in 1990 the population was 672,971.
   (a) Find the equation of exponential growth.
   (b) Project the population in 2010.

6. For $f(t) = 500/(1 + 100 \cdot e^{-t})$, find $f(t)$ for $t = 0$, 1, 2, 3, 4, 5, 6, 7, 8, 9, 10, and sketch the curve over the interval $[0, 10]$.

7. For $f(t) = 1000/(1 + 100 \cdot e^{-t})$, find $f(t)$ for $t = 0, 1, 2, 3, 4, 5, 6, 7, 8, 9, 10$, and sketch the curve over the interval $[0, 10]$.

8. **(Biology, Fruit Flies)**   The population of *Drosophila melanogaster* (the fruit flies used in many genetic experiments) can be modeled by $N(t) = 1035/(1 + 71.45 \cdot e^{-0.1667t})$. Find $f(t)$ for $t = 0, 5, 10, 15, 20, 25, 30, 35, 40, 45$, and sketch the curve over the interval $[0, 45]$.

The next three problems all refer to the work done by Edwin Mansfield.[*] He investigated the spread of a new technology in 12 situations. In each case he was able to model the acceptance by a logistic curve of the form

$$p(t) = \frac{n}{1 + (n - 1)e^{-k(t-t_0)}}$$

*E. Mansfield, "Technical Change and the Rate of Imitation," *Econometrica* 29:4 (Oct. 1961).

where n is the number of companies, t_0 is the year the innovation began, and t is the year that $p(t)$ companies had adopted the innovation.

An adaptation of his figures yields the values shown in table 4.5.2.

TABLE 4.5.2

INNOVATION	n	t_0	k
Diesel locomotives on railroads	25	1925	0.252
Continuous mining machinery	17	1947	0.482
High-speed bottle fillers	16	1951	0.330

For each of the problems 9–11, refer to table 4.5.2.

9. **(Logistic, Spread of Technology)**   Sketch a graph of $p(t)$ for diesel locomotives.
   (a) Estimate the year by which half of the companies had adopted the innovation.
   (b) Estimate the year the innovation had been adopted by "essentially all" the industry.

10. **(Logistic, Spread of Technology)**   Sketch a graph of $p(t)$ for continuous mining machinery.
    (a) Estimate the year by which half of the companies had adopted the innovation.
    (b) Estimate the year the innovation had been adopted by "essentially all" the industry.

11. **(Logistic, Spread of Technology)**   Sketch a graph of $p(t)$ for high-speed bottle filling machines.
    (a) Estimate the year by which half of the companies had adopted the innovation.
    (b) Estimate the year the innovation had been adopted by "essentially all" the industry.

Problems 12–16 all use the logistic curve model $f(t) = M/(1 + S \cdot e^{-ct})$.

12. Suppose $f(10) = 12$ and $f(15) = 24$.
    (a) Find values of S and c such that $f(t) = 100/(1 + Se^{-ct})$.
    (b) Estimate the value of t such that $f(t)$ has the maximum rate of increase.

13. Suppose $M = 1000$, $f(0) = 5$, and $f(5) = 75$.
    (a) Find the equation of the logistic curve.
    (b) Find t so that $f(t) = 200$.
    (c) Find the value of t for which $f(t) = (1/2)M$.

14. Suppose $M = 500$, $f(0) = 1$, and $f(5) = 25$.
    (a) Find the equation of the logistic curve.
    (b) Find t so that $f(t) = 100$.
    (c) Find the value of t for which $f(t) = (1/2)M$.

15. *(Logistic, Spread of Disease)*   There are 550 susceptible students at Rydell High School. The school nurse found five of them had flu on day 0, and by day 5 the number was 65. Assume the curve is logistic.
    (a) Estimate the number of students who have or have had the flu by day 10.
    (b) Estimate the earliest time at which essentially all of the susceptible students will have had the flu.

16. *(Logistic, Spread of Rumor)*   There are 5000 people in Littleville, where everyone gossips. At noon Jake starts the "rumor" that the mayor is resigning. By 3 P.M. 50 people have heard the rumor. Assume the curve is logistic.
    (a) Estimate the number of people who will have heard the rumor by 6 P.M.
    (b) Estimate the earliest time at which essentially everyone in Littleville will have heard the rumor.

17. *(Eventual Total Response)*   Suppose the cumulative total response from a mail order offer on day 1 is 2047, by day 2 is 3745, and by day 3 is 5153. Estimate the eventual total response.

18. *(Eventual Total Response)*   Suppose the LaPlata Mall sends out 5000 coupons in the mail. The cumulative total number that have been cashed by day 2 is 830, by day 5 is 1298, and by day 8 is 1440. Estimate the eventual total response.

19. *(Radioactive Decay, Carbon-14 Dating)*   An attendance scroll, allegedly from Plato's Academy, has been offered to you for sale. You use a carbon-14 dating procedure and find that the amount of ^{14}C is 90% of the original amount. How old is the scroll? (The half-life of ^{14}C is 5730 years.)

20. *(Radioactive Decay, Carbon-14 Dating)*   A beam from a cliff dwelling at Mesa Verde is dated as having been cut in 1250 A.D. What percentage of its original ^{14}C would it contain in 1990? (The half-life of ^{14}C is 5730 years.)

21. *(Stress Testing with Thallium)*   During a stress test, a person exercises strenuously (usually on a treadmill) while closely monitored. The test lasts for approximately 12 minutes, depending on the person's health. As closely as possible to one minute from the conclusion of the test, thallium may be injected into a vein. Thallium is radioactive with a half-life of 73 hours. Muscle tissue absorbs the thallium, but plaque and lesions do not. Thus, by a series of photos, taken two hours apart, problems with blood flow to the heart can be detected.

    If the thallium were all retained in the body (it is not), how much of the original dose would be present after 24 hours?

22. *(Biochemical, Drug Levels)*   Tagamet, often prescribed for patients with ulcers, has a half-life of two hours.
    (a) How much of an initial dose is present five hours after the peak plasma level?
    (b) How long does it take to get down to only 10% of the peak plasma level?

23. *(Biochemical, Drug Levels)*   Penicillin V has a half-life of 30 minutes.
    (a) How long does it take to get down to only 10% of the peak plasma level?
    (b) How much of the peak plasma level is present after 90 minutes?

Graphing Calculator Problems

24. *(Eventual Total Response)*   Suppose a bag of packets of $20 bills falls from an airplane into an Oregon river. After thirty days, 20 packets have been found floating downstream. After fifty days, 25 packets have been found. Using a bounded exponential model, estimate the eventual total number of packets that will be found.

25. Use formula (4.5.6) to estimate ranges for a screen showing $f(t) = 50/(1 + 3e^{-0.03t})$.

Writing and Discussion Problems

26. *(Logistic Model, Black and White Televisions)*
In 1992 the company Technology Futures Inc.
reported the information in the following
table about American households, giving the
percentage having only black and white television sets.

Year	1957	1977	1982	1992
Percentage	99	24	12	1

(a) Discuss how this phenomenon can be
modeled by a modified logistic function.
(b) Using the data points (1977, 24) and
(1982, 12), find an equation for a modified
logistic function.
(c) Using the function from (b), estimate the
number of households having only black
and white TVs in 1967.
(d) Estimate the year during which the maximum rate of households were losing their
status of having only a black and white
television.
(e) Discuss whether the function in part (b) is
a good approximation. In particular, how
does it compare with the known values in
1957 and 1992?

27. *(Recycled Newsprint Capacity in North
America)* According to the Newspaper Association of America, the following table gives the
capacity that existed for producing recycled
newsprint in North America. The values are in
millions of tons.

Year	1989	1990	1991	1992	1993	1994
Capacity	2.1	2.4	3.4	6.9	8.7	9.1

(a) Plot these points.
(b) What sort of model seems to fit the data?
(c) If a logistic model is used, estimate the
ultimate capacity.

28. Why does the logistic function $f(t) = M/(1 + S \cdot e^{-ct})$ never reach the value M?

29. Explain why the rate of increase of the number
of people who have heard a rumor is proportional to both the number who have heard it and
also the number who have not heard it.

Enrichment Problems

30. Suppose growth is exponential and that in 10
years the population has increased by a total of
x percent. Find a formula for the rate of exponential growth, k, in terms of x.

31. Suppose $f(t) = M/(1 + S \cdot e^{-ct})$. By finding $f''(t)$
show that $f''(t) = 0$ when $t = (\ln S)/c$.

32. *(Eventual Total Response)* Suppose that
$f(t) = K - ae^{-bt}$ and j is some positive fixed
number. If $f(t) = K - ae^{-bt} = A$, $f(t + j) = K - ae^{-b(t+j)} = B$, and $f(t + 2j) = K - ae^{-b(t+2j)} = C$,
show that $K = (A \cdot C - B^2)/(A + C - 2B)$.

SOLUTIONS TO PRACTICE EXERCISES

1. Solving $632{,}910 = 565{,}000e^{10k}$, we have
$\ln(632{,}910/565{,}000) = 10k$, so $k = 0.0113$.

2. (a) By substitution, $f(0) = 3000/(1 + S \cdot e^{-c \cdot 0})$
$= 3000/(1 + S \cdot 1) = 3000/(1 + S)$. Hence, with
$f(0) = 10$, we have $1 + S = 300$, so $S = 299$.
Using this value for S, we know $f(t) = 3000/(1 + 299e^{-c \cdot t})$. Substituting $f(2) = 25$
yields $25 = 3000/(1 + 299e^{-c \cdot 2})$. This reduces
to $e^{-2c} = 119/299 = 0.3980$. Converting to an
expression with natural logarithms, $-2c = \ln(0.398)$, so that $c = 0.4607$. Hence, $f(t) = 3000/(1 + 299e^{-0.4607t})$.

(b) Using formula (4.5.5), we have $f(t) = 1500$
when $t = (\ln 299)/0.4607 = 12.4$.
(c) Using formula (4.5.6), we have $t = (2 \cdot \ln S)/c = (2 \cdot \ln 299)/0.4607 = 24.8$.

3. For $A(t) = 2 \cdot e^{-0.0577623t}$, we have $A(t) = 1.2$
when $2 \cdot e^{-0.0577623t} = 1.2$. Dividing by 2 and
taking the natural logarithm of each side, we
have $-0.0577623t = \ln(1.2/2) = \ln 0.6$. Solving
for t gives $t = 8.8436$ years.

To find the rate at 8.8436 years, use
$A'(t) = (2 \cdot e^{-0.0577623t}) \cdot (-0.0577623)$ and find
$A'(8.8436) = -0.06931$ pounds per year.

PRESENT VALUE AND ELASTICITY OF DEMAND

This section consists of two major topics, either of which can be studied independently of the other. The first topic is *present value.* Hereby we consider the present value of a future payment of money. For instance, suppose you buy a U.S. savings bond with a face value of $100 that has a stated rate of interest of 5.8% and is payable in five years. What amount of money is that bond worth today?

Another situation of present value concerns an asset that changes value over time, usually increasing. This topic was first studied by Martin Faustmann, who, in 1849, published a model for harvesting a forest in such a manner as to optimize its total value over perpetuity.

The second topic of the section is *elasticity of demand.* This concerns demand functions and how changing prices affect demands and revenues. For instance, suppose a rise in the price of gasoline only slightly reduces consumption. Then total revenue will probably increase because revenue is the product of the quantity of a commodity sold times the commodity's price. On the other hand, suppose an airline raises fares and ridership declines dramatically. Then the total revenue will probably decrease.

The topic of elasticity of demand is not required for subsequent sections of the book and may be omitted without loss of continuity.

Present Value

A common obligation is to pay some amount of money at a future date. Home mortgages, car loans, even state lotteries may establish a series of payments stretching out 20 years or more. Clearly, a promise to pay $10 million 20 years from now is worth a lot less than $10 million today. An indication of how much less can be seen by calculating what $10 million would grow into over 20 years.

Suppose we invest $10 million at 6% annual interest with continuous compounding. Recall that $P(t) = P_0 \cdot e^{it}$ is the formula for the value at time t of an account established with P_0 dollars, paying an annual interest rate of i (written decimally) and with continuously compounding interest.

Thus, if $P_0 = 10$ and $i = 0.06$, then $P(20) = 10 \cdot e^{0.06 \cdot 20} = 10 \cdot e^{1.2} = 10 \cdot 3.3201 = \$33.201$ million after 20 years. That is considerably more than $10 million. How much would we have to invest in order to generate just $10 million? This leads us to Example 1.

EXAMPLE 1   Finding a necessary initial deposit to fulfill a goal

Determine the amount of money we would have to invest at an annual interest rate of 6% with continuous compounding in order to create $10 million after 20 years.

SOLUTION

To find this, we solve the equation $P(t) = P_0 \cdot e^{it}$ for P_0. We have $i = 0.06$, $t = 20$, and $P(20) = 10$. Substituting these values gives us the equation $10 = P_0 \cdot e^{0.06 \cdot 20} = P_0 \cdot e^{1.2}$.

We have $P_0 = 10/e^{1.2} \simeq \$3.01194$ million.

The value of 3.01194 million is called the *present value* of $10 million, payable 20 years from now, assuming an annual interest of 6% with continuous compounding.

The general situation involves finding P_0 by dividing each side of $P(t) = P_0 \cdot e^{it}$ by e^{it}. The following definition uses the fact that $1/e^{it} = e^{-it}$.

DEFINITION

Present value, $P_v(t)$, and discount rate

The *present value* of P dollars, due t years in the future, with i the annual interest rate and continuous compounding, is $P_v(t) = P \cdot e^{-it}$. The rate i is the *discount rate*.

EXAMPLE 2   Present value

What is the present value of $4 million payable in five years? The annual interest rate is 6% compounded continuously.

SOLUTION

Use $P_v(t) = P \cdot e^{-it}$ with $P = 4$, $i = 0.06$, and $t = 5$. Substituting these values gives $P_v(5) = 4 \cdot e^{-0.06 \cdot 5} = 4 \cdot e^{-0.3} = 4 \cdot 0.7408 = \$2.9632$ million.

PRACTICE EXERCISE 1

What is the present value of $50,000 payable in 10 years? The annual interest rate is 8% compounded continuously.

Zero-Coupon Bonds

A special category of "present value" problems are *zero-coupon bonds.* A zero-coupon bond pays no interest during the life of the bond, and instead pays the entire amount at the maturity date. Thus, the original purchase price of such a bond is the present value of the face value of the bond at the stated annual interest rate, compounded continuously. Such bonds are a favorite investment when capital gains are taxed at a lower rate than ordinary income, or when an investor wishes to defer taxable income. Some types of savings bonds issued by the United States government are in this category.

EXAMPLE 3   Zero-coupon bonds

A zero-coupon bond with a face value of $10,000 is payable at the end of five years. If current annual interest rates are 8.4% with continuous compounding, what should be the price of the bond?

SOLUTION

Present value is $P_v = 10,000 \cdot e^{-0.084 \cdot 5} = 10,000 \cdot e^{-0.42} = \$6570.47.$

Frequently the buyer of a bond does not know, or think of, the interest rate. Rather, the buyer of a bond only knows the asking price, the face value, and the time until the payment. This information is sufficient to determine the implied annual interest rate (that is, the discount rate).

EXAMPLE 4   Interest rates determined by present value

A zero-coupon bond with a face value of $10,000 is payable at the end of 10 years. It is offered at $7000. Assuming continuous compounding, what is the implied annual interest rate?

SOLUTION

Use the formula $P_v(t) = P \cdot e^{-it}$ with $P_v(10) = 7000$, $t = 10$, and $P = 10,000$.

Then $7000 = 10,000 \cdot e^{-i \cdot 10}$. Dividing by 10,000 gives $0.7 = e^{-10 \cdot i}$. This should be converted to the equivalent equation with natural logarithms, $\ln 0.7 = -10i$.

Solving for i, we have $i = (\ln 0.7)/(-10) = -0.3567/(-10) = 0.03567$. This is a decimalized version of interest, so the rate is 3.567%.

Present Value of an Appreciating Asset

Suppose an asset is anticipated to increase in value over time. For example, fine wine increases in price over time. So does a growing plantation of pulpwood. However, in both of these situations there is a time at which the continuing increase in value slows and, in fact, for both of these the asset eventually begins to lose value. This type of asset is called an *appreciating asset*. The graph of a typical appreciating asset is shown in figure 4.6.1.

Obviously, there comes an optimal time to sell the wine or harvest the pulpwood. We use present value to determine that time.

Suppose that $W(t)$ is the dollar worth of an asset at time t, and assume the current annual interest rate (called again the *discount rate*) is i. Then the present value of future sale is

$$P_v(t) = W(t) \cdot e^{-it}$$

Examples of graphs of $P_v(t)$ and $W(t)$ are shown in figure 4.6.2. The present value at time t is less than $W(t)$ because the value of $W(t)$ has to be discounted to bring the value back to present-day dollars.

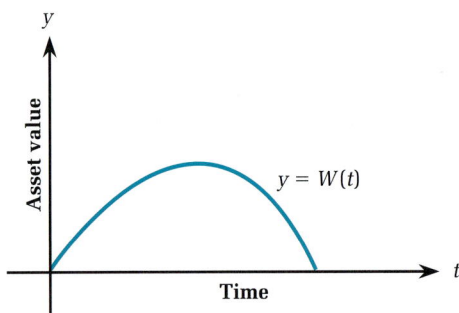

FIGURE 4.6.1

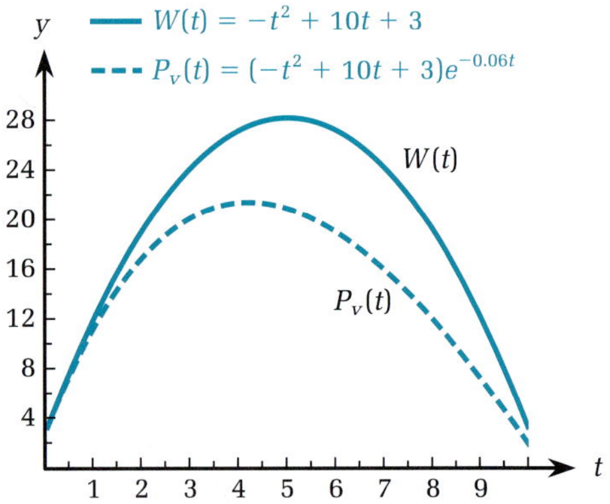

$W(t) = -t^2 + 10t + 3$
$P_v(t) = (-t^2 + 10t + 3)e^{-0.06t}$

FIGURE 4.6.2

In an effort to maximize $P_v(t)$ we search for critical numbers of $P_v(t)$ and we check any values of t that are endpoints of the time period. The critical numbers occur if we find a time b such that $P'_v(b) = 0$ or does not exist.

By the Product Rule for derivatives, $P'_v(t) = W(t) \cdot (e^{-it} \cdot (-i)) + e^{-it} \cdot W'(t)$. To solve $P'_v(t) = 0$, factor out the term e^{-it} to give $e^{-it} (W(t) \cdot (-i) + W'(t)) = 0$.

$e^{-it} \neq 0$ for any t whatsoever, so we have $W(t) \cdot (-i) + W'(t) = 0$. Solving this for i gives us the following result.

MAXIMUM VALUE OF AN APPRECIATING ASSET

Suppose the worth of an appreciating asset is given by $W(t)$, and $P_v(t)$ is the present value of $W(t)$ with a discount rate of i. If there is a maximum value for $P_v(t)$, then it occurs for t such that $t = 0$, t is the end of the time period (if such a time exists), or $W'(t)/W(t) = i$.

Because $W'(t)/W(t)$ is the relative rate of change, this result shows that when the relative rate of change is equal to the interest rate we are at a critical point. As long as the relative rate of change of the asset's value is greater than the interest rate, then we should hold on to the asset. When the relative rate of change is less than the interest rate, then we should sell the asset and invest the proceeds elsewhere.

EXAMPLE 5   Appreciating asset

Suppose the value of a bottle of wine is given by $W(t) = 2t$ and the discount rate is $i = 0.08$.

(a) In order to maximize present value, how many years should go by before the bottle is sold?

(b) What is the maximum present value?

SOLUTION

(a) $W'(t) = 2$. Thus, $W'(t)/W(t) = 2/(2t) = 1/t$. The discount rate i is 0.08. Solving $0.08 = 1/t$ gives $t = 1/0.08 = 12.5$ years. This is a maximum because $1/t > 0.08$ for $t < 12.5$ and $1/t < 0.08$ for $t > 12.5$.

(b) The maximum present value is $P_v(12.5) = 2 \cdot 12.5 \cdot e^{-0.08 \cdot 12.5} = 25 \cdot e^{-1} = 9.20$.

EXAMPLE 6   Appreciating asset

Suppose the value of a commodity is given by $W(t) = -t^2 + 10t + 3$, where $W(t)$ is in dollars and t is in years. The discount rate is 0.06.

(a) In order to maximize present value, how many years should go by before the commodity is sold?

(b) What is the maximum present value?

SOLUTION

Setting $i = W'(t)/W(t)$ gives $0.06 = (-2t + 10)/(-t^2 + 10t + 3)$. This is equivalent to solving $-0.06t^2 + 0.6t + 0.18 = -2t + 10$. Collecting terms on one side gives the quadratic equation $-0.06t^2 + 2.6t - 9.82 = 0$. Using the quadratic formula, we have

$$t = \frac{-2.6 \pm \sqrt{(2.6)^2 - 4(-0.06)(-9.82)}}{2(-0.06)} = \frac{-2.6 \pm \sqrt{6.76 - 2.3568}}{-0.12}$$

$$= \frac{-2.6 \pm \sqrt{4.4032}}{-0.12} = \frac{-2.6 \pm 2.0984}{-0.12} = 4.18 \text{ or } 39.15.$$

The time $t = 39.15$ cannot be correct because $W(39.15) = -(39.15)^2 + 10(39.15) + 3 < 0$. For $t = 4.18$ we have $W(4.18) = -(4.18)^2 + 10 \cdot 4.18 + 3 = -17.47 + 41.8 + 3 = 27.33$, and $P_v(4.18) = (-(4.18)^2 + 10 \cdot 4.18 + 3) \cdot e^{-0.06 \cdot 4.18} = 27.33 \cdot 0.7782 = 21.27$.

This answer means we should hold the commodity for 4.18 years, then sell it for $27.33 at that time. The future receipt of $27.33 has a present value of $21.27, as compared to the present immediate sales value (at $t = 0$) of $3.

These particular functions for $W(t)$ and $P_v(t)$ are used for the graphs in figure 4.6.2.

There are functions $W(t)$ such that the maximum present value continues to increase indefinitely. For these, if there is no limit on the time we can hold the asset, then we should never sell. On the other hand, there are some assets that we should sell right away, for their present values will only fall. Examples of each situation are presented in the problems at the end of this section.

PRACTICE EXERCISE 2

Suppose the value of a bottle of wine is given by $W(t) = -t^2 + 10t$, where $W(t)$ is in dollars and t is in years. The discount rate is 0.06.

(a) How many years should go by before the bottle is sold in order to maximize present value?

(b) What is the maximum present value?

When solving problems involving appreciating assets we may not have a formula for $W(t)$. However, we may be able to use empirical data to estimate $W'(t)$ and $W(t)$. If so, then we can estimate a solution to $i = W'/W$. For more information on this topic, see the paper by Robert H. Lamberson, "The Faustmann Model of Optimal Forest Rotation," *UMAP Journal* 1:3 (1980), pp. 27–34.

Elasticity of Demand

Total revenue is determined by both the quantity sold and the price of each item. However, the price and the quantity are related by what we have called a demand function. A change in price affects the demand. A small change in price may have very little effect on demand or it may have a significant effect. To help analyze these changes, economists have introduced the term *elasticity of demand*. This is the ratio of the relative change of the quantity sold with respect to the relative change of price. In order to better understand this concept and how it relates to maximizing revenue, we need to introduce some notation.

> ## RELATIVE CHANGE OF PRICE AND QUANTITY
>
> Suppose x units of a commodity are sold at a price p. If the price is changed and a new number of units are sold at the new price, then let Δx stand for the change in quantity sold and Δp stand for the change in price.
>
> The relative changes in quantity and price are $\Delta x/x$ and $\Delta p/p$, respectively.

EXAMPLE 7   Relative change in price and quantity

Suppose a clothing store is selling coats. Priced at \$39, nine coats are sold. Priced at \$56, eight are sold.

(a) Find the relative change of quantity.

(b) Find the relative change of price.

SOLUTION

(a) The number of coats fell from nine to eight, so the relative change of quantity is $-1/9$.

(b) The price rose from \$39 to \$56, an increase of \$17. The relative change of price is $17/39$.

In Example 7, an increase in price had little effect on demand. To provide perspective, the *ratio* of the relative change of quantity over the relative change in price is $(-1/9)/(17/39) = -39/153$.

It is certainly possible that a price increase has a considerable effect on demand. For example, suppose the store was selling three coats at \$111. Suppose they raised the price to \$116, and at the new price they sell only two coats. With this price increase, they again lost the sale of one coat, but let us recalculate the relative changes in quantity and price and the ratio of those changes.

The relative change in demand is $-1/3$, the relative change in price is $5/111$, and the ratio of the first to the second is $(-1/3)/(5/111) = -111/15$.

What can be said of revenues? In Example 7, the price increase caused an increase in revenue from $\$39 \cdot 9 = \$351$ to $\$56 \cdot 8 = \$448$. In the \$111 to \$116 situation, the revenue falls from $\$111 \cdot 3 = \$333$ to $\$116 \cdot 2 = \$232$. It appears that if the change in price causes too great a change in demand, then raising the price will cause the revenue to fall. On the other hand, if the change in price has little effect on the change in demand, then raising the price will increase revenue.

Let us consider these topics by using calculus. We assume the demand function is a differentiable function. It might be that quantity is expressed as a function of price or the other way around. It makes no difference to us because we will develop formulas for each situation.

Assume that the demand function, given in the form of the quantity, x, is a function of price, p. The ratio of the relative change of the quantity sold with respect to the relative change of price is $\frac{\Delta x/x}{\Delta p/p}$. This can be rewritten as $\frac{\Delta x}{x} \cdot \frac{p}{\Delta p} = \frac{p}{x} \cdot \frac{\Delta x}{\Delta p}$.

We assume x is a differentiable function, so as Δp approaches 0, the ratio $\Delta x/\Delta p$ approaches dx/dp.

Hence, the ratio of the relative changes approaches $(p/x) \cdot dx/dp$. If we had begun by assuming that the price, p, was a function of the quantity, x, then a similar argument shows that the ratio of the relative changes is $(p/x) \cdot 1/(dp/dx)$, which is exactly the same expression.

Demand functions generally decrease and so the values of dx/dp and dp/dx are negative. In order to produce a positive number, economists use the following definition with its additional minus sign.

DEFINITION

Elasticity of demand

The *elasticity of demand is represented by E and is given by*

$$E = -\frac{p}{x} \cdot \frac{dx}{dp} = -\frac{p}{x \cdot \frac{dp}{dx}}$$

EXAMPLE 8   Elasticity of demand if quantity is given as a function of price

Consider a demand function for coats given by $x = \sqrt{120 - p}$, where x is the number of coats sold if the price is p dollars. Find the elasticity of demand at
(a) $p = 39$ (b) $p = 111$.

SOLUTION

The derivative $dx/dp = (1/(2\sqrt{120 - p})) \cdot (-1) = -1/(2\sqrt{120 - p})$. Hence

$$E = -p/x \cdot dx/dp = (-p/(\sqrt{120 - p}) \cdot [-1/(2\sqrt{120 - p})]$$
$$= p/[2(120 - p)].$$

(a) At $p = 39$, $E = 39/[2(120 - 39)] = 39/162 \approx 0.24$.

(b) At $p = 111$, $E = 111/[2(120 - 111)] = 111/18 \approx 6.17$.

Economists prefer to express profits, costs, and so forth in terms of quantity sold. Hence, the demand function is usually expressed as price in terms of quantity. We can determine elasticity of demand if that is the case.

EXAMPLE 9   Elasticity of demand if price is given as a function of quantity

Suppose a demand function is given by $p = -x^2 + 120$, where x is the quantity and p is the price of those items. Find the elasticity of demand at
(a) $x = 9$ (b) $x = 3$.

SOLUTION

$$\frac{dp}{dx} = -2x, \text{ so we have } E = -\frac{p}{x \cdot \frac{dp}{dx}} = -\left(\frac{-x^2 + 120}{x \cdot (-2x)}\right)$$

$$= \frac{x^2 - 120}{-2x^2} = \frac{x^2}{-2x^2} - \frac{120}{-2x^2} = -\frac{1}{2} + \frac{60}{x^2}$$

(a) At $x = 9$, we have $E = -1/2 + 60/9^2 = -1/2 + 60/81 \approx 0.24$.

(b) At $x = 3$, we have $E = -1/2 + 60/3^2 = -1/2 + 60/9 \approx 6.17$. ●

The answers in the last two examples are identical because the demand functions are the same. The functions $x(p) = \sqrt{120 - p}$ and $p(x) = -x^2 + 120$ are simply two versions of the same demand curve. For both, if the price is $39, the demand is for 9 units. That is, $x(39) = \sqrt{120 - 39} = \sqrt{81} = 9$, and $p(9) = -9^2 + 120 = -81 + 120 = 39$.
We can compute E regardless of the form of the demand function.

EXAMPLE 10   Elasticity of demand if price is a function of quantity

Find the elasticity of $p = 100\, e^{-x}$.

SOLUTION

We need to find dp/dx and evaluate the expression for E.

$$\frac{dp}{dx} = 100e^{-x} \cdot \frac{d}{dx}(-x) = -100e^{-x}$$

$$E = -\frac{p}{x \cdot \frac{dp}{dx}} = -\frac{100e^{-x}}{x \cdot (-100e^{-x})} = \frac{1}{x}$$

●

PRACTICE EXERCISE 3

Find the elasticity of $p(x) = 100 - 2x$.

EXAMPLE 11   Elasticity of demand if quantity is given as a function of price

Find the elasticity of $x(p) = \dfrac{10}{p^2 + 1}$.

SOLUTION

The quantity is given as a function of price, so we need to find dx/dp. Using the Quotient Rule on $x(p)$, we have

$$\frac{dx}{dp} = \frac{(p^2 + 1) \cdot 0 - 10 \cdot 2p}{(p^2 + 1)^2} = \frac{-20p}{(p^2 + 1)^2}$$

Substituting dx/dp into the form for E, we have

$$E = -\frac{p}{x} \cdot \frac{dx}{dp} = \frac{-p}{\dfrac{10}{p^2 + 1}} \cdot \frac{-20p}{(p^2 + 1)^2} = \frac{2p^2}{p^2 + 1}$$

Consider E in the form $-\dfrac{dx/x}{dp/p}$. The terms dx/x and dp/p are relative changes and can be thought of as percentage increases or decreases. If we know E and the percentage change in either price or quantity, then we can determine the percentage change in the other.

EXAMPLE 12   Determining percentage changes

Suppose $E = 2.5$ and the quantity is increased by 5%. Find the associated decrease in price.

SOLUTION

An increase in quantity of 5% means $dx/x = 0.05$. Substituting in the formula $E = -(dx/x)/(dp/p)$, we have $2.5 = -0.05/(dp/p)$. Solving for dp/p gives us $dp/p = -0.05/2.5 = -0.02$.

The negative sign indicates a decrease in the price. Converting to percentage, we learn the price would decrease by only 2%.   ●

Maximizing Revenue Using Elasticity

We now show that revenue is maximized if elasticity of demand is 1. To begin, suppose price is given as a function of quantity. Consider the revenue function, $R(x) = x \cdot p$, and the marginal revenue, $R'(x) = x \cdot dp/dx + p \cdot 1$.

Revenue is increasing whenever $R'(x) > 0$. Substituting for $R'(x)$ we have $x \cdot dp/dx + p > 0$, which is equivalent to $p > -x \cdot dp/dx$.

Dividing each side by $-x \cdot dp/dx$ (which is positive and thus leaves the inequality unchanged), we have $\dfrac{p}{-x \cdot dp/dx} > 1$. Notice that the left side is the elasticity of demand, E.

We have shown that for $E > 1$, $R'(x) > 0$, so that increasing x will increase R. Similarly, we can show that for $E < 1$, increasing x will decrease R. Thus, if there is a maximum revenue, it will occur when $E = 1$.

The following terms are used to describe the various possibilities for E.

DEFINITION

Elastic, inelastic, unit elasticity

Suppose E is elasticity of demand.

(a) The demand is *elastic* when $E(x) > 1$.

(b) The demand is *inelastic* when $E(x) < 1$.

(c) The demand has *unit elasticity* when $E(x) = 1$.

In summary, the situation is as shown in table 4.6.1.

TABLE 4.6.1

	TYPE OF ELASTICITY	PRODUCTION	PRICE
To Increase Revenue	elastic $E > 1$ inelastic $E < 1$	increase decrease	decrease increase
To Maximize Revenue	unit elasticity $E = 1$		

EXAMPLE 13

Suppose a demand function is $p = 100e^{-x}$.

(a) Find values of x for which the demand function is elastic and for which it is inelastic.

(b) Find the maximum revenue possible.

SOLUTION

This is the demand function of Example 10. There we found $E = 1/x$. $1/x > 1$ for $x < 1$, so the demand is elastic for $x < 1$. Similarly, the demand curve is inelastic for $x > 1$. The maximum revenue would be at $x = 1$, and there we have $R(1) = 1 \cdot 100\, e^{-1} = 36.788$.

As a check on what we have done, let us find $R(x)$ at some elastic x-values, and see if $R(x)$ is increasing. Using $x = 0.5$ and $x = 0.7$, we have

$$R(0.5) = (0.5) \cdot 100\, e^{-0.5} = 50 \cdot 0.6065 = 30.33$$
$$R(0.7) = (0.7) \cdot 100\, e^{-0.7} = 70 \cdot 0.4966 = 34.76$$

The values $R(x)$ are increasing, as expected.

PRACTICE EXERCISE 4

For the demand function $p(x) = 100 - 2x$, find values of x for which the demand function is elastic and for which it is inelastic. Find the maximum revenue possible.

4.6 PROBLEMS

Foundations

The problems of this section require the basic skills illustrated by the following:

Find derivatives of the functions in problems 1–5.

1. $p(x) = \dfrac{50}{x + 1}$ **2.** $p(x) = \sqrt{24 - x}$

3. $p(x) = 500e^{-0.02x}$ **4.** $f(x) = e^x \sqrt{x}$

5. $g(x) = e^x x^2$

6. Solve the equation $e^x x^2 - e^x 2x = 0$.

Exercises

Present Value
In problems 7–10, find the present value of each of the following amounts, payable after the given number of years. Assume continuous compounding at the annual interest rates (discount rates) given. (Remember to change rates from percentages to decimalized interest rates.)

7. $100,000, 10 years, 5%

8. $1,000,000, 5 years, 10%

9. $2,000,000, 50 years, 8%

10. $10,000,000, 25 years, 11%

11. *(Zero-Coupon Bonds)* The Golden Rose is issuing a zero-coupon bond for a face amount of $1000 and a continuous compounding annual interest rate (discount rate) of 8%. What is the present value of the bond if it has a redemption date 10 years from now?

12. *(Zero-Coupon Bonds)* Recalculate the present value of the bond in problem 11 if the interest rate is raised to 9.5%.

13. *(Implied Interest of Zero-Coupon Bonds)* Suppose a zero-coupon bond has a face value of $10,000, due eight years from now. It is selling at $7200. What is the implied rate of interest? Assume continuous compounding.

14. *(Implied Interest of Zero-Coupon Bonds)* Suppose a U.S. savings bond has a face value of $10,000, due eight years from now. It is selling at $6700. What is the implied rate of interest? Assume continuous compounding.

Present Value of an Appreciating Asset
In problems 15–24, $W(t)$ is the dollar price t years into the future and i is the discount interest rate. Assume continuous compounding.
(a) Find how long to hold the asset before selling in order to realize the maximum present value.
(b) Find the maximum present value of the asset.

15. $W(t) = 4t$, $i = 0.08$ **16.** $W(t) = 3t$, $i = 0.1$

17. $W(t) = 1 + 2t$, $i = 0.04$

18. $W(t) = 3 + 2t$, $i = 0.03$

19. $W(t) = \sqrt{t}$, $i = 0.09$ **20.** $W(t) = t^2$, $i = 0.08$

21. $W(t) = -t^2 + 8t$, $i = 0.12$ over the time interval $[0, 8]$

22. $W(t) = -t^2 + 20t$, $i = 0.10$ over the time interval $[0, 20]$

23. $W(t) = -t^2 + 10t$, $i = 0.12$ over the time interval $[0, 2]$

24. $W(t) = -t^2 + 8t$, $i = 0.16$ over the time interval $[0, 2]$

25. Suppose the value of Nonclassico Wine is $W(t) = 2 + 1/(t + 1)$ dollars, where t is in years. For $i = 0.08$, why should the wine be sold immediately? (Hint: Show that $W'(t)/W(t)$ is negative.)

26. Suppose the value of Neoclassico Wine is $W(t) = 2 + 3^t$ dollars, where t is in years. For $i = 0.08$, why does the wine have an infinite present value and so should never be sold? (Hint: Show that $P_v(t) = W(t) \cdot e^{-0.08t}$ is always increasing by showing it has a positive derivative. When finding the derivative, it may be helpful to know that because $3 = e^{1.0986}$ we are able to rewrite 3^t as $3^t = (e^{1.0986})^t = e^{1.0986t}$.

Elasticity of Demand
For each demand function in problems 27–44, find
(a) the elasticity of demand,
(b) the values of x and p for which the demand is elastic (so that increasing production and decreasing prices will increase total revenues),
(c) the value(s) of x that produce the maximum revenue, and
(d) the maximum revenue.

27. $x(p) = \dfrac{100}{p^2 + 1}$ **28.** $x(p) = \dfrac{500}{2p^2 + 1}$

29. $x(p) = \dfrac{200 - p}{2}$ **30.** $x(p) = 100 - p$

31. $p(x) = 200 - 2x$ **32.** $p(x) = 300 - x$

33. $x(p) = 50/p$ **34.** $x(p) = 200/p$

35. $p(x) = \dfrac{100}{x}$ **36.** $p(x) = \dfrac{500}{x}$

37. $p(x) = \dfrac{50}{x + 1}$ **38.** $p(x) = \dfrac{500}{x + 2}$

39. $p(x) = \sqrt{24 - x}$ **40.** $p(x) = \sqrt{150 - x}$

41. $x(p) = 24 - p^2$ **42.** $x(p) = 150 - p^2$

43. $p(x) = 500e^{-0.02x}$ **44.** $p(x) = 150e^{-0.03x}$

45. Suppose for a particular quantity and price the elasticity is $E = 2.4$.
(a) If the quantity sold increases by 2%, what is the percentage decrease in the price?
(b) If the price increases by 4%, what is the percentage decrease in quantity sold?

46. Suppose for a particular quantity and price the elasticity is $E = 0.8$.
(a) If the quantity sold increases by 1.6%, what is the percentage decrease in the price?
(b) If the price increases by 2%, what is the percentage decrease in quantity sold?

Writing and Discussion Problems

47. What are some examples of commodities that have elastic and inelastic demand functions?

48. By conducting a survey of 10 people, attempt to approximate a demand curve for 1/2-gallon containers of ice cream. Over what range of prices is the demand elastic? inelastic?

49. Suppose an asset has cycles of increasing and decreasing worth. If those can be predicted, what are the best times at which the product should be bought and sold? Express your answer in terms of the change of the relative value of the product.

Enrichment Problems

50. The number of tickets sold for a concert is $x = 24{,}000/p - 500$, where p is in dollars. Currently the ticket price is \$12 and 1500 tickets are being sold.
   (a) Is this an elastic or an inelastic situation?
   (b) If ticket prices are increased, will total revenues increase?
   (c) Suppose the concert arena holds 2500 people. What price will maximize total revenue and what is that maximum total revenue?

51. The shuttle charges \$2 and currently has 12,000 riders. Suppose the number of riders is related to price by the formula $x = 6000 \cdot \sqrt{6 - p}$ riders, with p in dollars.

   (a) Is this an elastic or an inelastic situation?
   (b) If the charge increases, will total revenue increase?

Problems 52–54 require the following definition.

DEFINITION   Elasticity of cost

Suppose $C(x)$ is the total cost of producing x units.
   The *elasticity of cost*, E_c, is obtained by considering the ratio $(\Delta C/C)/(\Delta x/x) = (\Delta C/\Delta x) \cdot (x/C)$.
   As $\Delta x \to 0$, this expression goes to $(C' \cdot x)/C$. Hence, the *elasticity of cost*, $E_c = (xC')/C$.

---

52. Show that E_c is the marginal cost divided by the average cost.

53. Find E_c at $x = 25$ if the cost function is $C(x) = \sqrt{x} + 1/(x + 2) + 10$.

54. If $E_c = 1/2$ and the production increases by 3%, estimate the percentage increase in total cost.

55. *(Lottery Winnings)*   Suppose a lottery claims the winner receives \$10 million, but the actual payments consist of one million dollars today and then an additional one million dollars at the end of each of the next nine years. Using a discount rate of 8% and assuming continuous compounding, find the actual present value of the winnings.

SOLUTIONS TO PRACTICE EXERCISES

---

1. The present value of \$50,000 payable in 10 years with an interest rate of 8% compounded continuously is $P = 50{,}000 \cdot e^{-0.08 \cdot 10} = 50{,}000 \cdot e^{-0.8} = 50{,}000 \cdot 0.449329 = \$22{,}466.45$.

2. The worth function is $W(t) = -t^2 + 10t$, where $W(t)$ is in dollars and t is in years. The discount rate is 0.06. Thus, $P_v(t) = (-t^2 + 10t) \cdot e^{-0.06 \cdot t}$.
   Starting from $i = W'(t)/W(t)$, we have $0.06 = (-2t + 10)/(-t^2 + 10t)$. By cross-multiplying, we have $-0.06t^2 + 0.6t = -2t + 10$, which gives $0 = 0.06t^2 - 2.6t + 10$.
   Using the quadratic formula on this equation, we find $t = 39.07$ and 4.27. The first value gives a negative worth and certainly is not the maximum. By the First Derivative Test,

or by checking nearby values, we find $P_v(4.27) = \$18.94$ is the maximum present value.

3. For the demand function $p = 100 - 2x$, the elasticity is $E(x) = -(100 - 2x)/[x \cdot (-2)] = (50 - x)/x$.

4. This is the demand function of Practice Exercise 3, so the elasticity is $E(x) = (50 - x)/x$. The demand is elastic when $E > 1$.
   Solving $(50 - x)/x > 1$, we multiply each side by x to obtain $50 - x > x$. Adding x to each side we have $50 > 2x$, and thus $25 > x$. The demand is inelastic for $x > 25$, and it has unitary elasticity at $x = 25$. The maximum revenue is at $x = 25$, and is given by $R(25) = 25 \cdot 50 = 1250$.

CHAPTER 4 REVIEW

Define or discuss:

1. Base
2. Exponent power
3. Rules of exponents
4. Interval compounding of interest
5. Definition of e
6. Exponential function
7. Continuous compounding of interest
8. Population functions
9. Double declining balance depreciation
10. Chain Rule for exponential functions
11. Ebbinghaus memory model
12. Logarithm of a to the base b
13. Equivalent equations using logarithms and exponents
14. Common logarithm
15. Natural logarithm
16. Properties of logarithms
17. Logarithmic functions and their properties
18. Rule of 70, Rule of 72
19. Newton's Law of Cooling (or Heating)
20. Chain Rule for the derivative of $\ln |f(x)|$

21. Relative rate of change
22. Derivative of $\log_b |f(x)|$
23. Derivative of $b^{f(x)}$
24. Logarithmic differentiation
25. Exponential growth
26. Sigmoid curve
27. Malthus and Verhulst
28. Model of logistic growth
29. Logistic function
30. Maximum rate of increase of a logistic function
31. Significant duration of a logistic event
32. Eventual total response
33. Radioactive decay
34. Plasma drug levels
35. Present value
36. Discount rate
37. Zero-coupon bond
38. Present value of an appreciating asset
39. Maximum present value of an appreciating asset
40. Elasticity of demand
41. Elasticity of cost
42. Elastic, inelastic, and unit elasticity demands

REVIEW PROBLEMS FOR CHAPTER 4

Find derivatives of the functions in problems 1–20.

1. $f(x) = e^{3x}$
2. $f(x) = e^{-2x}$
3. $g(x) = e^{2x-3}$
4. $g(x) = e^{3x-1}$
5. $f(x) = x^2 \cdot e^x + x^3$
6. $f(x) = 3x \cdot e^x - 3x$
7. $f(x) = 2 \ln 3x$
8. $f(x) = 5 \ln 2x$
9. $h(x) = \ln(5x + 1)$
10. $h(x) = \ln(3 - 2x)$
11. $f(x) = x \cdot \ln x^2$
12. $f(x) = \dfrac{\ln \sqrt{x}}{x}$

13. $g(x) = \dfrac{\ln(x^2 + 1)}{x^2}$
14. $g(x) = (x^3 - 1) \ln 3x$
15. $f(x) = 2e^{x^2 - x}$
16. $f(x) = e^{\sqrt{x+1}}$
17. $f(x) = x \cdot \ln(x^3 - x)$
18. $f(x) = \ln(2x^2 + x)$
19. $f(x) = x \cdot e^x - e^x$
20. $f(x) = x \cdot \ln x - x$

Solve for x in problems 21–32.

21. $\ln 3x = 5$

22. $\ln 5x = -2$

23. $\ln(5x - 1) = 2$

24. $\ln(3 + 2x) = 5$

25. $e^{3x} = 4$

26. $e^{2x-1} = 6$

27. $5^{2x} = 8$

28. $2^{-3x} = 5$

29. $5e^{0.05x} = 2e^{0.1x}$

30. $2e^{2x} = 5e^x$

31. $\ln(x^2) = 8$

32. $\ln(\sqrt{x}) = 4$

Graph each of the functions given in problems 33–42 over the indicated intervals. Be careful to find and label any maximum or minimum values of any sort (relative or absolute).

33. $f(x) = e^{-x+1}, (-\infty, \infty)$

34. $g(x) = e^{0.5x}, (-\infty, \infty)$

35. $f(x) = 2 + 3^x, [-1, 5]$

36. $f(x) = \left(\dfrac{1}{3}\right)^{-x}, [-2, 3]$

37. $f(x) = e^{2x} - 4x, [0, 1]$

38. $f(x) = \dfrac{2x}{e^x}, [0, 2]$

39. $f(x) = x \cdot \ln 3x, (0, 3]$

40. $f(x) = x \cdot \ln \sqrt{2x}, (0.3]$

41. $g(x) = \dfrac{1}{2x} + \ln x, \left[\dfrac{1}{4}, 2\right]$

42. $g(x) = \dfrac{2}{x^2} + \ln x, [1, 3]$

43. The company finally bought Jake a bicycle. It cost $75 and the company is going to use double declining balance depreciation over 20 years. After Jake has ridden that bicycle 15 years and retires, what will be the book value of that bicycle?

44. A new candy-wrapping machine costs $25,000 and has its value determined by double declining balance depreciation of 20% over five years. Find its value at the end of four years.

45. Suppose $20,000 is invested at an annual rate of 12%. Find the value after five years if interest is compounded
(a) semiannually and   (b) continuously.

46. Suppose $25,000 is invested at an annual rate of 9%. Find the value after 10 years if interest is compounded
(a) annually and   (b) continuously.

47. Frank can invest his money at 12.1% compounded semiannually or 12% compounded quarterly. What would be the value of $1000 at the end of one year under each investment strategy?

48. The population of Africa in 1990 was approximately 661 million and was growing at an annual rate of 2.9%. Assuming this growth occurs continuously and will continue to do so in the future, estimate the population of Africa in the year 2025.

49. The population of Hong Kong in 1990 was approximately 5.7 million and was growing at an annual rate of 0.8%. Assuming this growth occurs continuously and will continue to do so in the future, estimate the population in the year 2025.

50. Assume continuous compounding and find the present value of $1,000,000 due in 10 years, with an interest discount rate of $i = 0.08$.

51. Assume continuous compounding and find the present value of $50,000 due in 25 years, with an interest discount rate of $i = 0.10$.

52. Suppose $f(x) = 3 - e^{0.5t}$.
(a) Find t such that $f'(t) = 0.3$.
(b) Suppose $t \geq 0$. What are the possible values of $f'(t)$?
(c) Sketch a graph of $f(t)$ for $t \geq 0$.

53. For the logistic curve $f(t) = M/(1 + S \cdot e^{-ct})$, let $M = 5000$, $f(0) = 4$, and $f(10) = 250$.
(a) Find the equation of the logistic curve.
(b) For what value of $f(t)$ is the rate of increase greatest? (Be careful—there is an easy and a difficult way to answer this. Hint: We may not need to find t.)
(c) When is $f(t) = 4000$?
(d) Estimate the significant duration of the event. That is, find a value of t for which $f(t)$ is essentially M.

54. *(Ebbinghaus Memory Model)* Suppose $P(t) = 75e^{-0.25 \cdot t} + 25$ is the percentage of items remembered after t days.
(a) Find t such that $P(t) = 40$.
(b) Find $P'(3)$ and give an interpretation of that value.

55. *(Cost of an Election)* The cost of running for the U.S. Senate (and winning) varies considerably for different states. One estimate of the cost is $C(t) = 6 \cdot e^{0.07t}$ million dollars, where t is the number of years after 1986.
(a) Estimate the cost for a successful campaign in 1996.
(b) In terms of dollars per year, what is the rate of increase of the cost of a successful campaign in 1996?

56. *(Cost of a First-Class Postage Stamp)* A first-class stamp was 3 cents in 1948 and 25 cents in 1988. Check that an exponential annual growth rate of 5.3% would have predicted the 1988 cost. Assume the annual growth rate continues at 5.3%.
(a) Estimate the cost of a first-class postage stamp in 2018.
(b) When would the cost be $5.00?

57. *(Eventual Total Response)* Suppose the response from an advertisement in the Sunday paper is 3947 by day 2; 5512 by day 3; and 6855 by day 4. Find an estimate for the eventual total response.

58. *(Drug Levels)* The beta-blocker Nadolol has a half-life of 22 hours.
(a) Find the equation that gives the blood plasma level at t hours, when the peak was P_0.
(b) How long until only 10% of the peak level is present?

59. *(Radioactive Decay)* Cobalt-60 is a radioactive isotope that has a half-life of 5.3 years. Starting with an amount P_0, the amount of cobalt-60 present after t years is given by $P(t) = P_0 \left(e^{ct} \right)$.
(a) Find c.
(b) How many days until only 10% of P_0 is present?

60. Nita estimates her recent painting, *The Lonely Tree,* will have a value of $W(t) = \$(100 + 25t)$ at t years from today. The discount rate is 8% on money.
(a) How many years should go by before she sells the painting?
(b) What is the maximum present value of the painting? (Assume continuous compounding.)

61. Suppose a demand function is given by $x(p) = 27 - p^2$.
(a) Find the elasticity of demand.
(b) For what values of x is the demand elastic? inelastic? of unit elasticity?
(c) What p produces the maximum revenue and what are the quantity and the revenue for that value of p?

62. Suppose a demand function is given by $x(p) = \sqrt{588 - p}$.
(a) Find the elasticity of demand.
(b) For what values of p is the demand elastic? inelastic?
(c) What p produces the maximum revenue and what are the quantity and the revenue for that value of p?

63. Suppose a demand curve is given by $p(x) = 60e^{-0.05x}$.
(a) Find the value of x that will maximize total revenue.
(b) Find the associated price and the maximum total revenue.

64. The Chocolate Works finds the demand function for chocolate roses is $p(x) = -2x^3 + 27$.
(a) Find the elasticity of demand.
(b) For what values of x is the demand elastic? inelastic?
(c) What x produces the maximum revenue and what are the price and the revenue for that value of x?

65. Suppose at some particular price the elasticity, E, is 2.2. If the price increases by 2%, what is the expected percentage decrease in quantity?

Enrichment Problems

66. Hermosa Security Bank offered an IRA paying 8% annual interest, compounded semi-annually. Mary bought $2000 on January 1, 1984, and additional amounts of $2000 on each of January 1, 1985 and January 1, 1986. Assume she made no additional contributions.
    (a) What is the total value of her IRA account on January 1, 1992?
    (b) When will she have $25,000 in her account (give month and year)?

Find the derivatives of the functions in problems 67–71.

67. $f(x) = 10^{2x}$

68. $f(x) = 2^{-3x}$

69. $f(x) = \log_3(x^2 + 3x)$

70. $f(x) = 2x^{3x+1}$

71. $g(x) = \ln(x^2)^{3x+5}$

72. **(Elasticity of Cost)**   Let $C(x) = 2000(1 - e^{-0.02x})$ be the total cost of producing x units. Find the elasticity of cost at $x = 60$.

CHAPTER 5

INTEGRATION

5.1 Antiderivatives

5.2 The Area Under a Curve and Riemann Sums

5.3 Definite Integrals and the Fundamental Theorem of Calculus

5.4 Area Between Two Curves

5.5 Applications of Integrals to Business and Economics

Chapter 5 Review

FIGURE 5.1.1

THE FIRST FOUR CHAPTERS LAID THE FOUNDATION FOR AND THEN DEVELoped the major ideas and applications of *differential calculus.* Therein we discussed derivatives and their interpretations as slopes of tangent lines and as rates of change.

Searching for extreme values, we paid particular attention to points at which the slopes of the tangent lines were zero or undefined. Because these slopes were given by derivatives, we developed many formulas to help us calculate derivatives.

Now, we shift our attention to *integral calculus.* A dominant—indeed an almost overwhelming—application is the area under a curve. There is a rich profusion of applications that can be related to such areas.

To help us appreciate why an area would be of interest to us, consider the following. Suppose we receive revenue at the rate of $100 a day for 30 days. Over that time our total revenue would be $3000, the result of multiplying the rate of revenue by the time over which that revenue was coming in. Figure 5.1.1 shows a graph of the rate of revenue as a function of time. The area between the graph and the x-axis, and between $x = 0$ and $x = 30$, is a rectangle. That area, which is the product of the width times the height, is $30 \cdot 100 = 3000$. Hence, the area represents the total revenue.

Suppose, however, that the rate of revenue is not a constant, but varies over time. For example, the graph in figure 5.1.2 may represent receipts on a toll bridge. We can still interpret the area under the curve as the total revenue. However, that area is not nearly as easy to determine as was the area in figure 5.1.1.

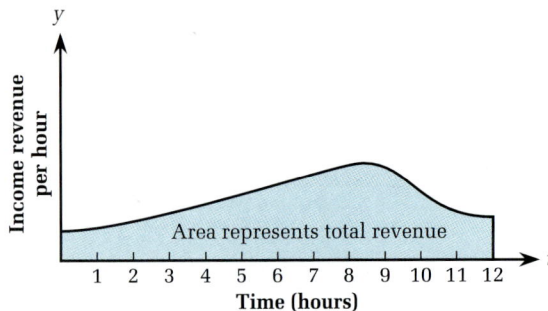

FIGURE 5.1.2

There are several techniques for evaluating the area. The first section of this chapter provides the necessary foundation for one particularly effective method. It uses what is called the Fundamental Theorem of Calculus, and this requires us to learn about antiderivatives.

5.1   ANTIDERIVATIVES

To this point in the book, we have started with a function and learned (and used) its derivative. Now we want to reverse the process. Like the detectives J. B. Fletcher and Sherlock Holmes, we want to go from a clue (the derivative) back to the sort of function that leaves that clue. For example, suppose the derivative of some unknown function is 2. Can we find one function with that derivative? If there are more than one, can we find all of them? As we may recall, if $F(x) = 2x$, then $F'(x) = 2$. On the other hand, for $G(x) = 2x + 3$, it is also true that $G'(x) = 2$. We can see that these are certainly different functions, as $F(0) = 0$ and $G(0) = 3$.

In figure 5.1.3 we have the graphs of several functions of the form $2x + C$ for various values of C. All of these functions have a derivative of 2. Each graph has 2 as its constant slope. Because C can be any real number, there actually are an infinite number of functions whose derivative is 2. Thus, knowing just the derivative of a function does not allow us to assert that there is only one function with that derivative.

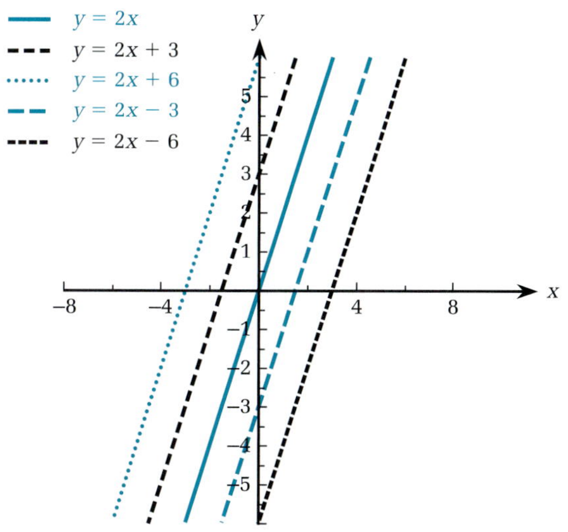

Legend:
- $y = 2x$
- $y = 2x + 3$
- $y = 2x + 6$
- $y = 2x - 3$
- $y = 2x - 6$

FIGURE 5.1.3

Antiderivatives

The next definition provides a name for a function whose derivative is specified.

DEFINITION

Antiderivative

Suppose $F(x)$ and $f(x)$ are functions and $F'(x) = f(x)$. Then $F(x)$ is an *antiderivative* of $f(x)$.

(A remark on notation: it is customary to use upper-case letters to stand for antiderivatives.)

For emphasis, notice that the definition states that $F(x)$ is *an* antiderivative of $f(x)$. It does not say that $F(x)$ is *the* antiderivative of $f(x)$. For $f(x) = 2$, we found that there are an infinite number of antiderivatives. Any function of the form $2x + C$ is an antiderivative of $f(x) = 2$.

Are there any functions other than ones of the form $2x + C$ that are antiderivatives of 2? The answer is given by the following, much more general, result.

THEOREM 5.1   ALTERNATE ANTIDERIVATIVES

If $F'(x) = G'(x)$, then $F(x) = G(x) + C$, for some constant C. In words: any two antiderivatives of a function can only differ by a constant of addition.

Theorem 5.1 results from the following line of thought. Let $H(x) = F(x) - G(x)$. The derivative of $H(x)$ is given by $H'(x) = F'(x) - G'(x)$. If $F'(x) = G'(x)$, then $F'(x) - G'(x) = 0$, so $H'(x) = 0$. If $H(x)$, or indeed any function, has a zero derivative for all x-values, then the output of $H(x)$ never varies and it must be just a constant. We can name the constant C.

Knowing that $H(x) = C$ and $H(x) = F(x) - G(x)$, we have $C = F(x) - G(x)$, which can be rewritten as $F(x) = G(x) + C$. That is the conclusion of Theorem 5.1.

By Theorem 5.1, we know that if we can find any antiderivative of some function, then we know all the antiderivatives because the others differ from the first one only by the addition of various values of a constant. Now, can we find the first antiderivative? That is, given a function $f(x)$, can we find some function that is an antiderivative of $f(x)$?

The process of finding antiderivatives depends on our knowing how to find derivatives. It is somewhat like long division; we need to know the multiplication tables to do long division.

EXAMPLE 1   Generating antiderivative formulas by knowing derivatives

Find the derivatives of each of the following. Then, state a general antiderivative formula in each case.
(a) $3x^2$ (b) e^{5x} (c) $x(\ln x) - x$

SOLUTION

(a) The derivative of $3x^2$ is $3 \cdot 2x = 6x$. Thus, one antiderivative of $6x$ is $3x^2$, and so a general antiderivative of $6x$ is $3x^2 + C$ for any constant C.

(b) The derivative of e^{5x} is $5e^{5x}$. Thus, one antiderivative of $5e^{5x}$ is e^{5x}, and so a general antiderivative of $5e^{5x}$ is $e^{5x} + C$ for any constant C.

(c) The derivative of $x(\ln x) - x$ is $[x \cdot 1/x + (\ln x) \cdot 1] - 1 = 1 + \ln x - 1 = \ln x$. Thus, one antiderivative of $\ln x$ is $x(\ln x) - x$, and so a general antiderivative is $x(\ln x) - x + C$ for any constant C.

Indefinite Integrals

We introduced symbols for the derivative concept so we could avoid having to express so much in words. The same is now done for the concept of antiderivative.

NOTATION AND DEFINITIONS FOR ANTIDERIVATIVES

$\int f(x)\,dx$ means a general antiderivative of $f(x)$ with respect to x.

The symbol "$\int$" is called the *integral sign*, the function $f(x)$ is called the *integrand*, and the symbol dx is called the *differential of* x. The whole expression $\int f(x)\,dx$ is an *indefinite integral*.

In writing $\int f(x)\,dx = F(x) + C$, the C is called the *constant of integration*.

The expression $\int f(x)\,dx$ is read aloud as the "indefinite integral of f of x with respect to x" or as the "antiderivative of f of x with respect to x."

Always writing the "$+C$" continually reminds us of the fact that any two antiderivatives differ by a constant. Hence, by adding various values of C to some particular antiderivative, we are sure to find every antiderivative.

WARNING

Do not omit writing the dx term. It is essential. It asserts what variable the antiderivative is being taken with respect to.

Rules for Integrals

The Power Rule for Derivatives states $\dfrac{d}{dx}x^n = nx^{n-1}$ for any real number n. In order to find an antiderivative of a power of x, we first find the following derivative.

$$\frac{d}{dx}\left(\frac{1}{n+1}\cdot x^{n+1}\right) = \frac{1}{n+1}\cdot\left(\frac{d}{dx}x^{n+1}\right)$$

$$= \frac{1}{n+1}\cdot\left((n+1)x^{(n+1)-1}\right) = x^n$$

Thus, we know one function whose derivative is x^n. This lets us produce the first of our integral formulas.

POWER RULE FOR INTEGRALS

For any real number $n \neq -1$, $\quad \int x^n\,dx = \dfrac{x^{x+1}}{n+1} + C.$

In words, the Power Rule for Indefinite Integrals says: "the general antiderivative for a power of x is found by *raising the power of x by 1 and dividing by the new power*. To this then add the constant C, the *constant of integration*."

The Power Rule for Integrals does not apply to $\int x^{-1} \, dx$. We could not divide by zero, which is what $n + 1$ would be if n were -1. Later we will introduce the formula for $\int x^{-1} \, dx$.

EXAMPLE 2 The Power Rule for Integrals

Find each of the following:
(a) $\int x^3 \, dx$ (b) $\int x^5 \, dx$ (c) $\int x^{2.3} \, dx$ (d) $\int x^{-3} \, dx$
(e) $\int dx$

SOLUTION

(a) For $\int x^3 \, dx$ we can use the Power Rule with $n = 3$. We then
have $\int x^3 \, dx = \dfrac{1}{3+1} \cdot x^{3+1} + C = \dfrac{x^4}{4} + C.$

(b) Using the Power Rule with $n = 5$, we have $\int x^5 \, dx = x^{5+1}/(5+1) = x^6/6 + C.$

(c) Using the Power Rule with $n = 2.3$, we have $\int x^{2.3} \, dx = x^{2.3+1}/(2.3+1) = x^{3.3}/3.3 + C.$

(d) Using the Power Rule with $n = -3$, we have $\int x^{-3} \, dx = x^{-3+1}/(-3+1) = x^{-2}/(-2) + C.$

(e) We can write dx as $1 \cdot dx$. Then use $1 = x^n$ with $n = 0$. By the Power Rule, $\int dx = \int x^0 \, dx = x^{0+1}/(0+1) + C = x/1 + C = x + C.$

Part (d) of the last example demonstrates that we need to be quite careful if $n < 0$ and to actually add $+1$ to the power of x.

WARNING

PRACTICE EXERCISE 1

Find (a) $\int x^2 \, dx$ (b) $\int x^4 \, dx$ (c) $\int x^{-2} \, dx$ (d) $\int x^{2.5} \, dx.$

Given $\int f(x) \, dx$, if we can write $f(x) = x^n$ for some n, then we should do so.

EXAMPLE 3 Rewriting the integrand and using the Power Rule

Find the following indefinite integrals:
(a) $\int (1/x^3)\, dx$ (b) $\int (1/\sqrt{x})\, dx$

SOLUTION

(a) Rewrite the quotient $1/x^3$ as x^{-3} and use the Power Rule to find $\int (1/x^3)\, dx = \int x^{-3}\, dx = x^{-2}/(-2) + C = -1/2x^2 + C.$

(b) The $\sqrt{x}$ can be expressed as $x^{0.5}$. Hence $\int (1/\sqrt{x})\, dx = \int (1/x^{0.5})\, dx = \int x^{-0.5}\, dx.$ By the Power Rule,

$$\int x^{-0.5}\, dx = \frac{x^{-0.5+1}}{-0.5 + 1} + C = \frac{x^{0.5}}{0.5} + C = 2x^{0.5} + C = 2\sqrt{x} + C$$

PRACTICE EXERCISE 2

Find $\displaystyle\int \frac{1}{x^2}\, dx.$

Each result we have about derivatives can be restated as a result about integrals. Thus, from $\dfrac{d}{dx} k \cdot f(x) = k \cdot \dfrac{d}{dx} f(x)$ we have the following rule.

CONSTANT MULTIPLE RULE FOR INTEGRALS

$\int k \cdot f(x)\, dx = k \cdot \int f(x)\, dx$ for any real number k.

EXAMPLE 4 Using the Constant Multiple Rule and the Power Rule

Find the following:
(a) $\int 5x^3\, dx$ (b) $\int (5/(3x^2))\, dx$ (c) $\int 5\, dx$

SOLUTION

(a) $\int 5x^3\, dx = 5 \int x^3\, dx$ by the Constant Multiple Rule for Integrals
$= 5 \cdot (x^4/4) + C$ by the Power Rule.

(b) First rewrite using negative exponents:

$$\int \frac{5}{3x^2}\, dx = \int \frac{5}{3}x^{-2}\, dx$$

$$= \frac{5}{3} \int x^{-2}\, dx \text{ by the Constant Multiple Rule}$$

$$= \frac{5}{3} \cdot \frac{x^{-1}}{-1} + C \text{ by the Power Rule}$$

$$= -\frac{5}{3} \cdot \frac{1}{x} + C = -\frac{5}{3x} + C$$

(c) $\int 5\, dx = 5 \int dx = 5x + C.$

Notice in part (c) we could have said $\int 5\, dx = 5(x + C) = 5x + 5C$. However, because C is simply being used as an arbitrary constant, whenever we have an antiderivative formula involving the product of C with another constant, then we simply change the name of the constant. We could use a new name, such as D, but the tradition is not to do so.

From the formula $\frac{d}{dx}(f(x) \pm g(x)) = \frac{d}{dx}f(x) \pm \frac{d}{dx}g(x)$, we have the Sum or Difference Rule for Integrals.

SUM OR DIFFERENCE RULE FOR INTEGRALS

$$\int (f(x) \pm g(x))\, dx = \int f(x)\, dx \pm \int g(x)\, dx$$

EXAMPLE 5   Using the Sum, Difference, Constant Multiple, and Power rules

Find $\int (2x^5 - 3x + \sqrt{x})\, dx.$

SOLUTION

In order to use the Power Rule, rewrite $\sqrt{x}$ as $x^{1/2}$ or $x^{0.5}$. Then

$$\int (2x^5 - 3x + \sqrt{x})\, dx$$

$$= \int 2x^5\, dx - \int 3x\, dx + \int x^{0.5}\, dx \text{ by the Sum and Difference rules}$$

$$= 2 \cdot \frac{x^6}{6} - 3 \cdot \frac{x^2}{2} + \frac{x^{1.5}}{1.5} + C \text{ by the Power and Constant Multiple rules}$$

$$= \frac{x^6}{3} - \frac{3}{2}x^2 + \frac{x^{1.5}}{1.5} + C$$

Notice that again only one constant of integration, C, is used even though there were three integrals. That is because if we added together three constants C_1, C_2, and C_3, one for each of the three integrals, then we could collect those together into one constant, C.

PRACTICE EXERCISE 3

Find   (a) $\int (4x^2 - 3x)\, dx$ (b) $\int \dfrac{3}{\sqrt{x}}\, dx.$

We may need to do some algebraic simplification before we can find the indefinite integral. In the next example, we have to expand an expression in order to give us powers of x.

EXAMPLE 6   Using algebraic simplification

Find   (a) $\int (x^2 + 3)^2\, dx$ (b) $\int \dfrac{2x^3 + 1}{x^2}\, dx.$

SOLUTION

(a) We first must square the expression $(x^2 + 3)$ so that we have powers of x.

$$\int (x^2 + 3)^2\, dx = \int (x^4 + 6x^2 + 9)\, dx = \frac{x^5}{5} + 2x^3 + 9x + C$$

(b) $(2x^3 + 1)/x^2$ is an improper fraction. Divide both terms in the numerator by x^2 and use the Sum Rule.

$$\int \frac{2x^3 + 1}{x^2}\, dx = \int \frac{2x^3}{x^2}\, dx + \int \frac{1}{x^2}\, dx$$

$$= \int 2x\, dx + \int x^{-2}\, dx$$

$$= x^2 + \frac{x^{-1}}{-1} + C$$

$$= x^2 - \frac{1}{x} + C$$

PRACTICE EXERCISE 4

Find $\int (2x + 1)^2\, dx.$ (Hint: Square the binomial first.)

If we had thought the answer to the last practice exercise was $(1/3)(2x + 1)^3$, we could check by finding the derivative.

$$\frac{d}{dx} \frac{1}{3}(2x + 1)^3 = \frac{1}{3} \cdot 3(2x + 1)^2 \cdot 2 \qquad \text{(Remember to use the Chain Rule on } 2x + 1.\text{)}$$

This derivative is not equal to $(2x + 1)^2$, so we realize $(1/3)(2x + 1)^3$ cannot be the antiderivative.

Antiderivative Formulas

Because every derivative formula corresponds to an antiderivative formula, let us list the pairings we know in table 5.1.1.

TABLE 5.1.1

DERIVATIVE FORMULA	ANTIDERIVATIVE FORMULA
1. $\frac{d}{dx} kx = k$	$\int k\, dx = kx + C$
2. $\frac{d}{dx} x^n = nx^{n-1}$	$\int x^n\, dx = \frac{x^{n+1}}{n + 1} + C \ (n \neq -1)$
3. $\frac{d}{dx} e^x = e^x$	$\int e^x\, dx = e^x + C$
4. $\frac{d}{dx} e^{kx} = k \cdot e^{kx}$	$\int e^{kx}\, dx = \frac{1}{k} \cdot e^{kx} + C$
5. $\frac{d}{dx} \ln|x| = \frac{1}{x}$	$\int \frac{1}{x}\, dx = \ln|x| + C$

Notice that (1) is a special case of (2) and that (3) is actually a special case of (4), but we state these anyhow because they occur so frequently. Furthermore, we see that (5) takes care of the case $\int x^{-1}\, dx$, which is not covered in (2).

All of these formulas are stated in terms of a variable x, but we must realize that x is not the only variable that we will meet. If we encounter expressions such as $\int 2t\, dt$ or $\int 2u\, du$, then we still can use formulas in Table 5.1.1 to have $\int 2t\, dt = t^2 + C$ and $\int 2u\, du = u^2 + C$.

If we have more than one variable in the integrand (the function we are trying to find the antiderivative of), the variable in the expression $d\_$ tells us what variable we need to be taking the antiderivative with respect to. For example, assume that x, u, and z are three completely independent variables. Then $\int 2ux\ dx = ux^2 + C$, whereas $\int 2ux\ du = xu^2 + C$ and $\int 2ux\ dz = 2uxz + C$. In the first, we take the antiderivative with respect to x, whereas in the second the antiderivative is with respect to u and in the last, the antiderivative is taken with respect to z.

Some more practice with antiderivatives is found in Example 7.

EXAMPLE 7   Finding antiderivatives

Find the following and check by taking derivatives.
(a) $\int 3.2\ dx$ (b) $\int 3e^x\ dx$ (c) $\int e^{2x}\ dx$ (d) $\int 7e^{-2x}\ dx$
(e) $\int (3/t)\ dt$ (f) $\int (2/(3x))\ dx$

SOLUTION

(a) $\int 3.2\ dx = 3.2x + C$, using (1). Check: $\frac{d}{dx}(3.2x + C) = 3.2 + 0 = 3.2$

(b) $\int 3e^x\ dx = 3 \int e^x\ dx = 3e^x + C$, using (3). Check:
$\frac{d}{dx}(3e^x + C) = 3 \cdot e^x + 0 = 3e^x$

(c) $\int e^{2x} = (1/2)e^{2x} + C$, using (4). Check: $\frac{d}{dx}[(1/2)e^{2x} + C] = (1/2)(2e^{2x}) + 0 = e^{2x}$

(d) $\int 7e^{-2x}\ dx = 7 \cdot (-1/2)e^{-2x} + C = -(7/2)e^{-2x} + C$, using (4) and the Constant Multiple Rule. Check: $\frac{d}{dx}[-(7/2)e^{-2x} + C] = -(7/2)(-2e^{-2x}) + 0 = 7e^{-2x}$

(e) $\int (3/t)\ dt = 3 \int (1/t)\ dt = 3 \ln |t| + C$, using (5) and the Constant Multiple Rule. Check: $\frac{d}{dx}(3 \ln |t| + C) = 3 \cdot 1/t + 0 = 3/t$

(f) $\int (2/(3x))\ dx = (2/3) \int (1/x)\ dx = (2/3) \cdot \ln |x| + C$, using (5) and the Constant Multiple Rule. Check: $\frac{d}{dx}[(2/3) \ln |x| + C] = 2/3 \cdot 1/x + 0 = 2/3x$

PRACTICE EXERCISE 5

Find each of the following and check by taking derivatives.
(a) $\int e^{-x}\ dx$ (b) $\int 2e^{5x}\ dx$ (c) $\int (2/x)\ dx$ (d) $\int (3/(2x))\ dx$

Antiderivatives with Initial Conditions

Some situations require an antiderivative that takes on a specified value. For instance, at the start of this section we asked for a function such that its derivative was 2. We found there were an infinite number of such functions. However, all of them were of the form $F(x) = 2x + C$. Had we required that the function also take on a specified value of y for some specified value of x, then there would have been only one such function. Graphically, we would have been looking among all of the parallel lines represented by $y = 2x + C$ for the one line that passed through a specified point.

Such a specified value is called an *initial condition*, even though it may not be the value of $F(0)$.

EXAMPLE 8   An antiderivative with initial condition

Find the function $F(x)$ such that $F'(x) = 2$ and $F(1) = 5$.

SOLUTION

$F'(x) = 2$, so we know $F(x)$ is of the form $F(x) = 2x + C$. To find C, we substitute 1 for x and solve $F(1) = 5$. From $F(1) = 2 \cdot 1 + C = 5$, we have $2 + C = 5$. Thus, $C = 5 - 2 = 3$, and the function is $F(x) = 2x + 3$. The graph of $y = 2x + 3$ is shown in figure 5.1.3 (p. 487).

We used $F(x)$ as the name of the antiderivative in Example 8, for, as we mentioned earlier, capital letters are often used for antiderivative functions. We will see why in the next section. However, we could have used $f(x)$ or any other name for the function.

EXAMPLE 9   An antiderivative with initial condition

Suppose $\dfrac{d}{dx} f(x) = x^2 - 1$ and $f(1) = 5$. Find $f(x)$.

SOLUTION

In figure 5.1.4 we see some of the many functions whose derivative is $x^2 - 1$. All such functions are given by the antiderivative $\int (x^2 - 1)\, dx = x^3/3 - x + C$. The particular function we want must be such that $f(1) = 5$. That is, if we substitute $x = 1$ into $f(x)$, then we need to get 5. Using the general antiderivative we need $f(1) = 1^3/3 - 1 + C = 5$ and so $C = 5 + 1 - 1/3 = 5 + 2/3 = 17/3$.

Thus, $f(x) = x^3/3 - x + 17/3$.

$$y = \frac{x^3}{3} - x$$

$$y = \frac{x^3}{3} - x - 1$$

$$y = \frac{x^3}{3} - x + 1$$

$$y = \frac{x^3}{3} - x + 2$$

$$y = \frac{x^3}{3} - x + \frac{17}{3}$$

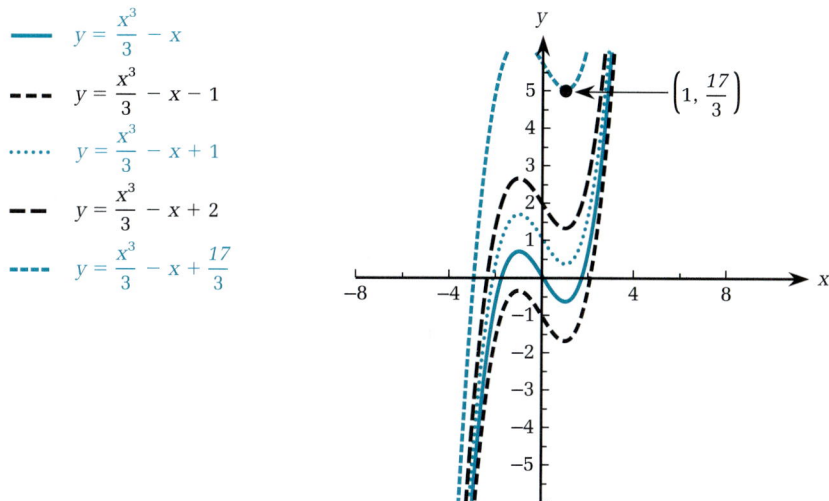

FIGURE 5.1.4

PRACTICE EXERCISE 6

Suppose $\dfrac{d}{dx}f(x) = 3x + 5$ and $f(2) = 10$. Find $f(x)$.

5.1 PROBLEMS

Foundations

The problems of this section require the basic skills illustrated by the following:

Write problems 1–5 in the form x^n for a real number n.

1. $\sqrt{x}$ 2. $\dfrac{1}{x^3}$ 3. $\sqrt[3]{x}$

4. $\sqrt[5]{x^2}$ 5. $\dfrac{1}{x^{-2}}$

6. Write $\dfrac{3}{e^{2x}}$ in the form ae^{bx}.

7. Solve for C if $f(x) = 3x + C$ and $f(2) = 1$.

Exercises

For problems 8–31, find antiderivatives.

8. $\int -2 \, dx$ 9. $\int 3 \, dx$

10. $\int x^4 \, dx$ 11. $\int x^6 \, dx$

12. $\int x^{-4} \, dx$ 13. $\int x^{-5} \, dx$

14. $\int x^{1.3} \, dx$ 15. $\int x^{2.5} \, dx$

16. $\int (2x + 7) \, dx$ 17. $\int (3x - 2) \, dx$

18. $\int (-t + 1) \, dt$ 19. $\int (2 - t) \, dt$

20. $\int (x^2 - 2x + 5) \, dx$ 21. $\int (x^2 + 3x - 4) \, dx$

22. $\int (t^2 + t) \, dt$ 23. $\int (t^2 + 4t) \, dt$

24. $\int (8y^4 - 3y + 2) \, dy$ 25. $\int (10y^4 + 3y - 1) \, dy$

26. $\int e^{3x}\,dx$

27. $\int e^{4x}\,dx$

28. $\int 5e^{2x}\,dx$

29. $\int -3e^{2x}\,dx$

30. $\int e^{-2x}\,dx$

31. $\int e^{-y}\,dy$

For problems 32–47, rewrite the expressions as powers or in the form e^{kx} and find antiderivatives.

32. $\int 3\sqrt{x}\,dx$

33. $\int 5\sqrt{x}\,dx$

34. $\int (x^3 + \sqrt{x})\,dx$

35. $\int (3x^2 - \sqrt{x})\,dx$

36. $\int \sqrt[3]{5x}\,dx$

37. $\int \sqrt[3]{2x}\,dx$

38. $\int \dfrac{2}{x^3}\,dx$

39. $\int \dfrac{3}{x^2}\,dx$

40. $\int \dfrac{x^3 + 1}{2x^2}\,dx$

41. $\int \dfrac{x^3 - 2}{x^2}\,dx$

42. $\int \dfrac{2}{x}\,dx$

43. $\int \dfrac{5}{x}\,dx$

44. $\int \dfrac{2x^2 + 3}{x}\,dx$

45. $\int \dfrac{x^2 - 2}{x}\,dx$

46. $\int \dfrac{3}{e^{2x}}\,dx$

47. $\int \dfrac{2}{e^{2x}}\,dx$

For problems 48–51, rewrite the expressions and find antiderivatives.

48. $\int (3x + 5)^2\,dx$

49. $\int (2x - 3)^2\,dx$

50. $\int \left(x + \dfrac{1}{x}\right)^2\,dx$

51. $\int \left(1 - \dfrac{1}{x}\right)^2\,dx$

For problems 52–57, find $f(x)$ given the initial condition.

52. $f'(x) = 3x + 1$ and $f(4) = 12$

53. $f'(x) = 2x - 1$ and $f(3) = 10$

54. $f'(x) = e^{2x}$ and $f(0) = 4.5$

55. $f'(x) = \dfrac{1}{x}$ and $f(1) = 3$

56. $f''(x) = 6$, $f'(1) = 8$ and $f(1) = 4$

57. $f''(x) = 4$, $f'(1) = 0$ and $f(1) = -1$

Writing and Discussion Problems

58. The final form of an answer quite often avoids our writing negative powers of x. What might be some reasons for this? Is it true or false that an expression such as x^{-2} has less meaning for most of us than $1/x^2$? Why?

59. Why is it wrong to write $\int 3x^2 = x^3$?

Enrichment Problems

60. Find $\int \left(2x + \dfrac{1}{x^2}\right)^2\,dx$.

61. Find $\int (e^x - e^{-x})^2\,dx$.

62. Assume x, u, and z are three completely independent variables and find each of the following:
(a) $\int x^2 u\,dx$ (b) $\int x^2 u\,du$ (c) $\int x^2 u\,dz$

SOLUTIONS TO PRACTICE EXERCISES

1. (a) $\displaystyle\int x^2\,dx = \dfrac{x^3}{3} + C$ (b) $\displaystyle\int x^4\,dx = \dfrac{x^5}{5} + C$

(c) $\displaystyle\int x^{-2}\,dx = \dfrac{x^{-1}}{-1} + C = -\dfrac{1}{x} + C$

(d) $\displaystyle\int x^{2.5}\,dx = \dfrac{x^{3.5}}{3.5} + C$

2. $\displaystyle\int \dfrac{1}{x^2}\,dx = \int x^{-2}\,dx = \dfrac{x^{-1}}{-1} + C = -\dfrac{1}{x} + C$

3. (a) $\displaystyle\int (4x^2 - 3x)\,dx = 4 \cdot \dfrac{x^3}{3} - 3 \cdot \dfrac{x^2}{2} + C$

(b) $\displaystyle\int \dfrac{3}{\sqrt{x}} = 3\int x^{-1/2}\,dx = 3 \cdot \dfrac{x^{1/2}}{\dfrac{1}{2}} + C =$

$\qquad 6x^{1/2} + C = 6\sqrt{x} + C$

4. $\displaystyle\int (2x + 1)^2\,dx = \int (4x^2 + 4x + 1)\,dx =$

$\qquad \dfrac{4}{3}x^3 + 2x^2 + x + C$

5. (a) $\displaystyle\int e^{-x}\,dx = \frac{e^{-x}}{-1} + C = -e^{-x} + C$ Check:

$$\frac{d}{dx}(-e^{-x} + C) = -(e^{-x} \cdot (-1)) + 0 = e^{-x}$$

(b) $\displaystyle\int 2e^{5x}\,dx = \frac{2}{5}e^{5x} + C$ Check:

$$\frac{d}{dx}\left(\frac{2}{5}\,e^{5x} + C\right) = \frac{2}{5}(e^{5x}5) + 0 = 2e^{5x}$$

(c) $\displaystyle\int \frac{2}{x}\,dx = 2\int \frac{1}{x}\,dx = 2\ln|x| + C$ Check:

$$\frac{d}{dx}(2\ln|x| + C) = 2\frac{1}{x} + 0 = \frac{2}{x}$$

(d) $\displaystyle\int \frac{3}{2x}\,dx = \frac{3}{2}\ln|x| + C$ Check:

$$\frac{d}{dx}\left(\frac{3}{2}\ln|x| + C\right) = \frac{3}{2} \cdot \frac{1}{x} + 0 = \frac{3}{2x}$$

6. Suppose $\dfrac{d}{dx} f(x) = 3x + 5$ and $f(2) = 10$.

Then $f(x) = (3/2) \cdot x^2 + 5x + C$. Evaluating the formula, $f(2) = (3/2) \cdot 4 + 5 \cdot 2 + C$. Because $f(2) = 10$, $C = -6$, and $f(x) = (3/2) \cdot x^2 + 5x - 6$.

5.2 THE AREA UNDER A CURVE AND RIEMANN SUMS

This section begins our study of the area of a region. We are interested in finding areas because many situations exist for which the area has a meaningful interpretation. For instance, at the beginning of this chapter we saw that area can represent the total revenue from a cash flow. Naturally, before we can interpret an area, we need to be able to find it.

Area

There are well-known formulas for areas of simple regions, as shown in figure 5.2.1.

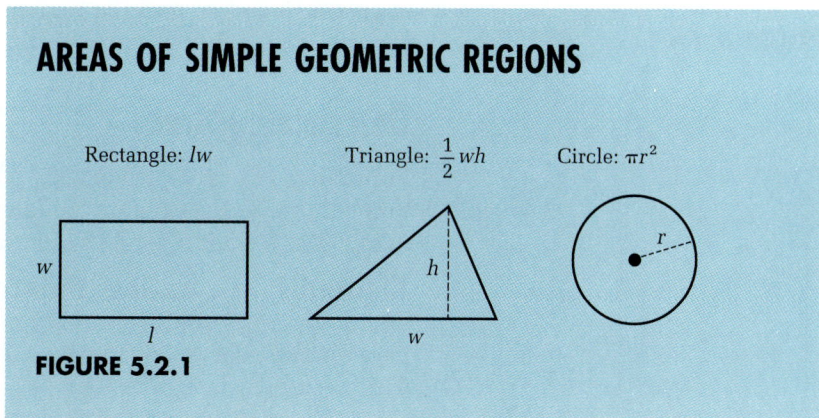

AREAS OF SIMPLE GEOMETRIC REGIONS

Rectangle: lw Triangle: $\frac{1}{2}wh$ Circle: πr^2

FIGURE 5.2.1

EXAMPLE 1   Areas of simple regions

For each of the following functions, find the area of the region bounded by the graph of the function, the x-axis, and the values of x at the start and end of the interval.

(a) $g(x) = 2x + 1$ over $[0, 2]$ gives a triangle on top of a rectangle, as shown in figure 5.2.2. The rectangle has width 2 and height 1, and so has an area of 2. The triangle has width 2 and a height of 4 and so has an area of $1/2 \cdot 2 \cdot 4 = 4$. Thus, the total area is $2 + 4 = 6$.

(b) $k(x) = \sqrt{9 - x^2}$ over $[0, 3]$ is one-fourth of a circle of radius 3. Hence the area is $1/4 \cdot \pi \cdot 3^2 = (9/4)\pi$. See figure 5.2.3.

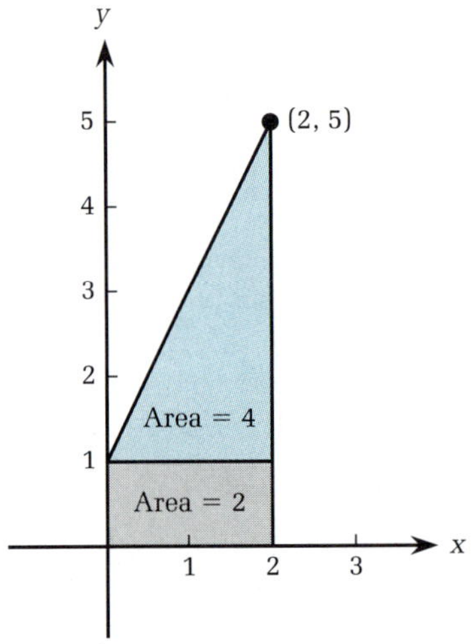

FIGURE 5.2.2

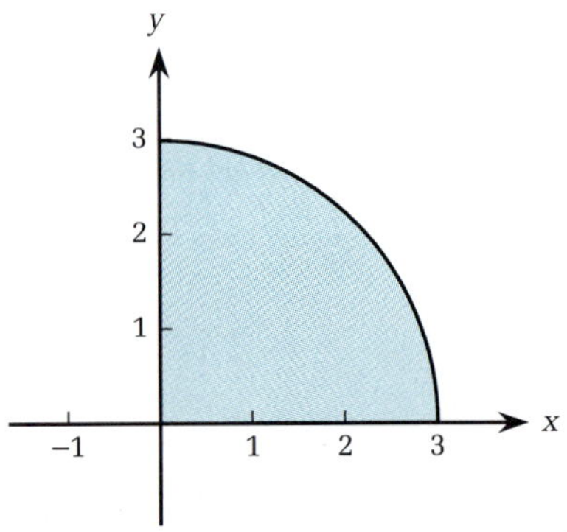

FIGURE 5.2.3

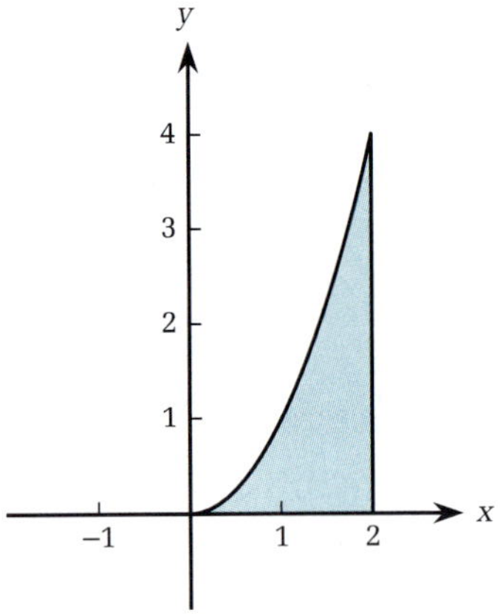

Most regions have more complicated boundaries. For instance, suppose $f(x) = x^2$ and the region is bounded by $f(x)$, the x-axis, and the line $x = 2$. This region is shown in figure 5.2.4.

FIGURE 5.2.4

Determining Area

Following are several methods which are commonly used to determine the area of a region which is bounded by the x-axis and the graph of a function $f(x)$, over an interval $[a, b]$.

Method I—Counting Squares. One method of estimating the area is to graph the region very carefully on graph paper with many very small squares and to try counting the squares within the region. This is shown in figure 5.2.5.

This method has the difficulty of our having to decide whether to count a square if the boundary passes through that square.

Should we count only the squares that are completely below the curve? Should every square that is touched by the curve be counted? Should we try to decide on a basis of how much of the square is below the curve?

Regardless of our decision on which squares to count and which not to count, there is a vast number of squares to count if we use fairly fine graph paper.

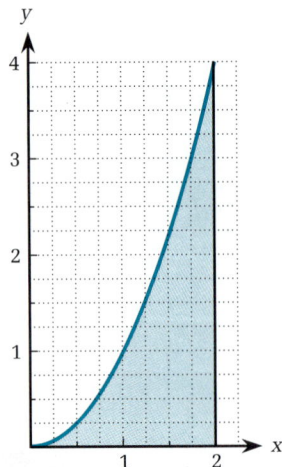

FIGURE 5.2.5

Method II—Proportional Measurements.

A second method is to trace out the curve quite carefully on a sheet of metal, such as a metal cookie sheet. Then cut out the region and weigh it. We can find out how much cookie sheets weigh per square foot, so we can translate our weight into an equivalent area.

Method II is often used in calculating the number of gallons of paint needed to paint various parts of a house. That is, you measure the number of square feet and divide by the stated coverage per gallon. A more exotic use of Method II is to calculate the number of flower blossoms needed for each part of a decked-out float in the Rose Bowl Parade in Pasadena, California.

There are situations that require Method II, but in general cookie sheets have a much better use, so save them and use this technique only as a last resort.

Method III—Graphing Calculators or Computers.

Access to a graphing calculator provides us the ability to do a vast number of computations quite quickly. Some calculators have the ability to graph a function and then show an approximation for the area under the curve and give a numerical value for that area. There are several methods used for these approximations, but they all use some sort of sum. Let us consider how these sums can be created and evaluated.

Method IV—Approximating Sums.

We now concentrate on sums that approximate an area. Suppose $f(x)$ is defined over $[a, b]$ and we partition $[a, b]$ into n subintervals. In doing so we let $x_0 = a$, and choose $x_1, x_2, x_3, \ldots, x_n$ so that $x_0 < x_1 < x_2 < \ldots < x_n = b$. This notation is used so that x_i is the right-hand endpoint of the ith subinterval.

The most straightforward partition is to choose the x_i so that each subinterval is the same width. The width of $[a, b]$ is $b - a$, so equal subintervals means each is $(b - a)/n$ wide. Therefore, $x_0 = a$, $x_1 = a + (b - a)/n$, $x_2 = a + 2[(b - a)/n]$, $x_3 = a + 3[(b - a)/n]$, and so on until we reach $x_n = a + n[(b - a)/n] = a + (b - a) = b$. In figure 5.2.6 we see the result of choosing five subintervals.

FIGURE 5.2.6

Let us assume $f(x) \geq 0$ over the interval $[a, b]$. To approximate the area under the graph of $f(x)$ over the interval $[a, b]$, suppose we use rectangles. Partition $[a, b]$ as above and let the base of each rectangle be one of those subintervals on the x-axis. Suppose the height of each rectangle is determined by evaluating $f(x)$ somewhere in the subinterval. Then we multiply each $f(x)$ by the width of its subinterval to find the area of the rectangle. Adding together all of these areas will provide an approximation to the area of $f(x)$ over $[a, b]$.

One possible choice is to use the midpoints of the subintervals.

MIDPOINT APPROXIMATION SUM

Suppose $f(x)$ is defined and $f(x) \geq 0$ over $[a, b]$. Furthermore, assume $[a, b]$ has been partitioned into n equal subintervals. Let the midpoint of the ith subinterval be denoted by x_i^*. The sum

$$f(x_1^*) \cdot [(b - a)/n] + f(x_2^*) \cdot [(b - a)/n] + \cdots + f(x_n^*) \cdot [(b - a)/n]$$

approximates the area of the region under $y = f(x)$, above the x-axis, and between $x = a$ and $x = b$.

Figure 5.2.7 is the graph of a function, together with the approximating region of rectangles that would be produced by using heights at the midpoints of five subintervals.

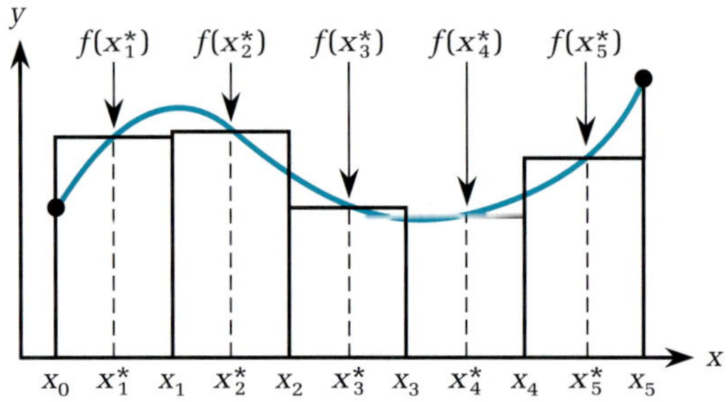

FIGURE 5.2.7

EXAMPLE 2   Midpoint approximation sum

Use five equal subintervals and the midpoint approximation sum to estimate the area under $f(x) = x^2$ above the x-axis, from $x = 1$ to $x = 2$.

SOLUTION

Dividing $[1, 2]$ into five equal subintervals means that each subinterval is $(2 - 1)/5 = 1/5 = 0.2$ long. Hence $x_0 = 1$, $x_1 = 1.2$, $x_2 = 1.4$, $x_3 = 1.6$, $x_4 = 1.8$, and $x_5 = 2$.

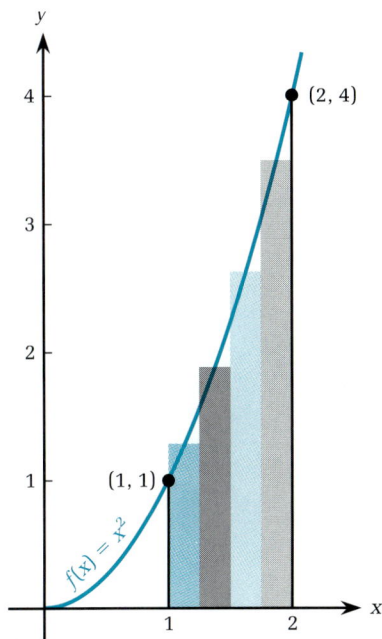

FIGURE 5.2.8

The midpoints of these subintervals are $x_1^* = 1.1$, $x_2^* = 1.3$, $x_3^* = 1.5$, $x_4^* = 1.7$, and $x_5^* = 1.9$. See figure 5.2.8.

The midpoint approximation is $f(x_1^*) \cdot [(b - a)/n] + f(x_2^*) \cdot [(b - a)/n] + \cdots + f(x_n^*) \cdot [(b - a)/n]$

$$= (1.1)^2(0.2) + (1.3)^2(0.2) + (1.5)^2(0.2) + (1.7)^2(0.2) + (1.9)^2(0.2)$$
$$= (1.21)(0.2) + (1.69)(0.2) + (2.25)(0.2) + (2.89)(0.2) + (3.61)(0.2)$$
$$= 0.242 + 0.338 + 0.45 + 0.578 + 0.722 = 2.33.$$

PRACTICE EXERCISE 1

Use four equal subintervals and the midpoint approximation to estimate the area under $f(x) = x^2$ above the x-axis from $x = 0$ to $x = 2$.

Partitioning the interval $[a, b]$ into many very small subintervals both increases the accuracy of the midpoint approximation sum and increases our desire to use a computer to do the computations. In each of the approximations shown in figures 5.2.9–5.2.11 we use the function $f(x) = x^3 - 3x^2 + 2x + 2$ over $[0, 3.2]$ but with progressively smaller subintervals.

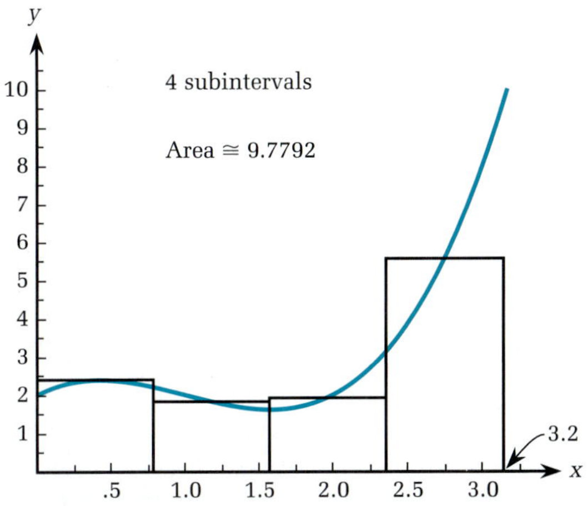

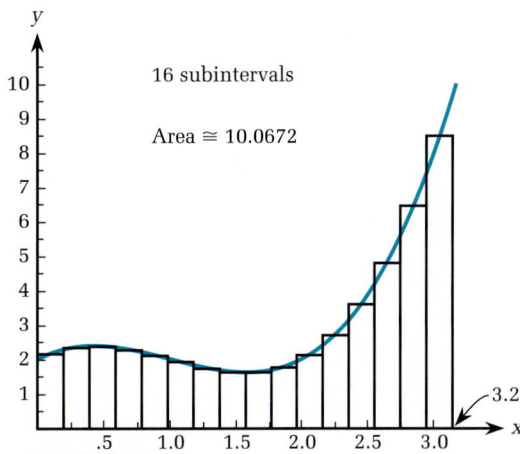

FIGURE 5.2.9

FIGURE 5.2.10

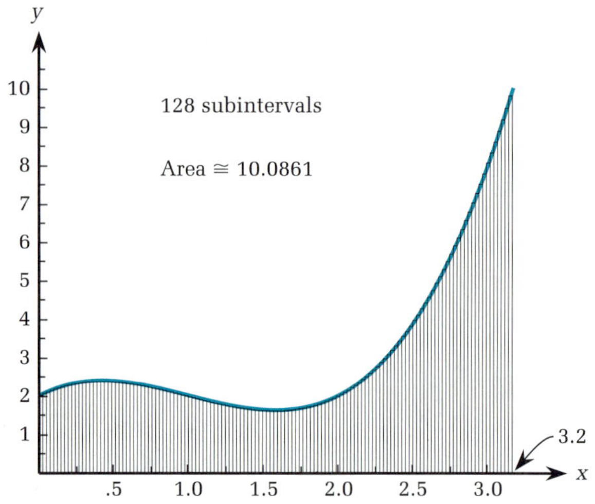

FIGURE 5.2.11

Definite Integrals as Riemann Sums

Bernhard Riemann (1826–1866) was one of those who showed that the partitioning of $[a, b]$, as well as the selection of the point in each subinterval at which to evaluate the function, can be done in a variety of ways, all of which lead to the same limiting value as the widths of the subintervals go to zero. In his honor, these approximating sums are referred to as *Riemann sums.* We have been using "midpoint" sums, but we could evaluate $f(x)$ at left-hand or right-hand endpoints of each subinterval, or anywhere we like.

The concept of area has been discussed for a function that was positive. Actually, a sum such as the midpoint approximation sum can exist whether a function is positive or negative; or even sometimes positive and sometimes negative. We have been using positive functions so that we can interpret the sums.

In the following definition, we do not assume the function is positive.

DEFINITION

Definite integral

Suppose $f(x)$ is defined over the interval $[a, b]$ and is such that no matter how the values x_i^* are chosen within the subintervals,

$$(5.2.1) \quad \lim_{n \to \infty} \left(f(x_1^*) \cdot \left(\frac{b - a}{n} \right) + f(x_2^*) \cdot \left(\frac{b - a}{n} \right) + \cdots + f(x_n^*) \cdot \left(\frac{b - a}{n} \right) \right) \text{ exists.}$$

Then the limit in (5.2.1) is called the *definite integral* of $f(x)$ from a to b, and it is represented by the symbol $\int_a^b f(x) \, dx$.

The term "dx" tells us what variable the definite integral is "with respect to" and so it is very important to write it.

The symbol $\int_a^b f(x)\,dx$ is read as "the definite integral from $x = a$ to $x = b$ of $f(x)$ with respect to x." The value $x = a$ is said to be the *lower limit* and the value $x = b$ is said to be the *upper limit* of the definite integral. Notice that this use of the word "limit" is in the sense of endpoints, so it is a different use than in the definition of the definite integral being a limit of sums.

Definite integrals look much like indefinite integrals but they are different. Definite integrals give us numerical values, in contrast to indefinite integrals, which produce antiderivative functions.

We have not yet found $\int_a^b f(x)\,dx$ by actually finding a limit using expression (5.2.1), nor do we intend to do so. There is a great deal of technical manipulation required to evaluate the limit and we receive very limited benefits from successfully doing so at this stage of our study. Either we will be satisfied with an approximation (perhaps using a computer or calculator) or we will use another method of evaluation. The next section of this chapter details the most famous method of evaluation, the Fundamental Theorem of Calculus.

Properties of Definite Integrals

The following properties are presented and illustrated by figures 5.2.12 and 5.2.13. The illustrations do not constitute proofs, but they do indicate the reasonableness of the results.

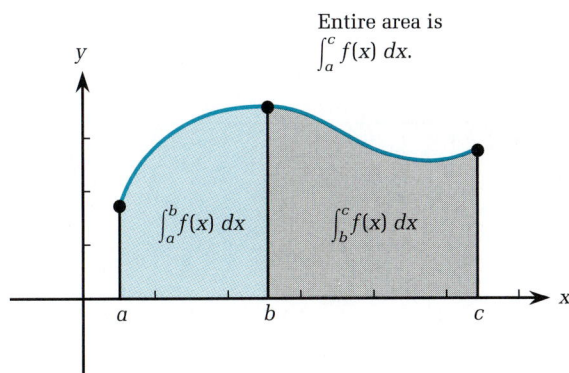

FIGURE 5.2.12

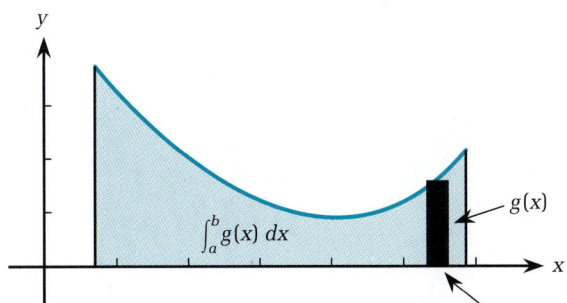

FIGURE 5.2.13

I. $\int_a^b f(x)\ dx + \int_b^c f(x)\ dx = \int_a^c f(x)\ dx.$

Consider figure 5.2.12.

II. $\int_a^b (f(x) \pm g(x))\ dx = \int_a^b f(x)\ dx \pm \int_a^b g(x)\ dx.$

Consider figure 5.2.13.

As we think about each term of the approximating sum, we realize that if $f(x)$ is positive over some particular subinterval, then the term is the area of the rectangle. If $f(x)$ is negative over a specific subinterval, the value of $f(x_i^*) \cdot [(b-a)/n]$ is negative, so it represents the negative of the area of the rectangle.

We may think of the approximating sum building a value by going along from a to b, adding together positive and negative numbers.

In figures 5.2.14–5.2.16, we see such an approximation for $f(x) = (x-1)^2 - 1$ over $[0, 4]$. In each pair of graphs we see the approximating rectangles in the upper portion. In the lower is a graph of the partial total to that x-value.

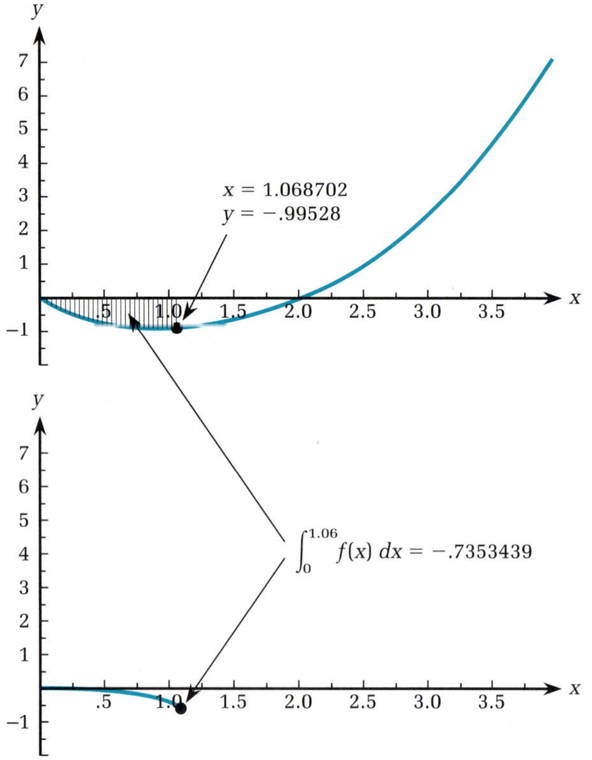

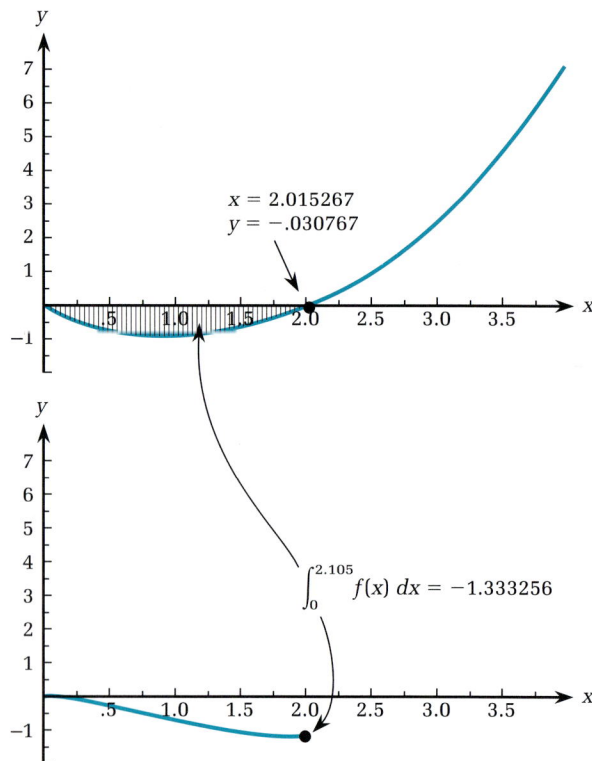

FIGURE 5.2.14

For the left figure:
$x = 1.068702$
$y = -.99528$
$\int_0^{1.06} f(x)\ dx = -.7353439$

FIGURE 5.2.15

For the right figure:
$x = 2.015267$
$y = -.030767$
$\int_0^{2.105} f(x)\ dx = -1.333256$

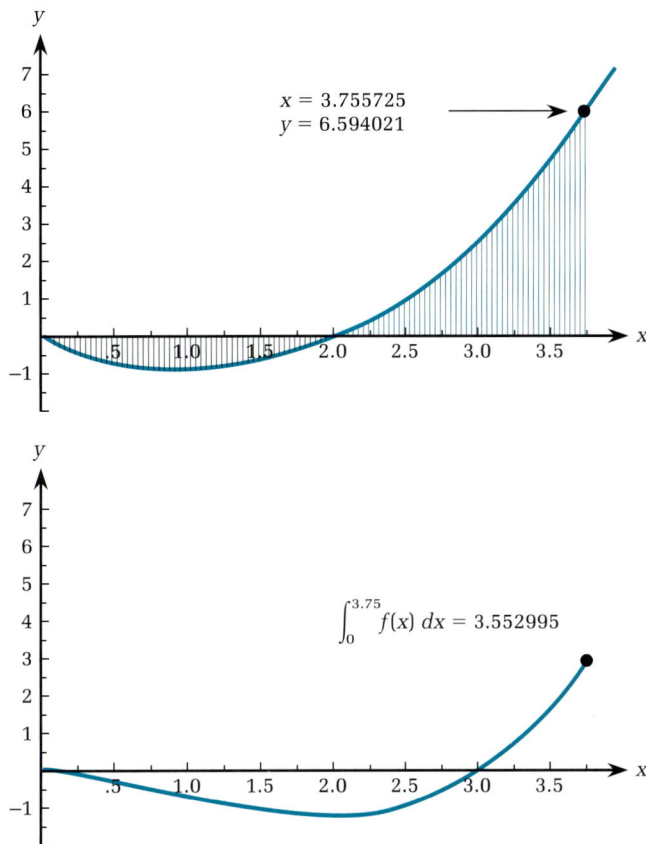

y
7
6
5
4
3
2
1
−1
x = 3.755725
y = 6.594021
.5
1.0
1.5
2.0
2.5
3.0
3.5
x
y
7
6
5
4
3
2
1
−1
$\int_0^{3.75} f(x)\,dx = 3.552995$
.5
1.0
1.5
2.0
2.5
3.0
3.5
x

FIGURE 5.2.16

In general, if $f(x)$ is continuous over the interval $[a, b]$, then $\int_a^b f(x)\,dx$ represents the net area (with positive area being measured above and negative area below the x-axis) between the graph of $y = f(x)$, the x-axis, and the lines $x = a$ and $x = b$. See figure 5.2.17.

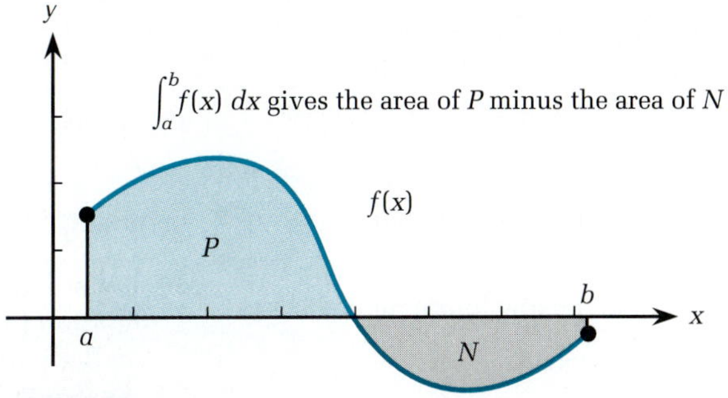

y
$\int_a^b f(x)\,dx$ gives the area of P minus the area of N
f(x)
P
N
a
b
x

FIGURE 5.2.17

EXAMPLE 3   Using the properties of the definite integral

Suppose $\int_a^b f(x)\, dx = 2$, $\int_a^b g(x)\, dx = 5$, and $\int_b^c f(x)\, dx = -3$. Find $\int_a^b (f(x) + g(x))\, dx$ and $\int_a^c f(x)\, dx$.

SOLUTION

$\int_a^b (f(x) + g(x))\, dx = \int_a^b f(x)\, dx + \int_a^b g(x)\, dx = 2 + 5 = 7$

$\int_a^c f(x)\, dx = \int_a^b f(x)\, dx + \int_b^c f(x)\, dx = 2 - 3 = -1$.

5.2 PROBLEMS

Foundations

The problems in this section require the basic skills illustrated by the following:

Suppose the interval $[0, 2]$ is to be equally subdivided into four subintervals.

1. How long is each subinterval?

2. Find the right-hand endpoint of each subinterval.

3. Find the midpoint of each subinterval.

Exercises

4. Find the area of the region bounded by $y = 5x - 2$, the x-axis, and the lines $x = 1$ and $x = 3$.

5. Find the area of the region bounded by $y = 2x + 1$, the x-axis, and the lines $x = 1$ and $x = 4$.

In problems 6–11 use the midpoint approximation sum to estimate the areas under the function, above the x-axis, and between the endpoints of the given intervals.

6. $f(x) = x^2 + x$ over $[0, 2]$ using four equal subintervals

7. $g(x) = \sqrt{x + 1}$ over $[0, 3]$ using four equal subintervals

8. $f(x) = \dfrac{1}{x^2 + 1}$ over $[0, 2]$ using four equal subintervals

9. $f(x) = x \cdot \ln x$ over $[1, 2]$ using four equal subintervals

10. $g(x) = \sqrt{9 - x^2}$ over $[0, 3]$, using six equal subintervals

11. $f(x) = 1/\sqrt{4 - x^2}$ over $[0, 2]$, using four equal subintervals

Find the values of the definite integrals in problems 12–19 assuming that

$$\int_a^b f(x)\, dx = 6, \quad \int_a^b g(x)\, dx = 5,$$

$$\int_b^c f(x)\, dx = 3, \quad \int_b^c g(x)\, dx = -2.$$

12. $\displaystyle\int_a^b (f(x) + g(x))\, dx$ 13. $\displaystyle\int_a^b (g(x) - f(x))\, dx$

14. $\displaystyle\int_a^c f(x)\, dx$ 15. $\displaystyle\int_a^c g(x)\, dx$

16. $\displaystyle\int_a^c (f(x) - g(x))\, dx$ 17. $\displaystyle\int_a^c (g(x) - f(x))\, dx$

18. $\displaystyle\int_a^b (2f(x) + 3g(x))\, dx$

19. $\displaystyle\int_a^c (3f(x) - 2g(x))\, dx$

Graphing Calculator Problems

20. If your graphing calculator can do numeric integration (which is what approximations are called), find $\int_0^3 \sqrt{9 - x^2}\, dx$ and compare your answer with the value for the area of a quarter-circle of radius 3.

21. If your calculator is programmable, write a program for the midpoint approximation and use it to estimate $\int_0^3 \sqrt{9 - x^2}\ dx$ using 24 equal subintervals and compare your answer with the value for the area of a quarter-circle of radius 3.

Writing and Discussion Problems

22. Suppose $\int_a^b f(x)\ dx > 0$. Can you conclude $f(x) > 0$ over $[a, b]$? Justify your answer.

23. Suppose $\int_a^c f(x)\ dx > 0$ for any c in the interval $[a, b]$. Can you conclude $f(x) > 0$ over $[a, b]$? Justify your answer.

Enrichment Problems

24. Find the area of the region bounded by $y = 5x - 2$, the x-axis, and the lines $x = 0$ and $x = 1$.

25. Find the area of the region bounded by $y = x + 1$ and $y = \sqrt{1 - x^2}$.

26. Show that $\int_a^a f(x)\ dx = 0$ by using Property I and the properties of real numbers.

27. Show that if $f(x) = 0$ then $\int_a^b f(x)\ dx = 0$ for any a and b whatsoever.

28. Use problem 27 and Property II to show that $\int_a^b (-f(x))\ dx = -\int_a^b f(x)\ dx$.

SOLUTION TO PRACTICE EXERCISE

1. The partition points are $x_0 = 0$, $x_1 = 0.5$, $x_2 = 1$, $x_3 = 1.5$, and $x_4 = 2$. The midpoints are $x_1^* = 0.25$, $x_2^* = 0.75$, $x_3^* = 1.25$, and $x_4^* = 1.75$. The midpoint approximating sum is
$f(x_1^*) \cdot [(b - a)/n] + f(x_2^*) \cdot [(b - a)/n] + \cdots + f(x_n^*) \cdot [(b - a)/n]$

$$= (0.25)^2(0.5) + (0.75)^2(0.5) + (1.25)^2(0.5) + (1.75)^2(0.5)$$

$$= (0.0625)(0.5) + (0.5625)(0.5) + (1.5625)(0.5) + (3.0625)(0.5)$$

$$= 0.03125 + 0.28125 + 0.78125 + 1.53125$$

$$= 2.625.$$

| 5.3 | # DEFINITE INTEGRALS AND THE FUNDAMENTAL THEOREM OF CALCULUS |

In Section 5.2 we found that definition of the definite integral $\int_a^b f(x)\ dx$ involved the existence and determination of

$$\lim_{n \to \infty} \left(f(x_1^*) \cdot \left(\frac{b - a}{n} \right) + f(x_2^*) \cdot \left(\frac{b - a}{n} \right) + \ldots + f(x_n^*) \cdot \left(\frac{b - a}{n} \right) \right)$$

We approximated this limit by midpoint approximation sums. We did not actually evaluate the limit of the sum for two reasons. First, the evaluation requires a great deal of skill at handling complex mathematical formulas for sums. Second, and much more important, for many functions there is an alternate method that creates the same value but is much simpler to carry out. The method, which uses results called the Fundamental Theorem of Calculus, was first used by Sir Isaac Newton in England and by Gottfried Wilhelm Leibniz in Europe.

Fundamental Theorem of Calculus, the Beginning

The fundamental theorem of calculus, discovered by both Newton and Leibniz, provides a connecting link between derivatives and integrals.

Newton

MATHEMATICAL INNOVATORS
Gottfried Wilhelm Leibniz (1646–1716) and Sir Isaac Newton (1642–1727)

Newton's book *Principia Mathematica* (published in 1687) remains one of the most important works in all of mathematics. In it, the concepts he developed laid the foundation for much of the material we now call calculus, an amazing invention of the human mind. For the first time, ideas based on infinite processes blossomed and ripened into rich and useful results. Newton's mathematics, which was simultaneously and independently being developed by Leibniz in Germany, revolutionized thought in all of science.

The seventeenth century was a tumultuous time throughout England and Europe. Leibniz was born into a chaotic central Europe. The Holy Roman Empire had degenerated into an ineffective and semi-riotous coalition of hundreds of small, separate states, each of them extremely touchy about its independence and many of them with private armies. England too was in upheaval. Revolutions, plagues, and the great London fire all disrupted Newton's life.

On the continent, France became the center of Europe. It was to Paris that Leibniz went as the representative of Mainz, one of the German states. There he attempted to persuade Louis XIV, the Sun King, to avoid further aggression into central Europe. Although unsuccessful as a court diplomat, his Paris days were not wasted, for he spent his time studying and developing his mathematical ideas.

Leibniz seemed to be interested in everything. He "invented" the psychological concept of the subconscious. He suggested that barometric readings could predict weather. He worked with microscopes and porcelain. Leibniz even reorganized the Harz silver mines and supervised the mint in Hanover, the city where he lived the last forty years of his life, serving the Dukes of Brunswick-Lüneberg.

Sir Isaac Newton, a professor at Cambridge University in England, wrote on many topics in science. For instance, he was one of the first to analyze the properties of light, motion, and gravity. However, in addition to his scientific contributions he was an innovator in business and industry. It was he, in his capacity as director of the Royal Mint of England, who introduced the process of "milling" coins. Milling is the process of putting ridges around the rims of a coin. It is still done on edges of coins, such as dimes and quarters, that traditionally were made of a precious metal. Before Newton, in an early form of white-collar crime, swindlers would shave off a little of the precious metal all around the smooth edge of a coin and then try to pass off the rest of the coin at "face" value. The shavings were then melted down to make new "coins." Putting ridges around the rims made such shaving immediately obvious to even the most unsuspecting person.

FUNDAMENTAL THEOREM OF CALCULUS (PART A)

Suppose $f(x)$ is continuous over the interval $[a, b]$ and $F(x)$ is an antiderivative of $f(x)$ there. Then $\int_a^b f(x)\, dx = F(b) - F(a)$.

The Fundamental Theorem of Calculus provides a computational method to find areas by integrals. We can call this *Method V.* This method gives an *exact* area, not merely an approximation.

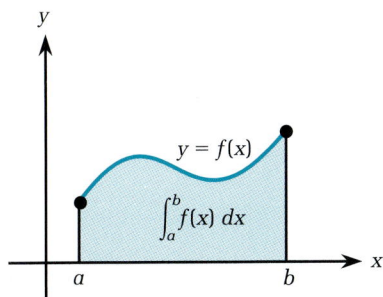

$$\int_a^b f(x)\, dx$$

FIGURE 5.3.1

THE AREA UNDER A FUNCTION THAT IS POSITIVE OVER [a, b]

Suppose $f(x)$ is a continuous function that is positive on $[a, b]$ and $F'(x) = f(x)$ there. Then the area below $f(x)$, above the x-axis, and between $x = a$ and $x = b$, as shown in figure 5.3.1, is given by

$$\int_a^b f(x)\, dx = F(b) - F(a)$$

Choosing an Antiderivative

To use the Fundamental Theorem (Part A) we need an antiderivative. It does not make any difference which one we use. The reason is that in Section 5.1 we found that any two antiderivatives of $f(x)$ could only differ by (at most) a constant. That is, if $F'(x) = f(x)$ and $G'(x) = f(x)$, then there is some constant C such that $G(x) = F(x) + C$. Thus, $G(b) - G(a) = (F(b) + C) - (F(a) + C) = F(b) + C - F(a) - C = F(b) - F(a)$.

As a result, we may use any antiderivative of $f(x)$ and evaluate that antiderivative at a and b. Given that fact, we choose an antiderivative that is as simple as possible. For example, if $f(x) = x^2$, we choose $F(x) = x^3/3$ rather than $F(x) = x^3/3 + 1$ or $F(x) = x^3/3 - 2$.

EXAMPLE 1   Determining the area under a positive function, choosing an antiderivative

Find the area between $y = x^2$ and the x-axis, between $x = 0$ and $x = 3$.

SOLUTION

To use the Fundamental Theorem of Calculus (Part A) we need to find an antiderivative of $f(x) = x^2$. We found in Section 5.1 that $\int x^2\, dx = (x^3/3) + C$, so we use $F(x) = (x^3/3)$. Then $\int_0^3 x^2\, dx = F(3) - F(0) = (3^3/3) - (0) = 9$ by the Fundamental Theorem of Calculus (Part A). This is an exact area. See figure 5.3.2.

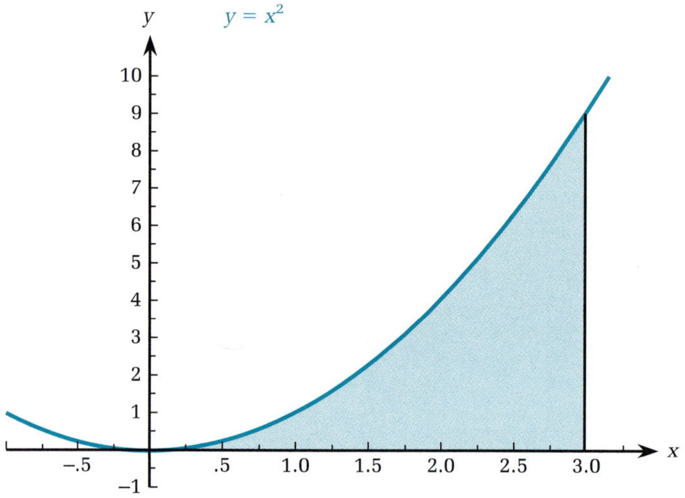

FIGURE 5.3.2

It is convenient to use the notation $F(x)|_{x=a}^{x=b}$ to stand for $F(b) - F(a)$. If there is no doubt as to the variable, then $F(x)|_{x=a}^{x=b}$ can be abbreviated as $F(x)|_a^b$.

EXAMPLE 2

Using the notation $F(x)|_a^b$, find $\int_2^4 x^2\, dx$.

SOLUTION

$\int_2^4 x^2\, dx = (x^3/3)|_2^4$. This means the antiderivative used is $x^3/3$ and its value at 2 should be subtracted from its value at 4. Thus, $(x^3/3)|_2^4 = (4^3/3) - (2^3/3) = (64/3) - (8/3) = 56/3 \approx 18.6667$.

EXAMPLE 3

Find $\int_0^2 (x^3 + 2)\, dx$.

SOLUTION

The Sum Rule for Integrals states that $\int (f(x) + g(x))\, dx = \int f(x)\, dx + \int g(x)\, dx$, so an antiderivative for $(x^3 + 2)$ is the sum of antiderivatives of x^3 and of 2. These can be $x^4/4$ and $2x$, respectively. Hence, an antiderivative of $\int (x^3 + 2)\, dx$ is $(x^4/4) + 2x$.

$$\int_0^2 (x^3 + 2)\, dx = [(x^4/4) + 2x]\big|_0^2 \text{ using the antiderivative and the}$$
$$\text{Fundamental Theorem (Part A)}$$
$$= [(2^4/4) + 2 \cdot 2] - [(0^4/4) + 2 \cdot 0]$$
$$= (16/4) + 4 = 4 + 4 = 8$$

PRACTICE EXERCISE 1

Find the area between $y = x^2 + 1$ and the x-axis from $x = 1$ to $x = 3$.

We know a definite integral assigns positive values to areas of a region above the x-axis and negative values to areas of a region below the x-axis. Example 4 shows what happens when the function crosses the x-axis on the given interval.

EXAMPLE 4   Interpreting a definite integral in terms of area

Find each of $\int_{-1}^0 (x^3/2)\, dx$, $\int_0^2 (x^3/2)\, dx$, and $\int_{-1}^2 (x^3/2)\, dx$, and give an interpretation in terms of areas.

SOLUTION

On the graph in figure 5.3.3 we can see $x^3/2$ is negative for $x < 0$ and positive for $x > 0$. Hence, the integral assigns a negative value to the area of the region to the left of $x = 0$ and a positive value to the area of the region to the right of $x = 0$.

$$\int_{-1}^0 \frac{x^3}{2}\, dx = \frac{x^4}{8}\bigg|_{-1}^0 = \frac{0}{8} - \left(\frac{(-1)^4}{8}\right) = 0 - \frac{1}{8} = -\frac{1}{8}$$

$$\int_0^2 \frac{x^3}{2}\, dx = \frac{x^4}{8}\bigg|_0^2 = \frac{2^4}{8} - \frac{0^4}{8} = \frac{16}{8} - 0 = 2$$

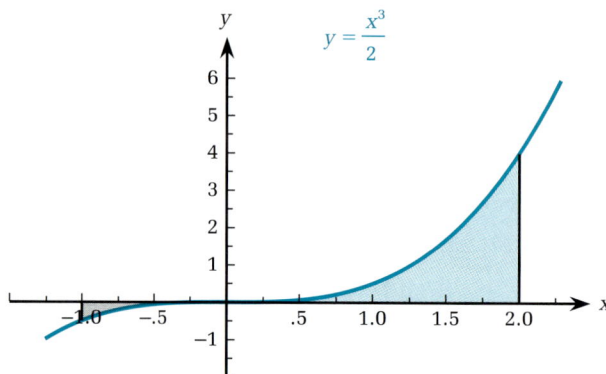

FIGURE 5.3.3

The value of the integral from -1 to 2 is the net value of these two regional values.

$$\int_{-1}^{2} \frac{x^3}{2}\, dx = \frac{x^4}{8}\bigg|_{-1}^{2} = \frac{2^4}{8} - \frac{(-1)^4}{8} = \frac{16}{8} - \frac{1}{8} = \frac{15}{8} = 1\frac{7}{8} = 1.875$$

EXAMPLE 5 Using the Fundamental Theorem of Calculus

Evaluate (a) $\int_{1}^{4}\left(3x - \frac{1}{x}\right) dx$ (b) $\int_{0}^{2}(e^{2x} - e^{-x})\, dx$

SOLUTION

(a) From the formulas in Section 5.1, $\int 3x\, dx = (3/2)x^2 + C$ and $\int (1/x)\, dx = \ln |x| + C$. Combining these, but not using C because we are working with a definite integral, we have

$$\int_{1}^{4}(3x - 1/x)\, dx$$

$$= [(3/2)x^2 - \ln x]\big|_{1}^{4} \text{ (We can eliminate the absolute}$$
$$\text{value sign because } x > 0.)$$

$$= [(3/2)(4)^2 - \ln 4] - [(3/2)(1)^2 - \ln 1]$$

$$= 24 - \ln 4 - 3/2 \approx 21.1137$$

(b) $\int_{0}^{2}(e^{2x} - e^{-x})\, dx$

$$= [(e^{2x}/2) - (e^{-x}/(-1))]\big|_{0}^{2}$$

$$= [(e^{2x}/2) + e^{-x}]\big|_{0}^{2}$$

$$= [(e^{2\cdot2}/2) + e^{-2}] - [(e^{0}/2) + e^{-0}] \text{ by substituting the limits}$$

$$= e^4/2 + e^{-2} - 1/2 - 1 \text{ (Recall that } e^0 = 1.)$$

$$= e^4/2 + e^{-2} - 3/2 \approx 27.2991 + 0.1353 - 1.5 = 25.9344$$

WARNING

1. Because of the possible mistakes with minus signs when working with $F(b) - F(a)$, it is a good idea to use many parentheses and group terms quite carefully.

2. If the antiderivative has been successfully found, be careful not to continue writing the integral sign, $\int$. As long as you write $\int$, you are expressing the need to find an antiderivative of the function.

PRACTICE EXERCISE 2

Evaluate $\displaystyle\int_0^5 (x^2 - e^x)\, dx$.

We encounter definite integrals arising from a wide variety of applications. For any integral, we first attempt an evaluation by the Fundamental Theorem of Calculus. The applicability of the theorem does not depend on the origin of the integral.

Determining Total Revenue from Rate of Revenue

Let us return to the model with which we opened this chapter: the total revenue from a toll bridge. In this model, revenue is equal to the product of the rate of revenue multiplied by the time over which that rate is received. Imagine a Riemann sum, with narrow rectangles representing small units of time and with heights being the rates of revenue over these times. As the limit is taken, with the intervals of time going to 0, we have a definite integral. The total revenue is equal to the area under the curve, which represents the rate of revenue (see figure 5.3.4).

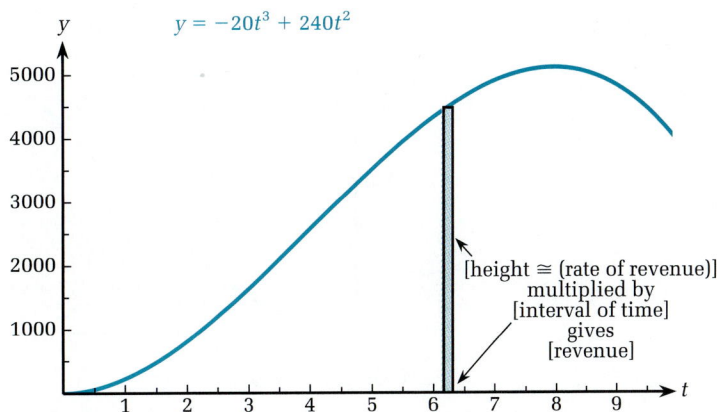

$y = -20t^3 + 240t^2$

[height $\cong$ (rate of revenue)] multiplied by [interval of time] gives [revenue]

FIGURE 5.3.4

EXAMPLE 6   Toll bridge collections

A simplified model for collections at a particular toll bridge is given by the formula $f(t) = -20t^3 + 240t^2$, where t is the hour of the day and $f(t)$ is the rate at which money is being collected per hour. The model is for midnight $(t = 0)$ to 10 A.M. $(t = 10)$.

How much revenue has been collected by 10 A.M.?

SOLUTION

The graph for $f(t)$ is shown in figure 5.3.4. The area under the curve represents the total revenue collected. By the Fundamental Theorem of Calculus, this area can be found by

$$\int_0^{10} f(t)\ dt = \int_0^{10} (-20t^3 + 240t^2)\ dt = \left(-20 \cdot \frac{t^4}{4} + 240 \cdot \frac{t^3}{3}\right)\Bigg|_0^{10}$$

$$= \left(-5t^4 + 80t^3\right)\Bigg|_0^{10}$$

$$= (-5 \cdot 10^4 + 80 \cdot 10^3) - 0 = 30{,}000 \qquad \bullet$$

Let us use the preceding example to develop the graph of $R(x)$, the total revenue up to time x. From a beginning time of $t = 0$, up to a time $t = x$, the total revenue, represented by $R(x)$, is the integral of the rate of revenue (the marginal revenue) from 0 to x. Suppose the marginal revenue is $f(t) = -20t^3 + 240t^2$. Then $R(x) = \int_0^x (-20t^3 + 240t^2)\ dt = \left(-20 \cdot \frac{t^4}{4} + 240 \cdot \frac{t^3}{3}\right)\Bigg|_0^x = -5x^4 + 80x^3$.

Figures 5.3.5 and 5.3.6 are the graphs of $f(t)$ and $R(x)$, respectively, both over [0, 10].

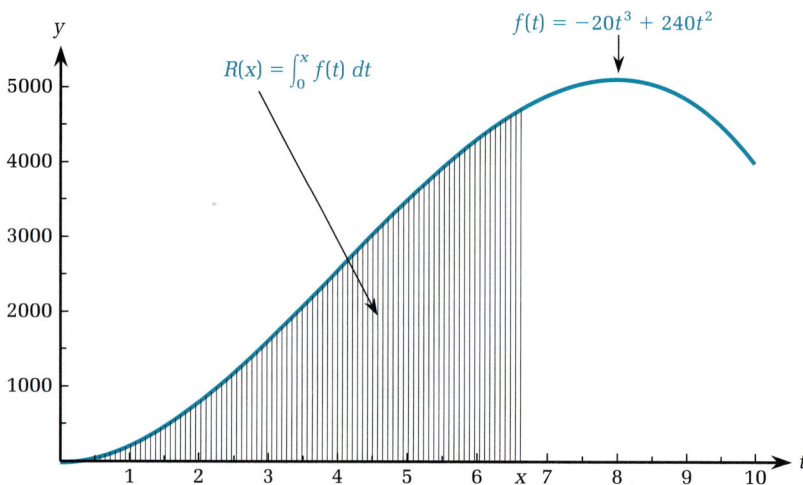

FIGURE 5.3.5

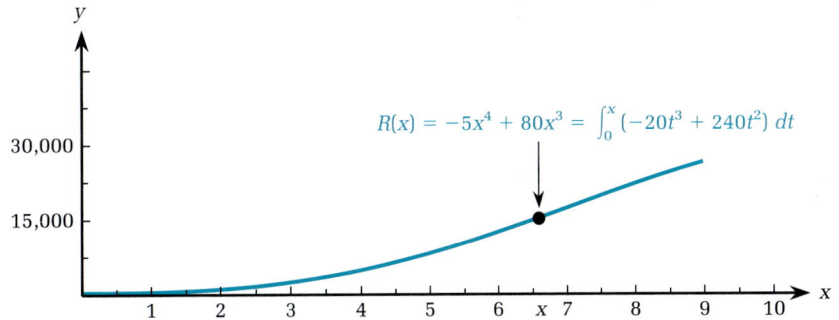

FIGURE 5.3.6

Observe that if $R(x) = \int_0^x (-20t^3 + 240t^2)\, dt$ we obtain $R(x) = -5x^4 + 80x^3$. If we take the derivative with respect to x we have $R'(x) = -20x^3 + 240x^2$.

Thus, we have $\dfrac{d}{dx} \displaystyle\int_0^x (-20t^3 + 240t^2)\, dt = -20x^3 + 240x^2$.

Fundamental Theorem of Calculus, the Conclusion

The preceding discussion may lead us to guess (correctly) that it is always true that the derivative with respect to x of the definite integral $\int_a^x f(t)\, dt$ is $f(x)$, that is, the integrand function evaluated at x.

FUNDAMENTAL THEOREM OF CALCULUS (PART B)

If $f(t)$ is continuous over $[a, b]$ and $a < x < b$, then

$$\frac{d}{dx}\left(\int_a^x f(t)\, dt \right) = f(x).$$

This formula, which we will not prove in this book, is truly a cornerstone of calculus. An outline of a partial proof of this result is contained in the enrichment problems at the end of this section.

Part B is what allows a proof of Part A of the Fundamental Theorem of Calculus. This is because if we assume, in Part A, that $\dfrac{d}{dx} F(x) = f(x)$ and now we know from Part B that $\dfrac{d}{dx} (\int_a^x f(t)\, dt) = f(x)$, then by Section 5.1, the functions $F(x)$ and $\int_a^x f(t)\, dt$ can differ only by a constant C. That is, $\int_a^x f(t)\, dt = F(x) + C$ for some C. To find C we let $x = a$, which gives $\int_a^a f(t)\, dt = F(a) + C$. But $\int_a^a f(t)\, dt = 0$, so we have $0 = F(a) + C$, or $C = -F(a)$. Thus, $\int_a^x f(t)\, dt = F(x) - F(a)$.

By letting $x = b$, we have $\int_a^b f(t)\,dt = F(b) - F(a)$. Finally, we use the fact that $\int_a^b f(t)\,dt = \int_a^b f(x)\,dx$ because t is only a dummy variable. Thus, we have Part A, $\int_a^b f(x)\,dx = F(b) - F(a)$.

Applications

It would be delightful if every time we had a rate function to graph, we could find a function whose graph approximates the curve and for which we could find an antiderivative. That will not be the case. We may have to use one of the methods in Section 5.2, rather than using the Fundamental Theorem of Calculus. However, that makes no difference to the concept we have been exploring. The area under the curve of the rate function represents the total amount of whatever quantity whose rate of change has been measured by the rate function.

Notice in Example 7 that the definite integral has a variable as one of its limits.

EXAMPLE 7   Advertising, 800 numbers

TABLE 5.3.1

w	$f(w)$	ACTUAL
0	0	0
1	114	100
2	394	400
3	934	950
4	772	750

In the March issue of a national magazine, a western resort town listed an 800 number for information. Readers were to call within four weeks. The actual rate of responses after w weeks is shown in table 5.3.1, along with values for $f(w) = 70e^w - e^{2w} - 69$, a function that closely approximates the actual response rate. Figure 5.3.7 is a graph of $f(w)$.

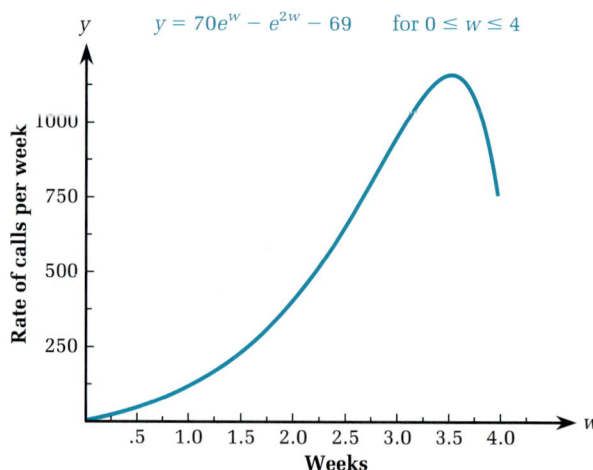

FIGURE 5.3.7

Estimate the total number of calls after t weeks.

SOLUTION

Let $C(t)$ be the total number of calls after t weeks.

$$C(t) = \int_0^t (70e^w - e^{2w} - 69) \, dw = \left(70e^w - \frac{e^{2w}}{2} - 69w \right) \Big|_0^t$$

$$= \left(70e^t - \frac{e^{2t}}{2} - 69t \right) - \left(70e^0 - \frac{e^0}{2} - 0 \right)$$

$$= 70e^t - \frac{e^{2t}}{2} - 69t - 70 + \frac{1}{2} \quad \text{(Remember that } e^0 = 1.\text{)}$$

$$= 70e^t - \frac{e^{2t}}{2} - 69t - 69.5$$

5.3 PROBLEMS

Foundations

The problems of this section require the basic skills illustrated by the following:

In problems 1–4 find the following indefinite integrals.

1. $\int x \, dx$

2. $\int x^3 \, dx$

3. $\int e^{3x} \, dx$

4. $\int \frac{2}{x} \, dx$

5. Solve for x if $x^2 - 3x - 10 = 0$.

Exercises

In problems 6–23 evaluate the definite integrals.

6. $\int_0^2 5x \, dx$

7. $\int_0^3 2x \, dx$

8. $\int_0^2 (x^2 - x) \, dx$

9. $\int_0^2 (x^2 + x) \, dx$

10. $\int_0^1 (4x^2 - x^3) \, dx$

11. $\int_{-2}^0 (x^2 + 3x^3) \, dx$

12. $\int_0^2 e^x \, dx$

13. $\int_0^1 e^x \, dx$

14. $\int_1^b \frac{1}{x} \, dx \quad (b > 1)$

15. $\int_a^5 \frac{1}{x} \, dx \quad (a > 0)$

16. $\int_a^b \left(2x + \frac{1}{x} \right) dx \quad (0 < a < b)$

17. $\int_1^b \left(x - \frac{1}{x} \right) dx \quad (b > 1)$

18. $\int_1^4 2\sqrt{x} \, dx$

19. $\int_4^9 4\sqrt{x} \, dx$

20. $\int_0^3 x^{2.1} \, dx$

21. $\int_0^2 2x^{1.5} \, dx$

22. $\int_1^1 e^{-x^2} \, dx$

23. $\int_3^3 \frac{1}{x^2 + 1} \, dx$

In problems 24–29 find the area of the region between the function and the x-axis over the given interval.

24. $f(x) = 2x^2$ over $[1, 4]$

25. $g(x) = x^2 + 2$ over $[0, 3]$

26. $f(x) = e^{2x} - 1$ over $[0, 3]$

27. $f(x) = 2e^{3x} - 1$ over $[0, 2]$

28. $f(x) = x + \frac{1}{x}$ over $[0.5, 2]$

29. $f(x) = 2x + \frac{1}{x}$ over $[0.5, 4]$

In problems 30–31 find the area of the region between the function and the x-axis.

30. $f(x) = x^2 - 3x - 10$ **31.** $g(x) = x^2 - x - 6$

32. Suppose the rate of revenue (marginal revenue) is given by $f(x) = x^3 - 9x^2 + 108$ when x hundred items are produced and there is no revenue for zero items. Find the total revenue for $x = 5$.

33. Suppose the rate of profit (marginal profit) is given by $g(x) = x^3 - 15x^2 + 500$ when x hundred items are produced and the "profit" of zero items is -10. Find the total profit for $x = 8$.

34. Suppose the rate at which ice melts in a picnic cooler is $g(t) = 1 + 0.1\sqrt{t}$ pounds an hour, where t is the number of hours. Estimate the number of pounds of ice needed for eight hours of cooling.

Writing and Discussion Problems

35. Evaluate $\int_3^1 x^2\, dx$ by the Fundamental Theorem of Calculus. The value is negative. Can this be interpreted as an area?

Enrichment Problems

36. Suppose $F'(x) = f(x)$ for all x and $f(x)$ is continuous. Use the Fundamental Theorem of Calculus to show that $\int_a^b f(x)\, dx + \int_b^c f(x)\, dx = \int_a^c f(x)\, dx$ for any values a, b, and c.

37. Find $\dfrac{d}{dx}\displaystyle\int_0^{x^2} e^t\, dt$. Propose a theorem about

    $\dfrac{d}{dx}\displaystyle\int_a^{g(x)} f(t)\, dt.$

38. **(Fundamental Theorem of Calculus, Part B)**
    Let $f(t)$ be continuous over $[a, b]$ and $f(t) \geq 0$ there.

    Define $A(x) = \int_a^x f(t)\, dt$. On the graph shown in figure 5.3.8 notice that $A(x + h) - A(x)$ is the rightmost shaded area.

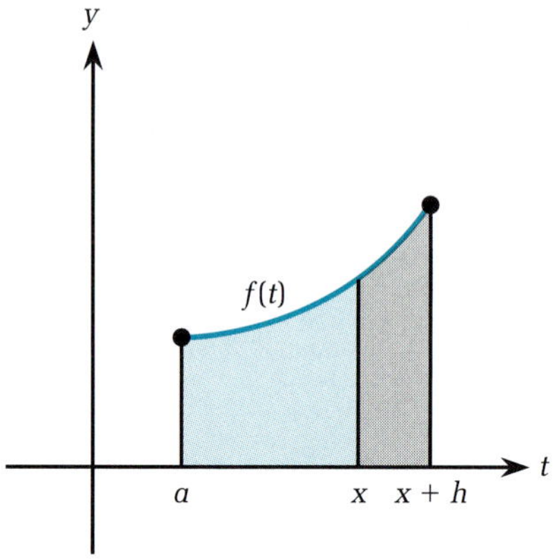

FIGURE 5.3.8

When h is small, an approximate value for $f(t)$ is $f(x)$. Give reasons for the following steps in the argument (for one step you may want to use a modified version of problem 36 above) that $A'(x) = f(x)$.

$$
\begin{aligned}
A'(x) &= \lim_{h \to 0} \frac{A(x + h) - A(x)}{h} \\
&= \lim_{h \to 0} \frac{\int_a^{x+h} f(t)\, dt - \int_a^x f(t)\, dt}{h} \\
&= \lim_{h \to 0} \frac{\int_x^{x+h} f(t)\, dt}{h} \\
&= \lim_{h \to 0} \frac{f(x) \cdot h}{h} \\
&= f(x)
\end{aligned}
$$

SOLUTIONS TO PRACTICE EXERCISES

1. The area between $y = x^2 + 1$ and the x-axis from $x = 1$ to $x = 3$ is given by

$$
\begin{aligned}
\int_1^3 (x^2 + 1)\, dx &= \left. \left(\frac{x^3}{3} + x \right) \right|_1^3 \\
&= \left(\frac{3^3}{3} + 3 \right) - \left(\frac{1^3}{3} + 1 \right) \\
&= 12 - \frac{4}{3} \\
&= \frac{32}{3} \approx 10.67
\end{aligned}
$$

2.
$$
\begin{aligned}
\int_0^5 (x^2 - e^x)\, dx &= \left. \left(\frac{1}{3}x^3 - e^x \right) \right|_0^5 \\
&= \left(\frac{1}{3}(5)^3 - e^5 \right) - \left(\frac{1}{3}(0)^3 - e^0 \right) \\
&= \frac{125}{3} - e^5 + 1 \quad (\text{Remember } e^0 = 1.) \\
&\approx -105.7465
\end{aligned}
$$

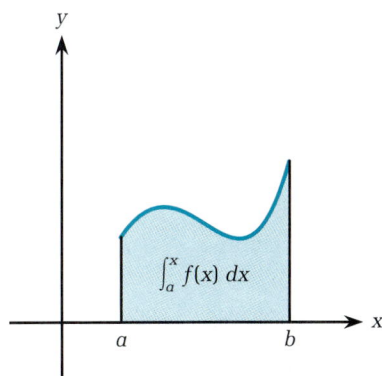

FIGURE 5.4.1

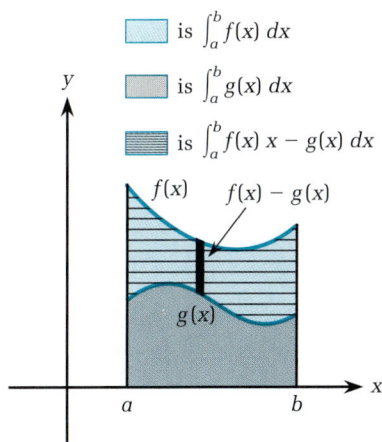

FIGURE 5.4.2

AREA BETWEEN TWO CURVES

This section and the next present a variety of applications, each of which depends on interpreting the area between two curves. We have defined and calculated the area of a region bounded by the graph of a function $y = f(x)$ and the x-axis over an interval $[a, b]$, as shown in figure 5.4.1.

Suppose a region is bounded by the graphs of two functions. Let these be represented by $f(x)$ and $g(x)$. Assume both of these are continuous over $[a, b]$. The reason for the assumption of continuity is to be certain both $\int_a^b f(x)\,dx$ and $\int_a^b g(x)\,dx$ actually exist.

Area of a Region Bounded by Any Two Continuous Functions

Let us first consider the case of two nonintersecting functions that have only positive values.

Suppose $f(x) \geq g(x) \geq 0$ over $[a, b]$. The region bounded by the curves of $y = f(x)$, $y = g(x)$, and $x = a$ to $x = b$ is indicated in figure 5.4.2.

The area of the region between the curves can be calculated by subtracting the area of the region below $y = g(x)$ from the area of the region below $y = f(x)$. This is given by the following.

AREA BETWEEN NONINTERSECTING CURVES OF POSITIVE FUNCTIONS

Suppose $f(x) \geq g(x) \geq 0$ over $[a, b]$. The area of the region bounded by the curves of $y = f(x)$, $y = g(x)$, and $x = a$ to $x = b$ is

$$(5.4.1) \quad \int_a^b f(x)\,dx - \int_a^b g(x)\,dx = \int_a^b (f(x) - g(x))\,dx$$

EXAMPLE 1   Area of region between two nonintersecting curves of positive functions

Find the area of the region bounded by $y = f(x) = x^2 + 1$ and $y = g(x) = x$ over $[1, 2]$.

SOLUTION

We need to check that $f(x) \geq g(x) \geq 0$ over $[1, 2]$. This is most easily done by graphing both functions, using the techniques of Chapter 3. Their graphs are shown in figure 5.4.3.

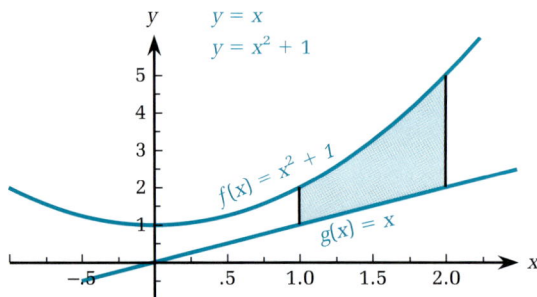

FIGURE 5.4.3

The area of the region is

$$\int_1^2 (f(x) - g(x))\, dx = \int_1^2 \left((x^2 + 1) - (x) \right) dx$$

$$= \int_1^2 (x^2 - x + 1)\, dx$$

$$= \left(\frac{x^3}{3} - \frac{x^2}{2} + x \right) \Bigg|_{x=1}^{x=2}$$

$$= \left(\frac{2^3}{3} - \frac{2^2}{2} + 2 \right) - \left(\frac{1^3}{3} - \frac{1^2}{2} + 1 \right)$$

$$= \frac{8}{3} - 2 + 2 - \frac{1}{3} + \frac{1}{2} - 1$$

$$= \frac{7}{3} - \frac{1}{2} = \frac{2 \cdot 7}{6} - \frac{1 \cdot 3}{6} = \frac{11}{6} \approx 1.8333$$

PRACTICE EXERCISE 1

Find the area of the region between $y = f(x) = -x^2 + 25$ and $y = g(x) = x + 5$ for $0 \leq x \leq 4$.

The special case of $g(x) = 0$ gives the area under $f(x)$ and above the x-axis, the situation addressed in Sections 5.2 and 5.3.

Suppose $f(x)$ and $g(x)$ are continuous functions such that $f(x) \geq g(x)$ over $[a, b]$, but not necessarily such that $g(x) \geq 0$. Formula (5.4.1) still gives the area of the region between the curves. The easiest way to see this is to realize that if some constant amount is added to both $f(x)$ and $g(x)$, then the region between them is shifted vertically, but its area is unchanged. So, we could add a constant C to $g(x)$

such that $g(x) + C$ is positive. Then we can use the result we already have for the area between two positive, nonintersecting curves. The area between $f(x) + C$ and $g(x) + C$ is the same as the area between $f(x)$ and $g(x)$.

We now consider the case of functions whose graphs intersect. If $y = f(x)$ intersects $y = g(x)$ for x-values in $[a, b]$ then the integral $\int_a^b [f(x) - g(x)]\, dx$ gives the net area (measuring area as positive if $f(x) \geq g(x)$ and negative if $f(x) \leq g(x)$). Thus, if we want to find the actual area of a region bounded by the two functions over the interval $[a, b]$, then we must locate each intersection and determine between intersections which is the greater function.

EXAMPLE 2   Finding the area of a region determined by functions that intersect

Find the area of the region bounded by $f(x) = -x^2 + 8x$ and $g(x) = 2x + 5$ over $[0, 5]$.

SOLUTION

The first thing we should do is *graph both functions,* which is done in figure 5.4.4.

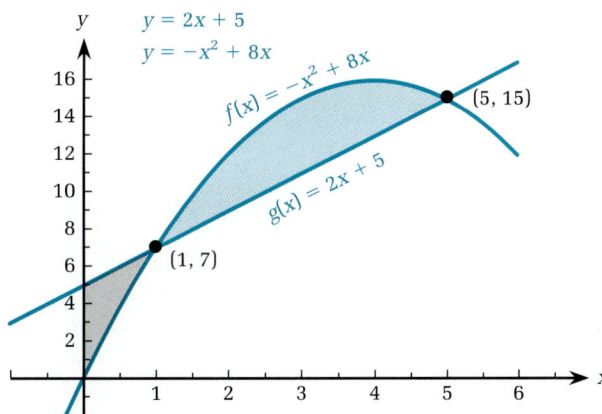

FIGURE 5.4.4

A graph shows which function is the greater. Furthermore, as always, a graph helps us determine whether our answers seem reasonable. If they do not, we need to discover why.

The graph of $f(x) = -x^2 + 8x$ is a parabola. At $x = 4$, it has a maximum value of $y = 16$ and it crosses the x-axis at $x = 0$ and $x = 8$. The equation of $g(x) = 2x + 5$ is a line, with y-intercept 5 and slope 2.

At $x = 0$, the value of $f(0) = 0$ is less than the value of $g(0) = 5$. However, at $x = 4$, the value of $g(4) = 2 \cdot 4 + 5 = 8 + 5 = 13$ is less than the value of $f(4) = 16$. Thus, the two curves must cross somewhere between 0 and 5. To find all crossings, we solve $f(x) = g(x)$. $-x^2 + 8x = 2x + 5$ is equivalent to $x^2 - 6x + 5 = 0$. To solve this, we can factor and find $(x - 5)(x - 1) = 0$. Thus, $f(x) = g(x)$ if $x = 5$ or $x = 1$.

In order to find the area of a region, we need to know which function is greater. Over the interval $[0, 1]$, we see $g(x) \geq f(x)$, but over the interval $[1, 5]$ we have $f(x) \geq g(x)$.

The area of the region between the curves is

$$\int_0^1 \left(g(x) - f(x) \right) dx + \int_1^5 \left(f(x) - g(x) \right) dx$$

$$= \int_0^1 \left((2x + 5) - (-x^2 + 8x) \right) dx + \int_1^5 \left((-x^2 + 8x) - (2x + 5) \right) dx$$

Each of these definite integrals needs to be evaluated separately. They cannot be combined into one integral because they have different integrands and different limits.

$$\int_0^1 \left((2x + 5) - (-x^2 + 8x) \right) dx = \int_0^1 (2x + 5 + x^2 - 8x) \, dx$$

$$= \int_0^1 (x^2 - 6x + 5) \, dx$$

$$= \left(\frac{x^3}{3} - 3x^2 + 5x \right) \Big|_0^1$$

$$= \left(\frac{1}{3} - 3 + 5 \right) - 0$$

$$= \frac{1}{3} + 2 = \frac{7}{3} \approx 2.3333$$

$$\int_1^5 [(-x^2 + 8x) - (2x + 5)] \, dx = \int_1^5 [-x^2 + 6x - 5] \, dx$$

$$= \left(-\frac{x^3}{3} + 3x^2 - 5x \right) \Big|_1^5$$

$$= \left(-\frac{5^3}{3} + 3 \cdot 5^2 - 5 \cdot 5 \right)$$

$$- \left(-\frac{1^3}{3} + 3 \cdot 1^2 - 5 \cdot 1 \right)$$

$$= -\frac{125}{3} + 75 - 25 + \frac{1}{3} - 3 + 5$$

$$= 52 - \frac{124}{3} = \frac{32}{3} \approx 10.6667$$

The area of the entire region is therefore $\dfrac{7}{3} + \dfrac{32}{3} = \dfrac{39}{3} = 13$.

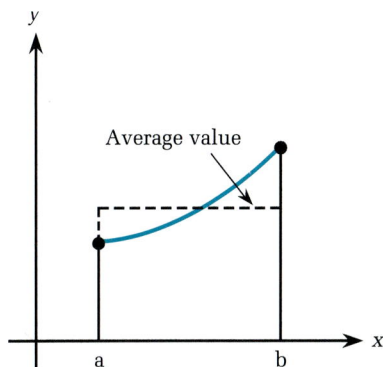

Average value

FIGURE 5.4.5

Average Value

The average velocity and the average monthly telephone bill are only two uses of the concept of "average." Public utility companies usually offer their customers some form of "level billing" rate so that heating and cooling bills can be spread out equally over the entire year.

Using the idea of area, we say that the average value of $f(x)$ is the height of that rectangle whose base lies on $[a, b]$ and whose area is equal to $\int_a^b f(x)\, dx$. This is shown in figure 5.4.5.

AVERAGE VALUE OF $f(x)$ OVER $[a, b]$

The *average value* of $f(x)$ over the interval $[a, b]$ is $\dfrac{\int_a^b f(x)\, dx}{b - a}$.

EXAMPLE 3   Average value

Find the average value of $f(x) = -x^2 + 8x$ over $[0, 6]$.

SOLUTION

$\int_0^6 (-x^2 + 8x)\, dx = [-(x^3/3) + 4x^2]|_0^6 = [-(6^3/3) + 4(6)^2] - 0 = -(216/3) + 144 = 72$.

Since $b - a = 6 - 0 = 6$, we have the average value is $72/6 = 12$. Figure 5.4.6 shows the graphs of both $y = f(x)$ and $y = 12$, which is the average value.

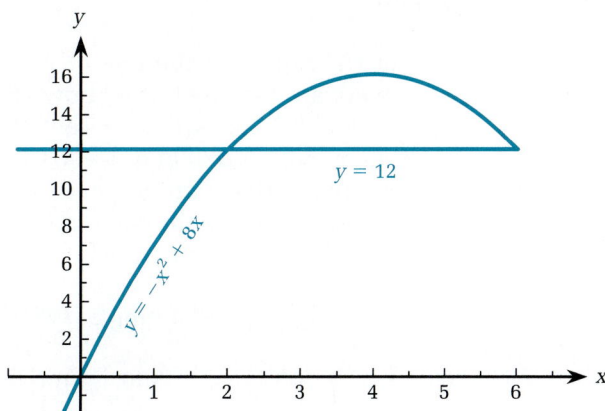

FIGURE 5.4.6

Notice the curves intersect at $(2, 12)$. We can think of the area in the region above $y = -x^2 + 8x$ and below $y = 12$ over the interval

[0, 2] as being equal to the area above $y = 12$ and below $y = -x^2 + 8x$ from $x = 2$ to $x = 6$.

Exponential Growth of Rate of Consumption

As background for the next application, we need to realize that equation (4.5.1), which was about the exponential growth of populations, can be modified to create the following formula.

EXPONENTIAL GROWTH OF RATE OF CONSUMPTION

Suppose a rate of consumption of a commodity is presently r_0 (measured in some units of quantity per some unit of time) and is changing by $100a$ percent (per unit of time). Then the rate is growing exponentially and the rate at time t is given by

(5.4.2) $$r(t) = r_0 e^{at}$$

EXAMPLE 4   Determining water conservation

The average water consumption in Mirage Lake City is 400 million gallons per year. There are no water meters, and without meters the city engineer estimates demand will grow exponentially at an average of 2% per year. However, if water meters are installed, the engineer estimates the average water consumption will drop immediately to 300 million gallons per year and demand will grow thereafter at an average of only 1% per year. How many million gallons would be saved over 20 years if water meters are installed?

SOLUTION

Let $r(t)$ represent the rate of water consumption assuming no water meters are installed. We know $r(0) = 400$ and $a = 0.02$, so that $r(t) = 400e^{0.02t}$. Let $s(t)$ represent the rate of consumption with water meters. $s(0) = 300$ and $a = 0.01$, so we have $s(t) = 300e^{0.01t}$. These two alternatives are graphed in figure 5.4.7. The area between the curves represents the total volume of water saved over 20 years.

This area is given by

$$\int_0^{20} (400 \cdot e^{0.02t} - 300 \cdot e^{0.01t})\, dt$$

$$= \left(\frac{400}{0.02} e^{0.02t} - \frac{300}{0.01} e^{0.01t} \right) \Bigg|_0^{20}$$

$$= (20{,}000 \cdot e^{0.4} - 30{,}000 \cdot e^{0.2}) - (20{,}000 \cdot e^0 - 30{,}000 \cdot e^0)$$

$$= 29{,}836.49 - 36{,}642.08 - 20{,}000 + 30{,}000$$

$$= 3194.41 \text{ million gallons}$$

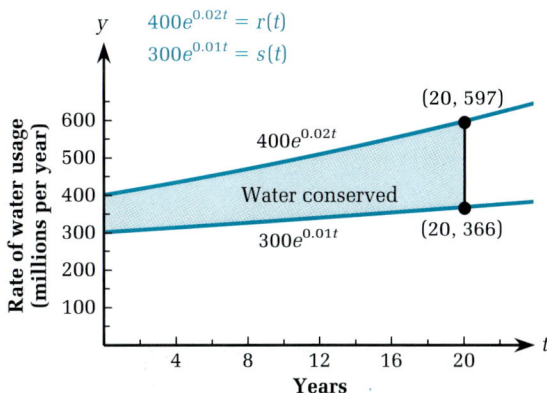

FIGURE 5.4.7

EXAMPLE 5   Reducing total health costs in the United States

In 1987, the rate of expenditure on health in the United States was 500 billion dollars yearly, increasing exponentially at 8% per year. If the increase could be cut to 7% per year, what would be the total savings in dollars for the period 1987–2007?

SOLUTION

The two rate functions are $500e^{0.08t}$ and $500e^{0.07t}$. The savings would be

$$\int_0^{20} (500e^{0.08t} - 500e^{0.07t})\, dt$$

$$= 500 \int_0^{20} (e^{0.08t} - e^{0.07t})\, dt$$

$$= 500 \left(\left(\frac{e^{0.08t}}{0.08} - \frac{e^{0.07t}}{0.07} \right) \Big|_0^{20} \right)$$

$$= 500 \left(\left(\frac{e^{0.08 \cdot 20}}{0.08} - \frac{e^{0.07 \cdot 20}}{0.07} \right) - \left(\frac{e^{0.08 \cdot 0}}{0.08} - \frac{e^{0.07 \cdot 0}}{0.07} \right) \right)$$

$$= 500 \left(\left(\frac{e^{1.6}}{0.08} - \frac{e^{1.4}}{0.07} \right) - \left(\frac{e^0}{0.08} - \frac{e^0}{0.07} \right) \right)$$

$$= 500(61.9 - 57.9 - (12.5 - 14.3)) = 500 \cdot 5.8$$

$$= 2900 \text{ billion dollars}$$

5.4 PROBLEMS

Foundations

The problems of this section require the basic skills illustrated by the following:

Find antiderivatives for problems 1–3.

1. $\int x^3\, dx$ 2. $\int e^{3x}\, dx$ 3. $\int \sqrt{2x}\, dx$

4. Convert 5% growth into an exponential decimal form of growth.

5. Solve for x such that $-x^2 = x - 2$.

Exercises

For problems 6–11, find the area between $f(x)$ and $g(x)$ over the interval indicated.

6. $f(x) = x^2, g(x) = x^3$ over [0, 1]

7. $f(x) = \sqrt{x}, g(x) = x$ over [0, 1]

8. $f(x) = x + 6, g(x) = x^2$ over [0, 3]

9. $f(x) = x + 2, g(x) = \frac{1}{2}x$ over [0, 2]

10. $f(x) = -x^2 + 10, g(x) = e^x$ over [0, 1]

11. $f(x) = e^{3x}, g(x) = x + 1$ over [0, 2]

In problems 12–15, find the average value over the indicated interval.

12. $f(x) = 2x + 3$ over [0, 4]

13. $g(x) = -3x + 5$ over [0, 2]

14. $g(x) = x^2 + 3$ over [−1, 2]

15. $f(x) = -2x^2 + 5x$ over [−1, 2]

16. *(Health Costs)*   In 1987 the total expenditure on health in the United States was 500 billion dollars annually, increasing exponentially at 8% per year. If the increase could be cut to only 5%, what would be the total savings in dollars for the period 1987–2007?

17. *(Electric Utility Power Purchases)*   Mid Range Power sells 50 megawatt-hours per year and expects usage to increase exponentially at 5% per year. It signs a contract promising to buy megawatt-hours on that assumption over the next 10 years. Suppose that usage increases at only 2% per year. How many excess megawatt-hours would Mid-Range Power find it had bought?

18. *(Landfill Disposal Costs of Low-Level Radio-active Waste)*   Low-level radioactive waste (LLRW) consists of waste contaminated with short-lived isotopes having half-lives of less than 90 days. Much of this waste is generated by hospitals performing medical tests. In 1991, the landfill disposal cost of one 55-gallon drum of LLRW averaged $600 in the United States. Costs for such disposal were estimated to increase ex-

ponentially by 25% each year over the period 1991–2001.[*]

Suppose a hospital generates 500 barrels of LLRW per year.
(a) Estimate the hospital's disposal cost for the period 1991–2001.
(b) Some hospitals have found that by careful separation of LLRW, they need to store only 20% of the number of barrels disposed of previously. Estimate the total savings for this hospital for the period 1991–2001 if this program is successful.

19. *(Savings on Landfill Disposal Costs of Low-Level Radioactive Waste)*   Refer to problem 18. In 1991, Monsanto Chemical disposed of 135 barrels of LLRW.[†] The Oak Ridge Laboratories has a super-compactor that can reduce wastes to only 10% of their original volume. Suppose Monsanto subjected their LLRW to the compactor and suppose costs of disposal rise from the 1991 cost of $600 a barrel at an exponential increase of 25% each year for the next 10 years. What would Monsanto expect to spend on landfill disposal of its LLRW for the period 1991–2001?

20. *(Inventory)*   Greg's Bakery uses flour at the rate of one ton per month. Through a slight mistake, Greg signs a two-year contract with the flour supplier on the expectation that demand will increase by 12% per month rather than 12% per year. Assuming he uses flour in a continuous manner, how much extra flour did he contract for? (Hint: Use 24 months and 1% per month for the needed flour, 12% per month for the contracted flour.)

Writing and Discussion Problems

21. If $f(x)$ is continuous over [a, b], then for some x-value, call it c, $\int_a^b f(x)\, dx = f(c) \cdot (b - a)$. Find such a value of c for $f(x) = -x^2 + 15$ over the interval [0, 3]. State this result in words, using the phrase "average value of $f(x)$."

22. *(Cavalieri's Principle)*   The statement that $\int_a^b (f(x) + g(x))\, dx = \int_a^b f(x)\, dx + \int_a^b g(x)\, dx$ illustrates a principle named in honor of Bonaventura Cavalieri (1598–1647). Basically, his

---

[*]*The Scientist*, 5:8, April 15, 1991.
[†]Ibid.

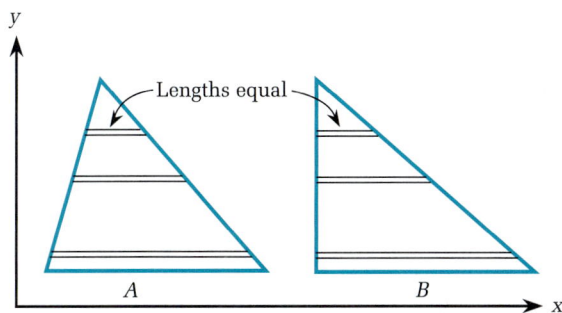

FIGURE 5.4.8

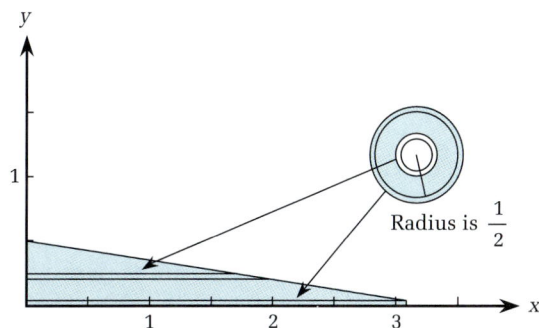

FIGURE 5.4.9

principle asserted that if an area is cut up into very thin sections, then their reassemblage will have the same area.

For example, the two triangles in figure 5.4.8 have the same area because thin horizontal sections of triangle A can be slid over so that each (almost) fits into triangle B.

Cavalieri's Principle predated the creation of calculus, but it anticipated some of its techniques and results. Explain how Cavalieri's Principle can be used to show that for a circle of radius r, the area is given by $A = \pi r^2$ (see figure 5.4.9). (Hint: Peel the circle's layers and arrange them to form, at least roughly, a triangle.)

Enrichment Problems

23. **(Recycling of Lubricating Oil)** In 1991, about one-half of all lubricating oil was being recycled. One western state had annual sales of 18 million gallons and recovery of 9 million gallons. Suppose the fraction of lubricating oil recycled could be increased to $[1 - (e^{-0.05t})/2]$, where t gives the number of years after 1991.

Assume annual sales of oil remain at the 18 million gallon level in the period 1991–2001.
   (a) Find the number of gallons that would be recycled in the period 1991–2001.
   (b) How many of these gallons would be "additional" beyond the 90 million gallons recycled under existing policies?

24. Find the area between the curves $f(x) = x^2$ and $g(x) = x$ between $x = -1$ and $x = 1$.

25. Find the area between the curves $f(x) = 2x$ and $g(x) = x^3$ between $x = -1$ and $x = 1$.

SOLUTION TO PRACTICE EXERCISE

1. Graph $f(x) = -x^2 + 25$ and $g(x) = x + 5$ for $0 \le x \le 4$ (see figure 5.4.10). The area is

$$\int_0^4 [(-x^2 + 25) - (x + 5)]\, dx$$

$$= \int_0^4 (-x^2 - x + 20)\, dx$$

$$= \left(-\frac{x^3}{3} - \frac{x^2}{2} + 20x \right) \Big|_0^4$$

$$= \left(-\frac{4^3}{3} - \frac{4^2}{2} + 20 \cdot 4 \right) - 0$$

$$= -\frac{64}{3} - 8 + 80 \approx 50.67$$

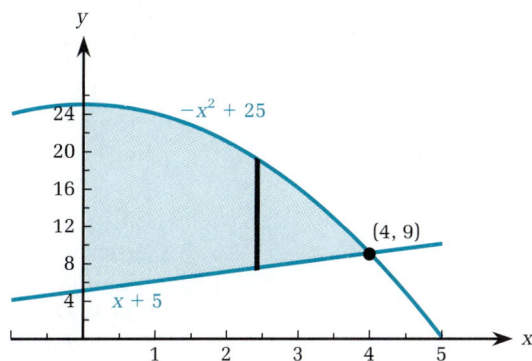

FIGURE 5.4.10

5.5   APPLICATIONS OF INTEGRALS TO BUSINESS AND ECONOMICS

In Section 4.6 we found that the present value of a future payment is less than the amount of the payment, because the money is not available until a future date. In that section, we calculated the present value of one single future payment. In this section, we determine the present value of a continuous stream of money over some interval of time. This is the capital value of an asset and is used in determining the value of a machine or even an entire company.

Capital values are used when a "leveraged buyout" is arranged. In such a transaction very little, if indeed any at all, of the purchaser's own money is used. The concept of the leveraged buyout has reshaped the U.S. economy since its introduction by Jerry Kohlberg in the mid-1970s. His firm, Kohlberg, Kravis, Roberts and Company, was founded in 1976 with $120,000. In 1987, they earned $75 million in fees alone in conjunction with the $26 billion takeover of RJR Nabisco.

Additional applications in this chapter are consumer's and producer's surpluses (of which the first measures what buyers do not have to spend and the second measures what producers realize in "extra" revenue), the Lorenz curve (which describes how goods are distributed), and the Gini Index (which is a measure of how uniformly goods are distributed).

Capital Value of an Asset

Suppose we have an enterprise that is going to produce revenue into the future. Examples might include a machine that makes aluminum cans, a toll bridge producing tolls, or an employee of a company. Let us discuss the problem in terms of the toll bridge, keeping in mind that the formulation is the same for any income-producing asset.

Suppose a toll bridge will produce revenue for the next 20 years at the rate of $700,000 per year. As we saw in Section 4.6, $700,000 produced t years from now is not worth $700,000 in today's dollars. The value of the bridge in today's dollars is far less than $20 \cdot 700,000 = \$14$ million. To dramatize this point, suppose we invested $14 million at 8% annual interest. We could draw out $1,120,000 every year without even touching the principal of that $14 million.

We need to discount future earnings to their present value. To find the present value of receiving a lump sum of P dollars t years from today at $r\%$ annual interest, we convert the percentage into a decimal value $i = r/100$, assume continuous compounding, and use the formula that the present value is $P \cdot e^{-it}$ in today's dollars.

We want to consider a continuous stream of revenue, so instead of one lump sum, we are going to receive revenue in a continuous flow.

EXAMPLE 1   Capital value of a continuous money flow

Find the capital value of an asset that produces continuously at the rate of $700,000 per year for the next 20 years. Use an annual interest rate of 8%.

SOLUTION

The amount of income is the product of the rate of income multiplied by the duration of time. The graph in figure 5.5.1 shows that t years from now we have $700,000 \cdot dt$ dollars, which flow in during a short period of time, dt. The graph in figure 5.5.2 shows the present value of that same amount of money, its value shrunken by the 8% interest discount rate to a present value of only $700,000 \cdot dt \cdot e^{-0.08t}$ dollars.

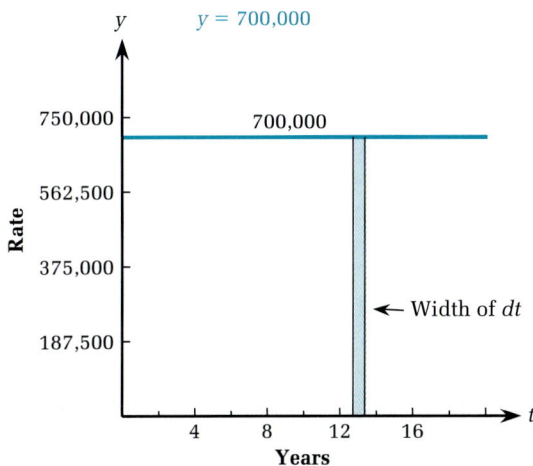

FIGURE 5.5.1

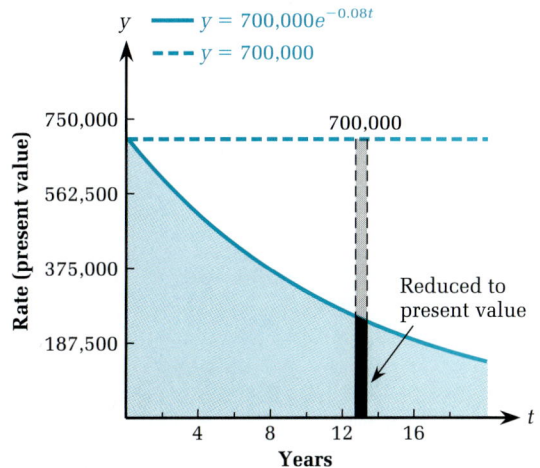

FIGURE 5.5.2

The area under the curve in figure 5.5.2 represents the total present value.

$$\int_0^{20} 700,000 \cdot e^{-0.08t} \, dt = 700,000 \left(\frac{e^{-0.08t}}{-0.08} \bigg|_0^{20} \right)$$

$$= 700,000 \left(\frac{e^{-0.08 \cdot 20}}{-0.08} - \frac{e^{-0.08 \cdot 0}}{-0.08} \right)$$

$$= 700,000 \cdot (-2.5237 + 12.5) = 6,983,410$$

Thus, the present value of the asset is $6,983,410, and the books of a company should list that capital value.

The continuous money flow was at a constant rate in Example 1, but we can adapt our method to find the capital value for a flow that is not constant. However, until the next chapter we do not have the necessary formulas and techniques to find complicated antiderivatives, so the examples in this section all use constant flows.

CAPITAL VALUE OF AN ASSET

Let $P(t)$ be the rate at which an asset generates money at time t and let i be the annual interest rate expressed decimally. The capital value of the asset from time $t = a$ to time $t = b$ is

$$\int_a^b P(t) \cdot e^{-it} \, dt$$

PRACTICE EXERCISE 1

Suppose a machine stamps out $3 million worth of cans a year. Using an annual interest rate of 7%, find the capital value of the machine in the next 10 years.

Traditionally a bond was issued based on a physical object, such as a truck, a press, or other capital item. However, in recent years the practice has arisen of issuing a bond based on the expected income from a product or service. The expected income has been capitalized and a bond issued based on that capital. Such a speculative bond is often called a "junk" bond.

Using the $6,983,410 capitalization of our asset in Example 1, we might decide to issue $6 million in "junk" bonds and try to persuade the bridge owners to sell the bridge for $6 million. If successful, we can acquire the property with none of our own money. It is clear that the estimates of the income flow of the asset, as well as interest rates, are critical in determining the capital book value of the asset, and the collapse of many enterprises demonstrates the great opportunities for manipulations.

Consumer's and Producer's Surplus

The axiom that "markets determine prices" describes an open market in which buyers and sellers are not forced to pay prices set artificially high or low. In such a market there is an equilibrium price. We will consider that case, but what follows applies to any situation in which there is a unified pricing structure and all buyers pay, and all sellers sell, at that one price.

As an example, the Organization of Petroleum Exporting Countries (OPEC) has tried for years to set a price for oil that would be uniform and totally independent of whether the production cost of the oil was $0.50 a barrel or $15 a barrel. The same situation exists in several other commodities, such as cocoa. On the other hand, many national governments artificially set the selling prices of consumer goods. Examples are tortillas in Mexico and sugar in the United States.

Consider the graph of a supply function in figure 5.5.3 and the graph of a demand function in figure 5.5.4.

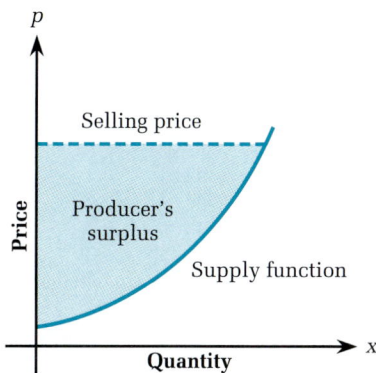

FIGURE 5.5.3

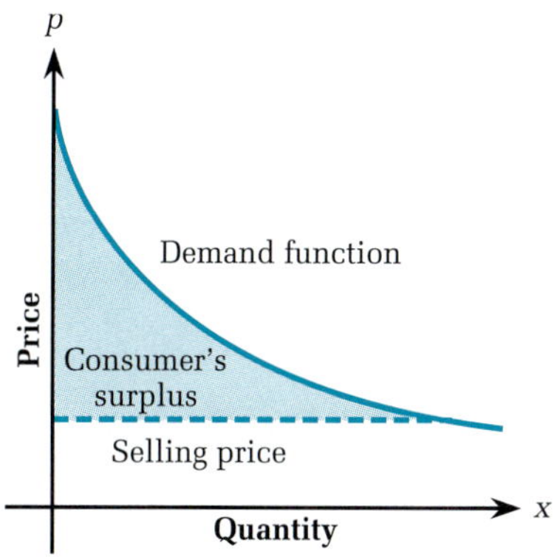

FIGURE 5.5.4

The shaded area in figure 5.5.3 represents dollars paid to suppliers who would have supplied the goods at a lower price. For example, if OPEC sets the price at $20 and some oil would have come on the market at $10, the supplier of that $10 oil gets a "surplus" of $10 a barrel.

The shaded area in figure 5.5.4 represents dollars that consumers did not have to pay for the goods they received. For example, if someone is willing to pay $2 a gallon for gasoline, but everyone who buys gasoline pays only $1.25 a gallon, then that consumer who would have paid $2 receives a "surplus" of $0.75 a gallon on gasoline.

DEFINITION

Consumer's surplus and producer's surplus

The *consumer's surplus* is the area of the region between the demand curve and the selling price. The *producer's surplus* is the area of the region between the supply curve and the selling price.

Demand and supply functions may be expressed as price being a function of quantity or as the quantity being expressed as a function of price. Whichever form is used, we need to know both the quantity and the price at which the transaction is occurring. The quantity values are used in the limits of integration, and the price values occur in the integrand.

EXAMPLE 2   Consumer's surplus for a known quantity

Suppose the demand function for concert tickets is $p_d(x) = -0.0005x^2 + 16$, as graphed in figure 5.5.5.

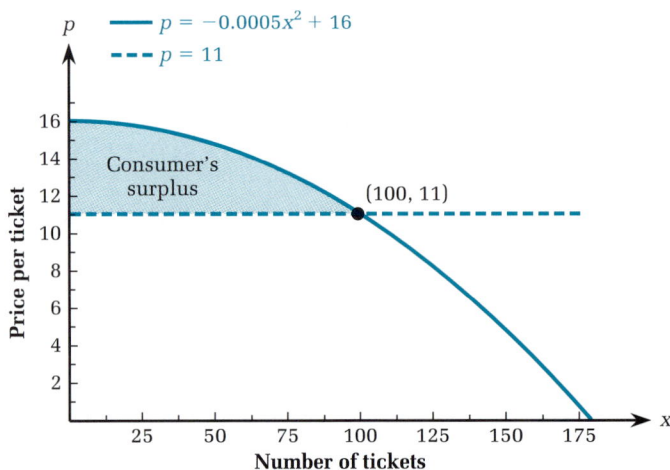

FIGURE 5.5.5

(a) Find the ticket price corresponding to $x = 100$.

(b) Find the consumer's surplus when $x = 100$.

SOLUTION

(a) Evaluating $p_d(x) = -0.0005x^2 + 16$ for $x = 100$ gives $p_d(100) = -0.0005(100)^2 + 16 = -5 + 16 = 11$ dollars per ticket.

(b) The consumer's surplus is the area of the region bounded by $y = 11$, which is the selling price, and the demand curve, $y = p_d(x)$, over the interval from $x = 0$ to $x = 100$. This area is given by the definite integral

$$\int_0^{100} (p_d(x) - 11)\ dx = \int_0^{100} ((-0.0005x^2 + 16) - 11)\ dx$$

$$= \int_0^{100} (-0.0005x^2 + 5)\ dx$$

$$= \left(-0.0005\left(\frac{x^3}{3}\right) + 5x\right)\Bigg|_0^{100}$$

$$= \left(-0.0005\left(\frac{100^3}{3}\right) + 5 \cdot 100\right)$$

$$- \left(-0.0005\left(\frac{0^3}{3}\right) + 5 \cdot 0\right)$$

$$= -166.67 + 500 = 333.33$$

10.6% of the aggregate income. So adding these two groups together we find the lowest 40% of all families receive a total of 15.2% of the aggregate income.

A graph displaying the data in table 5.5.1 is shown in figure 5.5.9. Such a graph is called a Lorenz curve in honor of the American statistician Max Otto Lorenz.

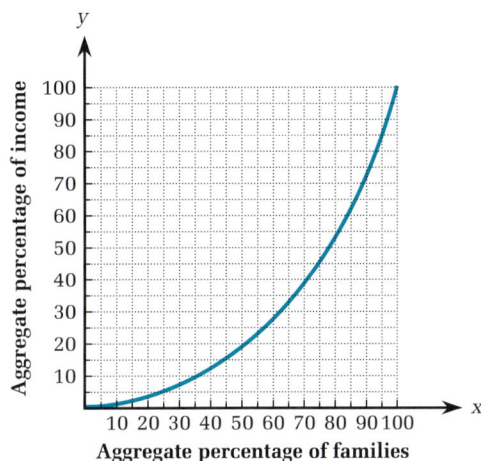

FIGURE 5.5.9

DEFINITION

Lorenz curve

Suppose $y = f_1(x)$ is that percent of a resource held by that x percent of the population that individually hold the least of the resource. The graph of y is a *Lorenz curve.*

In plotting the data, the shape of the curve between the data points is not exact, but with more information we could refine the graph.

PRACTICE EXERCISE 3

Produce a Lorenz curve for the data in the following table on the population of Mozzarellaville.

Percentage of People	0	20	40	60	80	100
Percentage of All Pizza Eaten	0	2	6	15	40	100

If we examine a Lorenz curve, several questions come to mind.

1. How uniformly is the resource distributed?

2. Has the uniformity changed over time?

3. Is one particular individual (person, family, country, or whatever) getting its "fair" share?

If everyone received the same amount, the Lorenz curve would be a straight line because the percentages of resource and percentages of people would be the same all the time, as shown in the table below, which yields the Lorenz curve in figure 5.5.10.

Percentage of Population	0	20	40	60	80	100
Percentage of Resource	0	20	40	60	80	100

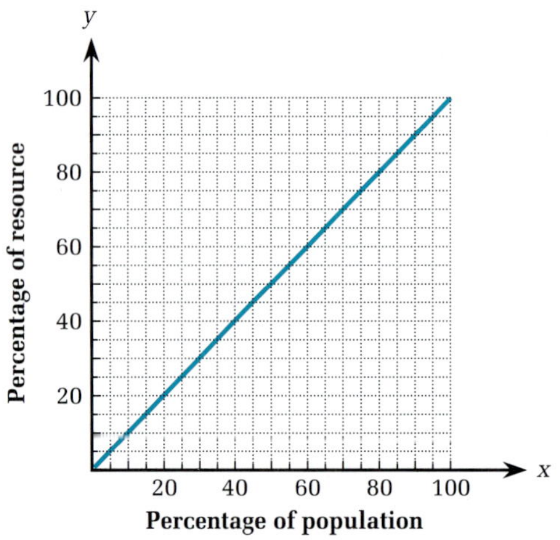

FIGURE 5.5.10

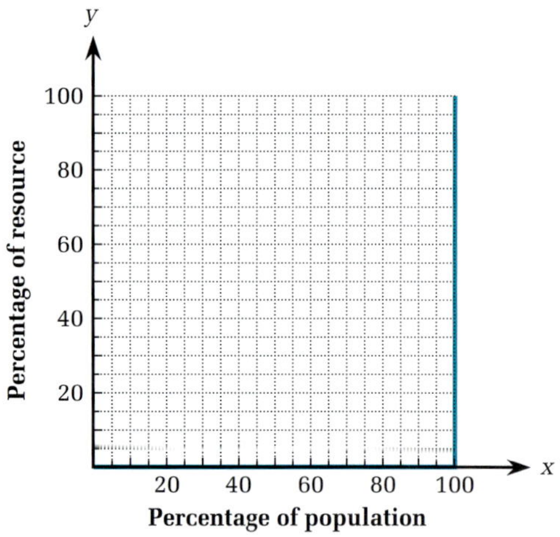

FIGURE 5.5.11

If one person had all the resource, the percentage of resource would be zero for all of the population except that one person, for whom the percentage would be 100. This is shown in the table below, which yields the Lorenz curve in figure 5.5.11.

Percentage of Population	0	20	40	60	80	100
Percentage of Resource	0	0	0	0	0	100

One measure of uniformity is given by letting a square of length 1 on each side be used to frame the graph of the Lorenz curve and then measuring the area between the Lorenz curve and the diagonal line representing a uniform distribution (see figure 5.5.12). The maximum size of this area is one-half (this occurring for a totally concentrated distribution). In order to create a convenient index number that takes on values from 0 to 1, the area is doubled. This number is called the *Gini Index,* in honor of Corrodo Gini (1884–1965). His 1912 book, *Variabilita e Mutabilita,* helped to create the discipline of sociology.

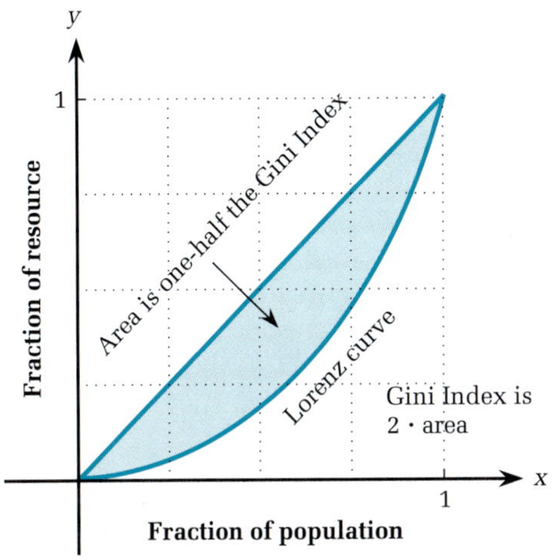

FIGURE 5.5.12

Because the area under the uniform distribution is 1/2, it often is easier to determine the Gini Index by counting the area *below* the Lorenz curve and subtracting that number from 1/2 before doing the doubling.

The typical Lorenz curve does not correspond to a neat mathematical equation. Thus, we are often forced into using other ways to find the Gini Index area, rather than using the Fundamental Theorem of Calculus. However, the following definition always applies.

DEFINITION

Gini Index

If $f_l(x)$ is the Lorenz curve for the distribution of a resource, then $2 \int_0^1 (x - f_l(x)) \, dx$ is the *Gini Index.*

EXAMPLE 6   Gini Index

Find the Gini Index for the Lorenz curve $f_l(x) = x^2$.

SOLUTION

$2 \int_0^1 (x - x^2) \, dx = 2(x^2/2 - x^3/3)|_0^1 = 2(1/2 - 1/3) - 2 (0 - 0) = 1/3 \simeq 0.3333.$

PRACTICE EXERCISE 4

Find the Gini Index for the Lorenz curve $f_l(x) = x^3$.

The Gini Index is a snapshot in time. The index itself does not depend on time. However, a collection of these snapshots, taken over some period of time, gives a moving picture. If the Gini Index is increasing over time, then the distribution is becoming less uniform. On the other hand, a sequence of decreasing Gini Index numbers over time means the distribution is becoming more uniform.

Lorenz curves may be used to compare salaries in different professions and for a variety of other applications. For further information, see *Economics* by P. A. Samuelson and W. Nordhaus (New York: McGraw-Hill), 12th edition, 1986.

Fair Share

Suppose at some point on the Lorenz curve the derivative is 1. Then the rate at which the share of the goods is increasing is exactly the same as the rate of increase of the share of the population. This can be thought of as achieving *fair share.*

EXAMPLE 7   Fair share

For the Lorenz curve $f_l(x) = x^{3/2}$, find the x-value for which "fair share" occurs.

SOLUTION

$f_l'(x) = 1$, where $(3/2)x^{1/2} = 1$. Hence $x^{1/2} = 2/3$, which yields $x = 4/9$.

5.5 PROBLEMS

Foundations

The problems of this section require the basic skills illustrated by the following:

1. Find the present value of a payment of $P(t)$ dol-lars, payable t years from today. Use continuous compounding and an annual interest rate of 6%.

2. Find $\int e^{0.06t} \, dt$.

3. Solve $-x + 30 = \sqrt{x}$. (Hint: Think of this as a quadratic equation in terms of $\sqrt{x}$.)

Exercises

4. **(Capital Value of an Asset)**   Find the capital value of an asset that produces continuously at the rate of $100,000 per year in the next 10 years. Use an annual interest rate of 6%.

5. **(Capital Value of an Asset)**   Find the capital value of an asset that produces continuously at the rate of $50,000 per year in the next 20 years. Use an annual interest rate of 7%.

6. **(Consumer's and Producer's Surplus at Equilibrium)**   A demand curve is given by $p_d(x) = -2x^2 + 100$, and a supply curve by $p_s(x) = 4x + 4$.
   (a) Find the equilibrium point.
   (b) Find the consumer's surplus at equilibrium.
   (c) Find the producer's surplus at equilibrium.

7. **(Consumer's and Producer's Surplus at Equilibrium)**   A demand curve is given by $p_d(x) = -x^2 + 13$, and a supply curve by $p_s(x) = x + 1$.
   (a) Find the equilibrium point.
   (b) Find the consumer's surplus at equilibrium.
   (c) Find the producer's surplus at equilibrium.

8. **(Consumer's and Producer's Surplus)**   A demand curve is given by $p_d(x) = -x^2 + 50$, and a supply curve by $p_s(x) = 3x + 10$.
   (a) Find the consumer's surplus at $x = 3$, $p = 41$.
   (b) Find the producer's surplus at $x = 4$, $p = 22$.
   (c) Find the equilibrium point.
   (d) Find the consumer's surplus at equilibrium.
   (e) Find the producer's surplus at equilibrium.

9. **(Consumer's and Producer's Surplus)**   A demand curve is given by $p_d(x) = -x + 30$, and a supply curve by $p_s(x) = \sqrt{x}$.
   (a) Find the consumer's surplus at $x = 2$, $p = 28$.
   (b) Find the producer's surplus at $x = 1$, $p = 1$.
   (c) Find the equilibrium point.
   (d) Find the consumer's surplus at equilibrium.
   (e) Find the producer's surplus at equilibrium.

10. **(Lorenz Curve and Gini Index)**
   (a) Graph the Lorenz curve and find the Gini Index for $f(x) = e^{0.693x} - 1$.
   (b) For what x is the "fair share" found?

11. **(Lorenz Curve and Gini Index)**
   (a) Graph the Lorenz curve and find the Gini Index for $f(x) = x^{1.5}$.
   (b) For what x is the "fair share" found?

12. A published formula for the Lorenz curve for wealth in the United States in 1985 was $f(x) = x^{5.6}$. Find the Gini Index.

13. **(Lorenz Curve and Gini Index, Navaho Rugs)**   At a weaver's Navaho rug auction in western New Mexico, 300 rugs sold for a total of $25,000. The following table summarizes the sales.

Percentage of Aggregate Rugs	0	20	40	60	80	100
Percentage of Aggregate Sales	0	8	23	40	60	100

   Use the graph of the Lorenz curve for these data, shown in figure 5.5.13, to estimate the Gini Index by counting portions of squares.

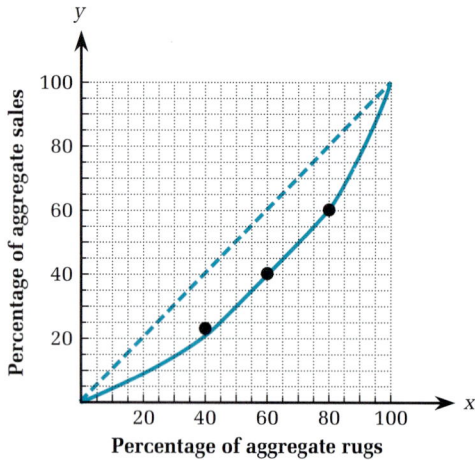

FIGURE 5.5.13

The data for United States family incomes based on aggregate percentages of aggregate income is given in table 5.5.2, which should be used for problems 14 and 15.

14. For the United States family incomes in 1970,
    (a) sketch a Lorenz curve

(b) estimate the Gini Index by counting portions of squares.

15. For the United States family incomes in 1980,
    (a) sketch a Lorenz curve
    (b) estimate the Gini Index by counting portions of squares.

16. A 1989 study of the income of 970 lawyers in Colorado revealed an average income of $68,680, distributed as shown in table 5.5.3.
    (a) Sketch a Lorenz curve.
    (b) Estimate the Gini Index by counting portions of squares.

Enrichment Problem

17. Suppose Fun-Time Computer Services is considering the purchase of a new computer, which costs $300,000. The new computer would reduce the cost of each concert ticket sale from $0.50 to $0.05. Suppose 100,000 ticket sales are processed each year. Using an annual interest rate of 7%, how many years must the computer be used before it has repaid its initial cost?

TABLE 5.5.2

PERCENTAGE OF AGGREGATE FAMILIES	0	20	40	60	80	95	100
Percentage of Aggregate Income (1970)	0	5.5	17.7	35.3	59.1	84.4	100
Percentage of Aggregate Income (1980)	0	5.1	16.7	34.2	58.5	84.8	100

TABLE 5.5.3

Percentage of Lawyers	9.2	12.8	19.3	33.9	59.6	74.8	84	93.6	97.4	100
Percentage of Aggregate Income	1	1.8	4.2	11.9	31.2	47.2	59.6	77.6	87.3	100

Source: *Denver Post*, November 11, 1991

SOLUTIONS TO PRACTICE EXERCISES

1. The machine stamps out $3 million worth of cans a year. Using an annual interest rate of 7%, the capital value of the machine over the next 10 years is $\int_0^{10} 3 \cdot e^{-0.07t} \, dt = [-3 \cdot (e^{-0.07t}/0.07)]|_0^{10}$ $= (-42.8571 \cdot e^{-0.7}) - (-42.8571) = 21.5749$ million dollars.

2. The demand function $p_d(x) = -x + 9.39$ and the supply function $p_s(x) = e^x$ have an approximate equilibrium for $x = 2$.
    (a) The equilibrium price is $p = 7.39$ (substituting into either p_s or p_d).
    (b) The graph of this situation is shown in figure 5.5.14.

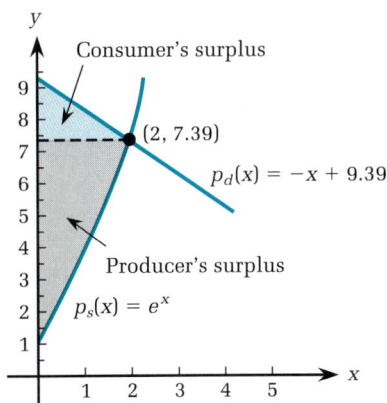

FIGURE 5.5.14

The producer's surplus at equilibrium is
$\int_0^2 (7.39 - e^x) \, dx = (7.39x - e^x)|_0^2 =$
$(14.78 - 7.389) - (0 - 1) = 8.391$.

The consumer's surplus at equilibrium is
$\int_0^2 [(-x + 9.39) - 7.39] \, dx =$
$\int_0^2 (-x + 2) \, dx = [-(x^2/2) + 2x]|_0^2 =$
$(-2 + 4) - (0) = 2$.

3. The Lorenz curve for the data on the population of Mozzarellaville is shown in figure 5.5.15.

4. $2 \int_0^1 (x - x^3) \, dx = 2[(x^2/2) - (x^4/4)]|_0^1 =$
$2[(1/2) - (1/4)] = 2(1/4) = 0.5$.

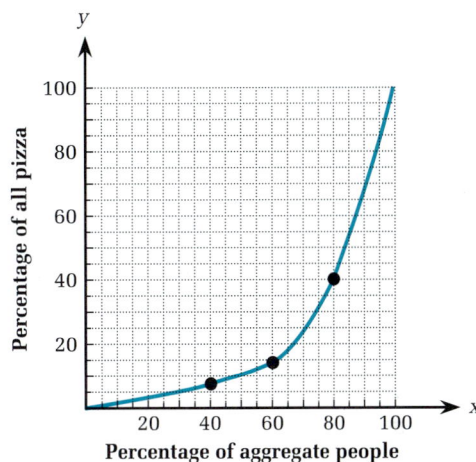

FIGURE 5.5.15

CHAPTER 5 REVIEW

Define or discuss:

1. Antiderivative
2. Constant of integration
3. Alternate antiderivative
4. Integrand
5. Indefinite integral
6. Power Rule for Integrals
7. Constant Multiple Rule for Integrals
8. Sum or Difference Rule for Integrals
9. Antiderivative formulas
10. Area
11. Methods of determining area
12. Average value

13. Midpoint approximation sum
14. Riemann sum
15. Definite integral
16. Properties of the definite integral
17. Fundamental Theorem of Calculus (Parts A and B)
18. Area of a region between curves
19. Exponential growth of rate of consumption
20. Capital value of an asset
21. Consumer's surplus and producer's surplus (in general and at equilibrium)
22. Lorenz curve
23. Gini Index
24. Fair share

REVIEW PROBLEMS FOR CHAPTER 5

1. Estimate $\int_0^2 [10/(x^2 + 4)] \, dx$ by using four equal subintervals and the technique of midpoint approximation.

2. Estimate $\int_1^9 \sqrt{2x} \, dx$ by using four equal subintervals and the technique of midpoint approximation.

3. Estimate $\int_{-2}^{0} [x/(x^2 + 1)]\ dx$ by using four equal subintervals and the technique of midpoint approximation.

Find antiderivatives for problems 4–12.

4. $\int 4x^2\ dx$

5. $\int \left(3\sqrt{x} + \dfrac{2}{x}\right) dx$

6. $\int \left(x^{-2} + \dfrac{3}{x^3}\right) dx$

7. $\int \left(6 + \dfrac{4}{e^{2x}}\right) dx$

8. $\int e^{-2x}\ dx$

9. $\int (y^3 - 2y)\ dy$

10. $\int \dfrac{x^2 + 1}{x}\ dx$

11. $\int \dfrac{u^3 + 2u}{u^2}\ du$

12. $\int (e^{2x} + x)\ dx$

Evaluate problems 13–16.

13. $\int_{1}^{4} (x^3 - x)\ dx$

14. $\int_{1}^{4} \sqrt{4x}\ dx$

15. $\int_{-2}^{-1} \left(u^2 + \dfrac{1}{u^2}\right) du$

16. $\int_{2}^{3} \dfrac{1}{x\sqrt{x}}\ dx$

For problems 17–20, find the area of the region bounded by the following pairs of curves and the vertical lines determined by the ends of the interval.

17. $f(x) = \dfrac{1}{x} + x,\ g(x) = 0$ over $\left[\dfrac{1}{2}, 2\right]$

18. $f(x) = x^3 + x,\ g(x) = 2x - 1$ over $[0, 2]$

10. $f(x) = o^x + 1,\ g(x) = -2$ over $[0, 3]$

20. $f(x) = 3x + 18,\ g(x) = x^2$ over $[-3, 6]$

21. Find the average value of $x^2 - 2x$ over $[0, 2]$.

22. Using an annual interest rate of 9%, find the capital value of an asset that is continuously generating income at the rate of $20,000 per year for a period of 10 years.

23. Using an annual interest rate of 12%, find the capital value of a parking lot that is continuously producing income at the rate of $2500 per month for 10 years.

24. Suppose $p_d(x) = 100 - (x^3/100)$ for $0 \le x \le 20$ is a demand function.
    (a) Sketch a graph of $y = p_d(x)$.
    (b) Find the quantity sold if the price is 90.
    (c) Find the consumer's surplus if the price is 90.

25. Let a demand function be $p_d = 35 - x^2$ and a supply function be $p_s = x + 5$.
    (a) Find the consumer's surplus at $p = 26$, $x = 3$.
    (b) Find the equilibrium price and quantity.
    (c) Determine the consumer's surplus at equilibrium.
    (d) Determine the producer's surplus at equilibrium.

26. Let a demand function be $p_d = -2x^2 + 86$ and a supply function be $p_s = 7x + 1$.
    (a) Find the equilibrium price and quantity.
    (b) Determine the consumer's surplus at equilibrium.
    (c) Determine the producer's surplus at equilibrium.

27. Consider the Lorenz curve $f(x) = (x^3 + x)/2$.
    (a) Find the Gini Index.
    (b) Find the x-coordinate for the fair share.

28. Estimate the Gini Index for the Lorenz curve graphed in figure 5.5.16.

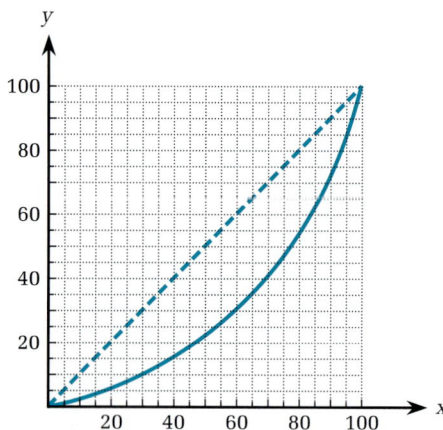

FIGURE 5.5.16

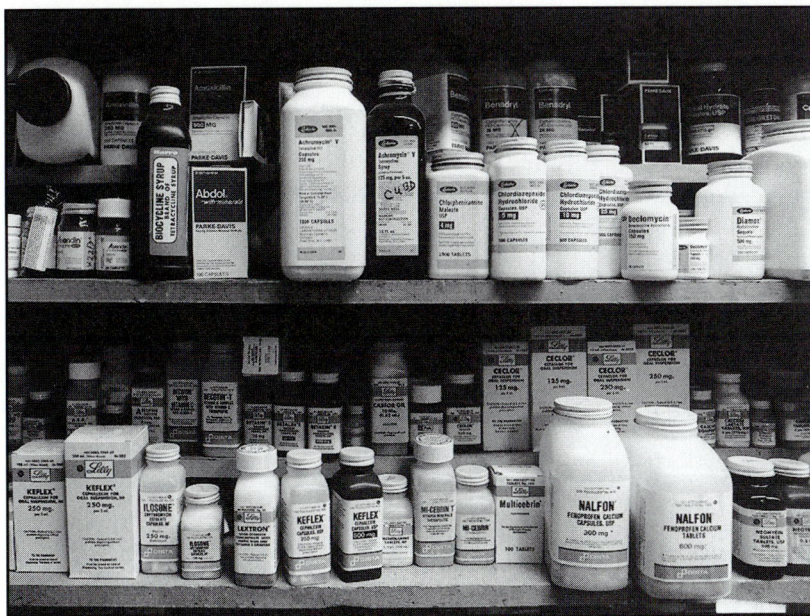

CHAPTER 6

TECHNIQUES OF INTEGRATION

THIS CHAPTER IS DIVIDED INTO THREE GENERAL TOPICS. FIRST, WE PRESENT several methods of finding antiderivatives. These methods are called *techniques of integration*. Many of these techniques exist, but we concentrate on only a very few of the more important ones.

Our second topic is the evaluation of definite integrals when we do not have an antiderivative for the integrand. The midpoint approximation sum of Section 5.2 is one such method of evaluation. We discuss several others in this chapter.

The last topic of this chapter concerns definite integrals that have infinite limits of integration or have integrands that are not continuous functions over an interval [a, b]. These are *improper integrals.* The applications that use such improper integrals need to be compared to similar applications with separated events, which are handled by *infinite geometric series.*

For an example of an improper integral, consider the present value of an asset. For instance, suppose $100 is being paid continuously each year and the annual interest rate is 5%. We have seen that the present value of this asset over 20 years is $\int_0^{20} 100e^{-0.05x}\,dx$. If we want the present value of this asset and we assume the money is coming in forever, then we would like to assign a value to $\int_0^{\infty} 100e^{-0.05x}\,dx$, which is an improper integral. See figure 6.1.1 for a visual representation of this case.

$y = 100e^{-0.05x}$

$\int_0^{20} 100e^{-0.05x}\,dx$ is revenue for 20 years.

$\int_0^{\infty} 100e^{-0.5x}\,dx$ is eternal revenue.

on out →

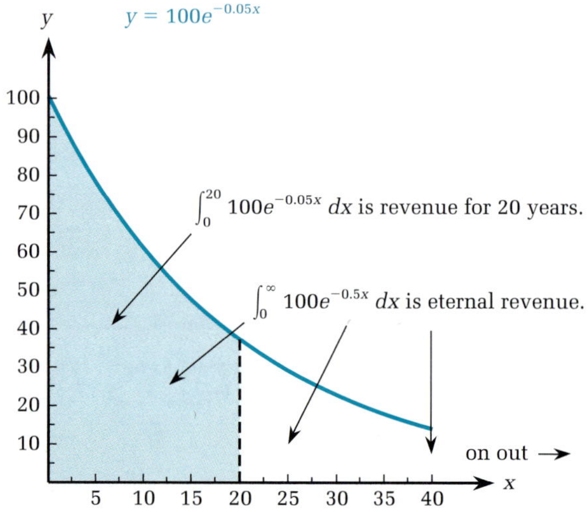

FIGURE 6.1.1

An example of an infinite geometric series occurs if we attempt to measure the total financial effect on a local economy of an initial expenditure of $100. If three-fourths of every dollar spent remains within the local economy to be respent, then the total effect would be

$$100 + \frac{3}{4}(100) + \frac{3}{4}\left(\frac{3}{4}(100)\right) + \frac{3}{4}\left(\frac{3}{4}\left(\frac{3}{4}(100)\right)\right)$$
$$+ \cdots = 100 + 75 + 56.25 + 42.19 + \cdots$$

See figure 6.1.2 for a visual representation of this case.

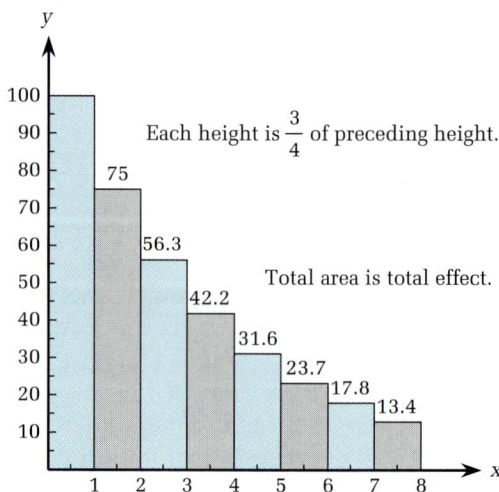

Each height is $\frac{3}{4}$ of preceding height.

Total area is total effect.

FIGURE 6.1.2

6.1   INTEGRATION BY SUBSTITUTION

Evaluating definite integrals can be done quite easily if we can find an antiderivative of the integrand, for then we can use the Fundamental Theorem of Calculus. Thus, finding antiderivatives becomes important in calculus. Table 6.1.1 summarizes the antiderivatives that we know from Chapter 5.

TABLE 6.1.1

$\int k\,dx = kx + C$	$\int x^n dx = \dfrac{x^{n+1}}{n+1} + C \quad (n \neq -1)$
$\int \dfrac{1}{x}\,dx = \ln\lvert x\rvert + C$	$\int e^{ax}\,dx = \dfrac{e^{ax}}{a} + C$
$\int (f(x) \pm g(x))\,dx = \int f(x)\,dx \pm \int g(x)\,dx$	$\int k \cdot f(x)\,dx = k \cdot \int f(x)\,dx$

These formulas are adequate for simple integrands, as found in $\int x^2\,dx$, $\int (3/x)\,dx$, or $\int 2e^{3x}\,dx$. However, we need to be able to find antiderivatives for many functions.

An integrand that seems to be quite complex can sometimes be transformed into a form that appears much simpler by a process known as *substitution*. This transformation helps us recognize a complicated integrand as being a simple integrand in disguise.

For instance, $\int 6(x^3 + 2)^2 x^2\,dx$ appears complex, whereas we easily recognize $\int u^2\,du$ and evaluate it as $(u^3/3) + C$. The first of these expressions can be made to look like the second, and we learn just how to do so in this section.

u-substitution

To introduce the idea of *u*-substitution, let us start with an example.

EXAMPLE 1

Find $\int (2x + 3)^4\,dx$.

SOLUTION

If we multiply out $(2x + 3)^4$, we have a sum of powers, which we can integrate one at a time by the methods of Section 5.1. Doing so involves quite a bit of effort, but it would succeed. However, let us try an alternate approach and see if it is easier.

Let us replace some seemingly complex term involving the variable x by a new variable u. For our integrand, we consider the term $(2x + 3)^4$ and say to ourselves, if we substitute u for $2x + 3$, then $(2x + 3)^4$ would be u^4. What is the effect on the rest of the integrand?

Making the substitution, we have

$$\int (2x + 3)^4\, dx = \int u^4\, dx, \text{ where we have substituted } u = 2x + 3.$$

WARNING

$\int u^4\, dx \neq (u^5/5) + C$. Notice there are two different variables involved in $\int u^4\, dx$.

We need to replace the dx by an equivalent term involving du. $\int u^4\, dx$ is the antiderivative of a function u *with respect to x,* but we want to be able to find the antiderivative of that function u *with respect to u.* We need an expression involving du.

If $u = 2x + 3$, then the derivative of u with respect to x is $du/dx = 2$. Multiplying by dx and dividing by 2 we have $du = 2\, dx$, so $du/2 = dx$. (For the moment, do not worry about how we can multiply by dx. The algebraic manipulation will be discussed shortly.)

If dx is replaced by $du/2$ in our example, we have

$$\int u^4\, dx = \int u^4\, \frac{du}{2} = \int \frac{1}{2} \cdot u^4\, du$$

This is a familiar form and we know its antiderivative.

$$\int u^4\, \frac{du}{2} = \frac{1}{2} \cdot \frac{u^5}{5} + C = \frac{u^5}{10} + C$$

With the antidifferentiation completed, we then substitute $2x + 3$ for u and have $\int (2x + 3)^4\, dx = (1/10)(2x + 3)^5 + C$.

We can, and should, check our result by differentiation. The derivative of $(1/10)(2x + 3)^5 + C$ is $(1/10) \cdot 5(2x + 3)^4 \cdot 2 = (2x + 3)^4$, and this is indeed our original integrand. ●

Now, a word about "multiplying" by dx. In Example 1, as we first go from $du/dx = 2$ to $du = 2\, dx$ and then on to $du/2 = dx$, we are using differentials. These were introduced in Section 3.7 and used there for approximations. Just like the early investigators of calculus, we quickly discover the need for such manipulations when finding antiderivatives.

Whenever we let $u = f(x)$, so that $du/dx = f'(x)$, we can use $du = \dfrac{du}{dx} \cdot dx = f'(x) \cdot dx$, or any convenient equivalent algebraic equality. Any variable name could be used for the substitution, but it is so common to use u that this substitution is called *u-substitution.* It is a very valuable technique for finding antiderivatives. Example 2 illustrates its use.

EXAMPLE 2   Using *u*-substitution

Find $\int (x^3 + 1)^5 3x^2 \, dx$.

SOLUTION

As for Example 1, we could multiply out $(x^3 + 1)^5$. However, if we let $u = x^3 + 1$, then $du/dx = 3x^2$, so $du = 3x^2 \, dx$, which lets us rewrite the integral as $\int u^5 du$. We recognize this integral and know $\int u^5 \, du = (u^6/6) + C$. Substituting for u gives $\int (x^3 + 1)^5 3x^2 \, dx = [(x^3 + 1)^6/6] + C$.

We can check by differentiating: $\dfrac{d}{dx} \left(\dfrac{(x^3 + 1)^6}{6} \right) =$

$\dfrac{6(x^3 + 1)^5}{6} \cdot 3x^2 = (x^3 + 1)^5 3x^2$, which is indeed our integrand.   ●

Students following along with an example such as Example 2 often ask, "What happens to the $3x^2$ term?" It appears to them that it simply vanishes. Their difficulty arises from not truly grasping the idea that du involves much more than simply being a replacement for dx. These differentials are not merely cosmetic additions but vital players in the process. We needed the product of $3x^2$ and dx to give us du.

Always be careful to replace dx when replacing the other terms involving x. The need to replace the dx term gives us yet another reason for making sure that whenever we write an integral, there is a d-something term in it.

u-substitution does for integration what the Chain Rule does for derivatives. We connect formulas together by function composition and hence must memorize fewer formulas. Table 6.1.2 shows this connection.

TABLE 6.1.2

CHAIN RULE	*u*-SUBSTITUTION	
$\dfrac{d}{dx} g(u(x)) = g'(u(x)) \cdot u'(x)$	$\int g'(u(x)) \cdot u'(x) \, dx$	
	$= \int g'(u(x)) \, du$	$\dfrac{du}{dx} = u'(x)$, so $du = u' \, dx$
	$= g(u(x)) + C$	

Expressing *u*-substitution in this form may lead us to believe that the application of *u*-substitution follows some simple mechanical, deterministic pattern. It does not. There may be several different choices for u that are workable. On the other hand, there may be no choice at all that succeeds.

To show the variety of choices, suppose in Example 2 that we had chosen $u = x^3$ instead of $x^3 + 1$. Then $\int (x^3 + 1)^5 3x^2 \, dx = \int (u + 1)^5 \, du$, and we might recognize this to be $[(u + 1)^6/6] + C$, which, when the x^3 is substituted back for u, yields $[(x^3 + 1)^6/6] + C$, as we found before.

If we had not recognized $\int (u + 1)^5 \, du$, we could have attempted to use substitution on it. Of course, we would need a new variable name, such as w, but there is no limit to the number of times we may use substitution. In the same way, there was no limit to the number of levels for functions being differentiated by using the Chain Rule. We will discuss choices for u-substitution shortly.

PRACTICE EXERCISE 1

Find $\int (x^2 - 5)^4 x \, dx$.

Table 6.1.3 expresses the basic integration formulas in terms of u. In each case, we should realize u may well be a function of some other variable.

TABLE 6.1.3

$\int k \, du = ku + C$	$\int u^n \, du = \dfrac{u^{n+1}}{n + 1} + C \quad (n \neq -1)$
$\int u^{-1} \, du = \int \dfrac{1}{u} \, du = \ln |u| + C$	$\int e^u \, du = e^u + C$
$\int (f(u) \pm g(u)) \, du = \int f(u) \, du \pm \int g(u) \, du$	$\int k \cdot f(u) \, du = k \cdot \int f(u) \, du$

To emphasize the use of $u = f(x)$, suppose we consider the formula $\int u^n \, du = \dfrac{u^{n+1}}{n + 1} + C$. Writing this using $f(x)$, we have $\int [f(x)]^n f'(x) \, dx = \dfrac{[f(x)]^{n+1}}{n + 1} + C$.

The following two examples use u-substitution. When choosing u, we often select a power or, as in Example 3, a denominator.

EXAMPLE 3   Using u-substitution to replace a denominator

Find $\int \dfrac{x}{x^2 + 1} \, dx$.

SOLUTION

Let $u = x^2 + 1$. Then $du/dx = 2x$.

By taking the first step in transforming the integrand, we have $\int \dfrac{x}{x^2 + 1}\, dx = \int \dfrac{x}{u}\, dx$. A substitution is still needed for $x\, dx$.

From $du/dx = 2x$ we have $du = 2x\, dx$, so $x\, dx = du/2$. Continuing with the u-substitution,

$$\int \frac{x}{x^2 + 1}\, dx = \int \frac{du/2}{u} = \frac{1}{2} \int \frac{1}{u}\, du = \frac{1}{2} \ln |u| + C$$

Replacing u by $x^2 + 1$, we get $\dfrac{1}{2} \ln |x^2 + 1| + C$.

We may remove the absolute value sign because $x^2 + 1 > 0$. In conclusion, we have

$$\int \frac{x}{x^2 + 1}\, dx = \frac{1}{2} \ln(x^2 + 1) + C$$

We can check our result by verifying that the derivative of $(1/2) \ln(x^2 + 1) + C$ is indeed $x/(x^2 + 1)$. ●

EXAMPLE 4   Using *u*-substitution with an exponential function

Find $\int xe^{3x^2+1}\, dx$.

SOLUTION

Here the exponential power is a good candidate for u because e^u has a simple antiderivative.

Let $u = 3x^2 + 1$. Then $du/dx = 3 \cdot 2x$, so $du = 6x\, dx$. As a first step in transforming the integrand, we have $\int xe^{3x^2+1}\, dx = \int xe^u\, dx$.

Because we need the x and dx terms replaced, we rewrite $du = 6x\, dx$ as $du/6 = x\, dx$, and the u-substitution process continues as $\int xe^u\, dx = \int \dfrac{e^u}{6}\, du = \dfrac{1}{6} \int e^u\, du = \dfrac{1}{6} e^u + C$. Replacing u by $3x^2 + 1$ gives us the final answer as $\int xe^{3x^2+1}\, dx = (1/6)\, e^{3x^2+1} + C$.

Checking by differentiation, we have $\dfrac{d}{dx}\left(\dfrac{e^{3x^2+1}}{6} + C\right) = \dfrac{1}{6} \cdot e^{3x^2+1} \cdot \dfrac{d}{dx}(3x^2 + 1) = \dfrac{1}{6} e^{2x^2+1} \cdot 6x$, which equals our original integrand. ●

PRACTICE EXERCISE 2

Find $\int xe^{-x^2}\,dx$.

Each time we face an integral that is not in a simple form, we should try u-substitution. Choosing u is not an exact process. Sometimes the entire development is scrubbed and we have to start over. It may happen that substitution does us no good, or at least so it appears, as shown in Example 5.

EXAMPLE 5   Showing that u-substitution does not always succeed

For $\int \sqrt{1 - x^2}\,dx$, $x > 0$, show the effect of u-substitution if $u = 1 - x^2$.

SOLUTION

If $u = 1 - x^2$, then $du/dx = -2x$ and so $du = -2x\,dx$.
Replacing $\sqrt{1 - x^2}$ by $\sqrt{u}$ gives $\int \sqrt{u}\,dx$. We need to substitute for the dx. From $du = -2x\,dx$ we have $dx = du/(-2x)$. To express x itself in terms of u we can solve $u = 1 - x^2$ for x as $x = \sqrt{1 - u}$.

With the substitution complete we have $\displaystyle\int \sqrt{u} \cdot \frac{du}{-2\sqrt{1 - u}}$,

which is not simpler at all.

A substitution may be successful for one integral and unsuccessful for another. For example, the substitution $u = 1 - x^2$, which we just saw failed to simplify $\int \sqrt{1 - x^2}\,dx$, does succeed in simplifying $\int x\sqrt{1 - x^2}\,dx$. For some integrals, no substitution reveals to us a simpler integrand.
We are only illustrating the idea of u-substitution, hence the integrals given in this section can all be successfully found by the use of an algebraic, exponential, or logarithmic substitution, often with an exponent or denominator being replaced. For each integral, try to pick u so that du will be most of the rest of the integrand, except possibly for constant multipliers.

EXAMPLE 6   Choosing u for u-substitution

Find $\int 5x^2 e^{x^3 + 2}\,dx$.

SOLUTION

If we try $u = x^3 + 2$ then $du/dx = \dfrac{d}{dx}(x^3 + 2) = 3x^2$ has the same power of x as $5x^2$. That there is a different constant term in a product is not important: Derivatives (and antiderivatives) of the product of a constant and a function are the product of the constant and the derivative (or antiderivative) of the function. In other words, a constant multiplier does not affect the success of a substitution.

For $u = x^3 + 2$, $du = 3x^2\ dx$, and so we have $\dfrac{du}{3} = x^2\ dx$.

Notice the constant 5 is of no use in the substitution. Whenever there is a constant in a product and the constant cannot be immediately used, then write the constant outside the integral, as in $\int 5x^2 e^{x^3+2}\ dx = 5 \int e^{x^3+2}(x^2\ dx)$. Using our substitution gives

$$5 \int \frac{e^u}{3}\ du = \frac{5}{3}e^u + C.$$

Finally, substituting back for u we have $(5/3)e^{x^3+2} + C$.
Checking our work by differentiating, we have

$$\frac{d}{dx}\left(\frac{5}{3}e^{x^3+2} + C\right) = \frac{5}{3}\ e^{x^3+2} \cdot \frac{d}{dx}(x^3 + 2)$$

$$= \left(\frac{5}{3}\ e^{x^3+2}\right)3x^2 = 5x^2 e^{x^3+2}$$

The following guidelines are drawn from our examples.

USING u-SUBSTITUTIONS

1. Keep in mind the basic integral formulas. We are working toward being able to use these.

2. Try to pick u so that du will be as much as possible of the rest of the integrand.

3. Do not worry about constants that appear in products.

4. Always be careful to replace dx while replacing the other x terms.

5. If there is a constant in a product and the constant cannot be immediately used, then write the constant outside the integral.

6. Check that all the original integrand is replaced.

In our examples we have been checking our antiderivative, whether found by using u-substitution or any other technique, by differentiating the "answer." If we have $\int f(x)\,dx$ and believe $F(x)$ is an antiderivative, then we should be able to check by showing that $F'(x) = f(x)$.

Suppose we find that $F'(x) \neq f(x)$ but instead $F'(x) = k \cdot f(x)$ for some constant k. Then actually $\int f(x)\,dx = [F(x)]/k + C$.

For instance, suppose we believe that $\int x\,(e^{3x^2})\,dx = e^{3x^2}$, but upon checking we find $\dfrac{d}{dx}(e^{3x^2}) = 6x\,(e^{3x^2})$. Then the correct indefinite integral is $\int xe^{3x^2}\,dx = (e^{3x^2}/6) + C$.

WARNING

This process of dividing by a term works only for constants. For example, suppose we had $\int e^{3x^2}\,dx$ and we guess an antiderivative is e^{3x^2}. Upon checking, we find $\dfrac{d}{dx}e^{3x^2} = 6xe^{3x^2}$. The correct antiderivative cannot be found by simply dividing e^{3x^2} by $6x$. This can be shown by the Quotient Rule, which yields

$$\frac{d}{dx}\frac{e^{3x^2}}{6x} = \frac{6x(6xe^{3x^2}) - 6(e^{3x^2})}{(6x)^2} \neq e^{3x^2}$$

In other words, $\int e^{3x^2}\,dx \neq e^{3x^2}/(6x)$.

Let us continue to demonstrate u-substitutions. So far we primarily have been replacing a polynomial by u and determining the effect on the integrand. There are other possibilities, and for each integral we should think about various candidates for u-substitution.

EXAMPLE 7   Letting $u = \ln x$ for u-substitution

Find $\displaystyle\int \frac{\ln x}{x}\,dx$.

SOLUTION

This integral appears simple in form. Nonetheless, unless we know the antiderivative, we need to try something and u-substitution may help.

Let $u = \ln x$. Then $du/dx = 1/x$, so $du = (1/x)\,dx$. We see $\dfrac{1}{x}\,dx$ is a part of the integrand, so it can be replaced by du.

Substituting for u and du produces $\int u\,du = u^2/2 + C$. Putting $\ln x$ back in for u results in $\displaystyle\int \frac{\ln x}{x}\,dx = \frac{(\ln x)^2}{2} + C$.

Differentiating our answer yields $\dfrac{d}{dx}\left(\dfrac{1}{2}(\ln x)^2 + C\right) =$

$(1/2)(2\ln x)\cdot(1/x) = \dfrac{\ln x}{x}$.

PRACTICE EXERCISE 3

Find $\int (e^{\sqrt{x}}/\sqrt{x})\,dx$.

Definite Integrals with Substitution

Suppose we have a definite integral and attempt a solution by u-substitution. As we begin this process, the integral is written for the original variable and original limits, but then there is a time when the integral is being written for the substituted variable. During that time it is critical to indicate what new limits are being used for the substituted variable. Or we may explicitly continue using the limits for the original variable as long as we are clear in our notation.

EXAMPLE 8   Changing limits during u-substitution

Evaluate $\displaystyle\int_0^2 2xe^{x^2}\,dx$.

SOLUTION

Using $u = x^2$ and $du = 2x\,dx$, the integral becomes $\int e^u\,du$, but the limits on u are no longer going from 0 to 2. If we said they were and we went on to evaluate $\displaystyle\int_0^2 e^u\,du = e^u\Big|_0^2 = e^2 - e^0 = e^2 - 1 = 6.389$, we would be wrong. We have to indicate that it is x that goes from 0 to 2. We can do this in one of two ways.

We can write $\displaystyle\int_{x=0}^{x=2} e^u\,du$ and keep writing the limits in terms of the x-values until we return to writing the antiderivative in terms of x, or we can change the limits from being those of the first variable to those of the new variable.

Using the first method we have $\displaystyle\int_0^2 2xe^{x^2}\,dx = \int_{x=0}^{x=2} e^u\,du = e^u\Big|_{x=0}^{x=2}$. This is the antiderivative, but it is still x that goes from 0 to 2. By substituting back $u = x^2$, we get $e^{x^2}\Big|_0^2$. We are now back to x, so we do not need to indicate the limits are on x. Evaluating this, we get $e^{2^2} - e^{0^2} = e^4 - 1$.

Using the second method, we first find what limits on u correspond to the original limits on x.

If $u = x^2$ and $0 \leq x \leq 2$, then by squaring we have $0 \leq x^2 \leq 4$. Hence u has the limits $0 \leq u \leq 4$ and we can write the integral as

$$\int_0^2 2xe^{x^2}\,dx = \int_0^4 e^u\,du = e^u\,\Big|_0^4 = e^4 - e^0 = e^4 - 1, \text{ as in the first}$$

method. Using either method we have $e^4 - 1 \approx 53.5982$.

It does not make any difference which method we use. The important point is to be certain we do not mix up the different limits and evaluate u at the limits for x.

PRACTICE EXERCISE 4

Evaluate $\int_0^1 x^2(x^3 + 1)^5\,dx$.

Someone who works a great many integration problems develops the ability to use u-substitution without writing down all the steps. For example, he or she might look at Practice Exercise 4 and say, "The antiderivative must be something similar to $(x^3 + 1)^6$, which by the Chain Rule has the derivative $6(x^3 + 1)^5 \cdot 3x^2$. This is fairly close to what is needed, but it does have a $6 \cdot 3 = 18$ that is unwanted. Constants can be compensated for in a product, so the actual antiderivative is $(1/18)(x^3 + 1)^6$."

Having said that, it is still important to emphasize that until we feel quite familiar with all the complications, it is worthwhile for us to write down explicitly all the steps when using u-substitutions.

6.1 PROBLEMS

Foundations

The problems of this section require the basic skills illustrated by the following:

In problems 1–4, find an expression for du in terms of x and dx.

1. $u = 2x + 3$ **2.** $u = 3x^2 + 1$

3. $u = 1 + \ln x$ **4.** $u = 1 + e^x$

Find the antiderivatives of problems 5–8.

5. $\int x^3\,dx$ **6.** $\int 5x^2\,dx$

7. $\int 3e^{4x}\,dx$ **8.** $\int \dfrac{2}{x}\,dx$

Exercises

Find the integrals in problems 9–32.

9. $\int (x - 2)^6\,dx$ **10.** $\int (x + 3)^6\,dx$

11. $\int (2x + 3)^5\,dx$ **12.** $\int (3x - 2)^5\,dx$

13. $\int \dfrac{3}{(x + 2)^2}\,dx$ **14.** $\int \dfrac{2}{(3x - 1)^3}\,dx$

15. $\int \sqrt{2x + 1} \, dx$

16. $\int \sqrt{5x - 2} \, dx$

17. $\int 2x(x^2 - 5)^3 \, dx$

18. $\int 6x(3x^2 + 1)^5 \, dx$

19. $\int \dfrac{3}{x + 3} \, dx$

20. $\int \dfrac{1}{x - 2} \, dx$

21. $\int \dfrac{x}{2x^2 + 1} \, dx$

22. $\int \dfrac{x}{x^2 - 3} \, dx$

23. $\int t\sqrt{t^2 + 4} \, dt$

24. $\int 2t\sqrt{t^2 - 3} \, dt$

25. $\int e^{2t+1} \, dt$

26. $\int e^{-2t+3} \, dt$

27. $\int 2t \cdot e^{3t^2} \, dt$

28. $\int t \cdot e^{t^2} \, dt$

29. $\int \dfrac{e^x}{1 + e^x} \, dx$

30. $\int \dfrac{e^x}{e^x - 1} \, dx$

31. $\int \dfrac{1}{x \ln x^2} \, dx$ (Hint: Use a property of logarithms to rewrite $\ln x^2$, and then use $u = \ln x$.)

32. $\int \dfrac{1}{x \ln \sqrt{x}} \, dx$ (Hint: Use a property of logarithms to rewrite $\ln \sqrt{x}$, then use $u = \ln x$.)

In problems 33–36 use the given u-substitution to solve each definite integral two ways.
(a) Find the limits on u corresponding to the given limits on the original variable.
(b) Evaluate the antiderivative by returning to values of the original variable.
(c) Evaluate the antiderivative by using the limits on u.

33. $\int_0^5 \dfrac{4}{x + 1} \, dx, \, u = x + 1$

34. $\int_0^3 \dfrac{2x}{x^2 + 1} \, dx, \, u = x^2 + 1$

35. $\int_0^4 t \cdot e^{t^2} \, dt, \, u = t^2$

36. $\int_0^2 e^{4t} \, dt, \, u = 4t$

Check that $\dfrac{d}{dx}(x \ln x - x) = \ln x$ and use it to evaluate problems 37 and 38.

37. $\int_1^4 \ln x \, dx$

38. $\int_1^6 \ln x \, dx$

Problems 39 and 40 are based on data from the American Veterinary Medical Association.

39. *(Expenditures on Cats)* The amount spent on veterinary fees by cat owners averaged \$44.59 per household in 1983 and \$63.87 in 1987. Assume that the expenditures are increasing exponentially and will continue to do so at 9% per year.

Use $A(t) = \$63.87 \cdot e^{0.09t}$ per year for the average amount spent t years from 1987. Estimate the total amount spent by the average household with cats on cat veterinary expenses for the 13 years to the year 2000.

40. *(Expenditures on Dogs)* The amount spent on veterinary fees by dog owners averaged \$74.51 per household in 1983 and \$86.57 in 1987. Assume that the expenditures are increasing exponentially and will continue to do so at 3.75% per year.

Use $A(t) = \$86.57 \cdot e^{0.0375t}$ per year for the average amount spent t years from 1987. Estimate the total amount spent by the average household with dogs on dog veterinary expenses for the 13 years to the year 2000.

Enrichment Problems

41. To show another reason for always adding the constant C when we take an antiderivative, suppose we left C off and were trying to figure out the seeming paradox in the following two ways of finding $\int (x + 3)^2 \, dx$.

By substitution, with $u = x + 3$, we have
$\int (x + 3)^2 \, dx = \int u^2 \, du = u^3/3 = (x + 3)^3/3 = (x^3 + 9x^2 + 27x + 27)/3 = (x^3/3) + 3x^2 + 9x + 9$. Yet, $\int (x + 3)^2 \, dx = \int (x^2 + 6x + 9) \, dx = x^3/3 + 3x^2 + 9x$. With $x^3/3 + 3x^2 + 9x + 9 = x^3/3 + 3x^2 + 9x$, we seem to be saying $9 = 0$. What's wrong?

42. Find $\int 1/(x + \sqrt{x}) \, dx$. (Hint: $1/(x + \sqrt{x}) = \dfrac{1}{\sqrt{x}} \left(\dfrac{1}{\sqrt{x}} + 1 \right)$, and try $u = \sqrt{x} + 1$.)

SOLUTIONS TO PRACTICE EXERCISES

1. To find $\int (x^2 - 5)^4 x \, dx$ let $u = x^2 - 5$.
   Then $du/dx = 2x$, which means $du = 2x \, dx$,
   so $du/2 = x \, dx$, which yields $\int u^4 \cdot (du/2) =$
   $1/2 \int u^4 \, du = (1/2) \cdot (u^5/5) + C = (u^5/10) + C =$
   $[(x^2 - 5)^5/10] + C$.

2. To find $\int xe^{-x^2} \, dx$, let $u = -x^2$. Then
   $du/dx = -2x$.
      We transform the integrand $xe^{-x^2} \, dx$ into
   $xe^u \, dx$. We need to replace the $x \, dx$ as well.
   From $du/dx = -2x$, we have $x \, dx = -du/2$.
   Then $\int xe^{-x^2} \, dx = \int e^u (-du/2) = -1/2 \int e^u \, du$
   $= -(1/2)e^u + C = -(1/2)e^{-x^2} + C$.

3. To find $\int (e^{\sqrt{x}}/\sqrt{x}) \, dx$ let $u = \sqrt{x}$. Then $du/dx =$
   $\dfrac{1}{2\sqrt{x}}$ which means $2 \, du = dx/\sqrt{x}$, yielding
   $$\int e^{\sqrt{x}} \cdot \frac{dx}{\sqrt{x}} = \int e^u 2 \, du = 2 \int e^u \, du =$$
   $2e^u + C = 2e^{\sqrt{x}} + C$.

4. Evaluate $\int_0^1 x^2(x^3 + 1)^5 \, dx$. Using the method in
   which we return to x for the evaluation, let
   $u = x^3 + 1$. Then $du/dx = 3x^2$, which means
   $$\frac{1}{3} \, du = x^2 \, dx \text{ giving } \int_{x=0}^{x=1} u^5 \cdot \frac{1}{3} \, du =$$
   $$\frac{1}{3} \int_{x=0}^{x=1} u^5 \, du = \frac{1}{3}\left(\frac{u^6}{6}\Big|_{x=0}^{x=1}\right). \text{ Now substitute back}$$
   for u to get $\dfrac{1}{3}\left(\dfrac{(x^3 + 1)^6}{6}\Big|_0^1\right) = \dfrac{1}{3}\left(\dfrac{2^6}{6} - \dfrac{1^6}{6}\right) =$
   $\left(\dfrac{1}{3}\right) \cdot \left(\dfrac{63}{6}\right) = 3.5.$
      Using the method in which we change
   from the limits for x to the limits for u, we
   start off the same way, letting $u = x^3 + 1$ and
   having $(1/3)du = x^2 \, dx$. Because $0 \le x \le 1$, we
   have $0 \le x^3 \le 1$, which means $1 \le x^3 + 1 =$
   $u \le 2$, and the integrals are $\int_1^2 u^5 \cdot (1/3)du =$
   $(1/3) \int_1^2 u^5 \, du = (1/3)[(u^6/6) \,|_1^2] = \dfrac{1}{3}\left(\dfrac{2^6}{6} - \dfrac{1^6}{6}\right)$
   $= (1/3) \cdot (63/6) = 3.5$, as before.

6.2   # INTEGRATION BY PARTS

In our search for antiderivatives we often use integral tables and various computer programs. Some tables list hundreds of antiderivatives. Nonetheless, no matter how extensive the table, it is assumed every user knows several elementary techniques of integration. We have seen the technique of u-substitution. Because of it, no table lists $\int x \cdot e^{x^2} \, dx$ because the substitution $u = x^2$ transforms $\int x \cdot e^{x^2} \, dx$ into $\int (1/2)e^u \, du$.

Both $\int e^u \, du = e^u + C$ and $\int a f(x) \, dx = a \int f(x) \, dx$ would be in the usual table. Hence, we have

$$\int x \cdot e^{x^2} \, dx = \int (1/2)e^u \, du \text{ by } u\text{-substitution}$$

$$= (1/2) \int e^u \, du = (1/2)e^u + C \text{ by using the table}$$

or simply knowing the formulas

$$= (1/2)e^{x^2} + C \text{ returning to } x \text{ as the variable.}$$

A second technique of integration that every user of tables is assumed to know is integration by parts, the topic of this section.

Integration by Parts

The technique of *integration by parts* comes from thinking about the Product Rule for derivatives, which is

$$\frac{d}{dx}(u \cdot v) = u \cdot \frac{d}{dx}v + v \cdot \frac{d}{dx}u.$$

Suppose we write this in differential form as $d(uv) = u \cdot dv + v \cdot du$. If we take antiderivatives of both sides we have $\int d(uv) = \int u \cdot dv + \int v \cdot du$. As we found in Section 5.1, $\int d(uv) = uv + C$. Hence, $uv + C = \int u \cdot dv + \int v \cdot du$. Rearranging terms and recognizing that the existence of a constant, C, is implied in any indefinite integral, we have the formula for integration by parts.

INTEGRATION BY PARTS

$$\int u \, dv = uv - \int v \, du$$

The formula for integration by parts, which appears to be so simple and is the result of such a minor bit of rearranging, provides us an amazingly powerful technique. The use of this technique requires us to choose u and dv so that their product is the original integrand. First, let us look at an example and then we will discuss in detail different choices for the u and dv terms.

EXAMPLE 1   Integration by parts

Find $\int x \cdot e^x \, dx$.

SOLUTION

Suppose we choose $u = x$ and $dv = e^x \, dx$. Then $u \, dv = x \cdot e^x \, dx$, and we have $\int x \cdot e^x \, dx = \int u \, dv$.

The formula for integration by parts is $\int u \, dv = uv - \int v \, du$, so we need to pause and find both v and du. We know u, so finding du is a matter of taking its derivative. From $u = x$ we have $du/dx = 1$, so that $du = dx$.

Obtaining v from dv requires finding an antiderivative. In our case, $\int dv = \int e^x \, dx = e^x + C$. At this stage of the process we are not through taking antiderivatives and our final answer needs only one constant of integration, so we can omit the C. Thus we use $v = e^x$.

Then $uv - \int v \, du = x \cdot e^x - \int e^x \, dx = x \cdot e^x - e^x + C$ because $\int e^x \, dx = e^x + C$. Notice that if there is no further antidifferentiation to do, then at last we write the constant of integration as part of our answer.

As always, we should check our work by taking the derivative of our solution to see if we get the integrand we started off with:

$$\frac{d}{dx}(x \cdot e^x - e^x + C) = \frac{d}{dx}(x \cdot e^x) - \frac{d}{dx}e^x + \frac{d}{dx}C = (x \cdot e^x + e^x \cdot 1)$$
$$- e^x + 0 = x \cdot e^x + e^x - e^x = x \cdot e^x, \text{ as desired.}$$

Choosing How to Express $\int f(x) \, dx$ in the form $\int u \, dv$

The choices for u and dv are not automatic. Suppose we had made a different selection for $\int x \cdot e^x \, dx$ and chosen $u = e^x$ and $dv = x \cdot dx$.

These choices would give $du = e^x \, dx$ and $v = x^2/2$. Then $\int x \cdot e^x \, dx = \int u \cdot dv = u \cdot v - \int v \cdot du$ by the formula for integration by parts.

By substituting for u, v, and du, we have $\int x \cdot e^x \, dx = e^x \cdot (x^2/2) - \int (x^2/2) \cdot e^x \, dx$. This does not appear promising. We now would have to integrate $\int (x^2/2) \cdot e^x \, dx$, a more formidable integrand than our original problem.

A third choice for u and dv could have been $u = x \cdot e^x$ and $dv = dx$. These choices give $du = (x \cdot e^x + e^x \cdot 1) \, dx$ and $v = x$. Using the formula, we have $\int x \cdot e^x \, dx = \int u \cdot dv = u \cdot v - \int v \cdot du = x \cdot e^x \cdot x - \int x \cdot (x \cdot e^x + e^x) \, dx = x^2 e^x - \int x^2 \cdot e^x \, dx - \int x e^x \, dx$. This also is more difficult than our original integral. Hence, our original choice was best. Following are general guidelines for using integration by parts.

GUIDELINES FOR CHOOSING u AND dv WHEN USING INTEGRATION BY PARTS:
$$\int u \cdot dv = u \cdot v - \int v \cdot du$$

1. Choose the largest, most complicated, dv that can be integrated to find v. Do not use a constant of integration in writing v.

2. Try to choose u so that du is "simpler" than u. (An example of "simpler" is having x being raised to a smaller positive power.)

3. Be sure that $\int v \cdot du$ is no more complicated than $\int u \cdot dv$.

4. Check that the choices for u and dv account for each term in the entire integrand.

Choosing u so that du is "simpler" than u is not as essential as choosing dv so that we can find v. The next example illustrates this point.

EXAMPLE 2   Choosing u and dv

Find $\int x \cdot \ln x \, dx$.

SOLUTION

If we try to use $dv = \ln x \, dx$, then we need to know $\int \ln x \, dx$ and we may not know this integral. Thus, even though picking $dv = x \, dx$ means the power of x is going to increase, we are forced into trying this choice.

It is worthwhile to list the choices for u and dv, along with the du and v. Doing so highlights the function for which we are finding the derivative or the antiderivative.

$$
\begin{array}{lll}
u = \ln x & \text{gives} & du = \dfrac{1}{x} \, dx \\[2em]
dv = x \, dx & \text{gives} & v = \dfrac{x^2}{2}
\end{array}
$$

This choice certainly produces a more complicated v than we had for dv and this concerns us, as it should. However, it turns out that the resulting integral is one we can solve.

$$
\begin{aligned}
\int x \cdot \ln x \, dx &= \int (\ln x)(x \cdot dx) \\
&= \ln x \cdot \frac{x^2}{2} - \int \frac{x^2}{2} \cdot \frac{1}{x} \, dx \text{ by integration by parts} \\
&= \frac{x^2}{2} \cdot \ln x - \int \frac{1}{2} x \, dx \\
&= \frac{x^2}{2} \cdot \ln x - \frac{x^2}{2 \cdot 2} + C \\
&= \frac{x^2}{2} \cdot \ln x - \frac{x^2}{4} + C
\end{aligned}
$$

Check by taking the derivative,

$$
\begin{aligned}
\frac{d}{dx}\left(\frac{x^2}{2} \cdot \ln x - \frac{x^2}{4} + C \right) &= \left(\frac{x^2}{2} \cdot \frac{1}{x} + x \cdot \ln x \right) - \frac{2x}{4} \\
&= \frac{x}{2} + x \cdot \ln x - \frac{x}{2} = x \cdot \ln x
\end{aligned}
$$

Example 3 demonstrates that even though an integrand may not appear to be a product, nonetheless integration by parts may be useful.

EXAMPLE 3   Integration by parts

Find $\int \ln x \, dx$.

SOLUTION

Because dv must contain dx and we do not want dv to be the entire integrand, so we let $dv = dx$ and let $u = \ln x$. Then

$u = \ln x$	gives	$du = \dfrac{1}{x} dx$
$dv = dx$	gives	$v = x$

The formula for integration by parts gives us

$$\int \ln x \, dx = (\ln x) \cdot x - \int x \cdot \frac{1}{x} \, dx$$

$$= x \cdot \ln x - \int dx$$
$$= x \cdot \ln x - x + C$$

PRACTICE EXERCISE 1

Find $\int x e^{2x} \, dx$.

Repeated Use of Integration by Parts

An integral may require that the technique of integration by parts be applied several times, as shown in Example 4.

EXAMPLE 4   Needing two uses of integration by parts

Find $\int x^2 e^{2x} \, dx$.

SOLUTION

We can find the antiderivative of either e^{2x} or x^2, so our choice of u and dv is guided by considering whether du is simpler than u. Finding that $\frac{d}{dx}e^{2x} = 2e^{2x}$ and $\frac{d}{dx}x^2 = 2x$, we see that the derivative of e^{2x} is no simpler than e^{2x} itself, whereas $2x$ is simpler than x^2.

$u = x^2$	gives	$du = 2x\,dx$
$dv = e^{2x}\,dx$	gives	$v = \dfrac{1}{2}e^{2x}$

We have $\int x^2e^{2x}\,dx = x^2\left(\dfrac{e^{2x}}{2}\right) - \int\left(\dfrac{e^{2x}}{2}\right)2x\,dx =$

$\dfrac{x^2e^{2x}}{2} - \int xe^{2x}\,dx$

Although this expression still involves an integral, we continue on because the new integral is simpler than the original. We need a second application of integration by parts on $\int xe^{2x}\,dx$. This is the integral of Practice Exercise 1. By letting $dv = e^{2x}\,dx$ and $u = x$, we found that $\int xe^{2x}\,dx = (1/2)xe^{2x} - (1/4)e^{2x} + C$.

Hence, $\displaystyle\int x^2e^{2x}\,dx = \dfrac{x^2e^{2x}}{2} - \dfrac{xe^{2x}}{2} + \dfrac{e^{3x}}{4} + C =$

$(1/2)x^2e^{2x} - (1/2)xe^{2x} + (1/4)e^{2x} + C$.   ●

Notation Convention.   Notice in the last step of Example 4 we have continued our notation convention of always writing the constant of integration as $+C$, even though it may have seemed more appropriate to write it in that last step as $-C$. The C is standing for a constant that can be positive or negative. Hence, the convention is to always use $+C$.

Integration by Parts for Definite Integrals

Integration by parts may be used for a definite integral. The formula involves evaluating the product $(u \cdot v)$ at a and at b.

INTEGRATION BY PARTS FOR DEFINITE INTEGRALS

$$\int_a^b u\,dv = (u \cdot v)\,\Big|_a^b - \int_a^b v\,du$$

EXAMPLE 5   Integration by parts, biomedical drug level

In an earlier example, we saw that thallium had a half-life of 73 hours and may be injected into patients undergoing a treadmill stress test. In fact, the thallium is eliminated from the body faster by excretion than by radioactive decay. Functions that describe the rate of excretion generally are of the form $f(t) = te^{-kt}$, where k is a positive constant and t is in hours. For $k = 1.5$ find $\int_0^4 te^{-1.5t}\, dt$.

SOLUTION

Let $u = t$ and $dv = e^{-1.5t}\, dt$. Then $du = dt$ and $v = e^{-1.5t}/(-1.5)$. Using integration by parts, we have

$$\int_0^4 te^{-1.5t}\, dt = t \cdot \frac{e^{-1.5t}}{-1.5}\Big|_0^4 - \int_0^4 (e^{-1.5t}/(-1.5))\, dt$$

Evaluating the first part and finding the antiderivative of the second, we have

$$\left(4 \cdot \frac{e^{-6}}{-1.5} - 0\right) - \left(\frac{e^{-1.5t}}{2.25}\Big|_0^4\right) = -0.0066 - \left(\frac{e^{-6}}{2.25} - \frac{e^0}{2.25}\right)$$

$$= -0.0066 - 0.0011 + 0.4444$$

$$= 0.4367$$

EXAMPLE 6   Integration by parts, capital value of an asset

Suppose an asset is generating revenue at the rate of $f(t) = 1000t + 10$ dollars per year. Using an annual interest rate of 9%, what is the capital value of the asset over the next five years?

SOLUTION

From Section 5.5, we see that the capital value of this asset is given by $\int_0^5 (1000t + 10)e^{-0.09t}\, dt$. To evaluate this we need to find both $1000\int_0^5 te^{-0.09t}\, dt$ and $10\int_0^5 e^{-0.09t}\, dt$. To find $\int_0^5 te^{-0.09t}\, dt$, let $u = t$ and $dv = e^{-0.09t}\, dt$. Then $du = dt$ and $v = e^{-0.09t}/(-0.09)$.

$$\int_0^5 te^{-0.09t}\, dt = t \cdot \frac{e^{-0.09t}}{-0.09}\Big|_0^5 - \int_0^5 \frac{e^{-0.09t}}{-0.09}\, dt$$

$$= t \cdot \frac{e^{-0.09t}}{-0.09}\Big|_0^5 + \int_0^5 \frac{e^{-0.09t}}{0.09}\, dt$$

$$= -5 \cdot \frac{e^{-0.09 \cdot 5}}{0.09} - 0 \cdot \frac{e^{-0.09 \cdot 0}}{0.09} + \left[\left(\frac{e^{-0.09t}}{-(0.09)0.09}\right)\Big|_0^5\right]$$

$$= -5 \cdot \frac{e^{-0.45}}{0.09} - 0 + \left[-\left(\frac{e^{-0.09 \cdot 5}}{(0.09)^2}\right) + \left(\frac{e^{-0.09 \cdot 0}}{(0.09)^2}\right)\right]$$

$$= -35.4238 - 78.7195 + 123.4568 = 9.3135$$

$$\int_0^5 e^{-0.09t}\, dt = \frac{e^{-0.09t}}{-0.09}\Big|_0^5 = \frac{e^{-0.45}}{-0.09} - \frac{e^0}{-0.09}$$

$$= -7.0848 + 11.1111 = 4.0263$$

Combining these, $\int_0^5 (1000t + 10)e^{-0.09t}\, dt =$
$1000 \cdot 9.3135 + 10 \cdot 4.0263 = 9353.76.$

6.2 PROBLEMS

Foundations

The problems of this section require the basic skills illustrated by the following:

1. Find $\dfrac{d}{dx}e^{-3x}$ and $\int e^{-3x}\, dx$.

2. If $\int x \ln 2x\, dx = \int u\, dv$ and $dv = x\, dx$, then find $\dfrac{d}{dx}u$.

Exercises

Resolve problems 3–14 using integration by parts.

3. $\displaystyle\int xe^{3x}\, dx$ 4. $\displaystyle\int xe^{2x}\, dx$

5. $\displaystyle\int 2xe^{-3x}\, dx$ 6. $\displaystyle\int 3xe^{-2x}\, dx$

7. $\displaystyle\int -xe^{-2x}\, dx$ 8. $\displaystyle\int -3xe^{-x}\, dx$

9. $\displaystyle\int x \ln 2x\, dx$ 10. $\displaystyle\int 3x \ln x\, dx$

11. $\displaystyle\int x^2 \ln x\, dx$ 12. $\displaystyle\int x^2 \ln 2x\, dx$

13. $\displaystyle\int \ln(x - 2)\, dx$ 14. $\displaystyle\int \ln(x + 3)\, dx$

Evaluate problems 15–18 by using integration by parts for definite integrals.

15. $\displaystyle\int_0^2 xe^{-x}\, dx$ 16. $\displaystyle\int_0^2 2xe^x\, dx$

17. $\displaystyle\int_{2.5}^3 \ln(x - 2)\, dx$ 18. $\displaystyle\int_{-1}^0 \ln(x + 3)\, dx$

19. **(Biomedical Drug Levels)**   Refer to Example 5 and find the total amount of a drug excreted over six hours if the rate of excretion is $f(t) = te^{-2t}$.

20. **(Capital Value of an Asset)**   Find the capital value over 10 years of an asset that is producing revenue at the rate of $100t + 200$ dollars at time t. Use an annual interest rate of 6%.

Enrichment Problems

Some formulas are given as *reduction formulas* because without presenting the final answer, the formula shows a way of reducing the exponent of x, and thus presenting a simpler problem. An example of a reduction formula is:

$$\int x^m e^x\, dx = x^m e^x - m \int x^{m-1} e^x\, dx$$

21. Solve $\int x^3 e^x\, dx$ by repeatedly using the reduction formula above.

22. Use integration by parts to derive the reduction formula above.

23. Use integration by parts to derive the reduction formula

$$\int (\ln x)^m\, dx = x(\ln x)^m - m \int (\ln x)^{m-1}\, dx$$

24. Use the reduction formula in problem 23 to find $\int (\ln x)^3\, dx$.

Problems 25–28 require the knowledge of $\int a^x\, dx$ and $\int \log_a x\, dx$, which can be found by using the optional material in Section 4.4.

25. $\int x\, 2^x\, dx$ 26. $\int 2x\, 3^x\, dx$

27. $\int \log x\, dx$ 28. $\int \log(2x + 1)\, dx$

SOLUTION TO PRACTICE EXERCISE

1. To find $\int xe^{2x}\, dx$ let $u = x$ and $dv = e^{2x}\, dx$. Then

$u = x$	gives	$du = dx$
$dv = e^{2x}\, dx$	gives	$v = \dfrac{e^{2x}}{2}$

$$\int xe^{2x}\, dx = x \cdot \frac{e^{2x}}{2} - \int \frac{e^{2x}}{2}\, dx$$
$$= x \cdot \frac{e^{2x}}{2} - \frac{e^{2x}}{4} + C$$

6.3

INTEGRAL TABLES

Tables of antiderivatives are called *integral tables* and written in the form of indefinite integrals. Traditionally, such tables appeared in a printed form, such as the short integral table in this section. However, tables now are being produced also in electronic form in what are called *symbolic algebra systems*. These automatically provide exact antiderivatives for some functions.

Regardless of the form of an integral table, there will be functions omitted. In part this is because not all functions have antiderivatives in terms of ordinary functions. The quest for antiderivatives occupied some of the most creative minds from the beginnings of calculus in the late seventeenth century until the nineteenth century. They succeeded with many functions, but for some functions there seemed to be no antiderivative. Finally the French mathematician Joseph Liouville (1809–1882) demonstrated that indeed there were some functions without any ordinary functions whatsoever for antiderivatives. One of these is that function so familiar to students of statistics, $(1/\sqrt{2\pi})e^{-x^2/2}$, whose graph is the bell-shaped normal distribution curve shown in figure 6.3.1.

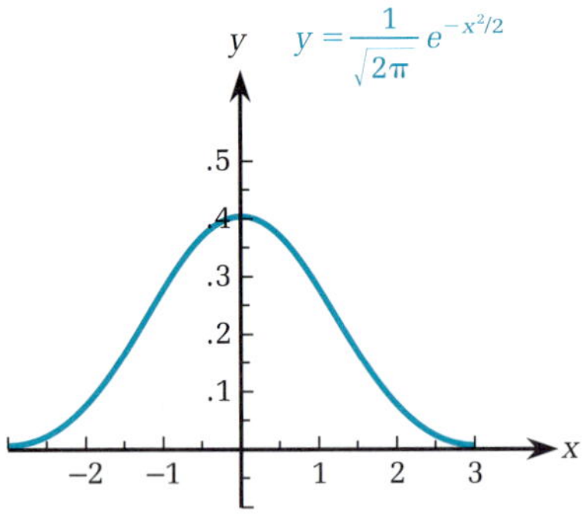

$y = \dfrac{1}{\sqrt{2\pi}} e^{-x^2/2}$

FIGURE 6.3.1

A Short Integral Table

The following is a sample of entries in an integral table. These are some we would find after we had found the easy formulas, such as $\int u^n \, du = u^{n+1}/(n+1) + C \ (n \neq -1)$.

Each of these formulas is stated in terms of the variable u. In part this is done to remind us immediately of u-substitution as a technique of integration. To remind us that integration by parts is also a valuable technique of integration, it is listed in most integral tables.

#1. $\displaystyle \int \frac{1}{u(au + b)} \, du = \frac{1}{b} \ln \left| \frac{u}{au + b} \right| + C$

#2. $\displaystyle \int \frac{u}{au + b} \, du = \frac{u}{a} - \frac{b}{a^2} \ln |au + b| + C$

#3. $\displaystyle \int \frac{u}{\sqrt{au + b}} \, du = \frac{2au - 4b}{3a^2} \sqrt{au + b} + C$

#4. $\displaystyle \int \sqrt{u^2 + a^2} \, du = \frac{u}{2} \sqrt{u^2 + a^2} + \frac{a^2}{2} \ln |u + \sqrt{u^2 + a^2}| + C$

#5. $\displaystyle \int \sqrt{u^2 - a^2} \, du = \frac{u}{2} \sqrt{u^2 - a^2} - \frac{a^2}{2} \ln |u + \sqrt{u^2 - a^2}| + C$

#6. $\displaystyle \int \frac{1}{\sqrt{u^2 + a^2}} \, du = \ln |u + \sqrt{u^2 + a^2}| + C$

#7. $\displaystyle \int \frac{1}{\sqrt{u^2 - a^2}} \, du = \ln |u + \sqrt{u^2 - a^2}| + C$

#8. $\displaystyle \int \frac{1}{a + be^{cu}} \, du = \frac{u}{a} - \frac{1}{ac} \ln |a + be^{cu}| + C$

Customizing an Entry of an Integral Table

An integral table presents formulas, but usually we must customize them for our particular problem.

For example, rather than listing the special formula of $\int x^3 \, dx = (x^4/4) + C$, the table gives $\int u^n \, du = u^{n+1}/(n+1) + C$ (for $n \neq -1$). To use the table we must substitute x for u and 3 for n.

The following examples are done by using the short integral table given above. The numbers of the integrals are the numbers of that table.

EXAMPLE 1   Customizing an integral table

Find $\int \dfrac{1}{x(x+2)} \, dx$.

SOLUTION

From the short integral table, we have

$$\boxed{\#1} \qquad \int \frac{1}{u(au+b)} \, du = \frac{1}{b} \cdot \ln \left| \frac{u}{au+b} \right| + C \qquad \text{(for } b \neq 0\text{)}$$

Using $a = 1$, $b = 2$, $\boxed{\#1}$ gives $\displaystyle \int \frac{1}{x(x+2)} \, dx = \frac{1}{2} \cdot \ln \left| \frac{x}{x+2} \right| + C.$

We can check by taking the derivative.

$$\frac{d}{dx} \left[\frac{1}{2} \cdot \ln \left| \frac{x}{x+2} \right| + C \right] = \frac{1}{2} \cdot \frac{1}{\dfrac{x}{x+2}} \left(\frac{(x+2) \cdot 1 - x \cdot 1}{(x+2)^2} \right)$$

$$= \left(\frac{x+2}{2x} \right) \left(\frac{2}{(x+2)^2} \right)$$

$$= \frac{1}{x(x+2)}$$

If our integrand is very similar to the integrand in the table, but it differs by having a product or divisor consisting of a constant, then we may factor out the constant and proceed. Remember, factoring can be done through an integral sign *only* if the factor is a constant.

EXAMPLE 2   Factoring out constants

Find $\int \dfrac{2}{3x(4x-1)} \, dx$.

SOLUTION

$$\int \frac{2}{3x(4x - 1)} \, dx = \frac{2}{3} \int \frac{1}{x(4x - 1)} \, dx$$ gives us the form of $\boxed{\#1}$. Using $a = 4$ and $b = -1$, we have

$$\int \frac{2}{3x(4x - 1)} \, dx = \frac{2}{3} \cdot \left(\frac{1}{-1} \cdot \ln \left| \frac{x}{4x - 1} \right| \right) + C$$

$$= \frac{-2}{3} \ln \left| \frac{x}{4x - 1} \right| + C$$

We can check by taking the derivative, first using a property of logarithms.

$$\frac{d}{dx} \left(-\frac{2}{3} \ln \left| \frac{x}{4x - 1} \right| + C \right) = \frac{d}{dx} \left(-\frac{2}{3} (\ln |x| - \ln |4x - 1|) + C \right)$$

$$= -\frac{2}{3} \left(\frac{d}{dx} (\ln |x|) - \frac{d}{dx} \ln |4x - 1| \right) + 0$$

$$= -\frac{2}{3} \left(\frac{1}{x} - \frac{4}{4x - 1} \right)$$

$$= -\frac{2}{3} \left(\frac{(4x - 1) - 4x}{x(4x - 1)} \right)$$

$$= -\frac{2}{3} \left(\frac{-1}{x(4x - 1)} \right)$$

$$= \frac{2}{3x(4x - 1)}$$

EXAMPLE 3   Using an integral table

Find $\int \frac{4}{\sqrt{x^2 - 9}} \, dx$.

SOLUTION

From the integral table, we have

$\boxed{\#7}$ $$\int \frac{1}{\sqrt{u^2 - a^2}} \, du = \ln \left| u + \sqrt{u^2 - a^2} \right| + C.$$

Factor the 4 out of the integrand. Then we may let $u = x$ and $a = 3$, so $a^2 = 9$, and we have $\int (4/\sqrt{x^2 - 9}) \, dx = 4 \ln \left| x + \sqrt{x^2 - 9} \right| + C.$

An integral table can be used in the evaluation of a definite integral by the Fundamental Theorem of Calculus.

EXAMPLE 4   Using an integral table for a definite integral

Evaluate $\displaystyle\int_2^5 \frac{1}{\sqrt{x^2 + 4}}\, dx$.

SOLUTION

From the integral table, we have

$\boxed{\#6}$ $\displaystyle\int \frac{1}{\sqrt{u^2 + a^2}}\, du = \ln |u + \sqrt{u^2 + a^2}| + C.$

Using this with $u = x$ and $a = 2$, so as to make $a^2 = 4$, we have

$$\int_2^5 \frac{1}{\sqrt{x^2 + 4}}\, dx = (\ln |x + \sqrt{x^2 + 4}|) \Big|_2^5$$

$$= \ln |5 + \sqrt{5^2 + 4}| - \ln |2 + \sqrt{2^2 + 4}|$$

$$= \ln (5 + \sqrt{29}) - \ln (2 + \sqrt{8})$$

$$= \ln (5 + 5.3852) - \ln (2 + 2.8284)$$

$$= 2.3404 - 1.5745$$

$$= 0.7659$$

6.3 PROBLEMS

In problems 1–6, find values of a and b so that we can use the formulas of the short integral table.

1. $\dfrac{1}{x(x + 3)}$ **2.** $\sqrt{x^2 - 9}$ **3.** $\dfrac{2x}{x + 5}$

4. $\sqrt{9x^2 + 1}$ **5.** $\sqrt{9x^2 - 4}$ **6.** $\dfrac{5}{3x - x^2}$

Use the short integral table in this section to find the antiderivatives of problems 7–28.

7. $\displaystyle\int \frac{1}{x(x + 4)}\, dx$ **8.** $\displaystyle\int \frac{1}{x(x + 3)}\, dx$

9. $\displaystyle\int \frac{x}{x + 8}\, dx$ **10.** $\displaystyle\int \frac{x}{2x + 1}\, dx$

11. $\displaystyle\int \frac{x}{\sqrt{x + 1}}\, dx$ **12.** $\displaystyle\int \frac{x}{\sqrt{x + 3}}\, dx$

13. $\displaystyle\int \sqrt{x^2 - 9}\, dx$ **14.** $\displaystyle\int \sqrt{x^2 + 9}\, dx$

15. $\displaystyle\int \frac{4}{\sqrt{x^2 - 4}}\, dx$ **16.** $\displaystyle\int \frac{1}{\sqrt{x^2 - 5}}\, dx$

17. $\displaystyle\int \frac{1}{5 + 3e^{2x}}\, dx$ **18.** $\displaystyle\int \frac{2}{3 + 2e^{4x}}\, dx$

19. $\displaystyle\int_0^2 \frac{2x}{x + 5}\, dx$ **20.** $\displaystyle\int_{-1}^1 \frac{x}{2x - 5}\, dx$

21. $\displaystyle\int \sqrt{4x^2 - 5}\, dx$ **22.** $\displaystyle\int \sqrt{9x^2 + 1}\, dx$

23. $\displaystyle\int \frac{3}{\sqrt{4x^2 - 1}}\, dx$ **24.** $\displaystyle\int \frac{2}{\sqrt{9x^2 - 4}}\, dx$

25. $\sqrt{9x^2 - 4}\, dx$ **26.** $\displaystyle\int \sqrt{4x^2 + 9}\, dx$

27. $\displaystyle\int_{1/2}^2 \frac{5}{3x - x^2}\, dx$ **28.** $\displaystyle\int_0^2 \frac{4}{2x^2 - 5x - 3}\, dx$

Enrichment Problems

29. Formula $\boxed{\#1}$,

$$\int \frac{1}{u(au + b)} \, du = \frac{1}{b} \ln \left| \frac{u}{au + b} \right| + C$$

is derived by a technique called *partial fractions*. The following is an outline of how this works. Supply the requested information.

(a) From algebra, we know values of A and B exist such that $\dfrac{A}{u} + \dfrac{B}{(au + b)} = \dfrac{1}{u(au + b)}$. To find A and B, first find a common denominator and add together the left side. $\dfrac{A}{u} + \dfrac{B}{(au + b)} = \dfrac{A(au + b) + Bu}{u(au + b)}$. This must equal the right side, $1/[u(au + b)]$. The numerators must be equal. Why?

(b) Rearranging the terms, we find $(aA + B)u + bA = 1$. Then $aA + B = 0$ and $bA = 1$. Why? (Hint: Try letting $u = 0$ and $u = 1$.)

(c) Now solve for A and B in terms of a and b.

(d) The integrals $\displaystyle\int \frac{A}{u} \, du$ and $\displaystyle\int \frac{B}{au + b} \, du$ can both be found by using substitution and the formula $\int (1/u) \, du = \ln |u| + C$. Do so.

(e) Finally, use the property of logarithms, which states that $\ln r - \ln s = \ln (r/s)$, to obtain the form of the antiderivative in $\boxed{\#1}$.

6.4 NUMERICAL INTEGRATION

This section discusses methods for approximating $\int_a^b f(x) \, dx$. We first illustrate the methods with functions for which we do know an anti-derivative and thus can use the Fundamental Theorem of Calculus. That way we can check the closeness of the approximation. Then we use the methods of approximation on functions for which we do not have any antiderivative and so cannot use the Fundamental Theorem.

In Section 5.2, we discussed the midpoint approximation sum, one of the approximations that use rectangles. In addition to rectangular approximations, in this section we investigate two additional approximations: the Trapezoidal Rule (which uses trapezoidal sections) and Simpson's Rule (which uses parabolic sections).

Approximations by Rectangles

The definite integral was introduced in Section 5.2 by using a midpoint approximation sum. For that sum we divide the interval $[a, b]$ into n equally wide subintervals. We then find the midpoint of each subinterval and evaluate the function at that point. The sum is formed by taking those function values and multiplying them by the width of the corresponding subintervals. Thus, the midpoint approximation uses rectangles.

Example 1 is a review of that method.

EXAMPLE 1   Midpoint approximation sum

Find the midpoint approximation sum of $\left.\int_1^5 \dfrac{5}{x}\, dx\right.$ by using four equal subintervals. Compare the approximation with the exact value.

SOLUTION

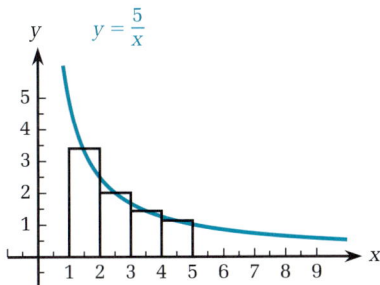

FIGURE 6.4.1

The graph of $f(x) = 5/x$ is shown in figure 6.4.1.

The four-interval partition consists of $x_0 = 1$, $x_1 = 2$, $x_2 = 3$, $x_3 = 4$, and $x_4 = 5$.

The midpoints are at 1.5, 2.5, 3.5, and 4.5, and we have

$$f(1.5) = \frac{5}{1.5} = 3.3333$$

$$f(2.5) = \frac{5}{2.5} = 2$$

Similarly, $f(3.5) = 1.4286$ and $f(4.5) = 1.1111$.

The midpoint approximation sum is thus $3.3333(1) + 2(1) + 1.4286(1) + 1.1111(1) \simeq 7.873$.

We can compare this with the exact value because we can use the Fundamental Theorem of Calculus on this definite integral:

$$\int_1^5 \frac{5}{x}\, dx = 5 \ln x \Big|_1^5 = 5 \ln 5 - 5 \ln 1 = 5 \ln 5 \simeq 8.0472$$

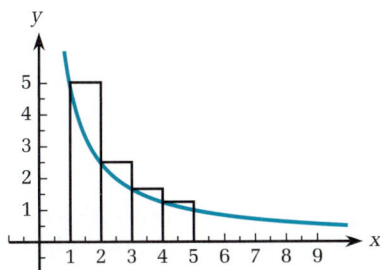

FIGURE 6.4.2

To use the midpoint approximation sum, we need to find the value of the function at the midpoint of each subinterval. If we already know the value of the function at each endpoint of the subintervals, then we could form a rectangular approximation for $\int_1^5 \dfrac{5}{x}\, dx$ by using left-hand endpoints or right-hand endpoints, as shown in figures 6.4.2 and 6.4.3.

Left-hand approximation value is 10.4167.

Right-hand approximation value is 6.4167.

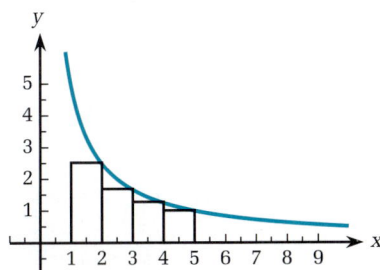

FIGURE 6.4.3

For $\int_1^5 \dfrac{5}{x}\, dx$, the left-hand estimation is too large because all the rectangles are above the curve. Similarly, the right-hand estimation is too small, because the rectangles are all below the curve. The midpoint approximation sum is closer to the actual value. However, it does require that we evaluate $f(x)$ at values between the endpoints of the subintervals. The next two methods of approximation allow us to achieve good approximations by using only values of $f(x)$ that are calculated at the endpoints of the subintervals.

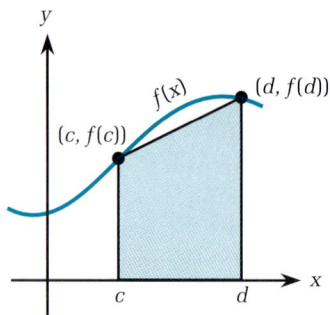

FIGURE 6.4.4

Trapezoidal Rule

The *Trapezoidal Rule* uses the average of two adjacent values to estimate an average value of the function over an interval.

The graph of $f(x)$ over $[c, d]$ is shown in figure 6.4.4. The area under the curve, above the x-axis, and over the interval $[c, d]$ can be approximated by the area of a trapezoid. This area is given by the following product of the average height multiplied by the width.

$$\text{Area} = \left(\frac{f(c) + f(d)}{2} \right) \cdot (d - c)$$

Suppose we subdivide the interval $[a, b]$ into n equal subintervals by the points $x_0 = a, x_1, x_2, x_3, \cdots, x_n = b$, as shown in figure 6.4.5.

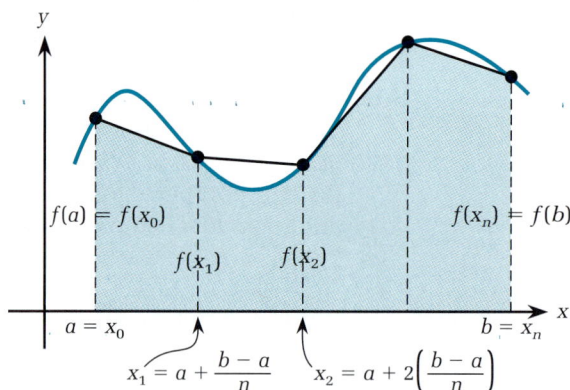

FIGURE 6.4.5

The entire interval has width $b - a$ and so the width of each of the n subintervals is $(b - a)/n$, which we represent by Δx.

Adding together the areas of all of the trapezoids gives an approximation for the area under the curve. Notice that in the following sum, the value of the function appears twice for all the division points except the endpoints. That is because each interior partition point is used twice, as a left-hand endpoint of the subinterval to its right and as a right-hand endpoint of the subinterval to its left.

The areas of trapezoids

$$= \frac{f(x_0) + f(x_1)}{2} \cdot \Delta x + \frac{f(x_1) + f(x_2)}{2} \cdot \Delta x + \cdots + \frac{f(x_{n-1}) + f(x_n)}{2} \cdot \Delta x$$

$$= \left[\frac{f(x_0) + f(x_1)}{2} + \frac{f(x_1) + f(x_2)}{2} + \cdots + \frac{f(x_{n-1}) + f(x_n)}{2} \right] \Delta x$$

$$= \left[\frac{f(x_0)}{2} + \frac{f(x_1)}{2} + \frac{f(x_1)}{2} + \frac{f(x_2)}{2} + \cdots + \frac{f(x_{n-1})}{2} + \frac{f(x_n)}{2} \right] \Delta x$$

$$= \left[\frac{f(x_0)}{2} + f(x_1) + f(x_2) + \cdots + f(x_{n-1}) + \frac{f(x_n)}{2} \right] \Delta x$$

TRAPEZOIDAL RULE

Suppose $f(x)$ is continuous over $[a, b]$, which is divided into n equal subintervals by $x_0, x_1, x_2, \cdots, x_n$. Then $\int_a^b f(x)\, dx$ exists and is approximated by

$$\left[\frac{f(x_0)}{2} + f(x_1) + f(x_2) + \cdots + f(x_{n-1}) + \frac{f(x_n)}{2}\right]\left(\frac{b-a}{n}\right)$$

EXAMPLE 2   Trapezoidal rule

Use the Trapezoidal Rule with $n = 4$ to estimate $\displaystyle\int_1^5 \frac{5}{x}\, dx$.

SOLUTION

For $a = 1$ and $b = 5$, each subinterval is of length 1. The values of $f(x) = 5/x$ are given in the following table.

x	1	2	3	4	5
$f(x) = \dfrac{5}{x}$	5	2.5	1.6667	1.25	1

Then $\left[\dfrac{f(x_0)}{2} + f(x_1) + f(x_2) + \cdots + f(x_{n-1}) + \dfrac{f(x_n)}{2}\right][(b-a)/n] =$
$2.5 + 2.5 + 1.6667 + 1.25 + 0.5 = 8.4167$. The graph showing this approximation is shown in figure 6.4.6.

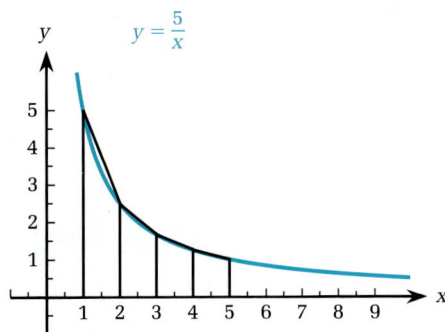

FIGURE 6.4.6

A second example of the Trapezoidal Rule follows.

EXAMPLE 3   Trapezoidal Rule

Use the Trapezoidal Rule with $n = 4$ to estimate $\int_1^2 \left(\dfrac{e^x}{x} - 2 \right) dx$.

SOLUTION

The graph of $f(x) = e^x/x - 2$ over $[1, 2]$ is shown in figure 6.4.7.

x	$\dfrac{e^x}{x} - 2$
1.00	0.7183
1.25	0.7923
1.50	0.9878
1.75	1.2883
2.00	1.6945

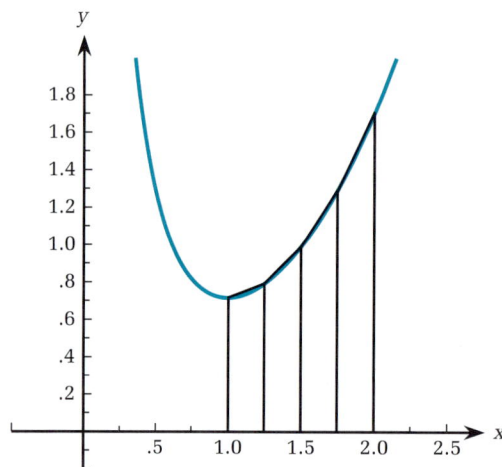

FIGURE 6.4.7

So $\int_1^2 \left(\dfrac{e^x}{x} - 2 \right) dx \approx [(1/2) \cdot 0.7183 + 0.7923 + 0.9878 + 1.2883 + (1/2) \cdot 1.6945] \cdot 0.25 = (0.3592 + 0.7923 + 0.9878 + 1.2883 + 0.8472)0.25 = 1.0687.$

The Trapezoidal Rule does not depend on whether the function is given by a formula. It can be used on data associated with any continuous function. In Example 4, we are given particular values of the function.

EXAMPLE 4   Using the Trapezoidal Rule with data

The data in the following table indicate the rate (in thousands of gallons per hour) of water being delivered by the Mirage Lake Water Co. Use the Trapezoidal Rule (with $n = 9$) to estimate the total amount of water delivered over the nine-hour period.

Hour	3 A.M.	4 A.M.	5 A.M.	6 A.M.	7 A.M.	8 A.M.	9 A.M.	10 A.M.	11 A.M.	Noon
Rate (gallons per hour)	20	25	35	45	110	200	95	75	65	80

SOLUTION

The data are graphed in figure 6.4.8.

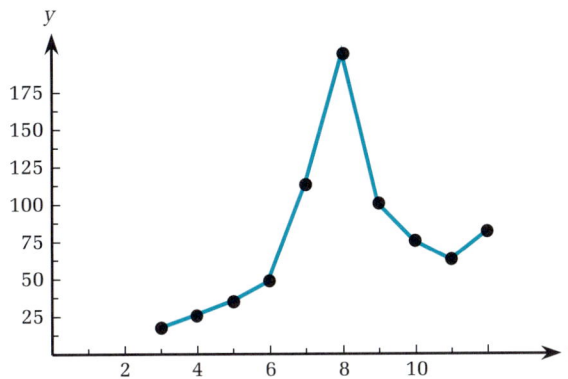

FIGURE 6.4.8

With $a = 3$, $b = 12$, and $n = 9$, we have $(12 - 3)/9 = 1$ for the width of each subinterval. By the Trapezoidal Rule, $\int_{3 \text{ A.M.}}^{\text{noon}}$ (rate) $dt \approx$ $[(1/2)20 + 25 + 35 + 45 + 110 + 200 + 95 + 75 + 65 + (1/2)80] \cdot 1$ $= 700$ (thousand gallons).

PRACTICE EXERCISE 1

Estimate $\int_{0}^{2} \sqrt{4 - x^2} \, dx$ by the Trapezoidal Rule with $n = 4$.

Error Estimate Using the Trapezoidal Rule

We saw in Section 5.2 that if $f(x)$ is a continuous function, then $\int_{a}^{b} f(x) \, dx$ exists and is exactly some real number. The approximations we use are just approximations, in the same way that 22/7 is an approximation of π or that 1.4 is an approximation of $\sqrt{2}$. We can see that if the number of subintervals is large, then the trapezoidal approximation is quite good. But, just how good is it?

The following result gives us a bound on how large an error occurs if we use the Trapezoidal Rule with n subintervals.

ERROR ESTIMATE USING THE TRAPEZOIDAL RULE

If B is a positive real number such that $|f'(x)| \le B$ for every value of x in $[a, b]$, then the maximum possible error resulting by using the Trapezoidal Rule as an approximation of $\int_a^b f(x)\, dx$ is no more than $\dfrac{B(b-a)^3}{12n^2}$.

This is the maximum possible error. The actual error usually is less.

EXAMPLE 5   Maximum possible error using the Trapezoidal Rule

Find the maximum possible error if we use the Trapezoidal Rule with $n = 4$ to estimate $\displaystyle\int_1^5 \frac{5}{x}\, dx$.

SOLUTION

The absolute value of the derivative of $5/x$ is given by $\left|-5/x^2\right| = 5/x^2$. We need to know how large this can be over the interval $[1, 5]$. Quotients are made large by having large numerators or by having small denominators. For this quotient, the numerator is constantly 5 and the denominator is small if x^2 is small. The smallest that x^2 can be over the interval $[1, 5]$ is 1. Thus, the absolute value of this derivative is always less than $5/1 = 5$, which is then the value B we use in the error estimate.

For $n = 4$ and $b - a = 4$, the maximum that the error can be is

$$\frac{B(b-a)^3}{12n^2} = \frac{5 \cdot 4^3}{12 \cdot 4^2} = 5/3 \approx 1.667.$$

Actually, this overstates the error. We know this because we know the exact value for $\displaystyle\int_1^5 \frac{5}{x}\, dx$ is $5 \cdot \ln 5$, which is about 8.0472. The Trapezoidal Rule approximation was 8.4167 (see Example 2). These differ by only 0.3695, just one-fifth of the bound on the approximation's error.

The real use of the error-bound formula is to determine how many partitions are needed to guarantee the error of approximation is small enough. For instance, if we know that using n subintervals for estimating $\displaystyle\int_1^5 \frac{5}{x}\, dx$ produces an error no more than $(5 \cdot 4^3)/(12n^2)$, then if we want the error to be no more than 0.001, we simply need to use an n such that $(5 \cdot 4^3)/(12n^2) < 0.001$. Solving this for n we have $\dfrac{5 \cdot 4^3}{12(0.001)} < n^2$, so $26{,}666.67 < n^2$ and $163.3 < n$. Of course, we do not intend to do a hand calculation with $n = 164$, but this value is the number of partitions we need to tell a calculator or computer to use to guarantee the accuracy we want.

Simpson's Rule

A more accurate, but more complicated, approximation than the Trapezoidal Rule is given by using segments of parabolas. Each segment is constructed to pass through three points on the graph whose x-coordinates are equally spaced. The area under each segment is found and the total of all the areas under the segments is used to approximate the area under the curve. The interval $[a, b]$ is divided up into an even number of subintervals and each adjacent pair of subintervals is used for one triple of points.

The rule is named Simpson's Rule in honor of Thomas Simpson (1710–1761), a successful textbook writer in England who popularized the rule, although it was not invented by him.

SIMPSON'S RULE

Suppose $f(x)$ is a continuous function over the interval $[a, b]$. Let $x_0, x_1, x_2, \cdots, x_n$ partition $[a, b]$ into n subintervals of equal width, n being even.

Then $\int_a^b f(x)\, dx$ exists and has an approximation given by

$$\left(\frac{b - a}{3n}\right)(f(x_0) + 4f(x_1) + 2f(x_2) + 4f(x_3) + 2f(x_4) +$$
$$\cdots + 2f(x_{n-2}) + 4f(x_{n-1}) + f(x_n))$$

Notice the coefficients of $f(x_0)$ and $f(x_n)$ are 1 and the others alternate between 2 and 4, starting with the 4.

EXAMPLE 6   Simpson's Rule

Use Simpson's Rule with $n = 4$ to approximate $\displaystyle\int_0^2 \sqrt{4 - x^2}\, dx$.

SOLUTION

For $n = 4$, the values of x_0 through x_4 and the associated values of $\sqrt{4 - x^2}$ are shown in the accompanying table. The graph of

$\sqrt{4 - x^2}$ is shown in figure 6.4.9. Notice in that figure the two parabolic segments, one from 0 to 1 and the other from 1 to 2, are also shown. Compare this with the Trapezoidal Rule estimation shown in the solution to Practice Exercise 1.

x_n	$\sqrt{4 - x^2}$
$x_0 = 0$	2
$x_1 = 0.5$	1.9365
$x_2 = 1.0$	1.7321
$x_3 = 1.5$	1.3229
$x_4 = 2.0$	0

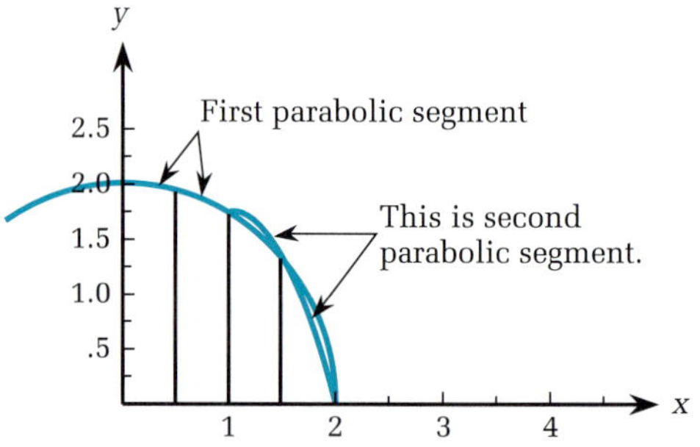

FIGURE 6.4.9

$$\int_0^2 \sqrt{4 - x^2} \, dx \simeq \frac{2}{3 \cdot 4}(2 + 4 \cdot 1.9365 + 2 \cdot 1.7321 + 4 \cdot 1.3229 + 0)$$

$$= 0.1667(2 + 7.746 + 3.4642 + 5.2916 + 0)$$

$$= 0.1667 \cdot 18.5018 = 3.0843$$

Compare this value with the answer to Practice Exercise 1. We know that $\int_0^2 \sqrt{4 - x^2}$ gives the area of one-fourth of a circle of radius 2. Thus, an exact answer would be $(1/4)(\pi \cdot 2^2) = \pi$.

We are using these approximations on an integral whose value is known to demonstrate the better accuracy of Simpson's Rule. ●

Practice Exercise 2 involves a function whose antiderivative we know and can determine, so as to give ourselves a check on our approximation.

PRACTICE EXERCISE 2

Use Simpson's Rule with $n = 4$ to approximate $\int_0^2 \sqrt{x} \, dx$.

We consider both the Trapezoidal and Simpson's Rules because if we are interested in estimating the errors involved in using these rules, then the error associated with the Trapezoidal Rule is much easier to estimate than the error involved with Simpson's Rule. The error associated with Simpson's Rule is discussed in the enrichment problems.

![calculator icon] Graphing Calculators and Symbolic Algebra Systems

Some graphing calculators and symbolic algebra systems will calculate approximations of definite integrals. These often use Simpson's Rule because applying Simpson's Rule is not really any more difficult than applying the Trapezoidal Rule and yet the accuracy is much greater.

A graphing calculator that does numeric integration may not allow you to decide the number of subintervals, instead simply choosing some subinterval width itself.

6.4 PROBLEMS

Foundations

The problems of this section require the basic skills illustrated by the following:

1. If the interval [1, 5] is to be divided into eight equal subintervals, find the coordinates of each of the endpoints of those subintervals.

2. Evaluate $f(x) = e^x/x$ at $x = 1.5$.

3. Evaluate $g(x) = e^{-x^2}$ at $x = 0.75$.

Exercises

Using four-place accuracy, as in the Examples, approximate the definite integrals in problems 4–11 using $n = 4$ by
(a) the Trapezoidal Rule (b) Simpson's Rule.

4. $\int_0^2 (2x + 1)\, dx$

5. $\int_0^4 x^2\, dx$

6. $\int_1^3 \dfrac{e^x}{x}\, dx$

7. $\int_1^5 \dfrac{\ln x}{x}\, dx$

8. $\int_0^1 e^{-x^2}\, dx$

9. $\int_0^2 e^{-x^2}\, dx$

10. $\int_0^2 \dfrac{1}{4 + x^2}\, dx$

11. $\int_0^1 \dfrac{1}{1 + x^2}\, dx$

12. The chart in table 6.4.1 gives the seasonally adjusted annual rate of sales of new homes (measured in thousands of units), as reported by the U.S. Department of Commerce for the 12-month period August 1, 1988, to July 31, 1989.* Use

the rates as midpoints of each month and use midpoint approximation to estimate the total annual sales of new homes over that period.

TABLE 6.4.1

Aug 15	705	Dec 15	670	Apr 15	603
Sept 15	682	Jan 15	690	May 15	642
Oct 15	719	Feb 15	620	June 15	641
Nov 15	640	March 15	558	July 15	739

13. **(Balance of Trade)** The values given in table 6.4.2 are reported by the U.S. Department of Commerce as annual rates of the balance of trade for the United States for 1982–1991.† Use the Trapezoidal Rule to estimate the total number of billions of dollars exported by the United States over the nine-year period, January 1, 1982, to January 1, 1991.

TABLE 6.4.2

1982	6.89	1987	−145.39
1983	−5.87	1988	−162.31
1984	−40.14	1989	−128.86
1985	−99.01	1990	−110.03
1986	−122.33	1991	−99.30

*Denver Post, August 30, 1989.

†*Denver Post*, March 13, 1991.

14. (Electricity Consumption) A chart for the rates of consumption of an electric utility is given below. The rate at a particular time is the number of megawatts per hour being recorded at that time. With $n = 6$, use
(a) the Trapezoidal Rule   (b) Simpson's Rule
to approximate the total consumption of megawatts during the day.

Time	Midnight	4 A.M.	8 A.M.	Noon
Rate	40	35	60	55

Time	4 P.M.	8 P.M.	Midnight	
Rate	50	65	42	

15. (Voter Turnout) A chart of the rate at which voters in one precinct were arriving to vote during an election is shown below. The rate is given as a number of voters per hour, although the poll watcher watched only a few minutes and multiplied the number seen arriving in that interval by an appropriate number to determine a rate per hour. With $n = 6$, use
(a) the Trapezoidal Rule   (b) Simpson's Rule
to estimate the total number of voters at that precinct from 6 A.M. to 6 P.M.

Time	6 A.M.	8 A.M.	10 A.M.	Noon
Rate	40	70	20	35

Time	2 P.M.	4 P.M.	6 P.M.	
Rate	16	40	60	

16. A sketch map of six building lots alongside a river is shown in figure 6.4.10. Estimate the total area by using
(a) the Trapezoidal Rule   (b) Simpson's Rule.

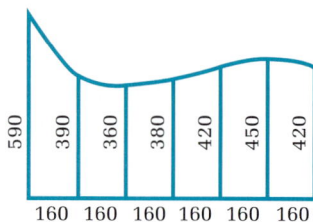

FIGURE 6.4.10

17. A sketch showing one side of a house built on a hillside is shown in figure 6.4.11. Using the seven measurements given, each five feet apart, estimate the number of square feet above ground on that side by using
(a) the Trapezoidal Rule   (b) Simpson's Rule.

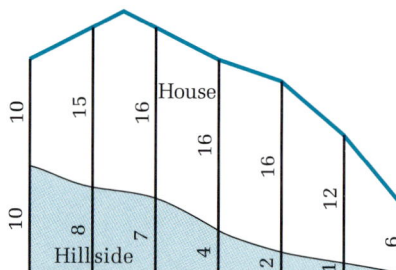

FIGURE 6.4.11

Writing and Discussion Problem

18. How does the existence of graphing calculators and computers affect our need to know about approximations? Is it good enough simply to instruct the calculator to use the most subintervals it can and be content with the resulting approximation?

Enrichment Problems

19. $f(x) = \sqrt{4 - x^2}$ is the equation of a semicircle with radius 2. Hence $\int_0^2 \sqrt{4 - x^2}\, dx$ is the area of one-fourth of a circle of radius 2. The area of a circle of radius r is πr^2, so the integral is $(1/4)(\pi \cdot 2^2)$, which is π. We found values for the integral of 2.9958 using the Trapezoidal Rule and 3.0843 using Simpson's Rule.

Another integral that equals π is $\int_0^1 \frac{4}{1 + x^2}\, dx$. For this integral, use Simpson's Rule with $n = 10$ and *all the accuracy of your calculator* to find an estimate for π.

20. Use the error estimate for the Trapezoidal Rule to find a bound on the error that might exist by using the Trapezoidal Rule with $n = 10$ on $\int_0^1 \frac{4}{1 + x^2}\, dx$ to estimate π. What happens when we try to use this bound on the error with our other integral, which estimated $\pi \simeq \int_0^2 \sqrt{4 - x^2}\, dx$?

SOLUTIONS TO PRACTICE EXERCISES

1. The table below shows the estimation of $\int_0^2 \sqrt{4 - x^2}\, dx$ by the Trapezoidal Rule with $n = 4$. The graph is shown in figure 6.4.12.

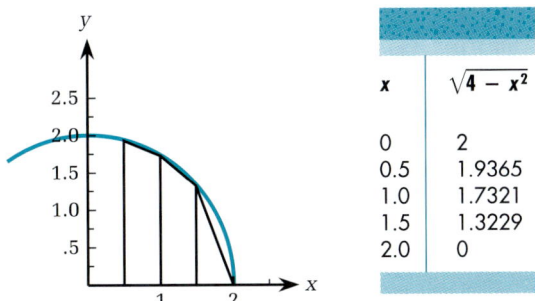

x	$\sqrt{4 - x^2}$
0	2
0.5	1.9365
1.0	1.7321
1.5	1.3229
2.0	0

FIGURE 6.4.12

Therefore, $\int_0^2 \sqrt{4 - x^2} \simeq$
$$\left[\frac{1}{2} \cdot 2 + 1.9365 + 1.7321 + 1.3229 + (1/2) \cdot 0\right] \cdot$$
$(2/4) = (1 + 1.9365 + 1.7321 + 1.3229) \cdot 0.5 =$
2.9958.

2. The following table shows values using Simpson's Rule with $n = 4$ to approximate $\int_0^2 \sqrt{x}\, dx$. The graph is shown in figure 6.4.13.

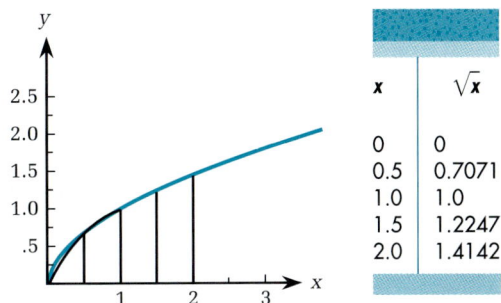

x	$\sqrt{x}$
0	0
0.5	0.7071
1.0	1.0
1.5	1.2247
2.0	1.4142

FIGURE 6.4.13

Therefore, $\int_0^2 \sqrt{x}\, dx \simeq [2/(3 \cdot 4)] \cdot$
$(0 + 4 \cdot 0.7071 + 2 \cdot 1 + 4 \cdot 1.2247 + 1.4142) =$
$0.1667 \cdot (2.8284 + 2 + 4.8988 + 1.4142) =$
$0.1667 \cdot 11.1414 \simeq 1.8573$. This integral can be evaluated using the Fundamental Theorem of Calculus. Doing so, we have
$$\int_0^2 \sqrt{x}\, dx = \int_0^2 x^{1/2}\, dx = \left. \frac{2}{3} x^{3/2} \right|_0^2 =$$
$$\frac{2}{3}(2^{3/2} - 0) \simeq 1.8856.$$

6.5   IMPROPER INTEGRALS AND INFINITE GEOMETRIC SERIES

Two types of processes continue indefinitely. There are those that consist of separate events and those with continuous activity. To illustrate the first, consider the cumulative impact of a construction project on a local economy. There is an initial spending, but then some of that initial money is respent in the community in the form of goods, housing, entertainment, and so forth. Some of the money spent in this second round of spending would again be respent in the local community for a third round of spending. These spending rounds continue on and on.

R. F. Kahn was one of the first to discuss the cumulative impact of initial expenditures. In 1931 he presented a case for government spending on public works as a recovery measure from the Great Depression. In Kahn's analysis, the rounds of successive spending occur at separate times. This is in contrast to a continuous process, which goes on without interruption. Examples of continuous processes include continuous money flow, oil-well production, or the total eventual sales of a book.

Infinite Geometric Series

We first need to review a topic from algebra.

DEFINITION

Geometric series

A *geometric series* has the form $a + ax + ax^2 + ax^3 + \cdots + ax^n$, where $a \neq 0$, $x \neq 0$, and the expression ax^n represents a general term of the series.

An example of a geometric series is $1 + (0.5) + (0.5)^2 + \cdots + (0.5)^n$. The value of this geometric series can be represented by a series of rectangles, as shown in figure 6.5.1.

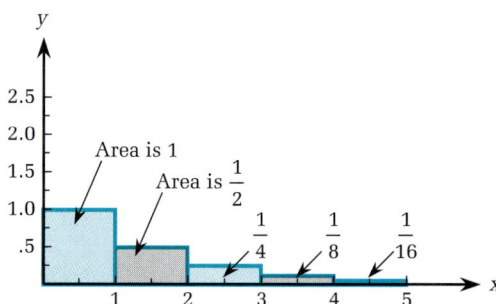

FIGURE 6.5.1

Now consider an *infinite geometric series,* which is often written to look like an infinite sum of numbers. For example, $1 + (0.5) + (0.5)^2 + (0.5)^3 + (0.5)^4 + \cdots + (0.5)^n + \cdots$, where the final three dots indicate that the adding never stops.

We cannot actually perform an infinite number of additions, no matter what brand of computer we use, but we are able to give meaning to some infinite geometric series. To do so, first realize that for any positive integer n we can add together n terms. Let us call such a partial sum S_n. As the number of terms increases, the values of the S_n may have a limit. If so, that value is said to be the value of the infinite series.

DEFINITION

Infinite geometric series, partial sum, converges

The expression $a + ax + ax^2 + ax^3 + \cdots + ax^n + \cdots$ is an *infinite geometric series* and $S_n = a + ax + ax^2 + ax^3 + \cdots + ax^n$ is the *nth partial sum.* If $\lim_{n \to \infty} S_n$ exists and is S, then we say the series *converges* to S.

A convergent infinite geometric series is evaluated by the following formula.

> ## THE SUM OF AN INFINITE GEOMETRIC SERIES
>
> (6.5.1) If $|x| < 1$, then $a + ax + ax^2 + ax^3 + \cdots + ax^n + \cdots = \dfrac{a}{1 - x}$.
>
> If $|x| \geq 1$, then the infinite geometric series does not converge to a value.

EXAMPLE 1   Evaluating an infinite geometric series

Find $1 + (0.5) + (0.5)^2 + (0.5)^3 + \cdots + (0.5)^n + \cdots$.

SOLUTION

We can use formula (6.5.1) because $|0.5| < 1$. For $a = 1$ and $x = 0.5$, we have $1/(1 - 0.5) = 1/0.5 = 2$. ●

Applications (Perpetual Annuities, Gambling, Drug Levels)

Let us apply formula (6.5.1) to an annual payment that goes on forever. This is a perpetual annuity.

EXAMPLE 2   Determining the value of a perpetual annuity

The King grants the Duke of Granado a rare vineyard on April 1. The Duke is so pleased that he promises he and his descendants will forever pay $1000 every year on April 1 to the King. The King's Keeper of the Metal knows that x dollars n years in the future has present value $(1 + i)^{-n}x$ dollars under annual compounding at a rate of interest i (written decimally). Using an annual interest rate of 6%, what is the present value of the Duke's promise to the King?

SOLUTION

The interest rate of 6% should be written as 0.06 in decimal form. The $1000 delivered a year from now is worth $(1.06)^{-1} \cdot 1000$ today. The $1000 delivered two years from now is worth $(1.06)^{-2} \cdot 1000$

today. This continues, year after year, and each year the $1000 delivered n years from now is worth $(1.06)^{-n} \cdot 1000$ today.

Because the payments go on forever (a perpetual annuity), the total present value is $(1.06)^{-1} \cdot 1000 + (1.06)^{-2} \cdot 1000 + \cdots + (1.06)^{-n} \cdot 1000 + \cdots = (1/1.06) \cdot 1000 + (1/1.06)^2 \cdot 1000 + \cdots + (1/1.06)^n \cdot 1000 + \cdots$

The King's Keeper recognizes she almost has an infinite geometric series of the form $a + ax + ax^2 + ax^3 + \cdots + ax^n + \cdots$ except for the fact that the term consisting of a itself is missing.

At this point, the King's Keeper has two options. First, she could rewrite the series as $\dfrac{1}{1.06} \left(1000 + \dfrac{1}{1.06} \cdot 1000 + \left(\dfrac{1}{1.06}\right)^2 \cdot 1000 + \cdots\right)$ and use formula (6.5.1) on the part in parentheses to give

$$\frac{1}{1.06}\left(\frac{1000}{1 - (1/1.06)}\right) = \frac{1}{1.06}\left(\frac{1000}{1 - 0.943396}\right) = \frac{1}{1.06}\left(\frac{1000}{0.056604}\right) =$$

$$\frac{1}{1.06}(17{,}666.60) = 16{,}666.60.$$

Second, she could find the result by using the formula for the infinite series directly, and then subtract 1000 from the initial result because the 1000 (which was the a term) was not present.

Using formula (6.5.1) with $x = 1/1.06 = 0.943396$ and $a = 1000$ gives $a/(1 - x) = 1000/(1 - 0.943396) = 1000/0.056604 = 17{,}666.60$.

Subtracting 1000 provides a final answer of $16,666.60 (as calculated before).

An alternative interpretation of our calculations in Example 1 is to realize that if $16,666.60 is invested today in an account that earns money at 6% compounded annually, then the account would generate an annual annuity payment of $1000 in perpetuity.

PRACTICE EXERCISE 1

Gopher Hole Golf Co. agrees to operate Gopher City's golf course. They will pay a nominal fee of $100 each year. Using an interest rate of 8%, annual compounding, and starting with the first payment of $100 today, what is the present value of the lease?

EXAMPLE 3   Calculating the multiplier effect of spending

Using the concept expressed by R. F. Kahn in the introduction to this section, find the cumulative impact of an initial expenditure of $1000 on a local economy. Assume that for each round of spending, 80% of what is spent is respent.

SOLUTION

If $1000 is spent on merchandise and 80% of what is spent is spent in turn, then $800 would be spent by those who received the first $1000. Of those receiving the $800, they in turn would spend $640 (which is 80% of $800) and so on. Thus, the total impact of an initial $1000 would be $1000 + (0.8) \cdot 1000 + (0.8)^2 \cdot 1000 + (0.8)^3 \cdot 1000 + \cdots = 1000/(1 - 0.8) = 1000/0.2 = \$5000$.

The *multiplier effect* is usually calculated by sociologists and economists as part of an Environmental Impact Statement for a federal project. In the case of Example 3, the newspapers would report that the impact on the local economy would be $5000.

PRACTICE EXERCISE 2

Find the total impact of an expenditure of $2 million if 55% of the money spent is going to be respent.

George Bernard Shaw, the British playwright, remarked of gambling, "Many lose so that few can win." Gambling has increased as a business venture. If we calculate the average return on a bet, then we can estimate the total number of times a bet of that size may be placed until the player is broke.

EXAMPLE 4   Estimating the number of betting rounds until bankruptcy

Suppose a casino pays out $0.80 on each dollar bet by a player. If a player begins with $1000 and places bets of $1 each, estimate the number of bets before the money is all gone.

SOLUTION

Think of this as consisting of rounds of betting. In round I the player bets all the $1000, $1 at a time, setting aside any winnings to be bet in round II. There will be $800 available for round II (80% of $1000). After round II, there will be $0.80 \cdot 800 = \$640$ available for round III. This continues, round after round, giving us exactly the same numbers as we had in Example 3. Hence, the sum is the same, or $5000.

Because each $1 bet is one of these $5000, we have found that the player will place an average of 5000 bets before being broke. (See problem 44 at the end of this section.)

From the field of medicine comes the following application. For a patient on a continuing drug therapy, such as a thyroid supplement or digitalis, the ideal drug is very long acting so that the blood serum level of the drug can be maintained in the effective range without oscillating severely back and forth between an ineffective level and a toxic level. For some patients on a drug such as digitalis, the effective level may be 75% of the lethal level.

EXAMPLE 5   Determining blood serum levels of a drug

For the cardiac-care drug digoxin, 60% of the drug is still present in the blood serum after one day. Suppose that one pill contributes an amount of k milligrams per liter to the blood serum level. Determine the long-term concentrations of digoxin in the bloodstream, both shortly before and shortly after taking a daily pill.

SOLUTION

After quite a while the amount is

$$k + (0.6) \cdot k + (0.6)^2 \cdot k + (0.6)^3 \cdot k + \cdots .$$

The $(0.6) \cdot k$ term results from digoxin taken yesterday and still present, the $(0.6)^2 k$ term (which is 60% of 60% of k) comes from digoxin taken two days ago but still present, and so forth.

Using the formula, the long-term concentration is

$$k/(1 - 0.6) = k/0.4 = 2.5k$$

shortly after taking the daily dosage and $1.5k$ shortly before the daily dosage.

The graph in figure 6.5.2 shows the concentration in the blood.

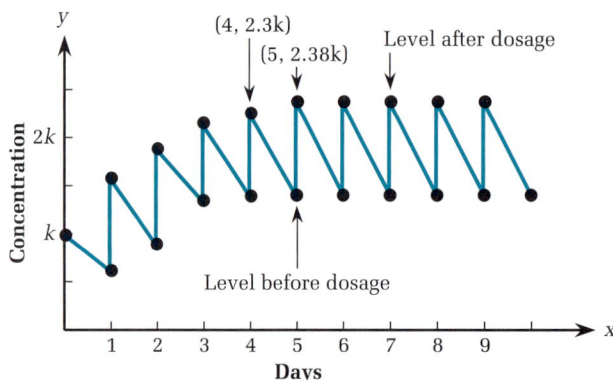

FIGURE 6.5.2

Improper Integrals

We began geometric series with a geometric interpretation in terms of rectangles, one rectangle for each term in the series. For an infinite geometric series, which converges to a number, the geometric interpretation is an infinite number of rectangles, which in total have a finite area as determined by the limit of the corresponding infinite geometric series. Hence, an infinite number of rectangles can have an associated finite area.

We now want to do something similar for definite integrals. Our development of a definite integral, $\int_a^b f(x)\,dx$, assumed $f(x)$ is continuous over the interval $[a, b]$ and that both a and b are actual, finite numbers. If $f(x) \geq 0$, we associated the area shown in figure 6.5.3 with the value of the definite integral.

Consider a function $f(x) \geq 0$, which is continuous on $[a, \infty)$ for some a. Can some number be assigned to the area of the region between $f(x)$ and the x-axis, with $x \geq a$ but with x being unbounded? See figure 6.5.4.

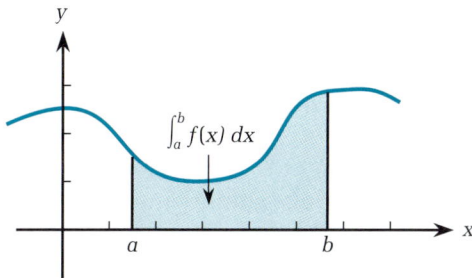

FIGURE 6.5.3

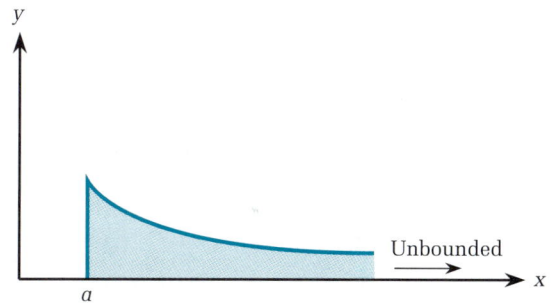

FIGURE 6.5.4

EXAMPLE 6   Assigning an area value to an unbounded region

Suppose we want to find the area of the region bounded by $f(x) = 1/x^2$, the x-axis, and the lines $x = 1$ and $x = b$. This area can be represented by the integral $\int_1^b (1/x^2)\,dx$, as shown in figure 6.5.5.

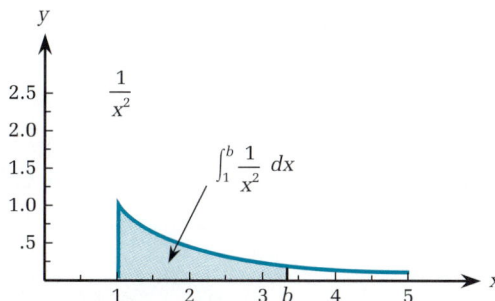

FIGURE 6.5.5

By the Fundamental Theorem, $\int_1^b \frac{1}{x^2}\,dx = \left.\frac{-1}{x}\right|_1^b = \left(\frac{-1}{b}\right) -$ $\left(\frac{-1}{1}\right) = \frac{-1}{b} + 1$. For instance, if $b = 5$, the area is $-(1/5) + 1 = 4/5 = 0.8$.

What if b is allowed to increase? As it does, so does the area. However, no matter how far b moves to the right, the area will never be larger than 1 because $1 - (1/b) < 1$ for $b > 0$. It will approach 1 and can be made as close to 1 as we want it to be. This is a limit concept and we have

$$\lim_{b\to\infty} \int_1^b \frac{1}{x^2}\,dx = 1$$

Thus, we can assign the value of 1 to the area of the region bounded by $f(x) = 1/x^2$, the x-axis, and for which $1 \le x$. ⬤

The value of $\lim_{b\to\infty} \int_1^b (1/x^2)\,dx$ is often represented by $\int_1^\infty (1/x^2)\,dx$, which is called an *improper integral*.

DEFINITION   Improper integrals, convergent and divergent integrals

For real numbers a, b and c, the following expressions are *improper integrals* and are defined by limits of definite integrals, provided those limits exist.

$$\int_a^\infty f(x)\,dx = \lim_{b\to\infty} \int_a^b f(x)\,dx$$

$$\int_{-\infty}^b f(x)\,dx = \lim_{a\to-\infty} \int_a^b f(x)\,dx$$

$$\int_{-\infty}^\infty f(x)\,dx = \int_{-\infty}^c f(x)\,dx + \int_c^\infty f(x)\,dx$$

If the limit does exist, the improper integral is said to be *convergent*.

If the limit does not exist, the improper integral is said to be *divergent*.

EXAMPLE 7   Evaluating an improper integral that converges

Evaluate $\int_1^\infty \frac{1}{x^3}\,dx$.

SOLUTION

$$\int_1^\infty \frac{1}{x^3}\,dx = \lim_{b\to\infty} \int_1^b \frac{1}{x^3}\,dx \text{ by the definition of the improper integral}$$

$$= \lim_{b\to\infty} -\frac{1}{2}x^{-2}\Big|_1^b \text{ by the Fundamental Theorem of Calculus}$$

$$= \lim_{b\to\infty}\left[-\frac{1}{2}b^{-2} - \left(-\frac{1}{2}\cdot 1^{-2}\right)\right] = \lim_{b\to\infty}\left(-\frac{1}{2}\cdot\frac{1}{b^2}\right) + \frac{1}{2}$$

As $b \to \infty$, the term $(1/2)\cdot(1/b^2) \to 0$, so the value of the improper integral converges to $1/2$.

EXAMPLE 8 Determining whether an improper integral does converge

Does $\displaystyle\int_1^\infty \frac{1}{\sqrt{x}}\,dx$ converge?

SOLUTION

$$\int_1^\infty \frac{1}{\sqrt{x}}\,dx = \lim_{b\to\infty}\int_1^b \frac{1}{\sqrt{x}}\,dx \text{ by the definition of the improper integral}$$

$$= \lim_{b\to\infty} 2\sqrt{x}\Big|_1^b \text{ by the Fundamental Theorem of Calculus}$$

$$= \lim_{b\to\infty} 2\sqrt{b} - 2\sqrt{1}$$

But this has no limit as $b \to \infty$ so the integral diverges.

An example of an application that involves an improper integral is money flowing in steadily. Thereby we have a *continuous* situation and we need an integral that goes on indefinitely. If we have a continuous flow of money, continuous compounding is used to discount the income to present values. Continuous money flows are sometimes referred to as *revenue streams*.

EXAMPLE 9 Present value of an infinite continuous money flow

Suppose a toll bridge generates money continuously. If that amount is A dollars per month and we use $i = 0.0075$ as the monthly discount of continuous compounding (9% per year divided by 12 and decimalized), then find the present value of the bridge over the indefinite future.

SOLUTION

The formula for present value is $\int_0^\infty A \cdot e^{-0.0075t}\, dt$, which is an improper integral.

$$\lim_{b \to \infty} \int_0^b A \cdot e^{-0.0075t}\, dt = \lim_{b \to \infty} \left(A \cdot \frac{e^{-0.0075t}}{-0.0075} \right)\Bigg|_0^b$$

$$= \lim_{b \to \infty} \left(\frac{A}{-0.0075 \cdot e^{0.0075b}} - \frac{A}{-0.0075 \cdot e^0} \right)$$

$$= 0 + \frac{A}{0.0075} = 133.3333\,A$$

PRACTICE EXERCISE 3

Find the present value of a revenue stream of $20 per month, being paid perpetually. Use an interest rate of 1% per month and continuous compounding.

In the last example, money flows constantly. When that is the case, the present value can be computed by annual payments or on a continuous basis. Whether the actual payment is done in a lump sum each month or as the money flows in does make a significant difference. Having the *use* of money is valuable. For just that reason, the commercial banks quote overnight rates of deposit for loans that literally are left on deposit for only a few hours.

When the payment varies over time, we need to be very careful in our analysis so we do not overstate (or understate) the present value. Use geometric series for lump sum payments. Use indefinite integrals for continuous revenue streams.

EXAMPLE 10   An application of an improper integral to determine book sales

Suppose the *rate* at which copies of the book, *A Π For All Seasons*, are selling is given by $500 \cdot e^{-0.02t}$ books a day, where t is in days. Estimate the total sales into the indefinite future.

SOLUTION

Total sales are given by the improper integral $\int_0^\infty 500 \cdot e^{-0.02t}\, dt$.

$$\lim_{b\to\infty} \int_0^b 500 \cdot e^{-0.02t} \, dt = \lim_{b\to\infty} \left(500 \cdot \frac{e^{-0.02t}}{-0.02} \right) \Big|_0^b$$

$$= \lim_{b\to\infty} \left(\frac{500}{-0.02 \cdot e^{0.02b}} - \frac{500}{-0.02 \cdot e^0} \right)$$

$$= 0 + \frac{500}{0.02} = 25{,}000$$

The last two examples are instances of the following general formula.

EXPONENTIAL FLOW RATE FORMULA

If P and k are arbitrary constants, then

(6.5.2) $$\int_0^\infty Pe^{-kt} \, dt = \frac{P}{k} \quad \text{provided} \quad k > 0$$

One interpretation of formula (6.5.2) is the next example of a perpetual money flow.

PRESENT VALUE OF CONTINUOUS MONEY FLOW IN PERPETUITY

Assume a continuous money flow of P dollars per time interval, a decimal interest rate of k for that time interval, and continuous compounding. Beginning at the present time and continuing indefinitely into the future

$$\boxed{Present\ Value = P/k}$$

The next example appears to be quite different, but nonetheless it uses the same improper integral.

EXAMPLE 11   Using an improper integral to estimate populations created by fish stocking

Fish in the Rainbow River do not reproduce. As a result, the state Game and Fish Division stock (to stock is to place in the river) 2000 eating-size fish a week in Rainbow River for sport fishers. The fish in each load are caught at an exponential rate, with the number uncaught after t weeks being $2000e^{-0.23t}$ (about half are caught in the first three weeks). Assume the river started with no fish in it. What is the eventual maximum fish population (that is, just after a stocking occurs)?

SOLUTION

The graph in figure 6.5.6 shows what is happening as the weeks go by. As we put ourselves on the bank of the river many weeks into the future, we see there are $2000e^{-0.23t}$ fish left from the 2000 stocked t weeks ago. This is true for each of the weeks. So the total is $\int_0^\infty 2000e^{-0.23t}\,dt$, which we have just seen is $P/k = 2000/0.23 = 8696$ fish.

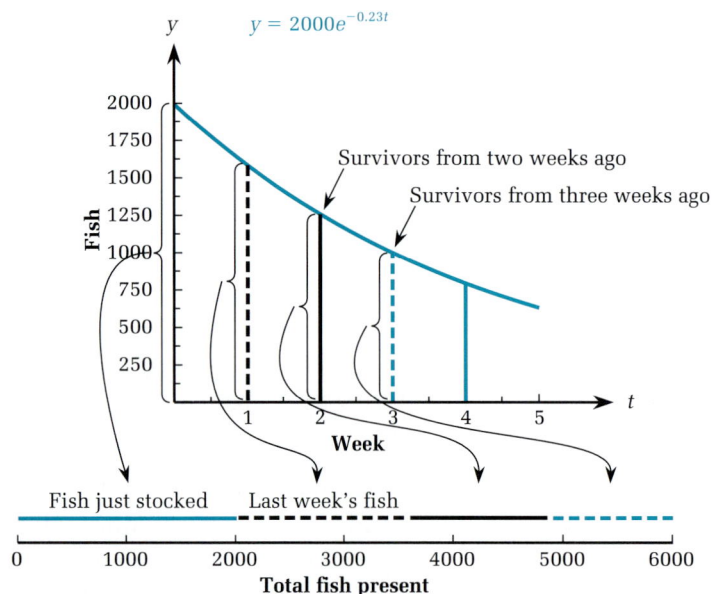

FIGURE 6.5.6

The next example considers an improper integral in which the values of x that are unbounded are negative values.

EXAMPLE 12   An improper integral with $-\infty$ as a lower limit

Evaluate $\int_{-\infty}^{2} e^{3x}\,dx$.

SOLUTION

$$\lim_{a \to -\infty} \int_a^2 e^{3x}\, dx = \lim_{a \to -\infty} \left(\frac{e^{3x}}{3} \bigg|_a^2 \right) = \lim_{a \to -\infty} \frac{e^6}{3} - \frac{e^{3a}}{3}$$

As a goes to $-\infty$, e^{3a} goes to 0 because $3a$ is becoming quite negative. Hence, the limit is $e^6/3 \approx 134.4763$.   ●

Sometimes we want the values of x to be unbounded both positively and negatively.

EXAMPLE 13   The normal curve, an improper integral of the form $\int_{-\infty}^{\infty} f(x)\, dx$

One of the most important improper integrals is

$$(1/\sqrt{2\pi}) \int_{-\infty}^{\infty} e^{-x^2/2}\, dx$$

The area under the entire improper integral is 1 and the graph of the function in the integral is called the *normal curve*. This curve, shown in figure 6.3.1, is used *extensively* in probability and statistics.   ●

6.5 PROBLEMS

Foundations

The problems of this section require the basic skills illustrated by the following:

1. Find the decimal equivalent of 75%.

2. Find $\int x^{-2}\, dx$.   3. Find $\int e^{2x}\, dx$.

4. Find $\int (1/x)\, dx$.

Exercises

Infinite Geometric Series
For problems 5–12, find the value of those infinite geometric series that converge.

5. $1 + \dfrac{1}{3} + \left(\dfrac{1}{3}\right)^2 + \left(\dfrac{1}{3}\right)^3 + \cdots + \left(\dfrac{1}{3}\right)^n + \cdots$

6. $1 + 0.75 + (0.75)^2 + (0.75)^3 + \cdots + (0.75)^n + \cdots$

7. $3 + 3(0.2) + 3(0.2)^2 + 3(0.2)^3 + \cdots + 3(0.2)^n + \cdots$

8. $5 + 5\left(\dfrac{1}{2}\right) + 5\left(\dfrac{1}{2}\right)^2 + 5\left(\dfrac{1}{2}\right)^3 + \cdots + 5\left(\dfrac{1}{2}\right)^n + \cdots$

9. $5\left(\dfrac{1}{2}\right)^3 + 5\left(\dfrac{1}{2}\right)^4 + 5\left(\dfrac{1}{2}\right)^5 + \cdots + 5\left(\dfrac{1}{2}\right)^n + \cdots$

10. $1 + \dfrac{3}{2} + \left(\dfrac{3}{2}\right)^2 + \left(\dfrac{3}{2}\right)^3 + \cdots + \left(\dfrac{3}{2}\right)^n + \cdots$

11. $5 + 5 \cdot 2 + 5 \cdot 2^2 + 5 \cdot 2^3 + \cdots + 5 \cdot 2^n + \cdots$

12. $2\left(-\dfrac{1}{2}\right) + 2\left(-\dfrac{1}{2}\right)^2 + 2\left(-\dfrac{1}{2}\right)^3 + \cdots + 2\left(-\dfrac{1}{2}\right)^n + \cdots$

13. **(Drug Levels)**   The drug nadolol (a beta-blocker used in cardiac therapy) persists in the blood serum according to the formula $A(t) = A_0 e^{-0.75t}$, where t is in days and A_0 is the maximum level from the dosage in one tablet. If a patient takes one tablet a day, what is the long-term level in the blood serum?

14. **(Multiplier Effect of Spending)**   A dam costing $500 million is proposed for a western river. Sociologists assume that 20% of the cost will flow into the local economy, and the local economy will respend locally 75% of whatever it takes in. Find the total effect of the dam on the local economy.

15. (Casino Gambling) A casino may have a slot machine that returns 90% of the money bet. If it accepts bets of one dollar, on average how many bets would be needed to reduce a player who started with $100 to having nothing?

16. (State Lotteries) A typical state lottery pays back an average of 50¢ of each dollar wagered. If a player buys 1000 $1 tickets and continually re-buys $1 tickets with his winnings, what is the average number of tickets he will buy before he is broke?

17. (Criminology, Repeat Offenders) The *rate of recidivism* is the percentage of criminals who will be rearrested and convicted for committing a crime after being found guilty of an earlier crime. If the recidivism rate is 72%, how many times would arrests and convictions be made on a group of 800 criminals? (Count the original arrest and conviction and assume these individuals live "forever" and have a tendency to relapse that is always 72%. Actually, the number of times a person is arrested affects their likelihood of being rearrested.)

Improper Integrals

For problems 18–35 decide whether these improper integrals converge or diverge. Evaluate the convergent ones.

18. $\int_2^\infty \frac{1}{x^2}\, dx$ **19.** $\int_3^\infty \frac{4}{x^2}\, dx$ **20.** $\int_2^\infty \frac{3}{\sqrt{x}}\, dx$

21. $\int_3^\infty \frac{2}{\sqrt{x}}\, dx$ **22.** $\int_1^\infty \frac{1}{x}\, dx$ **23.** $\int_2^\infty \frac{1}{x}\, dx$

24. $\int_2^\infty \frac{3}{x^4}\, dx$ **25.** $\int_1^\infty \frac{2}{x^3}\, dx$ **26.** $\int_2^\infty e^{-x}\, dx$

27. $\int_1^\infty e^{-2x}\, dx$ **28.** $\int_0^\infty e^{2x}\, dx$ **29.** $\int_2^\infty e^{1.5x}\, dx$

30. $\int_0^\infty e^{-x+2}\, dx$ **31.** $\int_2^\infty e^{-2x+1}\, dx$ **32.** $\int_{-\infty}^{-1} x^{-2}\, dx$

33. $\int_{-\infty}^{-1} x^{-3}\, dx$ **34.** $\int_{-\infty}^{-2} e^{2x}\, dx$ **35.** $\int_{-\infty}^{0} e^{2x}\, dx$

36. (Present Value) Suppose that starting at the present time for the first payment, $500 annual payments are to be made in perpetuity. Assume an annual interest rate of 6%.
(a) Find the present value with annual compounding of interest.
(b) Find the present value with continuous compounding of interest.

37. (Present Value) Suppose that starting one year from now, $500 annual payments are to be made in perpetuity. Assume an annual interest rate of 8%.
(a) Find the present value with annual compounding of interest.
(b) Find the present value with continuous compounding of interest.

38. (Present Value of a Revenue Stream) Find the present value of a continuous revenue stream of $5000 per year with continuous compounding of interest at an annual interest rate of 8%.

39. (Present Value of a Revenue Stream) Find the present value of a continuous revenue stream of $2000 per year with continuous compounding of interest at an annual interest rate of 12%.

40. (Fish Stocking) Black Lake, which is too warm for natural trout reproduction, is being stocked with 1500 trout a week. The number of trout surviving for t weeks from a particular batch is $A(t) = 1500\, e^{-0.1t}$. Estimate the maximum long-term population of trout in the lake (that is, just after a stocking takes place).

41. (Radioactive Buildup) Tritium, a naturally occurring radioactive form of hydrogen, is constantly being made by cosmic ray bombardment of Earth. It has a decay rate of 5.77% per year. If A tons of it are being made a year, what is the limiting value for the amount of tritium on Earth? (Tritium formerly could be used to date water samples, just as carbon-14 is used for organic materials. The tests of hydrogen bombs increased the levels of tritium by a factor of 100 and rendered it useless for dating.)

42. (Magazine Renewals) The rate at which subscriptions to *Ski Bum Tales* arrive is $1000e^{-0.06t}$ per week, where t is in weeks. Find the total number of subscriptions that will come in.

43. *(Oil-Well Production)*   In eastern New Mexico an oil well is producing at the rate of $300e^{-0.005t}$ barrels of oil per week, where t is in weeks. Find the total amount of oil the well would produce if it could run indefinitely.

Writing and Discussion Problem

44. Is Example 4 (about gambling) actually an infinite process? How could we justify using the result of that example?

Enrichment Problems

45. Justify formula (6.5.1) for the value of an infinite geometric series by letting $S_n = a + ax + ax^2 + ax^3 + \cdots + ax^n$.

    Multiply S_n term-by-term by x to get $x \cdot S_n$. Subtract S_n from $x \cdot S_n$ and show

    $$S_n = \frac{a(1 - x^{n+1})}{1 - x}.$$ Then find $\lim_{n \to \infty} S_n$. In taking the limit, explain the importance of the restriction that $|x| < 1$.

Find the values of the infinite geometric series in problems 46−50.

46. $1 + (0.3)^2 + (0.3)^4 + (0.3)^6 + \cdots$

47. $(0.3)^2 + (0.3)^4 + (0.3)^6 + (0.3)^8 + \cdots$

48. $(0.2)^2 + (0.2)^4 + (0.2)^6 + (0.2)^8 + \cdots$

49. $(0.1)^3 + (0.1)^6 + (0.1)^9 + \cdots$

50. $(0.2)^2 + (0.2)^5 + (0.2)^8 + \cdots$

51. *(Present Value of Variable Continuous Cash Flow)*   Suppose that money is flowing in at the rate of t dollars per year in year t. The present value of this perpetual cash flow, using a continuous compounding rate of 8%, is $\int_0^\infty t \cdot e^{-0.08t}\, dt$.

    Evaluate this integral. (Hints: From an integral table $\int xe^{ax}\, dx = \dfrac{xe^{ax}}{a} - \dfrac{e^{ax}}{a^2} + C$. We need the fact that $\lim_{x \to \infty} (x/e^{bx}) = 0$ for any $b > 0$. We do not have the technical tools to prove this convincingly, but it is true and may seem reasonable to us if we try several different values for x and b.)

52. Two appraisal firms, PrimePrice and LowBond, are evaluating a gold mine. Both agree the present rate of production is 90 ounces per week,

but they disagree on how fast the mine is being depleted. PrimePrice believes production in week t will be $90 \cdot e^{-0.02t}$, where t is measured in weeks, and LowBond's engineers assert the production will be $90 \cdot e^{-0.015t}$.

   (a) Find the *absolute total* production that each firm believes the mine can produce.
   (b) At a price of $400 an ounce, what is the difference between the value of gold in the two appraisals?
   (c) Suppose we assume a discount rate of 5.2%. Find the present values of both estimates. (Hint: Convert the 5.2% into a weekly decimalized interest rate.)

53. *(Perpetual Annuities)*   First, let's look at an example. Suppose an account earns 6%, but interest is compounded continuously. What amount would have to be invested in order to generate an annual annuity of $1000?

    To solve this, we first need to convert the 6% into the effective annual rate of interest. $e^{0.06} = 1.061836$, so the effective annual rate is 6.1836%. Thus, we would need (using the second method used in Example 2, where we subtract 1000 from the sum for the complete series) $(1000 + (1.061836)^{-1}1000 + (1.061836)^{-2}1000 + (1.061836)^{-3}1000 + \cdots) - 1000 =$

    $$\frac{1000}{1 - 0.9417645} - 1000 = \frac{1000}{0.058235} - 1000 =$$

    $17{,}171.67 - 1000 = \$16{,}171.67$.

    This is somewhat less than the amount needed when the interest rate is only paid annually, as would be expected because the effective interest rate is higher under continuous compounding.

    Show that the general formula for an annual annuity payment of A dollars per year, using continuous compounding of interest, is given by the following expression.

PERPETUAL ANNUITY UNDER CONTINUOUS COMPOUNDING

To generate a perpetual annual annuity payment of A dollars under an interest rate i (decimally expressed) and with continuous compounding requires an initial investment of $A[1/(e^i - 1)]$ dollars.

SOLUTIONS TO PRACTICE EXERCISES

1. Using an interest rate of 8%, annual compounding, and starting with a payment of $100, the present value of the lease is
$100 + (1.08)^{-1}100 + \cdots + (1.08)^{-n}100 + \cdots =$
$100 + 0.925926 \cdot 100 + \cdots + (0.925926)^n \cdot 100 =$
$100/(1 - 0.925926) = 100/0.074074 = \$1350.$

2. The total impact of an expenditure of $2 million if 55% of the money spent is going to be respent is $2 + 0.55 \cdot 2 + (0.55)^2 \cdot 2 + (0.55)^3 \cdot 2 + \cdots + (0.55)^n \cdot 2 + \cdots = 2/(1 - 0.55) = 2/0.45 = \$4.444444$ million.

3. The present value of a revenue stream of $20 each month paid perpetually, using an interest

rate of 1% per month and continuous compounding, is

$$\int_0^\infty 20e^{-0.01t} \, dt$$

$$= \lim_{b \to \infty} \int_0^b 20e^{-0.01t} \, dt$$

$$= \lim_{b \to \infty} \left(\frac{20e^{-0.01t}}{-0.01} \right) \Big|_0^b \quad \begin{array}{l} \text{by the Fundamental} \\ \text{Theorem of Calculus} \end{array}$$

$$= \lim_{b \to \infty} \left(\frac{20}{-0.01 \cdot e^{0.01b}} - \frac{20}{-0.01 \cdot e^0} \right)$$

$$= \frac{20}{0.01} = \$2000$$

CHAPTER 6 REVIEW

Define or discuss:

1. Integration by u-substitution

2. Change of limits for definite integrals when using u-substitution

3. Integration by parts

4. Reduction formulas

5. Capital value of an asset

6. Integral tables

7. Customizing an entry of an integral table

8. Numerical integration

9. Trapezoidal Rule

10. Simpson's Rule

11. Error estimates for Trapezoidal Rule

12. Geometric series (finite and infinite)

13. Sum of an infinite geometric series

14. Multiplier effect of spending

15. Convergence of a geometric series

16. Improper integrals

17. Convergence and divergence of an improper integral

REVIEW PROBLEMS FOR CHAPTER 6

Find the antiderivatives of problems 1–9.

1. $\int x(x^2 + 6)^8 \, dx$

2. $\int x(x^2 + 2)^4 \, dx$

3. $\int \frac{x^2}{2x^3 - 3} \, dx$

4. $\int x \cdot e^{x^2} \, dx$

5. $\int e^{x+1} \, dx$

6. $\int 2xe^{3x} \, dx$

7. $\int \frac{1}{x \ln x} \, dx$

8. $\int x \ln \sqrt{x} \, dx$

9. $\int \ln x^2 \, dx$

Evaluate the integrals in problems 10–12.

10. $\int_{-1}^1 xe^x \, dx$

11. $\int_1^2 \frac{2}{(3x - 1)^2} \, dx$

12. $\int_0^1 t\sqrt{t^2 + 1} \, dt$

13. Use the integral tables to find the following:

(a) $\int \dfrac{1}{x(3x-1)}\, dx$ (b) $\int \dfrac{1}{\sqrt{x^2-9}}\, dx$

(c) $\int \dfrac{5x}{\sqrt{5x+3}}\, dx$

14. Estimate $\int_0^2 \ln(x+1)\, dx$ by each of the following methods, using four equal subintervals for each:
(a) midpoint approximation
(b) Trapezoidal Rule
(c) Simpson's Rule

15. Estimate $\displaystyle\int_0^1 \dfrac{e^x}{x+1}\, dx$ by each of the following methods, using four subintervals for each:
(a) midpoint approximation
(b) Trapezoidal Rule
(c) Simpson's Rule

16. The rate of water use (measured in gallons per minute) is shown in the following table.

6 A.M.	7 A.M.	8 A.M.	9 A.M.	10 A.M.	11 A.M.	Noon
5	23	15	8	12	17	23

Estimate the total consumption of water from 6 A.M. to noon by
(a) the Trapezoidal Rule
(b) Simpson's Rule.

17. Find $(2/3) + (2/3)^2 + (2/3)^3 + (2/3)^4 + \cdots + (2/3)^n + \cdots$

18. Find $3(1/2)^2 + 3(1/2)^3 + 3(1/2)^4 + 3(1/2)^5 + \cdots + 3(1/2)^n + \cdots$

19. What is the present value of $200 paid at the end of every year and done so indefinitely? Use 5% as the annual interest rate.

20. Find the total economic impact on the local economy of a small town in Tennessee where a movie company spends $100,000 making a film. Assume 35% of every dollar spent in the town is respent in the town.

For problems 21–23, evaluate the improper integrals that exist.

21. $\displaystyle\int_0^\infty e^{x+1}\, dx$

22. $\displaystyle\int_0^\infty \dfrac{1}{x+1}\, dx$

23. $\displaystyle\int_2^\infty e^{-3x+1}\, dx$

24. Using an annual interest rate of 9%, find the capital value of an asset that is continuously generating income at the rate of $20,000 per year indefinitely into the future.

25. Using an annual interest rate of 12%, find the capital value of a parking lot that is continuously producing income at the rate of $2500 per month indefinitely into the future.

26. Suppose gold is being mined at the rate of $15e^{-0.003t}$ ounces per day.
(a) Find the amount of gold mined in 365 days.
(b) Find the day on which production has dropped down to 1 ounce.
(c) By evaluating an improper integral, estimate the *absolute total* amount of gold available in the mine.

27. Two appraisal firms, PrimePrice and LowBond, are evaluating an oil well. Both agree the present rate of production is 30 barrels per day, but they disagree on how fast the well is being depleted. PrimePrice believes production on day t will be $30 \cdot e^{-0.0015t}$, where t is measured in days, and LowBond's engineers assert the production will be $30 \cdot e^{-0.002t}$.
(a) Find the *absolute total* production that each firm believes the well can produce.
(b) At a price of $23 a barrel, what is the difference between the value of oil in the two appraisals?
(c) Using an annual discount rate of 7.2% and a year containing 360 days, find the present value of each estimated appraisal. (Hint: Convert the 7.2% into a daily, decimalized rate.)

CHAPTER 7

FUNCTIONS OF SEVERAL VARIABLES

WE HAVE ACHIEVED SIGNIFICANT RESULTS BY STUDYING FUNCTIONS OF A single variable. In this chapter we extend our ideas of derivatives and integrals to functions of two, or even more, variables. We want to find extreme values of such functions and determine a volume enclosed by their surfaces. Mathematical models using these ideas help us understand situations for which several variables are required.

The wind chill index is a well-known example of a function of several variables. When the winter weather is bad we are fascinated by, and sometimes positively delight in, knowing just how bad it is. For winds ranging in velocity from 4 to 45 miles per hour, the wind chill index (WCI) is

$$(7.0.1) \quad WCI = 91.4 - (0.475 + 0.304\sqrt{v} - 0.0203 \cdot v)(91.4 - F)$$

where v is the wind speed in miles per hour and F is the air temperature in degrees Fahrenheit. The WCI gives the temperature in degrees Fahrenheit at which calm air has the same chilling effect on exposed skin as the actual temperature and wind velocity.*

To illustrate the use of this formula, suppose the temperature is 20°F and the wind is blowing at 15 miles an hour. Then $WCI = 91.4 - (0.475 + 0.304\sqrt{15} - 0.0203 \cdot 15)(91.4 - 20) = -5°F$. Thus, the WCI is 25 degrees lower than the air temperature.

The WCI formula (7.0.1) is a function of two variables. If we wanted to consider all the sources of cold weather's discomfort, the WCI would have to be a function of several more variables. For instance, the humidity makes a difference in the effect of cold wind on bare skin. A WCI of $-5°$ F on the shores of Lake Michigan (with a relative humidity of 90%) is far more chilling than the same WCI in Las Vegas (with a relative humidity of 10%).

Models created for simulations, in business or in other topics, are usually functions of many variables. Consultants for the Italian city of Venice used more than a dozen variables to model the effect of a new sea barrier on protecting the city from flooding. Weather forecasters use hundreds of variables in an attempt to predict tomorrow's chance of rain. These are only a few of the many situations that need to consider and incorporate a great many variables.

In the following sections of this chapter, for the sake of simplicity of notation, most of our discussions and definitions are stated as they relate to functions of only two variables. Nonetheless, keep in mind that the ideas involved also relate to more than just two variables.

7.1 FUNCTIONS AND SURFACES OF TWO VARIABLES

Our understanding of functions of one variable has been greatly helped by graphs Similarly, a visual presentation of a function of two variables can be very helpful. Unfortunately, it is much more difficult to offer such visual presentations. Two-dimensional images of three-dimensional objects are difficult to produce. Cross-hatching, shifting perspective points-of-view, and color are just a few of the techniques used to provide an illusion of depth.

Figures 7.1.1 and 7.1.2 both represent the same surface, but from different perspectives. In figure 7.1.1 we can see there are two pits, but we might not recognize that there are two relative maximum values. In figure 7.1.2 the two peaks are clearly separated, and we might even guess which is higher, but we could miss the fact that there are two pits.

---

*For a history of this formula and a discussion of the topic, see W. Bosch and L. G. Cobb, "Windchill," *UMAP* 5:4 (1984), pp. 477–493.

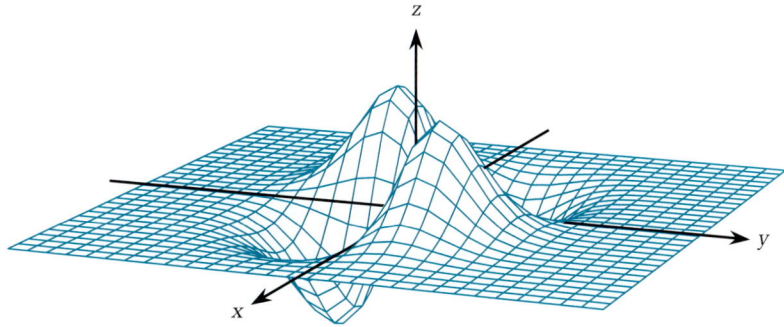

FIGURE 7.1.1

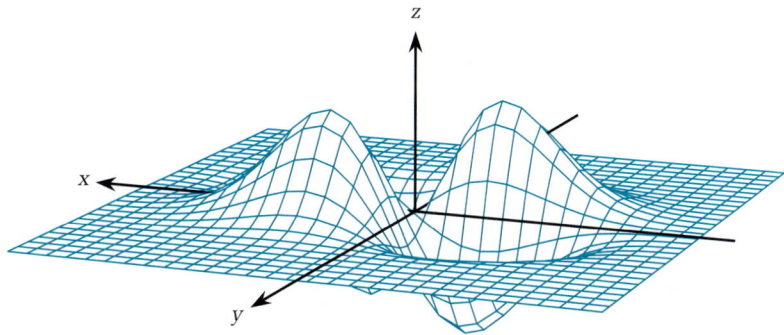

FIGURE 7.1.2

In this section we learn the terminology and the graphing conventions for functions of two variables. To actually create images of these surfaces most of us need a computer because recognizing and visualizing surfaces associated with functions of two variables are difficult skills. Ask any artist.

The Definition, Domain, and Range of $f(x, y)$

The concept of *function* given in Chapter 1 was expressed in terms of only one variable. Now, with more than one variable, we must generalize the concept of function.

First, let us recall what characterizes a function of one variable. It is a rule that produces ordered pairs of numbers. For each pair, the first number uniquely determines the second, often by an algebraic equation. The "uniquely determined" part means that if we know the first number of a pair (the "input" number), then there is only one possibility for the second number of the pair (the "output" number). The set of all possible first numbers is the *domain* and the set of all possible second numbers is the *range*.

The change required for a function of two variables is that the inputs must be pairs of ordered numbers.

DEFINITION

Function of two variables, domain, range

Suppose D is a subset of the set of all ordered pairs of real numbers and that for every pair (x, y) in D, there is associated a unique number $z = f(x, y)$. Then D is the *domain* of $f(x, y)$, which is a *function of two variables*. The *range* of $f(x, y)$ is the set of values of z produced by the function.

When we write $z = f(x, y)$ both x and y are *independent variables*, for they are unrelated. The output z is a *dependent variable*, for it depends on the particular values of x and y.

The following are examples of functions of two variables. The cost of gasoline for a car is a function of both the quantity *and* the kind of gasoline. The price of an airplane ticket is a function of both our class of travel *and* our destination. A monthly pepperoni pizza bill is a function of both the number *and* the sizes of pepperoni pizzas.

As with functions of one variable, the rule for functions of two variables is often given by an algebraic formula, as in Example 1. Part of the example asks us to find the domain. Unless we are told otherwise, we always assume the domain is the largest possible subset of all ordered pairs of real numbers.

EXAMPLE 1   Function of two variables, determining domain

Let $f(x, y) = \dfrac{xy}{x^2 + y^2}$. Find $f(1, 2)$ and $f(3, -1)$, and give the domain of $f(x, y)$.

SOLUTION

$$f(1, 2) = \frac{1 \cdot 2}{1^2 + 2^2} = \frac{2}{5} = 0.4$$

$$f(3, -1) = \frac{3 \cdot (-1)}{3^2 + (-1)^2} = \frac{-3}{10} = -0.3$$

The domain consists of all the ordered pairs of real numbers for which $f(x, y)$ is defined, that is, for which the denominator is not zero. The domain therefore consists of all pairs (x, y) except $(0, 0)$ because the denominator is zero only if both x and y are zero.

PRACTICE EXERCISE 1

Let $f(x, y) = xy + \sqrt{x}$. Find $f(1, 2)$ and $f(3, -2)$, and give the domain of $f(x, y)$.

Example 2 uses a function of two variables to describe an average value.

EXAMPLE 2   Modeling an average price in skiing

Hoot and Holler ski area charges \$26 for an all-day ticket and \$18 for an afternoon ticket. The all-day buyers always ski all day long. By 3 P.M., 150 all-day tickets and 50 afternoon tickets have been sold. At that time, what is the average price paid by the skiers on the hill?

SOLUTION

Let x stand for the number of all-day tickets sold and y stand for the number of afternoon tickets sold. The revenue is $26x + 18y$, so the average price is given by $A(x, y) = \dfrac{26x + 18y}{x + y}$. For $x = 150$ and $y = 50$, $A(150, 50) = \dfrac{26 \cdot 150 + 18 \cdot 50}{150 + 50} = 24$. ●

PRACTICE EXERCISE 2

Using the information of Example 2, find the average price paid by skiers on the hill on a day when Hoot and Holler sells 250 all-day tickets and 50 afternoon tickets.

Graphic Displays

A graphic display of a function is useful, but it is very difficult for us to produce one for a multivariate function. Nonetheless we need to learn about such graphic displays.

For a simple function of one variable, $y = f(x)$, we plot a few points and connect them in some manner. In the process, we try to make sure we plot all of the "interesting" points at which a maximum or minimum might occur. Doing so requires calculus, in particular checking for critical numbers for which the derivative is zero or does not exist.

Graphing functions of two variables also requires calculus to find "interesting points," but first we need to learn how to plot points.

The set $\{(x, y) : x$ and y any real numbers$\}$ corresponds to the set of points on a plane. For some, maybe even all, of these pairs of numbers there may be associated a third number. This third number can be thought of as being a height above or below the plane, which gives a function (the height) of two variables (the x- and y-coordinates).

DEFINITION

Surface

If $f(x, y)$ is a function of two variables, then the *surface* of $f(x, y)$ is defined as the set of points (x, y, z) that satisfy $z = f(x, y)$.

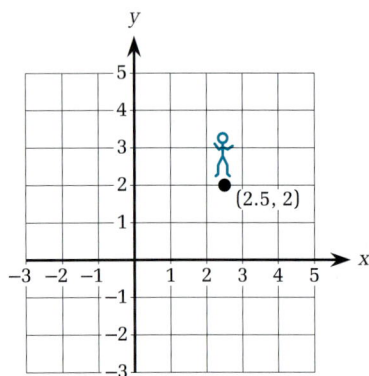

FIGURE 7.1.3

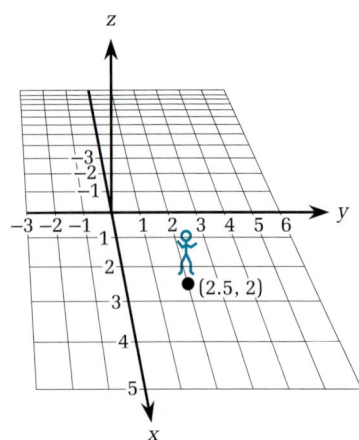

FIGURE 7.1.4

Right-Hand Coordinate System

First, let us establish axes for graphing three variables. Imagine we are walking on an immense checkerboard, as shown in figure 7.1.3. It is made of squares that are 1 unit on each side and go on forever.

We choose one point on the board as the origin. Then we leap up off the plane and locate a viewing point so that the board appears to be two-dimensional in a three-dimensional space. Such a view is shown in figure 7.1.4.

The x- and y-axes are on the plane. Positive x-values are measured toward us and positive y-values are toward the right.

A line perpendicular to the xy-plane is established as the z-axis, with positive values going up. This establishes a *right-hand coordinate system,* so named because if we imagine our right hand encircling the vertical axis, fingers going from the x-axis to the y-axis, then the thumb is pointing up along the vertical.

Think of returning to the xy-plane and walking around, looking straight up at the canvas of a circus tent, the rafters of a superdome, or the gilded ceiling of an opera house. If we stand at some point with particular x- and y-coordinates and look directly overhead, the ceiling at that point creates a "height" function. The collection of all those heights are the z-values of the function on whose x and y values we may stand. We may use the word "height" and think of a positive number, but actually the value of z may be positive or negative, as occurs in measuring the height and depth of dry land and the ocean floor relative to sea level.

A point on the surface is located by using an ordered triple of numbers (x, y, z). If we wish to emphasize that a triple represents a point, we may write $P(x, y, z)$ or something similar, just as we did for points on the graph of a function of one variable. The first coordinate is the x-coordinate, the second is the y-coordinate, and the last is the z-coordinate. Let us emphasize that the order of these three numbers is important.

Example 3 indicates by a dotted line the location of a given point above or below the plane. The dotted lines connect each of the given points with a point on the plane directly below or above the given point.

EXAMPLE 3   Locating a given point

Plot the points $A(2, 3, 2)$, $B(1, -2, 1)$, $C(-4, 4, 5)$, and $D(-5, -3, -2)$.

SOLUTION

See figure 7.1.5.

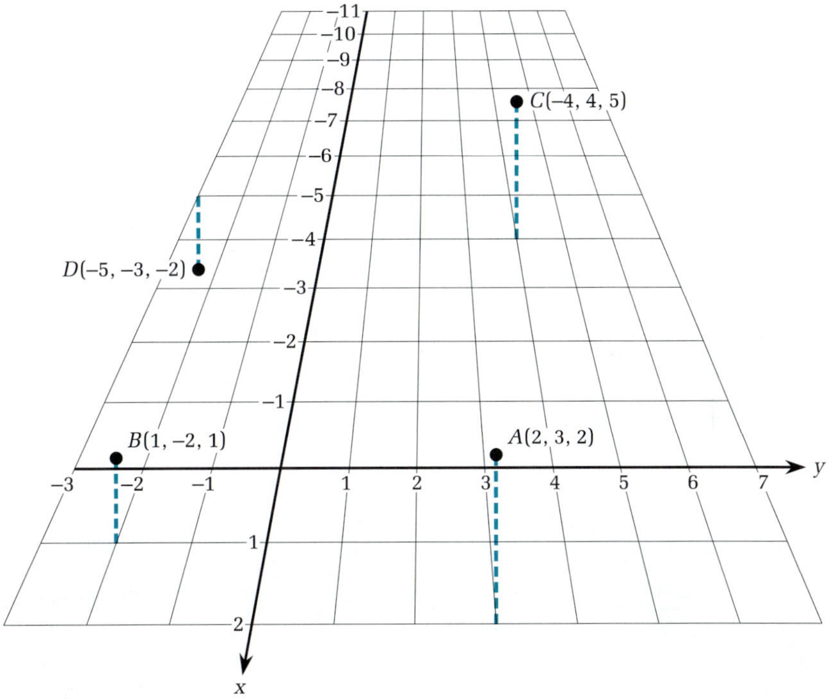

FIGURE 7.1.5

PRACTICE EXERCISE 3

Plot the points $A(2, 6, 4)$, $B(-2, 5, 2)$, and $C(0, -2, 3)$.

Level Curves and Other Traces

To help visualize a surface, we often find *traces*.

DEFINITION

Trace and level curve

The intersection of a plane with the surface $z = f(x, y)$ is a *trace* of the surface. If the plane is horizontal, then the trace is called a *level curve*.

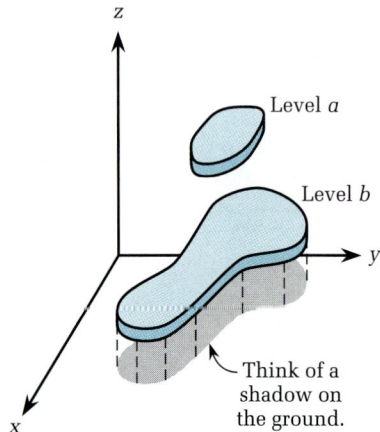

FIGURE 7.1.6

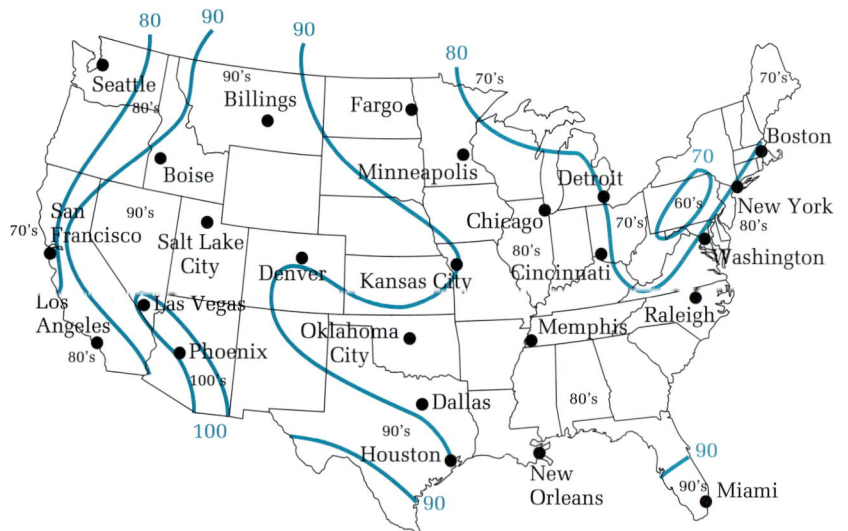

FIGURE 7.1.7

A topographic map shows elevations at various locations; these are horizontal traces, each parallel to the xy-plane, and so are level curves. The map shows the trace of each particular elevation collapsed onto a sheet of paper and labeled with the elevation of that level curve. This produces a family of level curves, and those locations at some specified elevation above sea level are shown as being connected by a curve. Figure 7.1.6 shows the topographic map and figure 7.1.7 shows the level curves spread out as they might appear on a hillside.

On a map showing forecast temperatures, locations that are forecast to have the same temperature are shown as being connected by curves called *isotherms*, as seen in figure 7.1.8. Boston, Washington D.C., and Cincinnati all have a forecast high of 80° F.

FIGURE 7.1.8

As an example of level curves in business, consider budget lines.

DEFINITION Budget lines

Suppose all of a person's income, I, is spent on only two goods, X and Y, with prices x_p and y_p, respectively. If x units of X and y units of Y are bought, $I = x_p \cdot x + y_p \cdot y$. Considering the income to be constant, y can be written as

$$y = \frac{I}{y_p} - \frac{x_p}{y_p} \cdot x$$

This is the equation of the *budget line.*

The slope of the budget line is $-x_p/y_p$, which is the ratio of the two prices.

Budget lines are examples of level curves for which the I is being graphed as a function of x and y, with a fixed x_p and y_p.

EXAMPLE 4 Budget lines

Ken is attending college and establishing his budget for the term. His favorite pizza, the Mt. Etna, costs \$15, and a movie costs \$5. Hence, if he buys x pizzas and goes to y movies, he would spend $I = 15x + 5y$ dollars. Find level curves for $I = 75$, $I = 100$, $I = 150$, and $I = 200$.

SOLUTION

To graph $15x + 5y = 75$, we can use any two points whose values satisfy the equation and connect them by a line. The x- and y-intercepts are easy to determine, so let us use them.

If $x = 0$, then $y = 15$, and if $y = 0$, then $x = 5$. Thus, $A(0, 15)$ and $B(5, 0)$ are two points on the budget line $I = 75$.

In graphing the budget line for $I = 100$, if $x = 0$, then $y = 20$. If $y = 0$, then $15x = 100$, which gives $x = 20/3 \simeq 6.6667$. While the points $C(0, 20)$ and $D(6.667, 0)$ determine the budget line $I = 100$, we (and Ken) realize $(6.6667, 0)$ is not a genuine choice because Ken must order a whole number of pizzas.

In a similar manner, for $I = 150$, we have $E(0, 30)$ and $F(10, 0)$, and for $I = 200$ we have $G(0, 40)$ and $H(13.3333, 0)$. See figure 7.1.9.

If we graph the surface of $f(x, y) = 15x + 5y$, we can see the location of two of these level curves, as shown in figure 7.1.10.

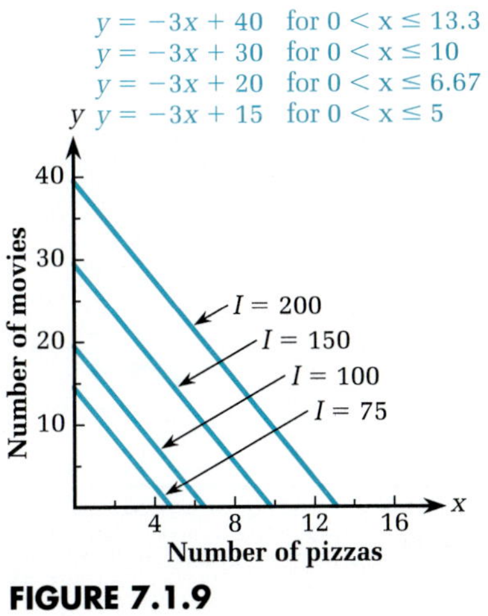

y = −3x + 40 for 0 < x ≤ 13.3
y = −3x + 30 for 0 < x ≤ 10
y = −3x + 20 for 0 < x ≤ 6.67
y = −3x + 15 for 0 < x ≤ 5
y
x
40
30
20
10
4
8
12
16
I = 200
I = 150
I = 100
I = 75
Number of movies
Number of pizzas

FIGURE 7.1.9

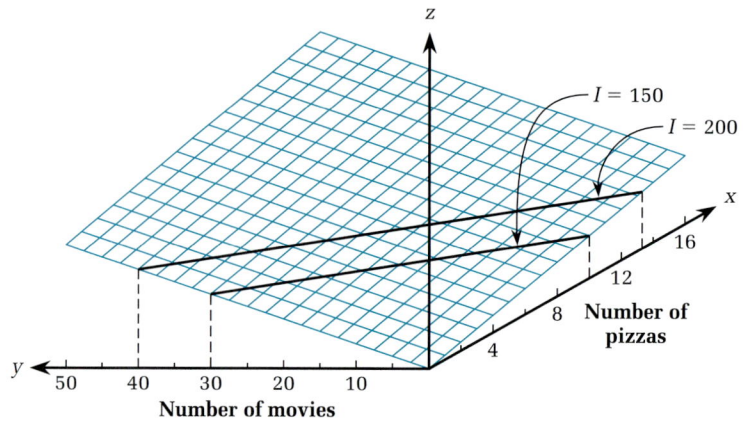

FIGURE 7.1.10

EXAMPLE 5   Level curves for yield management

An airline is offering tickets at two rates. For a 30-day advance pur-chase, nonrefundable ticket the price is $240. Otherwise the price is $540. Find level curves for total revenues of $50,000, $93,000, and $125,000.

SOLUTION

Let x represent the number of tickets at $240 and y represent the num-ber of tickets at $540. The total revenue is $R = 240x + 540y$. Solving for R being each of the level-curve values gives the lines shown in figure 7.1.11.

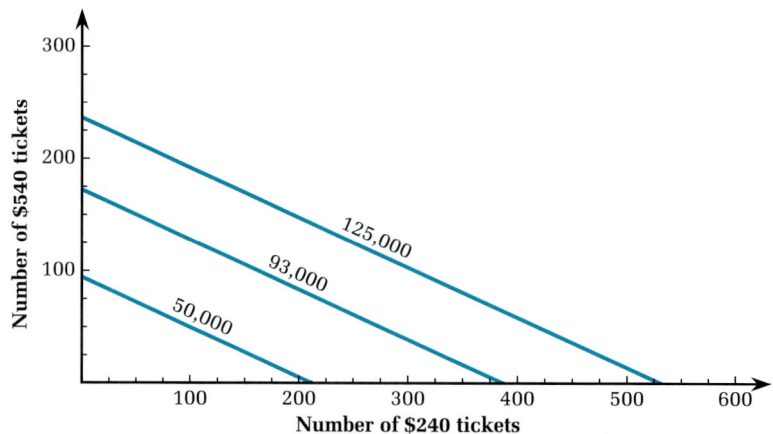

FIGURE 7.1.11

PRACTICE EXERCISE 4

A producer is thinking of taking a small group of college performers on a 10-week summer tour. Each performer will be paid the same amount of money. Fixed production costs are $5000 a week. Using as variables the number of performers and the salary to each, find level curves if the producer can raise $60,000, $70,000, or $80,000 for the entire season.

Cobb-Douglas Production Functions

Manufacturers realized at an early date that they could maintain production and reduce labor costs by installing machinery. Adam Smith (1723–1790) was a Scottish writer on moral and political economics, as well as being Commissioner of Customs and Salt Duties for Scotland. His famous book *Inquiry into the Nature and Causes of the Wealth of Nations,* published in 1776, opens with an example wherein 10 persons are said to be capable of producing 48,000 dressmaker's straight pins each day, whereas each person working alone and using only the simple tools of forge, hammer, and anvil may produce only a few such pins.

Such replacement of labor by capital continues today. During the 1970s and 1980s, tomato growers replaced an estimated 30,000 to 40,000 pickers by machinery that could strip the tomatoes from each plant in just seconds.

The first careful, quantitative analysis of this substitution of capital for labor was presented by Charles Cobb and Paul Douglas in 1928.*

DEFINITION

Cobb-Douglas model, isoquant, constant and increased returns to scale

The *Cobb-Douglas model* states that the amount of production is $P(x, y) = A \cdot x^b \cdot y^c$, where A is a constant, x is some measure of labor costs, and y is some measure of the capital expense (including machinery, materials, buildings, etc.).

A level curve for such a function is an *isoquant* (which means "same quantity").

If $b + c = 1$, then the situation is said to provide *constant returns to scale.*

If $b + c > 1$, then the situation is said to provide *increased returns to scale.*

In the examples of pins or tomatoes, the new machinery allows for an increase in production, sometimes called *scale economies.* For

*A history and general introduction of this topic is found in Robert Geitz, "The Cobb-Douglas Production Function," *UMAP* 2:2 (1981), pp. 73–96.

simplicity, let us assume constant returns to scale, so that $b + c = 1$ and we can write $P(x, y) = A \cdot x^b \cdot y^{1-b}$.

The model that Cobb and Douglas derived for the U.S. economy for the first quarter of the twentieth century was $P = 1.01x^{0.75}y^{0.25}$.

To find a level curve, we substitute a value for P and then solve for y as a function of x.

EXAMPLE 6 Finding a level curve for a Cobb-Douglas model

Find the level curve for $P = 100$ in the Cobb-Douglas model of the U.S. economy mentioned above.

SOLUTION

For $P = 1.01x^{0.75}y^{0.25}$, if $P = 100$ we have $100 = 1.01x^{0.75}y^{0.25}$.

We want y expressed as a function of x. To achieve this, first we isolate y as follows:

$$\frac{100}{1.01x^{0.75}} = y^{0.25} = y^{1/4}$$

To have y, rather than $y^{1/4}$, we need to raise each side to the 4th power:

$$\left(\frac{100}{1.01x^{3/4}}\right)^4 = y$$

Finally we have $y = \left(\frac{100}{1.01}\right)^4 x^{-3}$

$$= 96{,}098{,}034x^{-3}$$

A graph of this function is shown in figure 7.1.12.

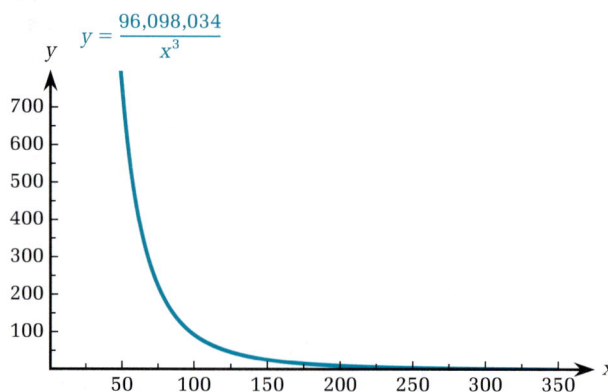

$$y = \frac{96{,}098{,}034}{x^3}$$

FIGURE 7.1.12

Cobb-Douglas models have been proposed that subdivide the economy into smaller sectors. In the enrichment problems are examples in which the labor force is subdivided into production and non-production workers for a number of industries.

Limits and Continuity

Functions of two variables can exhibit a wide variety of behavior as the values of x and y become close to a particular point $P(x_0, y_0)$. We do not have the time or space here to offer a worthwhile discussion of all the possibilities. Rather, we are simply going to assume the surfaces of the functions are smooth enough that they are connected and do not have any holes in them.

7.1 PROBLEMS

Foundations

The problems of this section require the basic skills illustrated by the following:

1. For what values of x is $\sqrt{x + 1} \geq 0$?

2. For what values of x and y is the denominator of $(x + 2)/(y - 1)$ not equal to 0?

3. If $f(x) = e^{-x^2}$, find $f(2)$.

Exercises

Evaluate the following functions at the indicated points and give the largest possible domain of each.

4. $f(x, y) = 3xy - y^2$ at $(0, 0)$, $(1, 2)$, $(-2, 3)$, $(-1, 2)$

5. $f(x, y) = -2xy + 2y^3$ at $(0, 0)$, $(2, 1)$, $(-3, 1)$, $(-1, 2)$

6. $f(x, y) = 2^x + 3y$ at $(0, 0)$, $(1, 3)$, $(-1, 2)$, $(-2, -1)$

7. $f(x, y) = 3^{-x} + y$ at $(0, 0)$, $(-1, 1)$, $(2, 1)$, $(-1, -2)$

8. $f(x, y) = y \cdot e^{-x}$ at $(0, 0)$, $(1, 2)$, $(-2, 1)$, $(-1, -1)$

9. $f(x, y) = y \cdot e^{-x^2}$ at $(0, 0)$, $(2, 1)$, $(1, 1)$, $(-2, 1)$

10. $f(x, y) = \sqrt{x + 1} + y$ at $(0, 0)$, $(3, 2)$, $(-1, 0)$, $(3, -1)$

11. $f(x, y) = \dfrac{x + 2}{y - 1}$ at $(0, 0)$, $(4, 3)$, $(5, 2)$, $(-1, -1)$

12. $f(x, y) = x^2 + \ln y$ at $(1, 1)$, $(-1, 1)$, $(2, e)$, $(2, 2)$

13. $f(x, y) = e^x - \ln y$ at $(1, 1)$, $(0, 1)$, $(1, e)$, $(-1, e)$

In problems 14 and 15, use formula (7.0.1).

14. Find the wind chill index for $40°$ F and a wind of 20 mph.

15. Find the wind chill index for $-10°$ F and a wind of 25 mph.

16. Plot the points $A(1, 0, 1/2)$, $B(-1, -1, 2)$, $C(3, 1, 3)$, and $D(1, -1, -2)$.

17. Plot the points $D(-1, 1, 0)$, $E(1, -2, 3)$, $F(0, -1, -2)$, and $G(1/2, 2, -1)$.

18. *(Price-Earnings Ratio)* The *price-earnings ratio*, $R(P, E) = P/E$, is defined as the ratio of the price per share of the stock divided by the earnings per share. Find the price-earnings ratio for a company earning \$1.27 per share and selling for \$20 per share.

19. *(Stock Yield)* The *yield* of a stock is the ratio of its dividends per share divided by the price per share. In 1989, Hershey Chocolate was paying a dividend of 70¢ per share and its stock sold for about \$30. Find the yield.

20. *(Future Value)* When p dollars are invested at an interest rate i, under continuous compounding, the money will grow to be worth $W(p, i) = p \cdot e^{it}$ in t years. Find $W(5000, 0.06)$.

21. *(Present Value)* When d dollars are due in t years and the discount rate (interest rate for continuous compounding) is i, the present value is $P(d, i) = d \cdot e^{-it}$. Find $P(5000, 0.06)$.

Graham Value for Stocks

Following the stock market crash of 1987, analysts once again sought a formula for the "true" value of a stock. One formula is the *Graham value,* first described in 1962. In that year the yield on AAA corporate bonds had been 4.4%.

The Graham value is calculated by letting E be the earnings per share, g be the long-term earnings growth estimate, and Y be the present yield on AAA corporate bonds. Then G, the Graham value, is

$$(7.1.1) \qquad G = \frac{E \cdot (8.5 + 2g) \cdot 4.4}{Y}$$

(g and Y are in percent, *not* in decimalized form.)

If the stock price is p, then the relative Graham value is $RG = G/p$. Analysts who use this method advise buying when $RG > 1$ (because the stock is undervalued) and selling when $RG < 1$ (because the stock would be overvalued). An account by Arbel, Carvell, and Postnicks asserted that many stocks returned to their Graham values following the 1987 crash.*

22. Use formula (7.1.1) to find the Graham value for a stock if the earnings are $2 a share, the long-term earnings growth estimate is 6%, and the yield on AAA corporate bonds is 9.5%.

23. Use formula (7.1.1) to find the Graham value for a stock if the earnings are $1.25 per share, the long-term earnings growth estimate is 14%, and the yield on AAA corporate bonds is 9.5%.

24. **(Ecology, Capture/Recapture)** When a population is difficult to count and yet mixes up well over time, such as fish in a lake, a common means of estimating the population is capture/recapture. The method involves catching a "bunch" of fish, marking each one, and then releasing them. After enough time has passed for thorough mixing of the marked and unmarked fish, another "bunch" is caught. The proportion of this bunch that is marked is assumed to be the same as the proportion of all the marked fish in the lake. Because we know how many fish are marked, we can estimate how many fish there are.

Let m be the number of marked fish, r be the number of marked fish in the second "bunch," t be the total number of fish in the second "bunch," and T be the estimated total number of fish in the lake. Our reasoning

about T yields $r/t = m/T$, which can be rewritten as $T = mt/r$.

Find T if 100 fish were caught and marked the first time, and then 75 were caught the second time, of which 10 were marked.

25. **(Forestry, Biology)** The height (in meters) of a lodgepole pine (*Pinus contorta*) can be estimated by $H = 3.7 + 0.08D - 3.2W$, where D is the stump diameter and W is the average width of growth rings, both measured in millimeters.† H is measured in meters. Find the height of a tree with a diameter of 210 millimeters and an average growth ring width of 0.7 millimeters.

26. Figure 7.1.13 shows a map of isotherms (curves connecting points of equal temperature).
    (a) Which of the following cities have the same forecast temperature as Denver: Boise, Houston, or Phoenix?
    (b) Which of the following cities have the same forecast temperature as Los Angeles: Phoenix, Dallas, or New Orleans?

27. **(Budget Curves)** At the start of the month, Pat realizes that the total she can spend on frozen pizzas is $T = np$ where n is the number of pizzas and p is the price of each. The possible prices for p are from $2 to $10. Find level curves for $T = \$25$, $T = \$50$, $T = \$75$, and $T = \$100$.

28. **(Budget Curves)** At the start of the month, Marie realizes that the total she can spend on delivered pizzas is $T = n(2 + p)$, where n is the number of pizzas, p is the price of the pizza ($4 \le p \le 12$), and there is a $2 delivery charge. Find level curves for $T = \$20$, $T = \$40$, $T = \$60$, and $T = \$90$.

29. **(Cobb-Douglas)** Suppose a Cobb-Douglas function is $P(x, y) = 1.2 \cdot x^{0.6}y^{0.4}$.
    (a) Find $P(25, 14)$.
    (b) Find a level curve for $P = 100$, giving y as a function of x.

30. **(Cobb-Douglas)** Suppose a Cobb-Douglas function is $P(x, y) = 23 \cdot x^{0.2} \cdot y^{0.8}$.
    (a) Find $P(20, 15)$.
    (b) Find a level curve for $P = 200$, giving y as a function of x.

31. The *DuBois and DuBois formula* for the surface area of a human being is $S \simeq 0.202W^{0.425}H^{0.725}$,

---

*Arbel, Carvell, and Postnicks, "The Smart Crash of Oct 19," *Harvard Business Review,* May–June 1988, pp. 124–136.

†P. Koch and J. Schlieter, "Spiral Grain and Annual Ring Width in Natural Unthinned Stands of Pine in North America," USDA Forest Service, Intermountain Research Station Paper INT-499, September 1991.

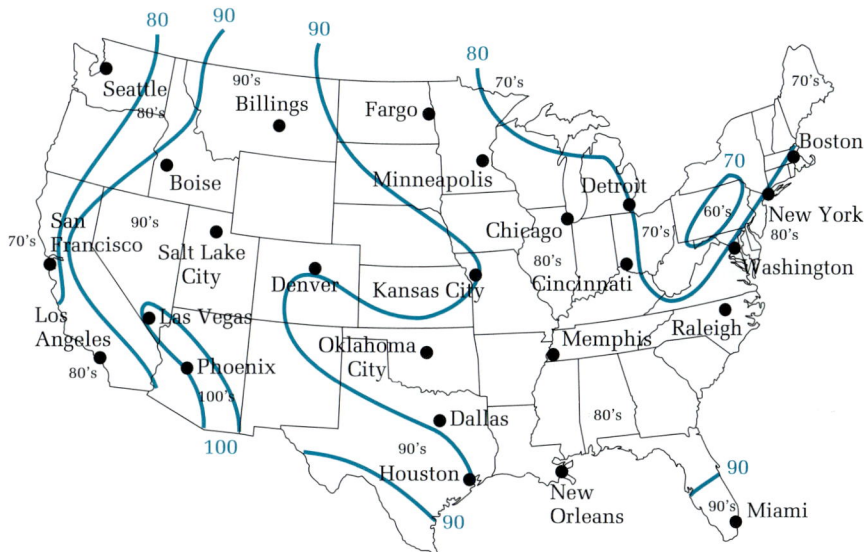

FIGURE 7.1.13

where S is measured in square meters, W is the human's weight in kilograms, and H is the human's height in meters.

(a) Find the surface area of a person who weighs 70 kilograms and is 1.6 meters tall.

(b) Find a level curve for $S = 1.8$, giving W as a function of H.

Writing and Discussion Problems

32. *(Perspective in Art)* Leonardo da Vinci (1452–1519) was one of the first to discover how to draw with perspective. Discuss *vanishing points* in drawing. How can you draw railroad tracks so they do not just look like a waterfall?

33. *(Readability, Flesch Index)* A widely used measure of the "readability" of material was developed in 1941 by Rudolph Flesch. The formula uses passage samples 100 words in length. If w is the average number of syllables in these 100-word samples and s is the average number of words in the sentences of the samples, then the *Flesch Index* is given by

$F = 206.84 - 0.85w - 1.02s.$

Find the *Flesch Index* for several passages and see how your findings compare with the psychologists' findings that scores in the 0–30 range indicate the material is quite difficult to understand, in the 35–45 range indicate material appropriate for college graduates, and scores in the 90–100 range indicate the

material is very easy to understand.[*] This index was applied to decide whether people who were arrested were able to understand their rights by reading a printed document.

Enrichment Problems

34. Find the domain of $f(x, y) = \sqrt{x + y}$.

35. *(Cobb-Douglas)* Let a Cobb-Douglas function have "constant returns to scale" and so be of the form $P(x, y) = A \cdot x^b \cdot y^{1-b}$.

(a) Show that if we double both x and y, then we will double P.

(b) Show that for any constant m, when we replace x by mx and y by my, then the production is also multiplied by m.

36. *(Cobb-Douglas)* Consider a Cobb-Douglas production function $P = b \cdot (L_1)^\alpha (L_2)^\beta (K)^\gamma$, where L_1 measures production workers, L_2 measures nonproduction workers, and K measures capital. Table 7.1.1 comes from work done by Moroney.[†] He gives values for various U.S. industry sectors. These figures are actual figures and hence the equation $\alpha + \beta + \gamma = 1$ is not always satisfied. One of the questions Moroney discussed was whether there were "constant returns to scale."

[*]G. H. Gugjonsson, "The 'Notice to Detained Persons,' PACE Codes, and Reading Ease," *Applied Cognitive Psychology* 5 (1991), pp. 89–95.

[†]J. Moroney, "Cobb-Douglas Production Functions and Returns to Scale in U.S. Manufacturing Industry," *Western Economic Journal* 6 (1967), pp. 39–51.

TABLE 7.1.1

INDUSTRY	α	β	γ
Food and Beverage	0.43882	0.07610	0.55529
Apparel	0.43705	0.47654	0.12762
Furniture	0.80154	0.10263	0.20458
Lumber	0.50391	0.14533	0.39170
Transportation Equipt.	0.74885	0.04103	0.23353
Paper and Pulp	0.36666	0.19732	0.42054

Find the production value for each industry if we invest 10 in L_1, 2 in L_2, and 20 in K.

SOLUTIONS TO PRACTICE EXERCISES

1. For $f(x, y) = xy + \sqrt{x}$, $f(1, 2) = 1 \cdot 2 + \sqrt{1} = 3$ and $f(3, -2) = 3(-2) + \sqrt{3} \approx -4.2679$. The domain of f consists of all $x \geq 0$, whereas y can be any real number.

2. With 250 all-day tickets at $26 and 50 afternoon tickets at $18, the average price paid per skier is $(26 \cdot 250 + 18 \cdot 50)/(250 + 50) = \$24.67$.

3. The points are plotted in figure 7.1.14.

4. During the summer, the fixed costs will be $50,000 (10 weeks at $5000 each week). Subtracting this from the given gross amounts and letting x be the number of performers and y be the salary of each, the level curves will be as shown in figure 7.1.15.

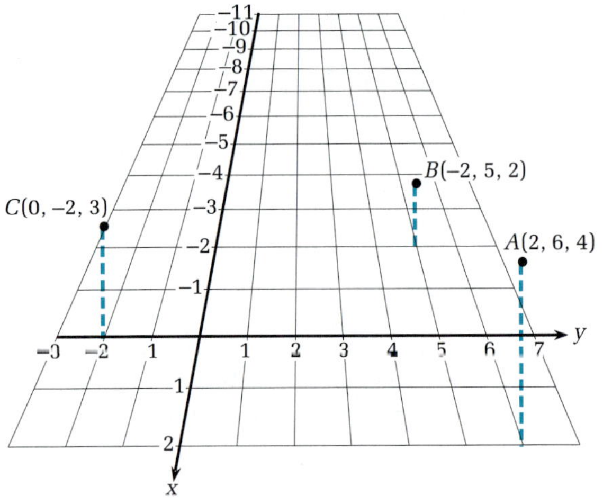

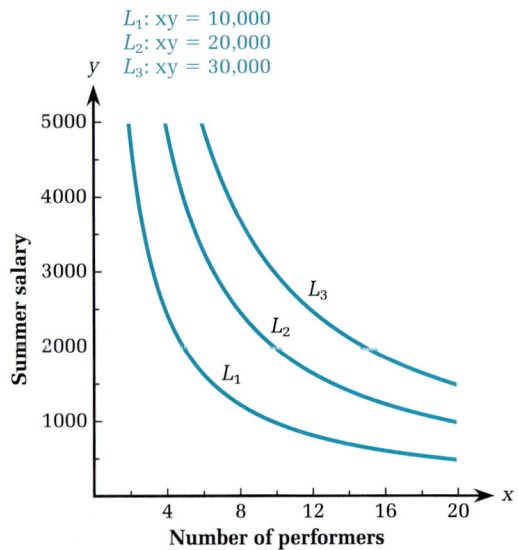

FIGURE 7.1.14

L_1: xy = 10,000
L_2: xy = 20,000
L_3: xy = 30,000

FIGURE 7.1.15

| 7.2 | ## PARTIAL DERIVATIVES |

Our study of the derivative of a function of one variable began with a focus on the rate of change of that function. This led us to the derivative and provided an important interpretation of it. Taking a similar approach for a function of two variables, there is a rate of change of the function with respect to each one of the variables. For instance, returning to our example of the wind chill index, suppose the temperature is 10°F and the wind is 15 miles per hour. We can ask, "At what rate would the wind chill index be dropping per each lower degree? Or per each increase in mile per hour of wind velocity?"

To find these answers we need a concept that quantifies the rate of change with respect to just one variable at a time. This brings us to the idea of a *partial derivative.*

Vertical Traces

Suppose we return to our imaginary walk about the *xy*-plane. Look straight up at a surface over us. Usually its height changes as we walk about, and we want to describe those changes. Although we might wander about in all sorts of rambling, random paths, what we actually will use are paths parallel to the *x*-axis or the *y*-axis.

For a path parallel to the *x*-axis, the points of the *xy*-plane all have the same *y*-values. That is, on such a path *y* is a constant and any change in the value of the function depends on changes in just the one variable *x*.

The surface given by $z = \dfrac{-5y}{x^2 + 2y^2 + 1}$ for $|x| \leq 4$, $|y| \leq 4$ is shown in figure 7.2.1.

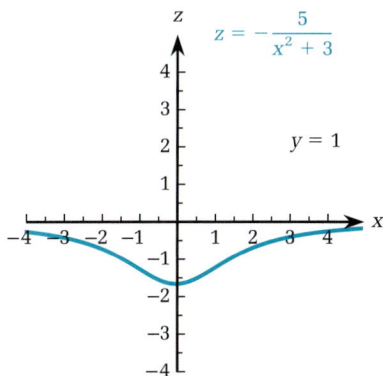

FIGURE 7.2.2

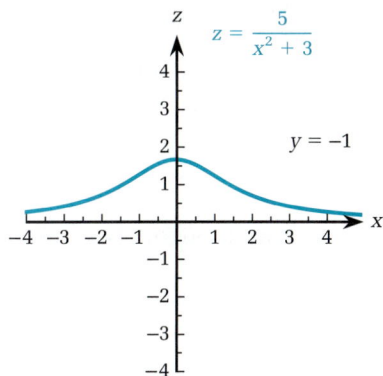

FIGURE 7.2.1

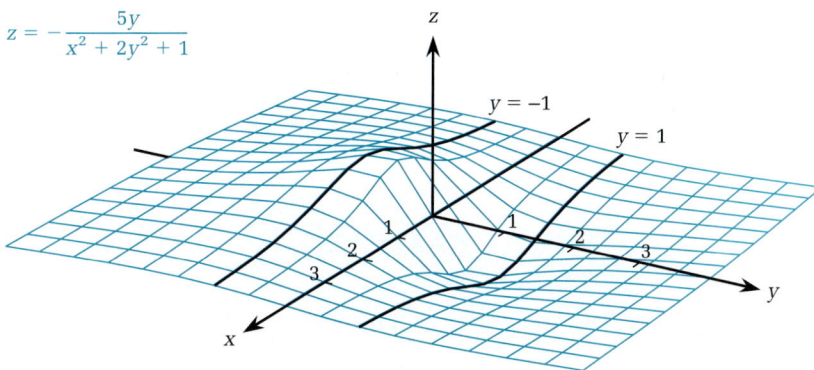

FIGURE 7.2.3

If we fix $y = 1$, then we have $z = \dfrac{-5 \cdot 1}{x^2 + 2 \cdot 1^2 + 1} = \dfrac{-5}{x^2 + 3}$, which is a function of *x* alone. The trace of this function is shown on the surface in figure 7.2.1. The graph of the vertical trace along $y = 1$ is shown in figure 7.2.2. Also shown on figure 7.2.1 is the vertical trace produced by $y = -1$. This is given by $z = \dfrac{(-5)(-1)}{x^2 + 2(-1)^2 + 1} = \dfrac{5}{x^2 + 3}$, shown in figure 7.2.3.

We need to realize that in figures 7.2.2 and 7.2.3 x increases from left to right, whereas in figure 7.2.1 x increases from back to front.

We can hold either x or y fixed for a vertical trace. Figure 7.2.4 shows the surface patterns for both $y = -1$ and $x = 1$.

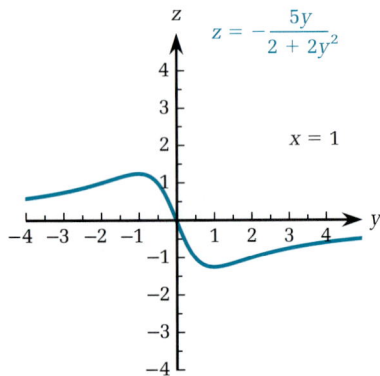

$$z = -\frac{5y}{x^2 + 2y^2 + 1}$$

$y = -1$

$x = 1$
$y = -1$
$z = 1.25$

FIGURE 7.2.4

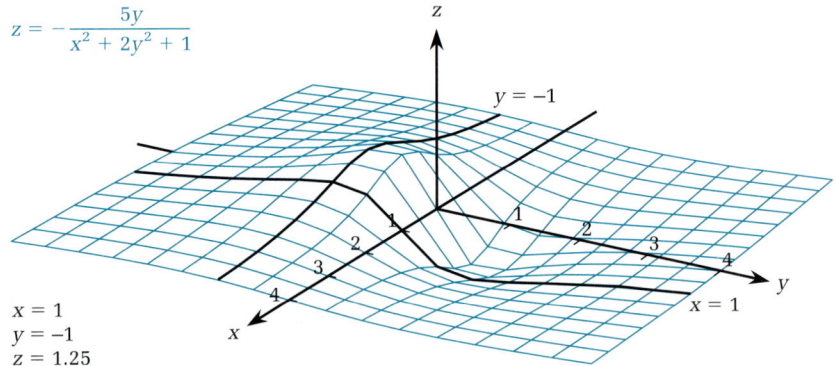

$$z = -\frac{5y}{2 + 2y^2}$$

$x = 1$

FIGURE 7.2.5

The equation of the trace for $x = 1$ is given by letting x be 1 in $z = \dfrac{-5y}{x^2 + 2y^2 + 1}$. Doing so gives $z = \dfrac{-5y}{1^2 + 2y^2 + 1} = \dfrac{-5y}{2 + 2y^2}$, as shown in figure 7.2.5.

Rates of Change

Now we find the rate of change for a multivariant function.

EXAMPLE 1   Determining rates of change

Suppose $z = f(x, y) = \dfrac{-5y}{x^2 + 2y^2 + 1}$. A sketch of this surface for $-4 \le y \le 4$, $-4 \le x \le 4$ is shown in figure 7.2.4.

Suppose we stand at $x = 1$ and $y = -1$. Then the height of the surface above you is

$$z = \frac{-5(-1)}{1^2 + 2(-1)^2 + 1} = \frac{5}{1 + 2 + 1} = \frac{5}{4} = 1.25$$

If we walk on a path parallel to the x- or y-axis and look up overhead, how is the surface changing?

SOLUTION

Suppose we walk parallel to the x-axis with y held at $y = -1$. All along this particular path in the xy-plane, we may substitute $y = -1$

in the formula for z to get $f(x, -1) = [-5(-1)]/[x^2 + 2(-1)^2 + 1] = 5/(x^2 + 3)$, which is graphed in figure 7.2.3.

From our study of derivatives, we know that as x increases, $\dfrac{5}{x^2 + 3}$ increases at the rate of $\dfrac{d}{dx}\left(\dfrac{5}{x^2 + 3}\right) = \dfrac{-5(2x)}{(x^2 + 3)^2} = \dfrac{-10x}{(x^2 + 3)^2}$.

Hence, at $x = 1$, a tangent line to the surface in the x-direction has slope $-10/4^2 = -5/8 \approx -0.625$.

We can confirm this by estimating the slope from figure 7.2.3.

Consider what happens if we start at the point $(1, -1)$ and begin walking along a line parallel to the y-axis. Then x is being held fixed at $x = 1$ and y is allowed to vary.

So $f(x, y) = \dfrac{-5y}{x^2 + 2y^2 + 1}$ would become $f(1, y) = \dfrac{-5y}{1^2 + 2y^2 + 1} = \dfrac{-5y}{2y^2 + 2}$.

As y changes, the rate of change of $f(1, y)$ is

$$\frac{d}{dy}\left(\frac{-5y}{2y^2 + 2}\right) = \frac{(2y^2 + 2)(-5) - (-5y)(4y)}{(2y^2 + 2)^2}$$

$$= \frac{-10y^2 - 10 + 20y^2}{(2y^2 + 2)^2} = \frac{10y^2 - 10}{(2y^2 + 2)^2}$$

At the point $(1, -1)$, this derivative is $\dfrac{10(-1)^2 - 10}{(2(-1)^2 + 2)^2} = 0$. Hence the tangent line is horizontal. Compare this result with figure 7.2.5.

Using this information, if we start at $x = 1$ and $y = -1$ and begin to move along the path $x = 1$, then the height of the surface is decreasing. This is true whether we have y increasing or decreasing. However, if we start at $x = 1$ and $y = -1$ and move along the path $y = -1$, then if x increases, the height decreases, whereas if x decreases the height increases.   ●

PRACTICE EXERCISE 1

For $f(x, y) = -3x^2 - 5(y - 2)^2 + 12$, we have $f(2, 1) = -5$.

Find the rate of change of $f(x, y)$ at $(2, 1)$ in the x-direction (that is, holding y fixed at 1).

Partial Derivatives

The process of proceeding parallel to the x-axis or the y-axis and determining the rate of change of the surface is one interpretation for the following definition.

DEFINITION

Partial derivatives

Suppose $f(x, y)$ is a function of two variables.

The *partial derivative with respect to x* is

$$\lim_{h \to 0} \frac{f(x + h, y) - f(x, y)}{h}$$ and is written as $f_x(x, y)$ or $\frac{\partial}{\partial x} f(x, y)$.

The *partial derivative with respect to y* is

$$\lim_{h \to 0} \frac{f(x, y + h) - f(x, y)}{h}$$ and is written as $f_y(x, y)$ or $\frac{\partial}{\partial y} f(x, y)$.

Each of these partial derivatives exists only if the appropriate limit exists.

Notation. We may abbreviate $f_x(x, y)$ and $f_y(x, y)$ as f_x and f_y, respectively. Furthermore, $\frac{\partial}{\partial x} f(x, y)$ and $\frac{\partial}{\partial y} f(x, y)$ may be written as $\frac{\partial f}{\partial x}$ and $\frac{\partial f}{\partial y}$, respectively.

WARNING

We cannot use the prime notation, such as f', because it does not indicate the variable for which we are taking the derivative.

When we first are learning to take partial derivatives, it is helpful to emphasize the variable in question. We can do this by using a separate color, or by putting a frame around the variable.

For example, if we are considering $f(x, y) = x^2y + 2x - 3y + 1$ to be a function of x, then we can write it as

$$f(\boxed{x}, y) = \boxed{x}^2y + 2 \boxed{x} - 3y + 1.$$

This framing emphasizes the fact that for the purpose of finding the derivative, all the unframed symbols are constants.

Finding a partial derivative is no more difficult than finding a derivative for a function of just one variable. What is difficult is learning to look at an expression such as $x^2y + 2x - 3y + 1$ and being able to think of an apparent variable as being momentarily fixed as a constant.

Suppose we think of that (momentarily fixed) y as being replaced by a. For $\boxed{x}^2a + 2\boxed{x} - 3a + 1$, to find the derivative with respect to $\boxed{x}$, we are able to determine the derivative to be $2\boxed{x} \cdot a + 2$, which is rewritten as $2a\boxed{x} + 2$. We recognize the a is merely some (unspecified) constant and hence the derivative of $\boxed{x}^2a$ is $2\boxed{x}a$. The derivative of $-3a$ is 0, just as the derivative of 1 is 0. All these derivatives are being found *with respect to x*.

In finding the derivative of $\boxed{x}^2y + 2\boxed{x} - 3y + 1$ with respect to x, we need to regard y as a constant, like a above. Doing so, we find the derivative with respect to x to be $2\boxed{x}y + 2$.

The derivative with respect to y is found by considering x held constant. For the same function, $f(x, y) = x^2y + 2x - 3y + 1$, we can indicate this is a function of y by writing

$$f(x, \boxed{y}) = x^2\boxed{y} + 2x - 3\boxed{y} + 1.$$

EXAMPLE 2   Finding partial derivatives

Suppose $f(x, y) = x^2y + 2x - 3y + 1$. Find f_x and f_y.

SOLUTION

In the following dual approach, we emphasize that in taking the derivative with respect to a variable, everything else in the expression is a constant.

Finding f_x Write $f(x, y)$ as a function of x.

These are all constants.

$$f(\boxed{x}, y) = \boxed{x}^2y + 2\boxed{x} - 3y + 1$$

Take the derivative with respect to x.
$f_x(\boxed{x}, y) = 2xy + 2$

Finding f_y Write $f(x, y)$ as a function of y.

These are all constants.

$$f(x, \boxed{y}) = x^2\boxed{y} + 2x - 3\boxed{y} + 1$$

Take the derivative with respect to y.
$f_y(x, \boxed{y}) = x^2 - 3$

EXAMPLE 3   Finding partial derivatives

Let $f(x, y) = x^2y^3 + e^{xy} + 3x^2 - \sqrt{y}$. Find $f_x(x, y)$ and $f_y(x, y)$.

SOLUTION

To find f_x, we first write f as $f(\boxed{x}, y) = \boxed{x}^2y^3 + e^{\boxed{x}y} + 3\boxed{x}^2 - \sqrt{y}$ to emphasize we are taking the partial derivative *with respect to x*.

Then $f_x = 2\boxed{x}y^3 + e^{\boxed{x}y} \cdot y + 3 \cdot 2\boxed{x} = 2xy^3 + e^{xy} \cdot y + 6x$.

To find f_y, write f as $f(x, \boxed{y}) = x^2\boxed{y}^3 + e^{x\boxed{y}} + 3x^2 - \sqrt{\boxed{y}}$. Then

$$f_y = x^2 \cdot 3\boxed{y}^2 + e^{x\boxed{y}} \cdot x - \frac{1}{2\sqrt{\boxed{y}}} = 3x^2y^2 + e^{xy} \cdot x - \frac{1}{2\sqrt{y}}. \quad \bullet$$

There are several notations for the numerical evaluation of a partial derivative. The simplest form is writing $f_x(1, 2)$ to indicate that we have evaluated the partial derivative with respect to x at the point $(1, 2)$. Alternative forms are $\frac{\partial}{\partial x}f\big|_{(1, 2)}$ or $\frac{\partial}{\partial x}f(x, y)\big|_{(x, y) = (1, 2)}$.

EXAMPLE 4   Evaluating partial derivatives

Suppose $f(x, y) = y \cdot e^{-x^2}$. Find $f_x(1, 2.5)$ and $f_y(1, 2.5)$.

SOLUTION

Starting with $f(\boxed{x}, y) = y \cdot e^{-\boxed{x}^2}$, the partial derivative with respect to x is $f_x = (y \cdot e^{-\boxed{x}^2}) \cdot (-2\boxed{x}) = -2xye^{-x^2}$. Evaluating f_x at $(1, 2.5)$ gives $f_x(1, 2.5) = -2 \cdot 1 \cdot 2.5 \cdot e^{-1^2} = -5e^{-1} = -1.8394$.

Starting from $f(x, \boxed{y}) = \boxed{y}e^{-x^2}$, the partial derivative with respect to y is $f_y = e^{-x^2}$. (There is no y left to frame.) Evaluating f_y at $(1, 2.5)$ gives $f_y(1, 2.5) = e^{-1} = 0.3679$. $\quad \bullet$

PRACTICE EXERCISE 2

For $f(x, y) = y^2 \cdot e^{-x}$, find $f_x(2, 1)$ and $f_y(2, 1)$.

Applications of Partial Derivatives

In Section 7.1 we discussed production functions that depended on quantity of labor (represented by x) and capital (represented by y). Recall that the term *marginal* relates to the derivative.

DEFINITION   Marginal productivity of labor and capital

Suppose $P(x, y)$ is a production function of labor, x, and capital, y.

The *marginal productivity of labor* is $P_x(x, y)$.
The *marginal productivity of capital* is $P_y(x, y)$.

We can interpret the marginal productivity of labor to mean that if labor is increased by one unit (in whatever units the labor is being measured), the production is increased (approximately) by $P_x(x, y)$. We similarly interpret marginal productivity of capital.

EXAMPLE 5   Finding marginal productivities using a Cobb-Douglas production function

For the production function $P(x, y) = 25x^{0.2}y^{0.8}$, x represents hours per week of labor and y represents thousands of dollars.

(a) Find the production using 300 hours per week and $40,000.

(b) Find the marginal productivity of labor and capital at (300, 40).

SOLUTION

$P(300, 40) = 25 \cdot 300^{0.2} \cdot 40^{0.8} \approx 1496.2779$.
    Finding the appropriate partial derivatives we have

$$P_x(\boxed{x}, y) = 25 \cdot 0.2\boxed{x}^{0.2-1}y^{0.8} = 5 \cdot x^{-0.8}y^{0.8} = 5(y/x)^{0.8},$$

which is marginal productivity of labor. Similarly,

$$P_y(x, \boxed{y}) = 25 \cdot x^{0.2} \cdot 0.8 \cdot \boxed{y}^{0.8-1} = 20(x/y)^{0.2},$$

which is the marginal productivity of capital.
    Having the marginal productivities, we can substitute the values of 300 and 40 for x and y. Thus $P_x(300, 40) = 5(40/300)^{0.8} \approx 0.9975$ and $P_y(300, 40) = 20(300/40)^{0.2} \approx 29.9256$.   ●

Suppose we have some particular amount of money being invested in labor and in capital. Would production increase more rapidly if additional resources were invested in labor or in capital expenditures? The change in productivity can be estimated by determining the products in the following definition.

DEFINITION   Changes in productivity with respect to labor and capital

Suppose $P(x, y)$ is a production function, with x invested in labor and y invested in capital.

The *change in productivity with respect to labor* is the product $(P_x \cdot (\text{change in } x))$.

The *change in productivity with respect to capital* is the product $(P_y \cdot (\text{change in } y))$.

EXAMPLE 6   Determining changes in productivity with respect to labor and capital

Suppose a production function is $P(x, y) = 300 \cdot x^{0.6} \cdot y^{0.9}$, where x represents millions of dollars spent on labor and y represents millions of dollars spent on capital equipment. If at present, $x = 5$ and $y = 1$, then would production increase more by spending an additional $1 million on labor or $200,000 on capital equipment?

SOLUTION

We need to determine $P_x \cdot (1)$ and $P_y \cdot (0.2)$ for $x = 5$ and $y = 1$. (The partial derivative with respect to capital is multiplied by 0.2 because y is in millions and 200,000 is 0.2 million.) First, find the partial derivatives:

$$P_x = 300 \cdot (0.6x^{0.6-1}y^{0.9}) = \left(\frac{180}{x^{0.4}}\right)y^{0.9}$$

$$P_y = 300 \cdot x^{0.6}(0.9y^{0.9-1}) = \left(\frac{270}{y^{0.1}}\right)x^{0.6}$$

Hence the evaluations of the partial derivatives are $P_x(5, 1) = (180/5^{0.4})1^{0.9} \simeq 94.56$ and $P_y(5, 1) = (270/1^{0.1})5^{0.6} \simeq 709.16$. Finally, the changes in productivity are $P_x \cdot (1) \simeq 94.56 \cdot 1 = 94.56$ and $P_y \cdot (0.2) \simeq 709.16 \cdot (0.2) = 141.83$.

In conclusion, production would increase more by spending $200,000 on capital equipment than by spending $1 million on labor.

Returning to our example of the wind chill index, we now can decide whether a change in temperature of one degree Fahrenheit or a change in wind velocity of one mile per hour would cause the WCI to change by a greater amount.

EXAMPLE 7   Determining changes in the wind chill index

We use the formula for wind chill index that is given by $WCI = 91.4 - (0.475 + 0.304\sqrt{v} - 0.0203v)(91.4 - F)$ to find $\partial(WCI)/\partial v$ and $\partial(WCI)/\partial F$ if F is 10° and v is 15 miles per hour.

SOLUTION

Keep in mind that v and F are independent variables; therefore, in taking $\partial(WCI)/\partial v$ we regard F as a constant.

$$\frac{\partial(WCI)}{\partial v} = 0 - \left(0.304\frac{1}{2\sqrt{v}} - 0.0203\right) \cdot (91.4 - F).$$

Evaluating this gives

$$\frac{\partial(WCI)}{\partial v}\bigg|_{(v,\ F)=(15,\ 10)} = -\left(0.304\frac{1}{2\sqrt{15}} - 0.0203\right) \cdot (91.4 - 10)$$

$$= -(0.0189)(81.4) = -1.54.$$

To find the partial derivative with respect to F we regard the velocity, v, as a constant.

$$\frac{\partial(WCI)}{\partial F} = 0 - (0.475 + 0.304\sqrt{v} - 0.0203v)(0 - 1)$$

Evaluating this gives $\partial(WCI)/\partial F|_{(v,\ F)=(15,\ 10)} = 0.475 + 0.304\sqrt{15} - 0.0203 \cdot 15 = 0.475 + 1.177 - 0.304 = 1.35$.

These partial derivatives can be interpreted as follows: If the wind velocity increases by one mile per hour, then the WCI decreases by 1.54. If the temperature increases by one degree, then the WCI increases by 1.35. Notice that our formulas correctly reflect the fact that increasing temperatures cause us to feel warmer, but increasing wind velocities cause us to feel colder.

Competitive and Complementary Products

Products may compete or complement each other in the market. If two products compete, then the growth of one accompanies the decline of the other. For example, if the number of passengers remains constant, then the growth of ridership on one airline must be accompanied by a loss of customers on another airline.

Products that complement each other will gain and/or lose market share in tandem. For instance, the sales of video tapes is increased if consumers spend more on VCR machines. As a result, tape manufacturers want increased sales of VCR machines whether they make the machines or not.

Consider how changing prices affect two products. If the two products are competitive, an increase in the price of either one will result in a decrease in that product's sales and thus an increase in the sales of the other. If products are complementary, an increase in price of either one will result in a decrease in sales of both.

DEFINITION

Competitive and complementary products

Suppose two products have prices a and b, and demand functions $A(a, b)$ and $B(a, b)$, respectively.

The products *compete* if both $\dfrac{\partial A}{\partial b} > 0$ and $\dfrac{\partial B}{\partial a} > 0$

(each increases sales if other raises price)

The products *complement* if both $\dfrac{\partial A}{\partial b} < 0$ and $\dfrac{\partial B}{\partial a} < 0$

(each loses sales if other raises price)

EXAMPLE 8   Determining competitiveness of products

Suppose demand for T-shirts is given by $T(x, y) = 2000 - 100x + 5y$ and for sweatshirts the demand is $S(x, y) = 100 - 4y + 3x$, where x is the price of a T-shirt and y is the price of a sweatshirt. Show that these products are competitive.

SOLUTION

$\partial T/\partial y = 5$, and $\partial S/\partial x = 3$. Because both partial derivatives are positive, the products are competitive.

In Example 8 we have a pair of products that compete. It is possible that any pair of products may compete with each other, complement each other, or do neither.

Second-Order Partial Derivatives

Just as we did for derivatives of functions of one variable, we can treat a partial derivative as a function and take its derivative. Because we are dealing with two variables, there are more choices concerning which second derivative we determine.

DEFINITION

Second-order partial derivatives

Suppose $z = f(x, y)$ is a function of two variables. The four possible second-order partial derivatives are (using our various notations):

$$f_{xx} = (f_x)_x = \frac{\partial}{\partial x}(f_x) = \frac{\partial}{\partial x}\left(\frac{\partial}{\partial x}f\right) \qquad f_{xy} = (f_x)_y = \frac{\partial}{\partial y}(f_x) = \frac{\partial}{\partial y}\left(\frac{\partial}{\partial x}f\right)$$

$$f_{yy} = (f_y)_y = \frac{\partial}{\partial y}(f_y) = \frac{\partial}{\partial y}\left(\frac{\partial}{\partial y}f\right) \qquad f_{yx} = (f_y)_x = \frac{\partial}{\partial x}(f_y) = \frac{\partial}{\partial x}\left(\frac{\partial}{\partial y}f\right)$$

WARNING

Notice that the subscript notation is read from left to right, but the "∂" notation works right to left. Hence, f_{xy} means we first find the partial derivative with respect to x, and using that function we take the partial derivative with respect to y. This is the same as $\frac{\partial}{\partial y}\left(\frac{\partial}{\partial x}f\right)$.

EXAMPLE 9   Finding second-order partial derivatives

For $f(x, y) = x^3y - 3xy^2$, find f_{xx}, f_{xy}, f_{yy}, and f_{yx}.

SOLUTION

As a reminder of which variable we are taking the partial derivative with respect to, we are going to frame the variable. (Upon completing this section, we will stop framing.)

For $f(\boxed{x}, y) = \boxed{x}^3y - 3\boxed{x}y^2$ we find $f_x = 3x^2y - 3y^2$. Then
$$f_{xx} = (3\boxed{x}^2y - 3y^2)_x = 3 \cdot 2xy + 0 = 6xy \text{ and}$$

$$f_{xy} = (3x^2\boxed{y} - 3\boxed{y}^2)_y = 3x^2 \cdot 1 - 3 \cdot 2y = 3x^2 - 6y.$$

Starting with $f(x, \boxed{y}) = x^3\boxed{y} - 3x\boxed{y}^2$, we find
$$f_y = x^3 - 3x \cdot 2y = x^3 - 6xy.$$

Then $f_{yx} = (\boxed{x}^3 - 6\boxed{x}y)_x = 3x^2 - 6y$
and $f_{yy} = (x^3 - 6x\boxed{y})_y = -6x.$

PRACTICE EXERCISE 3

Let $f(x, y) = x^2/y$. Find f_x, f_y, f_{xx}, f_{xy}, f_{yy}, and f_{yx}.

For many (but not all) functions, $f_{xy} = f_{yx}$. This can often be used as a check when we are finding second-order partial derivatives.

7.2 PROBLEMS

Foundations

The problems of this section require the basic skills illustrated by the following:

1. Find $\dfrac{d}{dx}e^{2x}$.

2. Find $\dfrac{d}{dy}e^{2y}$.

3. Find $\dfrac{d}{dy}5y^5$.

4. Find $\dfrac{d}{dy}4y^{0.5}$.

Exercises

For the functions in problems 5–14, find both f_x and f_y.

5. $f(x, y) = x^3y^2$

6. $f(x, y) = x^2y^4$

7. $f(x, y) = x^{-1}y^2$

8. $f(x, y) = x^3y^{-2}$

9. $f(x, y) = e^xy^3$

10. $f(x, y) = e^{2x}y^5$

11. $f(x, y) = e^{xy} + \ln(x)$

12. $f(x, y) = ye^{xy}$

13. $f(x, y) = \dfrac{x}{x^2 + y^2}$

14. $f(x, y) = \dfrac{xy}{x^2 - y^3}$

In problems 15 and 16, evaluate the partial derivatives as indicated.

15. For $f(x, y) = x^2e^{-y}$, find $f_x(1, 2)$ and $f_y(1, 2)$.

16. For $f(x, y) = \dfrac{\ln x}{x^2 + y^2}$, find $f_x(2, 2)$ and $f_y(2, 2)$.

17. Suppose a production function is given by $P(x, y) = 10x^{0.1}y^{0.9}$, where x is a measure of labor and y is a measure of capital.
   (a) Find the marginal productivity of labor.
   (b) Find the marginal productivity of capital.

For the production functions in problems 18–21 at the specified levels of labor and capital funding, determine whether a greater immediate increase in production occurs by increasing the labor or capital expenditures by the amount indicated.

18. $P(x, y) = x^2 + xy + y$.
   (a) $x = 3$, $y = 5$, change in x of 0.3 or a change in y of 0.2.
   (b) $x = 1$, $y = 5$, change in x of 0.1 or a change in y of 0.2.

19. $P(x, y) = \dfrac{x^2 + y^2}{xy}$.
   (a) $x = 1$, $y = 2$, change in x of 0.1 or a change in y of 0.1.
   (b) $x = 2$, $y = -3$, change in x of 0.2 or a change in y of 0.1.

20. $P(x, y) = 300x^{0.6}y^{0.9}$.
   (a) $x = 5$, $y = 7$, change in x of 0.2 or a change in y of 0.1.
   (b) $x = 2$, $y = 3$, change in x or a change in y of 0.1.

21. $P(x, y) = 200\left(\dfrac{x}{y}\right)$.
   (a) $x = 1$, $y = 2$, change in x of 0.1 or a change in y of 0.1.
   (b) $x = -3$, $y = 2$, change in x of 0.2 or a change in y of 0.1.

22. The Heads-Up Co. manufactures tennis balls and knows its weekly demand function is given by $q = 20,000p^{-2}x^{0.3}$, where p is the price in dollars for a can of three balls and x is the number of thousands of dollars spent weekly on advertising. Presently it is pricing its cans at $5 and is spending $2000 on advertising.
   (a) Determine the function, R, that gives the weekly revenue.
   (b) Find $R_x(5, 2)$ and $R_p(5, 2)$.
   (c) Using $R_x(5, 2)$ and $R_p(5, 2)$, decide whether revenues would be increased more by spending an additional $1000 on weekly advertising or by reducing the price of the cans to $4.50.

23. The following are pairs of demand functions for two products. In each case determine whether the products compete, complement, or do nei-

ther. In each case, a and b are the prices of the two products.
   (a) $A(a, b) = 200 - 3a + b$
       $B(a, b) = 100 - 2b + 5a$
   (b) $A(a, b) = 200 - 2a - b$
       $B(a, b) = 100 - a + 6b$
   (c) $A(a, b) = 50 - 5a - b$
       $B(a, b) = 25 - a - 2b$
   (d) $A(a, b) = (a/b)^{0.3}$
       $B(a, b) = (b/a)^{0.7}$

24. Suppose the revenue recorded by a ski area can be approximated by $R(x, y) = 4000x - xy^4 - 2x^3$, where x is the price of the lift ticket and y is the price of gasoline needed by potential skiers to reach the ski area. Both x and y are in dollars.
   (a) Find $R_x(25, 1)$.
   (b) Find $R_x(30, 1)$.

25. *(Alternate Energy, Windpower)* The power generated by a windmill is given by $P(r, v) = k \cdot \pi r^2 v^3$, where r is the radius of the windmill and v is the velocity of the wind.
   (a) Find P_r and P_v.
   (b) Evaluate $P_r(20, 15)$ and $P_v(20, 15)$.
   (c) Le Moulin Neuf is a windmill of radius 20 feet. The wind is blowing at 15 miles per hour. Find the relative rate of increase in power at this wind velocity as the velocity increases.

For problems 26–31, find all four second-order partial derivatives.

26. $f(x, y) = 3x^2y^3 - 2x^5y$

27. $f(x, y) = x^3y^{-1} + x^2y^3$

28. $f(x, y) = \dfrac{x + 1}{y}$

29. $f(x, y) = \dfrac{y}{x^2}$

30. $f(x, y) = e^{3x+2y}$

31. $f(x, y) = \dfrac{e^{2y}}{e^{3x}}$

Enrichment Problems

In this section we have discussed partial derivatives for functions of two variables. We can extend this concept to any number of variables. When we decide to take a partial derivative with respect to one of the variables, we hold all the others at fixed values.

For example, suppose we have $f(x, y, z) = x^2y + y^2z + z^2x$, a function of three variables. Then $f_x = 2xy + z^2$ when y and z are treated as constants.

For problems 32–35 find f_x, f_y, and f_z.

32. $f(x, y, z) = 3x^2y^2 + xz - 2y^2$

33. $f(x, y, z) = e^{xyz} + x^2 - y^2$

34. $f(x, y, z) = z \ln(xy)$

35. $f(x, y, z) = \dfrac{xyz}{x^2 + y^2 - z^2}$

36. *(Poiseuille's Law)* Investigating blood flow, and using tubes 0.03 to 0.14 millimeters in diameter, Jean Poiseuille (1797–1869) discovered the volume flowing in narrow tubes could be approximated by $Q = k[(D^4p)/L]$, where D is the diameter and L is the length of the tube, and p is the pressure difference between the ends of the tube.

(a) Find $\partial Q/\partial L$

(b) Find $\partial Q/\partial D$.

37. Show that any Cobb-Douglas function of the form $Q = Cx^\alpha y^{1-\alpha}$ satisfies the equation $x(\partial Q/\partial x) + y(\partial Q/\partial y) = Q$.

38. For a rectangular box of volume V, we have $V = lwh$, where l is the length, w is the width, and h is the height of the box. Solving this equation for l, we have $l = V/(wh)$, so that $\partial l/\partial w = -V/(w^2h)$.

(a) Solve the equation for w, in terms of V, l, and h, and find $\partial w/\partial h$.

(b) Solve the equation for h in terms of V, l, and w, and find $\partial h/\partial l$.

(c) Show that $\dfrac{\partial l}{\partial w} \cdot \dfrac{\partial w}{\partial h} \cdot \dfrac{\partial h}{\partial l} = -1$.

[The moral is: We cannot treat the partial derivatives on the left side in (c) as fractions.]

SOLUTIONS TO PRACTICE EXERCISES

1. Given $f(x, y) = -3x^2 - 5(y - 2)^2 + 12$. For $y = 1$ we have $f(x, 1) = -3x^2 - 5(-1)^2 + 12 = -3x^2 + 7$. The rate of change of $f(x, 1)$ at $(2, 1)$ in the x-direction (that is, holding y fixed at 1) is the derivative of $f(x, 1)$. This derivative is $-6x$, which at $(2, 1)$ is -12.

2. For $f(x, y) = y^2 \cdot e^{-x}, f_x(\boxed{x}, y) = y^2(-1)e^{-\boxed{x}}$ and $f_y(x, \boxed{y}) = 2\boxed{y} \cdot e^{-x}$. Hence $f_x(2, 1) = 1^2(-1)e^{-2} \approx -0.1353$ and $f_y(2, 1) = 2 \cdot 1 \cdot e^{-2} \approx 0.2707$.

3. To find $f_x, f_y, f_{xx}, f_{xy}, f_{yy}$, and f_{yx} for $f(x, y) = x^2/y$, we have $f_x = 2x/y$ and $f_y = -x^2/y^2$. Hence, $f_{xx} = 2/y, f_{xy} = -2x/y^2, f_{yy} = 2x^2/y^3$, and $f_{yx} = -2x/y^2$.

7.3 OPTIMIZATION

We are now ready to study extreme values for functions of two variables. If we can locate high and low points on a surface, then we can optimize the quantity represented by that surface. For example, we would like to find a maximum for a profit and a minimum for a cost.

This section contains many examples of surfaces. Because they were all created by a computer graphics package, it may seem reasonable to ask ourselves, "Why not have the graphics package find all the maximum and minimum values for us?"

The truth is, computers are not wise. They are fast and smart, but we have to supply the wisdom. In consequence, because there are an infinite number of values to check, we cannot simply tell the computer to look among all the x-values and all the y-values to find those that produce optimal z-values. We need to be more specific concerning the values the computer should be checking.

Furthermore, even if we are supplied with a two-dimensional portrayal of a surface, we cannot determine the values for z because of perspective. Without additional visual or arithmetic clues, we cannot tell just how high or low a surface point is.

Figure 7.3.1 shows the surface of $f(x, y) = \dfrac{-5y}{x^2 + 2y^2 + 1}$. We have investigated this function in Sections 7.1 and 7.2. We found that $f(1, -1) = 1.25$, but that does not appear to be the maximum value for $f(x, y)$. We will now find the maximum.

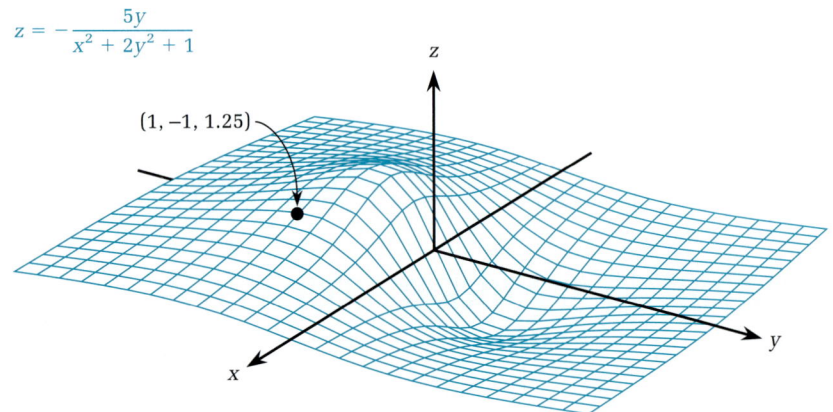

FIGURE 7.3.1

Relative Extremes

In Section 3.2 we defined relative extreme values of a function of one variable. We need to extend that definition to functions of two variables.

DEFINITION

Relative maximum and relative minimum for a function of two variables

Suppose $f(x, y)$ is a function of two variables. The value $f(a, b)$ is:

(a) a *relative maximum* if $f(a, b) \geq f(x, y)$ for all points (x, y) in some circular region around (a, b)

(b) a *relative minimum* if $f(a, b) \leq f(x, y)$ for all points (x, y) in some circular region around (a, b).

There is no guarantee the circular regions mentioned in the definition can be allowed to be very large. Consider the surface shown in figure 7.3.2. If the circle containing (a, b) becomes too large, then the values of $f(x, y)$ begin to create points on the backslope that are larger than the relative maximum at A, directly over (a, b).

$z = -y^3 - x^2 + 3y + 2$

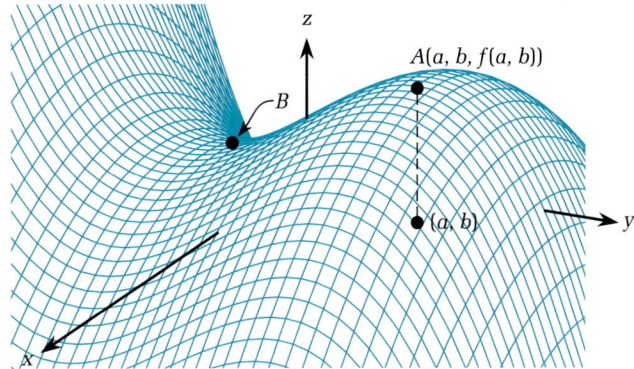

FIGURE 7.3.2

Suppose the surface in figure 7.2.1 represents a mountain. At any point on the surface we can measure slopes of vertical cross sections (vertical traces). When we are on the top of the mountain any cross-section trace would have a horizontal tangent. In particular, both f_x and f_y would be zero. This is true for any relative maximum or minimum, and we can use that fact as we did for a function of one variable.

For functions of one variable, we defined a "critical number" to be a value $x = a$ for which $f'(a) = 0$, or $f'(a)$ did not exist. The similar range of possibilities for functions of two variables is quite complex and beyond the scope of this book. Hence, our discussion will be restricted to those critical points for which both partial derivatives are zero.

DEFINITION

Critical point

The point (a, b) is a *critical point* for $f(x, y)$ if $f_x(a, b) = 0$ and $f_y(a, b) = 0$.

EXAMPLE 1   Finding critical points

Find the critical points for $f(x, y) = x^2 - 2x + y^2 - 4y + 6$.

SOLUTION

The partial derivatives are $f_x = 2x - 2$ and $f_y = 2y - 4$. These are zero if $2x - 2 = 0$ and $2y - 4 = 0$, which gives us $x = 1$ and $y = 2$. Hence $(1, 2)$ is the only critical point.

EXAMPLE 2   Finding critical points

Find the critical points for $f(x, y) = -y^3 - x^2 + 3y + 2$.

SOLUTION

The partial derivatives are $f_x = -2x$, and $f_y = -3y^2 + 3$. If $-2x = 0$, then $x = 0$. If $-3y^2 + 3 = 0$, then $y^2 = 1$, and $y = \pm 1$. Hence, the critical points are $(0, 1)$ and $(0, -1)$.

EXAMPLE 3   Finding critical points

Find the critical points for $f(x, y) = x^2 + 2xy - y^2 - 6x + 2y + 5$.

SOLUTION

$f_x = 2x + 2y - 6$ and $f_y = 2x - 2y + 2$. Setting each equal to zero gives a system of two equations, $2x + 2y - 6 = 0$ and $2x - 2y + 2 = 0$.

Solving the first equation for x gives $x = 3 - y$. Substituting for x in the second equation, we find $2(3 - y) - 2y + 2 = 0$, which yields $y = 2$. Substituting back for x, we have $x = 3 - 2 = 1$. Thus, the only critical point is $(1, 2)$.

PRACTICE EXERCISE 1

Find any critical points for $f(x, y) = x^2 + xy + y^2 - x - 2y + 4$.

There is no guarantee that we will find a relative maximum or minimum at a critical point. If we look back at figure 7.3.2 we see another point, *B*, where it seems quite possible that both partial derivatives are zero, but at which no relative maximum or minimum occurs. If we think of the surface as a terrain, we can imagine crossing a low point between two adjacent hills. This is called a pass. Another word used to describe the same situation is saddle, such as a leather horse-saddle on which the surface curves upward toward the head and tail, but downward on both sides. There are many other possibilities for the behavior of a surface at a critical point.

Let us focus our attention on finding relative extreme values. We need some way of sorting through all the critical points and determining which are associated with relative extreme values and which are not. One process for doing this is actually to find values of the function near the critical point. Often this is done by plotting selected values using a graphics package. A second, and often faster, test is given in the following result.

TEST FOR RELATIVE EXTREME VALUES FOR FUNCTIONS OF TWO VARIABLES

Suppose $z = f(x, y)$ is a function of two variables such that f_{xx}, f_{yy}, and f_{xy} all exist. If $f_x(a, b) = 0$ and $f_y(a, b) = 0$, then define the number

$$D = f_{xx}(a, b) \cdot f_{yy}(a, b) - [f_{xy}(a, b)]^2$$

(i)   If $D > 0$ and $f_{xx}(a, b) < 0$, then $f(a, b)$ is a *relative maximum*.

(ii) If $D > 0$ and $f_{xx}(a, b) > 0$, then $f(a, b)$ is a *relative minimum*.

(iii) If $D < 0$, then $f(a, b)$ is a *saddle point* (neither a relative maximum nor relative minimum).

(iv) If $D = 0$, this test gives no information.

This test appears to be, and actually is, quite complicated to state. However, it is fairly easy to apply, as the following examples show.

EXAMPLE 4   Applying the test for relative extreme values

Find any relative extreme values of $f(x, y) = x^2 - 2x + y^2 - 4y + 6$.

SOLUTION

This is the function in Example 1. There we found $f_x = 2x - 2$ and $f_y = 2y - 4$ and the only critical point occurred for $x = 1$ and $y = 2$.

To use the above test, we first find $f_{xx} = 2$, $f_{yy} = 2$, and $f_{xy} = 0$. Then $D = f_{xx}f_{yy} - f_{xy}^2 = 2 \cdot 2 - 0^2 = 4$. Because $D > 0$ and $f_{xx} > 0$, the test asserts there is a relative minimum at the point $(1, 2)$ and its value is $1^2 - 2 \cdot 1 + 2^2 - 4 \cdot 2 + 6 = 1$. See figure 7.3.3.

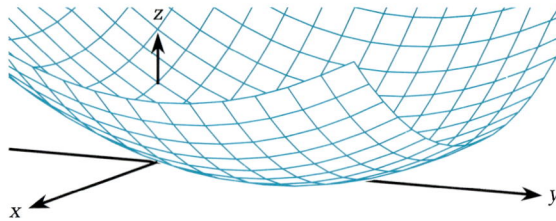

$$z = x^2 - 2x + y^2 - 4y + 6$$

FIGURE 7.3.3

EXAMPLE 5   Applying the test for relative extreme values

Find any relative extreme values of $f(x, y) = -y^3 - x^2 + 3y + 2$.

SOLUTION

This is the function in Example 2. There we found $f_x = -2x$, which equals zero if $x = 0$, and $f_y = -3y^2 + 3$, which equals zero if $y = \pm 1$. Thus, there are two critical points, $(0, 1)$ and $(0, -1)$.

Because $f_{xx} = -2$, $f_{yy} = -6y$, and $f_{xy} = 0$, we have $D = f_{xx}(a, b) \cdot f_{yy}(a, b) - [f_{xy}(a, b)]^2 = -2 \cdot (-6y) - 0^2 = 12y$.

At $(0, -1)$, the value $D = 12(-1) = -12 < 0$. Using the test for relative extreme values, we have a saddle-point value of $f(0, -1) = -(-1)^3 - 0^2 + 3 \cdot (-1) + 2 = 0$ at $(0, -1, 0)$.

At $(0, 1)$, the value $D = 12 \cdot 1 = 12 > 0$. Because $f_{xx} < 0$, by the test there is a relative maximum of $f(0, 1) = -1^3 - 0^2 + 3 \cdot 1 + 2 = 4$ at the point $(0, 1, 4)$.

The surface is shown in figure 7.3.2.

EXAMPLE 6   Applying the test for relative extreme values

Find any relative extreme values of $f(x, y) = x^2 + 2xy - y^2 - 6x + 2y + 5$.

SOLUTION

This is the function in Example 3. We found $f_x = 2x + 2y - 6$ and $f_y = 2x - 2y + 2$ to be zero at $(1, 2)$. The second derivatives are $f_{xx} = 2$, $f_{yy} = -2$, and $f_{xy} = 2$ for all values of x and y.

Thus, $D = 2 \cdot (-2) - (2)^2 = -4 - 4 = -8$. By the above test, the point $(1, 2, 4)$ is a saddle point. See figure 7.3.4.

$$z = x^2 + 2xy - y^2 - 6x + 2y + 5$$

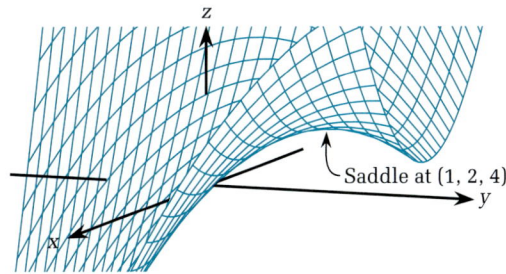

Saddle at (1, 2, 4)

FIGURE 7.3.4

PRACTICE EXERCISE 2

For $f(x, y) = x^2 + xy + y^2 - x - 2y + 4$, find any relative minimum or maximum values and where they occur. (Hint: This is the function in Practice Exercise 1.)

EXAMPLE 7   Determining a relative maximum

Suppose $f(x, y) = \dfrac{-5y}{x^2 + 2y^2 + 1}$. The surface of this function is shown in figure 7.3.1. Find any relative maximum.

SOLUTION

First, find critical points by solving $f_x = 0$ and $f_y = 0$.

$$f_x = \frac{(x^2 + 2y^2 + 1) \cdot 0 - (-5y)(2x)}{(x^2 + 2y^2 + 1)^2}$$

$$= \frac{10xy}{(x^2 + 2y^2 + 1)^2}$$

Hence if $x = 0$ or if $y = 0$, then $f_x = 0$.

$$f_y = \frac{(x^2 + 2y^2 + 1) \cdot (-5) - (-5y)(4y)}{(x^2 + 2y^2 + 1)^2}$$

$$= \frac{-5x^2 - 10y^2 - 5 + 20y^2}{(x^2 + 2y^2 + 1)^2}$$

$$= \frac{10y^2 - 5x^2 - 5}{(x^2 + 2y^2 + 1)^2}$$

If $x = 0$ and $f_y = 0$ then $10y^2 - 5 = 0$, so $y = \pm 1/\sqrt{2} \simeq \pm 0.7$.
If $y = 0$ and $f_y = 0$ then $-5x^2 - 5 = 0$, which has no solution.

Hence there are only two critical points, $(0, \pm 1/\sqrt{2})$. We need to decide which, if either, of these is a relative maximum. We could try using the extreme value test, but finding f_{xx}, f_{xy}, and f_{yy} appears to be quite a mess. In this type of situation a graphing utility comes in very handy, for we can tell immediately from the graph in figure 7.3.1 that there is a maximum at $(0, -1/\sqrt{2})$.

Knowing the exact coordinates for x and y, we can then find the maximum using $f(x, y)$:

$$f\left(0, -\frac{1}{\sqrt{2}}\right) = \frac{-5\left(-\dfrac{1}{\sqrt{2}}\right)}{0^2 + 2\left(-\dfrac{1}{\sqrt{2}}\right)^2 + 1}$$

$$= \frac{\dfrac{5}{\sqrt{2}}}{1 + 1} = \frac{5}{2\sqrt{2}} \simeq 1.7678$$

We could not have guessed that the maximum occurred at $(0, -1/\sqrt{2})$ from our drawing of the surface because we would not have known the critical point.

Applications

EXAMPLE 8   Maximizing profit

Suppose Frank is handpainting x sweatshirts and y T-shirts. The demand function for sweatshirts is $p_S(x) = 25 - 0.1x$ and the demand function for T-shirts is $p_T(y) = 12 - 0.2y$; both these give the dollar price of selling the respective number of items. The cost for manufacturing x sweatshirts and y T-shirts is $C(x, y) = 20x + 5y + 0.1xy$. Find the maximum profit and the number and price of sweatshirts and T-shirts that will produce that profit.

SOLUTION

Let $P(x, y)$ represent the profit from the sale of x sweatshirts and y T-shirts.

$$P(x, y) = x \cdot p_S(x) + y \cdot p_T(y) - C(x, y)$$
$$= x(25 - 0.1x) + y \cdot (12 - 0.2y) - (20x + 5y + 0.1xy)$$
$$= 25x - 0.1x^2 + 12y - 0.2y^2 - 20x - 5y - 0.1xy$$
$$= 5x + 7y - 0.1x^2 - 0.2y^2 - 0.1xy$$

For partial derivatives, we have $P_x = 5 - 0.2x - 0.1y$ and $P_y = 7 - 0.4y - 0.1x$. Hence, $P_{xx} = -0.2$, $P_{yy} = -0.4$, and $P_{xy} = -0.1$. To use the test, we find $D = P_{xx}P_{yy} - P_{xy}^2 = (-0.2)(-0.4) - (-0.1)^2 = 0.08 - 0.01 = 0.07 > 0$.

Because $D > 0$ and $P_{xx} < 0$, by the relative extreme value test, we have a maximum at any point for which $P_x = 0$ and $P_y = 0$.

Setting $P_x = 0$ and solving for y, we have $y = 50 - 2x$. Substituting this value for y in $P_y = 0$ yields $7 - 0.4(50 - 2x) - 0.1x = 0$, which simplifies to $7 - 20 + 0.8x - 0.1x = 0$. Solving for x, we have $x = 13/0.7 \simeq 18.6$. Returning to $y = 50 - 2x$, we find $y \simeq 50 - 2(18.6) = 12.8$.

The number of sweatshirts and T-shirts must be integer values. Hence, to find the maximum profit, the values for x that must be checked are 18 and 19, and the values for y are 12 and 13.

Substituting into $P(x, y) = 5x + 7y - 0.1x^2 - 0.2y^2 - 0.1xy$, we find $P(18, 12) = 91.2$, $P(18, 13) = 91.4$, $P(19, 12) = 91.3$, and $P(19, 13) = 91.4$.

Thus, Frank will have a maximum profit of $91.40. That profit occurs when he makes 18 sweatshirts and sells them for $p_S(18) = $23.20 each, and 13 T-shirts, which he sells at $p_T(13) = $9.40. Or, he could have the same profit by handpainting 19 sweatshirts and selling them at $23.10 each and selling 13 T-shirts at $9.40.

See figure 7.3.5.

$z = 5x + 7y - 0.1x^2 - 0.2y^2 - 0.1xy$

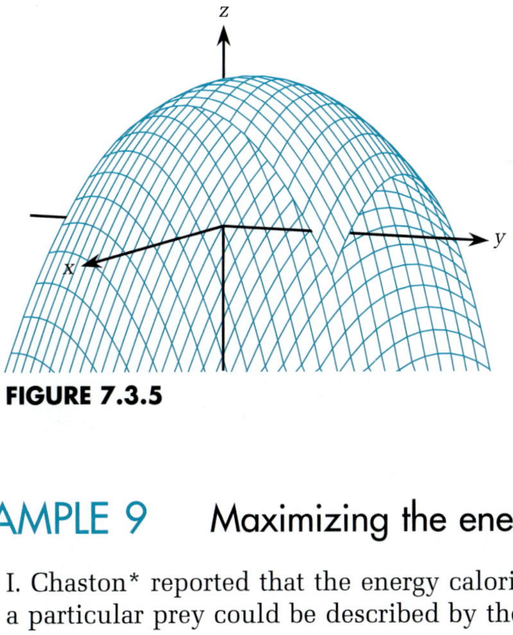

FIGURE 7.3.5

EXAMPLE 9   Maximizing the energy gain for birds

I. Chaston* reported that the energy calorie gain of a bird capturing a particular prey could be described by the function

$$E = 120P + 11SP - 0.5S^2P - 10P^2 - 30$$

where S is the energy expended in searching for prey and P is the energy of an individual prey, with both S and P measured in calories.

Find the calorie value for P that maximizes the energy gain, E.

*I. Chaston, *Mathematics for Ecologists,* London: Butterworths and Co., 1971, p. 40.

SOLUTION

Taking partial derivatives we have $E_P = 120 + 11S - 0.5S^2 - 20P$ and $E_S = 11P - 0.5(2S)P = 11P - SP$.

Solving $E_S = 0$, we have $(11 - S)P = 0$. Because $P \neq 0$, we find $S = 11$.

Substituting the value of 11 for S in E_P, and setting $E_P = 0$, we have $120 + 11(11) - 0.5(11)^2 - 20P = 0$.

Solving this for P gives us $P = [120 + 121 - 0.5(11)^2]/20 = 9.025$.

That this gives a maximum is determined by showing $E_{PP}E_{SS} - E_{PS}^2 > 0$ and $E_{PP} < 0$.   ●

7.3 PROBLEMS

Foundations

The problems of this section require the fundamental skills illustrated by the following:

1. Solve the system of equations: $2x + y = 0$ and $2y + x - 6 = 0$.

2. Solve the system of equations: $2x - 1 + y = 0$ and $2y + x = 0$.

3. Solve the system of equations: $3x^2 - 12 = 0$ and $-2y + 2 = 0$.

4. Solve the system of equations: $x^3 - 4x = 0$ and $2y - 2 = 0$.

Exercises

For problems 5–20, find any relative maximum or minimum values by using the relative extreme value test. Also list saddle points and points where the test is inconclusive.

5. $f(x, y) = x^2 - x + y^2 + xy$

6. $f(x, y) = x^2 + y^2 + xy - 6y + 1$

7. $f(x, y) = x^2 + 4y^2 + 2xy$

8. $f(x, y) = x^2 + y^2 + 4xy$

9. $f(x, y) = x^2 - xy + y^2 - 3y + 5$

10. $f(x, y) = x^2 - y^2$

11. $f(x, y) = x^3 - y^2 - 12x + 2y$

12. $f(x, y) = x^3 - y^2 - 27x + 4y$

13. $f(x, y) = 2x^3y - 2x + 16y - 5$

14. $f(x, y) = x^2 + xy + y^2 - 3x$

15. $f(x, y) = 8 \ln x - 5xy + y^2$

16. $f(x, y) = 2 \ln x - 3xy + y^2$

17. $f(x, y) = x^4 - 2x^2 + y^2 - 2y$

18. $f(x, y) = x^4 + 2x^2 - y^2 + 4y$

19. $f(x, y) = e^{(x-1)^2 + y^2}$

20. $f(x, y) = e^{-x^2 + 2y - y^2}$

21. **(Maximizing Profit)**   Suppose the demand function for light beer is $p_l(x) = 3 - 0.05x$, and for regular beer it is $p_r(y) = 2.75 - 0.2y$. The cost function is $C(x, y) = 2.25x + 2.50y + 0.01xy$.
    (a) Find the maximum profit possible.
    (b) Find the values for both x and y to produce that profit.

22. **(Maximizing Profit)**   The Dish DeLish ice-cream store has two grades of ice cream, Supreme and UltraRich. The demand function for Supreme is $p_s = 5 - 0.1x$, where x is in thousands of gallons and p_s is in dollars per gallon. The demand function for UltraRich is $p_u = 7 - 0.2y$, where y is in thousands of gallons and p_u is in dollars per gallon. The cost function is $C(x, y) = 4.5x + 4.75y - 0.1xy$.
    (a) Find the maximum profit possible.
    (b) Find the quantities (and prices) associated with the maximum profit.

23. **(Maximizing Profit)**   CF Zinc Co. has a smelter in Blanco and another in Fedwich. The cost of producing x units at Blanco is $C_1(x) = 0.02x^2 + 5x + 1000$, and the cost of producing y units at Fedwich is $C_2(y) = 0.05y^2 + y + 100$. All units produced sell for \$15. The profit function is $P(x, y) = 15(x + y) - C_1(x) - C_2(y)$.

(a) Find the number of units to produce at each smelter in order to maximize the total profit.

(b) What is the maximum total profit?

24. **(Maximizing Revenue of Competing Products)** (In this problem, the demand functions are given in the form of sales as a function of price.) Example 8 in Section 7.1 on competing products involved the sale of T-shirts and sweatshirts, where the demand functions were $T(x, y) = 2000 - 100x + 5y$ and $S(x, y) = 100 - 4y + 3x$, for which T is the number of T-shirts and S is the number of sweatshirts, when priced at x and y dollars, respectively.

In answering the following questions, assume for the sake of simplicity that we can sell fractions of T-shirts and sweatshirts.

(a) Find the prices and quantities of T-shirts and sweatshirts that maximize total revenue.

(b) Determine the maximum total revenue.

25. **(Biology, Hardy-Weinberg Law)**   There are four human blood types (disregarding Rh factors). These are determined by alternate forms (called *alleles*) of a gene. These alleles are named A, B, and O, and the blood types formed from these are A (by AA or AO), B (by BB or BO), O (by OO), and AB. The *Hardy-Weinberg Law* states that the proportion of humans in a population whose genes have two different alleles is given by the formula

$$P = 2pq + 2pr + 2rq$$

where p, q, and r are the proportions of the alleles A, B, and O, respectively, in the population. Show that $P \leq 2/3$. (Hint: Because $p + q + r = 1$, replace r by $1 - p - q$, then solve $P_p = 0$ and $P_q = 0$.)

26. A box is to be rectangular in shape and to have a volume of 2000 cubic inches. There is no top on the box. The sides of the box cost twice as much, per square inch, as does the bottom of the box. Find the dimensions of the least costly box.

Graphing Calculator Problems

27. To use a graphing calculator to search for extreme values, we could try the following process. Choose a fixed value for x, say $x = a$, graph $f(a, y)$ (using x as the variable name for the graphing calculator), and estimate the value of y that gives an extreme value. Suppose that is $y = b$. Go back and graph $f(x, b)$ and estimate the value of x that gives an extreme value. Use this as an improved guess for a and start the process over.

(a) Use the process, starting with $x = 1$, for the function $f(x, y) = -y^3 - x^2 + 3y + 2$. (This is the function in Example 5.)

(b) Discuss the possible ways this process can fail to locate extreme values.

Writing and Discussion Problems

28. Suppose a function is defined over a region given by $a \leq x \leq b$, $c \leq y \leq d$. If you have checked for all relative extreme values, how can you check for absolute extreme values?

29. For problem 24 we allowed fractions of T-shirts to be sold. If we require an integer number to be sold, then where do we check to find a solution?

30. Should students attempt to memorize the relative extreme value test for two variables?

Enrichment Problems

31. **(Finding Relative Extreme Values)**   The surface shown in figure 7.3.6 is given by

$z = (xy)/e^{\sqrt{x^2+2y^2}}$. Find any relative extreme values. (Hint: After taking the partial derivatives, you will find terms in which $e^{\sqrt{x^2+2y^2}}$ appears as a factor. Remember that $e^x \neq 0$ for any x.)

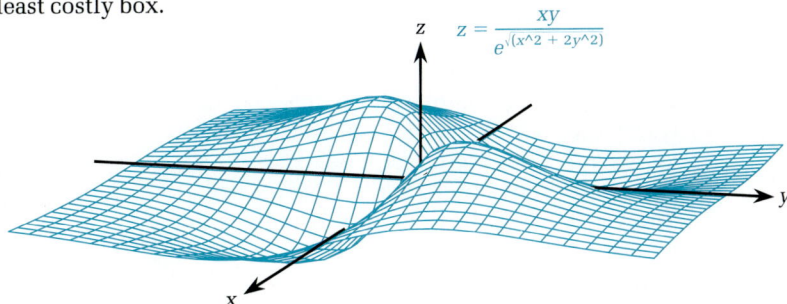

$$z = \frac{xy}{e^{\sqrt{x^2 + 2y^2}}}$$

FIGURE 7.3.6

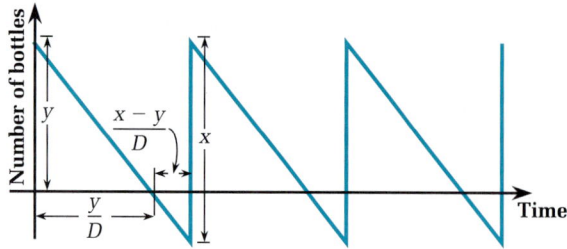

FIGURE 7.3.7

32. **(Economic Order Quantity, with Shortages Allowed)** In Section 3.6 we solved the EOQ problem of minimizing costs of ordering a commodity, for which we had a steady demand and for which no shortages were allowed.

In figure 7.3.7 we see a situation in which shortages are allowed. An example might be a merchant who orders French wines. Let us follow the costs involved in one cycle of ordering.

Let y be the stock on hand at the start of the cycle and D be the rate at which the wine is being sold. For the period of time y/D the merchant is selling wine from stock. From that time until the next shipment is delivered, the merchant is simply taking orders. Eventually the merchant places an order for a shipment of x bottles. The period of time when there is no wine on hand is $(x - y)/D$.

Let P be the cost of placing an order, S be the price per bottle for storage, and Q be the cost of not having the wine available for delivery. We continue with our earlier assumption that on average one-half of the total inventory is in stock all the time that any is in stock, and add the similar assumption that on average one-half of the difference between the

size of the inventory and the size of the shipment is not in stock all the time there is no stock.

The cost over one cycle is

$$P + \left(\frac{y}{D}\right)\left(\frac{y}{2}\right)S + \left(\frac{x - y}{D}\right)\left(\frac{x - y}{2}\right)Q.$$

The number of cycles per year is D/x, so the total cost of ordering for one year is

$$C(x, y) = \left[P + \frac{y^2}{2D}S + \frac{(x - y)^2}{2D}Q\right]\left(\frac{D}{x}\right)$$

$$= \frac{PD}{x} + \frac{y^2 S}{2x} + \frac{Q}{2x}(x - y)^2.$$

(a) Find $C_x(x, y)$ and $C_y(x, y)$.
(b) By solving $C_y = 0$, show that the minimum total annual cost occurs for
$$y = [Q/(S + Q)]x.$$
(c) By solving $C_x = 0$, show that the minimum total annual cost occurs for
$$x = \sqrt{\frac{2PD + y^2(S + Q)}{Q}}.$$
(d) Combining your results of parts (b) and (c), show that the minimum annual cost occurs
for $x = \sqrt{\dfrac{2PD(S + Q)}{SQ}}$ and $y =$
$$\sqrt{\frac{2PDQ}{S(S + Q)}}.$$
(e) Suppose the merchant is charged $40 to place an order, the demand is for 900 bottles per year, it costs $1 to store a bottle in inventory, and it costs $0.25 for each bottle on order and not in inventory. Find the optimal shipment size to minimize total annual cost. How many bottles are in inventory at the start of each cycle?

SOLUTIONS TO PRACTICE EXERCISES

1. To find the critical points for $f(x, y) = x^2 + xy + y^2 - x - 2y + 4$, set $f_x = 2x + y - 1 = 0$ and $f_y = x + 2y - 2 = 0$. From the first of these, $y = -2x + 1$. Substituting for y in the second, we have $x + 2(-2x + 1) - 2 = 0$, so $x - 4x + 2 - 2 = 0$, giving $x = 0$.

Substituting back in $y = -2x + 1$, we find $y = 1$. Thus, $(0, 1)$ is the only critical point.

2. $f(x, y) = x^2 + xy + y^2 - x - 2y + 4$ is the function of Practice Exercise 1, for which the only critical point is at $(0, 1)$. To use the relative extreme value test, we find $f_{xx} = 2, f_{yy} = 2$, and $f_{xy} = 1$, so that $f_{xx}f_{yy} - (f_{xy})^2 = 2 \cdot 2 - 1 = 3$. Because $D > 0$ and $f_{xx} > 0$, we have a relative minimum of $f(0, 1) = 1 - 2 + 4 = 3$ at $(0, 1, 3)$.

FIGURE 7.4.1

7.4 LAGRANGE MULTIPLIERS AND CONSTRAINED OPTIMIZATION

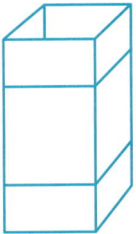

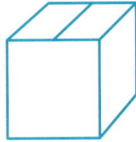

If we are seeking to maximize or minimize some quantity, there may be specific constraints that must be observed. For example, suppose we are packing cans in a box and every can must have some specified volume. Such a problem is called an *isoperimetric* problem. A typical problem of this form asks us to find a relative extreme value for a variable z, defined by some function $z = f(x, y)$ and subject to some constraint equation involving another function and expressed in the form $g(x, y) = 0$.

As an illustration, consider the problem encountered by Wildflower Mountain Honey. It packs gift boxes with four cylindrical jars of honey, each jar having a volume of 30 cubic inches. (This is somewhat more than 16 ounces of liquid.) The boxes are made of cardboard, with the flaps meeting in the middle of the top and bottom of the box and being glued together there, as shown in figure 7.4.1.

What values for the radius and height of the individual cans will minimize the total amount of cardboard used in making the box?

In 1755, Joseph Louis Lagrange, then only 19 years old, wrote of his discovery of a means of solving such problems. To illustrate the *method of Lagrange multipliers* we first consider an example that also can be worked by reducing the problem to a function of one variable. By solving this problem using both a familiar method and the new method, we are more confident of our ability to use Lagrange's method.

Lagrange Multipliers

The following is given for functions of two variables, but the result is correct for any number of variables.

LAGRANGE'S THEOREM

Any relative extreme value of the function $z = f(x, y)$, subject to the constraint condition $g(x, y) = 0$, will occur among those points (x, y) for which there exists a value of λ such that defining $L(x, y, \lambda)$ by $L(x, y, \lambda) = f(x, y) + \lambda \cdot g(x, y)$ we have each of the following:

$$L_x(x, y, \lambda) = 0$$
$$L_y(x, y, \lambda) = 0$$
$$L_\lambda(x, y, \lambda) = 0$$

provided all these partial derivatives exist.

The symbol λ is the Greek letter *lambda,* and in the definition of $L(x, y, \lambda)$, the λ is called a *Lagrange multiplier.* The following outlines the method of using Lagrange multipliers.

THE METHOD OF LAGRANGE MULTIPLIERS

1. If the constraint is not already in the correct form, rewrite it as $g(x, y) = 0$.

2. Define the function $L(x\ y, \lambda) = f(x, y) + \lambda \cdot g(x, y)$.

3. Determine the partial derivatives, L_x, L_y, and L_λ.

4. Establish and solve the system of equations $L_x = 0$, $L_y = 0$, and $L_\lambda = 0$.

5. Check among the solutions of the system in step 4 for any extreme values.

EXAMPLE 1   Maximizing volume

Asis Auction wants to construct a new exhibit and sales building. It is to be rectangular, have a flat roof, and be 10 feet high. Neglecting the costs of doors, windows, and floor, the total cost of sides and roof is to be $75,000. The front side costs $5 per square foot, and the other three sides cost $2 per square foot. The roof costs $25 per square foot. Find the floor area of the largest such building.

SOLUTION

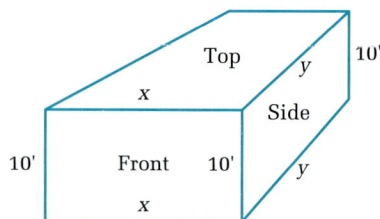

FIGURE 7.4.2

In figure 7.4.2, we see a sketch of the building. The quantity to be maximized is $A = f(x, y) = xy$.

The front side has area $x \cdot 10$, so its cost is $5 \cdot 10x = 50x$. The cost of the back side is $2 \cdot 10x = 20x$. The remaining two sides each have a cost of $2 \cdot 10y$. The cost of the roof is $25 \cdot xy$. Adding these costs together, we have a total cost of $70x + 40y + 25xy = 75,000$.

The method of Lagrange multipliers involves the following steps.

1. Rewrite the constraint as $g(x, y) = 70x + 40y + 25xy - 75,000 = 0$.

2. Define $L(x, y, \lambda) = f(x, y) + \lambda \cdot g(x, y) = xy + \lambda(70x + 40y + 25xy - 75,000)$.

3. Find $L_x = y + 70\lambda + 25\lambda y$, $L_y = x + 40\lambda + 25\lambda x$, and $L_\lambda = 70x + 40y + 25xy - 75,000$.

4. Solve the system of partials, each set equal to zero.
   From $L_x = 0$, we have $\lambda = -y/(70 + 25y)$.
   From $L_y = 0$, we have $\lambda = -x/(40 + 25x)$.
   We can use these two equations to eliminate λ. This is quite often possible when using Lagrange's method and should be tried when we are not interested in the value for λ. (Sometimes we very much do want to know λ, as we will discover later in an important application to business.)
   Eliminating λ, we get $\dfrac{-y}{70 + 25y} = \dfrac{-x}{40 + 25x}$, which gives us $-y(40 + 25x) = -x(70 + 25y)$.
   Simplifying, we have $-40y - 25xy = -70x - 25xy$, and so $y = 7x/4$.
   Substituting for y in $L_\lambda = 70x + 40y + 25xy - 75{,}000 = 0$, we get $70x + 40\left(\dfrac{7x}{4}\right) + 25x\left(\dfrac{7x}{4}\right) - 75{,}000 = 0$, which simplifies to $(175/4)x^2 + 140x - 75{,}000 = 0$.
   By the quadratic formula, this has the solution

   $$x = \frac{-140 \pm \sqrt{(140)^2 + 4 \cdot \dfrac{175}{4} \cdot 75{,}000}}{2 \cdot \dfrac{175}{4}}$$

   $$= \frac{-140 \pm \sqrt{13{,}144{,}600}}{\dfrac{175}{2}}$$

   $$\simeq \frac{-140 \pm 3625.5}{87.5}$$

   Because x must be positive, we have $x = (-140 + 3625.5)/87.5 = 39.8$.
   Using the fact that $y = (7/4)x$, we find $y = 69.7$.

5. There is only one solution in step 4. Lagrange's method does not, by itself, guarantee a maximum or minimum value; nonetheless, there surely is some optimal building size. Because we have found only one possibility, the area $A = xy = 39.8 \cdot 69.7 = 2774.1$ is truly a maximum area.

As we commented at the beginning of this example, one reason for its choice was our ability to check our work by representing A as a function of x. To do so we solve $70x + 40y + 25xy = 75{,}000$ for y in terms of x and substitute that value for y in the formula for area, $A = xy$.
From $70x + 40y + 25xy = 75{,}000$ we have $y = (75{,}000 - 70x)/(40 + 25x)$. We wish to maximize

$$A = xy = x\left(\frac{75{,}000 - 70x}{40 + 25x}\right) = \frac{75{,}000x - 70x^2}{40 + 25x}$$

To do so, find

$$A'(x) = \frac{(40 + 25x)(75{,}000 - 140x) - (75{,}000x - 70x^2)(25)}{(40 + 25x)^2}$$

and solve $A'(x) = 0$ for x by solving for the numerator being zero.

$$(40 + 25x)(75{,}000 - 140x) - 25(75{,}000x - 70x^2)$$
$$= 3{,}000{,}000 + 1{,}875{,}000x - 5600x - 3500x^2 - 1{,}875{,}000x$$
$$+ 1750x^2 = 3{,}000{,}000 - 5600x - 1750x^2.$$

By the quadratic formula, this is zero when

$$x = \frac{5600 \pm \sqrt{(5600)^2 - 4 \cdot (-1750)(3{,}000{,}000)}}{2 \cdot (-1750)}$$

$$= \frac{5600 \pm \sqrt{21{,}031{,}360{,}000}}{-3500}$$

$$= \frac{5600 \pm 145{,}022}{-3500}$$

Since x must be positive, we have $x = 39.8$, just as before.

Consider another application of the method, the design of the packing boxes for Wildflower Mountain Honey, the company mentioned in the introduction to this section.

EXAMPLE 2   Minimizing packing box material

Refer to the design of a packing box for jars as shown in figure 7.4.1. Let the height of each jar be h and the radius of each jar be r. The top and bottom flaps are each one jar wide, because when folded over, each covers the width of a jar. What values for the radius and height of the individual cans will minimize the total amount of cardboard used in making the box?

SOLUTION

In figure 7.4.3 we see that the amount of cardboard used for each side is calculated by [(jar height) + (top flap) + (bottom flap)] · (two jars wide) = $[h + 2r + 2r](4r) = (h + 4r)(4r) = 4rh + 16r^2$. There are four sides of the box, so the total amount of cardboard is

$$A(r, h) = 4(4rh + 16r^2) = 16rh + 64r^2$$

The constraint is that the volume of each jar is 30 cubic inches. The volume of a jar is the product of the area of the base multiplied by the height, so we have $\pi r^2 h = 30$.

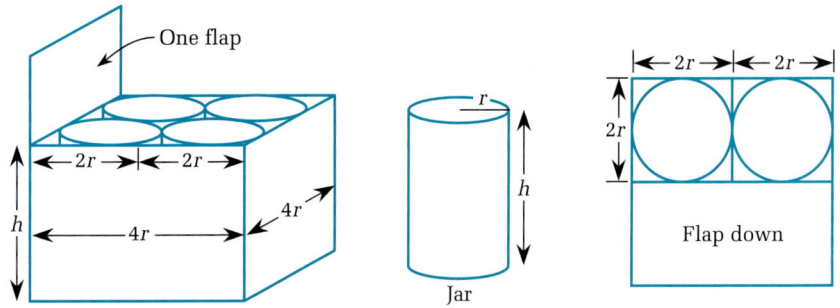

FIGURE 7.4.3

Now we use the method of Lagrange multipliers.

1. Rewrite the constraint in the form $g(r, h) = 0$. Because $\pi r^2 h = 30$, we have $\pi r^2 h - 30 = 0$.

2. Form $L(r, h, \lambda) = A(r, h) + \lambda\, g(r, h) = (16rh + 64r^2) + \lambda(\pi r^2 h - 30)$.

3. Determine the partial derivatives.

   (2.1) $\qquad\qquad\qquad L_r = 16h + 128r + 2\lambda\pi rh$

   (2.2) $\qquad\qquad\qquad L_h = 16r + \lambda\pi r^2$

   (2.3) $\qquad\qquad\qquad L_\lambda = \pi r^2 h - 30$

4. From formula (2.1), with $L_r = 0$, we get

   $$-\lambda = (16h + 128r)/(2\pi rh) = (8h + 64r)/(\pi rh).$$

   From formula (2.2), with $L_h = 0$, we get

   $$-\lambda = 16r/(\pi r^2) = 16/(\pi r).$$

   Setting these two values for $-\lambda$ equal, we have $(8h + 64r)/(\pi rh) = 16/(\pi r)$.

   Multiplying by πrh gives $8h + 64r = \dfrac{16}{\pi r}(\pi rh) = 16h$.

   Hence $h = 8r$. Substituting for h in $L_\lambda = 0$, we have $\pi r^2(8r) - 30 = 0$. Solving for r, we find $r^3 = 15/(4\pi)$, so that

   $$r = \sqrt[3]{\frac{15}{4\pi}} \simeq 1.0608.$$

   Because $h = 8r$, we have $h \simeq 8.4863$.

5. Because these values for r and h are the only possible solution to the problem of finding a minimum for the area, and we do believe there is a solution to the problem, we claim that the area given by $A(r, h) = 16rh + 64r^2$ has a minimum value of $A(1.0608, 8.4863) = 16(1.0608)(8.4863) + 64(1.0608)^2 \simeq 144.0363 + 72.0190 = 216.0553$.

   A realistic value for the area would be approximately 216 square inches.

There are two principal reasons we include Lagrange multipliers in our study. First, there may be constraints expressed so that it is difficult (or impossible) for us to solve for one variable in terms of the other. It was that situation which Lagrange and his contemporaries were trying to handle.

A second reason is shown in the following definition of a term from economics. In the definition, we should realize that solutions of $L_x = 0$ and $L_y = 0$ are defining λ as a function of x and y.

DEFINITION

Marginal productivity of money

Suppose $f(x, y)$ is a production function and $g(x, y) = 0$ is a constraint.

$$L(x, y, \lambda) = f(x, y) + \lambda g(x, y)$$

The *marginal productivity of money* at (x, y) is $-\lambda$, where λ is such that $L_x = L_y = 0$.

The following formula is from economics. (Remember that $-\lambda$ is positive.)

OPTIMAL INCREASE IN PRODUCTION

Assume $-\lambda$ is the marginal productivity of money, and N additional dollars are available. The optimal increase in production is given by $-\lambda N$.

Example 3 demonstrates how to find the marginal productivity of money and how to use it in a situation in which we change the constraint function, usually by allowing more resources to be devoted to the project.

EXAMPLE 3   Finding marginal productivity of money for a Cobb-Douglas production function and determining the optimal increase in production by additional allocation of resources

Suppose a Cobb-Douglas production function is $f(x, y) = 200x^{0.2}y^{0.8}$, where x represents units of labor and y represents units of capital.

Suppose each unit of labor is $80 and each unit of capital is $200. Assume the total expense for labor and capital is $40,000.

(a) Find the maximum production.

(b) Find the marginal productivity of money for that division of labor and capital.

(c) Find the optimal increase in production if an additional $5000 is allocated.

SOLUTION

(a) The constraint is $80x + 200y = 40,000$, so
$$g(x, y) = 80x + 200y - 40,000.$$
Define $L(x, y, \lambda) = 200x^{0.2}y^{0.8} + \lambda(80x + 200y - 40,000)$.
The partial derivatives are:

(3.1) $L_x = 200(0.2x^{-0.8})y^{0.8} + 80\lambda = 0$

(3.2) $L_y = 200x^{0.2}(0.8y^{-0.2}) + 200\lambda = 0$

(3.3) $L_\lambda = 80x + 200y - 40,000 = 0$

From equation (3.1), we have
$$\lambda = -[200(0.2x^{-0.8})y^{0.8}]/80 = -0.5(x^{-0.8})y^{0.8} = -0.5\ (y/x)^{0.8}.$$

From equation (3.2), we have
$$\lambda = -[200x^{0.2}(0.8y^{-0.2})]/200 = -0.8(x/y)^{0.2}.$$

Equating these two values for λ gives $0.5\left(\dfrac{y}{x}\right)^{0.8} = 0.8\left(\dfrac{x}{y}\right)^{0.2}$. To eliminate the decimal exponents, multiply each side by $\left(\dfrac{y}{x}\right)^{0.2}$. This gives $0.5\left(\dfrac{y}{x}\right)^{0.8}\left(\dfrac{y}{x}\right)^{0.2} = 0.8\left(\dfrac{x}{y}\right)^{0.2}\left(\dfrac{y}{x}\right)^{0.2}$, or $0.5\left(\dfrac{y}{x}\right)^{1} = 0.8$. Solving $0.5\left(\dfrac{y}{x}\right) = 0.8$ for y gives us $y = (0.8/0.5)x = 1.6x$. Substituting for y in equation (3.3), we have $80x + 200(1.6x) - 40,000 = 0$.

Solving for x, we have first $80x + 320x - 40,000 = 0$, and then $400x = 40,000$, so that $x = 100$.

Because $y = 1.6x$, we find $y = 1.6(100) = 160$.

Thus, the maximum production is $200(100)^{0.2}(160)^{0.8} = 200 \cdot 2.5119 \cdot 57.9824 = 29,129.2$.

(b) To find the marginal productivity of money at $x = 100$ and $y = 160$, we return to equation (3.1) and evaluate $\lambda = -0.5\ (160/100)^{0.8} = -0.7282$.

Hence the marginal productivity of money is 0.7282. The interpretation of the marginal productivity of money is that for each additional dollar available, the production will increase by 0.7282 units.

(c) The optimal increase in production is 0.7282(5000) = 3,641. This means that the total number of units that can be produced under optimal allocation of resources between labor and capital investment will yield a return of

29,129 (the production possible with $40,000)
+3,641 (the additional production of 0.7282 · 5000)
32,770 (the optimal production possible with $45,000)

WARNING

Suppose we have a constraint equation such as $10x + 20y = 2000$. We may be tempted to rewrite this as $x + 2y = 200$ before applying the method of Lagrange multipliers. *Do not do so!* Although the solutions for the extreme values are the same, the meaning of λ is no longer that of being the marginal productivity of money.

PRACTICE EXERCISE 1

Using the Cobb-Douglas production function of Example 3 and the costs therein, determine the optimal allocation of $45,000 between labor and capital so as to maximize total production. By determining the optimal allocation, check that the answer we obtained using the marginal productivity of money was correct.

Least Cost Rule in Economics

Let us reconsider the production function $f(x, y) = 200x^{0.2}y^{0.8}$ of Example 3. The *marginal productivity with respect to labor* is $f_x = 200(0.2x^{-0.8})y^{0.8}$, and the *marginal productivity with respect to capital* is $f_y = 200x^{0.2}(0.8y^{-0.2})$.

At the optimal allocation of $x = 100$ and $y = 160$, we find by substitution that $f_x = 200(0.2912902)$ and $f_y = 200(0.7282256)$.

The ratio $0.2912902/0.7282256 = 0.4$ is the same as the ratio of the unit cost of labor ($80) to the unit cost of capital ($200). This is always true. Thus, the following is stated as a law of economics.

LEAST COST RULE

At an optimal allocation of labor and capital, the ratio of their respective marginal productivities is equal to the ratio of their respective unit costs.

Lagrange's Method for a Nonlinear Constraint

So far, our examples have used Lagrange's method with a linear constraint. The following example shows its use with a nonlinear constraint.

EXAMPLE 4   Using Lagrange's method with nonlinear constraint

Maximize the production function $f(x, y) = x^{0.8}y^{0.4}$ provided that $x^2 + y^2 = 100$.

SOLUTION

Geometrically, we are choosing possible values for x and y that correspond to points on a circle of radius 10.

Following the method of Lagrange multipliers, we perform the following steps.

1. Rewrite the constraint as $x^2 + y^2 - 100 = 0$.

2. Define the function $L(x, y, \lambda) = x^{0.8}y^{0.4} + \lambda \cdot (x^2 + y^2 - 100)$.

3. Determine the partial derivatives.

$$L_x = 0.8x^{-0.2}y^{0.4} + 2\lambda x$$
$$L_y = 0.4x^{0.8}y^{-0.6} + 2\lambda y$$
$$L_\lambda = x^2 + y^2 - 100$$

4. Establish and solve the system of equations.

$$L_x = 0.8x^{-0.2}y^{0.4} + 2\lambda x = 0 \text{ gives us } \lambda = \frac{-0.8y^{0.4}}{x^{0.2} \cdot 2x} = \frac{-0.8y^{0.4}}{2x^{1.2}}$$

$$L_y = 0.4x^{0.8}y^{-0.6} + 2\lambda y = 0 \text{ gives us } \lambda = \frac{-0.4x^{0.8}}{y^{0.6}2y} = \frac{-0.4x^{0.8}}{2y^{1.6}}$$

Setting these two values of λ equal, we have $\dfrac{0.8y^{0.4}}{2x^{1.2}} = \dfrac{0.4x^{0.8}}{2y^{1.6}}$, which is equivalent to $0.8y^{0.4}(2y^{1.6}) = 2x^{1.2}(0.4x^{0.8})$, which simplifies to $2y^2 = x^2$.

The equation $L_\lambda = x^2 + y^2 - 100 = 0$ then gives us $2y^2 + y^2 - 100 = 0$. Hence $3y^2 = 100$ and so $y = \pm 5.7735$. Substituting back to find x, we have $x = \pm\sqrt{2}y = \pm 8.165$.

5. Because we want both x and y to be positive, the maximum will be $f(x, y) = (8.165)^{0.8}(5.7735)^{0.4} = (5.3649)(2.0164) = 10.82$.

EXAMPLE 5 Optimizing revenue

The El Casa Grande Hotel has two types of rooms. Type A is ordinary, whereas type B overlooks a beach. A tour agent wants a block of 100 rooms. The prices that a tour agent will pay are given by the following formulas, in which p_A is the price of each of the x number of A rooms and p_B is the price of each of the y number of the B rooms.

$$\text{For A rooms, } p_A = 200 - x + y$$
$$\text{For B rooms, } p_B = 200 - 2y^2 + 3x$$

Find the number of rooms of each type, and their prices, so that total revenue is maximized.

SOLUTION

The total revenue is given by

$$\begin{aligned}
R(x \cdot y) &= p_A \cdot x + p_B \cdot y \\
&= (200 - x + y)x + (200 - 2y^2 + 3x)y \\
&= 200x - x^2 + yx + 200y - 2y^3 + 3xy \\
&= 200x - x^2 + 4xy + 200y - 2y^3
\end{aligned}$$

The constraint is given by the need for 100 rooms, so that $x + y = 100$.

First, define $L(x, y, \lambda) = 200x - x^2 + 4xy + 200y - 2y^3 + \lambda(x + y - 100)$.

$$\text{Then } L_x = 200 - 2x + 4y + \lambda = 0$$
$$L_y = 4x + 200 - 6y^2 + \lambda = 0$$
$$L_\lambda = x + y - 100 = 0$$

Solving for λ in each of the first two of these and setting them equal to each other, we have $200 - 2x + 4y = 4x + 200 - 6y^2$, which can be rewritten $6x = 6y^2 + 4y$.

Solving for x we have $x = y^2 + (2/3)y$. Substituting this value for x in the equation for L_λ, we have $[y^2 + (2/3)y] + y - 100 = 0$, or $y^2 + (5/3)y - 100 = 0$.

This quadratic can be solved by the quadratic formula:

$$\begin{aligned}
y &= \frac{-\dfrac{5}{3} \pm \sqrt{\left(\dfrac{5}{3}\right)^2 - 4 \cdot 1 \cdot (-100)}}{2 \cdot 1} \\[2em]
&= \frac{-\dfrac{5}{3} \pm \sqrt{\dfrac{25}{9} + 400}}{2} \\[2em]
&= \frac{-1.6667 \pm 20.0693}{2}
\end{aligned}$$

The values for x and y must be positive, so we have $y = 9.2$. Substituting this value back into the equation for x yields $x = y^2 + (2/3)y = (9.2)^2 + (2/3)(9.2) = 84.6 + 6.1 = 90.7$.

Because an integer number of each type of room must be rented, we should check the revenue from renting 90 or 91 rooms of type A and 10 or 9 rooms of type B, respectively.

$$R(90, 10) = 200(90) - (90)^2 + 4(90)(10) + 200(10) - 2(10)^3$$
$$= \$13,500$$
$$R(91, 9) = 200(91) - (91)^2 + 4(91)(9) + 200(9) - 2(9)^3$$
$$= \$13,537$$

To maximize revenue there should be 91 rooms of type A and 9 rooms of type B. The type A rooms rent for $p_A = 200 - 91 + 9 = \$118$. The type B rooms rent for $p_B = 200 - 2(9)^2 + 3(91) = \$311$.

The method of Lagrange multipliers can be adapted to functions of three or even more variables and to situations in which there are several constraint functions. Let us solve a problem involving minimizing a function of three variables.

EXAMPLE 6   Lagrange multipliers for three variables

Figure 7.4.4 depicts a rectangular container to be made out of plastic. It has two compartments, a bottom, but no top. The thickness of all the plastic is uniform, so minimizing the area of plastic also minimizes the amount of plastic. The total volume of the container is to be 6000 cubic inches. Find the dimensions of a container made with least area.

SOLUTION

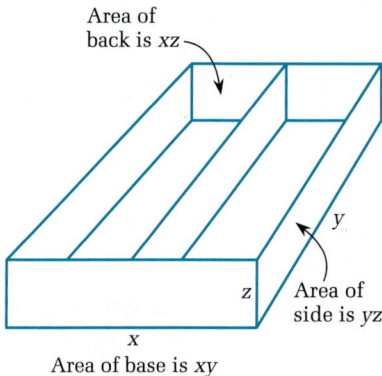

Area of back is xz

Area of side is yz

Area of base is xy

FIGURE 7.4.4

The total area is $A = xy + 2xz + 3yz$ and the volume is $V = xyz = 6000$. Our constraint is therefore $xyz - 6000 = 0$. Define $L = xy + 2xz + 3yz + \lambda(xyz - 6000)$.

There are three variables in the formula for A and four variables in L, so we need to solve a system of four partial derivatives.

$$(6.1) \qquad\qquad L_x = y + 2z + \lambda yz = 0$$
$$(6.2) \qquad\qquad L_y = x + 3z + \lambda xz = 0$$
$$(6.3) \qquad\qquad L_z = 2x + 3y + \lambda xy = 0$$
$$(6.4) \qquad\qquad L_\lambda = xyz - 6000 = 0$$

From equation (6.1) we have $\lambda = -\left(\dfrac{y + 2z}{yz}\right)$, and from equation (6.2) $\lambda = -\left(\dfrac{x + 3z}{xz}\right)$.

Setting these equal to each other and multiplying each by xyz we have $xy + 2xz = xy + 3yz$, which implies $2xz = 3yz$. Because $z \neq 0$, we can divide by z, which gives us $2x = 3y$. Solving for x in terms of y, we have $x = (3/2)y$.

From equation (6.3) we have $\lambda = -\left(\dfrac{2x + 3y}{xy}\right)$. Using this and the expression for λ from equation (6.2), we have $\dfrac{x + 3z}{xz} = \dfrac{2x + 3y}{xy}$.

Multiplying by xyz, we have $xy + 3zy = 2xz + 3yz$. Canceling the $3zy$ terms and then dividing by x ($x \neq 0$) gives us $y = 2z$. Rewriting z in terms of y, we have $z = (1/2)y$.

Using our relations between x, y, and z, we can substitute in equation (6.4) to get

$$\left(\frac{3}{2}y\right)y\left(\frac{1}{2}y\right) - 6000 = 0$$

$$\frac{3}{4}y^3 = 6000$$

$$y^3 = \frac{4}{3}(6000) = 8000$$

$$y = 20$$

Using this value for y and solving for x and z, we find $z = (1/2)(20) = 10$ and $x = (3/2)(20) = 30$.

The minimum total area is

$$A = 30 \cdot 20 + 2 \cdot 30 \cdot 10 + 3 \cdot 20 \cdot 10 = 600 + 600 + 600 = 1800.$$

7.4 PROBLEMS

Foundations

The problems in this section require the basic skills illustrated by the following:

1. Suppose $f(x, y) = 3x^2 - 2xy - y^3$. Find
   (a) f_x (b) f_y

2. Suppose $f(x, y) = xy^2$. Find   (a) f_x (b) f_y

Exercises

Find the indicated relative extreme values in problems 3–12 by the method of the Lagrange multiplier.

3. Maximum of $f(x, y)$, $= 3xy$ with constraint $x + 2y = 10$.

4. Minimum of $f(x, y) = x^2 + 2y^2$ with constraint $2x + y = 8$.

5. Minimum of $f(x, y) = x^2y$ with constraint $2x - y = 6$.

6. Maximum of $f(x, y) = x^2 - xy^2$ with constraint $x + y = 2$.

7. Maximum of $f(x, y) = x^2 - 2y^2$ with constraint $y - x^2 = 1$.

8. Minimum of $f(x, y) = x^2y$ with constraint $x^2 - y = 2$.

9. Minimum of $f(x, y, z) = x^2 + y^2 + z^2$ with constraint $x + y + z = 6$.

10. Maximum of $f(x, y, z) = x + y + z$ with constraint $x^2 + y^2 + z^2 = 12$.

11. Minimum of $f(x, y, z) = x^2 + 2y^2 + 3z^2$ with constraint $x = 1 + y + z$.

12. Maximum of $f(x, y) = x^{0.4}y^{0.6}$ with constraint $2x + y = 500$.

13. Rapper Gift Rapping Service has a Cobb-Douglas production function $f(x, y) = 6x^{0.2}y^{0.8}$, where x is the number of units of labor and y is the number of units of capital. Suppose each unit of labor is \$50 and each unit of capital is \$400. The total available resources are \$50,000.
    (a) Find the maximum level of production and the allocation to both labor and capital to produce that maximum level.
    (b) Find the marginal productivity of money for Rapper Gift Rapping.
    (c) Use the answer to part (b) to determine the increase in production if the total available resources increase by \$10,000.

14. MonoLog Records has a Cobb-Douglas production function $f(x, y) = 4x^{0.7}y^{0.3}$, where x is the number of units of labor and y is the number of units of capital. Suppose each unit of labor is \$150 and each unit of capital is \$50. The total available resources are \$15,000.
    (a) Find the maximum level of production and the allocation to both labor and capital to produce that maximum level.
    (b) Find the marginal productivity of money for MonoLog Records.
    (c) Use the answer to part (b) to determine the increase in production if the total available resources increase by \$10,000.
    (d) Show that at the optimal level of production, the Least Cost Rule is satisfied. This rule states that at an optimal allocation of labor and capital, the ratio of their respective marginal productivities is equal to the ratio of respective unit costs. In a more symbolic form,

$$\frac{\text{marginal productivity of labor}}{\text{marginal productivity of capital}} = \frac{\text{unit price of labor}}{\text{unit price of capital}}$$

15. The Screechway Wayside Motel has two types of rooms. Type A rooms directly face the freeway, whereas type B rooms are slightly quieter. A bicycle tour agent wants a block of 40 rooms. The prices that the tour agent will pay are given by the following formulas, in which x is the number of type A rooms and y is the number of type B rooms.

$$\text{For A rooms, } p_A = 100 - 2x + y$$
$$\text{For B rooms, } p_B = 200 - y^2 + 3x$$

Find the number of rooms of each type, and their prices, so that total revenue is maximized.

16. A paint company has two manufacturing plants. The cost of manufacturing x gallons of paint at plant A is given by $f(x) = 1000 + 2x + 0.001x^2$ and the cost of manufacturing y gallons of paint at plant B is given by $g(x) = 500 + y + 0.03y^2$. All paint sells for \$8 per gallon. The total number of gallons produced is given by the formula $x + 2y = 3500$.
    (a) Determine a function for total revenue.
    (b) Determine a function for total cost.
    (c) Determine a function for total profit.
    (d) Determine the production at each plant that will maximize the total profit.

17. The new gymnasium, Pecs-R-Us, is trying to decide how to spend its advertising budget. It can advertise on radio, in print media, or by flyers handed out at the mall. It believes that the number of inquiries in response to each form of advertising is $10x$, $20y^2$, and $5z^2$, when they run x, y, and z sets of ads on the radio, in print, and by flyers, respectively. The unit costs are \$1 for radio, \$24 for print, and \$14 for flyers. Total resources are \$1000. Find the optimal allocation of resources to produce the maximum number of inquiries.

18. Refer to Example 6. Suppose a container is to be made of plastic and have three parallel compartments, a bottom, but no top. For a volume of 216 cubic inches, find the dimensions that use the least surface area of plastic.

19. An ordinary 12-ounce can of soft drink or beer has a volume of approximately 22 cubic inches. Find the dimensions of such a cylindrical can having the least surface area.

20. An ordinary 12-ounce can of soft drink or beer has a volume of approximately 22 cubic inches. Suppose we are packaging six cans in a box made from a piece of cardboard. The flaps from the sides will just touch when folded over and they will be glued to the flaps from the ends, as shown in figure 7.4.5. Let r be the radius of each can and h be the height of each can. Using the constraint that $\pi r^2 h = 22$, find the dimensions for r and h for which the least cardboard is used in making the box.

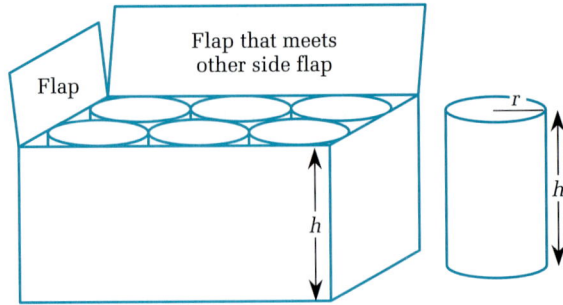

FIGURE 7.4.5

Enrichment Problems

21. Suppose a Cobb-Douglas production function is given by $P = f(x, y) = Ax^{\alpha}y^{1-\alpha}$ with constraint $ax + by = c$.
    (a) Show that the maximum of P occurs for $x = \alpha c/a$ and $y = [(1 - \alpha)c]/b$.
    (b) Use part (a) to show that for optimal production, the ratio of labor to capital is independent of the total resources available.

SOLUTION TO PRACTICE EXERCISE

1. The constraint is $80x + 200y = 45,000$, so $g(x, y) = 80x + 200y - 45,000$.
   Define $L(x, y, \lambda) = 200x^{0.2}y^{0.8} + \lambda(80x + 200y - 45,000)$.
   We have the following equations for the partial derivatives:

   (1) $L_x = 200(0.2x^{-0.8})y^{0.8} + 80\lambda = 0$

   (2) $L_y = 200x^{0.2}(0.8y^{-0.2}) + 200\lambda = 0$

   (3) $L_\lambda = 80x + 200y - 45,000 = 0$

   Only the third of these equations differs from those in Example 2, so the relation between x and y is the same, $y = 1.6x$.
   Substituting for y in equation (3), we have $80x + 200(1.6x) - 45,000 = 0$, or $80x + 320x - 45,000 = 0$, $400x = 45,000$, so that $x = 112.5$.
   From $y = 1.6x$, we find $y = 1.6(112.5) = 180$.
   The maximum production is $200(112.5)^{0.2}(180)^{0.8} = 32,770$.

7.5   LEAST SQUARES

We have found mathematical models in business, sociology, economics, biology, medicine, and many other areas. Given any situation, we ask ourselves two questions:

1. Do we have the correct type of model? (That is, do the data suggest we should use a linear, quadratic, logarithmic, exponential, or what type of model?)

2. Having decided on a particular type of model, what is the best fitting model of that particular type?

One of the most widely used measures of fitness relies on determining the sum of the squares of all the differences between the predicted and the actual values. Finding the least of these sums produces a best fit in some sense. Calculating these "best fits" is an interesting application of minimization of a function of several variables.

The computations done in this section require a significant amount of work using a hand-held calculator. In fact, most users of least sum of squares tests use a computer package or a graphing calculator, which automatically does the arithmetic for fitting by linear functions and perhaps even polynomial, exponential, and logarithmic functions.

Frankly, we are less interested in the actual computations than in deciding upon the formulas for the best-fit functions of a particular type. Thus, we should try to keep from getting bogged down in numerical details.

Least Squares Line

Let us reconsider our example, first introduced in Section 1.4, of the share of the daytime television audience held by the broadcast networks (ABC, NBC, CBS, and Fox).

1. Should we use a linear model, such as $f(x) = -2x + 78$?

2. If we use $f(x) = ax + b$, what values for a and b will give the least sum of squares of differences?

Table 7.5.1 gives the actual percentages and the percentages predicted by using the formula $f(x) = -2x + 78$, where x is the number of years after the 1977–1978 season. Also shown are the squares of the differences between the actual and the predicted values.

Let D stand for the sum of the squares of differences between the actual values and the predicted values using $f(x) = ax + b$. Because D is a function of a and b, we may write $D(a, b)$. From table 7.5.1, we find $D(-2, 78) = 8$. Is this the smallest D can be?

TABLE 7.5.1
Using $f(x) = -2x + 78$ to predict the audience share x years after 1977–1978

PROGRAMMING SEASON	x YEARS AFTER 1977–1978	ACTUAL SHARE	PREDICTED SHARE $f(x)$	DIFFERENCE	DIFFERENCE SQUARED*
1977–1978	0	78	78	0	0
1979–1980	2	75	74	1	1
1981–1982	4	68	70	2	4
1983–1984	6	65	66	1	1
1985–1986	8	61	62	1	1
1987–1988	10	57	58	1	1

*D = Sum of squares of differences = 8

When choosing to approximate a collection of data by a line, we can give a recipe for just how to choose the values of a and b so that the value of D is minimum. The line $y = ax + b$ for those particular values of a and b is called the *least squares line*, or the *regression line*.

We are first going to illustrate the technique of finding the least squares line by fitting a line to just three points. The reason for so few points is that the algebraic details are quite messy, even for only three points. After this three-point example, we present the general linear regression formulas, which are based on partial derivatives. It is helpful to see how partial derivatives are used in the linear situation because the ideas can be applied to nonlinear models, for which the formulas may not be as simple as we find for a linear model.

EXAMPLE 1   An illustration of the method of least squares

For the points $P(1, 1)$, $Q(2, 3)$, and $R(3, 4)$, find the equation of the line such that the sum of the squares of differences between the given y-coordinates and the corresponding y-coordinates on the line is least.

SOLUTION

The points are plotted in figure 7.5.1. Also shown is the line $y = ax + b$, which, although it looks like particular values of a and b have been chosen, is actually a general line, not a particular line. On the line, the y-coordinates given by the values of $x = 1, 2,$ and 3 would be as shown.

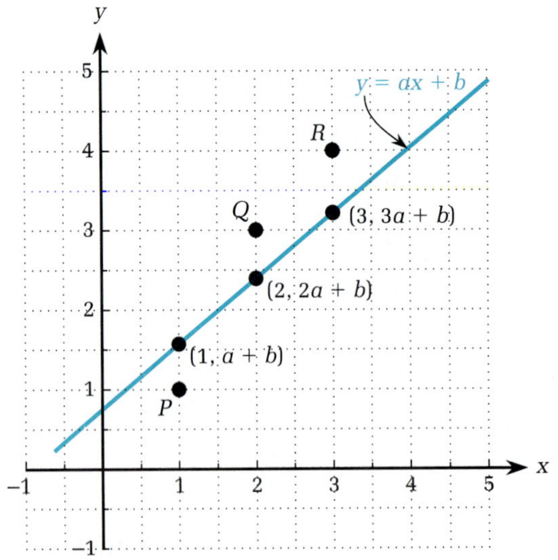

FIGURE 7.5.1

The squares of the differences between the y-values on the line and the y-values of the points are added together to give D, which we want to minimize.

$$D = [(a + b) - 1]^2 + [(2a + b) - 3]^2 + [(3a + b) - 4]^2$$

In order to follow the work, let us break this into three parts and expand each separately. Then we will add them all together to find D.

$$[(a + b) - 1]^2 = (a + b)^2 - 2(a + b) + (-1)^2$$
$$= a^2 + 2ab + b^2 - 2a - 2b + 1$$
$$[(2a + b) - 3]^2 = (2a + b)^2 - 2 \cdot 3(2a + b) + (-3)^2$$
$$= 4a^2 + 4ab + b^2 - 12a - 6b + 9$$
$$[(3a + b) - 4]^2 = (3a + b)^2 - 2 \cdot 4(3a + b) + (-4)^2$$
$$= 9a^2 + 6ab + b^2 - 24a - 8b + 16$$

Then the sum $D = 14a^2 + 12ab + 3b^2 - 38a - 16b + 26$.

To minimize D, we take both of the partial derivatives and set them equal to zero. $D_a = 28a + 12b - 38 = 0$ and $D_b = 12a + 6b - 16 = 0$, which produces two linear equations in two unknowns. One means of solving them simultaneously would be to divide the first by 2 and subtract the result from the second.

$$\begin{array}{ll} D_b & 12a + 6b - 16 = 0 \\ D_a/2 & \underline{14a + 6b - 19 = 0} \\ & -2a \qquad\quad +3 = 0 \end{array}$$

From this, $a = 3/2 = 1.5$, and we may substitute this value for a back into either of our equations to solve for b. Suppose we use the equation $12a + 6b - 16 = 0$. Then $12(1.5) + 6b - 16 = 18 + 6b - 16 = 2 + 6b = 0$ gives us that $b = -1/3$.

The value of D for the values $a = 3/2$ and $b = -1/3$ is $D = 14(3/2)^2 + 12(3/2)(-1/3) + 3(-1/3)^2 - 38(3/2) - 16(-1/3) + 26 \approx 0.1667$.

For these three points, the line $y = 1.5x - 1/3$ provides the best fit in the sense that it is the least squares line.

As we try for a linear fit using more data points, the algebraic details of solving for a and b become rather complicated. To state the conclusion succinctly, it is quite helpful to use *sigma notation*, whereby we express sums in a compact manner. The Greek letter, Σ, a capital sigma, is used because it is the letter in the Greek alphabet corresponding to our letter S and it reminds us that a sum is being expressed. The notation $\sum_{i=1}^{n} x_i$ means we take the sum $x_1 + x_2 + x_3 + \cdots + x_n$.

THE LEAST SQUARES LINE

The *least squares line* for the n data points (x_1, y_1), (x_2, y_2), (x_3, y_3), $\cdots$, (x_n, y_n) is given by $y = f(x) = ax + b$, where

$$a = \frac{\left(\sum_{i=1}^{n} x_i\right)\left(\sum_{i=1}^{n} y_i\right) - n\left(\sum_{i=1}^{n} x_i y_i\right)}{\left(\sum_{i=1}^{n} x_i\right)^2 - n\left(\sum_{i=1}^{n} x_i^2\right)}$$

and

$$b = \frac{\sum_{i=1}^{n} y_i - a\left(\sum_{i=1}^{n} x_i\right)}{n}$$

EXAMPLE 2   Finding the least squares line

Use the formulas to find the least squares line for the data given in Example 1.

SOLUTION

Set up the data as shown in table 7.5.2.

TABLE 7.5.2

x_i	y_i	x_i^2	$x_i y_i$
1	1	1	1
2	3	4	6
3	4	9	12
$\sum x_i = 6$	$\sum y_i = 8$	$\sum x_i^2 = 14$	$\sum x_i y_i = 19$

$$a = \frac{6 \cdot 8 - 3(19)}{6^2 - 3 \cdot 14} = \frac{48 - 57}{36 - 42} = \frac{-9}{-6} = 1.5$$

$$b = \frac{8 - (1.5) \cdot 6}{3} = \frac{8 - 9}{3} = -1/3$$

Hence the least squares line is $y = 1.5x - 1/3$, as we found directly by minimizing using partial derivatives.

PRACTICE EXERCISE 1

Fill in the table below and find the equation of the least squares line for the indicated data.

x_i	y_i	x_i^2	x_iy_i	
0	0.5			$a =$
1	2			
2	4			$b =$
$\sum x_i =$	$\sum y_i =$	$\sum x_i^2 =$	$\sum x_iy_i =$	

Most calculators have various keys labeled $\boxed{\Sigma y}$, $\boxed{\Sigma y^2}$, and $\boxed{\Sigma xy}$, whose use greatly simplifies the work of finding a least squares line. For those who need to calculate many such lines, a computer software package should be learned.

EXAMPLE 3   Least squares line, television audience share

Let us use the formulas for our TV audience share examples and find the least squares line. See table 7.5.3.

TABLE 7.5.3

x_i	y_i	x_i^2	x_iy_i
0	78	0	0
2	75	4	150
4	68	16	272
6	65	36	390
8	61	64	488
10	57	100	570
$\sum x_i = 30$	$\sum y_i = 404$	$\sum x_i^2 = 220$	$\sum x_iy_i = 1870$

$$a = \frac{30 \cdot 404 - 6 \cdot 1870}{(30)^2 - 6 \cdot 220} = \frac{12{,}120 - 11{,}220}{900 - 1320} = -2.1429$$

$$b = \frac{404 - (-2.1429)30}{6} = \frac{404 + 64.287}{6} = 78.0476$$

Table 7.5.4 shows the sum of squares using the equation $f(x) = -2.1429x + 78.0476$.

TABLE 7.5.4
Using $f(x) = -2.1429x + 78.0476$ to predict the audience share x years after 1977–1978

PROGRAMMING SEASON	x YEARS AFTER 1977–1978	ACTUAL SHARE	PREDICTED SHARE $f(x)$	DIFFERENCE	DIFFERENCE SQUARED*
1977–1978	0	78	78.0476	0.0476	0.002266
1979–1980	2	75	73.7618	−1.2382	1.5331
1981–1982	4	68	69.476	1.476	2.1786
1983–1984	6	65	65.1902	0.1902	0.0362
1985–1986	8	61	60.9044	−0.0956	0.00914
1987–1988	10	57	56.6186	−0.3814	0.1455

*Sum of squares of differences = 3.9048

Comparing the sums of squares of differences, we find $f(x) = -2.1429x + 78.0476$ provides a better linear approximation to the data than does the line $y = -2x + 78$, for which 8 was the sum of squares of differences.

We should not be misled into believing that $f(x) = -2.1429x + 78.0476$ is a model wherein all those digits truly have significance. All we are asserting is that among all possible linear models, this is the best one, where best means the smallest sum of the squares of differences between predicted and observed values. ●

Least Squares for Quadratic and Other Models

Suppose we decide a quadratic model would be appropriate for a set of data $\{(x_1, y_1), (x_2, y_2), (x_3, y_3), \cdots, (x_n, y_n)\}$.

Among all quadratic models, $y = ax^2 + bx + c$, if we use least sum of squares of differences to judge, we would choose the one for which the sum $S = \sum_{i=1}^{n} [(ax_i^2 + bx_i + c) - y_i]^2$ is smallest.

We do not intend to take partial derivatives with respect to three variables in this book because of the complicated manipulations involved. However, we state the conclusion of such work in the following form.

THE QUADRATIC WITH THE LEAST SUM OF SQUARES

For the points (x_1, y_1), (x_2, y_2), $\cdots$, (x_n, y_n), the least squares quadratic is $y = ax^2 + bx + c$ for which a, b, and c satisfy the system of equations

$$a\sum x_i^4 + b\sum x_i^3 + c\sum x_i^2 = \sum x_i^2 y_i$$
$$a\sum x_i^3 + b\sum x_i^2 + c\sum x_i = \sum x_i y_i$$
$$a\sum x_i^2 + b\sum x_i + cn = \sum y_i$$

where $\sum$ stands for $\displaystyle\sum_{i=1}^{n}$.

We can judge the faithfulness of any sort of model by minimizing sums of least squares of differences. Unfortunately, there are no techniques for finding the coefficients involved with each model. That is, we might guess that a drug level in the body is given by $y = e^{at} - e^{bt}$, but the determination of the best values for a and b from some particular data may be quite difficult. Once again, we emphasize the invaluable need for computers to solve these types of problems.

7.5 PROBLEMS

Exercises

1. Assume we have data points $(0, 1)$, $(1, 3)$, and $(3, 2)$. Find D, the sum of the differences squared, when the points are approximated by the following lines.
   (a) $y = x/2 + 1$
   (b) $y = ax + b$
   (c) For your function in part (b), find D_a and D_b.

2. Assume we have data points $(-1, 1)$, $(0, 3)$, and $(2, 4)$. Find D, the sum of the differences squared, when the points are approximated by the following lines.
   (a) $y = (4/3)x + 2$
   (b) $y = ax + b$
   (c) For your function in part (b), find D_a and D_b.

For problems 3–8, use the data points.
(a) Find the equation for the least squares line for the points.
(b) Calculate the sum of the squared differences.

3. $(1, 3)$, $(2, 5)$, $(4, 7)$

4. $(0, 2)$, $(1, 4)$, $(2, 5)$

5. $(0, 1)$, $(2, 3)$, $(3, 5)$, $(4, 4)$

6. $(-1, 2)$, $(1, 1)$, $(2, 0)$, $(3, -2)$

7. $(1, 1)$, $(1, 2)$, $(2, 2.5)$, $(3, 4)$

8. $(1, 0.5)$, $(2, 2)$, $(2, 4)$, $(3, 3)$

9. **(Least Squares, Aerobics)** A sports physical center measured the caloric output of four women doing aerobics for one hour. The number of calories and the weight of each person are given in the following table.

| Weight (in kilograms) | 57 | 48 | 53 | 51 |
| Calories in One Hour | 380 | 350 | 360 | 355 |

Find the least squares line $f(x)$ for calories per hour as a function of body weight.

10. **(Least Squares, TV Audience Share)** The percentages of the TV audience watching prime-time broadcast networks (CBS, ABC, NBC, and Fox) as reported by the Nielsen is shown in the table below.

Season Percentage	1977–1978 91	1979–1980 87	1981–1982 80
Season Percentage	1983–1984 75	1985–1986 73	1987–1988 66

Find the least squares line $f(x)$, wherein x measures the number of years past 1977–1978.

11. **(Least Squares, Library Holdings and Staffing)** The following table shows some selected library holdings and the number of staff at four major research libraries in 1990.*

	NUMBER OF VOLUMES (IN MILLIONS)	TOTAL STAFF (IN THOUSANDS)
Harvard University	11.8	1.1
University of Minnesota	4.6	0.5
University of Florida	2.9	0.44
University of Oregon	1.8	0.22

*Source: *Chronicle of Higher Education*, Mar. 20, 1991.

Find the least squares line $y = f(x)$, where x is the staff size and y the number of volumes.

12. **(Least Squares, Foundation Giving)** In the table below are listed foundations that gave the most money in grants in 1989, together with the total assets of those foundations in that year.*

FOUNDATION	ASSETS (IN BILLIONS)	GRANTS (IN MILLIONS)
Ford Foundation	5.8	218
Pew Charitable Trusts	3.3	137
W. K. Kellogg Foundation	4.2	106
Robert Wood Johnson Foundation	2.6	99
John D. and Catherine MacArthur Foundation	3.1	95

*Source: *Chronicle of Higher Education*, Dec. 5, 1991.

Find the least squares line $y = f(x)$, where x is assets and y is total grants.

Enrichment Problems

13. (a) Find the quadratic with the least sum of squares for the data of problem 11.
    (b) Compare the sums of squares of the quadratic model found in part (a) and the linear model, whose equation we found in problem 11.

SOLUTION TO PRACTICE EXERCISE

1. Fill in the following table and find the equation of the least squares line for the indicated data.

x	y	x^2	xy
0	0.5	0	0
1	2	1	2
2	4	4	8
$\sum x = 3$	$\sum y = 6.5$	$\sum x^2 = 5$	$\sum xy = 10$

$$a = \frac{3 \cdot 6.5 - 3 \cdot 10}{3^2 - 3 \cdot 5} = \frac{19.5 - 30}{9 - 15} = 1.75$$

$$b = \frac{6.5 - 1.75 \cdot 3}{3} = \frac{6.5 - 5.25}{3} \approx 0.4167$$

The line of least squares is therefore $y = 1.75x + 0.4167$.

7.6

TOTAL DIFFERENTIALS AND THEIR APPLICATIONS

The topic of this section is a generalization to several variables of the discussion of differentials presented in Section 3.7. Let us first quickly review differentials for a function of one variable.

For $y = f(x)$ we let Δx and $\Delta y = f(x + \Delta x) - f(x)$ represent an exact change in x and the corresponding exact change in y, respectively. In Section 3.7, the *differentials* of x and of $y = f(x)$ were represented by dx and dy and defined by:

$$dx = \Delta x \qquad \text{(the actual change in } x\text{)}$$
$$dy = \frac{dy}{dx} \cdot \Delta x \qquad \begin{array}{l}\text{(the product of the derivative times}\\ \text{the actual change in } x\text{)}\end{array}$$

In terms of rates, we can think of dy as

$$dy = \begin{array}{l}\text{(rate of change of } y \text{ per}\\ \text{unit change of } x\text{)}\end{array} \cdot \text{(actual change in } x\text{)}$$

For small values of dx, we found that dy was a good approximation for Δy. As illustrated in figure 7.6.1, we have

$$f(x + \Delta x) \simeq f(x) + dy = f(x) + \frac{dy}{dx}\Delta x$$

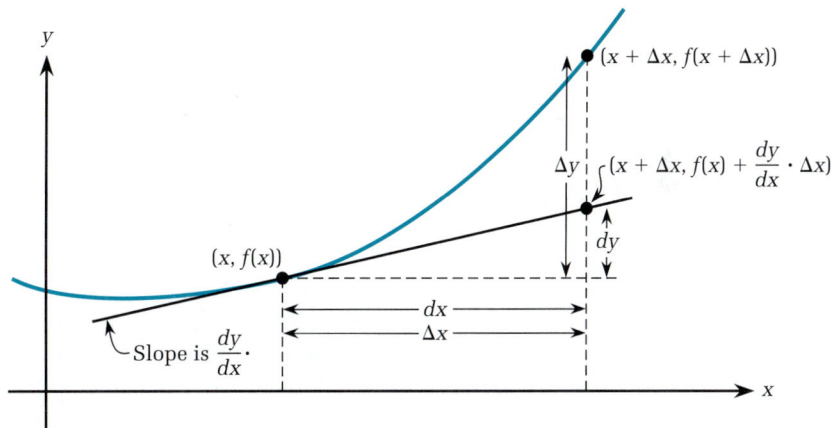

FIGURE 7.6.1

To illustrate a differential approximation, let $y = x^{3/2}$. We can easily compute by hand the value of y for $x = 4$ because $4^{3/2} = (\sqrt{4})^3 = 2^3 = 8$. But it is difficult to find $(4.05)^{3/2}$ by hand.

The derivative is $y' = (3/2)x^{1/2}$. To estimate $(4.05)^{3/2}$, let $x = 4$ and $\Delta x = 0.05$. We first evaluate y' at $x = 4$. This is $(3/2)(4)^{1/2} = (3/2) \cdot 2 = 3$. Then $dy = \frac{dy}{dx} \cdot \Delta x = 3 \cdot 0.05 = 0.15$. Hence, an estimate for $(4.05)^{3/2}$ is $4^{3/2} + 0.15 = 2^3 + 0.15 = 8 + 0.15 = 8.15$.

We now want to be able to find a similar approximation for a function of several variables. We continue our tradition of expressing our remarks in terms of a function of two variables, but similar statements can be made about any number of variables.

The geometric interpretation of dy was the change that occurs along the tangent line to a curve. For a function of two variables, the analogous idea is to find the change that occurs if we travel over a plane tangent to the surface of the curve, as shown in figure 7.6.2.

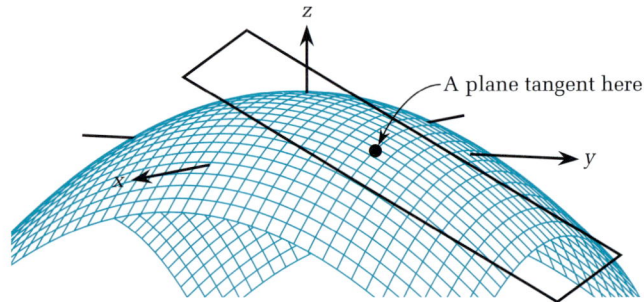

FIGURE 7.6.2

A natural use of this differential would be as an approximation. For example, what volume of paint is used to cover a cylinder 8 inches long, with a radius of 1 inch, to be covered with a 0.03-inch coating of paint? We can compute this by finding the difference between the volume of an 8-inch long cylinder with a radius of 1 inch and a cylinder 8.06 inches long with radius 1.03 inches. (Remember to paint both ends.)

The Total Differential

In the following, $f(x, y)$ is a function of two independent variables. The differentials dx and dy are equal to Δx and Δy, respectively, and these are also independent of each other.

DEFINITION

Total differential

Suppose $z = f(x, y)$ is a function of two variables, and both z_x and z_y exist.

The *total differential* is defined by

$$dz = \left(\frac{\partial}{\partial x} f(x, y)\right) \cdot dx + \left(\frac{\partial}{\partial y} f(x, y)\right) \cdot dy$$

Alternately expressed, $dz = z_x\,dx + z_y\,dy$.

Thinking in terms of rates, we could say that the rate of change of z is approximately $dz =$ (the rate of change of f per unit change in x, with y held constant) · (actual change in x) + (the rate of change of f per unit change in y, with x held constant) · (actual change in y).

EXAMPLE 1   Evaluation of a total differential

Suppose $z = x^2y + 3xy + y^3$. Find dz and evaluate the total differential for $x = 2$, $y = 3$, $dx = 0.05$, $dy = 0.2$.

SOLUTION

The partial derivatives are $z_x = 2xy + 3y$ and $z_y = x^2 + 3x + 3y^2$.

$$dz = \quad z_x \quad dx + \quad z_y \quad dy$$
$$dz = (2xy + 3y)dx + (x^2 + 3x + 3y^2)dy$$

Evaluating, $dz = (2 \cdot 2 \cdot 3 + 3 \cdot 3)0.05 + (2^2 + 3 \cdot 2 + 3 \cdot 3^2)0.2 = 21 \cdot 0.05 + 37 \cdot 0.2 = 1.05 + 7.4 = 8.45$.

Using dx to Approximate $f(x + \Delta x, y + \Delta y)$

The value of dz provides a good approximation for $\Delta z = f(a + dx, b + dy) - f(a, b)$ if dx and dy are fairly small.

EXAMPLE 2   Using dz to approximate $f(x + \Delta x, y + \Delta y)$

Consider a rectangular solid with a square base.

(a) Find the total differential of the volume.

(b) Using the total differential, estimate the change in volume if the base is 10 inches on a side, the height is 20 inches, and the base increases by 0.02 inches and the height decreases by 0.03 inches.

SOLUTION

(a) Let x be one side of the base and y be the height. Then the volume is given by $V = x^2y$, and $dV = V_x \cdot dx + V_y \cdot dy = 2xy\,dx + x^2\,dy$.

Evaluating this for the values given, we have $dV = 2(10)(20)(0.02) + 10^2(-0.03) = 8 - 3 = 5$. See figure 7.6.3.

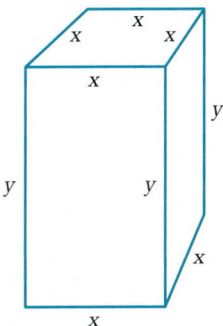

FIGURE 7.6.3

(For the sake of comparison, $V(10.02, 19.97) = (10.02)^2(19.97) = 2004.996$ and $V(10, 20) = 2000$, so the actual change in volume is $2004.996 - 2000 = 4.996$. This is very close to $dV = 5$.) ●

In this example we had a negative value for dy because y was decreasing. We need to be careful to consider the signs involved, so that increasing and decreasing changes are considered in an appropriate manner.

PRACTICE EXERCISE 1

Suppose $z = f(x, y) = x^2y + 3xy + 2y$. Find dz for $x = 2$, $y = -1$, $dx = 0.2$, and $dy = -0.3$.

EXAMPLE 3 Using the total differential to estimate a change in volume

Estimate the volume of paint used to cover a cylinder 8 inches long, with a radius of 1 inch, which is to be covered with a 0.03-inch coating of paint.

SOLUTION

The formula for the volume of a cylinder of radius r and height h is $V = \pi r^2 h$.

The total differential is $dV = V_r dr + V_h dh = 2\pi rh \cdot dr + \pi r^2 \cdot dh$.

Evaluating this for $h = 8$, $r = 1$, $dr = 0.03$, and $dh = 0.06$ (remember both ends), we have $dV = 2\pi \cdot 1 \cdot 8 \cdot 0.03 + \pi \cdot 1^2 \cdot (0.06) = 0.48\pi + 0.06\pi = 0.54\pi \approx 1.6965$ cubic inches. ●

Example 3 demonstrates the need to understand the interpretation of a formula. To go back to a function of one variable for a moment, recall that the volume of a cube of edge x is given by $V = x^3$. If we find $dV = 3x^2 dx$, then an interpretation of this is that three faces, each of area x^2 and of thickness dx, have been added to the existing volume. This is perfectly correct for a cube whose edge increased from x to $x + \Delta x$. However, if we wanted to estimate the additional volume created by adding a thickness dx all over a cube, we need six faces, not three. The edges of the coated cube increased from x to $x + 2dx$. The dV measured a change of volume, with three of the faces remaining in place. As a result, to find the volume created by coating, we would need to use $2dV$ or else use a dx that is twice the thickness of the coating.

Error Estimation

A second use of total differentials is in the estimation of errors. There is often some uncertainty about the various values of measurements, and these errors affect the use of those values.

EXAMPLE 4   Estimating a possible error in volume

The formula for the volume, V, of a conical pile of radius r and height h is $V = (1/3)\pi r^2 h$. Suppose we estimate the radius to be 15, with a possible error of 0.1, and the height to be 6, with a possible error of 0.05. Find an estimation for the possible error in the volume.

SOLUTION

The total differential is $dV = V_r dr + V_h dh$.

In order to allow for the worst possible situation, we take $|V_r| \cdot |dr| + |V_h| \cdot |dh|$ as a bound on the possible error. Absolute values are used to make sure we do not accidentally use positive and negative values of dr and dh, which incorrectly cancel out (maximum) errors.

For our function, we have $V_r = (2/3)\pi rh$ and $V_h = (1/3)\pi r^2$. Substituting our values for the variables gives us that the error is no worse than about

$$|V_r| \cdot |dr| + |V_h| \cdot |dh| = \left(\frac{2}{3}\pi \cdot 15 \cdot 6\right) \cdot 0.1 + \left(\frac{1}{3}\pi \cdot 15^2\right) \cdot 0.05$$

$$= 6\pi + 3.75\pi = 9.75\pi \approx 30.63$$

Thus, if we compute the volume of the pile to be $V = (1/3)\pi r^2 h = (1/3)\pi(15)^2 \cdot 6 = 450\pi \approx 1413.72$, we know this value is no more than (about) 30.63 too large or too small.   ●

In many instances, the relative error of measurement is of interest. Recall that

$$\text{relative error of measurement} = \frac{\text{error of measurement}}{\text{measurement}}$$

For Example 4, we calculated that for a conical pile of radius 15 and height 6, the volume of the pile is 1413.7. We do not know the error of measurement, but we can approximate it by the total differential. Using the estimate, our relative error is about $(30.63/1413.7) = 0.02167$.

Calculating the relative error for the radius and the height, we have $0.1/15 = 0.0067$ and $0.05/6 = 0.0083$, respectively.

Expressed in percentages, the relative percentage error in volume is 2.167%, whereas for the radius and height the relative percentage errors are 0.67% and 0.83%, respectively. This illustrates the cumulative nature of errors. It is quite difficult to have a low percentage of error in a finished product made of many parts, each of which has some error, even though each is a very small error.

7.6 PROBLEMS

Foundations

The problems of this section require the basic skills illustrated by the following:

1. For $f(x, y) = x^2 + 3xy + y^3$, find f_x and f_y.

2. For $f(x, y) = 2x^3 - 3xy + y^2$, find f_x and f_y.

3. For $f(x, y) = e^x + xe^{2y}$, find f_x and f_y.

Exercises

For problems 4–11, evaluate the total differentials for the functions and values indicated.

4. $f(x, y) = x^2 + 3xy + y^3$ at $(1, 2)$, for $dx = 0.1$ and $dy = 0.2$

5. $f(x, y) = 2x^3 - 3xy + y^2$ at $(1, 2)$, for $dx = 0.1$ and $dy = 0.2$

6. $f(x, y) = x^2y + \dfrac{x}{y}$ at $(2, 1)$, for $dx = 0.2$ and $dy = -0.1$

7. $f(x, y) = 2xy^2 - \dfrac{2y}{x}$ at $(2, 1)$, for $dx = -0.1$ and $dy = 0.2$

8. $f(x, y) = e^x + xe^{2y}$ at $(0, 1)$, for $dx = 0.1$ and $dy = 0.2$

9. $f(x, y) = e^{2x} - ye^x$ at $(1, 0)$, for $dx = 0.3$ and $dy = -0.1$

10. $f(x, y) = \ln\left(\dfrac{x + 2y}{xy}\right)$ at $(1, 1)$, for $dx = 0.2$ and $dy = -0.1$

11. $f(x, y) = \ln(x^2 + 3xy)$ at $(1, 0)$, for $dx = -0.1$ and $dy = 0.2$

12. *(Volume of Coating, Electroplating Gold Coins)* Suppose we are producing electroplated gold coins. The core has thickness 1/8 inch and radius 1 inch. The coating is 7-millionths of an inch thick, which is the legal minimum in the United States.
   (a) Estimate by differentials the volume of the coating on one coin. (Do not forget the bottom.)

(b) Estimate the number of coins that could be plated using one cubic inch of gold.

13. *(Volume of Coating, Storage Drums for Waste Disposal)* A company is going to spraycoat cylindrical drums with a fiberglass coating 1/4 cm thick. Each drum is 1 meter tall and has a radius of 0.25 meters.
   (a) Estimate by differentials the volume of the coating on one drum. (Do not forget the bottom.)
   (b) Estimate the number of drums that can be coated using 0.5 cubic meters of fiberglass.

14. *(Volume of Material in Manufacturing)* A manufacturer is producing plastic cylindrical drinking tumblers with walls 1/8 inch thick. Use differentials to estimate the volume of plastic in a glass of radius 1.5 inches and height 5 inches.

15. *(Volume of Material in Manufacturing)* A manufacturer is producing curbside pickup bins for materials to be recycled. Each bin is made of plastic and is rectangular of height 45 cm, width 40 cm, and length 75 cm. The walls of the bin are 1/4 cm thick. Assume the bin has no top. Use differentials to estimate the volume of plastic used to make each bin.

16. *(Wax to Cover the Human Surface Area)* The owner of Waxworks Alive! needs to know the surface area of a human "dummy." The formula of DuBois and DuBois for the surface area of a human body is $S = 0.202(W^{0.425})(H^{0.725})$, where S is in square meters, W is in kilograms, and H is in meters.

   A particular "dummy" is of a person who, when alive, was said to weigh 80 kilograms and be 1.8 meters tall. By using dS for an estimate, find how much error in S may occur if the measurements are off by 2 kilograms in weight and 0.05 meters in height.

17. The area of a triangle is given by $A = (1/2)bh$, for a base of b and a height of h. We measure one triangle as having $b = 200$ and $h = 50$. Suppose the possible errors in the measurements for base and height are 0.1 and 0.3, respectively.

(a) Find the possible error in the calculated area.

(b) Find the relative percentage errors in the base, the height, and the area.

18. The number of tickets sold each week at Waxworks Alive! is given by $T(x, y) = 500 + 10x^{0.3}y^{0.5}$, where x dollars are spent on newspaper ads and y dollars are spent on handout flyers. Currently $600 is being spent on newspaper ads and $400 is being spent on handout flyers. Estimate the effect on ticket sales by spending $20 more on newspaper ads and $25 more on flyers.

19. The weekly demand for tickets at a concert can be expressed as $d(x, p) = 10,000(x^{0.3}/p^2) + 500$, where p is the price of the ticket and x is the number of dollars spent on advertising. Suppose $x = \$20,000$ and $p = 10$. Estimate the effect of decreasing the ticket price by $1 and increasing the advertising budget by $1000.

20. **(Gathering Food in a Changing Environment)** In Section 7.3, Example 3, we found the energy gain of a bird to be given by $E = 120P + 11SP - 0.5S^2P - 10P^2 - 30$, when the bird spent S calories searching and the prey contained P calories. Estimate the effect on the energy gain if the calories spent searching increase from 11 to 12 and the calories of the prey decrease from 9 to 8.

Enrichment Problems

Evaluate the total differentials for the functions and values indicated in problems 21 and 22.

21. $f(x, y, w) = x^2y + yw + w^3$ at $(1, 1, 0)$ for $dx = 0.1, dy = 0.2, dw = -0.1$

22. $f(x, y, w) = xyw^2 + x^3w$ at $(1, 1, -1)$ for $dx = 0.2, dy = -0.1, dw = 0.2$

SOLUTION TO PRACTICE EXERCISE

1. For $z = x^2y + 3xy + 2y$ and for $x = 2, y = -1$, $dx = 0.2$, and $dy = -0.3$,

$$z_x = 2xy + 3y, \quad z_y = x^2 + 3x + 2$$

$$\begin{aligned}
dz &= z_x\,dx + z_y\,dy \\
&= (2xy + 3y)dx + (x^2 + 3x + 2)dy \\
&= (2 \cdot 2 \cdot (-1) + 3(-1))(0.2) + \\
&\quad (2^2 + 3 \cdot 2 + 2)(-0.3) \\
&= (-4 - 3)(0.2) + (12)(-0.3) \\
&= -1.4 - 3.6 = -5
\end{aligned}$$

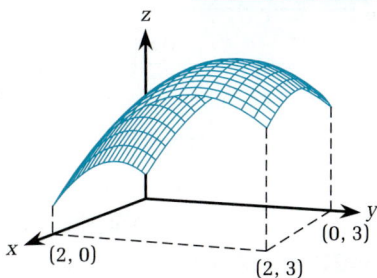

This is a cross section of the region under the surface $z = -x^2 - y^2 + 2x + 4y + 1$ for $0 \le x \le 2$ and $0 \le y \le 3$.

FIGURE 7.7.1

7.7 DOUBLE INTEGRALS AND APPLICATIONS

Definite integrals for functions of one variable were introduced in Sections 5.2 and 5.3, and we found two principal interpretations: the area of a region and the total quantity of something for which we knew the rate of change. In this section we extend the idea of a definite integral to functions of two variables and offer several interpretations.

One application is to determine the volume of a solid, such as shown in figure 7.7.1.

Suppose figure 7.7.1 represents the volume of oil located under a particular section of land. If we can determine this volume, we can determine the value of that oil and thus the value of the mineral rights.

Another example would be to let $f(x, y)$ represent the quantity of corn produced at the point (x, y) of an irrigated cornfield. Then the volume of the region under the surface defined by this function represents the total production of the cornfield.

A third example could consider the water pouring out of the end of a rectangular chute at an amusement park ride. The rate of flow is greater for water at the top of the chute. By knowing the rates of flow at each point in a cross section, we can determine the total flow through the chute.

These examples provide the flavor of definite integrals for a function of two variables.

Double Integrals

Suppose R is some bounded region of the xy-plane, as shown in figure 7.7.2.

If we graph R on a three-dimensional axis, it is as shown in figure 7.7.3.

Suppose the function $f(x, y) \geq 0$ is defined over all of R. We can estimate the volume of the solid under the surface $z = f(x, y)$ and over the region R in the following manner.

The entire region of R can be subdivided into small subregions. There are many ways that this could be done, but let us agree to use uniformly sized square subregions, or any part of such a subregion that lies in R. Let the collection of all these subregions be represented by P. Suppose that somewhere in each subregion of P we choose a point (x_P, y_P) at which to evaluate f. Next, let A_P be the area of the subregion with (x_P, y_P) in it, and multiply A_P by the value of $f(x_P, y_P)$ for that subregion. This can be thought of as giving the volume of a typical "tower." (Think of the skyscrapers in any major city.)

Finally, take the sum of all the products $A_P \cdot f(x_P, y_P)$. This can be represented as $\sum_P (A_P \cdot f(x_P, y_P))$, and is shown in figure 7.7.4.

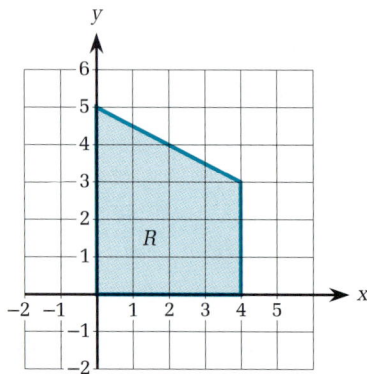

FIGURE 7.7.2

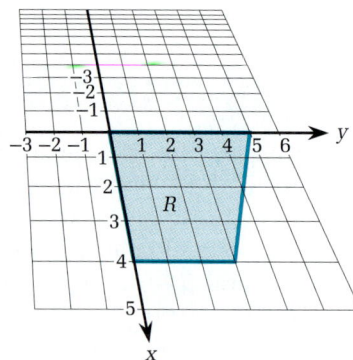

FIGURE 7.7.3

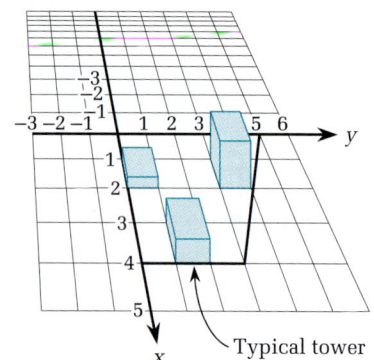

Typical tower

FIGURE 7.7.4

If we take partitionings of R that have smaller subregions, it would seem likely that the approximation for the volume would improve. In fact, we may even say that the volume is defined as the number that is the limit of such sums as the size of the largest subregion area goes to zero. Keep in mind that we are using square subregions (or portions thereof).

Using $f(x, y)$ as simply some function of two variables, not necessarily the height of some surface, lets us introduce a more general definition.

DEFINITION

Double integral

Suppose $f(x, y)$ is a function defined over a region R. Suppose P is some partition of R into subregions and that points (x_P, y_P) are chosen so that each point is in one subregion. If, as the sizes of the subregions go to zero, the various sums $\sum_P(f(x_P, y_P) \cdot (\text{area of the subregion}))$ go to some number, then that number is the *double integral of $f(x, y)$ over the region R* and is denoted as

$$\iint\limits_R f(x, y) \, dR$$

Iterated Integrals

The evaluation of a double integral is carried out by a process called iteration. This uses the Fundamental Theorem of Calculus and represents the region R in a special way.

EVALUATION OF A DOUBLE INTEGRAL BY AN ITERATED INTEGRAL FIRST WITH RESPECT TO y AND THEN WITH RESPECT TO x

Suppose R is a region described by an interval of values for x and two functions $g(x)$ and $h(x)$ such that $a \leq x \leq b$ and $g(x) \leq y \leq h(x)$. Suppose $\iint_R f(x, y) \, dR$ exists. Then

$$\iint\limits_R f(x, y) \, dR = \int_a^b \left(\int_{g(x)}^{h(x)} f(x, y) \, dy \right) dx$$

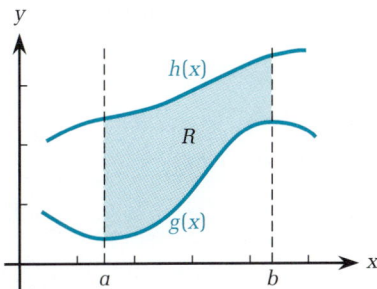

FIGURE 7.7.5

Figure 7.7.5 is a geometric interpretation of such a region R.

The evaluation of this iterated integral requires that we first regard the outer variable x as actually a constant that is somewhere between a and b and find a definite integral with respect to y, the inner variable. The effect of this is to find the area of one particular trace of the surface.

Having the inside integral represent traces, we go back to the x variable that we had held constant and find the definite integral obtained by integrating with respect to it. The cumulative effect is to add together all the volumes we would have by using each trace as one cross section.

A familiar example might be a loaf of sliced bread. We can find the area for one side of each slice. Assuming that area can be multiplied by the thickness of the slice to approximate the volume of a slice, the total volume of the loaf can be found by adding together all the volumes of the individual slices. For thin slices we obtain good approximations.

EXAMPLE 1   Evaluating a double integral by using an iterated integral

Suppose R is the region described by $0 \le x \le 2$ and $0 \le y \le x$, as shown in figure 7.7.6. If $f(x, y) = x^2 - 2xy + 3y^2$, find

$$\iint_R f(x, y) \, dR.$$

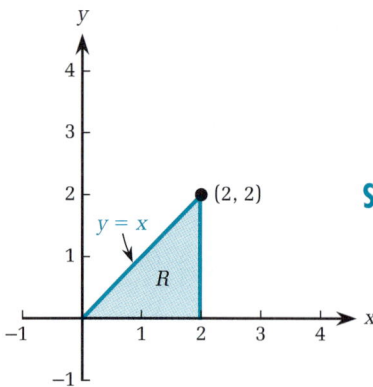

FIGURE 7.7.6

SOLUTION

We can use $\iint_R f(x, y) \, dR = \int_a^b \left(\int_{g(x)}^{h(x)} f(x, y) \, dy \right) dx$ by letting $a = 0$, $b = 2$, $g(x) = 0$, and $h(x) = x$.

Starting with $\int_0^2 \left(\int_0^x (x^2 - 2xy + 3y^2) \, dy \right) dx$ and using the Fundamental Theorem of Calculus, we need to realize that within the inside integral, the x is just a constant, some fixed number between 0 and 2. Hence, in finding an antiderivative with respect to y, we let x be just as much a constant as the 2 or the 3. Then

$$\int_0^2 \left(\int_0^x (x^2 - 2xy + 3y^2) \, dy \right) dx = \int_0^2 \left((x^2y - xy^2 + y^3) \Big|_{y=0}^{y=x} \right) dx$$

by finding the antiderivative with respect to y. Now we must evaluate it at 0 and at x. We thus get

$$\int_0^2 \left[(x^2 \cdot x - x \cdot x^2 + x^3) - (x^2 \cdot 0 - x \cdot 0^2 + 0^3) \right] dx =$$

$$\int_0^2 x^3 \, dx = \frac{x^4}{4} \Big|_{x=0}^{x=2} = \frac{2^4}{4} - \frac{0^4}{4} = \frac{16}{4} = 4.$$

It is possible that the region R may be described so that y is given to be in an interval and then the values for x lie between two functions of y. If this happens, then we define an iterated integral with the values for y on the outermost integral.

EVALUATION OF A DOUBLE INTEGRAL BY AN ITERATED INTEGRAL WITH RESPECT TO *x* AND THEN WITH RESPECT TO *y*

Suppose R is a region described by an interval of values for y and two functions $g(y)$ and $h(y)$ such that $a \leq y \leq b$ and $g(y) \leq x \leq h(y)$. Suppose that $\iint_R f(x, y)\, dR$ exists. Then

$$\iint_R f(x, y)\, dR = \int_a^b \left(\int_{g(y)}^{h(y)} f(x, y)\, dx \right) dy$$

The region R of Example 1 can be described by $0 \leq y \leq 2$, $y \leq x \leq 2 - y$. If we use this representation for R, then we have

$$\iint_R f(x, y)\, dR = \int_0^2 \left(\int_y^{2-y} f(x, y)\, dx \right) dy$$

$$= \int_0^2 \left(\int_y^{2-y} (x^2 - 2xy + 3y^2)\, dx \right) dy$$

$$= \int_0^2 \left[\left(\frac{x^3}{3} - x^2 y + 3y^2 x \right) \Big|_{x=y}^{x=2-y} \right] dy$$

But, notice that the next step of substituting y and $2 - y$ for x will create terms that are quite complicated.

It is often the case that different descriptions of the region R lead to iterated integrals of varying difficulties to solve.

Alternate Descriptions of Regions in a Plane

We have seen that the region described by $0 \leq y \leq 2$, $y \leq x \leq 2$ is the same as the region described by $0 \leq x \leq 2$ and $0 \leq y \leq x$. Both are shown in figure 7.7.7.

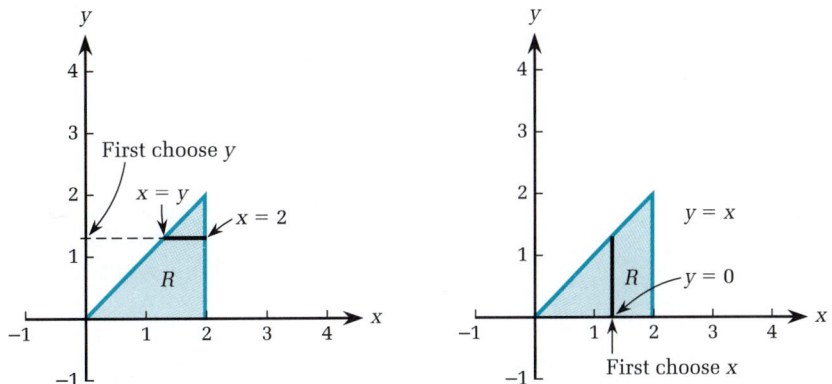

FIGURE 7.7.7

Given the job of describing a region in the plane, we consider whether it is better to use x as the initial variable and then describe the variation in y in terms of x or, alternately, to describe the region by first specifying y and then expressing the range of x values in terms of y. We are guided by two principles. First, which description seems easier in terms of the algebraic expressions? Second, which formulations of definite integrals are easier (or even at all possible) to evaluate?

Choosing an alternate description usually requires us to find an inverse function because we may be given y as a function of x and then be searching for x as a function of y, or conversely. Example 2 illustrates this.

EXAMPLE 2   Finding an alternate description of a region in the xy-plane

Suppose a region R is given by $0 \le x \le 4$, and for each value of x the values of y are given by $0 \le y \le \sqrt{x}$. Describe this region by first describing the values of y and then specifying the values of x in terms of functions of y.

SOLUTION

The region can be described by $R = \{(x, y) : 0 \le x \le 4, 0 \le y \le \sqrt{x}\}$. A sketch of the region is shown in figure 7.7.8.

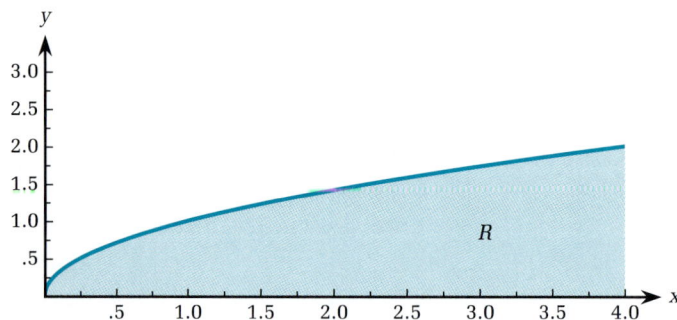

FIGURE 7.7.8

We see that the smallest value of y is 0 and the largest value of y occurs at $\sqrt{4} = 2$, giving us $0 \le y \le 2$. For any particular value of y, the values of x go from the x value on the curve of $y = \sqrt{x}$ up to an x-value of 4. To find the actual values of x, we solve $y = \sqrt{x}$ for x in terms of y. This gives $y^2 = x$. Hence the values for x go from $x = y^2$ to $x = 4$.

In summary, R can be described as follows:

$$\{(x, y) : 0 \le y \le 2, y^2 \le x \le 4\}$$

PRACTICE EXERCISE 1

Suppose the region R is described by $0 \le x \le 4$, $x/2 \le y \le 2$. Describe R by first giving an interval for y and then specifying the values of x in terms of functions of y.

Changing the Order of Integration

It makes no difference whether the values for x are used on the inner integral and first we take the antiderivative with respect to x, or the values for x are used on the outer integral and first we take the antiderivative with respect to y and evaluate that using the limits of y. The important thing is to keep in mind which limits accompany which variable.

EXAMPLE 3   Showing that changing the order of integration does not affect the value of the integral

Find the double integral of $f(x, y) = y + 2x$ over the region R shown in Example 2.

SOLUTION

A sketch of the region and the surface of $z = f(x, y)$ is shown in figure 7.7.9.

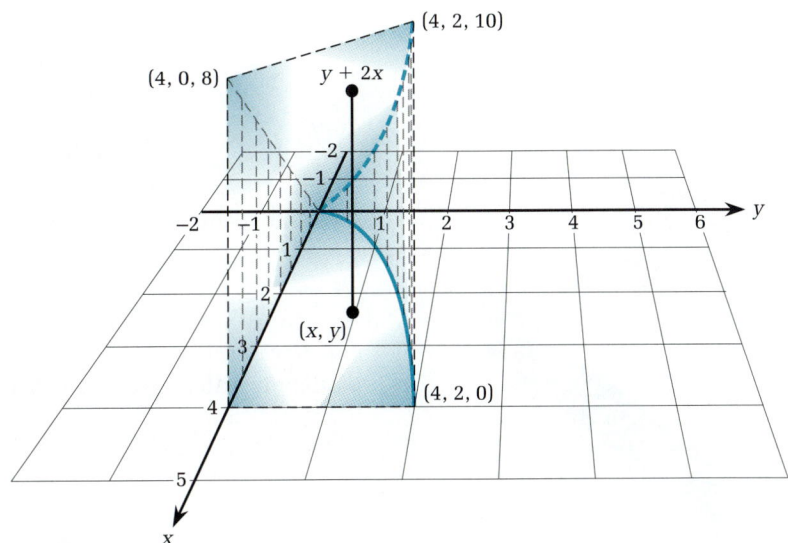

FIGURE 7.7.9

Using the description of R with x being the outside variable, we have $\int_0^4 \int_0^{\sqrt{x}} (y + 2x)\, dy\, dx$. Using the description of R with y being the outside variable, we have $\int_0^2 \int_{y^2}^4 (y + 2x)\, dx\, dy$. Evaluating the first of these, and remembering to treat x as a constant when finding the antiderivative with respect to y, we get the following:

$$\int_0^4 \int_0^{\sqrt{x}} (y + 2x)\, dy\, dx = \int_0^4 \left(\int_0^{\sqrt{x}} (y + 2x)\, dy \right) dx$$

$$= \int_0^4 \left[\left(\frac{y^2}{2} + 2xy \right) \Big|_{y=0}^{y=\sqrt{x}} \right] dx$$

$$= \int_0^4 \left[\left(\frac{(\sqrt{x})^2}{2} + 2x\sqrt{x} \right) - 0 \right] dx$$

$$= \int_0^4 \left(\frac{x}{2} + 2x^{1.5} \right) dx$$

$$= \frac{x^2}{2 \cdot 2} + 2\left(\frac{x^{2.5}}{2.5} \right) \Big|_{x=0}^{x=4}$$

$$= \frac{4^2}{4} + 2\left(\frac{4^{2.5}}{2.5} \right) - 0 = 29.6$$

Now we start over with the second formulation of the iterated integral.

$$\int_0^2 \int_{y^2}^4 (y + 2x)\, dx\, dy = \int_0^2 \left((yx + x^2) \Big|_{x=y^2}^{x=4} \right) dy$$

$$= \int_0^2 \left[(y \cdot 4 + 4^2) - (y \cdot y^2 + (y^2)^2) \right] dy$$

$$= \int_0^2 (4y + 16 - y^3 - y^4)\, dy$$

$$= 2y^2 + 16y - \frac{y^4}{4} - \frac{y^5}{5} \Big|_{y=0}^{y=2}$$

$$= 8 + 32 - 4 - \frac{32}{5} = 29.6$$

Thus we have the same value as before.

WARNING

Make certain that when we are finding antiderivatives in connection with an iterated integral that we regard an expression, such as x or y, as a constant unless we have d(that same variable).

Example 4 and its accompanying practice exercise present another instance of iterated integration and illustrate that reversing the order does not affect the value of the double integral.

EXAMPLE 4 Seeing that changing the order of integration does not affect the double integral's value

Find the double integral of $f(x, y) = x^2 y - x + 2y + 3$ over the region described by $0 \leq y \leq 3$, where for each value of y, the values for x are determined by $1 \leq x \leq 2$.

SOLUTION

Keep in mind that in finding an antiderivative with respect to x, y is just a constant. Then

$$\int_0^3 \int_1^2 (x^2 y - x + 2y + 3) \, dx \, dy$$

$$= \int_0^3 \left(\left(\frac{x^3}{3} y - \frac{x^2}{2} + 2yx + 3x \right) \Big|_{x=1}^{x=2} \right) dy$$

$$= \int_0^3 \left[\left(\frac{2^3}{3} y - \frac{2^2}{2} + 2y \cdot 2 + 3 \cdot 2 \right) - \left(\frac{1^3}{3} y - \frac{1^2}{2} + 2y \cdot 1 + 3 \cdot 1 \right) \right] dy$$

$$= \int_0^3 \left(\frac{13}{3} y + \frac{3}{2} \right) dy$$

$$= \left(\frac{13}{6} y^2 + \frac{3}{2} y \right) \Big|_{y=0}^{y=3}$$

$$= \frac{13}{6} (3^2) + \frac{3}{2} \cdot 3 = \frac{39}{2} + \frac{9}{2}$$

$$= \frac{48}{2} = 24$$

In the following practice exercise, the variable y is between 0 and 3, whereas x has 1 and 2 for its limits. Show the answer to be the same as that found in Example 4.

PRACTICE EXERCISE 2

Evaluate $\int_1^2 \int_0^3 (x^2 y - x + 2y + 3) \, dy \, dx$.

We have seen that the order of integration does not matter as far as the final value is concerned. Furthermore, we have seen an instance in which one choice of the order made a difference in the ease of using the Fundamental Theorem of Calculus, but we could use either order. However, following an introduction to the topic of volume, we see an example in which the order we choose is critical.

Volume and Other Applications

If the function $f(x, y) \geq 0$ over the region R, then $\int \int_R f(x, y)\, dR$ gives the volume under the surface of $z = f(x, y)$ and above the region R. If it happens that for some points in R the function is positive and for some points the function is negative, then $\int \int_R f(x, y)\, dR$ gives the *net* volume between the surface and the xy-plane. This is analogous to the situation we have with a function of one variable.

Let us evaluate the integral that arises from the introductory discussion of this section.

EXAMPLE 5   Finding volume by a double integral

Find the volume of the solid over $\{(x, y) : 0 \leq x \leq 2, 0 \leq y \leq 3\}$, the region above the xy-plane and below the surface of the function $f(x, y) = -x^2 - y^2 + 2x + 4y + 1$. See figure 7.7.1.

SOLUTION

The volume is $\int_0^3 \left(\int_0^2 (-x^2 - y^2 + 2x + 4y + 1)\, dx \right) dy$.

Taking the antiderivative with respect to x and using the Fundamental Theorem of Calculus, we get

$$\int_0^3 \left(\left(-\frac{x^3}{3} - y^2 x + x^2 + 4yx + x \right) \Bigg|_{x=0}^{x=2} \right) dy$$

$$= \int_0^3 \left[\left(-\frac{2^3}{3} - y^2 \cdot 2 + 2^2 + 4y \cdot 2 + 2 \right) - 0 \right] dy$$

$$= \int_0^3 \left(-\frac{8}{3} - 2y^2 + 4 + 8y + 2 \right) dy$$

$$= \int_0^3 \left(-2y^2 + 8y + \frac{10}{3} \right) dy$$

Now, taking the antiderivative with respect to y, and using the Fundamental Theorem of Calculus, we get

$$\left(-\frac{2}{3}y^3 + 4y^2 + \frac{10}{3}y \right) \Bigg|_{y=0}^{y=3} = \left(-\frac{2}{3} \cdot 3^3 + 4 \cdot 3^2 + \frac{10}{3} \cdot 3 \right) - 0$$

$$= -18 + 36 + 10 = 28 \qquad \bullet$$

In Example 6, we find that sometimes the order of integration is critical to our success in being able to find antiderivatives with which to use the Fundamental Theorem of Calculus.

EXAMPLE 6   Showing order of integration may be critical

Find the volume of the solid under the surface $f(x, y) = e^{x^2} + y$ and over the region of the xy-plane bounded by $y = x$, $y = 0$, $x = 1$.

SOLUTION

If we describe R whereby the y values are given in terms of x, then we would use $0 \leq x \leq 1$, and for each x we have $0 \leq y \leq x$. If we choose to express the x values in terms of y, we would have $0 \leq y \leq 1$, and for each y we have $y \leq x \leq 1$. These alternate characterizations are shown in figure 7.7.10.

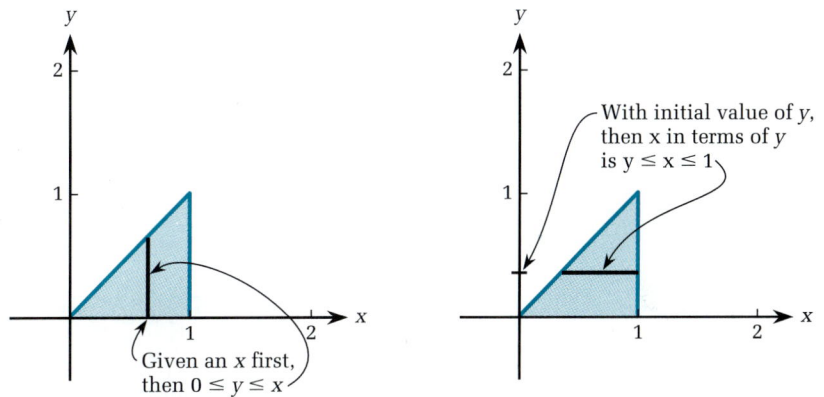

FIGURE 7.7.10

The integrals using the two descriptions are

(7.1)
$$\int_0^1 \int_0^x \left(e^{x^2} + y \right) dy \, dx$$

(7.2)
$$\int_0^1 \int_y^1 \left(e^{x^2} + y \right) dx \, dy$$

The integral of expression (7.1) appears simpler than that of (7.2) because both limits start at 0, but there is an even more compelling reason for using (7.1). We expect to use the Fundamental Theorem of Calculus to do our evaluation. If we start working with (7.2), we need to find an ordinary function whose derivative with respect to x is e^{x^2}. There is no such function. Thus, we must use expression (7.1).

Beginning with (7.1), we first write the integral in its iterated form.

$$\int_0^1 \int_0^x \left(e^{x^2} + y \right) dy\, dx = \int_0^1 \left(\int_0^x (e^{x^2} + y) dy \right) dx$$

$$= \int_0^1 \left[\left(e^{x^2} \cdot y + \frac{y^2}{2} \right) \Big|_{y=0}^{y=x} \right] dx$$

$$= \int_0^1 \left[\left(e^{x^2} \cdot x + \frac{x^2}{2} \right) - \left(e^{x^2} \cdot 0 + \frac{0^2}{2} \right) \right] dx$$

To integrate this, we let $u = x^2$ and use the result

$$\int e^u \cdot \frac{du}{2} = \frac{e^u}{2} + C$$

Then,

$$\left(\frac{e^{x^2}}{2} + \frac{x^3}{2 \cdot 3} \right) \Big|_{x=0}^{x=1} = \left(\frac{e^1}{2} + \frac{1}{6} \right) - \left(\frac{e^0}{2} + \frac{0}{6} \right)$$

$$= \frac{e}{2} + \frac{1}{6} - \frac{1}{2} = \frac{e}{2} - \frac{1}{3} \approx 1.0258$$

Applications of double integrals occur whenever we have a multivariant function whose values at some coordinates represent some rate of some quantity and we want to know the total quantity. Example 7 uses a cornfield, but it could be any situation in which we know the amount of a quantity located at each point.

EXAMPLE 7   Finding the total yield of a cornfield

Suppose a cornfield is rectangular, 1320 feet by 660 feet (a quarter mile by an eighth of a mile), with the long side next to an irrigation canal. The productivity of the field can be approximated by the formula $p = 0.003 - 0.000001x$, where p is in bushels per square foot and x is the number of feet from the canal. Find the total production from the cornfield.

SOLUTION

Let y be the distance down the field from one end and x be the distance from the canal. The double integral, taken over the entire field, will give the total production.

$0 \le y \le 1320$ and $0 \le x \le 660$, so we have

$$\int_0^{1320} \left(\int_0^{660} (0.003 - 0.000001x)\, dx \right) dy$$

$$= \int_0^{1320} \left(\left(0.003x - 0.000001 \cdot \frac{x^2}{2} \right) \Big|_{x=0}^{x=660} \right) dy$$

$$= \int_0^{1320} \left[\left(0.003 \cdot 660 - 0.000001 \cdot \frac{660^2}{2} \right) - 0 \right] dy$$

$$= \int_0^{1320} (1.98 - 0.2178) \, dy$$

$$= \int_0^{1320} 1.7622 \, dy$$

$$= 1.7622y \Big|_{y=0}^{y=1320} = 2326.1$$

Average Value

Values of some items are often expressed as averages. For instance, in Example 7, corn production would often be expressed by the number of bushels per acre. The field was 660 by 1320 feet, which is 871,200 square feet. There are 42,580 square feet in an acre, so the field of 871,200 square feet is 20.46 acres. To determine the average number of bushels per acre, we divide the total production by the number of acres and find that the average yield per acre is 113.69 bushels per acre.

The average value of a function is determined by dividing its integral over a region by the area of the region. In symbols, we may express this as follows.

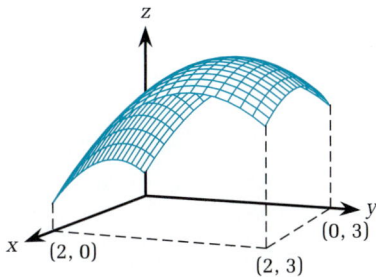

AVERAGE VALUE OF A FUNCTION OVER A REGION

The *average of f(x, y) over the region R* is given by

$$\frac{\displaystyle\iint_R f(x, y) \, dR}{\displaystyle\iint_R dR}$$

To illustrate, in Example 5 we found the total volume under $f(x, y) = -x^2 - y^2 + 2x + 4y + 1$ and above the region $R = \{(x, y) : 0 \le x \le 2, 0 \le y \le 3\}$ to be 28. R is a rectangular area 2 by 3 in size, so its area is $2 \cdot 3 = 6$. The average of $f(x, y)$ is then $28/6 \approx 4.6667$.

In figure 7.7.11 we see both the surface of $f(x, y)$ over R and a level surface whose height is the average of $f(x, y)$ over R. The volumes under both are equal.

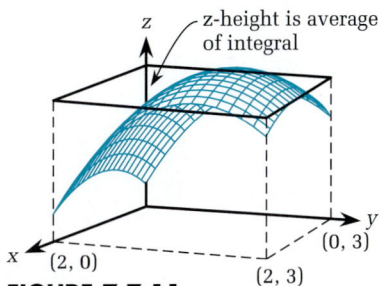

FIGURE 7.7.11

7.7 PROBLEMS

Foundations

The problems of this section require the basic skills illustrated by the following:

1. Find $\int_1^2 x^2 \, dx$.

2. Find $\int_1^2 x^2 \, dy$.

3. Find $\int_1^y dx$.

4. Find $\int_1^3 yx^2 \, dx$.

5. Find $\int_1^3 yx^2 \, dy$.

6. Find $\int_1^y x^2 \, dx$.

7. Describe the region

$$R = \{(x, y) : 0 \le x \le 4, 0 \le y \le x\}$$

so that $0 \le y \le 4$ and then x depends on y.

Exercises

Evaluate the integrals in problems 8–23.

8. $\int_0^1 \int_0^2 (x^2 y) \, dy \, dx$

9. $\int_0^2 \int_0^1 (xy^2) \, dy \, dx$

10. $\int_0^1 \int_1^2 (x^3 y + 3x - y) \, dy \, dx$

11. $\int_0^2 \int_1^3 (x^2 y^2 - 2y) \, dy \, dx$

12. $\int_0^2 \int_0^1 (x^2 y^3 + 2x + y) \, dx \, dy$

13. $\int_1^2 \int_1^3 (x^{-2} y + x + y^{-2}) \, dx \, dy$

14. $\int_1^2 \int_0^9 (x \sqrt{y}) \, dy \, dx$

15. $\int_0^9 \int_0^2 (y^2 \sqrt{x}) \, dy \, dx$

16. $\int_1^4 \int_1^2 \left(\frac{1}{xy}\right) \, dy \, dx$

17. $\int_1^2 \int_1^e \left(\frac{x}{y}\right) \, dy \, dx$

18. $\int_1^3 \int_0^2 (e^{x+3y}) \, dy \, dx$

19. $\int_{-1}^1 \int_0^1 (e^{2x+y}) \, dy \, dx$

20. $\int_1^2 \int_1^x (x^3 y + 1) \, dy \, dx$

21. $\int_0^1 \int_x^2 (xy^2 - 3) \, dy \, dx$

22. $\int_0^1 \int_y^1 (x^2 y + 2) \, dx \, dy$

23. $\int_0^1 \int_0^y (x^2 y) \, dx \, dy$

In problems 24–35, find the volume of the solid under the surface given by $x = f(x, y)$ and above the region in the xy-plane given by R.

24. $f(x, y) = 2x, R = \{(x, y) : 0 \le x \le 1, 0 \le y \le 2\}$

25. $f(x, y) = 3y + 1, R = \{(x, y) : 0 \le x \le 2, 0 \le y \le 1\}$

26. $f(x, y) = xy, R = \{(x, y) : 1 \le x \le 2, 0 \le y \le 1\}$

27. $f(x, y) = x^2 + y^2, R = \{(x, y) : 0 \le x \le 1, 0 \le y \le 2\}$

28. $f(x, y) = \sqrt{x + y}, R = \{(x, y) : 1 \le x \le 4, 0 \le y \le x\}$

29. $f(x, y) = \sqrt{xy}, R = \{(x, y) : 0 \le x \le 9, 0 \le y \le x\}$

30. $f(x, y) = \frac{1}{x}, R = \{(x, y) : y \le x \le 2y, 1 \le y \le 3\}$

31. $f(x, y) = \frac{1}{x}, R = \{(x, y) : y \le x \le y^2, 1 \le y \le 4\}$

32. $f(x, y) = 2x + y, R$ bounded by $y = x$ and $y = x^2$

33. $f(x, y) = x + 3y, R$ bounded by $y = x$ and $y = \sqrt{x}$

34. $f(x, y) = e^{x+y}, R$ bounded by $y = 0, y = 2x, x = 2$

35. $f(x, y) = x^3 y, R$ bounded by $y = \sqrt{x}, y = 0$, and $x = 4$

For problems 36–41, find the average value of the function $f(x, y)$ over the region R.

36. $f(x, y) = 2x + y, R = \{(x, y) : 0 \le x \le 1, 0 \le y \le 2\}$

37. $f(x, y) = x + 3y, R = \{(x, y) : 0 \le x \le 2, 0 \le y \le 2\}$

38. $f(x, y) = e^x + 1, R = \{(x, y) : 0 \le x \le 1, 0 \le y \le 1\}$

39. $f(x, y) = x + e^y$,
$R = \{(x, y) : 0 \le x \le 1, 0 \le y \le 1\}$

40. $f(x, y) = x^2 + y^2$,
$R = \{(x, y) : -1 \le x \le 1, 0 \le y \le 2\}$

41. $f(x, y) = 2x^2 - y^2$,
$R = \{(x, y) : 0 \le x \le 1, 0 \le y \le x\}$

42. *(Water Use by Amusement Park Ride)*   The rate at which water is flowing out the top of a rectangular opening, such as at a boat ride in an amusement park, depends on the depth of the water. Water at the top is flowing faster than at the bottom. The following information has converted a speed of approximately 2 miles per hour into gallons per square foot per second.
    Suppose the rate of water flow is given by $18 + 3y$ gallons per square foot per second, where y measures the depth of water flowing through a rectangular opening 4 feet wide and 2 feet deep, as shown in figure 7.7.12. Find the total number of gallons per second used by this ride.

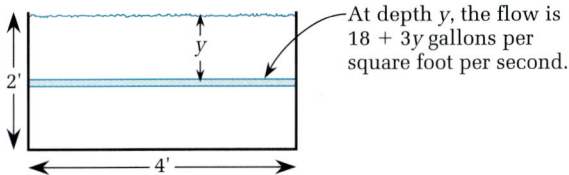

FIGURE 7.7.12

43. *(Averages, Landfill)*   A town wishes to close a sanitary landfill. They find that at present the height of the surface of the landfill above the water table can be approximated by $f(x, y) = 0.002x + 0.01y + 2$ feet, over a region defined by $R = \{(x, y) : 0 \le x \le 100, 0 \le y \le 200\}$, where x and y are measured in feet.

(a) If the town levels the landfill, how far above the water table would the surface be?

(b) If the Environmental Commission rules that the landfill must be at least 3.5 feet above the water table, how many (if any) additional cubic feet of material must be brought in?

44. A mineral deposit is of thickness $f(x, y) = 4xy/(x^2 + y)^2$ over the region $R = \{(x, y) : 1 \le x \le 3, 1 \le y \le 2\}$.

(a) Find the total volume of the deposit.

(b) Find the average thickness of the deposit.

Enrichment Problems

Evaluate the integrals in problems 45 and 46.

45. $\displaystyle\int_0^1 \int_0^y (y^2 e^{xy})\, dx\, dy$ **46.** $\displaystyle\int_0^1 \int_0^1 \int_0^2 x\, dx\, dy\, dz$

47. A college student who took calculus and was hired by the amusement park mentioned in problem 42 finds that the cross section of the ride need not be a rectangle, but can be a parabola instead. Suppose the region

$$R = \{(x, y) : -2 \le x \le 2, x^2/2 \le y \le 2\}$$

is the cross section of the ride's opening.

(a) Find the number of gallons per second that would flow through the opening.

(b) Assume water costs 7 cents per thousand gallons to pump back to the top of the ride. How much money would be saved in a 10-hour day by using the parabolic cross section in contrast to the rectangular cross section of problem 42?

SOLUTIONS TO PRACTICE EXERCISES

1. The region R described by $0 \le x \le 4$, $x/2 \le y \le 2$ can be described as $0 \le y \le 2$, $0 \le x \le 2y$.

2.
$$\int_1^2 \left(\int_0^3 (x^2 y - x + 2y + 3)\, dy \right) dx$$
$$= \int_1^2 \left(\left(x^2 \frac{y^2}{2} - xy + y^2 + 3y \right) \Big|_{y=0}^{y=3} \right) dx$$
$$= \int_1^2 \left[\left(x^2 \left(\frac{3^2}{2} \right) - x \cdot 3 + 3^2 + 3 \cdot 3 \right) - 0 \right] dx$$
$$= \int_1^2 \left(\frac{9}{2} x^2 - 3x + 18 \right) dx$$

$$= \left(\frac{9}{2} \cdot \frac{x^3}{3} - 3 \cdot \frac{x^2}{2} + 18x \right) \Big|_{x=1}^{x=2}$$
$$= \left(\frac{9}{2} \cdot \frac{2^3}{3} - 3 \cdot \frac{2^2}{2} + 18 \cdot 2 \right)$$
$$- \left(\frac{9}{2} \cdot \frac{1^3}{3} - 3 \cdot \frac{1^2}{2} + 18 \cdot 1 \right)$$
$$= 12 - 6 + 36 - \frac{3}{2} + \frac{3}{2} - 18 = 24$$

CHAPTER 7 REVIEW

Discuss or define:

1. Function of several variables
2. Domain and range of a function of several variables
3. Right-hand coordinate system
4. Surface
5. Level curves
6. Trace
7. Budget lines
8. Cobb-Douglas production function
9. Returns to scale
10. Limits and continuity for functions of two variables
11. Partial derivatives
12. Marginal productivity of labor and capital
13. Competing and complementary products (in terms of partial derivatives)
14. Second-order partial derivatives
15. Relative maximum and relative minimum
16. Critical point
17. Saddle point
18. Test for relative extreme values for functions of two variables
19. Lagrange's method
20. Method of Lagrange multipliers
21. Marginal productivity of money
22. Least Cost Rule in economics
23. Least squares test
24. Least squares line
25. Total differential
26. dz as an approximation to Δz
27. Error estimation by dz
28. Double integrals
29. Iterated integrals
30. Volume by a double integral
31. Describing regions in the plane
32. Changing order of integration
33. Average of $f(x, y)$ over a region

REVIEW PROBLEMS FOR CHAPTER 7

For each of the functions in problems 1–5, give the largest possible domain and evaluate the function at the indicated points.

1. $f(x, y) = 3x + y$ at $(0, 0)$, $(-2, 1)$, $(2, 2)$

2. $f(x, y) = 2x^2 + \dfrac{1}{xy}$ at $(1, 1)$, $(2, -1)$, $(1, 3)$

3. $f(x, y) = \dfrac{x^2 + y^2}{x + y}$ at $(1, 1)$, $(-1, 2)$, $(2, 0)$

4. $f(x, y) = \ln(x^2 + y^2)$ at $(1, 0)$, $(1, 2)$, $(2, -1)$

5. $f(x, y, z) = \dfrac{\sqrt{xyz}}{e^{x+y+z}}$ at $(0, 0, 0)$, $(0, 1, 0)$, $(1, 1, 1)$

6. For $f(x, y) = 3x^2 + xy + y^2$, find f_x and f_y.

7. For $f(x, y) = \dfrac{-2x^2 + 1}{y^2 - 1}$, find $\dfrac{\partial f}{\partial x}$ and $\dfrac{\partial f}{\partial y}$.

8. For $f(x, y) = e^{y/x}$, find f_x and f_y.

9. For $f(x, y, z) = -x^3 + yz^2$, find $f_x, f_y,$ and f_z.

10. Suppose a production function is given by $P(x, y) = (x^3 + 2y)/(xy)$ and currently we are at $(3, 2)$. Find the following increases in production.
    (a) Increase x by 0.2.
    (b) Increase y by 0.5.

For the functions in problems 11 and 12, find all the second-order partial derivatives.

11. $3x^3 - y^2$ 12. $xy^{-2} + \ln y$

The equation $z_{xx} + z_{yy} = 0$ is called the *Laplace equation*. Show that the functions in problems 13 and 14 satisfy this equation.

13. $z = x^2 - y^2$

14. $z = \ln[(x - a)^2 + (y - b)^2]$

Find the critical points for problems 15–17.

15. $f(x, y) = -x^2 - 2x - y^2 + 4y - 4$

16. $f(x, y) = e^{-x^2 - y^2 + 2x - 1}$

17. $f(x, y) = 3y^2 + 2xy + x^2 + 10y + 2x + 1$

18. Suppose a production function is given by $P(x, y) = 5x^{0.3}y^{0.7}$, where x is the measure of labor and y is the measure of capital.
(a) Find the marginal productivity of labor.
(b) Find the marginal productivity of capital.

Find all relative extremes of the functions in problems 19 and 20.

19. $f(x, y) = -x^2 - 2x - y^2 + 4y - 4$

20. $f(x, y) = x^2 + y^3 - 6xy$

21. The Pueblo Bonita Adobe Brick Co. has two plants. The cost of producing x tons of adobe bricks at Plant A is $f(x) = 5000 + 2x + 0.01x^2$, and the cost of y tons at Plant B is $g(y) = 1000 + 5y + 0.2y^2$. The bricks sell for $50 per ton, no matter where produced. Find how many tons to produce at each plant in order to maximize the total profit, and find that maximum profit.

22. Find the least squares line for each of the following sets of points.
(a) {(0, 2), (1, 4), (3, 5)}
(b) {(1, 2), (2, 0), (3, −1), (5, −2)}

23. Suppose a Cobb-Douglas production function is given by $f(x, y) = x^{0.85}y^{0.15}$ and that $2x + y = 70$.

(a) Maximize $f(x, y)$ by using the method of Lagrange multipliers.
(b) Find the marginal productivity of money at $x = 20$, $y = 30$.

24. Use the method of the Lagrange multiplier to find the minimum of $x^2 + xy + y^2$ with $x + 3y = 1$.

25. Use the method of the Lagrange multiplier to find the maximum of $-x^2 - 2y^2 + 5x + y$ with $x = y + 1$.

26. The Great Chipmunk Nut Co. is manufacturing a new sampler box of eight kinds of nuts. They will use a cylinder, divided into eight sectors, as shown in figure 7.8.1. There is to be a clear plastic top. Not considering the top and edge, what are the dimensions of a box that holds 50 cubic inches and is made of the least total surface area including the surface area of those dividers inside the box?

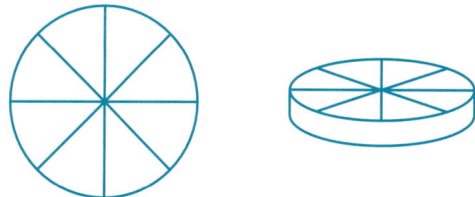

FIGURE 7.8.1

27. Find the total differential of $f(x, y) = x^2 + e^{2xy}$.

28. Suppose the sales of hot dogs can be modeled by
$$S(x, y) = 1000\left(\frac{x^{0.15}}{y - 60}\right), y \geq 70,$$ where x is the number of square feet in the vendor's sign and y is the Fahrenheit temperature.
(a) Find $S(20, 80)$.
(b) If x could be in error by 2 square feet and y could be in error by 5 degrees, use differentials to estimate the error in the value of $S(20, 80)$.

Evaluate the integrals in problems 29 and 30.

29. $\int_0^1 \int_0^3 (x^2 y + y) \, dy \, dx$

30. $\int_0^1 \int_0^x xy^2 \, dy \, dx$

Let the region R in problems 31 and 32 be given by $\{(x, y): 0 \le x \le 1, 0 \le y \le x\}$.

31. Find $\displaystyle\iint_R (3x + y)\, dy\, dx$

32. Find $\displaystyle\iint_R e^{x+y}\, dy\, dx$

33. Find the average of $f(x, y) = xy + x^2$ over the following regions.
  (a) $\{(x, y) : 0 \le x \le 1, 0 \le y \le 2\}$
  (b) $\{(x, y) : 0 \le y \le 4, 0 \le x \le \sqrt{y}\}$

SELECTED ANSWERS FOR CHAPTER 1

Section 1.1

1. 10/3 **3.** 4x **5.** 21.7% **7. (a)** 2
(b) 2/5 **(c)** 4 **(d)** 3 **(e)** 5 **(f)** 7/16
(g) $\sqrt{3} - (17/10) \approx +0.03205$ **(h)** $-2a$
9. $(-1, 3)$ **11.** $(-2, \infty)$ **13.** the numbers 5
and -2 **15.** $\{1, 3, 5, 7, 9\}$ **17.** $\{x : x \geq 8\}$
19. (a) 3 **(b)** 2 **(c)** 5 **(d)** 2

21. 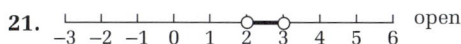 open
23. 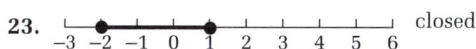 closed
25. 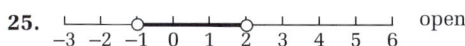 open
27. 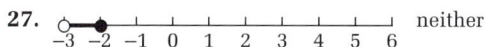 neither
29.  open
31. 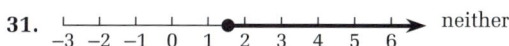 neither
33. 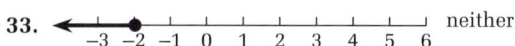 neither

35. $(-1/2, \infty)$ **37.** $(-\infty, 1]$ **39.** $(-3/4, \infty)$
41. $[1, 7/2]$ **43.** $[-1, 0]$ **45.** $[-2, 4]$
47. $(-4/3, 2)$ **49.** for no values **51.** $[0, 4/3]$
53. $(1.5, 3.1)$ **55.** 0.8 **57. (a)** 0.732
(b) 0.228 **59. (a)** 0 to 2 **(b)** completely evenly
distributed **(c)** completely isolated populations
61. 0.466, West **63.** 81.7 to 82.4

Section 1.2

1. $2 \leq x$ **3.** $4 + 4h + h^2$ **5.** $\mathbb{R}, -2, 4$ **7.** $\mathbb{R}$,
2, 1/2 **9.** $\mathbb{R}, 3, 5, 7$ **11.** $x \geq -1/2, \sqrt{3}, \sqrt{7}$
1
3. $\mathbb{R}$ except -2, 1/4, 1/2, not defined **15.** $\mathbb{R}$
except 0, 19/3, not defined, $10.4 + (1/5.2) \approx 10.5923$
17. $y \leq 2/3, \sqrt{2}, \sqrt{1/2}$, not defined
19. $-2 \leq x \leq 2, 2, 0$ **21.** $\mathbb{R}, 7, -1$ **23.** $\mathbb{R}, 8,$
-2 **25.** function **27.** not a function **29.** 4
31. (a) $0, 8, a^2 - 1, 8 + 6h + h^2, (a + h)^2 - 1$
(b) $2\sqrt{17} \approx 8.2462$ **(c)** $2a + h$ **33.** 1 **35.** 9
37. -1 **39.** $f(x) = \begin{cases} 29 & 0 \leq x \leq 1 \\ 29 + 23\lceil x - 1\rceil, & 1 < x \end{cases}$
43. $\sqrt{5} + \sqrt{10} \approx 5.3983$ **45.** $x^2 + (y - 3)^2 = 4$
47. $(x - 2)^2 + (y - 1)^2 = 9$ **55.** A to D to B to C is
85.6 miles.

Section 1.3

1. $x > -3$ **3.** $3x/(2 - 2x)$
5. (a) $x^2 - 4 - [x/(x + 3)]$ **(b)** $(x^2 - 4)[x/(x + 3)]$
(c) $[(x^2 - 4)(x + 3)]/x$ **(d)** -3.84
(e) $[x/(x + 3)]^2 - 4$ **(f)** 0 **(g)** $(x^2 - 4)/(x^2 - 1)$
(h) $\mathbb{R}$ except -3 **(i)** $\mathbb{R}$ except ± 1
7. (a) $f(g(x)) = (1 - x)/2$ has a domain $\mathbb{R}$ except -1.
[Note: -1 is not in the domain of $g(x)$, so it is not in
the domain of $f(g(x))$.] **(b)** $g(f(x)) = 2x/(2x - 1)$
has a domain $\mathbb{R}$ except 1/2 and 0. [Note: 0 is not in
the domain of $f(x)$ so it is not in the domain of
$g(f(x))$.] **9.** $f(x) = 1/x, g(x) = \pi x^2 + 1$
11. $4 + 1.071x$ **13.** $(3 - x)/5$ **15.** $1/(2x)$
17. $(2 + x)/(1 + x)$ **19.** no inverse
21. $\sqrt{(x - 1)/2}$ **23.** $\sqrt[3]{1 - x}$

Section 1.4

1. -5 **3.** $(-2x - 5)/3$ **5.** 2 **7.** $-5/2$
9. 4/5 **11.** -1

13.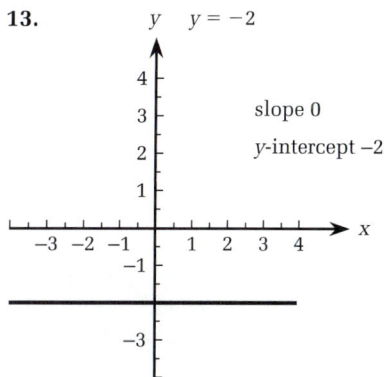

15.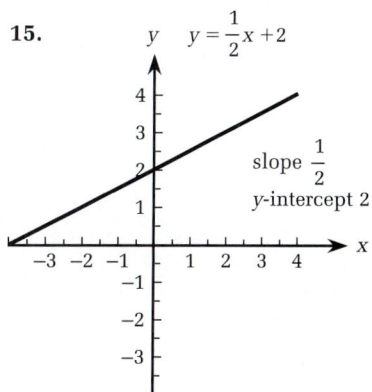

17.

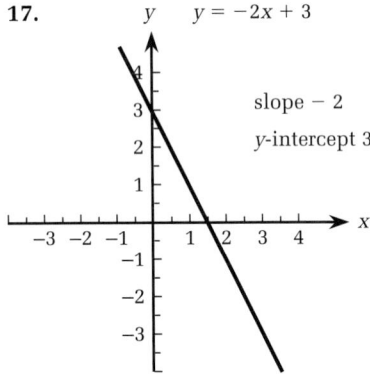

$$y \quad y = -2x + 3$$

slope -2

y-intercept 3

17.

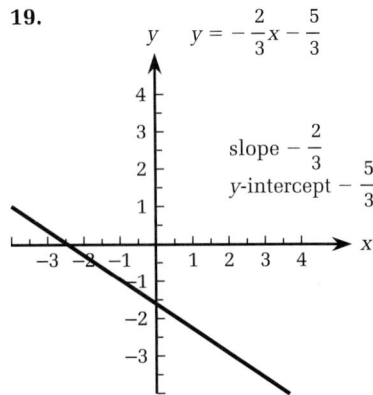

$$y \quad y = x^2 - 3x - 10$$

$\left(\dfrac{3}{2}, -\dfrac{49}{4}\right)$

19.

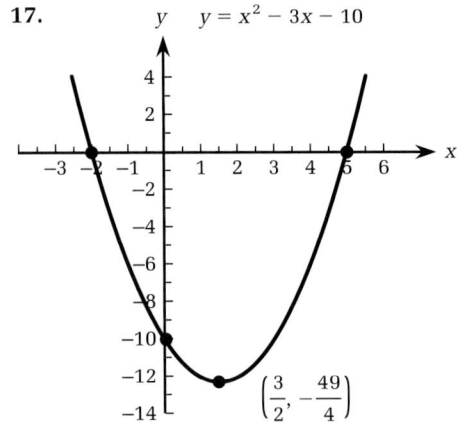

$$y \quad y = -\dfrac{2}{3}x - \dfrac{5}{3}$$

slope $-\dfrac{2}{3}$

y-intercept $-\dfrac{5}{3}$

19.

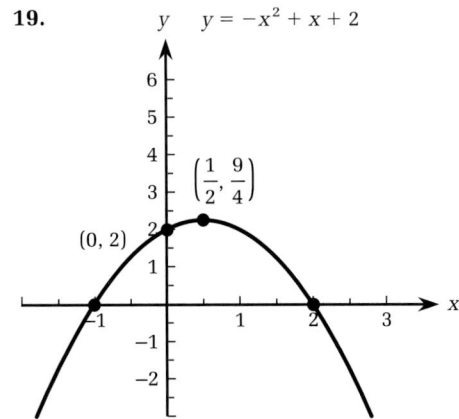

$$y \quad y = -x^2 + x + 2$$

$\left(\dfrac{1}{2}, \dfrac{9}{4}\right)$

$(0, 2)$

21. $y = -(1/2)x + 3$ **23.** $y = -2x + 9$
25. $y = -(3/2)x + (13/2)$ **27.** $y = (2/3)x + (1/3)$
29. 378 km   **31. (a)** $-40t + 450$ **(b)** \$210
33. $(5/9)(F - 32)$
35. $f(x) = \begin{cases} 30 + 2(x - 55) & \text{for } 55 < x \le 75 \\ 100 + 5(x - 75) & \text{for } x > 75 \end{cases}$
37. (a) Choose A   **(b)** 75 per hour
39. (a) i. 15 **ii.** 75 **iii.** \$31.50   **(b) i.** 54 **ii.** 270
iii. $-\$1.00$ **41. (a)** $865{,}000\ \text{ft}^3$ **(b)** 11 ft
43. (a) 0.45 sec   **(b)** 12   **45. (a)** C is 16, F is 24
(b) C is 19, F is 32   **(c)** about 1   **(d)** For C, $t = 23$;
for F, $t = 12.5$.   **47. (a)** 750   **(b)** 495   **(c) i.** 2
ii. 4.2 **iii.** 11.9   **(d)** 216   **49. (a)** 108
(b) \$3.60   **51. (a)** 17.5 meters   **(b)** 24.2 cm
59. (a) yes   **(b)** 87 cents

21.

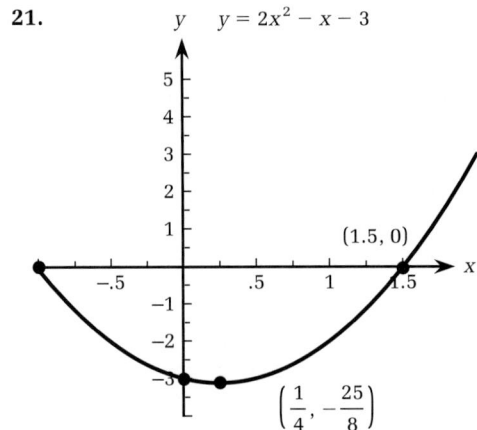

$$y \quad y = 2x^2 - x - 3$$

$(1.5, 0)$

$\left(\dfrac{1}{4}, -\dfrac{25}{8}\right)$

Section 1.5

1. $y = -(1/4)x + 8$ **3.** $1 \cdot 10 = 2 \cdot 5 =$
$(-1)(-10) = (-2)(-5)$ **5.** $(x - 3)(x + 2)$
7. $20, 2\sqrt{5}$ real   **9.** $5, -2$ **11.** $0, 2$
13. $4, -1$ **15.** $-1 \pm \sqrt{3}$

23.

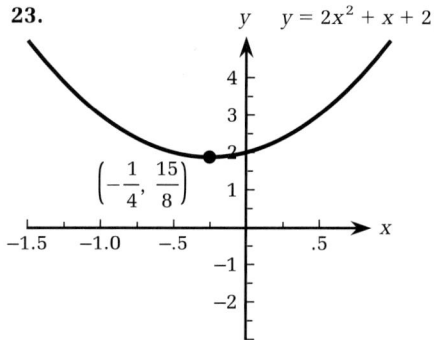

$y \quad y = 2x^2 + x + 2$

$\left(-\dfrac{1}{4}, \dfrac{15}{8}\right)$

25.

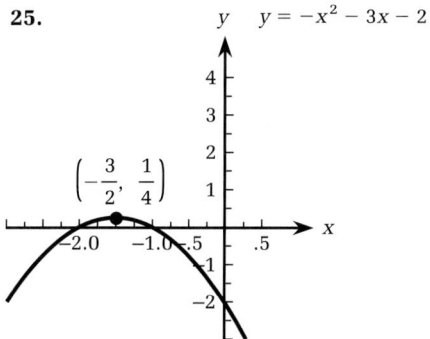

$y \quad y = -x^2 - 3x - 2$

$\left(-\dfrac{3}{2}, \dfrac{1}{4}\right)$

27. $0, 2, -2$

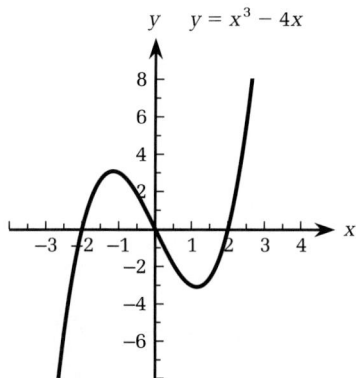

$y \quad y = x^3 - 4x$

29. $0, 3, -2$

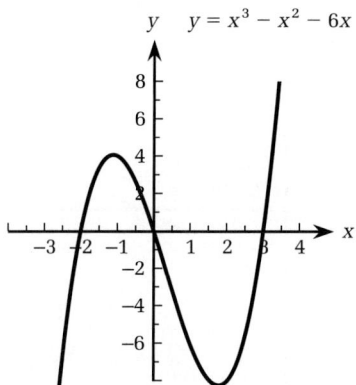

$y \quad y = x^3 - x^2 - 6x$

31. 1

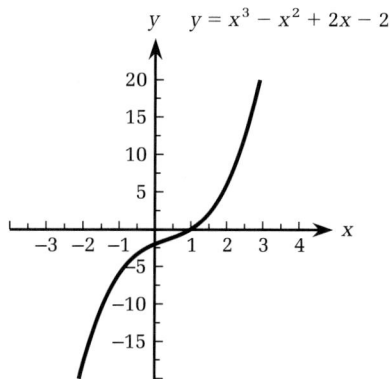

$y \quad y = x^3 - x^2 + 2x - 2$

33. $0, -2, 2$

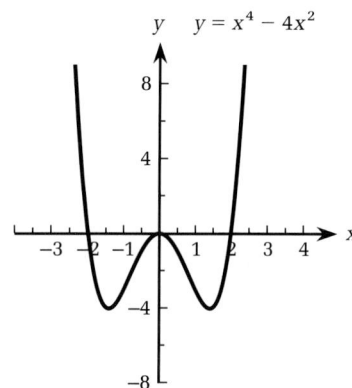

$y \quad y = x^4 - 4x^2$

35. $f(0) = -2, f(1) = 2$ **37.** $f(-1) = 4, f(1) = -6$
39. (a) $C(0) = 0; C(5) = 21.4; C(10) = 92.6;$
$C(15) = 183.7; C(20) = 174.6$
(b) $C(23) = 37.5 C(24) = -42.9$
(c)

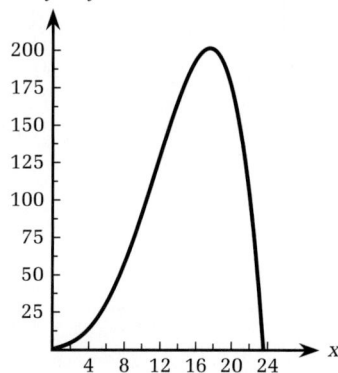

$y \quad y = -0.006x^4 + 0.140x^3 - 0.053x^2 + 1.79x$

41. (a) 7 new, 14 total   **(b)** 4900   **43. (a)** 90
(b) 324,000   **45. (a)** $p = -0.4x + 50$ **(b)** 35
sacks at \$36 each   **(c)** \$490   **47.** $t(100) = 204.75,$

$t(400) = 944.5$, $t(700) = 1872.5$
49. $-1, -0.47, 8.47$ **55.** $r < 75$
57. 500 discs, \$10 retail

Section 1.6

1. $-3/2$ **3.** no values   **5.** $(2x + 1)(2x - 1)$
7. $(3x^2 + 1)/x$ **9.** $(3x - 1)/(x - 1)$
11. $\mathbb{R}$ except $3, -1$

13.

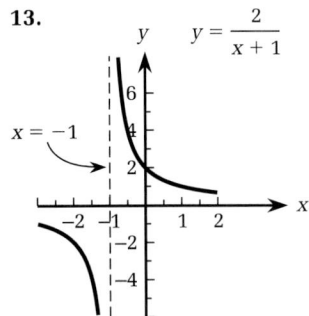

$y = \dfrac{2}{x + 1}$

$x = -1$

15.

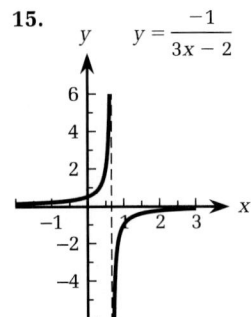

$y = \dfrac{-1}{3x - 2}$

17.

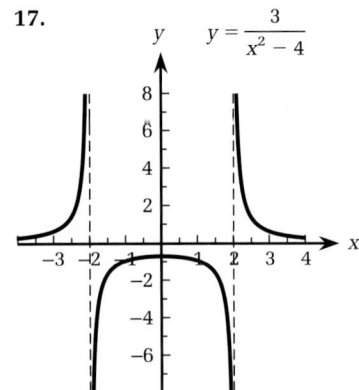

$y = \dfrac{3}{x^2 - 4}$

19.

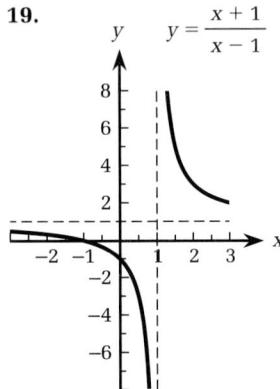

$y = \dfrac{x + 1}{x - 1}$

21.

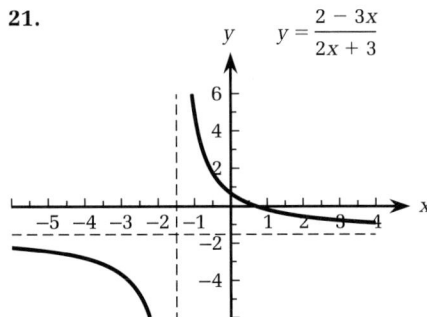

$y = \dfrac{2 - 3x}{2x + 3}$

23.

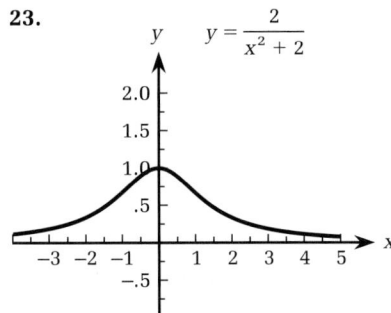

$y = \dfrac{2}{x^2 + 2}$

25.

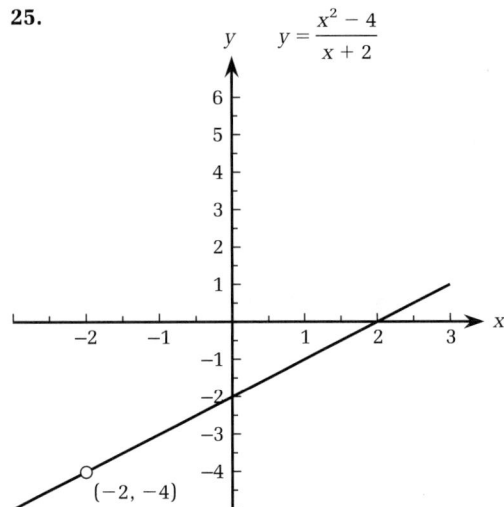

$y = \dfrac{x^2 - 4}{x + 2}$

$(-2, -4)$

27.

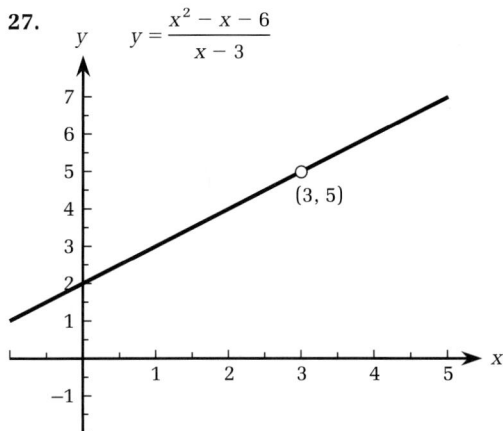

$$y = \frac{x^2 - x - 6}{x - 3}$$

(3, 5)

29.

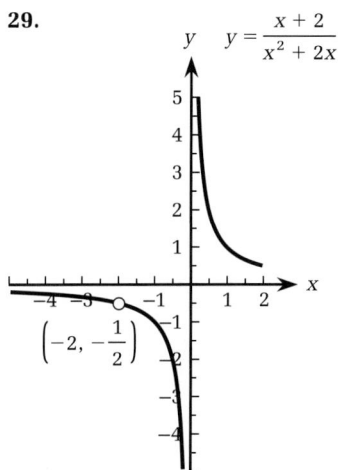

$$y = \frac{x + 2}{x^2 + 2x}$$

$\left(-2, -\frac{1}{2}\right)$

31.

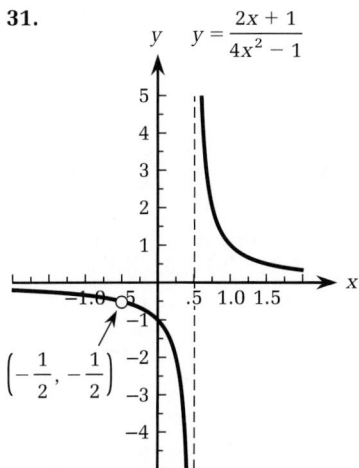

$$y = \frac{2x + 1}{4x^2 - 1}$$

$\left(-\frac{1}{2}, -\frac{1}{2}\right)$

33. (a) 4 million   **(b)** 90%

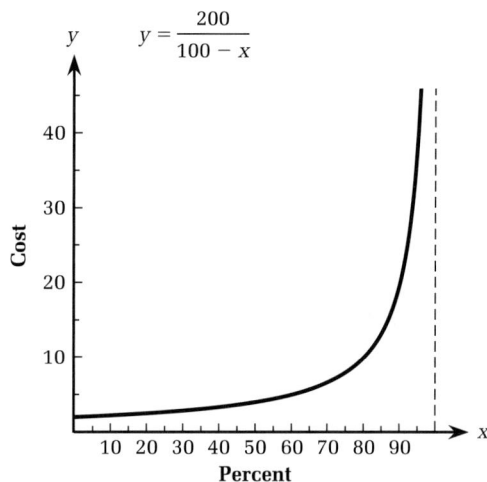

$$y = \frac{200}{100 - x}$$

Cost

Percent

35. (a) 8.67%   **(b)** 10.58%   **(c)** $470.59
37. (a) 3.29   **(b)** 2.38   **(c)** 0.37   **(d)** 0.29

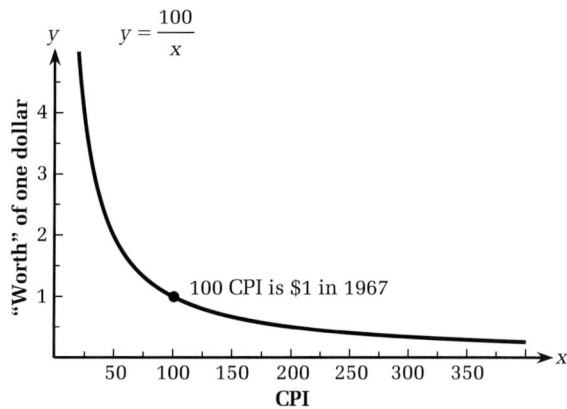

$$y = \frac{100}{x}$$

"Worth" of one dollar

100 CPI is $1 in 1967

CPI

39. (a) 192   **(b)** $(1125 + 25\,n)/(5 + n)$
41. H.A. of $y = 0$, V.A. of $x = \pm 2$

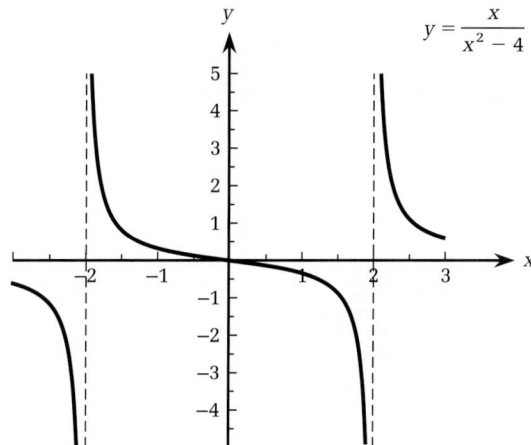

$$y = \frac{x}{x^2 - 4}$$

43. H.A. of $y = 1/2$, V.A. of $x = 0$, $x = 1$

$$y = \frac{x^2 - x - 6}{2x^2 - 2x}$$

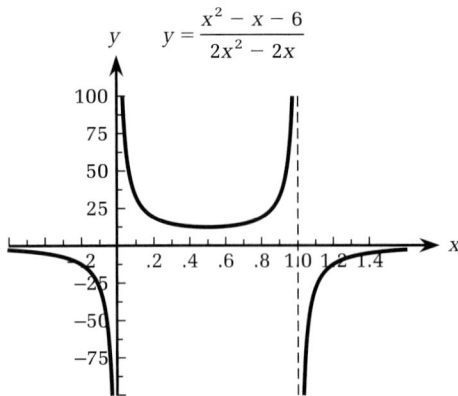

45. H.A. of $y = 1$, V.A. of $x = \pm 25$

$$y = \frac{x^2 - 2x - 4}{x^2 - 625}$$

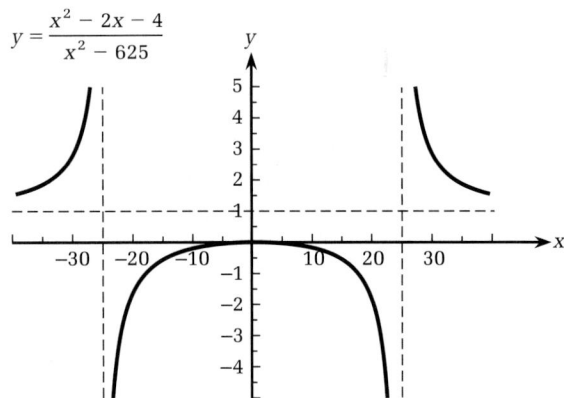

Section 1.7

1. 4   **3.** 3   **5.** 1/2   **7.** $2^5 = 32$
9. $-2^5 = -32$ **11.** $2^2 = 4$ **13.** $2^7 = 128$
15. $(2/(\;3))^2 = 4/9$ **17.** $2^{-5} = 1/32$ **19.** 8
21. 0.4   **23.** 1.5157   **25.** 3.6056   **27.** 3^{2x+1}
29. x^5 **31.** x^4 **33.** r^2 **35.** st^3 **37.** $x^2 y$
39. 1.5651   **41.** 2.9907

43.

$$y = x^{1.5}$$

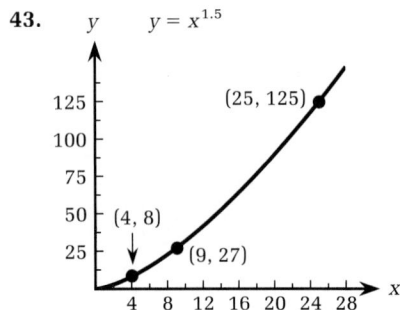

45. (a) 51.5   **(b)** 124.5   **47. (a)** 5305
(b) 58,000   **49.** 171

51.

$$y = 17.4(x^{0.25})$$

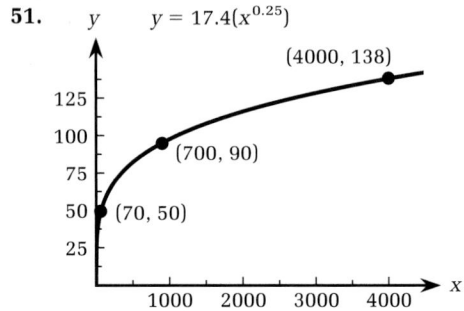

53. (a) 1.6   **(b)** 2.1

Chapter 1 Review

1. 3.2   **2.** 2.3   **3.** $2 - \sqrt{3}$ **4.** $4 - 2\sqrt{3}$
5. 1/15   **6.** 2/3   **7. (a)** real numbers greater than or equal to -3 and also less than 5   **(b)** real numbers greater than or equal to -2 and also less than 5   **(c)** real numbers greater than -1 and also less than 2.5

8. (a)

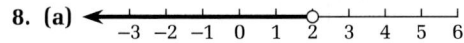

(b)

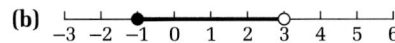

(c)

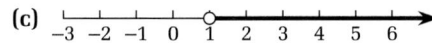

(d)

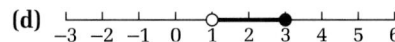

(e)   no such x exist

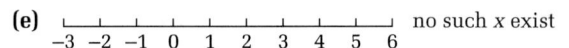

(f)

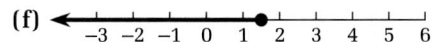

9. (a) $(-4, 6)$ **(b)** $(-3/2, 5/2)$ **(c)** $(-9/2, 3/2)$
(d) $[3/2, \infty)$ **(e)** $(2/3, 2]$ **(f)** $[-1, 2/3)$
10. (a) $\mathbb{R}$ except ± 1 **(b)** $x \geq -4$ **(c)** $x \geq 0$
(d) $\mathbb{R}$ except 0, 1   **11. (a)** $\sqrt{3}$ **(b)** 2
(c) $\sqrt{4 - a}$ **(d)** $\sqrt{3 - a}$ **12. (a)** 3   **(b)** not
defined   **(c)** $\dfrac{3}{a}$ **(d)** $\dfrac{3}{a + 1}$ **13. (a)** 2   **(b)** 1
(c) 3   **(d)** -3 **14.** 0.32   **15.** $4\sqrt{2} \approx 5.6569$
17. $x^2 + 2x + y^2 - 8y + 8 = 0$ **18. (a)** $(2.5, 2.5)$
(b) $\sqrt{10}/2$ **(c)** $x^2 - 5x + y^2 - 5y + 10 = 0$
19. (a) $20, 18, 20 - 2a^2, 20 - 2a^2 - 4ah - 2h^2$
(b) $2\sqrt{17} \approx 8.2462$ **(c)** $-4x - 2h$
20. $5 + \lceil x - 10 \rceil 0.5$ **21. (a)** 3/2   **(b)** 3
22. (a) $y = 2/3x + 13/3$ **(b)** 2/3   **(c)** 13/3
23. $y + 2x = 4$ **24.** $y = -2/5x + 3$
25. $y = -950t + 16,000$

26. $y = -14{,}000t + 80{,}000$ **27. (a)** $(x + 2)(x + 4)$
(b) $(2x + 1)(x - 6)$ **28. (a)** $-2/3, 1$ **(b)** $4, -3$
29. (a) $(1 \pm \sqrt{17})/4$ **(b)** $(-5 \pm \sqrt{29})/2$
30.

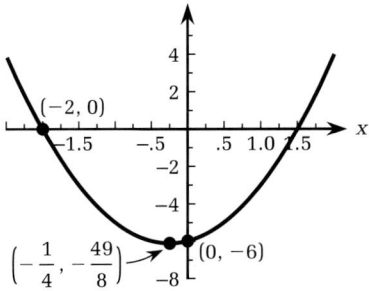

$y = 2x^2 + x - 6$

$(-2, 0)$

$\left(-\dfrac{1}{4}, -\dfrac{49}{8}\right)$ $(0, -6)$

31.

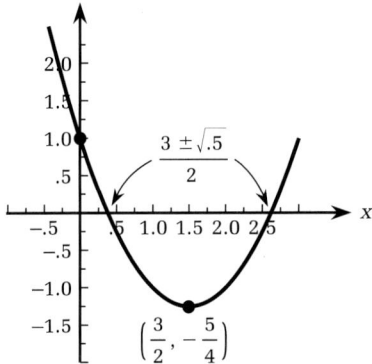

$y = x^2 - 3x + 1$

$\dfrac{3 \pm \sqrt{.5}}{2}$

$\left(\dfrac{3}{2}, -\dfrac{5}{4}\right)$

32. $b^2 - 4ac < 0$ **33. (a)** $P(x) = -x^2 - 2x + 8$
(b) $x = 2$ **34. (a)** $x(x - 3)(x + 2)$
(b) $(x + 2)(x - 2)(x + 1)$ **35. (a)** 46.5 **(b)** 66
36. (a) $0.015x$ **(b)** 45 **(c)** 2000 **(d)** $6g$
(e) $0.09x$ **(f)** 1667 **(g)** $\$200$
37. (a) $a = 0.942, b = -15.5$ **(b)** 551.6
(c) $13{,}790$ miles vs. $9{,}456$ miles
38. (a) $y = -0.05x + 159$ **(b)** $-0.05x^2 + 159x$
(c) 79 cents **(d)** 1600 **(e)** $\$1264.00$
39. (a) $n = 1000$ **(b)** 8
40. (a) $\dfrac{91}{9}$ **(b)** $\dfrac{10}{9}$ **(c)** $\dfrac{1}{3x^3 + x^2}$ **(d)** $\dfrac{3x + 1}{x^2}$

(e) $\dfrac{1}{(3x + 1)^2}$ **(f)** $\dfrac{3}{x^2} + 1$ **(g)** $\mathbb{R}$ except $-1/3$
41. (a) $2x^2 + \sqrt{x} + (1/x)$ **(b)** $(2x^2 + \sqrt{x})/x$
(c) $x(2x^2 + \sqrt{x})$ **(d)** $(2/x^2) + (1/\sqrt{x})$
(e) $1/(2x^2 + \sqrt{x})$ **42.** $x = \pm 1/2$
43. (a) $(x - 3)/2$ **(b)** $x/(x - 2)$
44. (a) $h(x) = x^2 + 2$ **(b)** $g(x) = \sqrt{x}$
45. (a) $3{,}000$ **(b)** 85% **46. (a)** x^2 **(b)** 9
(c) x^3y^2 **(d)** x^6/y^6 **(e)** 3 **47. (a)** 84.4485
(b) 125 **(c)** 4 **(d)** 2.7832 **48.** $x^{-4.6}\,y^{-6}$

49.

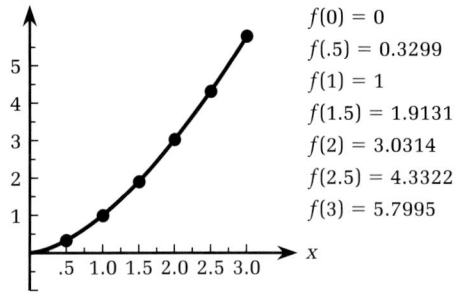

$y = x^{1.6}$

$f(0) = 0$
$f(.5) = 0.3299$
$f(1) = 1$
$f(1.5) = 1.9131$
$f(2) = 3.0314$
$f(2.5) = 4.3322$
$f(3) = 5.7995$

50.

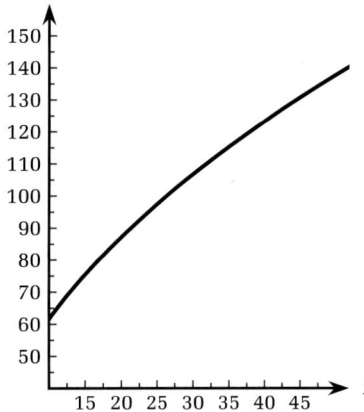

$v = 19.5\sqrt{(l)}$

SELECTED ANSWERS FOR CHAPTER 2

Section 2.1

1. (a) 2 **(b)** 5 **3. (a)** 1 **(b)** 1 **(c)** 0.5
(d) -1 **(e)** no limit **5. (a)** 1 **(b)** 1 **(c)** 0.5
(d) 0.5 **(e)** 0.5 **7.** 6 **9.** 6 **11.** -2
13. 4 **15.** 1/2 **17.** no limit **19.** 1
21. -1 **23.** no limit **25. (a)** 3 **(b)** 2
(c) no limit **27. (a)** 3 **(b)** 2 **(c)** no limit
29. (a) 10 for $0 \le x \le 1$, $10 + \lceil x - 1 \rceil$ for $x > 1$
(b) positive reals except integers

(c)

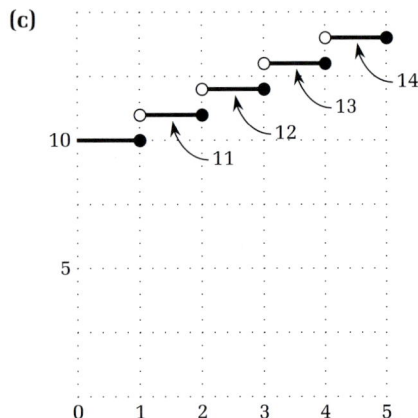

31. (a) $1.84x$ **(b)** 2.76 **(c)** 2.58 to 2.94
37. (a) 1.23 to 1.77 **(b)** 1.45 to 1.55

Section 2.2

1. 0, 6, 4 **3.** $(x - 3)(x + 3)$ **5.** $x - 4$ **7.** 8
9. -6.5 **11.** 35 **13.** -14 **15.** 4/5
17. 2 **19.** 4 **21.** -3 **23.** $(u^2 + a)/(a - 1)$
25. -6 **27.** 0 **29.** 5 **31.** 3 **33.** -3
35. 3/5 **37.** 4 **39.** 0 **41.** $-1/2$
43. $2x + 1$ **45.** $4x^2 + 1$ **47.** 1.5 **51. (a)** 0
(b and c) no limit **53.** 2 **55.** $2x - 5$
57. $\lim\limits_{x \to 0^-} f(x) = 1$, $\lim\limits_{x \to 0^+} f(x) = 0$

Section 2.3

1. $9 + 6h + h^2$ **3.** $3; 3; 19; 2x^2 + 4xh + 2h^2 + 1$
5. (a) 3 **(b)** 3 **7. (a)** 8 **(b)** 14 **9. (a)** 0
(b) 1.5 **11. (a)** 1/7 **(b)** 3.1429 **13.** 4
15. 3 **17. (a)** 7 **(b)** 8 **(c)** -20 **19. (a)** 3
(b) 12.5 **(c)** 14 **21. (a)** 2.2 **(b)** 7.1 **(c)** 7.1
(d) 13.6 **23. (a)** 5 **(b)** 16 **25. (a)** 1

(b) 1.9 **(c)** 0.1 **27. (a)** -0.19 **(b)** -0.095
(c) -0.04 **(d)** 1.6 **29. (a)** 38 **(b)** 37 **(c)** 50
(d) 1100 **31.** -90 **35.** **(a)** lost 20
(b) -10 **(c)** -12 **(d)** 5 **(e)** 60

Section 2.4

1. (a) $-1/4$ **(b)** $y = -(1/4)x + (13/4)$
3. (a) A, B, D **(b)** A, B, C **(c)** A, B, D **(d)** A, C
5. 3/8 **7.** -22.2 at A, -85.7 at B
9. -0.024 at A, -0.013 at B **11.** $y = 2x - 5$
13. $y = 2t$ **15.** $y = -6x - 2$ **17.** $y = 8x - 8$
19. $y = -2x + 4$ **21.** 2 **23.** $2t - 5$
25. $-2x$ **27.** $6x - 1$ **29.** $1/(2x^2)$
31. (a) 0.8 **(b)** 1.3

Section 2.5

1. $2x^{1/2}$ **3.** $2x^{-1/2}$ **5.** $x^{3/2}$ **7.** ± 2 **9.** 0
11. 0 **13.** $10x^9$ **15.** $21x^6$ **17.** $-24x^7$
19. $1/\sqrt{x}$ **21.** $0.5x^{-0.5}$ **23.** $-1.5x^{-1/2}$
25. $-6x^{-3}$ **27.** $4.8w^{1.4}$ **29.** $-3u^{-2.5}$
31. $3x^2 - 10x + 1$ **33.** $8x^3 + 12x - 3$
35. $-3/(2x^{3/2}) - 28x^3$ **37.** $(1/3)x^{-2/3} + 10/(x^3)$
39. $-3u^{-5/2} - 8u$ **41.** $-26x^{4.2} + 12x^2$
43. 20 **45.** 15.5 **47.** 34 **49.** ± 2 **51.** 7
53. (a) $0.7x$ **(b)** 14 **(c)** 14.35 **55.** $10x + 2$
57. (a) 0.05, 0.16 **(b)** 0.00019 **59.** 0.44
61. (a) $(\pi/2)d$ **(b)** 7π, area increases about 22 in.^{2}
per inch of d. **63.** 791 **65. (a)** $2.725/\sqrt{x}$
(b) 0.2725 miles per hour faster per additional foot
of height **67. (a)** $k/(2\sqrt{l})$ **(b)** The 4-ft shark's
maximum velocity is increasing twice as fast as that
of the 16-ft shark. **69. (a)** $0.18v - 1.29$ **(b)** 0.51
additional watts per additional meter per second
71. (a) $0.37m^{-2/3}$ **(b)** 0.59 additional meters per
additional kilogram **73. (a)** $f'(x) = 0.0098x -$
$0.2629 - (17.1793/x^2)$ **(b)** -0.1098 additional
minutes per additional year of age **75. (a)** $-2, 3, 1$
(b) 0.74 **(c)** 1 **79.** $18x$ **81.** $2x - 4$
83. $1/\sqrt{2x}$ **85.** 17.4

Section 2.6

1. (a) x^5 **(b)** x^{-1} **(c)** x^3 **(d)** $10x^{-1} + 8$
3. $5x^4 - 3x^2$ **5.** $-0.5x^{-1.5}$ **7.** $-2x^{-3}$
9. $4x - 5x^4$ **11.** $4x^3 - 4x$ **13.** $3x^2 - 4x - 3$

15. $8x^3 + 15x^2 + 16x + 18$
17. $(5/2)x^{3/2} + (1/2)x^{-1/2}$ **19.** $7.5x^{6.5} - 13x^{5.5}$
21. $4.5x^{3.5} - 10x^4$ **23.** $-4/(x-2)^2$
25. $(2x^2 + 6x)/(2x + 3)^2$ **27.** $(x^2 - 1)/(x^2 + 1)^2$
29. $(x^2 - 2x - 2)/(x-1)^2$ **31.** $-3x^{-4} + 16x^{-5}$
33. $(-x^2 - 2x + 4)/(x+1)^2$
35. $(6u^2 + 8u + 11)/(u+2)^2$ **37.** $(9x^2 + 4)/6x^{3/2}$
39. $(2x^{5/2} - 3x^2 + 2x^{1/2} - 1)/[2\sqrt{x}(x^2 - 1)^2]$
41. $y = (1/2)x - (1/2)$ **43. (a)** $200/(100 - x)^2$
(b) 0.08   **(c)** 2   **(d)** 0.08 m per additional
percent at 50%; 2 m per additional percent at 90%
45. $-56,000/(1000 + x)^2$, -0.01636% per
additional dollar in price   **47. (a)** $1.5 - (500/x)$
(b) $500/x^2$ **(c)** 0.0008889 dollars per additional
video   **49. (a)** $-0.01x^3 + 20x$
(b) 8   **(c)** $(-0.01x^3 + 20x)/(x+2)$
(d) $(-0.02x^3 - 0.06x^2 + 40)/(x+2)^2$ **51.** 70
years

Section 2.7

1. $6x + 12x^2$ **3.** $8x^5 - 40x^3$ **5.** 5
7. (a) $9x^2 + 12x + 5$ **(b)** $\mathbb{R}$ **(c)** $18x + 12$
9. (a) $3/(x^2 + x - 2)$ **(b)** $\mathbb{R}$ except $-2, 1$
(c) $[-3(2x + 1)]/(x^2 + x - 2)^2$ **11.** $24(1 + 3x)^7$
13. $10x(2 + x^2)^4$ **15.** $16(1 + x)(2x + x^2)^7$
17. $3/(2\sqrt{3x + 4})$ **19.** $2x/\sqrt{2x^2 - 3}$
21. $x^2(x^3 + 1)^{-2/3}$ **23.** $3.6(3x - 1)^{0.2}$
25. $0.5(x + 1)(x^2 + 2x)^{-0.75}$ **27.** $-20x(x^2 + 2)^{-11}$
29. $16x^{-3}(2x^{-2} + 1)^{-5}$
31. $x^4(x^2 - 1)^2(14x^3 - 5)$
33. $(x + 2)^4(7x^2 + 4x)$
35. $[(x^2 + 1)(6x^3 - 14x)]/(3x^2 - 2)^2$
37. $(-18x + 4)/(2x - 3)^3$
39. $(x - 4)^2(2x + 3)(10x - 7)$, 4, $-3/2$, 0.7
41. $y = 2.7x - 3.1$
43. $(-x^4 + 9x^2 + 2x)/(x^3 + 1)^2$
45. (a) $\dfrac{-2.01x + 0.9}{(x + 0.53)^3}$ **(b)** 2.8 at 0.1, -0.3 at 1
(c) cost increasing at 0.1, decreasing at 1
47. (a) $(410)/(2 + 2x)^2$ **(b)** 11.4%, 0.85%, 0.23%
49. $R(w) = 100\sqrt{w + 2} - (\sqrt{w + 2})^3$, $dR/dw = 14.9$,
to increase revenue add more workers
55. $(df/dg)(dg/dh)(dh/dx)$

Section 2.8

1. 0   **3.** $(x^2 + 2x)^2/\sqrt{2x - 1} +$
$2\sqrt{2x - 1}(x^2 + 2x)(2x + 2)$
5. (a) $(2 - y)/(1 + x)$ **(b)** 1/2   **7. (a)** $(x + 1)/y$
(b) 1   **9. (a)** $(10xy - 4x)/(3y^2 - 5x^2)$ **(b)** 4/5
11. (a) $(3 - 5x^4 - 3yx^2)/(x^3 - 1)$ **(b)** 5/2

13. (a) $(6\sqrt{xy} - 2y\sqrt{x} - y\sqrt{y})/(2x\sqrt{y} + x\sqrt{x})$
(b) 4/63
15. (a) $\dfrac{-\dfrac{(x^2 + 2y)^2}{\sqrt{2x - 1}} - 4x(x^2 + 2y)\sqrt{2x - 1}}{4(x^2 + 2y)\sqrt{2x - 1} - 7}$
(b) 5/3   **17.** $y = 1$ **19.** $y = -(1/4)x + 3/2$
21. -1.42

Section 2.9

1. $245x - 1.1x^2$ **3.** $0.003h^{1/2}$ **5. (a)** $-2x - 2$
(b) $1/(-2x - 2)$ **7. (a)** $(-2 - 2px)/x^2$
(b) $x^2/(-2 - 2px)$ **9. (a)** $-15/7$ **(b)** $-7/15$
11. (a) $-1/2$ **(b)** -2 **13.** 0.221%/min
15. (a) 0.00585   **(b)** 0.000745   **17. (a)** 210
(b) 24.5   **(c)** 185.5   **19.** $1.6\pi \approx 5$ **21.** 42.3
ft/sec   **23.** -0.00000575 ft/min   **25.** $dr/dt =$
0.95 ft/year, $dh/dt = 0.22$ ft/year

Chapter 2 Review

1. (a) 2   **(b)** 2   **(c)** 4   **(d)** 4   **(e)** 4   **(f)** 2   **(g)** 5
(h) no limit   **(i)** 3   **2. (a)** 1   **(b)** does not exist
(c) 5   **3. (a)** $-3/4$ **(b)** does not exist   **(c)** 5
4. (a) -1 **(b)** 1   **(c)** does not exist   **5. (a)** 0
(b) does not exist   **6.** 3   **7.** $3a + 2$ **8.** 2
9. $2a$ **10. (a)** \$2.10 to 2.70   **(b)** .75 lb to .92 lb
11. $-1/12$ **12.** $-1/4$ **13.** 5/2   **14. (a)** 12
(b) 0   **15.** 2   **16.** 2   **17. (a)** \$0.96 per year
(b) \$2.23 per year   **(c)** 6   **(d)** 20%   **18. (a)** 15
more cars per hour per car per mile   **(b)** 1 more car
per hour per car per mile   **(c)** -4.75 cars per hour
per car per mile   **(d)** 0   **19.** $2x + h + 3$
20. $(-4x - 2h)/(x^2(x + h)^2)$
21. $6/(\sqrt{3(x + h)} + \sqrt{3x})$ **22. (a)** $4x - 3$
(b) $-1/(x + 2)^2$ **(c)** $2 + 3/x^2$
23. $y = (1/6)x + (11/6)$
24. $y = -0.24x + 1.62$ **25.** $6x^2 - 1$
26. $-6x^{-3} - 5x^4$ **27.** $6x + (1/x^2)$ **28.** $2x +$
$3 - (2/x^3)$ **29.** $\sqrt{5}/(2\sqrt{t})$ **30. (a)** $9x^2 + 46x - 8$
31. $-3x^2 + 3x + 1$ **32.** $10.5x^{2.5} + 12x^2 - 3x^{-0.5}$
33. $5(2x^3 + x^{-2})^4(6x^2 - 2x^{-3})$
34. $-3(t^2 + t)^{-4}(2t + 1)$
35. $(-3x^2 + 10x + 9)/(x^2 + 3)^2$
36. $(-x^2 + 32x - 16)/(16 - x^2)^2$
37. $(3x^2 - x + 2)/(2\sqrt{x + 2}) + \sqrt{x + 2}(6x - 1)$
38. $(-2x + 3)/(2x^2\sqrt{x^2 - x})$ **39.** 2.6
40. $-5/32$ **41.** 64   **42.** $-6/25$ **43.** $-4/13$
44. 11   **45. (a)** $-3x^2 + 5$ **(b)** $1/(-3x^2 + 5)$
46 44.44   **47. (a)** 381.67   **(b)** 24.56
48. 3.25 additional profit per additional T-shirt
49. (a) $(3x^2 + 25)/(x^2 - 3x)$ **(b)** -4

50. (a) $R(x) = 0.09x - 0.003x^2$ **(b)** 0.03
51. -0.133 minutes per year of age   **52.** 0.02297
53. (a) $8.816m^{-0.24}$ **(b)** 3   **54. (a)** $0.36(\sqrt{v} - 9.7)$
(b) -1.33 watts per additional meter per second
55. (a) 1.25   **(b)** 31   **56. (a)** $t(19) = 32.2$,

$t(75) = 50$ **(b)** $t'(19) = 0.137$, $t'(75) = 0.8$
(c) At 19, it takes 0.13 minutes to collect an
additional pound, at 75 it takes 0.8 minutes.
(d) 48   **57.** 0.16 ft per minute
58. -176mm^2 per day

SELECTED ANSWERS FOR CHAPTER 3

Section 3.1

1. $(-2, 0]$ **3.** $[3, 7.5]$ **5.** $(1, 3)$ **7.** $(1.5, \infty)$
9. $\{x : -2 < x < 3.5\}$ **11.** $\{x : 1 \le x \le 5\}$
13. $\{x : 1 < x\}$ **15.** $\{x : 1 \le x \le 2\}$ **17.** $4/3$
19. $t > 1/3$ **21.** $\mathbb{R}$ **23.** $\mathbb{R}$ except -1
25. $x \ge -3/2$ **27.** $\mathbb{R}$ except $3/2$
29. $\mathbb{R}$ except $\pm\sqrt{3}$ **31.** $t \ge -1/2, t \ne 0$
33. $\mathbb{R}$ except integers

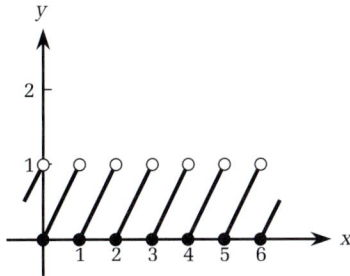

35. (a) $C(x) = \begin{cases} 10 & \text{for } x \le 4 \\ 10 + \lceil 2x - 8 \rceil & \text{for } x > 4 \end{cases}$
(b) $0 \le x \le 4$ and all $x > 4$ except $4.5, 5, 5.5, \cdots$

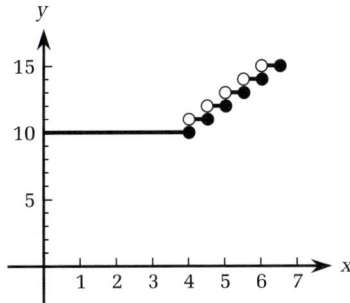

37. $f(0) = 1/3, f(2) = -1$ **39. (a)** $f(0) = -1$ and
$f(2) = 5$ **(b)** $f(1) = -1$, so there is a zero in $(1, 2)$
41. 4.8 **43.** Pos. $(-1, \infty)$; Neg. $(-\infty, -1)$
45. Pos. $(-1, 2)$ and $(3, \infty)$; Neg. $(-\infty, -1)$ and $(2, 3)$
47. Pos. on $\mathbb{R}$ **49.** Pos. $(-2, 0)$ and $(2, \infty)$;
Neg. $(-\infty, -2)$ and $(0, 2)$ **51.** increasing
55. $x < -1$ **57.** 1.32 **61.** Inc. $(-\infty, -3)$ and
$(1, \infty)$; Dec. $(-3, 1)$ **63.** Inc. $(-\infty, 0)$ or $(1, \infty)$; Dec.
$(0, 1)$

65. continuous except at 1, 2, and 3

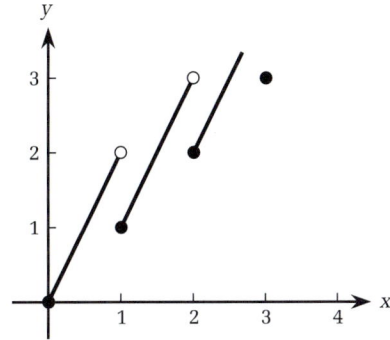

67. continuous except at 0.5, 1, 1.5, 2, 2.5

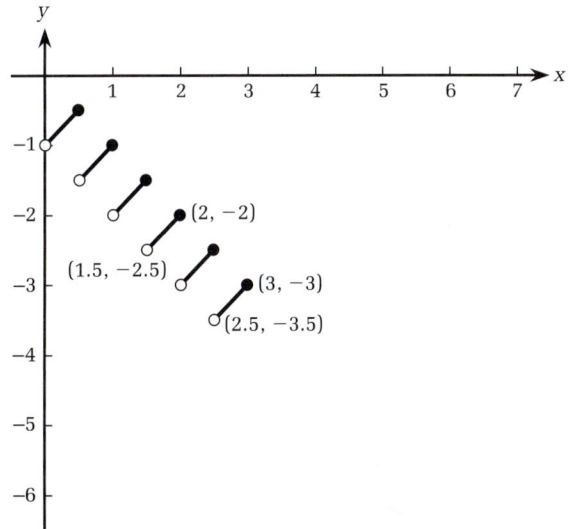

69. $c = 1/2$ **71.** Inc. for $x > 1$, Dec. for $0 < x < 1$

Section 3.2

1. (a) -24 **(b)** 6 **(c)** $-4x - 8$ **(d)** -2
3. $(x - \sqrt{3})^2(x + \sqrt{3})^2$ **5.** A abs min, C rel and
abs max, D rel min   **7.** A abs min, B rel max, C rel
min, E rel max, F rel min, G abs max   **9. (a)** B, C
(b) D **11. (a)** B, D **(b)** C, E **13.** $-1/10$
15. none   **17.** $-1/3$ **19.** $0, 4/3$ **21.** $1/4$
23. $4/3$ **25.** none   **27.** 1 **29.** none

31. $\sqrt{2}$ **33. (a)** none   **(b)** none   **(c)** $-17, 1$

$y = -2x^2 + 4x - 1$

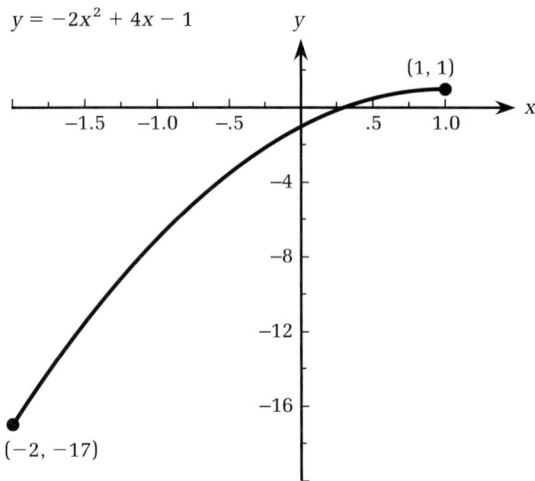

35. (a) $2/3$ **(b)** none   **(c)** $-8, 1$

$y = (3x - 2)^3$

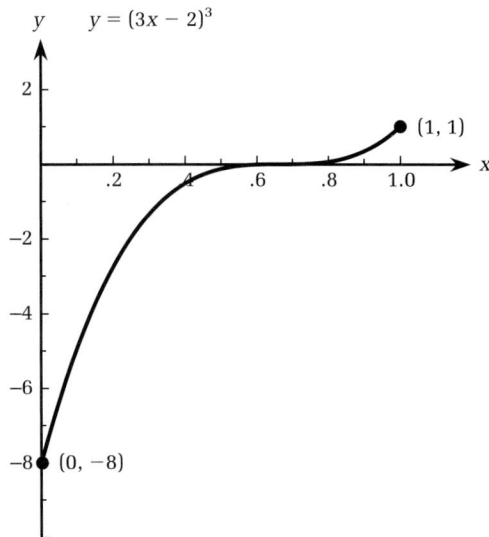

37. (a) 1 **(b)** -4 **(c)** $-4, 23$

$y = x^3 + 3x^2 - 9x + 1$

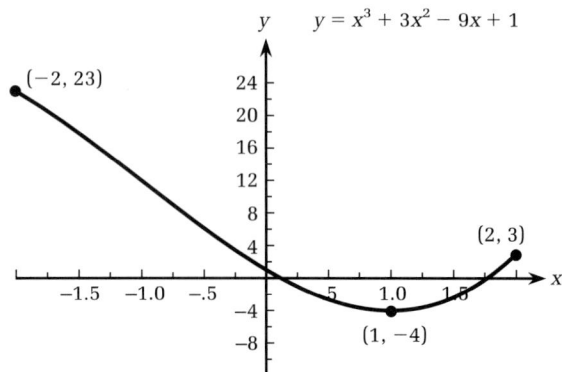

39. (a) $-1, 0, 1$ **(b)** $1, 2$ **(c)** $1, 10$

$y = x^4 - 2x^2 + 2$

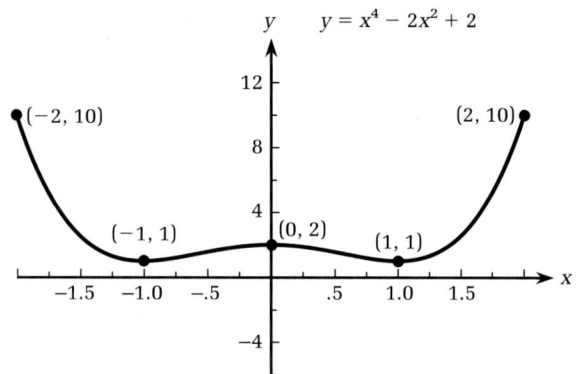

41. (a) 1 **(b)** none   **(c)** $-1, 27$

$y = (x^2 - 1)^3$

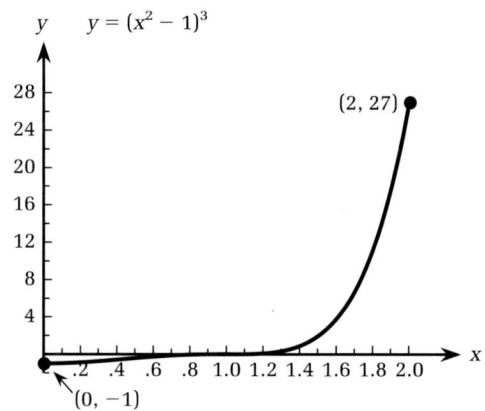

43. (a) 0 **(b)** 1 **(c)** $1/5, 1$

$y = \dfrac{1}{x^2 + 1}$

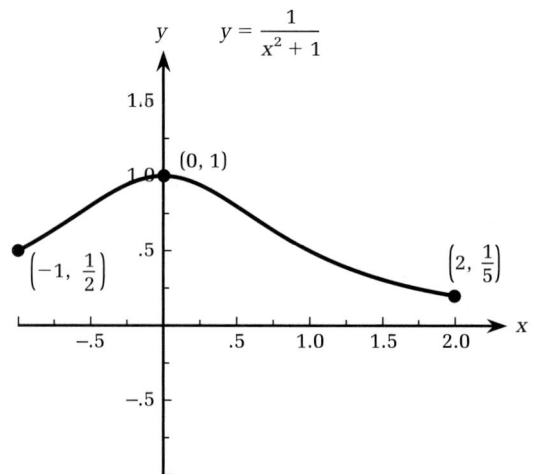

45. (a) $1/\sqrt{3}$ **(b)** 0.5699 **(c)** $0, 0.5699$

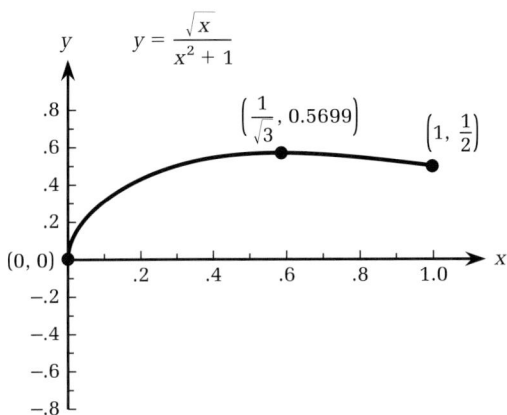

$$y = \frac{\sqrt{x}}{x^2 + 1}$$

47. (a) none **(b)** none **(c)** $0, 2\sqrt{2}$

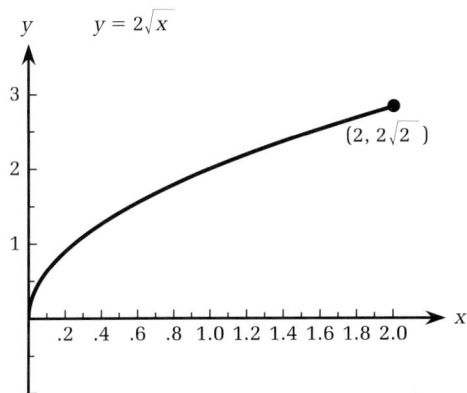

$y = 2\sqrt{x}$

49. (a) 0 **(b)** -1 **(c)** $-1, 0.5874$

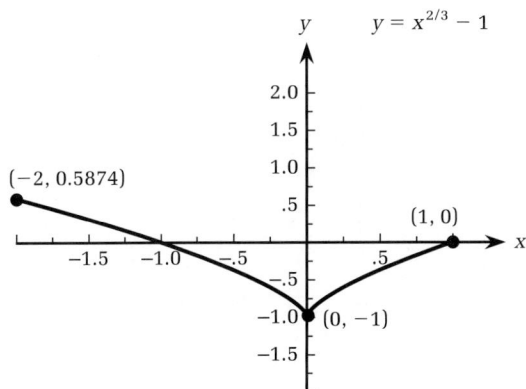

$y = x^{2/3} - 1$

57. (a) 1 **(b)** none **(c)** none
59. (a) $0, -1, 1$ **(b)** $2, -2$ **(c)** none
61. (a) ± 2 **(b)** none **(c)** none
63. (a) 0.4 **(b)** 2.6214 rel and abs max **(c)** 0 abs min **65.** shows $x^3 + 6x$ has no rel max/min (since $3x^2 + 6 > 0$ for all x)

Section 3.3

1. $-3x^{-4} + 4x$ **3.** $[4(1 - x^2)]/(x^2 + 1)^2$
5. $1/3, 2$ **7.** $-5 < x < 3$ **9.** 0
11. $6x^{-4} - 6x$ **13.** $19.53x^{1.1} - 4/(3x^3)$
15. $-(9/4)(3x - 2)^{-3/2}$
17. $(2x^3 + 1)^4(1224x^4 + 72x)$
19. $12 + (1/4)(x + 2)^{-3/2}$ **21.** 6
23. (a) CU $(2.5, 4)$, CD $(-2, 2.5)$ **(b)** 2.5
25. (a) CU $(-2, 1)$, CD $(1, 3)$ **(b)** 1 **27.** CU $\mathbb{R}$
29. CU $(0, \infty)$, CD $(-\infty, 0)$ **31.** CU $(5/6, \infty)$, CD $(-\infty, 5/6)$ **33.** CU on $(1/\sqrt{3}, \infty)$ and $(-\infty, -1/\sqrt{3})$, CD on $(-1/\sqrt{3}, 1/\sqrt{3})$
35. CU $(1/\sqrt{5}, \infty)$ and $(-1/\sqrt{5}, 0)$, CD $(-\infty, -1/\sqrt{5})$ and $(0, 1/\sqrt{5})$ **37.** CU $(-\infty, 0)$ and $(0, \infty)$ **39.** CU $(0, 12)$, CD $(12, \infty)$
41. (a) $2/3$ **(b)** AMax 7, R and AMin 5.37 **(c)** none

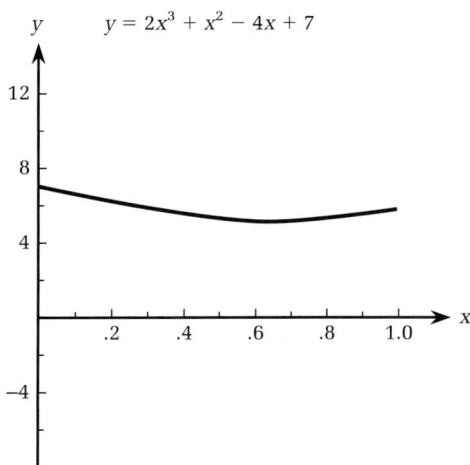

$y = 2x^3 + x^2 - 4x + 7$

43. (a) $-1/2, 1$ **(b)** RMax 2.75 at $x = -.5$, RMin and AMin -4 at $x = 1$, AMax 9 at $x = 2$ **(c)** $(0.25, -0.625)$

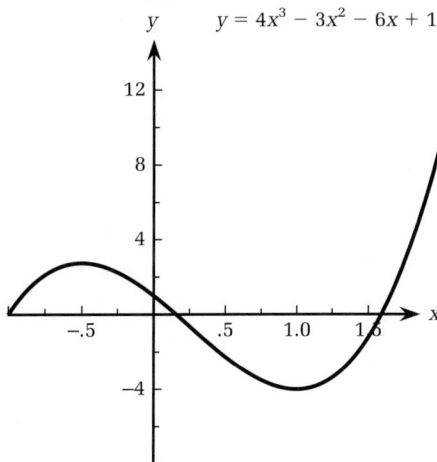

$y = 4x^3 - 3x^2 - 6x + 1$

45. (a) $2/3$ **(b)** AMin and RMin of 12 at $x = 2/3$, AMax 20 at $x = 2$ **(c)** none

$y \quad y = \dfrac{4}{x} + 9x \quad$ for $.5 \le x \le 2$

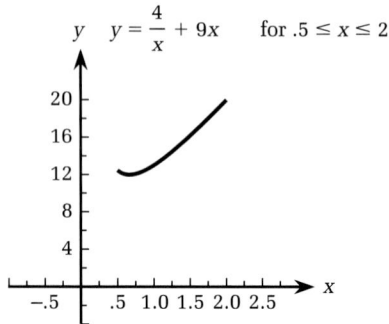

47. (a) 0 **(b)** RMin and AMin of -0.5 at $x = 0$, AMax of $7/13$ at $x = 3$ **(c)** $(\sqrt{4}/3, 0.125)$

$y \quad y = \dfrac{x^2 - 2}{x^2 + 4} \quad$ for $-1 \le x \le 3$

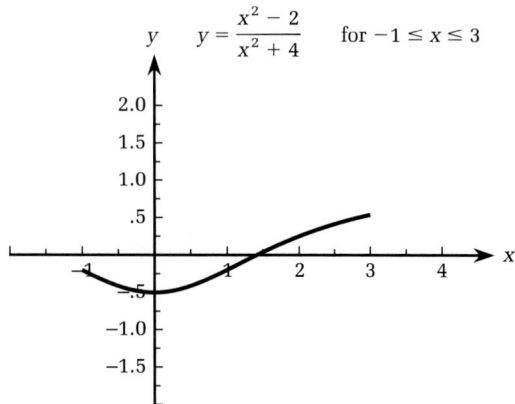

49. $180x^2 + 6$ **51.** $-6/(1 + x)^4$ **53. (a)** 15
(b) 10 **55. (a)** 3 **(b)** 9.125% **57.** 1947
59. 0 and 1.1006 **69.** $a = 1, b = 0, c = 5$

Section 3.4

1. (a) num: deg 2, lc 3; den: deg 2, lc 2 **(b)** num: deg 1, lc -2; den: deg 1, lc 1 **3.** 0 **5.** 2
7. dne **9.** 1 **11.** $-2/3$ **13.** ha: $y = 0$, va: $x = 2$ **15.** ha: $y = -1$, va: $x = 2$
17. ha: $y = -2$, va: $x = -1$ **19.** ha: $y = 0$, va: $x = \pm 2$ **21.** ha: $y = 0$, va: $x = \pm 1$
23. ha: $y = 0$, va: $x = 0, x = 1$ **25.** ha: $y = 0$, va: $x = -2, x = 0$ **27.** ha: $y = 1/3$, va: $x = 0$
29. ha: none, va: $x = -1/2$ **31.** ha: $y = 3/2$, va: $x = 0, x = 1/2$ **33.** ha: none, va: none

35. 10
37. ha: $y = 0$, va: $x = 2$

$y \quad y = \dfrac{3}{x - 2}$

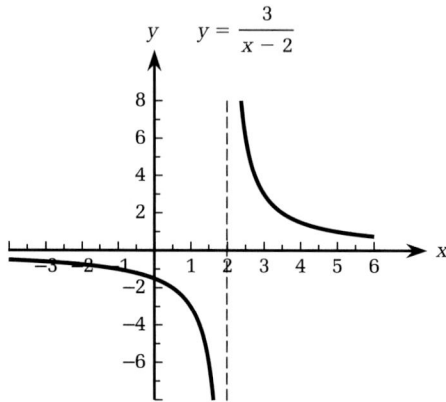

39. va: $x = 0$

$y \quad y = 4x + \dfrac{1}{x}$

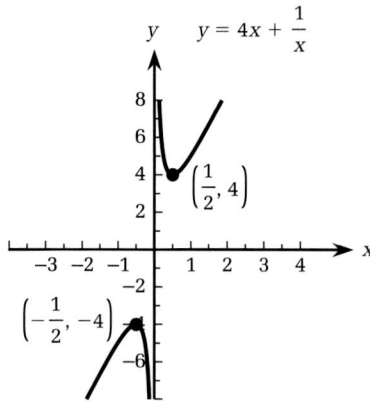

41. ha: $y = 2$, va: $x = 1$

$y \quad y = \dfrac{2x}{x - 1}$

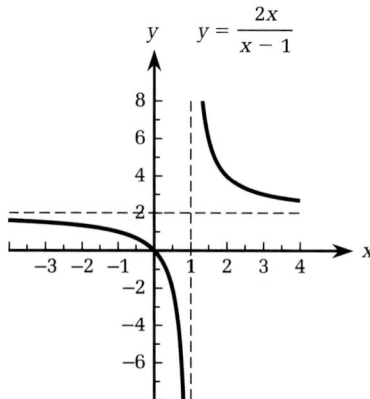

43. va: $x = 0$

$$y = 4x + \frac{1}{\sqrt{x}}$$

$\left(\frac{1}{4}, 3\right)$

45.

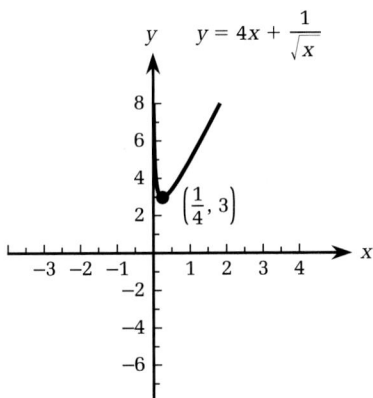

$$y = x + \frac{4}{x^2} \qquad \text{for } .5 \le x \le 3$$

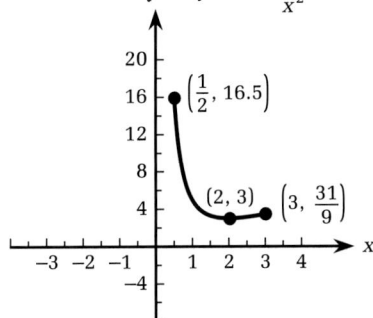

$\left(\frac{1}{2}, 16.5\right)$

$(2, 3)$ $\left(3, \frac{31}{9}\right)$

47.

$$y = \frac{5}{x^2} \qquad \text{for } 1 \le x \le 5$$

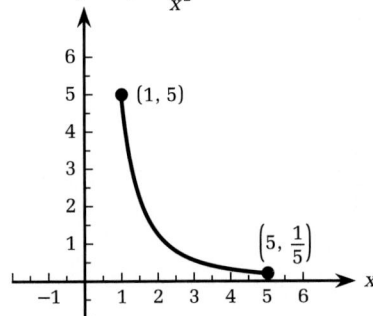

$(1, 5)$

$\left(5, \frac{1}{5}\right)$

49. ha: $y = 0$, va: $x = 0$ and $x = 2$

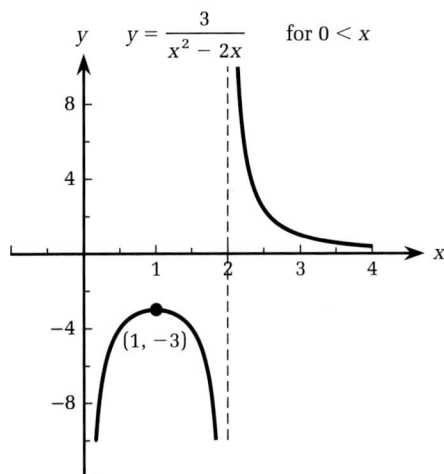

$$y = \frac{3}{x^2 - 2x} \qquad \text{for } 0 < x$$

$(1, -3)$

51.

$$y = \frac{1}{x^2 + 2} \qquad \text{for } -1 \le x \le 4$$

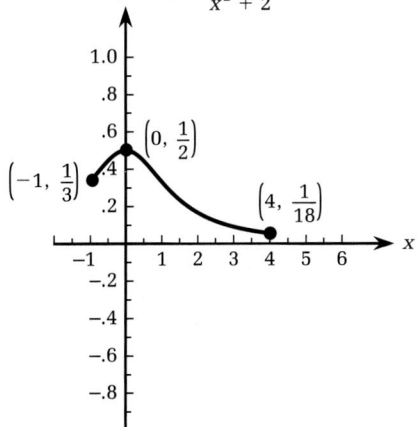

$\left(0, \frac{1}{2}\right)$

$\left(-1, \frac{1}{3}\right)$

$\left(4, \frac{1}{18}\right)$

53.

$$y = \frac{1}{x^2 - 4x + 5} \qquad \text{for } 1 \le x \le 3$$

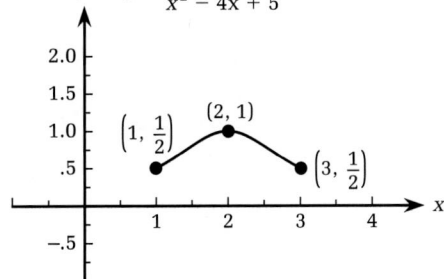

$(2, 1)$

$\left(1, \frac{1}{2}\right)$ $\left(3, \frac{1}{2}\right)$

55. $y = 3x^{2/3} - x$ for $0 \le x \le 10$

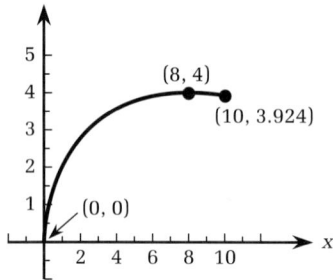

(8, 4)
(10, 3.924)
(0, 0)

57. $y = \dfrac{x}{2x^2 + 1}$ for $0 \le x \le 2$

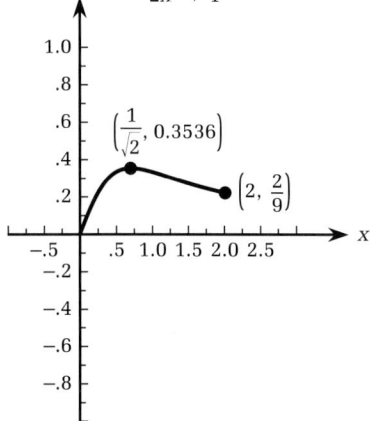

$\left(\dfrac{1}{\sqrt{2}}, 0.3536\right)$
$\left(2, \dfrac{2}{9}\right)$

59. $y = 3x^{1/3} - x$ for $-1 \le x \le 2$

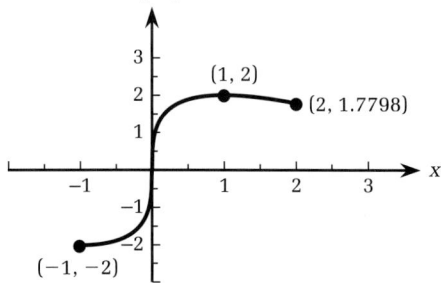

(1, 2)
(2, 1.7798)
(−1, −2)

63. Rel max at $(1.1711, 4.4067)$, Rel min at $(-0.9244, -1.3497)$ and $(2.4281, 168.7868)$
67. $\sqrt{x - 5}$ is not defined near 2.

69.

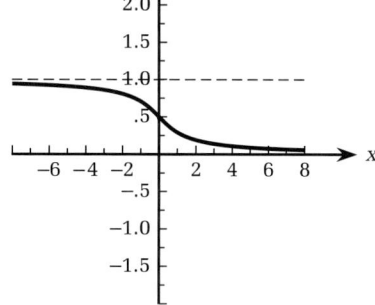

$y = \dfrac{x + 1 - \sqrt{x^2 + 1}}{2x}$

71. $y = 2x$

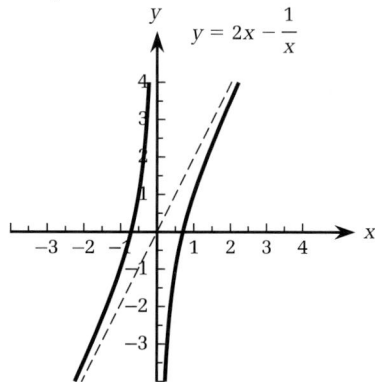

$y = 2x - \dfrac{1}{x}$

73. $y = x - 2$

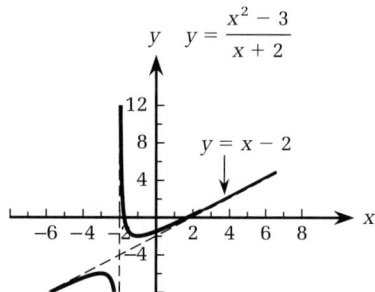

$y = \dfrac{x^2 - 3}{x + 2}$

$y = x - 2$

Section 3.5

1. (a) $4x + 2y$ **(b)** xy **3. (a)** 150 by 300
(b) 45,000 **5.** 200 by 150, area is 30,000
7. (a) 60 by 75 **(b)** 4500 **9. (a)** 20 by 35
(b) 700 **11. (a)** $15x - 0.1x^2$ **(b)** 75 **(c)** $7.50
(d) $562.50 **13.** $900 **15. (a)** $-(1/12)x +$
27.5 **(b)** $13.75 **(c)** $2268.75 **(d)** 165
17. (a) $-x^3 + 300x$ **(b)** 200 **(c)** 2000 **(d)** 10
19. (a) $24x - x^{3/2}$ **(b)** 19.5 **(c)** 256 **(d)** 8
(e) 2048 **21. (a)** $12.75 **(b)** 310 **(c)** $2402.50
23. $20.25 at $1.24 per box **25. (a)** 10
(b) 4500 **27. (a)** 840 **(b)** 10 **(c)** 84

29. 261   **31.** $a = 6, b = 6$ **33.** 108
35. (a) 1.57   **(b)** 67.6   **37.** $r \approx 1.585, h \approx 3.169$
39. $r = 140$, sidelength is 440   **(b)** 123,200

Section 3.6

1. $-96,000/x^2$ **3.** 50   **5. (a)** 50   **(b)** 10
(c) $500   **7. (a)** 10,000   **(b)** 20   **(c)** $40,000
9. (a) 500   **(b)** 10   **(c)** $500   **11. (a)** 25
(b) $150   **13. (a)** 84   **(b)** $21   **15.** 1789
17. (a) 128   **(b)** 4   **19. (a)** 156   **(b)** 156
21. 259   **25. (a)** 7071   **(b)** 28   **(c)** $56,568

Section 3.7

1. $2x - 2$ **3.** $dp/dx = (-20 - p)/(x + 2p)$
5. (a) 0.51   **(b)** 0.5   **7. (a)** 0.33333   **(b)** 0.4
9. (a) 0.04   **(b)** 0.04   **11. (a)** -0.1716
(b) -0.1667 **13.** $dy/dx = (2 - 2xy)/(x^2 - 3y^2)$
15. -4 **17.** -0.206452 **19.** 0.157 cm^3
21. (a) 103   **(b)** 1428   **23. (a)** 0.0005926
(b) 99.995   **25.** 2   **27.** 0.1008   **29. (a)** 12°
(b) 0.42   **31. (a)** $8\pi\text{ft}^2$ **(b)** $46.60

Chapter 3 Review

1. (a) inc $(-\infty, -\sqrt{2})$ and $(\sqrt{2}, \infty)$
(b) dec $(-\sqrt{2}, \sqrt{2})$ **2. (a)** inc $(-1/2, 2)$
(b) dec $(-\infty, -1/2)$ and $(2, \infty)$ **3.** RMax D,
AMax A, RMin B, F, AMin F, Inf C, E
4. RMax B, F, AMax F, RMin D, AMin A, Inf C, E
5. RMax C, AMax A, RMin B, AMin E, Inf D
6. RMax B, E, AMax E, RMin C, AMin G, Inf D, F

For 7–25 parts (b) and (c) are shown on the graphs.

7. (a) $-1/2$

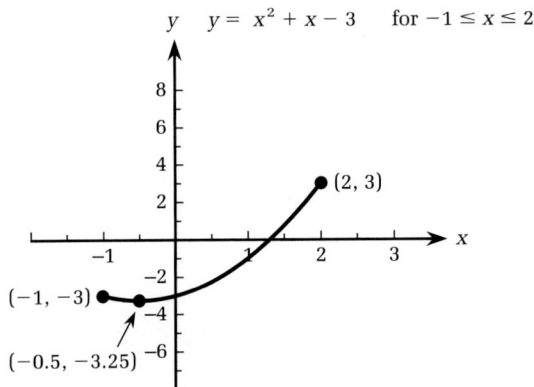

8. (a) $\dfrac{1}{4}$

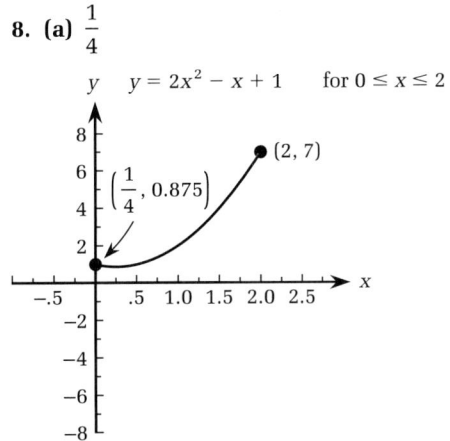

9. (a) -1

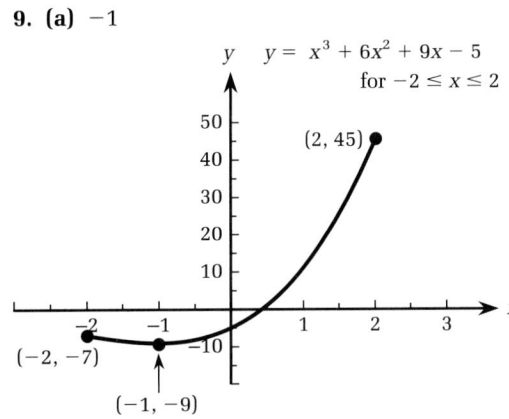

10. (a) 0 and 1/3

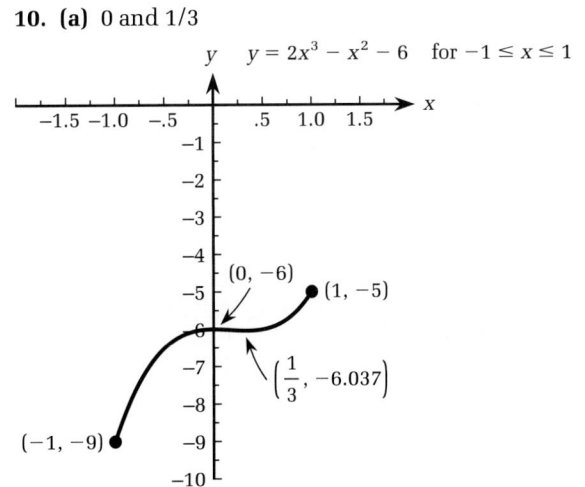

11. (a) 0 and 3

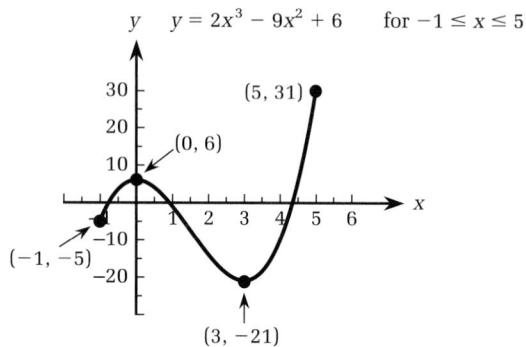
$y = 2x^3 - 9x^2 + 6$ for $-1 \le x \le 5$

12. (a) 6.25

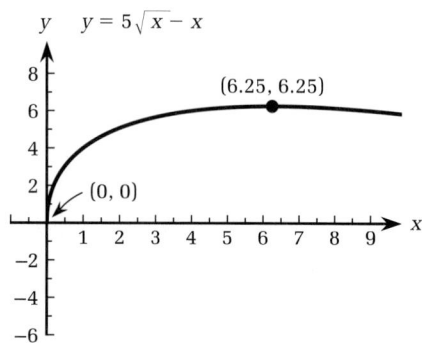
$y = 5\sqrt{x} - x$

13. (a) 1/27

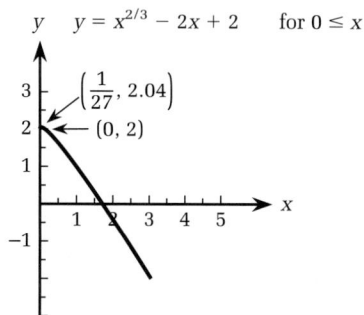
$y = x^{2/3} - 2x + 2$ for $0 \le x$

14. (a) 0.25

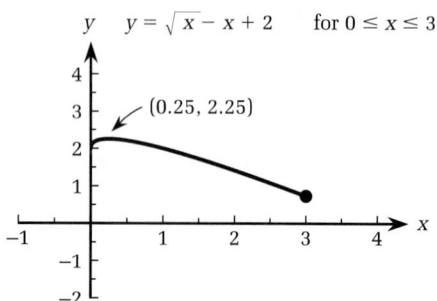
$y = \sqrt{x} - x + 2$ for $0 \le x \le 3$

15. (a) 2

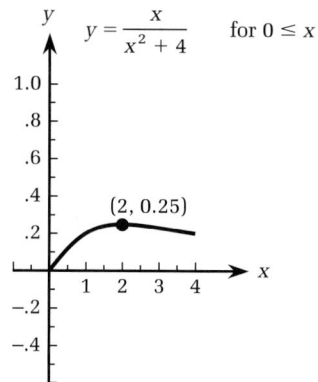
$y = \dfrac{x}{x^2 + 4}$ for $0 \le x$

16. (a) 1

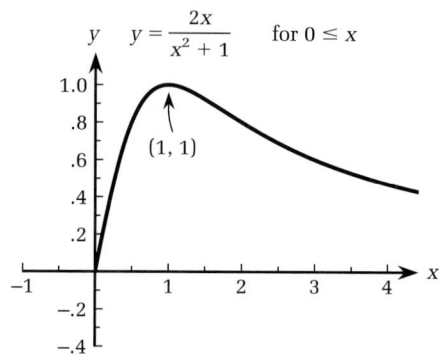
$y = \dfrac{2x}{x^2 + 1}$ for $0 \le x$

17. (a) 3

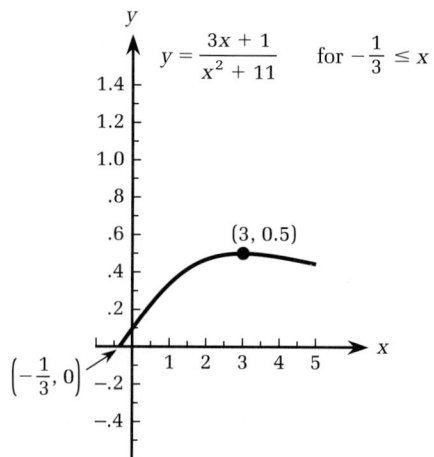
$y = \dfrac{3x + 1}{x^2 + 11}$ for $-\dfrac{1}{3} \le x$

18. (a) $\pm\sqrt{2}$

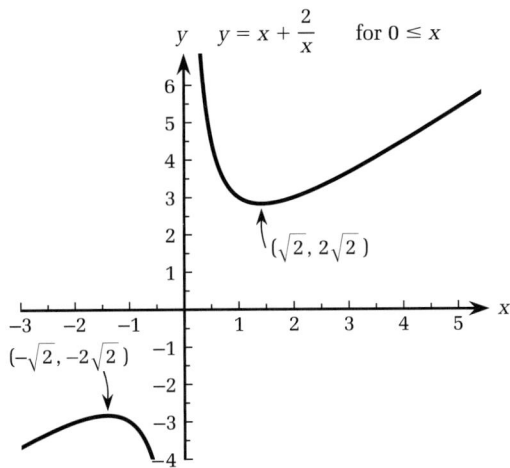

$y \quad y = x + \dfrac{2}{x} \qquad$ for $0 \le x$

$(\sqrt{2}, 2\sqrt{2})$

$(-\sqrt{2}, -2\sqrt{2})$

19. (a) 2

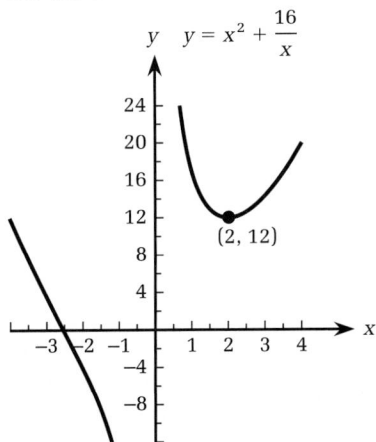

$y \quad y = x^2 + \dfrac{16}{x}$

$(2, 12)$

20. (a) 2/3

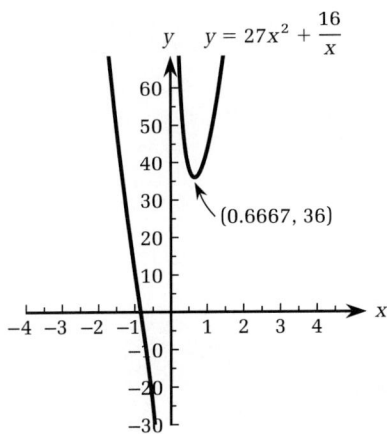

$y \quad y = 27x^2 + \dfrac{16}{x}$

$(0.6667, 36)$

21. (a) 0.5

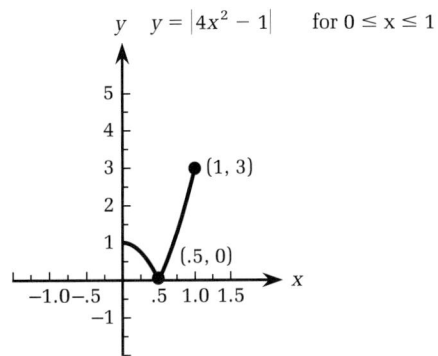

$y \quad y = |4x^2 - 1| \qquad$ for $0 \le x \le 1$

$(1, 3)$

$(.5, 0)$

22. (a) 1

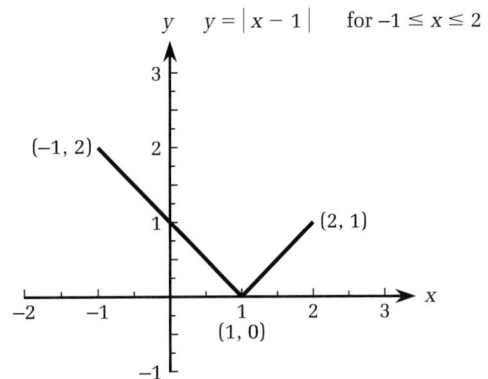

$y \quad y = |x - 1| \qquad$ for $-1 \le x \le 2$

$(-1, 2)$

$(2, 1)$

$(1, 0)$

23. (a) 5

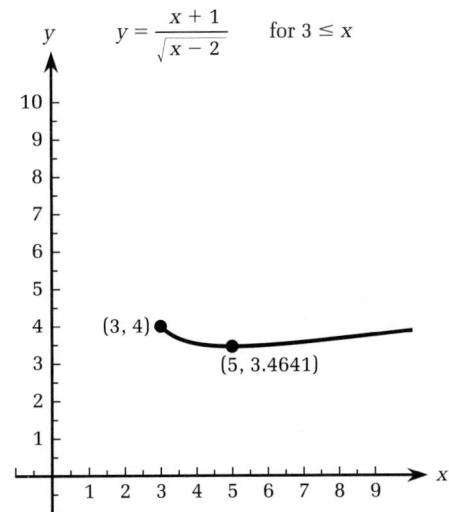

$y \quad y = \dfrac{x + 1}{\sqrt{x - 2}} \qquad$ for $3 \le x$

$(3, 4)$

$(5, 3.4641)$

24. (a) $\dfrac{8}{3}$

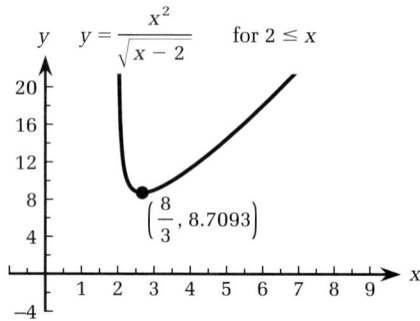

$y = \dfrac{x^2}{\sqrt{x-2}}$ for $2 \le x$

$\left(\dfrac{8}{3}, 8.7093\right)$

25. (a) 0 and 8

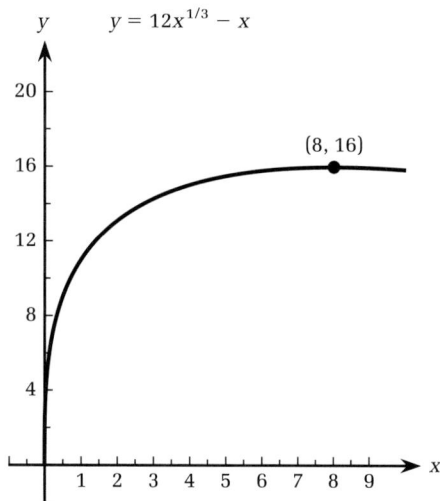

$y = 12x^{1/3} - x$

$(8, 16)$

26. (a) 0 and 8

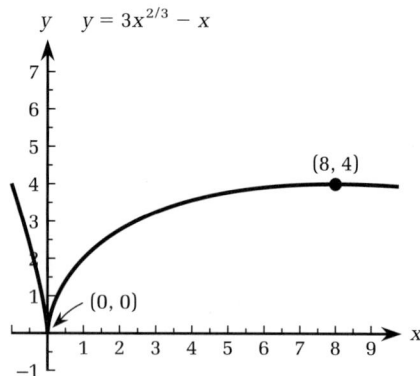

$y = 3x^{2/3} - x$

$(8, 4)$

$(0, 0)$

27. $(-2, -7)$ **28.** $(1/6, -325/54)$, about $(0.1666 - 6.0185)$ **29.** $(0,0), (2\sqrt{3}, \sqrt{3}/8)$ and $(-2\sqrt{3}, -\sqrt{3}/8)$ **30.** $(0,0), (\sqrt{3}, \sqrt{3}/2), (-\sqrt{3}, -\sqrt{3}/2)$ **31.** va $x = 1$; ha $y = 2$ **32.** ha $y = 0$ **33.** ha $y = 0$ **34.** va $x = 0$ **35.** va $x = \pm 1/2$; ha $y = -1/4$ **36.** va $x = \pm 2$; ha $y = 3$ **37. (a)**

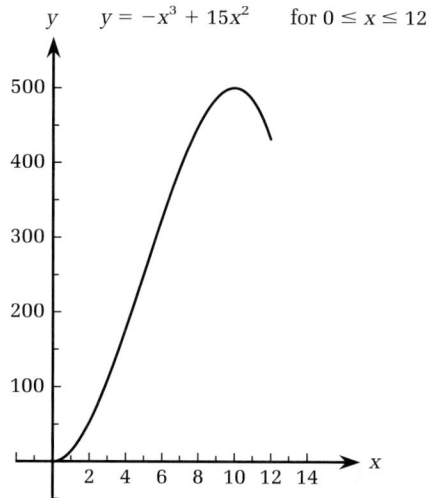

$y = -x^3 + 15x^2$ for $0 \le x \le 12$

(b) 500   **(c)** 10   **(d)** $(5, 250)$ **(e)** Rate of profit increase starts to decline.   **38. (a)** $t = 2$ **(b)** 95%   **39. (a)** 250 by 200   **(b)** 50,000   **40. (a)** $y''' = (-3/8)x^{-3/2} + 5x^4$ **(b)** $y^{(4)} = (9/16)x^{-5/2} + 20x^3$ **41.** $60x^2$ **42. (a)** 520,833   **(b)** 625 by 833.3   **(c)** 160 by 120   **43. (a)** 30,000   **(b)** 75 by 200/3   **(c)** 5000   **44. (a)** 450   **(b)** 45   **(c)** $20,250   **45. (a)** $100x/(x+1)$ **(b)** 85.25   **(c)** 19   **46. (a)** 25   **(b)** 15   **47. (a)** 12   **(b)** 5   **(c)** $300   **48. (a)** $\sqrt{3}$ mile from a point on shore directly opposite the well   **(b)** 5098   **49.** 135   **50.** 32   **51.** 5.1   **52.** 0.075   **53.** order 8   **54.** $\sqrt{2}\, c$

SELECTED ANSWERS FOR CHAPTER 4

Section 4.1

1. 32   **3.** 1/9   **5.** 2.4623   **7.** 4.1787
9. 0.04979   **11. (a)** 0.06   **(b)** 0.005   **(c)** 0.1
(d) -0.03 **13.** 6665.49   **15.** 28,470.38

17.

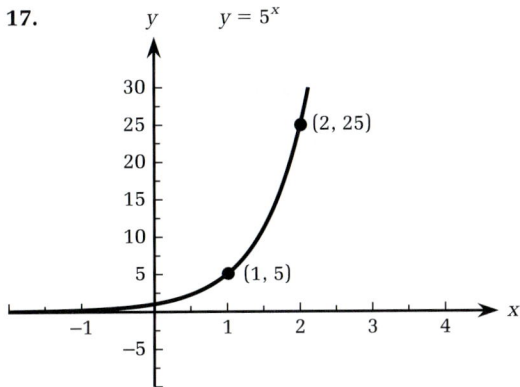

$y = 5^x$

19.

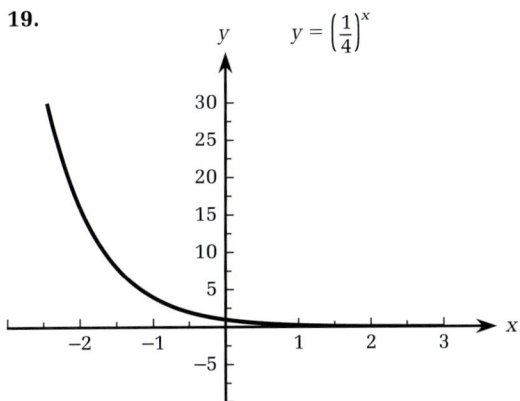

$y = \left(\frac{1}{4}\right)^x$

21.

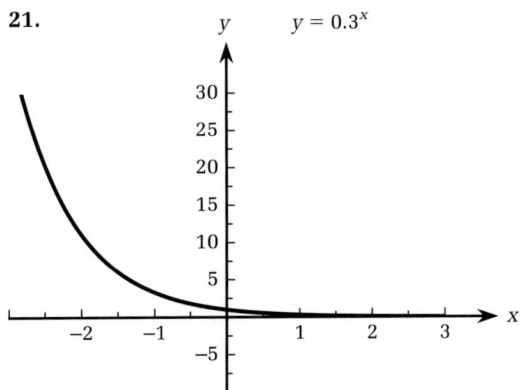

$y = 0.3^x$

23.

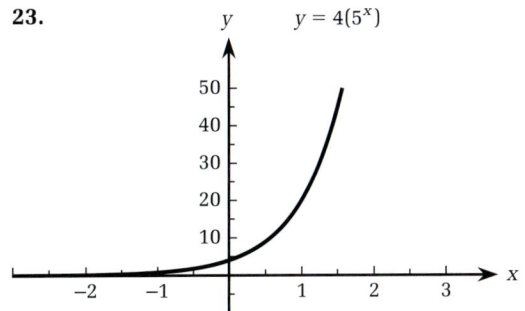

$y = 4(5^x)$

25.

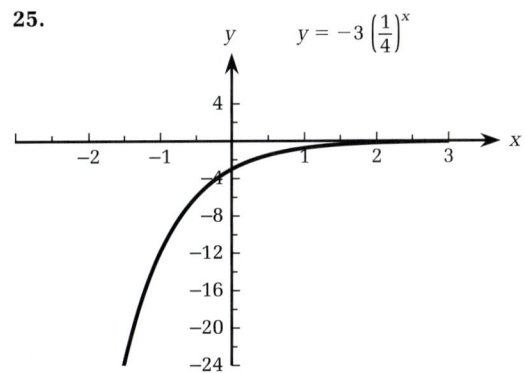

$y = -3\left(\frac{1}{4}\right)^x$

27.

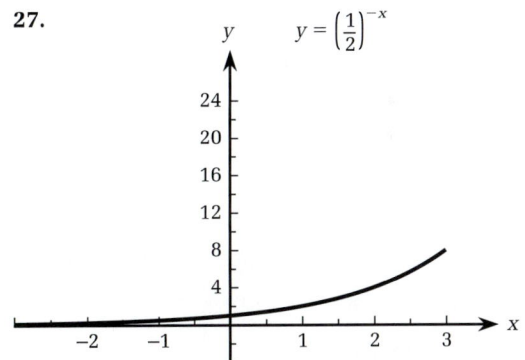

$y = \left(\frac{1}{2}\right)^{-x}$

29.

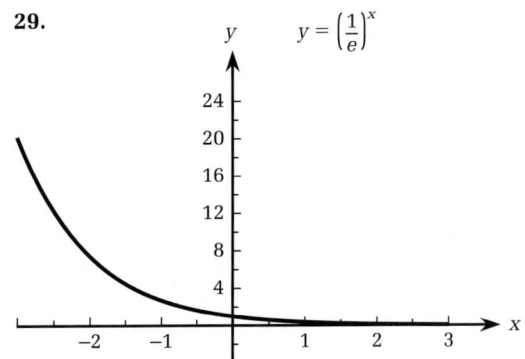

$y = \left(\frac{1}{e}\right)^x$

31.

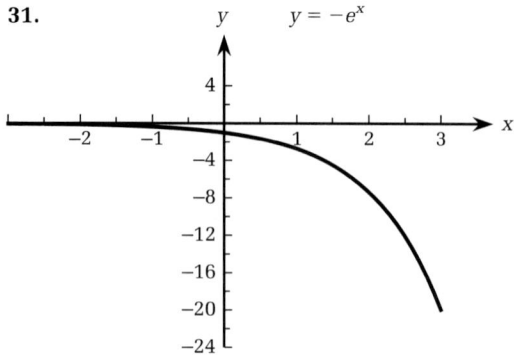

$y = -e^x$

33.

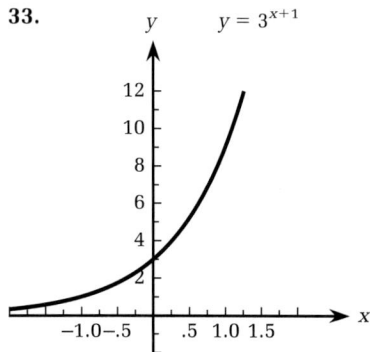

$y = 3^{x+1}$

35. 11,127.71   **37.** 8,853,650.10
39. 71,850.32   **41. (a)** 208.5   **(b)** 121.6
43. 8.2   **45.** 1678   **47.** 1.335 and -2.961
49. 2.72

55.

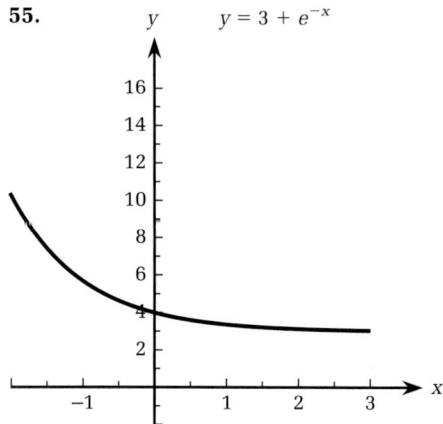

$y = 3 + e^{-x}$

57. (a) 8%   **(b)** 8.243%   **(c)** 8.3%   **(d)** 8.329%

Section 4.2

3. 0, 2   **5.** $2x - 5$ **7.** $R(x) = x \cdot p(x)$
11. $0.5e^{0.5x}$ **13.** $-6e^{-3x}$ **15.** $2 - 3e^{3x}$
17. $-2x + 2e^{2x}$ **19.** $3e^x(x + 1)$
21. $x^2 e^{-2x}(-2x + 3)$ **23.** $e^{2x}[2\sqrt{x} + 1/(2\sqrt{x})]$

25. $(3 - 3x)/e^x$ **27.** $e^{x^2-5x}(2x - 5)$
29. AMin -4.3891, AMax 2.8647

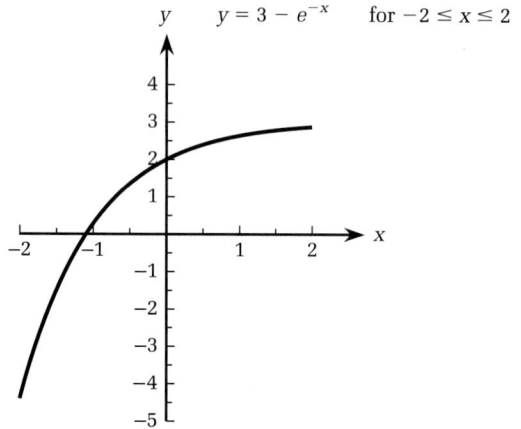

$y = 3 - e^{-x}$ for $-2 \le x \le 2$

31. AMax 27.2991, AMin and RMin 5.4366

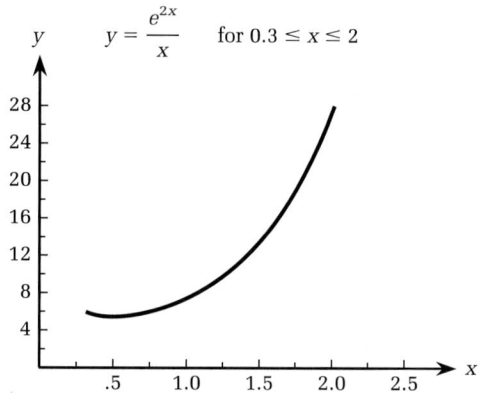

$y = \dfrac{e^{2x}}{x}$ for $0.3 \le x \le 2$

33. AMax and RMax of 1

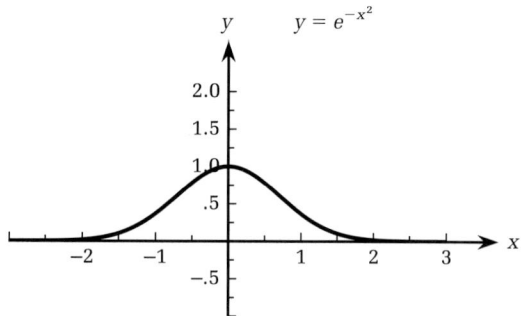

$y = e^{-x^2}$

35. (a) $3\frac{1}{3}$ **(b)** 36.7879   **(c)** 122.6
37. (a) 3.91 million, 27 thousand per year
(b) 4.78 million, 81.2 thousand per year
39. (a) $2.1e^{0.07t}$ **(b)** 17.149   **(c)** billion dollars
added in 1990   **41. (a)** $-27e^{-0.3t}$ **(b)** 30%
remembered by fifth day, 6% of original forgotten
during fifth day   **43. (a)** $19.8e^{-.2t}$ **(b)** 10.86

(c) After three weeks, the checker is learning about 11 new codes per week.

45.

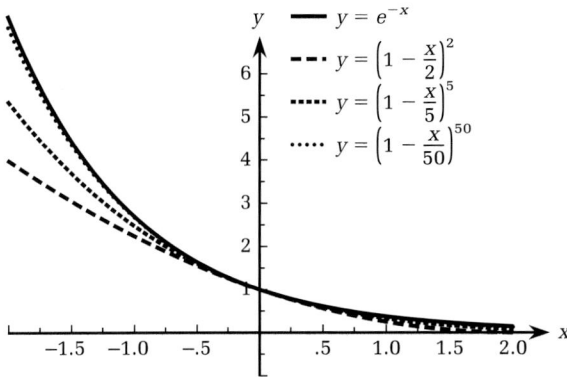

47. $-1.84 \underline{\text{and 1}}.15$ **49.** 12.0549

51. $\dfrac{x \cdot e^{\sqrt{x^2 + 1}}}{\sqrt{x^2 + 1}}$ **53.** $\dfrac{2e^{2x} + 1}{2\sqrt{e^{2x} + x}}$

55. RMin $(-0.4224, 0.4600)$

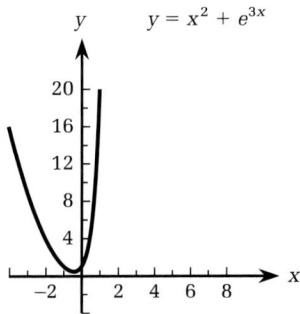

57. $x = \pm 1, y = \dfrac{1}{\sqrt{2\pi}} e^{-0.5} \approx 0.242$ **59.** $y = 0$

Section 4.3

1. 2^2 **3.** 2^{-3} **5.** 4^1 **7.** $4^{-3/2}$ **9.** $3^2 = 9$
11. $2^5 = 32$ **13.** $2^{-3} = 1/8$ **15.** $4^{3/2} = 8$
17. $10^{-3} = 0.001$ **19.** $100^{1.5} = 1000$
21. $e^{1.7918} = 6$ **23.** $\log_4 16 = 2$
25. $\log_3 (1/9) = -2$ **27.** $\log_{(1/5)} 25 = -2$
29.

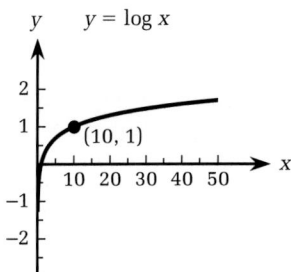

31.

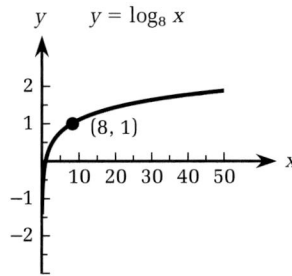

33.

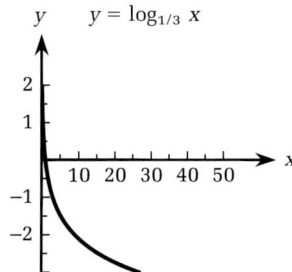

35. $\log 6 = 0.7782$ **37.** $\log 25 = 1.3979$
39. $\ln 2.3 = 0.8329$ **41.** $\ln 234.1 = 5.4557$
43. $\log_2 5 = 2.3219$ **45.** $\log_3 2 = 0.6309$
47. 8   **49.** $1/9$ **51.** $e^5/2 \approx 74.2066$
53. $e^2 - 1 \approx 6.3891$ **55.** $(\ln 10)/2 \approx 1.1513$
57. $2 + \ln 5 \approx 3.6094$ **59.** -998.577
61. $(\ln 4)/(\ln 3) \approx 1.2619$ **63.** $(\ln 3)/(\ln 0.5) \approx -1.5850$ **65. (a)** 14   **(b)** 14.4
67. (a) $\$18,221.19$ **(b)** 15.3   **69.** 3.2
71. 5.5   **73.** 17.5   **75.** $1/4$ **77.** 0.3589
81. 0.9   **83.** 16,173

Section 4.4

1. $(1/3)x$ **3.** $(1/2)x^{-4}$ **5.** $(1 + 2x^3)/2x$
7. $(1 - 2x^{3/2})/\sqrt{x}$ **9.** $1/(2x)$ **11.** $3/(3x + 2)$
13. $1 + \ln 2x$ **15.** $\dfrac{2x - (2x + 1) \ln (2x + 1)}{(2x + 1)x^2}$
17. $(3/x)[\ln (1/2)x]^2$ **19.** $(2x + 2)/(x^2 + 2x)$
21. $2e^{2x}/(e^{2x} - 1)$ **23.** $\dfrac{-1 - x^2}{x(x^2 - 1)}$
25. $2x + 4x \ln x$ **27.** $x^{-4} (2 - 6 \ln x)$
29. $\dfrac{3x^3 \ln x - 1 - x^3}{x(\ln x)^2}$ **31.** $2/(2x - 1)$
33. $-2/x^2$ **35.** 3.3863   **37.** $y = (1/e) x$
39. $-1/55$

41. RMin and AMin of $-1/(2e)$ at $x = e^{-1/2}$; AMax of 0 at $x = 1$

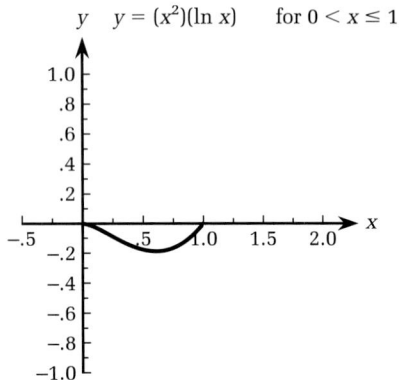

y $y = (x^2)(\ln x)$ for $0 < x \leq 1$

43. RMin of $e^2/4$ at $x = e^2$ **45. (a)** $(\lambda m)/e$ at $x = m/e$ **(b)** λ **47.** $\dfrac{3}{(3x - 1)\ln 2}$

49. $\dfrac{2x + 2}{(x^2 + 2x)\ln 10}$ **51.** $2^{3x} \cdot 3\ln 2$

53. $2^{x^2+1} \cdot 2x\ln 2$ **55.** $2^{-2x}(1 - 2x\ln 2)$

57. RMin of -0.4882 at $x = 0.6913$

59. domain of $x > -0.5671$ (approximately)

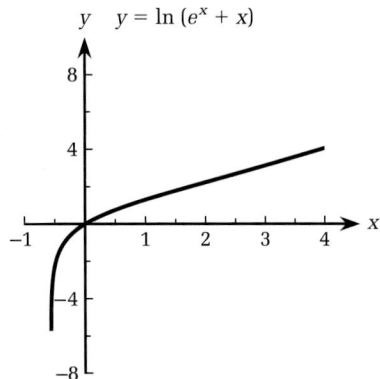

y $y = \ln(e^x + x)$

61. $x^{-x}(-1 - \ln x)$ **63.** $(3x + 1)^x[(3x/(3x + 1)) + \ln(3x + 1)]$ **65.** $x^{\sqrt{x}}\left(\dfrac{\sqrt{x}}{x} + \dfrac{\ln x}{2\sqrt{x}}\right)$

Section 4.5

1. 403.4288, 1635.9844   **3.** -0.8047
5. (a) $541,000 \cdot e^{0.02183t}$ **(b)** 1,041,398

7.

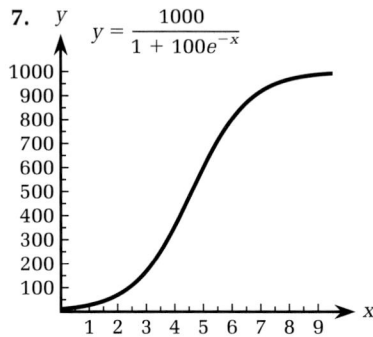

$y = \dfrac{1000}{1 + 100e^{-x}}$

9. (a) 1938   **(b)** 1950

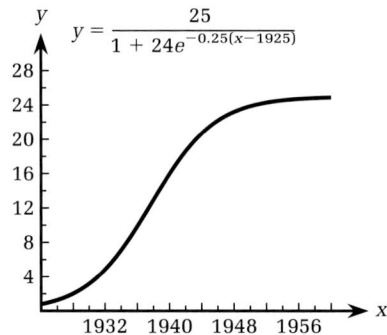

$y = \dfrac{25}{1 + 24e^{-0.25(x-1925)}}$

11. (a) 1959   **(b)** 1967

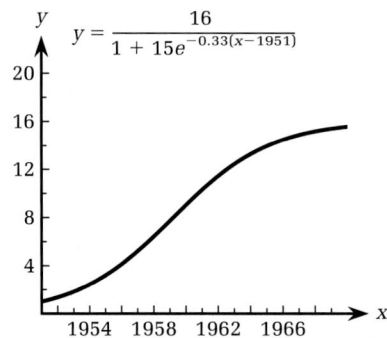

$y = \dfrac{16}{1 + 15e^{-0.33(x-1951)}}$

13. (a) $f(t) = 1000/(1 + 199e^{-0.5562t})$ **(b)** 7
(c) 9.5   **15. (a)** 364   **(b)** 17.5   **17.** 11,989
19. 870 yr   **21.** 80%   **23. (a)** 99.7 min
(b) 12.5%   **25.** $0 \leq x \leq 73, 0 \leq y \leq 50$
27. (c) a max of about 10

Section 4.6

1. $-50/(x + 1)^2$ **3.** $-10e^{-0.02x}$ **5.** $2xe^x + e^x x^2$
7. 60,653   **9.** 36,631   **11.** $449.33
13. 4.1%   **15. (a)** 12.5   **(b)** 18.39
17. (a) 24.5   **(b)** 18.77

19. (a) 5.55 **(b)** 1.43   **21. (a)** 3.09 **(b)** 10.47
23. (a) 2 **(b)** 12.59
27. (a) $200p^2/[100(p^2 + 1)]$ **(b)** $p > 1, x < 50$
(c) 50 **(d)** 50   **29. (a)** $p/(200 - p)$
(b) $p > 100, x < 50$ **(c)** 50 **(d)** 5000
31. (a) $(200 - 2x)/2x$ **(b)** $p > 100, x < 50$
(c) 50 **(d)** 5000   **33. (a)** 1 **(b)** none
(c) any $x > 0$ **(d)** 50   **35. (a)** 1 **(b)** none
(c) any $x > 0$ **(d)** 100   **37. (a)** $(x + 1)/x$
(b) none **(c and d)** no maximum
39. (a) $(48 - 2x)/x$ **(b)** $p > \sqrt{8}, x < 16$ **(c)** 16
(d) 45.25   **41. (a)** $2p^2/(24 - p^2)$ **(b)** $p > \sqrt{8}$,
$x < 16$ **(c)** 16 **(d)** 42.25   **43. (a)** $50/x$
(b) $p > 500/e, x < 50$ **(c)** 50 **(d)** 9197
45. (a) 0.83% decrease **(b)** 9.6% decrease
51. (a) inelastic **(b)** yes   **53.** $E_c(25) = 0.164$
55. \$7,162,394

Chapter 4 Review

1. $3e^{3x}$ **2.** $-2e^{-2x}$ **3.** $2e^{2x-3}$ **4.** $3e^{3x-1}$
5. $x^2 e^x + 2xe^x + 3x^2$ **6.** $3xe^x + 3e^x - 3$
7. $2/x$ **8.** $5/x$ **9.** $5/(5x + 1)$
10. $-2/(3 - 2x)$ **11.** $2 + \ln x^2$
12. $\dfrac{1/2 - \ln \sqrt{x}}{x^2}$ **13.** $\dfrac{2}{x^3 + x} - \dfrac{2 \ln (x^2 + 1)}{x^3}$
14. $x^2 - (1/x) + 3x^2 \ln 3x$ **15.** $2e^{x^2-x}(2x - 1)$
16. $e^{\sqrt{x+1}}/(2\sqrt{x + 1})$ **17.** $\left(\dfrac{3x^2 - 1}{x^2 - 1}\right) +$
$\ln (x^3 - x)$ **18.** $(4x + 1)/(2x^2 + x)$ **19.** xe^x
20. $\ln x$ **21.** $e^5/3 \approx 49.4711$
22. $e^{-2}/5 \approx 0.02707$ **23.** $(e^2 + 1)/5 \approx 1.6778$
24. $(e^5 - 3)/2 \approx 72.7066$ **25.** $(\ln 4)/3 \approx 0.4621$
26. $(1 + \ln 6)/2 \approx 1.3959$ **27.** $(\ln 8)/(2 \ln 5) \approx$
0.6460 **28.** $(\ln 5)/(-3 \ln 2) \approx -0.7740$
29. $20 \ln 2.5 \approx 18.3258$ **30.** $\ln 2.5 \approx 0.9163$
31. $e^4 \approx 54.5982$ **32.** $e^8 = 2980.9580$

33.

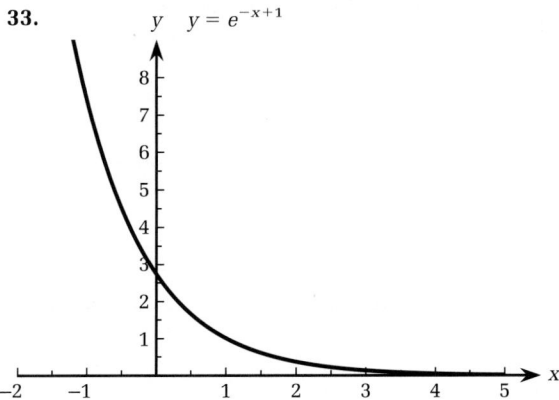
$y \quad y = e^{-x+1}$

34.

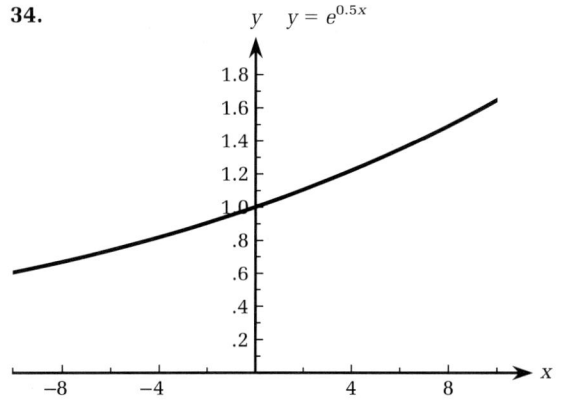
$y \quad y = e^{0.5x}$

35. AMin 2.3333, AMax 245

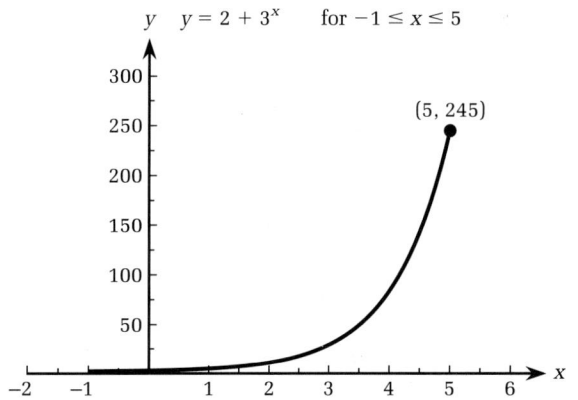
$y \quad y = 2 + 3^x \quad$ for $-1 \le x \le 5$
$(5, 245)$

36. AMin 0.037, AMax 9

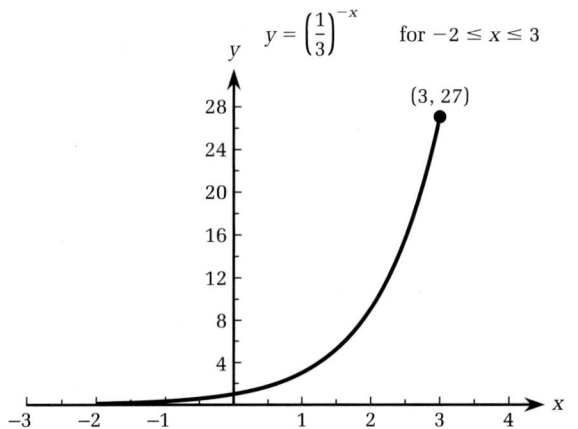
$y = \left(\dfrac{1}{3}\right)^{-x} \quad$ for $-2 \le x \le 3$
$(3, 27)$

37. AMax $e^2 - 4 \simeq 3.39$, A and RMin 0.614

$y \quad y = e^{2x} - 4x \quad$ for $0 \le x \le 1$

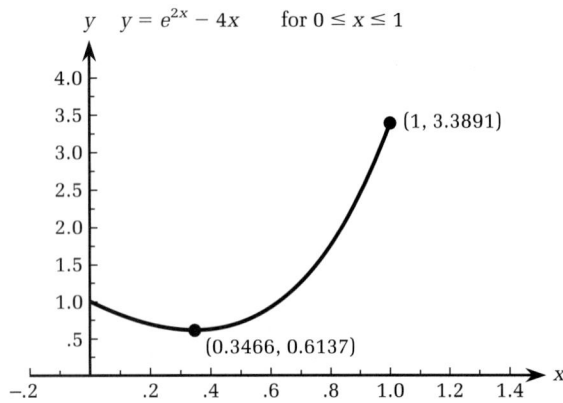

(1, 3.3891)

(0.3466, 0.6137)

38. AMin 0, A and RMax $\dfrac{2}{e} \simeq 0.7358$

$y \quad y = \dfrac{2x}{e^x} \quad$ for $0 \le x \le 2$

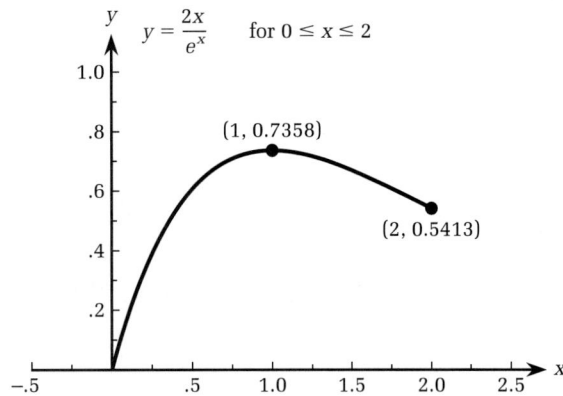

(1, 0.7358)

(2, 0.5413)

39. AMax 6.5917, A and RMin -0.1226

$y \quad y = x \ln (3x) \quad$ for $0 < x \le 3$

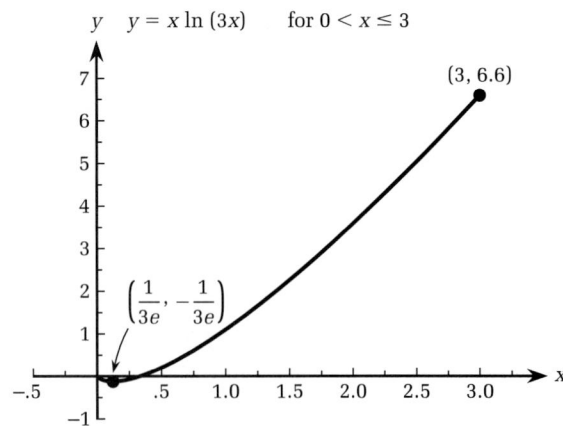

(3, 6.6)

$\left(\dfrac{1}{3e}, -\dfrac{1}{3e} \right)$

40. AMax 2.687, A and RMin -0.09197

$y \quad y = x \ln \sqrt{2x} \quad$ for $0 < x \le 3$

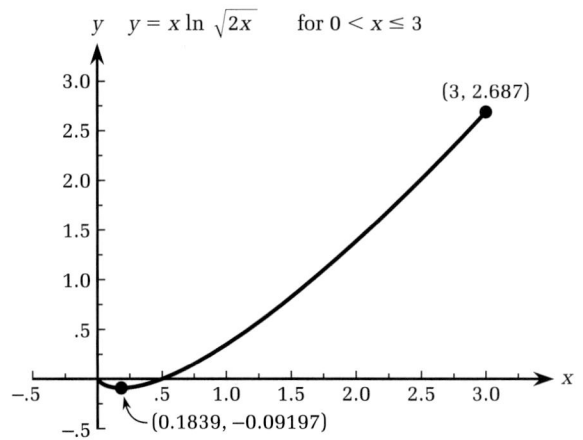

(3, 2.687)

(0.1839, -0.09197)

41. AMax 0.9431, A and RMin 0.3069

$y = \dfrac{1}{2x} + \ln (x) \quad$ for $\dfrac{1}{4} \le x \le 2$

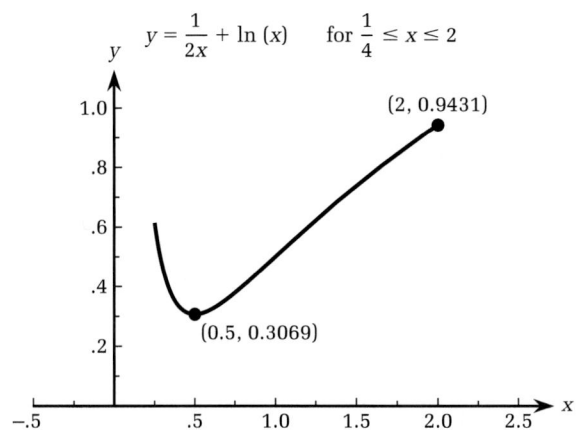

(2, 0.9431)

(0.5, 0.3069)

42. AMax 2, A and RMin 1.1931

$y \quad y = \dfrac{2}{x^2} + \ln (x) \quad$ for $1 \le x \le 3$

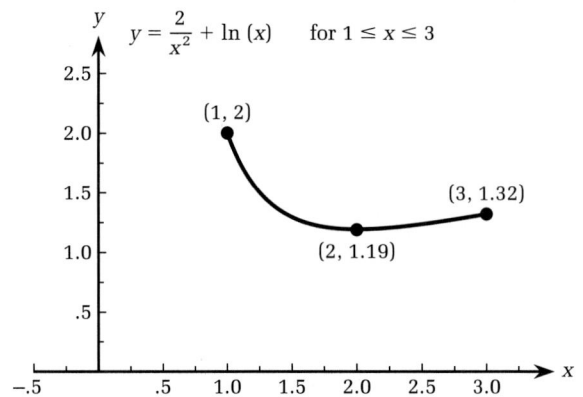

(1, 2)

(3, 1.32)

(2, 1.19)

43. 15.44   **44.** 3240   **45. (a)** 35,817
(b) 36,442   **46. (a)** 59,184   **(b)** 61,490
47. 1124.66 at 12.1%, 1125.51 at 12%   **48.** 1.82
billion   **49.** 7.5 million   **50.** 449,329
51. 4104.25   **52. (a)** 1.0217   **(b)** $f'(t) > 0$
(c)

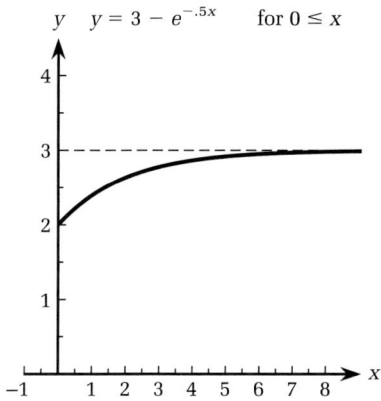

y $y = 3 - e^{-.5x}$ for $0 \le x$

53. (a) $f(t) = \dfrac{5000}{1 + 1249e^{-0.4186t}}$ **(b)** 2500

(c) 20.3   **(d)** 34   **54. (a)** 6.4   **(b)** forgetting at a
rate of 8.8% of original material per day

55. (a) 12.08 million   **(b)** 846 thousand
56. (a) \$1.23   **(b)** 2044   **57.** 14,980
58. (a) $P_0 e^{-0.0315t}$ **(b)** 73 hours
59. (a) -0.1308 **(b)** 17.6   **60. (a)** 8.5
(b) 158.32   **61. (a)** $E = 2p^2/(27 - p^2)$
(b) Inelastic for $x > 18$, Elastic for $x < 18$, Unit Elastic at $x = 18$ **(c)** $p = 3$, $x = 18$, $R = 54$
62. (a) $E = p/[2(588 - p)]$ **(b)** Elastic for $p > 392$,
Inelastic for $p < 392$ **(c)** $p = 392$, $x = 14$, $R = 5488$
63. (a) 20   **(b)** $p = 22.07$, $R = 441.40$
64. (a) $E = (2x^3 - 27)/(-6x^3)$ **(b)** Elastic for $x <$
1.5, Inelastic for $x > 1.5$ **(c)** $x = 1.5$, $p = 20.25$,
$R = 30.38$ **65.** 4.4%   **66. (a)** 10,411
(b) March of 2003   **67.** $10^{2x} (2 \ln 10)$

68. $2^{-3x}(-3 \ln 2)$ **69.** $\dfrac{2x + 3}{(x^2 + 3x) \ln 3}$

70. $2x^{3x+1} \left(\dfrac{3x + 1}{x} + 3 \ln x \right)$

71. $(6x + 10)/x + 6 \ln x$ **72.** 0.52

SELECTED ANSWERS FOR CHAPTER 5

Section 5.1

1. $x^{1/2}$ **3.** $x^{1/3}$ **5.** x^2 **7.** -5 **9.** $3x + C$
11. $(x^7/7) + C$ **13.** $(x^{-4}/-4) + C$
15. $(x^{3.5}/3.5) + C$ **17.** $(3/2)x^2 - 2x + C$
19. $2t - (t^2/2) + C$ **21.** $x^3/3 + (3/2)x^2 - 4x + C$
23. $t^3/3 + 2t^2 + C$ **25.** $2y^5 + (3/2)y^2 - y + C$
27. $e^{4x}/4 + C$ **29.** $-(3/2)e^{2x} + C$
31. $-e^{-y} + C$ **33.** $(10/3)x^{3/2} + C$
35. $x^3 - (2/3)x^{3/2} + C$ **37.** $(3/4) \cdot \sqrt[3]{2} \cdot x^{4/3} + C$
39. $3x^{-1} + C$ **41.** $x^2/2 + 2x^{-1} + C$
43. $5 \ln|x| + C$ **45.** $(x^2/2) - 2 \ln|x| + C$
47. $-e^{-2x} + C$ **49.** $(4/3)x^3 - 6x^2 + 9x + C$
51. $x - 2 \ln|x| - (1/x) + C$ **53.** $x^2 - x + 4$
55. $\ln|x| + 3$ **57.** $2x^2 - 4x + 1$ **61.** $e^{2x}/2 - 2x - (1/2)e^{-2x} + C$

Section 5.2

1. 0.5   **3.** 0.25, 0.75, 1.25, 1.75   **5.** 18
7. 4.6724   **9.** 0.6345   **11.** 1.3583   **13.** -1
15. 3   **17.** -6 **19.** 21   **25.** $\pi/4 - 1/2 \simeq$ 0.2854

Section 5.3

1. $x^2/2 + C$ **3.** $e^{3x}/3 + C$ **5.** $5, -2$ **7.** 9
9. 14/3   **11.** $-28/3$ **13.** $e - 1$ **15.** $\ln 5 - \ln a$ **17.** $[(b^2 - 1)/2] - \ln b$ **19.** 152/3
21. $(4/5)(2)^{2.5} \simeq 4.5255$ **23.** 0   **25.** 15
27. $(2e^6 - 8)/3 \simeq 266.2859$
29. $15.75 + \ln 8 \simeq 17.8294$ **31.** $125/6 \simeq 20.83$
33. 2454   **37.** $2xe^{x^2}$

Section 5.4

1. $x^4/4 + C$ **3.** $\sqrt{2}\,(2/3)x^{3/2} + C$ **5.** $-2, 1$
7. 1/6   **9.** 5   **11.** $[(e^6 - 1)/3] - 4 \simeq 130.1429$
13. 2   **15.** 1/2   **17.** 95.214   **19.** 362,313
23. (a) 109   **(b)** 19   **25.** 1.5

Section 5.5

1. $P(t) \cdot e^{-0.06t}$ **3.** 25   **5.** 538,145
7. (a) $(3, 4)$ **(b)** 18   **(c)** 4.5   **9. (a)** 2   **(b)** 1/3
(c) $(25, 5)$ **(d)** 312.5   **(e)** $125/3 \simeq 41.66$
11. (a) 0.2   **(b)** 0.44

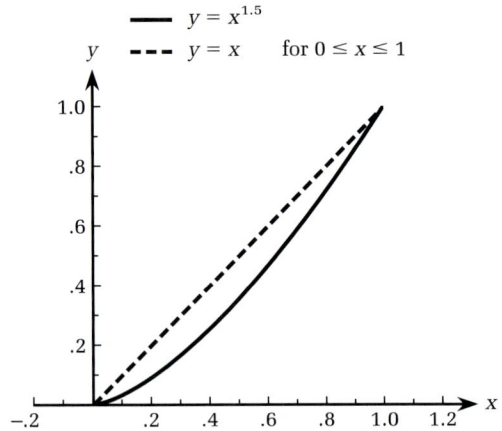

$y = x^{1.5}$
$y = x$ for $0 \le x \le 1$

13. $\simeq 0.26$ **15.** $\simeq 0.32$
17. 9 years

Chapter 5 Review

1. 3.9335   **2.** 24.5851   **3.** -0.8169
4. $(4/3)x^3 + C$ **5.** $2x^{3/2} + 2 \ln|x| + C$
6. $-x^{-1} - (3/2)x^{-2} + C$ **7.** $6x - 2e^{-2x} + C$
8. $-(1/2)e^{-2x} + C$ **9.** $y^4/4 - y^2 + C$
10. $x^2/2 + \ln|x| + C$ **11.** $(u^2/2) + 2 \ln|u| + C$
12. $e^{2x}/2 + x^2/2 + C$ **13.** 56.25   **14.** 28/3
15. 17/6   **16.** $-2/\sqrt{3} + \sqrt{2} \simeq 0.2595$
17. $2 \ln 2 + 15/8 \simeq 3.2613$ **18.** 4
19. $e^3 + 8 \simeq 28.0855$ **20.** 121.5   **21.** $-(2/3)$
22. 131,873   **23.** 174,701
24. (a)

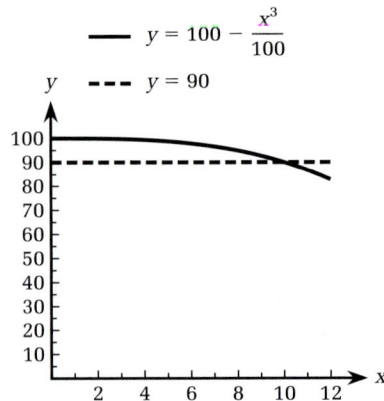

$y = 100 - \dfrac{x^3}{100}$
$y = 90$

(b) 10   **(c)** 75
25. (a) 18   **(b)** $10, 5$ units   **(c)** $250/3 \simeq 83.3$
(d) 12.5   **26. (a)** $36, 5$ units   **(b)** 166.67
(c) 87.5   **27. (a)** 0.25   **(b)** 0.577   **28.** 0.32

SELECTED ANSWERS FOR CHAPTER 6

Section 6.1

1. $du = 2\,dx$ **3.** $du = dx/x$ **5.** $x^4/4 + C$
7. $(3/4)e^{4x} + C$ **9.** $(x-2)^7/7 + C$
11. $(2x+3)^6/12 + C$ **13.** $-3/(x+2) + C$
15. $(1/3)(2x+1)^{3/2} + C$ **17.** $(x^2-5)^4/4 + C$
19. $3\ln|x+3| + C$ **21.** $(1/4)\ln|2x^2+1| + C$
23. $(1/3)(t^2+4)^{3/2} + C$ **25.** $(1/2)e^{2t+1} + C$
27. $(1/3)e^{3t^2} + C$ **29.** $\ln|1+e^x| + C$
31. $(1/2)\ln|\ln|x|| + C$ **33. (a)** $1 \le u \le 6$
(b and c) $4\ln 6$ **35. (a)** $0 \le u \le 16$ **(b and
c)** $(1/2)e^{16} - 1/2 \approx 4{,}443{,}054$ **37.** $4\ln 4 - 3$
39. $\$1576.87$

Section 6.2

1. $-3e^{-3x}, e^{-3x}/(-3) + C$
3. $(1/3)xe^{3x} - e^{3x}/9 + C$ **5.** $-(2/3)xe^{-3x} -$
$(2/9)e^{-3x} + C$ **7.** $(1/2)xe^{-2x} + (1/4)e^{-2x} + C$
9. $(x^2/2)\ln 2x - (1/4)x^2 + C$ **11.** $(x^3/3)\ln x -$
$x^3/9 + C$ **13.** $x\ln(x-2) - x - 2\ln(x-2) + C$
15. $1 - 3e^{-2} \approx 0.5940$ **17.** $-0.5 - 0.5\ln 0.5 \approx$
-0.1534 **19.** $-3.25e^{-12} + 1/4 \approx 0.25$
21. $x^3e^x - 3x^2e^x + 6xe^x - 6e^x + C$
25. $(x \cdot 2^x)/(\ln 2) - 2^x/(\ln 2)^2 + C$
27. $x\log x - x/\ln 10 + C$

Section 6.3

1. $a=1, b=3$ **3.** $a=1, b=5$ **5.** $a=2/3$
7. $(1/4)\ln|x/(x+4)| + C$ **9.** $x - 8\ln|x+8| + C$
11. $[(2x-4)/3]\sqrt{x+1} + C$
13. $(x/2)\sqrt{x^2-9} - (9/2)\ln|x+\sqrt{x^2-9}| + C$
15. $4(\ln|x+\sqrt{x^2-4}|) + C$
17. $x/5 - (1/10)\ln|5+3e^{2x}| + C$
19. $2(2 - 5\ln 7 + 5\ln 5) \approx 0.6353$
21. $2[(x/2)\sqrt{x^2-5/4} -$
$(5/8)\ln|x+\sqrt{x^2-5/4}|] + C$
23. $(3/2)[\ln|x+\sqrt{x^2-(1/4)}|] + C$
25. $3[(x/2)\sqrt{x^2-(4/9)} - (2/9)\ln|x +$
$\sqrt{x^2-(4/9)}|] + C$ **27.** $(5/3)\ln 10 \approx 3.8376$

Section 6.4

1. $\{1, 1.5, 2, 2.5, 3, 3.5, 4, 4.5, 5\}$ **3.** 0.5698
5. (a) 22 **(b)** $64/3$ **7. (a)** 1.2204 **(b)** 1.2757
9. (a) 0.8806 **(b)** 0.8818 **11. (a)** 0.7828
(b) 0.7854 **13.** -860.14 **15. (a)** 462
(b) 501 **17. (a)** 415 **(b)** 420 **19.** 3.1415926

Section 6.5

1. 0.75 **3.** $e^{2x}/2 + C$ **5.** 1.5 **7.** 3.75
9. 1.25 **11.** does not converge **13.** $1.333A_0$
15. 1000 **17.** 2857 **19.** $4/3$ **21.** does not
converge **23.** does not converge **25.** 1
27. $1/(2e^2)$ **29.** does not converge
31. $1/(2e^3)$ **33.** $-1/2$ **35.** $1/2$
37. (a) $\$6250$ **(b)** $\$6003.33$ **39.** $\$16{,}666.67$
41. $17.33A$ **43.** $60{,}000$ **47.** $9/91 \approx 0.0989$
49. $1/999 \approx 0.001001$ **51.** 156.25

Chapter 6 Review

1. $(x^2+6)^9/18 + C$ **2.** $(x^2+2)^5/10 + C$
3. $(\ln|2x^3-3|)/6 + C$ **4.** $(1/2)e^{x^2} + C$
5. $e^{x+1} + C$ **6.** $(2/3)xe^{3x} - (2e^{3x}/9) + C$
7. $\ln|\ln x| + C$ **8.** $(x^2/2)\ln\sqrt{x} - (x^2/8) + C$
9. $2(x\ln x - x) + C$ **10.** $2/e \approx 0.7358$
11. 0.2 **12.** $(1/3)2^{3/2} - 1 \approx 0.6095$
13. (a) $-\ln|x/(3x-1)| + C$
(b) $\ln|x+\sqrt{x^2-9}| + C$
(c) $[(10x-12)/15]\sqrt{5x+3} + C$
14. (a) 1.3026 **(b)** 1.2821 **(c)** 1.2953
15. (a) 1.1236 **(b)** 1.1289 **(c)** 1.1254
16. (a) 5340 **(b)** 5480 **17.** 2 **18.** 1.5
19. 4000 **20.** $153{,}846$ **21.** does not exist
22. does not exist **23.** $(e^{-5}/3) \approx 0.002246$
24. $222{,}222$ **25.** $250{,}000$ **26. (a)** 3327.3
(b) 902 days **(c)** 5000 **27. (a)** PP $20{,}000$,
LB $15{,}000$ **(b)** $\$115{,}000$ **(c)** PP $405{,}882$,
LB $313{,}636$

SELECTED ANSWERS FOR CHAPTER 7

Section 7.1

1. $x \geq -1$ **3.** $e^{-4} \approx 0.01832$ **5.** $0, -2, -8,$
20, all reals   **7.** $1, 4, 10/9, 1,$ all reals   **9.** $0,$
$e^{-4}, 1/e, e^{-4},$ all reals   **11.** $-2, 3, 7, -1/2, y \neq 1,$
x any real   **13.** $e, 1, e - 1, 1/e - 1, y > 0, x$ any
real   **15.** -59.4
17. $D(-1, 1, 0)$
$E(1, -2, 3)$
$F(0, -1, -2)$
$G(1/2, 2, -1)$

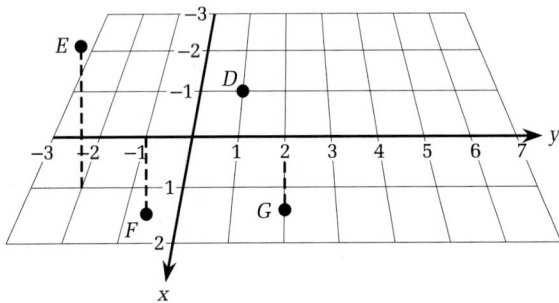

19. 0.0233 **21.** $5000e^{-0.06t}$ **23.** 21.13
25. 18.26 meters

27.

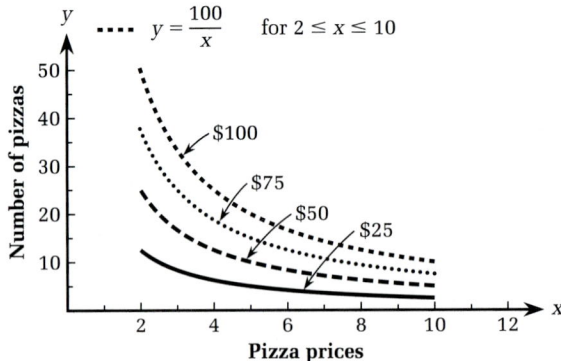

$y = \dfrac{25}{x}$ for $2 \leq x \leq 10$

$y = \dfrac{50}{x}$ for $2 \leq x \leq 10$

$y = \dfrac{75}{x}$ for $2 \leq x \leq 10$

$y = \dfrac{100}{x}$ for $2 \leq x \leq 10$

29. (a) 23.79 **(b)** $y = 63393.8x^{-3/2}$
31. (a) 1.7278 **(b)** $W = 172\,H^{-1.71}$

Section 7.2

1. $2e^{2x}$ **3.** $25y^4$ **5.** $f_x = 3x^2y^2, f_y = 2x^3y$
7. $f_x = -x^{-2}y^2, f_y = 2x^{-1}y$ **9.** $f_x = e^x y^3,$
$f_y = 3e^x y^2$ **11.** $f_x = ye^{xy} + (1/x), f_y = xe^{xy}$
13. $f_x = (y^2 - x^2)/(x^2 + y^2)^2, f_y = -2xy/(x^2 + y^2)^2$
15. $2e^{-2} \approx 0.2707, -e^{-2} \approx -0.1353$
17. $P_x = (y/x)^{0.9}, P_y = 9(x/y)^{0.1}$ **19. (a)** increase y
(b) increase x **21. (a)** increase x **(b)** increase x
23. (a) compete   **(b)** complement
(c) complement   **(d)** complement
25. (a) $P_r = 2k\pi rv^3, P_v = 3k\pi r^2 v^2$ **(b)** $P_r(20, 15) =$
$135{,}000k\pi, P_v(20, 15) = 270{,}000k\pi$ **(c)** 0.2
27. $f_{xx} = 6xy^{-1} + 2y^3, f_{xy} = -3x^2y^{-2} + 6xy^2, f_{yy} =$
$2x^3y^{-3} + 6x^2y, f_{yx} = f_{xy}$ **29.** $f_{xx} = 6y/x^4, f_{xy} =$
$-2/x^3, f_{yy} = 0, f_{yx} = f_{xy}$ **31.** $f_{xx} = 9e^{2y - 3x}, f_{xy} =$
$-6e^{2y - 3x}, f_{yy} = 4e^{2y - 3x}, f_{yx} = f_{xy}$
33. $f_x = yze^{xyz} + 2x, f_y = xze^{xyz} - 2y, f_z = xye^{xyz}$
35. $f_x = \dfrac{(x^2 + y^2 - z^2)yz - 2x^2yz}{(x^2 + y^2 - z^2)^2},$

$f_y = \dfrac{(x^2 + y^2 - z^2)xz - 2xy^2z}{(x^2 + y^2 - z^2)^2},$

$f_z = \dfrac{(x^2 + y^2 - z^2)xy + 2xyz^2}{(x^2 + y^2 - z^2)^2}$

Section 7.3

1. $x = -2, y = 4$ **3.** $x = \pm 2, y = 1$
5. RMin at $(2/3, -1/3, -1/3)$ **7.** RMin at $(0, 0, 0)$
9. RMin at $(1, 2, 2)$ **11.** Saddle at $(2, 1, -15),$
RMax at $(-2, 1, 17)$ **13.** Saddle at $(-2, 1/12, -1)$
15. Saddle at $(4/5, 2, -4 + 8\ln(4/5))$ **17.** RMin
at $(1, 1, -2)$, Saddle at $(0, 1, -1)$ and RMin at $(-1, 1,$
$-2)$ **19.** RMin at $(1, 0, 1)$ **21. (a)** 2.851
(b) $x = 7.456, y = 0.44$ **23. (a)** 250 units at
Blanco, 140 units at Fedwich   **(b)** 1130
31. RMax of $\sqrt{2}/e^2 \approx 0.1914$ at $x = \sqrt{2}, y = 1$ and at
$x = -\sqrt{2}, y = -1$; RMin of $-\sqrt{2}/e^2$ at $(-\sqrt{2}, 1)$ and
$(\sqrt{2}, -1)$

Section 7.4

1. (a) $6x - 2y$ **(b)** $-2x - 3y^2$ **3.** 37.5
5. -8 **7.** 2 **9.** 12 **11.** $66/121$
13. (a) 689.2 **(b)** 0.013784 **(c)** 137.8
15. $\$50$ for 30 type A, $\$190$ for 10 of type B, for
$\$3400$ total   **17.** 660 sets on radio, 6 of print, 14
of mall ads   **19.** $r \approx 1.52, h \approx 3.04$

Section 7.5

1. (a) 2.5   **(b)** $3b^2 - 12b + 8ab + 10a^2 - 18a + 14$
(c) $D_a = 8b + 20a - 18, D_b = 6b - 12 + 8a$
3. (a) $y = 1.286x + 2$ **(b)** 0.2857
5. (a) $y = 0.8857x + 1.2572$ **(b)** 1.8857
7. (a) $y = 1.2273x + 0.2273$ **(b)** 0.5455
9. $3.36x + 185.56$ **11.** $11.83x - 1.41$
13. (a) $5.0157x^2 + 4.8703x + 0.3890$ **(b)** for
quadratic, $D = 0.62$; for linear, $D = 1.22$

Section 7.6

1. $f_x = 2x + 3y, f_y = 3x + 3y^2$ **3.** $f_x = e^x + e^{2y}$,
$f_y = 2xe^{2y}$ **5.** 0.2   **7.** 1.15   **9.** 4.7053
11. 0.4   **13. (a)** 0.0049   **(b)** 101   **15.** 3337.5
17. (a) 32.5   **(b)** base 0.05%, height 0.6%, area
0.65%   **19.** 419   **21.** 0.3

Section 7.7

1. 7/3   **3.** $y - 1$ **5.** $4x^2$
7. $R = \{(x, y) : 0 \le y \le 4, y \le x \le 4\}$ **9.** 2/3
11. 64/9   **13.** 6   **15.** 48   **17.** 3/2
19. $(e^3 - e^2 - e^{-1} + e^{-2})/2 \simeq 6.232$
21. $-97/30 \simeq -3.2333$ **23.** 1/15   **25.** 5
27. 10/3   **29.** 162   **31.** $4\ln 4 - 3$
33. 19/60   **35.** 102.4   **37.** 4   **39.** $e - 1/2$
41. 5/6   **43. (a)** 3.1 feet   **(b)** an extra 8000
cubic feet   **45.** $e/2 \cong 1.3591$ **47. (a)** 108.8
(b) 11.2 gal/second saving; $28.22 per 10 hours

Chapter 7 Review

1. all reals, $f(0, 0) = 0, f(-2, 1) = -5, f(2, 2) = 8$
2. all reals except for $xy = 0, f(1, 1) = 3$,
$f(2, -1) = 7.5, f(1, 3) = 7/3$ **3.** all reals except
for $y = -x, f(1, 1) = 1, f(-1, 2) = 5, f(2, 0) = 2$
4. all reals except for both x and y being zero, $f(1, 0)$
$= 0, f(1, 2) = \ln 5, f(2, -1) = \ln 5$ **5.** all reals,
$f(0, 0, 0) = 0, f(0, 1, 0) = 0, f(1, 1, 1) = 1/e^3$
6. $f_x = 6x + y, f_y = x + 2y$
7. $\dfrac{\partial}{\partial x} f = -\dfrac{2y(-2x^2 + 1)}{(y^2 - 1)^2}, \dfrac{\partial}{\partial y} f = \dfrac{-4x}{y^2 - 1}$
8. $f_x = (-y/x^2)e^{y/x}, f_y = (1/x)e^{y/x}$
9. $f_x = -3x^2, f_y = z^2, f_z = 2yz$ **10. (a)** 0.5555
(b) -1.125 **11.** $f_{xx} = 18x, f_{yy} = -2, f_{xy} = f_{yx} = 0$
12. $f_{xx} = 0, f_{yy} = 6xy^{-4} - y^{-2}, f_{xy} = f_{yx} = -2y^{-3}$
15. $x = -1, y = 2$ **16.** $x = 1, y = 0$
17. $x = 1, y = -2$ **18. (a)** $P_x = 1.5(y/x)^{0.7}$
(b) $P_y = 3.5(x/y)^{0.3}$ **19.** RMax at $(-1, 2, 1)$
20. RMin at $(18, 6, -108)$ **21.** 2400 at A, 112.5 at
B, for profit of $54,131   **22. (a)** $0.9286x + 2.4286$
(b) $-0.9429x + 2.3429$ **23. (a)** 25.4473
(b) 0.1063   **24.** RMin at $(2.5, -0.5, 5.25)$
25. 16/3   **26.** radius 2.73″, height 2.14″
27. $(2x + 2ye^{2xy}) dx + (2xe^{2xy}) dy$ **28. (a)** 78
(b) 21   **29.** 6   **30.** 1/15   **31.** 7/6
32. $e^2/2 - e + 1/2$ **33. (a)** 5/6   **(b)** 2.8

PHOTO ACKNOWLEDGMENTS

Chapter 1

1, © Brian Parker/Tom Stack; **44,** © Leonard Lessin/Peter Arnold, Inc; **70,** © Lee Snider/Image Works; **108,** Courtesy of AeroVironment, Inc.

Chapter 2

127, © David McGlynn 1992/FPG Int. Inc.; **152,** Brian Parker/Tom Stack & Assocs.; **263,** Spencer Swanger/Tom Stack & Assocs.

Chapter 3

265, © Gösta Håkansson/Photo Researchers, Inc.; **281,** © David L. Brown/Tom Stack & Assocs.; **322,** © M. Antman/The Image Works; **354,** © David Hill/Photo Researchers, Inc.

Chapter 4

383, © David M. Doody/Tom Stack & Assocs.; **401,** Courtesy of WEB Service Company Inc.; **467,** Leonard Lessin/Peter Arnold, Inc.

Chapter 5

485, © Joseph Sohm/The Image Works; **501,** Joe Sohm/The Image Works; **543,** John Cancalosi/Tom Stack & Assocs.

Chapter 6

457, © Jack Spratt/The Image Works; **599,** © Sam C. Pierson, Jr./Photo Researchers, Inc.

Chapter 7

601, © Benjamin/The Image Works; **670,** © Mark Antman, The Image Works; **685,** © John Gerlach/Tom Stack & Assocs.

Index

FORMULAS

GEOMETRIC

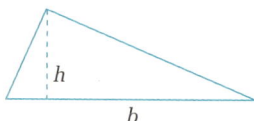

triangle

area: $\dfrac{1}{2}bh$

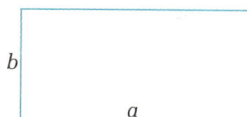

rectangle

area: ab

perimeter: $2a + 2b$

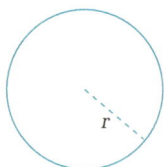

circle

area: πr^2

circumference: $2\pi r$

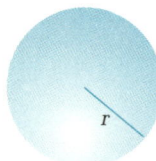

sphere

volume: $\dfrac{4}{3}\pi r^3$

surface area: $4\pi r^2$

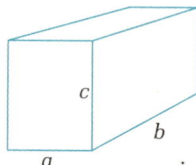

rectangular solid

volume: abc

surface area: $2ab + 2bc + 2ac$

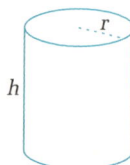

cylinder

volume: $\pi r^2 h$

surface area: $2\pi rh + 2\pi r^2$

For points $P(x_1, y_1)$ and $Q(x_2, y_2)$, the distance from P to Q is $\sqrt{(x_2-x_1)^2 + (y_2 - y_1)^2}$. The slope of the line joining them is $\dfrac{y_2 - y_1}{x_2 - x_1}$. Letting the line's slope be m, then the equation of that line is $y - y_1 = m(x - x_1)$.

ALGEBRAIC

$(a + b)^2 = a^2 + 2ab + b^2$ $(a - b)^2 = a^2 - 2ab + b^2$ $a^2 - b^2 = (a - b)(a + b)$

$ax^2 + bx + c = 0$ if $x = \dfrac{-b + \sqrt{b^2 - 4ac}}{2a}$ or $x = \dfrac{-b - \sqrt{b^2 - 4ac}}{2a}$ (quadratic formula)

$|x| = \begin{cases} x \text{ if } x \geq 0 \\ -x \text{ if } x < 0 \end{cases}$ For $a > 0$, if $|x| \leq a$ then $-a \leq x \leq a$.

EXPONENTIAL AND LOGARITHMIC

$a^0 = 1$ $a^{c+d} = a^c a^d$ $(a^c)^d = a^{cd}$ $a^{-c} = \dfrac{1}{a^c}$ $(ab)^c = a^c b^c$

$\ln 1 = 0$ $\ln e = 1$ $(e \approx 2.7182818285)$ $\ln a^c = c\,(\ln a)$ $\ln (ab) = \ln a + \ln b$ $\log_b a = \dfrac{\ln a}{\ln b}$